W. Fresenius K.E. Quentin
W. Schneider (Eds.)

# Water Analysis

## A Practical Guide to Physico-Chemical, Chemical and Microbiological Water Examination and Quality Assurance

Compiled by W. Schneider

With Contributions by
F.J. Bibo, H. Birke, H. Böhm, W. Czysz, H. Gorbauch,
H.J. Hoffmann, H.H. Rump, W. Schneider
with Additional Contributions by Other Experts

With 178 Figures and 55 Tables

Springer-Verlag Berlin Heidelberg New York
London Paris Tokyo

*Published by:*
Deutsche Gesellschaft für
Technische Zusammenarbeit
(GTZ) GmbH
6236 Eschborn 1, FRG



*Edited by:*
Professor Dr. Wilhelm Fresenius
Institut Fresenius GmbH
6204 Taunusstein-Neuhof, FRG

Professor Dr. Karl Ernst Quentin
Institut für Wasserchemie und Chemische Balneologie
der TU München
8000 München 70, FRG

Prof. Wilhelm Schneider, D. Sc.
Hoehenstraße 2
6209 Heidenrod 12

---

*Translated by:* Fachübersetzungen E. López-Ebri
Dipl.-Übersetzer A. Gledhill, B.A.
2800 Bremen 1, FRG

Ingenieurbüro für technische und naturwissenschaftliche
Übersetzungen Dr. W.-D. Haehl GmbH
Richard Holland B.A. (Hons.)
7000 Stuttgart 80, FRG

*Revised by:* T.J. Oliver B.A.
2090 Winsen/Luhe, FRG

---

ISBN 3-540-17723-X Springer-Verlag Berlin Heidelberg New York
ISBN 0-387-17723-X Springer-Verlag New York Berlin Heidelberg

Library of Congress Cataloging-in-Publication Data
Water analysis. Bibliography: p. Includes index. 1. Water–Analysis. 2. Water chemistry. 3. Water quality–Evaluation. I. Fresenius, Wilhelm, 1913-. II. Quentin, K.E. (Karl Ernst) III. Schneider, W. (Wilhelm)
TD380.W327 1988 628.1'61 87-28385
ISBN 0-387-17723-X (U.S.)


Printed in Germany



Printing: Druckhaus Beltz, Hemsbach/Bergstr.
Bookbinding: J. Schäffer GmbH & Co. KG., Grünstadt
2154/3145-543210

# Foreword

1977 saw the publication of "A Collection of Methods for Water Analysis", a three-volume work in ring-binder form compiled by W. Fresenius and W. Schneider for the GTZ (Deutsche Gesellschaft für Technische Zusammenarbeit), 6236 Eschborn 1, FRG. This publication was geared to the needs of a project in Algeria.

More recently, the editors were requested by the GTZ to produce, on the basis of the previous collection of the water analysis methods, which was published in French, an updated and revised version to be used in different partner countries and for publication in 1985/86. This was not only to take account of advances in water analysis and instrumental techniques, but also to include simple methods of analysis for use in the field, and methods suitable for use in laboratories with relatively unsophisticated equipment.

The approach envisaged by the GTZ was to divide up information on water, water supplies and water analysis into three broad groups, namely:

1. Simple modules on the physics and chemistry of water, water hygiene and water analysis capable of being understood and applied in practice by the layman using suitable chemicals and equipment (W. Schneider).
2. Information to supplement work by Rump-Krist, also commissioned by the GTZ, on water analysis methods which could be used in laboratories with simple equipment, particularly in the Third World (Verlag Chemie, Weinheim, FRG, 1986).
3. In addition to detailed instructions on sampling methods and on-the-spot analysis of water, the new work, "Water Analysis (A Practical Guide)", was also to provide a concise theoretical presentation of the various techniques of water analysis together with an indication of their relative importance, and to describe methods for use in water analysis laboratories with simple equipment or the latest in modern facilities. It was to include an account not only of up-to-date analysis methods, eg. tests for anthropogenic traces of organic and inorganic substances, but also biological approaches to water analysis. And finally it was to comment on the mathematical evaluation and weighting of the data obtained from water analysis, and to make it possible to arrive at a comparative appraisal of the findings in the light of new guidelines and rules on the quality of wastewater, surface water, ground water, drinking water etc.

One major consideration in this work has been the GTZ's desire to provide its counterparts and their analytical laboratories with summaries of analytical methods which have been tried and tested in practice, and were necessary to give theoretical and practical explanations of these methods, in order to permit laboratories with equipment of varying degrees of sophistication to handle the monitoring of the water quality. Information was also to be included to facilitate appraisal of the significance of analytical findings by comparison with legislation or guidelines from a number of different countries.

The editors and the individual contributors have amassed a broad range of experience of water analysis in virtually every part of the world. The choice of methods for determining the various parameters in water samples is the product of the experience and the results of individual laboratory work in Germany. This book thus embodies not only the author's and editor's practical experience with regard to feasibility of water analyses, but also the experience they have gained during periods of work overseas. Personal experience of the feasibility of the various methods in practice has been a key criterion in the selection of items for inclusion.

In the context of the International Water Decade 1981–1990 the GTZ felt there was a need for a comprehensive contribution of this kind to the field of water analysis and hence to the problem of assuring the quality of water, especially drinking water.

The editors and contributors would like to express their gratitude to Mr. Betz, Mr. Deichsel, Mr. Eichner, Mr. R.E. Fresenius, Mr. Frimmel, Mr. Golwer, Mr. Maushart, Mr. Scholz and Mr. Guckes for their contributions and advice, and to Mrs. Wienrich and Mrs. Bibo for their expert assistance. Our thanks are also due to the typists, Mrs. Wagner and Miss Haas, to Mrs. Fischer and Mrs. Krist for the drawings etc. We should particularly like to thank Mr. Kresse, head of GTZ's "Water Supply and Sanitation" Section, as well as Mrs. Zeumer and Mrs. Tazir of the GTZ Translation Service and the translators involved in producting the English text. Special thanks also go to Mr. T. Oliver for monitoring the English text and Mrs. M. Masson-Scheurer for the final type-script.

The editors and contributors hope that this work, "Water Analysis (A Practical Guide)", will provide water analysts in responsible positions with a valuable guide to assist them in their task of training staff to carry out local inspections and take samples, performing both simple and advanced analysis and assessing the significance and implications of analytical findings. If the book comes anywhere near achieving this ambitious purpose, it will be endorsement enough of its claim to have been produced by practitioners for practitioners.

Constructive criticism from colleagues is always welcome.

We should like to conclude with the hope that future water analysts, not to mention water technologists, will find this book useful in their studies at technical institutions and universities, especially in those countries involved in cooperation with the Federal Republic of Germany.

W. Fresenius
K.E. Quentin
W. Schneider

# Preface

This GTZ publication, "Water Analysis (A Practical Guide to Physico-Chemical, Chemical and Microbiological Water Examination and Quality Assurance)", belongs to the wide range of publications for the water sector within the scope of the International Decade for Drinking Water Supply and Sanitation.

The publication responds to the requirements for information and further training in the Federal Republic of Germany's partner countries in technical cooperation projects. It reflects the experience gathered in projects for monitoring and improving water quality, e.g. through establishing national environmental laboratories and water works laboratories for monitoring drinking water quality.

In the growing concern for our environment, maintaining water quality standards and protecting natural water resources have assumed key significance, particularly in newly industrialising countries and Third World conurbations.

Excessive strains on the environment, such as intensive farming, industry and centres of high population density, rapidly inflict serious damage both on the ecological balance and – a fact which often fails to be acknowledged due to a lack of monitoring methods – on human beings.

This publication provides those engaged in monitoring and environmental laboratories, waterworks laboratories and also in research and teaching with reliable and recognized techniques of day-to-day analysis and assessment, in order to develop sound monitoring systems and effective measures for protecting and improving water quality.

The GTZ wishes to thank the authors – in particular Prof. W. Schneider for his untiring efforts – and also the translators, and hopes that "Water Analysis" will be disseminated throughout the world.

Eschborn, September 1987

Dr. Ing. Klaus Erbel
Head of Division
GTZ "Hydraulic Engineering/
Water Resources Development"

# Table of Contents

# Literature

## I Basic Literature


Fachgruppe Wasserchemie in der GDCh (Ed) (1986) Deutsche Einheitsverfahren zur Wasser-, Abwasser- und Schlammuntersuchungen, VCH Verlagsgesellschaft, 6940 Weinheim, FRG

DIN, Deutsches Institut für Normung eV (1986) Beuth-Verlag, Berlin, FRG

ISO – International Organization for Standardization 1986 and Draft International Standards ISO Water Quality – Physical, Chemical, Biochemical, ISO, 1211 Geneva, Switzerland

American Public Health Assoc (Ed) (1979) Standard Methods for the Examination of Water and Wastewater. 16th Ed, Washington, DC, USA

ASTM (1984) Annual Book of ASTM Standards, Part 31: Water. American Society for Testing and Materials, Philadelphia, PA, USA

United States Environmental Protection Agency (1979) Methods for Chemical Analysis of Water and Wastes, USEPA, Cincinnati, OH, USA

World Health Organization (1984) Guidelines for Drinking-Water Quality, vol 1–3, WHO, 1211 Geneva, Switzerland

FAO (Ed) (1986) Codex Alimentarius, Methods of Analysis for Natural Mineral Waters. Food and Agriculture Organization of the United Nations, Rome, Italy

Suess MJ (Ed) (1982) Examination of Water for Pollution Control, vol 1–3 Pergamon Press, Oxford (World Health Organization), UK

Rodier J (1978) L'analyse de l'eau. Eaux naturelles, eaux résiduaires, eau de mer. 6me ed Dunod, Paris, France

Institut für Wasserwirtschaft (Ed) Ausgewählte Methoden der Wasseruntersuchungen, Vol 1 (1986), Vol 2 (1982). VEB G. Fischer Verlag, Jena, GDR

Hutton L (1983) Field Testing of Water in Developing Countries, Water Research Centre, Medmenham, UK

Sontheimer H, Spindler P, Rohman U (1980) Wasserchemie für Ingenieure, DVGW Forschungsstelle, Karlsruhe, FRG

Souci SW, Quentin KE (Ed) (1969) Handbuch der Lebensmittelchemie, Band VIII/1 und 2: Wasser – Wasser und Luft, Springer Verlag, Heidelberg-Berlin-New York, FRG

Fachgruppe Wasserchemie in der GDCh/Normenausschuß CNAW-Wasserwesen im DIN Jahrbuch „Vom Wasser", 1975–1986. Verlag Chemie, 6940 Weinheim, FRG

Cheeseman R, Wilson A (1978) Manual on Analytical Quality-Control for the Water Industry. Water Research Centre, Stevenage, UK

Förstner U, Wittman G, (1981) Metal Pollution in the Aquatic Environment, Springer Verlag, Berlin-Heidelberg-New York, USA

Stumm W, Morgan J (1981) Aquatic Chemistry. John Wiley & Sons, New York-Chichester

GTZ (Ed) Fresenius W, Schneider W (1979) Méthodes pour l'analyse des eaux, vol 1–3, GTZ, 6236 Eschborn, FRG

GTZ (Ed) Schneider W (1980) Technologie de l'eau potable. GTZ, 6236 Eschborn, FRG

GTZ (Ed) Fresenius W, Schneider W, Boehnke B, Poeppinghaus K (1984) Abwassertechnologie. Springer Verlag, Berlin-Heidelberg-New York-Tokyo, FRG

GTZ (1985) Water Modules, GTZ, 6236 Eschborn, FRG

GTZ (Ed) Rump HH, Krist H (1986) Manual for the Examination of Water, Wastewater and Soil, VCH Verlagsgesellschaft, Weinheim-Deerfield Beach FL, FRG/USA

Höll K (1986) Wasser, 7. Auflage. Walter de Gruyter, Berlin-New York, FRG

Huetter LA (1984) Wasser und Wasseruntersuchung, Verlag Diesterweg-Salle-Sauerlaender-Frankfurt-Berlin-München-Aarau, FRG

E. Merck. (1980) Die Untersuchung von Wasser, 6100 Darmstadt, FRG

## II Monographs

Welz B (1986) Atomic Absorption Spectrometry, 2nd Ed VCH Verlagsgesellschaft, Weinheim-Derfield Beach FL, FRG/USA

Snell FD (1978) Photometric and Fluorometric Methods of Analysis, vol 1–3. John Wiley & Sons, New York-Chichester

Kissinger PT, Heineman WR (Eds) (1984) Laboratory Techniques in Electroanalytical Chemistry. Marcel Dekker, New York-Basel, USA

Gasparic J, Churacek J (1978) Laboratory of Paper- and Thin-Layer Chromatography. Horwood – Wiley J, New York-London-Chichester, UK

Engelhardt H (1979) High Performance Liquid Chromatography. Springer-Verlag, Belin-Heidelberg-New York, FRG

Kraft G, Fischer J (1972) Indikation von Titrationen. Walter de Gruyter, Berlin-New York, FRG

Wang J (1985) Stripping Analysis, VCH Verlagsgesellschaft, Weinheim-Deerfield Beach FL, FRG/USA

Hachenberg H, Schmidt AP (1972) Gas Chromatographic Headspace Analysis. Heyden, London-New York-Rheine, UK/USA/FRG

Grob RL (1985) Modern Practice of Gas Chromatography, 2nd Ed John Wiley & Sons, New York-Chichester

Bertsch W, Jennings WG, Kaiser RE (1982) Recent Advances in Capillary Gas Chromatography. John Wiley & Sons, New York-Chichester

Seelmann-Eggebert W, Pfennig G, Muenzel H (1974) Chart of the Nuclides. Verlag Gersbach & Sons, Munich, FRG

Haberer K (1962) Radionuclide im Wasser. Thiemig Verlag, Munich, FRG

Handbook of Elektrode Technology, Orion Research – Colorameßtechnik, 7073 Lorch, FRG

## III Journals

Analyst RSC, London, UK

Analytical Chemistry, ACS, Washington DC, USA

Analytica Chemica Acta, Elsevier, Amsterdam, Nederland

Fresenius Zeitschrift für Analytische Chemie, Springer Verlag, Berlin-Heidelberg-New York-Tokyo, FRG

Atomic Spectroscopy, Norwalk CT, USA

Water Research, Pergamon Press, Oxford-New York, UK/USA

Journal American Water Works Association, New York, USA

Int J Environmental Analytical Chemistry, Gordon and Breach, New York-London-Tokyo

Gesundheitsingenieur, Oldenbourg Verlag, Munich, FRG

Journal for Water and Wastewater Research, VCH Verlagsgesellschaft, Weinheim, FRG

Gas- und Wasserfach – Wasser-Abwasser, Oldenbourg Verlag, München, FRG

Forum Städte-Hygiene, Patzer Verlag, Hannover-Berlin, FRG

Wasser, Luft und Betrieb, Vereinigte Fachverlage Krausskopf, 6500 Mainz, FRG

Labo, Hoppenstadt, 6100 Darmstadt 1, FRG

Zentralblatt für Bakteriologie, Mikrobiologie und Hygiene, G. Fischer Verlag, 7000 Stuttgart, FRG

GIT-Verlag, 6100 Darmstadt 11, FRG

International Labmate Newgate, Sandpit Lane St. Albans, Herts, UK

Water Research, Pergamon Preß, 6242 Kronberg, FRG

Applied Atomic Spectrometrie, Perkin Elmer Corp Analytic Instruments, Main Ave. Norwalk CT 06856, USA

# 1 Introduction, Sampling, Local Testing, etc.

## 1.1 General

All life on earth depends on water. The form in which the chemical compound $H_2O$ manifests itself is variously modified by its physical properties, its capacity to dissolve solid, liquid and gaseous substances and hence by its secondary chemical action and the fact that water provides a habitat for a wide variety of organisms.

These characteristics of water are an important factor to man, who uses the water for drinking or for technical purposes.

To be able to use the available water, man must test it. He must ascertain whether it can be used for the intended purpose or whether he must switch to another source of water. The simplest form of water analysis is local inspection and sensory examination. Modern methods of water analysis employ complex chemical and physico-chemical separation and determination techniques, in which readings are supplied by measuring instruments working on a variety of measuring principles, as well as microbiological techniques. Electronic data processing systems are used to evaluate the results of the analyses. Extensive automation of water analysis and the evaluation of results makes it possible to control water catchment, water treatment, water utilization, sewage treatment and water reclamation.

To some extent the simplest test methods, which still have their place in water analysis today and in many cases are actually indispensable, compete with the most modern methods of analysis.

The physico-chemical, chemical, radiochemical, bacteriological and biological analysis procedures that have been compiled in this collection of methods are based on experience which has shown that it is possible to work from these instructions without having to consult specialized literature. However, literature references are provided to supplement the analytical methods described.

Against this background the following guide to "Water Analysis" will endeavour to describe the broad field of water analysis.

Concise introductions to the theory of the methods discussed and the measurement procedures are provided (in Chapter 2) to make it easier to understand the instructions given for the methods selected.

### 1.1.1 Water resources

In principle, the water existing on earth can be divided into ground water and surface water.

In addition to ground water close to the surface and deep in the earth, ground water also includes spring water emerging spontaneously or by human intervention. **Ground water** comprises all water stagnating or moving below the earth's surface or even emerging at the earth's surface. Through this transition stage (spring), it becomes **surface water.** Surface water can be understood as water from precipitation and water from moving or standing waters, including lake and seawater.

The **hydrologic cycle,** in which ground water may be wholly or partially involved at various times, is governed by the evaporation of water and its condensation and return to the earth as precipitation (rain, fog, dew, snow and ice). By the accumulation of water from precipitation to **surface water,** with partial infiltration and as a result of direct percolation through the earth's layers, new **ground water** is formed.

As regards their physico-chemical properties and the substances dissolved in them, ground water and surface water can yield measurements which fluctuate between considerable extremes and make necessary a differentiation between the use of the different water resources.

Water from precipitation generally contains a low proportion of dissolved solids and, in absolute terms, of gases. Nevertheless, on account of its virtually non-existent buffering capacity, it can develop aggressive properties as a result of small quantities of dissolved solids or gases.

The pH of the precipitation (rain) water can fall as low as 4.2 due merely to the carbon dioxide in the air. If sulphur or nitrogen oxides are dissolved, pH values of four or even lower are possible, and if this happens, the $CO_2$ becomes unstable and is driven off.

In passing through the soil, the water absorbs soluble substances from the different geological levels. This results in a differing dissolution of inorganic substances and also, in rarer cases and to a lower degree, in a charging with soluble organic substances.

By means of the water's contact with soluble minerals below the surface, and also by means of processes of evaporation on the surface, water resources are created containing widely differing amounts of dissolved salts, culminating in highly concentrated salt water which, without further treatment, is suitable neither for human consumption nor for technical purposes.

In such cases, water analysis must supply information on the usability of the water; from the results of water analysis, the expert must be able to draw conclusions regarding direct application or the need for treatment plants.

Not only the physical properties and the chemical action of the water with dissolved substances undergo changes in the course of the hydrological cycle, but also its colonization with organisms. Above all, the presence of microorganisms is of interest for human use. For drinking purposes it is necessary to know whether the water contains bacteria or viruses, in particular those of a pathogenic or potentially pathogenic nature. This knowledge of the bacteriological properties of the water allows conclusions to be drawn on whether it is suitable for drinking purposes for humans and animals without treatment or only after pretreatment, or for bathing purposes, or whether such an application must be excluded. Many an epidemic has been due to the poor microbiological condition of the water.

Observations on intact or disturbed ecological conditions in the water of a surface water resource provide important information on infection. Plants and animals, microorganisms and macroorganisms living in the water are just as important for evaluation as physical and chemical parameters.

### 1.1.2 Water management

As it is fundamentally impossible to increase the water resources on the earth at will, careful water management, combined with expert water analysis and evaluation, and as far as possible perfect water reclamation, is essential.

Not only water experts are called upon to participate in this process of water management, but also the various water consumers in domestic households, industry and agriculture, not to mention administrative and public authorities. The latter can provide for water protection by issuing appropriate regulations and performing checks, and can also ensure that limit values are laid down and adhered to.

Water drainage areas, water catchment systems, water disposal and distributing plants should be monitored and protected by the authorities.

Monitoring of the quality of water for safe use in the case of ground water, spring water, surface water from rivers, ponds or lakes or from the sea for use as drinking, industrial or swimming pool water requires a wide range of continuous checks on hygienic, physical, chemical, bacteriological and biological factors.

For this purpose, cooperation between engineers and technicians, chemists, hygienists, doctors, biologists and bacteriologists, geologists and hydrogeologists is necessary.

This collection of methods for "Water Analysis" contains instructions for conducting the necessary local investigations, samplings and physicochemical, radiochemical, biological or bacteriological analyses. The expert user will be able to summarize from these methods the right range of analytical methods for each application.

### 1.1.3 Water analyses

The types of water analyses of varying complexity as they have been applied in practice are summarized briefly below; no value judgement is intended.

**Simple bacteriological analysis**, determining the total bacterial count per 1 ml, testing for Escherichia coli and coliform bacteria per 100 ml water in each case.

**Extended bacteriological analysis** with special investigations, e.g. for Salmonellae, Shigellae, Clostridia, anaerobic and similar microorganisms.

**Hygienic-chemical analysis.** This comprises, in addition to local inspection, a bacteriological analysis and testing for hygienically important chemical parameters, e.g. oxidizability, nitrogen compounds, iron and manganese, and also - where there are grounds for suspicion - determining toxic heavy metals or organic pollutants.

**Abbreviated chemical analysis.** These analyses can give a general outline of the chemical condition of the water, e.g. temperature of the water, appearance, pH value, electrical conductivity, oxidation reduction potential, "water hardness", iron and manganese, nitrogen compounds, chloride, sulphate, oxidizability.

**Extended chemical analysis.** Depending on the objective, additional analyses are performed on the same water, e.g. determining phosphate, silicic acid, calcium and magnesium, sodium and potassium, or organic substances.

**Comprehensive chemical analysis.** In addition to all quantitative measurements necessary to evaluate a water, this water analysis should also include tests for heavy metals, in particular toxic heavy metals, and where necessary quantitative measurements of these substances. Furthermore, tests should be conducted for indicators of organic contamination, such as phenols, surface-active agents, oil and grease-type substances, polycyclic aromatic hydrocarbons, halogenated hydrocarbons and for fuels or pesticides.

**Small-scale mineral water analysis.** In this type of analysis, the main components of a mineral water should be determined quantitatively and detailed bacteriological investigations should be conducted.

**Comprehensive mineral water analysis or medicinal water analysis.** In this analysis, all substances contained in the water should be determined, as far as they are qualitatively detectable and quantitatively recordable according to the latest advances in analytical chemistry. Tests should also be made to see whether anthropogenic substances, such as halocarbons, pesticides, etc., can be detected in the trace range. In addition, bacteriological investigations, tests for radioactivity and, where applicable, radiochemical analyses and gas analyses are necessary.

**Analyses of swimming pool water.** The analysis of swimming pool water should in particular provide information on the bacteriological, virological and hygienic-chemical quality of the water. Comparative investigations are therefore also necessary of the feed water on the one hand and the swimming pool water during use on the other. Tests to determine urea and $NH_4^+$ and/or $NO_3^-$, tests for worm eggs and microbiological analyses should be carried out in addition to the usual work.

**Industrial water analysis.** A distinction can be made between boiler feed water analysis, boiler water analysis and analysis of condensate. In addition to measuring pH, oxidation reduction potential and electrical conductivity, these three types of analysis also require the determination of traces of oxygen and carbon dioxide, but also quantitative analyses, e.g. for iron and manganese, copper, nitrogen compounds, "water hardness" and chloride, sulphate, phosphate and silicic acid.

**Analysis of the water to determine aggressiveness to metals and building materials.** In addition to local investigations with measurements of pH, electrical conductivity, oxidation reduction potential, oxygen and carbon dioxide, it is also necessary to determine the chloride content, the sulphate content, the proportion of magnesium and calcium, and, where necessary, metals. Organic acids, e.g. humic acids, should also be recorded.

**The analysis of a water as production water** must be geared to the requirements of the production plant.

**Analysis of the water for irrigation purposes.** The following are important: pH, electrical conductivity, oxidation reduction potential, sodium, potassium, calcium, magnesium, iron, manganese, chloride, sulphate, nitrogen compounds and boron, possibly also selenium and heavy metals, as well as pesticides (herbicides).

**Analysis of sewage.** In the routine analyses used to analyze sewage, processes to determine temperature, pH, settleable substances, chemical oxygen demand and biochemical oxygen demand are usual. More comprehensive analyses of sewage also record organic substances such as oil and grease-type substances, organic solvents, phenols, detergents, cyanides, heavy metals, pesticides or other pollutants which can in particular also inhibit chemical degradability. In addition to determining the 5-day biochemical oxygen demand (BOD), which in its relationship to chemical oxygen demand (COD) gives information on the biological degradability of organic substances contained in the water, toxicity tests are conducted in the sewage with bacteria, Daphniae or fish.

**Evaluation.** Following each water analysis, the expert should deliver a summarized evaluation of observations on site and of the results of analyses.

## 1.2 Local inspection and sampling of water

Information is given below on carrying out the **local inspection** and the **sampling of water.**

### 1.2.1 Sampling of water

General

The key points summarized below for water sampling are to be seen as suggestions designed to ensure that the information available to the evaluating body on the local conditions and the work conducted in sampling is as comprehensive as possible. The descriptions by the sampler should be so informative that a third party can gain a picture of the local conditions and the sampling point, and the conditions during the sampling process are reproducible.

**The following key points supplement each other and also partly overlap, so that if they are followed information can be gathered on different aspects.**

### 1.2.2 Local investigations

**Location of water resource** (description in relation to striking points in the terrain):

Coordinates (map 1 : 25 000)
Height amsl

**Location of sampling point** (relationship to the location of the water resource, description according to striking points in the surroundings):

Coordinates (map 1 : 25 000)
Height amsl

**Weather** on the day of sampling

Meteorological conditions, e.g. in the eight preceding days
Meteorological conditions in the last 4-week period

### 1.2.3 Geology and hydro-geology

a) Own observations
b) Information from existing documents

**Information on the local investigations on geology and hydro-geology**

Nature of the sampling points

Non-tapped cropping out of water at one particular point as a source or as an expanse of water as spring swamp. Tapped cropping out of water (adit, seepage line, drainage pipes) and wells (information on tapping, casing, sealing, top of well, etc.)

Sampling points in watercourses, lakes and pits with uncovered ground water (ground water pond)

**Topography**

Valley side, valley bottom, water bottom, shore area
Built-up area, forest, agriculturally used area, wasteland
Catchment area of the water resource: Description according to map
Size and extent: Area, depth, tributary waters, etc.
Borders between subterranean and above-surface catchment area

**Nature of the tributary waters**

Cropping out of water from cracks, caverns or pore spaces

Confined or non-confined ground water, moving or standing surface water

**Water quantity**

Water run-off from surface waters
Discharge from sources, adits, seepage lines and drainage pipes
Capacity of well pump rate (difference in ground water table before and after sampling)

**Hydro-geological conditions**

Nature and structure of the subsoil (types of rock, course of fissure zones)

Permeability of the subsoil (aquifers, underlying stratum of the ground water, water-bearing layers)

Cleansing capacity of the subsoil (chemical, physical and biological processes in the overlying strata and in the aquifer) and the surface waters (weedage, sediment load)

Flow direction and flow velocity of the ground water and the surface waters

Influencing the water resource by lowering or raising the ground water surface

Influence of anthropogenic sources of burden (building, industry, solid waste deposits, fertilizing, traffic, sewage disposal) on the water resource.

New formation of ground water by infiltration of water from precipitation and water from surface waters

**Ground water or water processes in hydro-geological context with gas resources:**

Influences due to pressure relief
Influencing due to creation of cones of depression

**Nature of the rocks**

Petrographic composition of the rocks (e.g. carbonate rocks, silicate rocks)

Degree of rock weathering (mechanical loosening of structure, chemical weathering, corrosion phenomena)

Changes to rock surfaces (separations, efflorescence, deposits, colonization with organisms)

Soil formation (nature and thickness of soil horizons, covering layers and plant growth)

### 1.2.4 Technical information on the water resource

Technical information on the development of the water resource

Technical information on tapping and casing (e.g. materials used such as metals, plastics, wood, stone, cement, transition filter sections, backfilling with gravel or sand, etc.)

Information on well head and sealing (type of construction)
Pump (suction pump, submerged-type pump, etc., pump material?)

**Water treatment** (plant present or not? Principle? - not present but necessary)

**Water utilization** (drinking water, swimming pool water, service water, e.g. for industry or for irrigation)

**Water disposal** (type and material, distances, slope, pumping station)

**Information on water run-off**, quantities of water produced and supplied

Pump tests (quantities of water delivered at different periods; where necessary, static water table and lowered water table at different pumping rates)

Other possibilities of water extraction in the surroundings:

a) from the same water resource (distance)
b) from other water resources (distance and geological conditions)

**Forms of water utilization in the surroundings** (e.g. springs, wells, waterworks, industry, irrigation, power plants, water-borne traffic)

### 1.2.5 Possibilities of environmental influences on the water resource

Sewage discharge locations in the surroundings (distances from the catchment area of the water resource to be investigated, distance from the sampling location)

Agriculture in the surroundings, where necessary distance and nature of agricultural utilization (seepage pits, dung heaps, fertilization - animal, artificial, plant protectives and herbicides, insecticides, etc.)

Surrounding industry, where necessary distance and type of industrial utilization

Roads in the surroundings

Canalization in the surroundings

Settlements in the surroundings (individual houses, individual farms, smaller settlements, larger villages or towns, other plants)

Rubbish tips/dumps, other waste tips

### 1.2.6 Hygienic considerations

Hygienic conditions surrounding the water resource

Hygienic conditions at the location of the water resource

Hygienic conditions at the sampling location
Hygienic conditions in the past in the region of the water resource
Have rehabilitation measures taken place in the past?
Are rehabilitation measures planned?
Has the water been disinfected, e.g. by chlorination, intermittent chlorination, filtering or other treatment?

Further additional information on the hygienic conditions in the region of the water resource, past and present.

## 1.3 Techniques of water sampling

### 1.3.1 General

In principle, water samples are to be taken such that

1. average samples are obtained
2. secondary contamination is excluded, and
3. the water samples taken, when analyzed in the laboratory, provide results which are representative of the water resource in question.

Any secondary, physico-chemical, chemical or biological change to the water should be prevented. If such a change cannot be prevented, the changeable substances should be determined analytically on site, or fixed so that they can be recorded in the laboratory. Information should accompany the sample for the analyst conducting investigations.

As far as possible, the sampler should be informed of the intended scope of investigation when taking samples to determine traces, so that he can use the appropriate bottles; otherwise instructions should be given on which bottles of which size he should fill with water on site, and for which investigations these are intended, and whether they should be fixed or preserved.

Water samples for determining phosphate may not be filled in bottles which have been rinsed with cleansing agents containing phosphate. The same applies to water samples to determine surface-active agents (detergents).

Preservation of samples may be necessary in individual cases. The best form of preservation of samples is a rapid investigation of the water sample after sampling. This should take place in the laboratory as far as possible no later than 2 days after sampling. During transport and until commencement of the investigation the water samples should be stored in cool conditions at around 4 °C.

The key points listed below are intended as suggestions and should be supplemented according to local conditions.

### 1.3.2 Bottle material

#### Glass bottles

Water samples are frequently taken in glass bottles. Glass bottles with ground-in stoppers of glass have proved particularly suitable. Glass bottles with cork stoppers, rubber stoppers or other forms of sealing are less suitable. The bottle glass should be of good hydrolytic quality. New bottles should be rinsed before use with diluted acid and subsequently washed out with distilled water (in order to exclude release of alkalis from new bottles to the water inside).

Glass bottles are particularly necessary for sampling if substances are to be determined which could be secondarily changed by plastics, or if changes in the concentration of substances contained in the water can occur by adsorption into plastics, e.g. certain heavy metals such as silver. Similar considerations naturally also apply to glass bottles. Here too, attention must be paid to wall adsorption. When determining traces of heavy metals, it is recommended that the water be acidulated directly after sampling, e.g. with hydrochloric acid or nitric acid, as acid solutions have less of a tendency to precipitate traces of heavy metals on the glass walls than neutral or alkaline solutions. The analyst should be informed that this has been done.

Attention should be paid to the danger of breakage in glass bottles. Gas-tight filling is necessary for gas chromatographic headspace analyses. Special bottles must be used with a lateral tube and a septum, or with an appropriate stopper with septum, so that gas samples can be taken from the gas space with an appropriate syringe without opening the bottles in the laboratory.

Water samples for determining oil and grease-type substances should be filled in glass bottles and special care taken that the glass bottles, together with ground surfaces and stoppers, have been rendered grease-free beforehand with the same solvent as will be used later to extract the water.

Glass vessels (5 to 10 litres) should be used for samples for special investigations, e.g. to determine polycyclic aromatic hydrocarbons, extractable substances, halogenated hydrocarbons, etc.

Glass bottles are also necessary for special investigations, e.g. to determine oxygen, and also usually for bacteriological analysis.

**Plastic bottles**

It is possible in many cases to use plastic bottles for taking water samples. These have the advantage of not breaking when they are completely filled, as can be the case with glass bottles after temperature changes. Care must be taken to use a plastic material which is as inert as possible with respect to the water used. Polyethylene bottles have proved suitable. Plastic bottles in the usual round bottle shape, but also plastic bottles of rectangular shape which can be particularly easily stacked, are common.

Plastic vessels should be preferred to glass bottles for determining silicate and borate, as possible release of these substances from the glass to the water can then be excluded.

Disposable plastic bottles can be useful when taking sewage samples. The cost of cleaning the used bottles is frequently higher than the price of new bottles.

Sterilizable plastic vessels, e.g. made of polytetrafluoroethylene (PTFE), can also be used for taking water samples for microbiological analysis.

**Labelling the bottles**

The bottles containing the water samples should be labelled so as not to be mistaken. Either tags or adhesive labels can be used. The following minimum information should be contained on the label:

Sampling location, designation of the water resource, sampling date and time.

The sample designation and the sampling documents can be coded on sampling for electronic data processing. Sampling records should also be prepared simultaneously on site. The general data on sampling should here be recorded in plain language, and all measured results of the local investigations of the water, of the quantities of water taken, of the method of filling, and, where necessary, fixing and sample preservation, should be noted.

### 1.3.3 Techniques of sampling

Air-free filling is necessary when taking certain water samples, e.g. for determining oxygen (see Instructions). The same applies to further special determining processes.

In general, the bottles should be prerinsed at least five times with the water to be sampled and then filled so that a gas bubble of approximately bean size remains in the bottle. Only in few exceptional circumstances is it necessary to fill the bottle completely, e.g. for determining radioactivity. In such cases, care should be taken to protect water in glass bottles from heat or, for example, to ensure that the water is able to expand via the stopper and a water-filled rubber tube, as otherwise the bottles will burst. In the case of short-lived radionuclides, determination is necessary on site.

When producing **composite samples** care should be taken to measure the water and to ensure that secondary contamination, and influences from increases in temperature or from atmospheric oxygen, are as far as possible excluded or taken into consideration in evaluation.

If water samples contain settleable substances, it may be appropriate according to task to separate the supernatant settled water from the sediment in order to produce composite samples, and to use the clarified water also to produce composite samples. If during transport to the laboratory precipitations resettle out of the composite sample, these should be homogenized or dissolved before analysis and included in the investigation, as they have only formed from out of the water at a later date. Settleable substances precipitated on site could be filled separately into bottles and examined separately in the laboratory. A note should be made of the settling time that elapsed and the temperature at which the settleable substances were precipitated.

Batch sampling

A basic distinction can be made between **batch sampling** (individual samples) and **continuous sampling** (time-proportional or flow-proportional).

Batch sampling, i.e. taking of individual samples (random samples), characterizes only the water flowing or contained in a vessel at the precise moment of sampling. The random sample is only appropriate if any changes in the composition of the water to be investigated can be expected to take place very slowly.

Equipment

For **taking water samples** from deeper water layers there are sampling devices which can be opened at prescribed water depths, thereby allowing the water to be taken from a certain level, e.g. sampling devices with evacuated glass ampoules. There are also sampling devices which are lowered in an open state to certain depths, with the water flowing through the sampling device as it is lowered. The devices are sealed at the desired depth by a falling weight and can bring the water collected at the relevant depth to the surface in the closed system. After sterilization, these devices can also be used for taking water samples for bacteriological analyses (e.g. Ruttner sampler).

Continuous sampling

Continuous sampling can be in the form of time-proportional or flow-proportional sampling.

Time-proportional sampling

Equal quantities of water are sampled and combined to form composite samples at certain intervals (e.g. every 15 minutes or at hourly intervals) within a longer period. This sample taking can be done by hand or automatically.

Flow-proportional sampling

Given that the quantity of water running off per time unit is known, individual samples of the water are taken at certain intervals by means of a flow

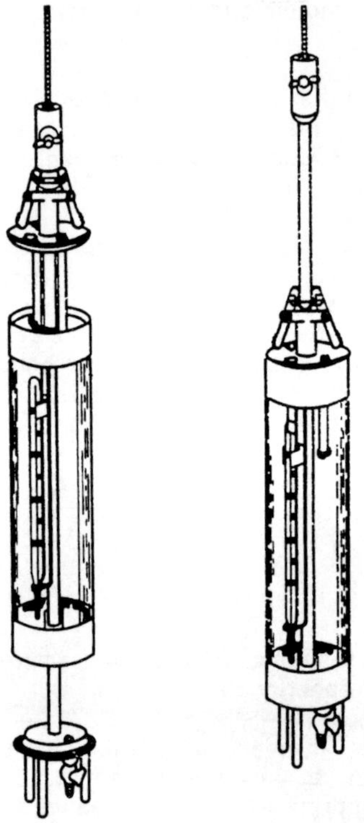

Figs. 1 and 2. Ruttner sampling bottle. The device cuts through the water with its caps open without intermixing with air. The device is closed automatically by actuation of a falling weight. Inside there is a thermometer graduated in 0.1 °C

diversion switch for a constant time. These individual samples are equal in quantity and are combined into composite samples.

In continuous sampling it is possible to determine the average concentration of substances contained in the water by mixing a fairly large number of individual samples.

**Sampling equipment**

For taking samples of water by hand, porcelain, plastic or stainless steel casseroles have proved effective for drawing the individual samples. Stainless steel buckets are very suitable for preparing composite samples. Graduated stainless steel buckets with a known capacity are used for discharge measurement with a stopwatch (10-litre, 15-litre, or 20-litre buckets). If samples are to be taken from surface water, suitable samplers with extension handles up to several metres long are necessary. Bamboo rods with a jaw grip for inserting the bottles have come into common use. In this case too, the bottles should be rinsed at least five times with the water to be analyzed.

For continuous sampling of water, e.g. from surface waters or from sewage pipes, there is a wide range of semi-automatic and automatic equipment.

The following schematic diagram (Steinecke-Dietrich) shows a reliable sampling device.

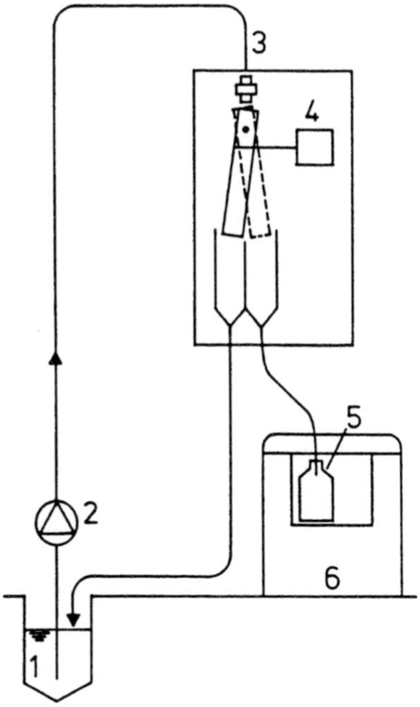


Fig. 3. 1) = Sampling point; 2) = MOHNO pump; 3) = Flow diversion switch; 4) = Drive unit; 5) = Collecting vessel; 6) = Refrigerated cabinet

This schematic drawing shows a unit for automatic sampling from a sewer duct.

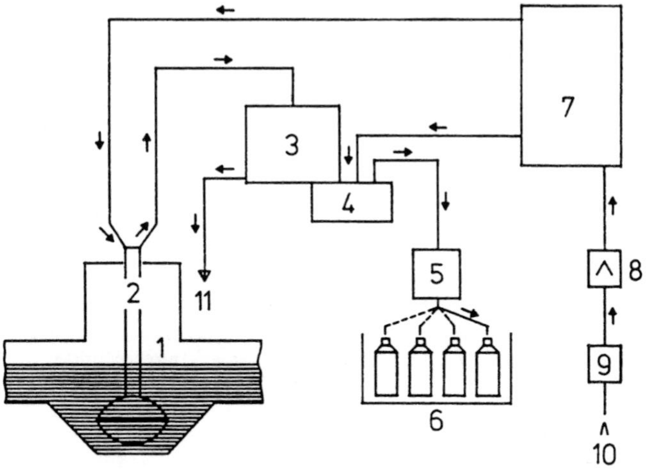


Fig. 4. 1) = Sewer duct; 2) = Sewage pump; 3) = Collecting vessel; 4) = Proportioning pump; 5) = Distribution unit; 6) = Bottle filler; 7) = Control unit; 8) = Pressure reducer; 9) = Air filter; 10) = Compressed air; 11) = To sewer

## 1.4 Water sampling in practice

### 1.4.1 Local investigations

In the local investigations as many measurements as possible must be taken, especially for variable parameters. The appearance of the water sample and any changes in it during the sampling must be described. Turbidity, colouring, escaping gases and any odour must be recorded. The odour must be described as accurately as possible, e.g. sourish, putrid, sewage-like, aromatic, fishy, etc. If there is no danger of epidemic the taste of the water should also be established on the spot, and attention should be drawn in particular to any off-taste, e.g. from hydrogen sulphide, organic compounds, dissolved metals such as iron and manganese or dissolved carbon dioxide. The tester must endeavour to record proper measurement results on site even under field and difficult conditions. Data concerning geology, situation of the localities, tapping and casing, ground water lowering, etc. are necessary. Biological indicators must be determined and flora and fauna recorded.

### 1.4.2 Sampling for microbiological analysis of water

An adequate number of sterile vessels, preferably glass vessels which can be sealed bacteria-tight, must be carried. When taking the samples all secondary microbiological contamination from the sampler himself or from the environment must be eliminated. In general 3 · 0.25 litre of water must be filled into sterile vessels for each sampling point. For special bacteriological investigations, e.g. for testing for Salmonellae, for sulphite-reducing Clostridia or hydrogen-sulphide-forming organisms and other micro-organisms, it may be necessary to take samples of several litres of water under sterile conditions (see Chapter 5).

Besides the data on local conditions and the local inspection, an analysis of drinking water should always give indications of hygiene problems and include a microbiological water analysis. Appropriate samples for bacteriological analysis must therefore be taken. To do this, the sampling point must be disinfected e.g. by flaming, and the water samples for the bacteriological analyses are to be taken in sterile vessels under sterile conditions after an appropriately long run-off time of about 5 minutes. If this is not possible under the prevailing conditions, drawn samples using a sterile sampler must be taken in sterile bottles.

If in exceptional cases no sterilized vessels are available, glass bottles can be sterilized with boiling water. However, in this case care must be taken to ensure that reinfection is avoided during cooling. As a precaution, several parallel samples must be taken under the most sterile conditions possible in the bottles sterilized in this way under field conditions and then analyzed separately. If the water to be tested contains disinfectants, e.g. chlorine, these must be eliminated by adding sterile sodium thiosulphate solution. Further details of bacteriological analyses will be found in Chapter 5.

### 1.4.3 Sampling for physico-chemical analyses

Samples for physico-chemical analysis will seldom present problems in drinking water sampling. Measurements of temperature, pH, electrical conductivity and oxidation reduction potential must be taken.

Attention must be paid here to changes caused by access or escape of gases. For example, oxidation of dissolved bivalent iron can occur through access of oxygen. The water will become turbid and the oxidation reduction potential will be shifted. The same applies to the measurement of electrical conductivity, when precipitation of e.g. calcium carbonate, iron hydroxide or silicic acid occurs through release of gas. The sampling procedure in question must be matched to the intended measurement procedures so that secondary changes in the water before measurement can be avoided, or can if necessary be reversed or taken into account in the calculations in the laboratory.

### 1.4.4 Sampling for chemical analyses (see also 1.3)

The choice of bottle material and bottle size to suit the intended purpose of the analysis is important. The concentration of any disinfectants such as chlorine, chlorine dioxide or ozone must be measured immediately after sampling.

Variable components e.g. $Fe^{2+}$, $H_2S$ or $HS^-$, phenols, cyanides, etc. must be fixed at the place of sampling. $O_2$ and $CO_2$ must be measured at the time of sampling.

For other measures see the special section on water preservation (1.7).

### 1.4.5 Water sampling for gas analyses

For these analyses a particularly complex sampling technique and a careful approach are required so that on the one hand dissolved gases or freely rising gases cannot escape, and on the other hand no secondary change in the gas composition can take place due to the air. Besides the glass

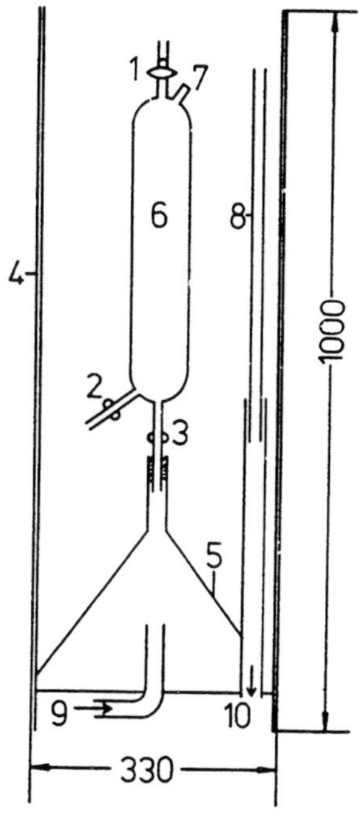


Fig. 5. Device for collecting the freely rising gases. 1, 2, 3) = gas-tight ground stopcocks; 4) = gas-water collecting vessel made of PVC; 5) = gas collecting funnel with lead rings for weight; 6) = gas collecting flask; 7) = lateral tube at top with silicone membrane; 8) = tele-scopic overflow pipe for the water; 9) = gas-water inlet; 10) = water drainage (from Fresenius, Fresenius, Schneider)

sampling and gas collection device with lateral tube which has already been discussed, glass bottles with a tube and septum in the stopper have also proved effective. In this case, it is a simple matter to take gas samples using a syringe for gas chromatographic analysis of the so-called free gases. The amount of these gases can also be determined e.g. by filling with a known quantity of water. In addition the dissolved gases in a second sample are driven off, e.g. under vacuum at increased temperature (50 - 80 °C), measured and analyzed.

For extracting gases from gas-retaining water samples see the authors' tried and tested device (Fig. 5).

### 1.4.6 Radioactive inert gases

Discontinuous sampling for measuring radioactivity due to the radioactive inert gas "radon 222" can be carried out relatively easily following the instructions below.

Some special points arise from the nature of the radioactive parent substance radium and the daughter radon, which is an alpha-emitting inert gas. Precautions must be taken when sampling to prevent the gas escaping from the water prior to measurement and, on the other hand, to prevent its enrichment by any change in the composition of the gas above the water.

In addition, consideration must be given to the fact that the decay products of radon are solids which are themselves again subject to radioactive decay. The vessels with radioactive water samples are therefore contaminated to a greater or lesser extent, so that before they are used again the radioactivity should have subsided and the vessels must be thoroughly cleaned. If the water contains dissolved radium 226, with the result that the radon 222 activity is constantly being replenished, precautionary measures are also necessary here since the radium 226 is easily separated from the water with precipitates (e.g. by adsorption on iron hydroxide as a non-isotopic carrier). Consequently any insoluble residues present must also be taken into account. In the case of basin or pond tapping of the water resource an increase in the concentration of radium 226 in the sinter of the source can occur.

In view of the relatively short half-life of radon 222 (approx. 3.8 days) the water sample must go for radioactivity measurement as quickly as possible. The time which elapses between sampling and the first measurement must be noted.

For determining what is known as the residual activity, water samples are filled into bottles which can be closed airtight, so that they can be analyzed for their radon content in the laboratory at a later stage, possibly after radioactive equilibrium has been established. For this, 1-litre or 2-litre glass bottles with double-bore rubber stoppers are used.

Through the holes, two glass tubes bent at right angles outside the vessel are inserted, one of these glass tubes ending immediately under the rubber stopper and the second reaching down to just above the base of the bottle (cf. illustration). After air-free filling the open outer ends of the two glass tubes are closed off with a rubber tube which is completely filled with water. Water samples can be taken from these bottles for measuring the radon 222 with a suitable device, e.g. a fontactoscope or an emanometer.

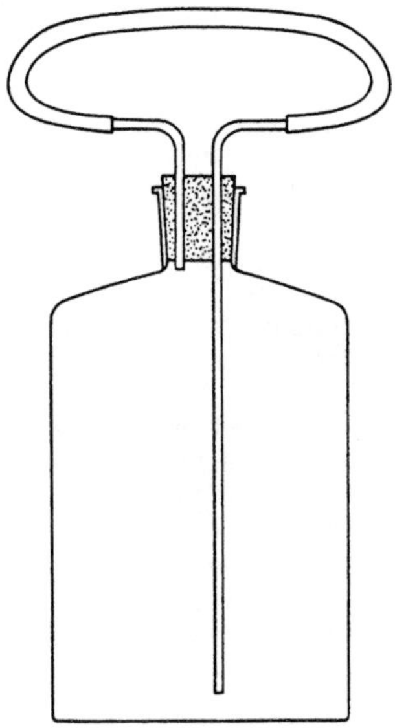

Fig. 6. Sampling bottle for determining residual activity

## 1.5 Special instructions for sampling water of various origins and for various investigation purposes

### 1.5.1 Surface water

Local measurements should be conducted, e.g. on flow velocity and discharge rate. Details are necessary on transverse profile, longitudinal profile and sampling depths. Where necessary, the sampler must be given instructions on the conditions for taking surface water samples, e.g. sampling from travelling surge, observation of stratification, production of composite samples, distance of the sampling location from the bank, sampling depth and similar instructions. The sampler should describe the local conditions, in particular the appearance of the surface water, and where possible give information on microflora and microfauna and also on macroorganisms. Bio-indicators provide important information on water quality. Assimilation and dissimilation processes in plants play their part, e.g. in determining the oxygen and carbon dioxide contained in a water. Investigations should be conducted into the discharge of sewage. Indications suggesting influences on or eutrophication of the water should be noted on site. Water samples should be taken for bacteriological, microscopic and chemical analyses. Local measurements should be taken of temperature, pH, electrical conductivity, oxidation reduction potential, oxygen and carbon dioxide ($CO_2$) content, turbidity, colouring, settleable substances, etc.

### 1.5.2 Ground water

Ground water samples should nearly always be taken from appropriate wells or boreholes; uncovered ground water will only exist in rare cases. According to the scope of investigation planned, samples should also be taken here for bacteriological, physico-chemical and chemical analyses. The taking of water samples may be necessary for trace analysis and for examinations for special indicators of pollution, such as cyanides, phenols or other organic substances. It can also be necessary to take samples for microscopic analyses. In addition to the usual local investigations and measurements, information is also necessary on whether the water samples were taken during a pump test and on the pumping rate and the depression at which the samples were taken in comparison to the static water table. The record must show whether the sample was taken from moving ground water or static ground water. Where possible, information should be provided on the flow velocity of the ground water and in general on the hydro-geological conditions of the ground water. The possible sampling locations for taking ground water samples may be classified roughly as follows:

1. Uncovered ground water
2. Springs, slope springs, boundary springs, etc.
3. Prospecting holes
4. Artesian wells (confined-water wells)
5. Shallow wells (dug wells, basin wells, driven wells)
6. Deep wells (drilled wells)
7. Horizontal wells (adits)

### 1.5.3 Salt water

This heading covers the sampling and investigation of seawater, lake water, brackish water, water from salt lakes, mines or mineral water. The general principles for taking water samples apply here to the bacteriological, physico-chemical and chemical analyses. It may be necessary on account of the high salt content to take larger water samples for certain trace analyses or for special investigations, which can then be examined in the laboratory after depletion of the greater part of the dissolved salts. Care should be taken that the bottles are rinsed with distilled water on the outside and particularly at the stopper after sealing, as otherwise salts may dry at the bottle neck and falsify the results of subsequent investigations. Before rinsing the bottle neck, the ground-in stopper should be checked for correct seating so that distilled water does not enter the bottle and dilute the water sample. Water samples should also be taken for microscopic analyses. On-site investigations are necessary for algae or other bio-indicators, and indications of eutrophication should be recorded. Secondary observations, such as water-borne traffic or bathing activity, sewage discharge, fish stocks, lagoon formation, should be noted.

### 1.5.4 Bathing water

The instructions described in the section on surface water apply in principle to bathing water as well. In addition, bathing water investigations will be necessary in swimming pools (open-air or indoor), and also in bath-tubs. A distinction must be made between public baths and baths only serving special purposes, e.g. cleansing baths in tub form or spas. A distinction must also be made between salt water and fresh water baths, and between treated bathing water and natural bathing water in open bathing waters.

In addition to the number of visitors, frequency of visitors and overall appearance of the swimming baths, the local inspections and investigations should also include hygienic inspections, e.g. so-called klatsch preparations. Details of water temperature, water appearance, weather conditions and date and time of sampling should be recorded. In the case of disinfected bathing water, the chlorine content and/or the ozone content and oxidation reduction potential should be measured together with the pH. In addition to evaluating the apparent quality and the hygienic conditions in and around the baths, water samples should also be taken for bacteriological water investigations, including testing for worm eggs. Samples should also be taken for the usual physico-chemical and chemical analyses and for the analysis of specific indicators of pollution, e.g. urea, nitrogen compounds or overall organic parameters such as chemical oxygen demand or 5-day biochemical oxygen demand. Water samples should also be taken of the fresh or uncontaminated feed water at the same time as sampling the used bathing water from the bath.

The sampler should be given instructions on the nature of the investigation required so that he can adapt his work on site to suit requirements.

### 1.5.5 Industrial water

A whole range of different aspects have to be observed when taking samples of industrial water, depending on the application. The following possible uses exist:

Boiler water
Boiler feed water (high-pressure or low-pressure boiler water)
Condensate
Industrial process water
Process water for factories in the foodstuffs industry
Process water for breweries
Process water for mineral water and lemonade plants
Cooling water and similar fields of application
Water for mixing concrete
Irrigation water in agriculture
Expanses of water for fish cultivation

Ground water or surface water must be checked for corrosiveness in relation to concrete and/or metals. Bacteriological investigations are necessary in the case of process water for foodstuffs factories or for the drinks manufacturing industry, and appropriate samples should therefore be taken. For the other fields of utilization, bacteriological investigations on service water will be less frequent. Here, however, attention must be paid to the fact that questions of aggressiveness or the formation of precipitations can frequently have microbiological causes. The local investigations should be conducted according to the usual sampling procedure; in particular, it is necessary to determine pH and electrical conductivity, and also the presence of oxygen or carbon dioxide. Special samples for testing corrosiveness in relation to metals or concrete should be mixed on site, or tests conducted using the pH-difference method. In addition to taking water samples for the usual physico-chemical or chemical analyses, sampling may also be necessary for microscopic investigations, e.g. when the water is used as cooling water.

### 1.5.6 Irrigation

Specific measured values on site are specially important for the use of a water resource for agriculture as irrigation water, e.g. determining electrical conductivity at the existing water temperature. On-site measurements of the hydrogen carbonate and carbonate content are also necessary. Furthermore, water samples should be taken for quantitative determination of the sodium, calcium, magnesium and boron content.

The values for electrical conductivity measured on site can provide information on the possibility of utilizing water as irrigation water for agriculture. Thus, in the case of electrical conductivities of up to 250 µS $cm^{-1}$ at 20 °C, no salinization of the soil need be feared. Such water is generally suitable. In the case of electrical conductivities of up to 750 µS $cm^{-1}$ at 20 °C, the water can as a rule still be tolerated.

Electrical conductivities of 750 to 2250 µS $cm^{-1}$, by contrast, are high enough to have a salinizing effect on the soil with continuous use, and salt-tolerant plants should be used for cultivation. A high degree of salination is to be expected from waters with electrical conductivities of up to 5000 µS $cm^{-1}$. Even higher conductivities usually render the water unusable even for salt-tolerant plants. An evaluation of the water quality for irrigation is also possible on the basis of the ratio of sodium to calcium + magnesium. In this connection see the diagram below.

In field 1 of the diagram, the water can be deemed suitable for general use; in field 2, only a small danger of alkalization exists. In field 3, the high proportion of sodium entails a danger of alkalization for most soils. This danger is reduced in gypsiferous soils; they must, where necessary, be fertilized with calcium and organic fertilizer. In field 4 of the diagram, the water is in general no longer suitable for irrigation purposes or only suitable in conjunction with the simultaneous use of calcium compounds.

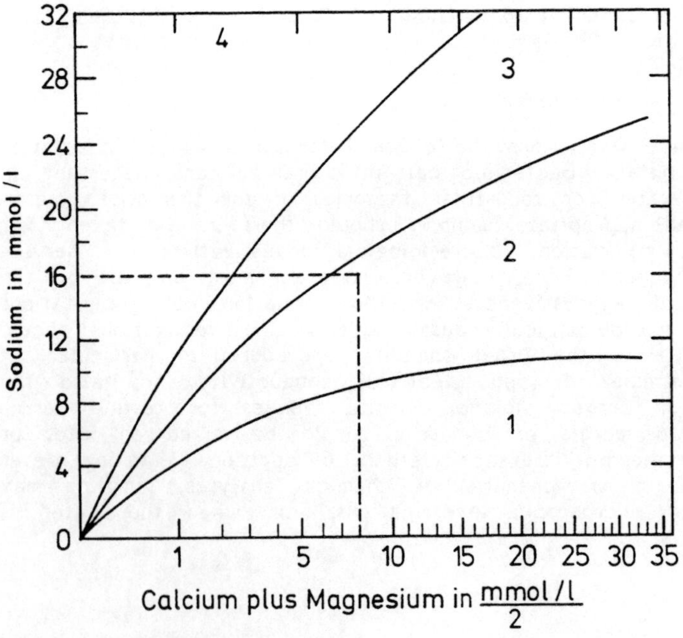


Fig. 7. Evaluation of water quality according to sodium content; 1) = Na low; 2) = Na avarage; 3) = Na high; 4) = Na very high

A certain estimation of suitability can also be made by examining the ratio of carbon dioxide compounds to alkaline earths. The ratio in mmol/l for the following compounds should be calculated according to the formula given:

$$(CO_3^{2-} + HCO_3^-) - (Ca^{2+} + Mg^{2+})$$

If these values are smaller than 1.25 mmol/l the water is probably suitable; between 1.25 and 2.50 mmol/l it is conditionally suitable; and over 2.50 mmol/l it is not suitable. See the section on boron regarding the importance of the boron content for irrigation water.

### 1.5.7 Expanses of water devoted to fish cultivation

In fish cultivation waters, which must generally be light and clear because the majority of fish species would otherwise migrate, the oxygen content plays a particularly important role. The oxygen demand by individual fish species varies (Salmoniformes: at least 5 mgl/l; carp: approx. 3.5 mgl/l; goldfish: 0.75 mgl/l). It should be noted that the oxygen content falls when the water is heated. A lack of oxygen can thus arise for some fish species at higher water temperatures in summer.

In addition, the pH is of importance. The compatibilities of individual fish species also vary in this respect. In general, expanses of water devoted to fish cultivation should have a pH ranging between six and eight. Fish cultivation waters should further be free of substances toxic to fish (e.g. heavy metals and cyanides) from sewage. Hydrogen sulphide is a strong poison for fish; processes of putrefaction must therefore be avoided in expanses of water devoted to fish cultivation. Local measurements should be taken and samples prepared for the laboratory in glass bottles; the analysis should be effected rapidly in such cases.

### 1.5.8 Waste water

Examples of water locations:

Domestic waste water

Industrial waste water

Waste water from critical locations, e.g. hospitals, chemical factories producing toxic materials, metal foundries and galvanization plants, etc.

Waste water with inflammable organic solvents, oils and greases, phenols or chlorine compounds

Seepage water from solid waste (rubbish dumps, etc.)

Clarified waste water from sewage plants

Waste water from surface channels

As a basic rule, **flow-proportional sampling** is recommended; flow-proportional composite samples can be produced from the individual samples taken, where necessary, with automatic sampling devices. In the case of **time-proportional sampling**, no account is taken of the quantity of water.

Nevertheless, this must be regarded as a better solution than a single or multiple random sample. Production of time-proportional composite samples is also possible, using automatic sampling devices where applicable. The composite samples can be taken with or without settleable substances. If the water contains many settleable substances, it is more favourable to remove these substances before producing the composite samples and to investigate them separately in the laboratory. After removing the settleable substances, the water from the composite samples should be used for laboratory tests. If secondary turbid substances have formed in the water, these should be included in analyses.

When taking the individual samples, which may for example be made at 15-minute or hourly intervals, it is useful to measure the temperature and pH of the waste water and to describe its appearance in each case.

For example, 3-hour or 4-hour composite samples or even daily composite samples can also be produced. The composite samples should be cooled (approx. 4 °C) so that secondary changes are avoided as far as possible.

The settleable substances should be determined immediately after sampling or after producing a composite sample, so that secondary precipitations during transport are not included in the record as settleable substances. It is also useful to measure the following physico-chemical parameters of individual samples on site:

Temperature
pH
Electrical conductivity
Oxidation reduction potential
Oxygen content

Waste water samples for determining hydrogen sulphide or its compounds should be derived from the individual samples, and zinc acetate or cadmium acetate added.

It is important to preserve samples of waste water, particularly for determining cyanide, phenol, organic compounds and nitrogen compounds, etc. (e.g. with chloroform, by alkalization or with mercuric chloride) (see 1.7).

Samples for determining the 5-day biochemical oxygen demand (BOD) or the chemical oxygen demand (COD) should be frozen if not examined immediately.

For special purposes, it may be necessary to test for certain toxic substances which can interfere with biological clarification. In addition to composite samples, random samples should then also be tested in suspicious circumstances. In particular, the possibility of cyanides, surges of acid or alkaline waste water, and also organic substances and heavy metals should be borne in mind here.

In the case of waste water discharge locations, it is useful to have water samples for conducting a fish test or a bacteria test for toxicity (see special section on fish test and bacteria test in Chapter 5).

Attention should be given during waste water sampling to the presence of organic solvents which enter the vapour phase and lead to explosions. Where necessary, water samples should be taken for gas chromatographic headspace analysis.

Safety precautions

The necessary safety precautions for the sampler should be emphasized. Securing equipment is always necessary when entering channels, shafts, wells, galleries, etc. A check should be made beforehand for toxic or ignitable gases. The sampler should be secured to a safety rope and backed up by a second person. Safety regulations should always be observed in all cases of water sampling.

## 1.5.9 Sludge

It may be necessary, in addition to sludge-containing water, to take samples of the sludge itself.

Examples

Waste water sludge
Sewage sludge
Digested sludge
Neutralization sludge and other industrial sludge
Sludge from surface waters
Sediments from surface water
Sludge and deposits from wells
Deposits from pipe systems

In this form of sampling, particular care should be taken to homogenize the cross-section taken, as sludge can frequently be of very heterogeneous composition. If this is the case, many individual samples (10 to 100) should be taken, mixed and the test sample taken from this homogenized raw sample.

Sufficient quantities of sludge should be taken so that a separation of sludge and water for separate investigation in the laboratory is possible. The sludge taken can be investigated immediately. Extracts with rain water or with seepage water containing carbon dioxide are also possible, however, and these allow one to determine which critical substances can be remobilized by water.

If it is necessary to determine the water content of the sludge, the samples must be packed in watertight vessels, preventing loss of water, so that the water content can be determined in the laboratory.

Special samples may be necessary for bacteriological and microscopic investigations, in addition to local measurements, such as pH (insertion electrode), temperature, appearance and observations on gas formation and odour. In general, sample quantities of sludge of the order of at least several hundred grams are recommended, and in special cases up to several kilograms. Digestion tests can be conducted on the samples taken, whereby precautions should be taken against hydrogen sulphide release or gas formation (methane or carbon dioxide). Where necessary, the appropriate gas analyses are to be made. In order to be able to judge the degree of disinfection of sludge, in particular sewage sludge, in conjunction with solid waste after the rotting process, special bacteriological investigations are necessary. For this purpose, the samples must be taken in sterile wide-necked glass bottles.

When taking samples of sludge, it is of equally decisive importance that the sampler carefully records all local conditions and observations to prevent the chemist or biologist who is conducting and evaluating the analysis from drawing false conclusions.

### 1.6 Local investigations (see also 1.7)

It is recommended that appropriate forms be issued depending on the type of individual analysis, as guidance not only for the laboratory but also for the sampler.

A summary of the local investigations generally to be conducted is given below. In the case of individual elements, fixation of the constituent to be determined is necessary and appropriate reference is made to the section of investigation.

Temperature of the water (to 0.1 °C with a calibrated mercury thermometer)

Temperature of the air

Barometric pressure

Height amsl

Appearance

Colour

Turbidity

Depth of visibility

Settleable substances in ml/l

Gases released

Odour

Taste

Measure pH electrometrically on site

Measure electrical conductivity electrometrically on site at the same time as temperature

Measure redox potential electrometrically excluding atmospheric oxygen and secondary redox processes

Measure $Cl_2$ on site

Measure $ClO_2$ on site

Measure $O_3$ on site

Density, (fill pycnometers without gas losses on site)

Measure oxygen on site electrometrically or chemically according to Winkler method

Determine carbon dioxide on site volumetrically, or fix the total carbon dioxide in a water sample by means of CaO, $Ba(OH)_2$ or NaOH and determine the total $CO_2$ in the laboratory, and distribute the hydrogen carbonate and/or carbonate components (titration on site) by recalculation in terms of free, dissolved $CO_2$.

Hydrogen sulphide: fix on site with zinc acetate or cadmium acetate

Bivalent iron: fix on site with 2,2'-bipyridyl

Silicic acid: hydrolysis and decomposition on site by addition of the water to a mixture of hydrofluoric acid and perchloric acid; determine photometrically in the laboratory; insoluble or colloidal silicates should first be removed from the water by filtration.

Aggressive carbon dioxide:
Either measure pH on site immediately and after saturation with lime, or determine in the laboratory (marble dissolution test according to Heyer method) after addition of calcium carbonate

Short-lived radionuclides

Bio-indicators

### 1.6.1 Preservation of water samples

General

The best method of preserving the water samples taken is to provide for analysis as soon as possible. The water samples taken should be stored in the dark and cooled to around 4 °C (cooling box with ice or refrigerator in the vehicle) for transport to the analytical laboratory.

Attempts should be made to ensure that samples taken arrive at the laboratory for analysis on the following day, or at the latest the day after that. If immediate analysis of the water samples in the analytical laboratory is not possible, these should be kept in cold storage at around 4 °C until analysis, which must take place as soon as possible. Decisions should be made from a specialist point of view on which analyses should be conducted immediately and which can be left for a later date.

Cooling or preservation of samples is particularly important when taking samples of waste water. Attention should be paid to microbiological reactions after sampling, which should be kept to a minimum by cooling the samples. Thus sulphate reduction with formation of hydrogen sulphide can be caused by desulphurizing vibrios, with consequent triggering of other secondary processes, e.g. iron precipitation with accompanying precipitation of other heavy metals. Organic substances may degrade microbiologically during transport thereby changing the COD or the 5-day BOD.

Some general information is included below for selected inorganic or organic substances contained in the water which experience has shown to be easily changeable. Further information is to be found under the individual descriptions of methods.

### 1.6.2 Inorganic substances contained in the water

Dissolved, bivalent iron: Fixing directly at the time of sampling with 2,2'-bipyridyl forming a red-coloured complex which can be analyzed photometrically in the laboratory.

Silicic acid ($SiO_2$): Decomplexing or depolymerization directly after sampling by adding the water sample to a digestion mix of perchloric and hydrofluoric acid. In order to be able to distinguish between dissolved and undissolved silicic acid, it is recommended that one sample be added directly to the acid mix and a second water sample after filtration, for example through a membrane filter. If the perchloric acid/hydrofluoric acid mix prepared in the plastic bottle has been weighed, any desired quantity of water can be added on site and determined by measuring the weight difference in the analytical laboratory.

Hydrogen sulphide $H_2S$, hydrogensulphide ($HS^-$) or sulphide ($S^{2-}$): To determine these, the water sample is added to a prepared solution containing zinc ions as excess zinc acetate. Here too, the quantity of the water can be determined by measuring the difference in weight.

Nitrogen compounds, such as ammonium, nitrite and nitrate, or samples to determine so-called organically fixed nitrogen: Transport under cool conditions for analysis as soon as possible. It is also possible to add approximately 1 ml chloroform per litre of water to prevent microbiological reactions. Analysis is to be recommended as soon as possible, at the latest after cooled transport of 1 to 2 days, particularly when determining nitrite content.

Heavy metals: For use in analysis according to the flame atom absorption method, the sample should be acidified using hydrochloric acid of reagent purity immediately after being drawn, and the samples should be taken in glass bottles and transported under refrigerated conditions. The quantity of hydrochloric acid should be noted for a blank experiment.

For determining heavy metals using the so-called furnace technique (graphite tube cuvette), acidification of the water sample with nitric acid is to be recommended, and here too provision should be made for a blank experiment. If heavy metals are to be expected in the trace range, the water samples should be taken with acidification in a polytetrafluoroethylene bottle (Teflon bottle).

To record metal traces e.g. for arsenic and selenium using the so-called hydride process, acidification with hydrochloric acid is recommended; the rest of the procedure is similar.

To determine the quantity of mercury, the water sample is stabilized in glass bottles with a solution of nitric acid and potassium dichromate until analysis.

Settleable substances in terms of volume are determined directly after sampling in a sedimentation funnel. If settleable substances are to be recorded in terms of weight, the water should be filtered via a membrane filter directly after sampling or after a sedimentation period of, for example, 2 hours, and the settleable substances can then be determined in the laboratory.

### 1.6.3 Organic substances contained in the water

Dissolved organic carbon (DOC) or total organic carbon (TOC):

a) Transport in refrigerated condition, or
b) deep-frozen at -15 to -20 °C

Chemical oxygen demand (COD):

a) Transport in refrigerated condition, or
b) deep-frozen at -15 to -20 °C

Acidification can also be effected with sulphuric acid directly after sampling, and before deep-freezing.

5-day biochemical oxygen demand (BOD):

a) Transport in refrigerated condition, or
b) deep-frozen at -15 to -20 °C.

Phenols: Preservation by adding 5 ml 35 % hydrochloric acid and 1 g copper sulphate per litre of water. Transport in refrigerated condition.

Surface-active agents (detergents):

a) Transport in refrigerated condition, or
b) deep-frozen at -15 to -20 °C.

Chlorinated hydrocarbons: Transport in refrigerated condition in specially cleansed, gas-tight sealed glass bottles.

Hydrocarbons: Transport in refrigerated condition in specially cleansed glass bottles.

Polycyclic aromatic hydrocarbons: Transport in refrigerated condition in specially cleansed glass bottles.

Cyanides: Water samples to determine cyanide must be kept chilled and dark for transport. They should be alkalized with sodium hydroxide solution (1 molar) on site and adjusted to a pH of around 8.
1 ml chloroform and 1 ml of a 10 % solution of hydrochloric acid and stannous chloride ($SnCl_2$) are added and where necessary the pH readjusted to 8 with sodium hydroxide solution using pH paper. After adding 10 ml solution of zinc/cadmium sulphate (10 %) per litre of water, the sample is poured into glass bottles for despatch.

### 1.6.4 Radioactivity

Measuring short-lived radionuclides, e.g. radon 222 which has a half-life of 3.8 days, should take place directly at the sampling location. Samples to be transported to the laboratory for measurement must be taken free of air bubbles in glass bottles, with compensation for pressure effected by means of a hose filled with water led through a double-bored stopper (cf. 1.4.6).

Date of sampling, time of sampling and appearance of the water samples should also be recorded on the label for measurements of individual radionuclides in the laboratory. This is particularly necessary for short-lived radionuclides in order to make it possible to recalculate activity back to the time of sampling.

## 1.7 Local analyses (see also 1.6)

### 1.7.1 Sensory examination on site

The **odour test** should be conducted before the taste test, as the sense of smell is more sensitive and also has a considerable influence on taste. Non-smokers are to be preferred as testers. Plastic bottles are useless for testing because of the disturbing influence of their own odour. Wide-necked flasks made of glass are most suitable; they should be filled approximately two thirds full of the water and shaken, and, if the water at ambient temperature has no odour, heated to approx. 40 - 60 °C and retested.

The odour should be characterized as accurately as possible both with respect to intensity and nature. The following five stages are commonly used to define the **intensity of the odour**: "Very weak" (only discernible by an experienced expert); "Weak" (also discernible by the layman when compared with reference sample); "Distinct" (discernible for any water consumer); "Strong" (unpleasant); "Very strong" (repulsive).

The following designations are used for the **nature of the odour**:

**Odours of general nature**

"Metallic" (e.g. iron-containing deep water)
"Earthy" (occasionally caused by Schizophyceae)
"Dankly mouldy" (e.g. stale water)
"Peaty" (marsh water)
"Musty-putrid" (impure water)
"Sewage-like" (considerably contaminated water or waste water)
"Faecal" (considerably contaminated water and/or waste water)
"Fishy" (e.g. caused by Bacillariophyta)
"Smelling of fish oil" (e.g. caused by Flagellata)
"Seaweed-like" (e.g. caused by Flagellata or algae)

**Odours of chemical nature**

| | | |
|---|---|---|
| e.g. smelling of: | Hydrogen sulphide; | Tarry substances; |
| | Mineral oils; | Phenols; |
| | Chlorophenols; | Chlorine; |
| | Ammonia; | Soap; |
| | Acetic acid; | Butyric acid. |

**Hydrogen sulphide and free chlorine** can mask other odours. Dechlorination can be effected for example by adding stoichiometric quantities of sodium thiosulphate. If hydrogen sulphide is present, sufficient zinc acetate is added to the water sample (formation of barely soluble zinc sulphide) and the sample retested.

## Taste test

A taste test on site for orientation purposes may only be carried out with extreme care. In the case of "repulsive" odour, the tester should refrain from this test. He should also refrain from this test if he suspects contamination with bacteria.

If no distinct taste is noticeable when tasting the water at normal temperature (a metallic taste usually occurs as after-taste), it should be heated to approximately 30 - 40 °C in a covered glass beaker before tasting.

The following distinction can be made according to the nature and strength of the taste:

1. No particular taste
2. Insipid
3. Soft
4. Mineral
5. Hard
6. Sour
7. Bitter
8. Salty
9. Metallic-astringent
10. Alkaline
11. Musty
12. Tasting of $H_2S$ or sulphides
13. Tasting of organic compounds, e.g. mineral oils, pharmaceuticals, fuels, phenols, etc.

The intensity of the taste can be characterized by adding "weak", "distinct" or "strong".

## Odour threshold value

Determining the odour threshold value for a water (or waste water) is always "subjective". It is nevertheless by no means worthless.

The odour threshold value is defined as the degree of dilution obtained in the water to be examined, using absolutely odourless water produced by filtration through activated carbon.

Formula

$$G = \frac{U + V}{U}$$

where:

G = the non-dimensional figure for the odour threshold value (dilution factor)
V = the volume of the diluting water (ml) and
U = the volume of the water to be investigated (ml) in the mixture in which an odour is only just discernible.

An odour-free glass flask of diluting water serves for comparison. The determining process can be carried out with water at 20 °C or 60 °C.

A pretest determines initially the approximate range of the odour threshold value, by filling 200 ml, 20 ml, 2 ml and 0.2 ml of the sample each into a test flask and topping up the three latter measures to 200 ml. Beginning with the diluting water, the flask is shaken, the stopper opened, the contents sniffed and the flask immediately resealed. The analogous process is applied to the remaining samples.

When the range of perception has been narrowed in this manner, samples are produced in which the dilution at which an odour was only just discernible in the pretest is diluted still further. Commencing with the most diluted sample, the samples are tested as far as that mixture in which an odour is still discernible with any certainty.

### 1.7.2 Colouring

Testing for colouring must be carried out on site as the water may change while the samples are being transported. Colouring gives information on possible contamination. Humic substances cause a yellow to brown colouring (marsh water). The yellowish colouring caused by iron content changes to a yellowish-brown turbidity when standing open to the air. The intrinsic colour of the water can be neglected, as pure natural water in thin layers is practically colourless.

Colorimeter cylinders, as far as possible with a plane-parallel base, are used for field determination. In routine practice, however, water bottles made of transparent light glass also fulfil the same purpose.

The colorimeter cylinder or transparent bottle is filled with the water sample after removing undissolved substances by sedimentation or filtration. The sample is then observed in diffused light against a white background. The sample is classified as

"colourless"
"very slightly coloured"
"slightly coloured"
"coloured"
"strongly coloured".

and the following colour shade is added

"yellowish"
"yellowish-brown"
"brownish"
"yellowish-green"

If a more accurate designation of the colouring is required, it is necessary to proceed with reference solutions for which appropriate series of dilutions must be produced. Colorimeter cylinders (e.g. 2.5 cm inside diameter and marked at 50 ml) made of colourless glass and with a plane-parallel base are used.

Waters with yellowish to brownish colouring are used with standard solutions of potassium hexachloroplatinate and cobalt (II) chloride (**platinum cobalt chloride reagent**):

Standard solution (1 ml = 0.5 mg Pt):

1.246 g $K_2PtCl_6$ of reagent purity and 1.00 g $CoCl_2 \cdot 6\ H_2O$ of reagent purity are dissolved in distilled water, containing 100 ml HCl of reagent purity, (1.19 g/ml) per litre, and topped up to 1 litre.

For example, a series with 0.5 - 1.0 - 1.5 to 10 ml of the standard solution, each topped up with distilled water to 50 ml, is prepared for visual colour comparison. In this series, the colour of the sample is classified in the colorimeter tube from above against a white background with diffused daylight. In cases of higher colour intensity, the samples are diluted appropriately. If the colour intensities lie below 1 mg Pt/l, for example, colorimeter tubes with 200 ml content are used and dilution series prepared with 0.1 - 1 mg Pt/l.

The results are given in mg/l Pt/Co. Any pretreatment of the sample (e.g. filtration) must be indicated.

Determination of the colouring can also be carried out with the aid of a **colorimeter** or photometer. The standard solutions should be prepared analogously and the measurements taken at 420 nm (violet filter).

### 1.7.3 Transparency and turbidity

Testing waters for transparency and turbidity is done for orientation purposes. The transparency depends on the colour and any turbidity of the water. The turbidity is caused by suspended or colloidally dissolved inorganic and/or organic substances. Apart from sludge particles, silicic acid, ferric and aluminium hydroxide, organic colloids, bacteria and plankton are possible. The determining process should be carried out immediately after sampling, and certainly not more than 24 hours afterwards at the latest.

Transparency

a) After taking a sample in a colorimeter cylinder with plane-parallel base disc (2.5 cm inside diameter and cm gradation), the transparency is determined by ascertaining the readability of a standard script (black letters of 3.5 mm height and 0.5 mm stroke thickness), viewing vertically in diffused daylight, and expressing this in cm (average value from several "readings").

b) The scale for the depth of visibility is that depth of water at which a white Secchi disc with an edge length or diameter of 20 cm, fixed to a chain or a rod, is just discernible when immersed in the water. Up to 1 m, the values are given in cm; over 1 m depth, the values are given in 0.1 m gradations rounded down.

For the simple visual test for **turbidity**, a colourless clear glass bottle of 1 litre is filled two thirds full with the water sample, which is shaken well and observed against a black and then a white background (possible Tyndall effect). The following are examples of the distinctions made:

"clear"
"opalescent"
"weakly turbid"
"strongly turbid"
"opaque".

A measurement can also be made by comparing the turbidity of the water with that of a series of dilutions of a **kieselguhr standard suspension** (1.00 g and 0.10 g $SiO_2$/l), which can be evaluated both visually and photometrically (nephelometrically) (scale in mg/l $SiO_2$).

## Local measurements

### 1.7.4 Temperature

#### Temperature of the air

**Equipment**

Calibrated mercury thermometer graduated in 0.5 °C.

**Method**

Measure with a carefully dried thermometer (otherwise too low a reading is given as a result of latent heat) close to the sampling location at 1 m height above the water or the ground. If the sun is shining, measurements must be taken in the shade and screened from reflected radiation of heat (light-coloured house walls or rocks).

#### Temperature of the water

**Equipment**

Calibrated mercury thermometer graduated in 0.1 °C. Usual measuring ranges: -5 to +30 °C, -5 to +60 °C, -5 to +100 °C, or electrical thermometer and/or thermocouple.

(Special thermometer: Maximum thermometer (particularly for hot springs and/or waters); well thermometer (the sphere of mercury remains below the water surface on withdrawing the sampling beaker).

**Method**

Immerse the thermometer in the water to reading depth and wait until the reading is constant (approx. 1 min.). If it is not technically possible to measure directly, the water sample is taken in a vessel containing at least 1 litre which must itself be at the same temperature as the water (with tap water, run tap approx. 5 mins.), and the process followed as described above.

As the solubility of gases is a function of the temperature, the reading should be to 0.1 °C.

### 1.7.5 Hydrogen-ion activity (pH) (cf. also Chapter 2)

**Introduction**

In chemically pure water, water molecules split up to form $H_3O^+$ (hydronium ion) and $OH^-$ (hydroxyl ion) according to the following formula:

$$2H_2O \rightleftharpoons H_3O^+ + OH^-$$

In its old form (merely taking into consideration the proton activity $H^+$) this reads:

$$H_2O \rightleftharpoons H^+ + OH^-$$

The hydrogen ion is carrier of the acid reaction. The pH is used as measuring unit; this is defined according to the following equation

$$pH = -\log a_{H^+}$$

as the negative logarithm of molar hydrogen-ion activity, corresponding in large dilutions approximately to the molar hydrogen-ion concentration ($c_{H^+}$).

The concentration of hydrogen ions in chemically pure, neutral water is $1 \cdot 10^{-7}$ g/l, corresponding to the dissociation constant of the water $K_W = 10^{-14}$ at 25 °C deducible from the principle of mass action;

$$[H^+] + [OH^-] = K_W \qquad (K_W = 10^{-14})$$

$$[H^+] = \frac{K_W}{[OH]} = \frac{10^{-14}}{10^{-7}} = 10^{-7} \text{ g/l; } OH^- = 10^{-7} \text{ g/l}$$

The conversion of the fraction $(H^+) = 10^{-7}$ into the negative logarithm pH 7 is in the interests of simplification.

| | | |
|---|---|---|
| pH 7 | = | neutral |
| pH>7 to 14 | = | alkaline range |
| pH 0 to <7 | = | acid range |

**a) Electrometric determination of pH**

**General**

The pH is determined electrometrically by measuring the difference in potential between the measuring electrode (glass electrode) and the reference electrode with known potential (saturated calomel electrode used instead of the normal hydrogen electrode).

The glass electrode has a spherical glass membrane which is immersed in the solution of unknown pH to be examined. The membrane is filled with a solution of known pH. If a different concentration and/or activity of hydrogen ions to that of the interior solution exists at the exterior surface of the glass membrane, a corresponding phase-boundary potential $\Delta E$ in contact with the outside solution is produced at the thin-walled glass membrane. This phase-boundary potential obeys the Nernst equation:

$$\frac{RT}{F} \cdot \ln \frac{a'_{H^+}}{a_{H^+}} = \Delta E \text{ (Volt)}$$

where:

| | | |
|---|---|---|
| $a_{H^+}$ | = | Activity of hydrogen ions in the water sample |
| $a'_{H^+}$ | = | Activity of hydrogen ions in the reference solution |
| R | = | General gas constant |
| T | = | Absolute temperature (K) |
| F | = | Faraday constant (96493 Coulomb/val) |

so that the glass electrode behaves practically like a hydrogen electrode (in which the phase-boundary potential is measured between the gaseous $H_2$ located on the platinum black of the Pt-electrode and the $H^+$ ions ($H_3O^+$) contained in the solution).

The saturated calomel electrode used as reference electrode has, in comparison to the defined normal hydrogen ion electrode (hydronium ion concentration 1 g ion/l to $H_2$ at 18 °C, 1000 mbar), a potential of +0.250 V, which should be subtracted from the measured potential difference to determine E (already allowed for in the pH meter).

If the filling of the glass electrode is connected conductively with the reference electrode also immersed in the water sample, the electromotive force (e.m.f.) of this cell can be determined by current-free voltage measurement. Because of the high internal resistance of the glass electrode, an instrumentation amplifier is required to indicate the voltage. Buffer solutions with defined pH serve for calibration and checking.

The e.m.f. in the measuring cell changes in direct proportion to the pH value by 58 mV for pH = 1 (20 °C). As a temperature change of 10 or 20 °C in the water sample results in a change in potential of approx. 2 or approx. 4 mV and thus in a pH indication error of approx. 0.03 or approx. 0.06 pH units, temperature compensation is not necessarily required in the normal temperature range, given that the mean reproducible accuracy of pH measurement with glass electrodes is 0.05 - 0.1 pH.

### Equipment

pH meter (battery, rechargeable or mains equipment)

Single-probe measuring cell (measuring and reference electrode)

Measuring range pH 1 to pH 10. Measurements can also be made in the range above pH 10 with the aid of so-called alkali-stable electrodes.

### Disturbances

The polarizability is low, disturbances due to colour, turbidity, oxidation and reduction agents practically do not occur. New glass electrodes which have been stored in dry conditions must be allowed to steep immersed in water for several days before use until they no longer show any movement in potential. Their sensitivity is affected by oily substances. In such cases, the electrodes must be carefully cleaned with soap or detergent solution after every measurement. They should then be rinsed with distilled water, HCl solution and again with distilled water. They should be subsequently recalibrated with standard buffer solutions.

### Measurement

The pH meter is calibrated with buffer solutions of known pH according to the instructions accompanying each appliance and the gradient of the reading corrected where necessary.

Ready-made buffer solutions can be used for calibration or buffer solutions produced according to the following instructions:

pH 2.0: 6.71 g potassium chloride (KCl) is dissolved in 1 litre 0.01 m hydrochloric acid.

pH 4.62: Mix 200 ml 1 m acetic acid, 100 ml 1 m sodium hydroxide solution and 700 ml bi-distilled water.

pH 6.4: Solution A. Dissolve 21.008 g citric acid and 200 ml 1 m sodium hydroxide solution in bi-distilled water to 1 litre.

Solution B. 0.1 m sodium hydroxide solution. Mix 54.4 ml of Solution A with 45.6 ml of Solution B.

pH 7.0: Solution A. Dissolve 9.078 g potassium dihydrogenorthophosphate ($KH_2PO_4$) in bi-distilled water to 1 litre.

Solution B. Dissolve 11.88 g disodium monohydrogenorthophosphate ($Na_2HPO_4 \cdot 2\ H_2O$) in bi-distilled water to 1 litre.

Mix Solutions A and B in the ratio 2 : 3.

pH 9.0: Solution A. Dissolve 12.40 g boric acid and 100 ml 1 m sodium hydroxide solution in bi-distilled water to 1 litre.

Solution B. 0.1 m hydrochloric acid.

Mix 8.5 parts of Solution A with 1.5 parts of Solution B.

For the actual process of measurement itself, the measuring cell is inserted into the water sample in the measuring vessel. $CO_2$ losses should be avoided as far as possible. The temperature of the sample is then measured, and the temperature compensation set on the meter. The measured value is read off after the reading has remained constant for at least 1 min.

Scale of the results

Depending on the sensitivity of the equipment, the pH is given to an accuracy of up to 0.05 pH units. Below pH 1, above pH 12 or in concentrated saline solutions, the reading can be limited to an accuracy of 0.1 pH units. pH measurement is problematical in distilled water or in waters with very low proportions of dissolved substances (e.g. rain water). In such cases, one can only arrive at approximate values.

b) Colorimetric pH measurement

Certain pigments or pigment mixtures show a colour dependent on pH in aqueous solutions, e.g. litmus turns red in the acid range and blue in the alkaline range.

By using so-called universal indicators, measurement of pH is possible by colorimetric comparisons in gradations for example of 0.5 pH units. Measurement is particularly simple if these indicator mixtures are applied to paper strips or inserted in plastic tubes. It is sufficient to immerse them for a short period in the water sample to be measured and subsequently compare the colour with the pH scale.

### 1.7.6 Electrical conductivity

General

The electrical conductivity of water is based on the presence of ions. It can be regarded as a non-specific yardstick for the content or the concentration of dissolved dissociable substances in water.

A simple continuous method for determining the content of dissociable substances, particularly in the case of measurements to be repeated regularly at specified intervals, is important not only for systematic but also for intermittent checking on water and its content of dissolved mineral substances.

In the structure of an electrical field in water, the anions migrate to the positively charged anode, the cations migrate to the negatively charged cathode. At constant temperature, the electrical conductivity of a given water is a function of its concentration of ions.

Electrical conductivity is expressed as the reciprocal of electrical resistance in ohm (Ω), in relation to a water cube of edge length 1 cm at 20 °C (specific electrical conductivity). It is given in Siemens ($S = \frac{1}{\Omega}$) per cm ($S \cdot cm^{-1}$).

Good distilled water should have values below $0.3 \times 10^{-6}\ S \cdot cm^{-1}$ (i.e. $<0.3\ \mu S \cdot cm^{-1}$).

The reference temperature or any conversion to a different temperature from the measuring temperature should always be indicated, as the results measured are dependent on temperature.

The following designations are usual in practice:

$$1\ S \cdot cm^{-1} = 10^3\ mS \cdot cm^{-1} = 10^6\ \mu S \cdot cm^{-1}.$$

The resistance of electrolytes, and thus electrical conductivity, cannot be measured with direct current on account of the polarization of electrodes and the additional resistances thus arising (which falsify the result). For this reason high frequency (>1000 Hz) alternating current is always used. Measurement is based on 25 °C. A factor f, to be taken from the following Table (from DIN 38404 Part 8, DEV-15, issue 1985) is used to convert the measuring temperature to other temperatures.

### Equipment

Conductivity meter (resistance meter with bridge fed with alternating current)

Conductivity cell or flow cell

Thermometer -10 °C to +50 °C (graduation in 0.05°).

### Method

The meter and conductivity cell are rinsed several times in the water to be examined. The vessel is then filled. When water and equipment have reached the same temperature, this is read off on the thermometer, followed by the resistance or the (directly indicated) conductivity on the meter.

Rinsing and measurement are repeated until the values of two consecutive measurements do not deviate by more than 2 % from their mean value.

Temperature correction factor for converting electrical conductivity values measured in natural waters at temperature t to the reference temperature of 25 °C

| t | f25 | | | | | | | | | |
|---|---|---|---|---|---|---|---|---|---|---|
| °C | 0.0 | 0.1 | 0.2 | 0.3 | 0.4 | 0.5 | 0.6 | 0.7 | 0.8 | 0.9 |
| 0 | 1.918 | 1.912 | 1.906 | 1.899 | 1.893 | 1.887 | 1.881 | 1.875 | 1.869 | 1.863 |
| 1 | 1.857 | 1.851 | 1.845 | 1.840 | 1.834 | 1.828 | 1.822 | 1.817 | 1.811 | 1.805 |
| 2 | 1.800 | 1.794 | 1.788 | 1.783 | 1.777 | 1.772 | 1.766 | 1.761 | 1.756 | 1.750 |
| 3 | 1.745 | 1.740 | 1.734 | 1.729 | 1.724 | 1.719 | 1.713 | 1.708 | 1.703 | 1.698 |
| 4 | 1.693 | 1.688 | 1.683 | 1.678 | 1.673 | 1.668 | 1.663 | 1.658 | 1.653 | 1.648 |
| 5 | 1.643 | 1.638 | 1.634 | 1.629 | 1.624 | 1.619 | 1.615 | 1.610 | 1.605 | 1.601 |
| 6 | 1.596 | 1.591 | 1.587 | 1.582 | 1.578 | 1.573 | 1.569 | 1.564 | 1.560 | 1.555 |
| 7 | 1.551 | 1.547 | 1.542 | 1.538 | 1.534 | 1.529 | 1.525 | 1.521 | 1.516 | 1.512 |
| 8 | 1.508 | 1.504 | 1.500 | 1.496 | 1.491 | 1.487 | 1.483 | 1.479 | 1.475 | 1.471 |
| 9 | 1.467 | 1.463 | 1.459 | 1.455 | 1.451 | 1.447 | 1.443 | 1.439 | 1.436 | 1.432 |
| 10 | 1.428 | 1.424 | 1.420 | 1.416 | 1.413 | 1.409 | 1.405 | 1.401 | 1.398 | 1.394 |
| 11 | 1.390 | 1.387 | 1.383 | 1.379 | 1.376 | 1.372 | 1.369 | 1.365 | 1.362 | 1.358 |
| 12 | 1.354 | 1.351 | 1.347 | 1.344 | 1.341 | 1.337 | 1.334 | 1.330 | 1.327 | 1.323 |
| 13 | 1.320 | 1.317 | 1.313 | 1.310 | 1.307 | 1.303 | 1.300 | 1.297 | 1.294 | 1.290 |
| 14 | 1.287 | 1.284 | 1.281 | 1.278 | 1.274 | 1.271 | 1.268 | 1.265 | 1.262 | 1.259 |
| 15 | 1.256 | 1.253 | 1.249 | 1.246 | 1.243 | 1.240 | 1.237 | 1.234 | 1.231 | 1.228 |
| 16 | 1.225 | 1.222 | 1.219 | 1.216 | 1.214 | 1.211 | 1.208 | 1.205 | 1.202 | 1.199 |
| 17 | 1.196 | 1.193 | 1.191 | 1.188 | 1.185 | 1.182 | 1.179 | 1.177 | 1.174 | 1.171 |
| 18 | 1.168 | 1.166 | 1.163 | 1.160 | 1.157 | 1.155 | 1.152 | 1.149 | 1.147 | 1.144 |
| 19 | 1.141 | 1.139 | 1.136 | 1.134 | 1.131 | 1.128 | 1.126 | 1.123 | 1.121 | 1.118 |
| 20 | 1.116 | 1.113 | 1.111 | 1.108 | 1.105 | 1.103 | 1.101 | 1.098 | 1.096 | 1.093 |
| 21 | 1.091 | 1.088 | 1.086 | 1.083 | 1.081 | 1.079 | 1.076 | 1.074 | 1.071 | 1.069 |
| 22 | 1.067 | 1.064 | 1.062 | 1.060 | 1.057 | 1.055 | 1.053 | 1.051 | 1.048 | 1.046 |
| 23 | 1.044 | 1.041 | 1.039 | 1.037 | 1.035 | 1.032 | 1.030 | 1.028 | 1.026 | 1.024 |
| 24 | 1.021 | 1.019 | 1.017 | 1.015 | 1.013 | 1.011 | 1.008 | 1.006 | 1.004 | 1.002 |
| 25 | 1.000 | 0.998 | 0.996 | 0.994 | 0.992 | 0.990 | 0.987 | 0.985 | 0.983 | 0.981 |
| 26 | 0.979 | 0.977 | 0.975 | 0.973 | 0.971 | 0.969 | 0.967 | 0.965 | 0.963 | 0.961 |
| 27 | 0.959 | 0.957 | 0.955 | 0.953 | 0.952 | 0.950 | 0.948 | 0.946 | 0.944 | 0.942 |
| 28 | 0.940 | 0.938 | 0.936 | 0.934 | 0.933 | 0.931 | 0.929 | 0.927 | 0.925 | 0.923 |
| 29 | 0.921 | 0.920 | 0.918 | 0.916 | 0.914 | 0.912 | 0.911 | 0.909 | 0.907 | 0.905 |
| 30 | 0.903 | 0.902 | 0.900 | 0.898 | 0.896 | 0.895 | 0.893 | 0.891 | 0.889 | 0.888 |
| 31 | 0.886 | 0.884 | 0.883 | 0.881 | 0.879 | 0.877 | 0.876 | 0.874 | 0.872 | 0.871 |
| 32 | 0.869 | 0.867 | 0.865 | 0.864 | 0.863 | 0.861 | 0.859 | 0.858 | 0.856 | 0.854 |
| 33 | 0.853 | 0.851 | 0.850 | 0.848 | 0.846 | 0.845 | 0.843 | 0.842 | 0.840 | 0.839 |
| 34 | 0.837 | 0.835 | 0.834 | 0.832 | 0.831 | 0.829 | 0.828 | 0.826 | 0.825 | 0.823 |
| 35 | 0.822 | 0.820 | 0.819 | 0.817 | 0.816 | 0.814 | 0.813 | 0.811 | 0.810 | 0.808 |

## Evaluation

The electrical conductivity K expressed in $S \cdot cm^{-1}$ is calculated according to the following formula:

$$K = \frac{F \cdot C}{R_t}$$

$R_t$ = Electrical resistance of the water at temperature t, measured in ohm
C = Resistance capacity (cell constant) of the meter in $cm^{-1}$
F = Temperature factor (Table)

The measured value is usually read off directly from the equipment in $S \cdot cm^{-1}$, $mS \cdot cm^{-1}$ or $\mu S \cdot cm^{-1}$.

The result is given rounded to two decimal places when read off for example in $mS \cdot cm^{-1}$.

## Calibration of conductivity meters

The conductivity meters on the market have a specific resistance capacity or cell constant C (in $cm^{-1}$) resulting from the distance between the electrodes l and the electrode area q:

$$C = \frac{l}{q}$$

The resistance capacity or cell constant C by which the measured value must be multiplied to obtain the value of an electrical conductivity K is generally given by the manufacturing firm. It is nevertheless necessary to check this value from time to time.

In such cases, or if the resistance capacity of a cell is unknown, this is determined with the aid of solutions whose electrical conductivity is known.

### Method

The cell, which has been previously rinsed repeatedly with bi-distilled water or fully desalinated water at 25 ± 0.1 °C and subsequently rinsed with 0.01 m KCl solution (see below), is filled with this 0.01 m KCl solution. The measuring bridge is fed with an alternating current of 1000 Hz and the resistance R read off. This process is repeated until the values for two consecutive measurements do not deviate by more than 2 % from their mean value.

This whole procedure is repeated with 0.1 m KCl solution.

### Calculation

The resistance capacity or cell constant C is given by the formula:

$$C = K_{25} \cdot R_{25}$$

C = Resistance capacity of the cell in $cm^{-1}$
$K_{25}$ = Electrical conductivity of the solution at 25 °C in $S \cdot cm^{-1}$
$R_{25}$ = Electrical resistance of the solution at 25 °C in ohm

The values determined for the two solutions (0.01 m KCl and 0.1 m KCl) after multiple measurements for C are averaged and the mean value given to 0.01 mS $cm^{-1}$.

**Producing the calibrating solutions**

0.1 m KCl solution: Top up 7.456 g potassium chloride of reagent purity (dried for 2 hrs. at 105 °C) to 1000 ml with bi-distilled water at 25 °C. This solution has an electrical conductivity of

$$0.01295 \text{ S} \cdot \text{cm}^{-1}, \text{ or } 12.95 \text{ mS} \cdot \text{cm}^{-1}, \text{ at } 25\ °\text{C}.$$

0.01 m KCl solution: 100 ml of the 0.1 m KCl solution is topped up to 1000 ml with bi-distilled water at 25 °C. Electrical conductivity of this solution

$$= 0.001421 \text{ S} \cdot \text{cm}^{-1}, \text{ or } 1.42 \text{ mS} \cdot \text{cm}^{-1}, \text{ at } 25\ °\text{C}.$$

The bi-distilled (or fully desalinated) water used to produce the KCl solutions must previously have been completely degasified by introducing nitrogen (dissolved carbon dioxide would increase the conductivity). The calibrating solutions should be stored in **air-tight** sealed glass bottles.

### 1.7.7 Redox potential

**General**

A redox (oxidation reduction) reaction is a process in which electrons are exchanged between a reduction agent and an oxidation agent, for example in an aqueous solution.

In a redox process, the electrons "migrate" under the influence of a difference in potential (expressed in V or mV) from the reduction agent (electron donor) to the oxidation agent (electron acceptor).

**Electrometrical determination of the redox potential (redox voltage)**

The redox potential is determined electrometrically by measuring the difference in potential between the measuring electrode (platinum electrode) and the reference electrode. A saturated calomel electrode is used as reference electrode instead of the normal hydrogen electrode ($E°_{St}$ = 0). The potential difference between the saturated calomel electrode and the normal hydrogen electrode at 25 °C is 241.0 mV.

The concurrent tendency to release electrons (reduction agent) or absorb electrons (oxidation agent) is measured, expressed as e.m.f. (equilibrium redox potential in mV).

If the two electrodes immersed in the water sample are joined, the e.m.f. in this cell can be measured, using current-free voltage measurement. It is necessary to determine the pH at the same time on account of the dependence of the redox potential on pH.

As the chemical constitution of the participants in the equilibrium, the overall electro-chemical reaction and its changes dependent on pH are completely unknown in the majority of cases, the values are either related to pH = 0 ($E_{00}$) or to pH 7 ($E_{07}$) and the stoichiometric factor is set at n = 1 so that the change in potential of 59.1 mV per pH unit at 25 °C should be taken into consideration.

The redox voltage plays a similar part in redox reactions to the pH in acid-base reactions. In the same way as the concentrations of all acids and bases contained in an aqueous solution can be clearly determined by the pH, the redox voltage in a solution determines the concentrations of all reversibly reducible or oxidizable oxidation and reduction agents contained in the solution.

It is possible to refer to the **redox voltage in a solution** because all corresponding redox pairs contained in a state of equilibrium in the solution must have the same redox potential (the same quantity of electrons must be released as are absorbed per time unit). Otherwise, transfers of electrons, i.e. redox reactions, would have to take place until this state of equilibrium is reproduced.

Method

**Pre-treatment of the electrodes**

If new electrodes are used, they should be degreased by washing in an approximately 10 % solution of surface-active agents and rinsing in ethanol. The platinum electrode is then immersed for 3 minutes in aqua regia at 70 - 90 °C, subsequently well rinsed in distilled water, reduced for at least 10 minutes in a 10 % solution of sodium sulphite, and again rinsed thoroughly in distilled water.

If measurements are taken with already worked electrodes or if contamination of the electrode becomes noticeable in the form of creep of the final value, they should also be degreased as above, but almost boiling nitric acid (1 + 1) is used instead of the aqua regia. The electrode is immersed for 5 minutes in this nitric acid. It is also possible, however, to use a cold chromic acid solution, but the electrode must be immersed in this case for approximately 30 minutes in order to achieve complete oxidizing cleaning of the surface. It is then rinsed in distilled water, reduced for at least 10 min. in a 10 % solution of sodium sulphite, and rerinsed in distilled water. In several cases, it has proved suitable to clean the platinum ring of the electrode with talcum powder.

**Measuring the redox potential**

If a special measuring vessel is used, the sample water should be allowed to flow through the measuring vessel fitted with the electrodes for approximately 10 minutes. The inlet and discharge tubes are then connected and pressure-free measurements taken.

If a glass beaker is used for measurement, the electrodes connected to the meter are immersed in the sample and the measured value read off when the reading has not changed for at least 5 minutes (establishment of equilibrium). Before repeating measurements, the platinum electrode is rinsed in distilled water. If the measured value takes longer than 5 minutes to reach equilibrium, the platinum electrode is contaminated and must be recleaned.

**Measuring equipment**

Measuring electrode: Platinum electrode

Reference electrode: Saturated calomel electrode

pH meter with mV display, accuracy ± 5 mV

Aqua regia:
3 parts by volume of hydrochloric acid (HCl) (1.19 g/ml) and 1 part by volume of nitric acid ($HNO_3$) (1.43 g/ml).

Solution of chromic acid:
Dissolve 5 g $K_2Cr_2O_7$ in 500 ml sulphuric acid (1.84 g/ml).

Nitric acid:
Mix 1 part by volume of nitric acid ($HNO_3$) (1.42 g/ml) with 1 part by volume of distilled water.

Solution of sodium sulphite ($Na_2SO_3$), 10 % in distilled water.

Measuring vessel with three vertical ground sleeves to receive the two electrodes and a thermometer, with two connecting nozzles situated opposite each other and just above the base of the vessel for feeding or discharging the sample water.

Redox single-probe measuring cell: When using a redox single-probe measuring cell, attention should be paid to the instructions of the manufacturing firm. When treating such equipment with acid, considerable care must be taken not to immerse it so far in the acid that the (acid-sensitive) diaphragms come into contact with the acid.

If redox single-probe measuring cells are used to measure redox potential, these should be calibrated in similar manner to the single-probe measuring cells for determining pH. Calibration should, for example, be conducted with the aid of equi-molar quantities of corresponding redox pairs (e.g. $K_3(Fe(CN)_6)$ and $K_4(Fe(CN)_6)$.

**Notes:**

When determining the redox potential in low-oxygen or oxygen-free samples, measurements should be made with continuous flow of the water sample (i.e. not in an open glass beaker) in order to prevent oxygen penetrating into measuring equipment with ground sleeve inserts.

In routine determinations with a meter reading accuracy of ± 5 mV, no temperature correction is generally necessary. The potential values dependent on temperature should be inserted in the calculation according to the Table given below.

## Calculation

Redox voltage related to pH 0:

$$E_{00} = E + pH \cdot f + E_{GKE}$$

Redox voltage related to pH 7:

$$E_{07} = E + pH \cdot f - 7 \cdot f + E_{GKE}$$

E = Potential measured against the saturated calomel electrode (GKE) at temperature t

Temperature-dependence of hydrogen (f) and saturated calomel electrode ($E_{GKE}$) (Chateau/Pouradier)

| t (°C) | f(mV) | $E_{GKE}$ (mV) | t (°C) | f(mV) | $E_{GKE}$ (mV) |
|---|---|---|---|---|---|
| 10 | 56.1 | 251 | 18 | 57.7 | 246 |
| 11 | 56.3 | 250 | 19 | 57.9 | 245 |
| 12 | 56.5 | 249 | 20 | 58.1 | 244 |
| 13 | 56.7 | 249 | 21 | 58.3 | 244 |
| 14 | 56.9 | 248 | 22 | 58.5 | 243 |
| 15 | 57.1 | 248 | 23 | 58.7 | 243 |
| 16 | 57.3 | 247 | 24 | 58.9 | 242 |
| 17 | 57.5 | 246 | 25 | 59.1 | 241 |

$E_{GKE}$ = Potential of the saturated calomel electrode (GKE) at temperature t (see Table)
f = E/pH (RT/MF) at temperature t (see Table)
pH = Measured pH of sample solution

Specimen calculation (after U. Hässelbarth):

$$E = -180 \text{ mV at } 10\text{ °C and pH } 6.8$$
$$E_{00} = -180 + 6.8 \cdot 56.1 + 251 = +452 \text{ mV}$$
$$E_{07} = -180 + 6.8 \cdot 56.1 - 7 \cdot 56.1 + 251 = +59 \text{ mV}$$

### 1.7.8 Oxygen (cf. also Section 3.6)

The concentration of oxygen in water which is in solubility equilibrium with the surrounding atmosphere at given temperature and pressure (also dependent on height amsl) is referred to as **oxygen saturation** (oxygen saturation concentration). Concentrations below this value are referred to as **oxygen deficit,** concentrations larger than the saturation value as oxygen supersaturation.

The **current concentration of oxygen** is that oxygen contained in the water received when determining after sampling. The oxygen contained in the water is necessary for the life of animal and plant organisms. This particularly applies to the metabolic behaviour of microorganisms which cause degradation of contaminants in the water. Dissolved oxygen is consumed in this process of aerobic degradation of contaminants in the water. A fall in the normal oxygen content thus indicates contamination or the presence of oxygen-consuming substances.

**a) Electrometrical detection (on site)**

The polarographic oxygen-measuring cell, covered with membranes and working on the so-called Clark principle, consists of two electrodes arranged so as to be insulated, which are joined by a liquid or paste-like electrolyte. The electrode compartment is separated from the measuring medium (water, waste water) by a membrane allowing $O_2$ to permeate.

With a constant electrode surface area, membrane thickness and electrolyte concentration, the diffusion current produced in the measuring cell is essentially only dependent on temperature, over and above the partial pressure of the oxygen. On the one hand, the diffusion of oxygen through the membrane of the measuring cell is dependent on temperature, insofar as when this rises (falls) the cell current indicated for the same $O_2$ concentration increases (decreases). Furthermore, the solubility of $O_2$ (e.g. in water) is also dependent on temperature. It is therefore not possible to dispense with temperature compensation in the case of an oxygen-measuring electrode. This correction can either be taken into account in the equipment by means of a thermistor combined with the cathode, or - with more simple equipment - it can be calculated.

## Equipment and chemicals

Oxygen membrane electrode (single-probe measuring cell). The construction of such an electrode varies with the manufacturer. The most important part of such an electrode is the measuring head (Fig. 8).

The cathode usually consists of pure gold, the anode of a metal less precious than gold; silver is very often used. Polyethylene, polypropylene or Teflon are suitable as material for the membrane. A paste-like mass, which has recently replaced the previously common KOH/KCl electrolytes, serves as electrolyte. The remaining parameters for the electrode, such as velocity of approach, polarization DC voltage, zero (residual) current, temperature compensation, settling time for the measured value, gradient or sensitivity and measuring accuracy, should be taken from the special instructions accompanying each piece of equipment.

Oxygen zero solution: A 3 - 5 % sodium sulphite solution prepared with warm water at approximately 60 °C is filled into a bottle with a long narrow neck, sealed and allowed to stand for 24 hours at room temperature.

Oxygen saturation solution: Iced water whose temperature is reduced to approximately 0 °C is stirred with an agitator so as to be well aerated (10 - 15 minutes). The saturation value remains constant at the temperature of the iced water.

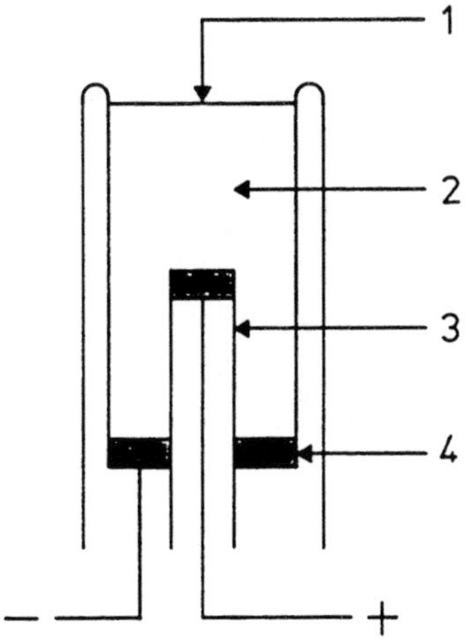


Fig. 8. Schematic diagram of a measuring head; 1) = Membrane film; 2) = Electrolyte compartment; 3) = Cathode; 4) = Anode (ring-shaped)

Calibration

The oxygen zero point is calibrated by immersing the measuring head in the oxygen zero solution, which must be completely free of air bubbles. After approximately 15 minutes, the zero value has been reached and this is then set at "zero mg $O_2/l$" by means of a knob on the equipment.

To determine the saturation calibration point, the measuring head is immersed in the oxygen saturation solution which is then well stirred. After approximately 3 minutes, a constant measured value must be recorded, which is then adjusted to the air saturation value with a ("gradient") knob on the equipment. (The partial pressure for a gas in a liquid saturated with this gas corresponds to the partial pressure of the gas above this liquid: Air saturation value = water saturation value).

Calibration can also be done in air. For this purpose, the membrane film and its surroundings are carefully dried and the electrode exposed for 5 - 10 minutes to the air. The calibrating procedure is then the same as described above. Admittedly, a calibration error of $\pm$ 3 % must be accepted in this method.

**Determining process**

To determine the oxygen content, the single-probe measuring cell is immersed in the sample in the measuring vessel and the necessary velocity of approach produced by actuating an agitator. In the case of equipment without temperature compensation, the temperature of the sample is measured simultaneously. Air bubbles adhering to the electrode should be removed by shaking briefly. The measured value is read off after the value indicated has remained constant for at least one minute. The values read off on the measuring equipment are recorded directly in mg/l $O_2$. The results are rounded down to 0.1 mg/l.

Reference should be made to the instructions issued by the manufacturer and accompanying each piece of equipment on storing the electrodes, preparing them for calibration and measurement and on check measurements to identify faults or indications of wear and tear.

**b) Determination of dissolved oxygen after Winkler (titrimetrically)**

In 1914, L.W. Winkler published in the Fresenius Journal of Analytical Chemistry a process for the iodometric determination of oxygen in water which is still valid today.

**Principle**

Upon the addition of sodium hydroxide solution, oxygen dissolved in water combines with manganese II ions to form higher-grade manganese oxides, which are dissolved on acidification as manganese III ions and oxidize iodide ions to iodine. The iodine thus formed is titrated with sodium thiosulphate. The consumption of sodium thiosulphate solution used for adjustment is a measure of the oxygen content in the water.

**Reaction equation**

$$2\ Mn^{3+} + 2\ I^- \longrightarrow 2\ Mn^{2+} + I_2$$

$$I_2 + 2\ S_2O_3^{2-} \longrightarrow 2\ I^- + S_4O_6^{2-}$$

Measurement should take place at the site of sampling, and strict attention should be paid to the working instructions on account of the continuous presence of atmospheric oxygen. A determination limit of approximately 0.2 mg/l can be achieved.

If disturbing substances are absent, this process is also suitable as a calibrating method, e.g. as a check on the electrometric determination of oxygen.

**Equipment**

Sampling tube made of rubber or plastic

Glass bottles with ground and bevelled glass stoppers, content approximately 100 to 150 ml. The quantity of water contained is fixed by weighing.

Measuring pipette 1 ml, graduated

**Reagent solutions**

Manganese II chloride:
800 g $MnCl_2 \cdot 4\ H_2O$ in 1 litre distilled water

Sodium hydroxide solution (containing potassium iodide and sodium azide):
Dissolve 360 g NaOH, 200 g KI and 5 g $NaN_3$ (the latter to eliminate disturbance by nitrite) with distilled water to 1 litre (paying attention to protective measures, e.g. protective goggles). The solution is to be filtered through glass wool.

Phosphoric acid ($H_3PO_4$):
1.70 g/ml

Sodium thiosulphate solution:
0.01 m
Titration against 0.01 m potassium iodate solution. 1 ml 0.01 m sodium thiosulphate solution corresponds to 0.08 mg oxygen ($O_2$).

Indicator solution:
20 g zinc chloride is dissolved in 100 ml water and heated to boiling point. 4 g starch (made into a paste with water) is stirred into this solution and heated until dissolved. Subsequently, the mixture is diluted with water to approximately 500 ml, 2 g zinc iodide dissolved in the mixture and the solution filled up to 1000 ml with distilled water. It is stored in a brown bottle.

**Sampling and Measuring**

The water sample to be analyzed is filled into the glass bottle so that any influence by atmospheric oxygen on the oxygen contained in the water can as far as possible be precluded. Where possible, the sample should be taken from flowing water by means of a tube reaching to the base of the glass bottle. The bottle is filled from the bottom upwards by laminar inflow of the water into the bottle. The water is allowed to overflow until the contents of the bottle have been renewed two or three times.

If the water sample is to be taken from a surface water, a sampling stopper should be used. This stopper fits the neck of the glass bottle and has two bore holes. A glass tube extends through one bore hole from the top edge of the stopper almost as far as the base of the bottle. A glass tube is inserted in the second bore hole, extending some 4 to 5 cm above the stopper and ending at the bottom end of the stopper. For sampling, the glass tube extending to the base is closed with the thumb and the bottle immersed in the water so that the second glass tube is above the water surface. The thumb is now removed and the water to be analyzed flows into the bottle. When the bottle is full, the sampling plug is removed underwater and the bottle completely filled.

0.5 ml of the manganese-II-solution is pipetted into the brim-full bottle, followed by 0.5 ml of the sodium hydroxide solution containing potassium iodide and sodium azide. Corresponding quantities of the water sample overflow. The bottle is now sealed, ensuring that no air bubbles are allowed to enter, and shaken. Depending on the oxygen content, precipitates of higher-grade hydrated manganese oxides are formed, which are allowed to settle for approximately one minute, after which 2 ml phosphoric acid solution is added. The bottle is sealed, and after ten minutes the contents of the bottle are transferred into a 300 ml flask and the bottle rinsed out with distilled water. The iodine released is titrated until coloured faintly yellow with 0.01 m sodium thiosulphate solution. 1 ml zinc iodide-starch solution is then added as indicator and titrated until the transition phase from blue to colourless.

Calculation

$$\text{mg/l } O_2 = \frac{a \cdot F \cdot 80}{V - V_R}$$

a = Consumption of 0.01 m sodium thiosulphate solution in ml
F = Factor of the 0.01 m sodium thiosulphate solution
V = Content of the oxygen bottle in ml
$V_R$ = Added reagent solutions in ml (without $H_3PO_4$)

The results are rounded down to 0.1 ml/l $O_2$.

### 1.7.9 Ozone

Ozone is increasingly used in the treatment of water as drinking water, mainly for disinfection and to oxidize organic and inorganic substances contained in the water. Ozone does not form any chlorine-containing haloforms such as $CHCl_3$, $CHCl_2Br$ or $CHClBr_2$, but does form $CHBr_3$ (bromoform) through the oxidation of the bromide. If therefore ozoning of the water by filtration with activated carbons is preceded by safety chlorination, corresponding halogenated organic compounds can scarcely be present in the water. The redox potential of ozone corresponds to the following equation:

$$O_3 + 2\,H^+ + 2e^- = H_2O + O_2 \qquad + 2.07 \text{ Volt}$$

This value for the redox potential shows clearly that ozone is one of the strongest oxidation agents. For sufficiently diluted solutions, the Henry-Dalton law applies to the rate of solubility.

More important for the practice of water technology is, however, the question of the rate of dissolution. It is basically true that a high concentration of ozone in the initial gas has an increasing effect on the concentration of the ozone dissolved in water. If a reaction with the substances contained in the water is suspected, the influence of the transport processes into the centre of the liquid is reduced and the rate of dissolution apparently increases.

Ozone, as a metastable molecule, cannot be stored or accumulated. It is therefore necessary to produce ozone at its location of use. Its decomposition rate is of considerable importance and depends on the following factors.

a) The concentration and nature of the dissolved salts
b) The presence of organic substances and their structure
c) The concentration of hydrogen ions
d) The temperature.

Dissociation of ozone is favoured by high pH values, e.g. pH >10. This is due to the formation of OH radicals, whereas it is essentially the $O_3$ molecule which is present in neutral and acid conditions. In order to keep the low-effect dissociation of ozone as low as possible, it is necessary to precisely adjust such a verifiable dissolved content of ozone as is necessary to solve the intended task of oxidation. This statement of principle depends heavily on the application in question. Depending on whether OH radicals and/or $O_3$ molecules are present, the rate of reaction with the substances contained in the water in question may differ.

## Quantitative determining process

Iodometric method, colorimetric and physical methods are above all used to determine ozone in water. In the latter case, the UV extinction at 258 nm is used, but considerable disturbances may occur, owing to the extinction of existing natural or anthropogenic substances contained in the water in this range of wavelengths.

The iodometric method and two colorimetric processes are described below.

### a) Determining ozone with the KI method

The following reaction:

$$O_3 + 2\,I^- + H_2O \quad = \quad I_2 + O_2 + 2\,OH^-$$

takes place stoichiometrically only in neutral solution. In the case of unbuffered solutions, a further reaction according to the following equation:

$$I_2 + 2\,OH^- \quad = \quad I^- + IO^- + H_2O$$

and disproportionation according to the following equation:

$$3\,IO^- \quad = \quad 2\,I^- + IO_3^-$$

take place. The iodine released is titrated with thiosulphate solution. (Starch solution is used as indicator. Change from blue to colourless.)

$$I_2 + 2\,S_2O_3^{2-} \quad = \quad S_4O_6^{2-} + 2\,I^-$$

b) **Determining ozone with diethyl phenylenediamine (DPD)**

This method is not specific and generally refers to oxidation agents. It has already been described in detail in determining chlorine so that it need only be mentioned briefly here. An ozone disc is to be used. If potassium iodide is added before the reagent, or potassium iodide and DPD reagent are added simultaneously, the colour reaction corresponding to the ozone equivalents is the result. Ozone apparently reacts first with iodide to iodine, which then quantitatively forms a red dye with DPD. The DPD reagent should be present in excess of the quantity of iodine for the colour reaction so that the reaction can proceed quantitatively in the way described.

c) **Determining ozone with indigo trisulphonate**

This method is successfully used if direct spectrophotometric determining of ozone is disturbed by UV-absorbing substances on account of UV adsorption at 258 nm, or if this method does not offer sufficient sensitivity.

Like DPD, indigo trisulphonate is not a specific reagent to ozone. How ever, it is only decolourized by such reactive oxidation agents as chlorine, chlorine dioxide and ozone. Other substances contained in the water, such as chlorite, chlorate and hydrogen peroxide do not interfere. Chlorine can be masked by adding malonic acid.

It proves advantageous that the reagent has a high molar coefficient of absorption at 600 nm. Self-colouring in natural waters does not interfere in this spectral range. A further favourable effect is that the oxidation products formed show only a very slow further reaction with excess ozone. For this reason, even local excess does not lead to uncontrolled consumption of oxidation agents. The stock solution is stored in a brown bottle and keeps for approximately 4 months.

**Equipment**

Spectrophotometer or filter photometer for measuring in the 600 nm range.

Several 100-ml measuring flasks

Pipettes, 10 ml, 100 ml

Indigo reagent:
0.62 g potassium indigo trisulphonate dissolved in 1 litre 0.5 molar phosphoric acid of reagent purity

Diluted phosphoric acid:
20 ml $H_3PO_4$ (1.71 g/ml) diluted to 1 litre with distilled $H_2O$.

**Method**

The method used in practice for determining ozone must take into consideration the concentration to be expected. In the case of 0.1 to 10 mg/l $O_3$, 1 ml of the indigo reagent and 10 ml of the diluted phosphoric acid should be prepared. In the case of 0 to 0.1 mg/l $O_3$, 0.1 ml of the indigo reagent should be used. The measuring flask with the reference solution is filled with ozone-free water. Enough sample water is added to the flask for the measuring sample for the quantity of ozone to lie in the range 0.01 to 0.04 mg absolute. This is subsequently filled to the mark and mixed. This should be measured in a 5-cm measuring cell.

Care should be taken that photometric measuring should be carried out as far as possible within one hour.

In order to determine the calibration curve, standard solutions are produced of ozone in distilled water which has been acidified with $H_3PO_4$ to a pH less than 2 to stabilize the ozone. The standard solutions are measured at 258 nm on account of their UV absorption. This should be carried out directly before adding the indigo reagent.

The mass per unit volume in the standard solution is calculated according to the following formula:

$$P_S = A_S : A \cdot B$$

where:

$P_S$ = mass per unit volume of ozone in the standard solution in mg/l
$A_S$ = spectral absorption in the standard solution at 258 nm
$A$ = specific spectral absorptiveness of the ozone in the water; it is $0.0604\ cm^{-1} \cdot (mg/l)^{-1}$
$B$ = optical length of path for the UV measuring cell (cm).

The mass per unit volume of ozone in the water sample is calculated according to the following equation:

$$P_x = F \cdot \frac{(A - A_o)}{B \cdot V_{max.}/V_{O_3}}$$

$P_x$ = Concentration of ozone in the water sample (mg/l)
$F$ = Calibration factor in mg/l; as a rule it is 0.417 per mg/l ozone
$A_o$ = Spectral absorption of the blank sample
$A$ = Spectral absorption of the measuring solution
$B$ = Length of path for the measuring cell (cm)
$V_{max}/V_{O_3}$ = Ratio of the final volume of the measuring solutions (100 ml) to the volume of the water samples used.

### 1.7.10 Chlorine

There are various methods of chlorinating water, for example with gaseous chlorine or with such hypochlorite preparations as sodium hypochlorite, calcium hypochlorite or chlorinated lime, but also using other chlorine compounds such as chlorine dioxide.

The added chlorine can exist in the water to be analyzed in various forms, e.g. as chlorine ($Cl_2$), as hypochlorite ion ($ClO^-$) or as hypochlorous acid ($HClO$).

Chlorine can react with any nitrogen compounds which may be present in the water, thus causing chloramines to be produced, which also display an oxidizing effect, even if this is only a weaker one. In a water with added chlorine one therefore speaks of "total chlorine", i.e. the total of free available chlorine (that is, of those chlorine compounds existing as hypochlorite ion and hypochlorous acid) and of those compounds which contain combined available chlorine (e.g. chloramines) as ingredients with an active oxidizing effect.

As a general rule, chlorine is determined colorimetrically, using a suitable comparator. These determining processes can be conducted very simply on site, and the comparators for visual colour comparison are common practically throughout the world.

The following are used as reagents for colorimetric testing on site:

N,N-diethyl-p-phenylenediamine reagent (DPD)

and

o-tolidine.

In recent years, colorimetric determination of the different available chlorine compounds with DPD has come more and more to the fore. Differentiation between the so-called free available chlorine and the combined available chlorine is possible with this reagent, namely by the additional use of potassium iodide. If o-tolidine is used, only "total chlorine" can be determined colorimetrically.

a) Working instructions for the determination of "free available chlorine" with DPD

   A normal commercial DPD reagent tablet without potassium iodide additive is placed with tweezers in the cuvette of a normal commercial comparator, and a measured quantity of the water to be analyzed, e.g. 1, 2 or 5 ml, is added.

   When the tablet has completely dissolved in the water, the cuvette is filled up to the mark with the water to be analyzed, and the solution mixed thoroughly with a glass rod.

   The reference cuvette is filled with the water sample to be analyzed without adding reagents. Both cuvettes are then placed in the comparator and adjusted until the colours are the same, with the aid of the colour comparator chart. The content of free available chlorine can then be read off. The values are normally read off in mg/l, taking into consideration the quantity of water involved.

b) In order to determine the "total chlorine", analysis is conducted on the lines of the instructions above. However, a DPD reagent tablet with potassium iodide additive is used.

   If the readings lie outside the colour range of the comparator, the water sample to be analyzed must be diluted appropriately with distilled water and the tests repeated. The dilution factor should be taken into consideration during conversion.

c) If these instructions are followed, the difference between total chlorine and free available chlorine represents the combined available chlorine.

If DPD is employed for determination of chlorine, it should be noted that in order to determine the so-called free available chlorine, a red dye is formed in the presence of chlorine, whereas for detection of the so-called "total chlorine" and/or "combined available chlorine" this reaction does not occur until iodide ions are added.

d) If **o-tolidine** is employed for determining total chlorine, a yellow dye is formed, incorporating the total available chlorine. If normal commercial test sets and normal commercial comparators are used, attention should be paid to the suppliers' instructions.

e) For checking and/or for calibration, it is, for example, possible to use appropriately diluted sodium hypochlorite solutions with a known chlorine content. A normal commercial sodium-hypochlorite solution indicates approximately 13 to 14 % active chlorine; the exact content can be determined iodometrically in the laboratory.

f) It should be noted that all reactions are non-specific, i.e. they do not respond to chlorine, etc., itself, but to the oxidation capacity of a substance.

### 1.7.11 Determination of chlorine dioxide ($ClO_2$) and chlorite ($ClO_2^-$)

Chlorine dioxide is frequently employed for disinfecting drinking water, as this prevents or noticeably reduces the formation of so-called haloforms or trihalomethanes.

Chlorine dioxide forms scarcely any chloramines with nitrogen compounds and has a considerable long-term effect in the water grid, even with higher pH values up to pH 9. A notable disadvantage is the formation of chlorite. In addition, attention is to be paid to the formation of chlorite-chlorate, depending on the pH value. Chlorite itself would seem critical to health, and no more than 0.1 mg/l of chlorite and 0.1 to 0.4 mg/l of chlorine dioxide should be present in drinking water.

Analysis of chlorine dioxide in all pH ranges is disturbed by the presence of chlorine; determination of chlorite is only disturbed in the acid pH field. Chlorine must therefore be excluded by the addition of potassium bromide and sodium formate before determination.

a) **Field method**

Principle: Colorimetric determination from a buffered solution with DPD and glycine.

**Equipment**

Comparator with colour wedge for chlorine dioxide, e.g.

Multicol and colour wedge No. 24 together with appropriate measuring cuvettes

Buffer solution

DPD reagent solution

Glycine tablets

Measurement

Both cuvettes are rinsed with the water to be analyzed. One cuvette is filled with the water to be analyzed as far as the top calibration mark, and a glycine tablet is dissolved in the water. The process of dissolving is aided by means of a glass rod to break up and mix the tablet.

Three drops of the buffer solution and two drops of the DPD reagent solution are added to the second cuvette. The content of the first cuvette with the water to be analyzed and the glycine additive are then transferred to the second cuvette with the buffer solution and the DPD reagent solution. This is then mixed and this cuvette employed as a measuring cuvette. Water to be analyzed as reference solution is filled up to the mark in this first cuvette which has also been rinsed with this water, without reagent additive.

Both cuvettes are placed in the comparator, e.g. in the Multicol apparatus, and the colour disk turned until the colours are equal. The value measured is multiplied by the factor 1.9, giving the content of mg $ClO_2$/l water (see also the sections on the determination of total chlorine, free available chlorine and combined available chlorine).

b) Determination of chlorine dioxide and chlorite

The instructions for analysis below follow the process devised by G. Hartung and K. E. Quentin, and the analysis should be conducted in

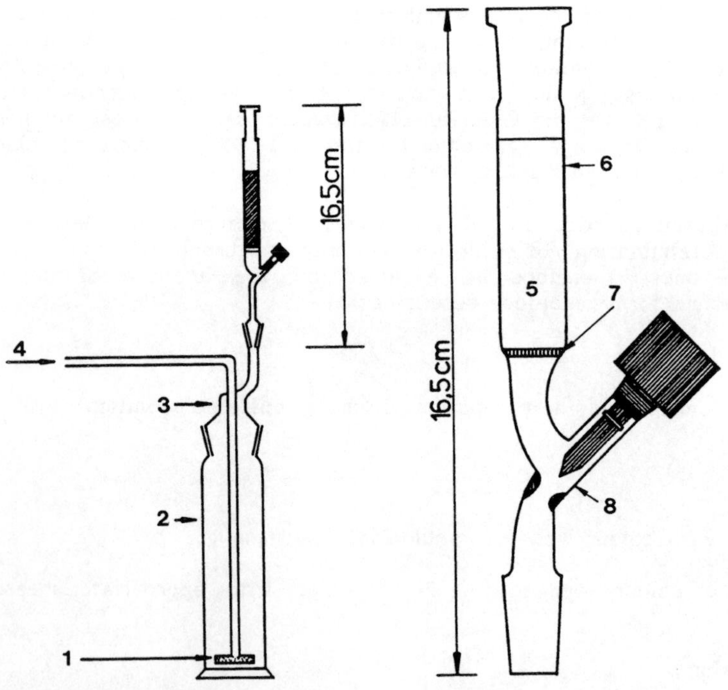


Figs. 9. and 10. 1) = Sintered plate; 2) = Wash bottle 250 or 1500 ml; 3) = Sintered plate insert; 4) = Nitrogen; 5) = KI solution; 6) = Column sleeve; diameter 1.7 cm; 7) = Filter plate; 8) = Screw cock

the laboratory, as far as possible immediately after the taking of the water samples.

Equipment

Spectrophotometer with 5-cm quartz cuvettes

Equipment for blowing out $ClO_2$, consisting of wash bottles of 250 ml and 1500 ml, column sleeve with filter plate and screw cock.

Sodium chlorite ($NaClO_2$), approximately 80 %

Sodium peroxide sulphate ($Na_2S_2O_8$)

Chloramine-T stock solution (1.4085 g/l corresponding approximately to 675 mg $ClO_2$/l)

Chloramine-T diluted solution (to be prepared freshly each day from the stock solution; the stock solution itself can be kept for approximately 2 weeks)

1st chloramine-T dilution:
10 ml of the stock solution to 1 litre,
corresponding to 6.75 mg $ClO_2$/l

2nd chloramine-T dilution:
1 ml of the stock solution to 1 litre,
corresponding to 0.675 mg $ClO_2$/l

Potassium iodide solution, 200 g/l

Phosphate buffer solution pH 7

Solution a:
$KH_2PO_4$ (68 g/l)

Solution b:
$Na_2HPO_4 \cdot 2H_2O$ (89 g/l)

3 parts of solution a) and 7 parts of solution b) give the phosphate buffer solution pH 7.

50 % sulphuric acid ($H_2SO_4$)

Potassium bromide solution (KBr) 70 g/l

Sodium formate solution (HCOONa) 40 g/l

Afterpurified nitrogen

Method

Three different process variants are given below for determining chlorine dioxide and two variants for determining chlorite.

**b1) Determination of chlorine dioxide at pH 7 (one oxidation equivalent)**

This process can be used in the range from 0.5 to approximately 0.01

mg/l $ClO_2$ using a 5-cm cuvette. The presence of chlorite does not cause problems, but chlorine must be eliminated beforehand. In order to remove chlorine, 1.5 ml phosphate buffer and 0.5 ml KBr solution are added to 50 ml of the water to be analyzed; after 1 min. 0.5 ml sodium formate solution is also added and allowed to stand for 15 min. If no chlorine is present, 2.5 ml phosphate buffer is merely added to the water to be analyzed. 0.5 ml potassium iodide solution is then added and mixed. Using a 5-cm cuvette, the extinction is now measured at 350 nm in contrast to a reference solution (50 ml double-distilled water + 2.5 ml phosphate buffer + 0.5 ml potassium iodide solution). The concentration of $ClO_2$ is read off from the calibration curve.

**b2) Determination of chlorine dioxide at pH 2.5 (five oxidation equivalents)**

This process can be used when no chlorite is present in the range from 0.1 to approximately 0.002 mg/l $ClO_2$, using a 5-cm cuvette. Chlorine must first be eliminated with potassium bromide and sodium formate. In order to remove chlorine, 1.5 ml double-distilled water and 0.5 ml KBr solution are added to 50 ml of the water to be analyzed; after 1 min. 0.5 ml sodium formate solution is added and allowed to stand for 15 mins. If no chlorine is present, 2.5 ml double-distilled water is simply added to the water to be analyzed. Approximately 3 drops of sulphuric acid are added in order to adjust to pH 2.5 with the aid of a pH meter. Then, 0.5 ml potassium iodine solution is added and mixed. The extinction is subsequently measured at 350 nm against a reference solution (52.5 ml double-distilled water, adjustment with approximately 3 drops $H_2SO_4$ to pH 2.5 + 0.5 ml potassium iodide solution), using a 5-cm cuvette. The concentration of $ClO_2$ is read off from the calibration curve.

**b3) Analysis after blowing out the $ClO_2$ from the water by means of nitrogen**

It is recommended that the chlorine dioxide be blown out of the water with nitrogen, if it is necessary to separate $ClO_2$ from chlorite and other disturbing substances contained in the water, or if very low concentrations of $ClO_2$ in the water require enrichment. Approximately 0.0002 mg/l $ClO_2$ can be determined.

In order to remove chlorine, 2.5 ml KBr solution is added to 250 ml of the water sample in a 250-ml extractor; 2.5 ml sodium formate solution is added after 1 min. and allowed to stand for 15 min. The additives are not necessary if chlorine is not present.

In order to blow out $ClO_2$ with nitrogen, the extractor is then sealed with the sintered plate insert, and the column sleeve with filter plate and screw cock, filled with 10 ml of freshly prepared KI solution of pH 2.5, is mounted.

A powerful flow of nitrogen is fed through the water sample and the KI solution above it for 3 hours to blow out the $ClO_2$. The extinction of the KI solution is subsequently measured at 350 nm against a reference solution (KI solution pH 2.5), using a 5-cm cuvette. The concentration of $ClO_2$ is read off from the calibration curve and converted as follows:

$$\text{ClO}_2\text{ content read off from the calibration curve} \cdot \frac{10}{53} \cdot \frac{1000}{250} = \text{mg/l ClO}_2$$

The determining processes should always be carried out three times. The water sample from which the $ClO_2$ has been expelled can be used for determining chlorite.

b4) **Determination of chlorite**

If the water to be analyzed also contains chlorine and chlorine dioxide in addition to chlorite, these must be removed. Chlorine is eliminated by potassium bromide and sodium formate, chlorine dioxide is removed by blowing out with nitrogen or by extraction in accordance with the previous sections.

**Determination at pH 2.5**

This process can be used in the range from 0.1 to approximately 0.002 mg/l chlorite (5-cm cuvette) if chlorine and chlorine dioxide are not present or have been removed.

2.5 ml double-distilled water is added to 50 ml of the water to be analyzed and adjusted to pH 2.5 with a number of drops of sulphuric acid. After adding 0.5 ml potassium iodide solution, measurements are taken at 350 nm after some 4 minutes. The quantity of chlorite in the water is determined from the calibration curve.

b5) **Determination of the chlorine dioxide formed from chlorite by oxidization:**

The indirect determination of chlorite by conversion into $ClO_2$ by means of sodium peroxide sulphate is necessary if the concentration of chlorite is below 0.02 mg/l and determination in accordance with the previous sections is not possible.

Calculation of the results of analysis in the iodometric process is always based on the oxidation equivalents of the substances involved per mole.

Where simple conversions do not achieve this objective, the calibration curves produced with chloramine-T must be produced for each determination variant, taking into consideration the volume of water and the volume of the reagent solutions. The extinction values for the iodine solutions resulting from every measuring process then serve to indicate

Molecular weights and oxidation equivalents

| Substance | Molecular weight | Oxidation equivalents per mole | Oxidation equivalents in grams |
|---|---|---|---|
| $I_2$ | 253.809 | 2 | 126.904 |
| Chloramine-T (trihydrate) | 281.696 | 2 | 140.848 |
| $ClO_2$ pH 7 | 67.457 | 1 | 67.457 |
| $ClO_2$ pH 2.5 | 67.457 | 5 | 13.491 |
| $ClO_2^-$ pH 7 | 67.457 | - | - |
| $ClO_2^-$ pH 2.5 | 67.457 | 4 | 16.864 |

the concentration of chlorine dioxide and/or chlorite, with the aid of the appropriate calibration curve. Readings for $ClO_2$ and/or $ClO_2^-$ in the 0.1 to 0.01 mg/l range should be given to two decimal places, readings below 0.01 mg/l to a maximum of three decimal places.

c) **Photometric determination of chlorine dioxide using chlorophenol red**

A variant of the photometric method for determining chlorine dioxide in water using chlorophenol red (CPR method) was described by I.J. Fletcher and P. Hemmings in ANALYST, June 1985. (Further literature on this method is also listed there.) They show how it is possible to optimize the procedure for determining chlorine dioxide in water and eliminate the effect of free available chlorine when using chlorophenol red by adding sodium cyclamate solution. A phosphate buffer is used to adjust to pH 7, after which a practically undisturbed reaction in the presence of thioacetamide permits the photometric determination of chlorine dioxide.

Equipment

Photometer, measuring wavelength 520 nm

4- or 5-cm cuvette

Sodium cyclamate solution 1 %

Thioacetamide solution 0.25 %

Phosphate buffer solution, pH 7:
Dissolve 35.2 g potassium dihydrogenorthophosphate and 27.2 g disodium hydrogenorthophosphate to 1 litre.

Chlorophenol red solution:
Dissolve 0.1436 g chlorophenol red in 100 ml of 0.01 m sodium hydroxide and top up with distilled water to 1 litre.

For use the latter solution is diluted to one-tenth strength with distilled water.

Method

c1) 5 ml of sodium cyclamate solution is measured into a 100-ml glass vessel.

c2) 50 ml of water sample is added and mixed.

c3) Immediately after mixing, 2 ml of buffer solution is added, followed without delay by 4 ml of the dilute chlorophenol red solution. 2 ml of the thioacetamide solution is then added.

c4) After mixing, the result is measured at 570 nm in 4- or 5-cm cuvettes against distilled water.

c5) A blind test is to be performed in parallel.

Calibration is done using dilute chlorine dioxide solutions of known content. The stock solution is prepared by dissolving approximately

600 mg of gaseous chlorine dioxide in 1 litre water by feeding it into a suitable absorption system. The solution is to be stored in dark glass bottles at approximately 4 °C. In each case the content is determined by iodometric titration of this solution. From this stock solution fresh dilutions are to be prepared every day, suiting the concentrations to the measuring range of the water samples to be analyzed.

### 1.7.12 Carbon dioxide ($CO_2$), titrimetric determining process on site (cf. also Section 3.6)

Carbon dioxide refers here to the so-called free and dissolved $CO_2$.

"Free" is here used in contrast to $CO_2$ existing as hydrogen-carbonate ion or carbonate ion and "dissolved" is intended to distinguish it from the $CO_2$ rising in gaseous state.

Direct titration of $CO_2$ with sodium hydroxide solution and conversion to sodium hydrogen carbonate has proved suitable as a determining process on site. The end-point of titration is at pH 8.3 and can either be determined electrometrically with the aid of a glass electrode and a pH meter or visually by using phenolphthalein. The process is suitable for direct determination up to a concentration of some 200 mg/l of free dissolved $CO_2$. If higher values are present, the water must be appropriately diluted.

If very hard water is to be tested, i.e. with contents of calcium and magnesium ions exceeding 28 mmol/l, or if the water contains more than 3 mg/l of dissolved iron, 2 ml of a 50 % potassium sodium tartrate solution should be added to the titration solution (e.g. 0.1 m sodium hydroxide solution). This should be titrated with a suitable burette up to pH 8.3 (measured electrometrically) or to the point at which phenolphthalein changes from colourless to red.

It is recommended that a preliminary titration should always be performed in order to determine approximate consumption values. The largest proportion of the sodium hydroxide solution, and where necessary this solution together with the added potassium sodium tartrate, should then be added for the second measuring process at once and then adjusted drop-by-drop to the end-point. (Cf. also Section 3.2)

### 1.7.13 p- and m-values (acid-base consumption, $HCO_3^-$ and $CO_3^{2-}$)

The consumption of acid in a water is understood as the quantity of a strong acid in mmol/l consumed in titration until specific pH values are achieved or specific indicators are converted. If titration takes place electrometrically to pH 8.3 or using phenolphthalein as indicator, the consumption of acid, i.e. the so-called p-value is determined.

If titration is continued to pH 4.3 or with the additional use of methyl orange or a corresponding mixed indicator which is converted in this range, the consumption of acid is determined as m-value.

Similarly, the consumption of base in an acid water represents the titration value with sodium hydroxide solution.

If titration is conducted electrometrically to pH 4.3 or using methyl orange or a mixed indicator which changes colour in a corresponding pH range, the consumption of base is determined as the negative m-value. If titration takes place electrometrically to pH 8.3 or using phenolphthalein as indicator, the consumption of base is measured as the negative p-value.

The consumption of base in a natural water is essentially caused by dissolved carbon dioxide ($CO_2$). Humic acids or other weak organic acids can also play a part. The pH of a water can in such cases lie in the range 4.3 to 4. If mineral acids are present, the pH will fall below 4, so the pH measured in a water already gives an indication of the measured values to be expected.

**Determining the p- and m-values**

Approximately three drops of phenolphthalein solution are added to 100 ml of the water sample. If the solution is coloured red, it should be titrated with 0.1 m hydrochloric acid until the colour disappears. The millilitres of 0.1 m hydrochloric acid consumed correspond to the p-value and thus in essence to carbonate (ml 0.1 m hydrochloric acid · 30 corresponds to mg $CO_3^{2-}$/l).

0.1 ml of an indicator is now added to the solution titrated in this manner; the indicator changes colour approximately in the region of pH 4.3 and the titration process is then continued with 0.1 m hydrochloric acid. When using a mixed indicator (e.g. following the Mortimer process), the change of colour is from bluish-green via grey to red. The millilitres of 0.1 m hydrochloric acid consumed correspond to the m-value and can be calculated in terms of hydrogen carbonate (ml 0.1 m hydrochloric acid · 61 corresponds to mg $HCO_3^-$/l).

In the case of acid waters, titrations should be carried out using 0.1 m sodium hydroxide solution to the negative m-value (pH 4.3) and to the negative p-value (pH 8.2). (Cf. also Section 3.2)

## 1.7.14 Corrosive carbonic acid

The concept of corrosive carbonic acid is in fact non-specific. It would be more correct to refer to the corrosiveness of the water which can be caused by inorganic or organic acids giving the water a pH in the acid range, or for example for amphoteric metals to a corrosiveness which is caused by a pH in the alkaline range, e.g. larger than pH 8.

Finally, corrosiveness can exist in neutral ranges as a result of the dissolved substances contained in water, e.g. sizeable concentrations (e.g. greater than 200 mg/l) of chloride, sulphate (greater than approx. 100 mg/l), or nitrate (greater than approx. 100 mg/l)

As a rule, however, the corrosiveness of a natural water is determined by the carbonic acid. Free, dissolved carbonic acid which can exist in the water as physically dissolved $CO_2$ but also as $H_2CO_3$, can give the water corrosive properties towards metals and constructional materials (e.g. cement). From the point of view of development, waters are described as lime-corroding if they can dissolve calcium carbonate and thus also attack such constructional materials and corrode metals. If a water is supersaturated with calcium carbonate, it can be described as lime-precipitating and the formation of coatings in pipelines must be evaluated positively and heavy incrustation negatively.

Processes for estimating the corrosiveness of a water dependent on carbonic acid, based on the work of G. Axt, A. Grohmann, U. Hässelbarth. J. Halopeau and D. Meier have been published in the "German Standard Processes for the Analysis of Water, Waste Water and Sludge" (publishers: Verlag Chemie GmbH, Weinheim/Bergstrasse), issued by the Hydro-Chemical Group of Experts in the Gesellschaft Deutscher Chemiker. The principles of two tried-and-tested methods of on-site measurement and/or preparation for laboratory tests are described below.

a) Lime solvent power of a water according to Heyer

This measured value still retains a certain importance today for rapid in formation in the context of the possible corrosiveness of a water. The water samples must be treated on site so that final determination is possible in the laboratory. This means that the water sample to be analyzed is filled into a glass bottle with a rubber tube and allowed to overflow until the bottle is filled with water, free of bubbles and the contents of the bottle have been replaced several times. After the rubber tube has been carefully removed, 1 g powdered calcium carbonate is added to the bottle which is completely filled with water and the bottle sealed with the ground glass stopper so as to be as far as possible free of bubbles. The temperature of the water on site should be measured. After the full bottle prepared in this way has been transported to the analytical laboratory, the bottle seal is carefully opened, a magnetic stirring rod inserted and the bottle with the water to be analyzed placed in a thermostatic system corresponding to the temperature of the water on site, i.e. at the location of sampling.

Subsequently, the contents are stirred for two hours and then 100 ml taken from the clear supernatant liquid and titrated with 0.1 m hydrochloric acid to the m-value (either methyl orange or mixed indicator or electrometrically pH 4.3).

This measured value is related to the m-value of the water to be analyzed as measured on site.

The difference in measured value for the so-called m-value on site and the m-value after this pretreatment is one variable for the proportion of so-called "lime-corroding carbonic acid" in a water. Approximately 1 ml 0.1 m hydrochloric acid of excess measured value corresponds to a solvent power of some 50 mg/l calcium carbonate.

b) In accordance with DIN 38404, Part 10, it is possible to measure the calcium carbonate saturation by the rapid pH test.

The pH of a water changes if the water comes into contact with a precipitate of calcium carbonate. If the water is lime-corroding, the pH rises. If the water is lime-precipitating, the pH falls.

A measuring vessel of the type shown in the drawing below is to be recommended for analyses.

Apart from this measuring vessel and its accessories, a single-probe glass electrode measuring cell and a pH meter are necessary. In addition, a thermometer, calcium carbonate, and hydrochloric acid with a concentration of 1 mmol/l are required.

The water to be analyzed is fed into the measuring vessel by means of a rubber or plastic tube reaching to the bottom of the vessel until the pH measured by the single-probe measuring electrode remains constant and the temperature does not fluctuate either.

The tube should then be carefully removed from the measuring system and the pH and temperature read off. 4 g calcium carbonate is then added, care being taken that the measuring head of the single-probe measuring cell is surrounded by sludge for the pH measurement. Several minutes after adding calcium carbonate, e.g. after 2 or 3 minutes, the pH is again read off on the pH-meter. The time interval should be noted.

Evaluation

If the latter value measured for pH **after** adding calcium carbonate in the experimental arrangement described lies above the pH value measured before

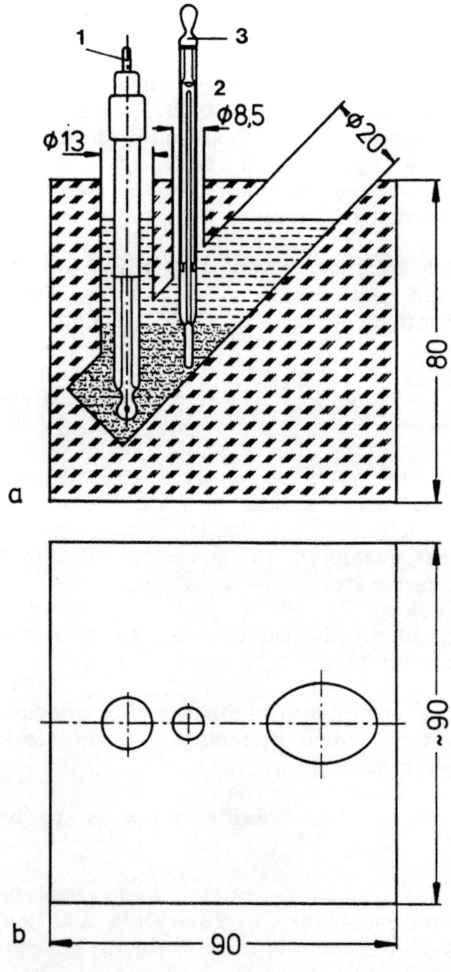


Figs. 11 a and b. a 1) = pH glass electrode measuring cell; 2) = Dimensions in mm; 3) = Thermometer; (DIN 38 404, Part 10); b Analytical vessel. Round or rectangular block for the rapid pH test

according to the same method, a water exists which can be expected to tend towards corrosiveness to lime, i.e. it will attack unprotected metals and constructional materials on a lime or cement basis.

If the pH measured according to this experimental arrangement lies below the original pH, the sample comprises a water which can be expected, for example, to tend to precipitate calcium carbonate in pipelines and vessels.

If the water to be analyzed lies in the range of the so-called lime/carbonic acid equilibrium, i.e. approximately in the range of calcium carbonate saturation, the difference in pH before and after adding calcium carbonate is smaller than 0.04.

The process should be stated when giving the results, and a record made of how large the rise or fall in pH was; indication should also be made of the temperature at which the measurements were taken and an evaluation given on whether the water should be classified as lime-corroding or lime-precipitating.

### 1.7.15 Settleable substances

The settleable substances in a water sample may be determined according to volume and according to weight. At the place of sampling it is customary to carry out determination according to volume. If the settleable substances are also to be determined according to weight, the water sample should be filtered after a settling time of 2 hours as described below. After rinsing with distilled water and drying in the laboratory, the mass of the residues on the filter should be determined by weighing.

It is necessary to determine the settleable substances immediately after the sample is taken in order to avoid the secondary formation of deposits in the water sample during transport. The volume of the settleable substances, or their mass, is increased for example by oxidation processes in the presence of air which cause the oxidation of such elements as dissolved iron. This secondary reaction causes an increase in the settleable substances. For a number of years the method using an Imhoff sediment cone has proved valuable in the determination of settleable substances. This method has also been adopted for the German Standard Processes for Waste Water and Sludge Analysis and by DIN. The determination of settleable substances by volume is possible down to a proportion by volume of about 0.1 ml/l. An Imhoff sediment cone is used, as shown in the diagram below.

The sediment cone holds a volume of 1 litre of water. It is conical in shape and the lower part is graduated. The quantity of water under investigation is transferred to the sediment cone immediately after sampling and left for 2 hours. During this settling time of 2 hours, the sediment cone should be turned sharply, for example every 30 minutes. This procedure is intended to cause the sedimentation of settleable substances adhering to the glass wall. After 2 hours the volume of the settled substances is read off and noted. Up to 2 ml/l the volume of settleable substances should be rounded off to 0.1 ml/l. If the proportion is higher, greater than 2 ml/l, the results should be rounded off to 0.5 ml/l.

If the settleable substances are also to be determined by weight, a water sample, taken in the same way, should be filtered locally through a weighed paper filter contained for transport in a Petri dish. The filter should

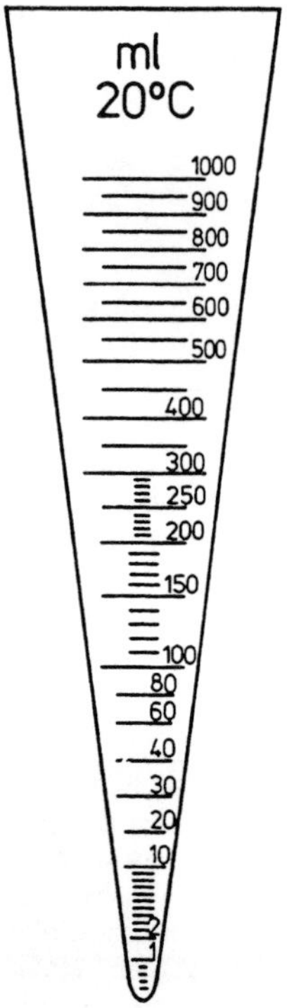


Fig. 12. Imhoff cone

previously be dried in the laboratory for 2 hours at 105 °C, and weighed. After filtration of 1 litre of the water sample, the residue on the filter is rinsed three times with distilled water and the filter returned to the Petri dish. In the laboratory, the filter is again dried for 2 hours at 105 °C and weighed, this time together with the settleable substances which were isolated locally by filtration. The difference between the two weights yields the mass of settleable substances in mg/l.

### 1.7.16 The investigation of wastes and sludges in the context of water pollution

General remarks

Waste products such as household or bulky refuse, industrial wastes, galvanic sludges, sewage sludges and similar substances may influence the quality of surface water and ground water.

To ensure harmless deposition, therefore, on the one hand the composition of the waste is of significance but on the other hand it is also important to know what soluble components may lead to pollution of the water under certain conditions. The deposition of waste substances always leads to the formation of seepage water loaded with soluble inorganic and organic substances from the waste. In order to judge whether harmless deposition is possible, knowledge of the substances soluble in water under practice-related conditions is required.

Problems are caused by the inhomogeneity of the wastes to be examined. An average sample, or at least an approximate average sample, must be taken from the material to be deposited. If the nature of the material allows, the sample should be reduced by thorough mixing. If the material is highly inhomogeneous, as in the case of refuse, it may be necessary to take a larger average sample, for example 100 kg, separating the sample into individual groups by fractionated sifting and recording the percentage constituents. In order to determine the elutability the samples should then be weighed in according to the fractionation. In the analysis of waste products it is necessary, if possible, to use uncrushed or only coarsely ground material as a basis and to select large weighed samples for the individual analyses. Experience has shown that weighed samples in the region of 1 to 10 kg are required in order to obtain reasonably reproducible results.

**Notes on the analysis of wastes**

The following investigation pattern for the homogenized sample may be considered as a guide.

Loss on drying at 105 °C
Total organic material as loss on ignition at 450 °C
Total carbon, determined by combustion analysis
Degradable organic matter determined by oxidation with potassium dichromate
Total nitrogen
Calculation of the carbon-nitrogen ratio
Total phosphorus
Qualitative spectral analysis or qualitative X-ray fluorescence analysis of the ignition residue for the determination of heavy metals
If relevant concentrations of hazardous heavy metals are suspected, these should then be determined quantitatively as total heavy metals, (The methods described here for water analysis may be used in modified form for the determination of heavy metals.)

**Further analyses which may be necessary for wastes**

Pesticides
Organohalogen compounds
Phenol-type substances
Cyanides
Extractable substances (oily and fatty substances)
Surface-active agents (detergents)
Sodium
Potassium
Calcium
Magnesium
Iron and manganese
Ammonium
Nitrite
Nitrate
Chloride
Sulphate
Total sulphur

In addition to these general analyses of the waste products themselves, **elution tests** are required for the assessment of water pollution caused by seepage water escaping from waste deposits.

An appropriately sized average sample or if necessary a fractionated average sample is likewise required in this case.

Generally speaking the material to be examined is not crushed but is examined in its original state as it would arrive for deposition. In exceptional cases, coarse grinding up to a maximum of 10 mm may be necessary.

The residues are extracted with 10 times the quantity of water. A number of variants of methods which have proven their worth in practice are suggested below. The water extracts, for their part, are then examined according to the general analysis procedures compiled in this collection of methods.

General methodological principles

Extract between 1 and 10 kg of the wastes one or more times with 10 times the quantity of rain water and/or seepage water (see below). Conduct the extractions at 20 °C. The wastes should remain in contact with the elution solution for 24 hours for each extract. During the first 8 hours, shake the preparation with the solid waste and 10 times the quantity of water for 10 minutes every hour. If larger quantities are taken, e.g. 5 kg of waste and 50 kg of water or 10 kg of waste and 100 kg of water, roll for 10 minutes every hour in appropriate plastic containers. The preparation may also be shaken or rolled continuously for 24 hours. The experimental procedure should be noted in the report.

After a 24-hour contact period, separate off the supernatant water, filter through a coarse filter and carry out the analysis.

If the residue is to be extracted a second or even third time so as to determine the extraction gradient, filter off the undissolved residues from the first elution and treat them again under analogous conditions, without further drying, with the same quantity of elution fluid for 24 hours.

Elution fluids

a) Rain water

Deionized or distilled water is generally used to simulate rain water.

b) Seepage water (such as may occur at waste disposal sites)

In this case rain water is used, or distilled water or deionized water which is almost saturated with carbon dioxide. Saturate with carbonic acid by passing $CO_2$ through water at about 4 °C in a plastic cask. Once the saturation process has begun, heat slowly overnight to 20 °C and then determine the content of free, dissolved carbon dioxide according to the procedure described in this collection of methods. Carbon dioxide concentrations in the region of about 1500 to 2000 mg/l of water are obtained if these test conditions are followed.

c) Dilute seepage water

Take a volume of the water obtained according to the method described under b), saturated with carbonic acid and dilute with distilled or deionized water to a concentration of free, dissolved carbon dioxide of about 500 mg/l.

**Investigation scheme** for the analysis of the water extracts of solid, liquid or sludge-like wastes.

Analyze separately the aqueous extracts obtained according to the methods described (rain water extract, seepage water extract with high carbon dioxide content and seepage water extract with low carbon dioxide content), and if appropriate determine the elution gradient.

The following may be determined in the aqueous extracts:
pH
Electrical conductivity at 25 °C
Redox potential in mV
Evaporation residue 180 °C
Ignition residue of the evaporation residue at 450 °C
m-value
p-value
So-called total hardness
So-called carbonate hardness
So-called non-carbonate hardness

Oxidizability with potassium permanganate
a) as potassium permanganate consumption
b) as oxygen consumption in mg $O_2$/l

Oxidizability with potassium dichromate (chemical oxygen demand, COD) as mg $O_2$/l

Biochemical oxygen demand (BOD) as mg $O_2$/l

Chloride
Sulphate
Phosphate
Nitrate
Nitrite
Ammonium

Qualitative spectral analysis of the evaporation residue in order to confirm the presence of toxic heavy metals in relevant concentrations. If appropriate, the heavy metals detected must be determined quantitatively in the seepage water extracts.

Cyanides
a) Total cyanide
b) Easily liberated cyanide

Phenols
a) Total phenols
b) Phenols volatile with water vapour

Organic solvents

Surface-active agents (detergents)

Oily and fatty substances extractable with petroleum ether
Polycyclic aromatic substances
Organohalogen compounds
Pesticides
Further analyses, depending on the origin of the wastes and suspicion of particular harmful substances.

In the case of the sludges containing water, take the first water extract by separating off this water by filtration or centrifuging. Subsequently extract the sludges a second or third time as described above.

The investigation results thus obtained can be used as a basis for assessing whether and to what extent pollution of the ground water or surface water may occur as a result of elution of the wastes. At the same time it is possible to predict whether alkaline-reacting substances may, in the course of the flow path, cause precipitation of e.g. iron, manganese, calcium, magnesium or heavy metals.

Using the relationship of COD to BOD as a basis, it is also possible to predict the expected biological degradability of dissolved organic substances.

Of course, the same demands cannot be made of seepage water from waste disposal sites as of drinking water. A comparison with the parameters allowed for waste water is conceivable. In certain cases higher loading of the seepage water may also be acceptable if the seepage water reaches the ground water and is retained there for a long period as a result of the slow flow rate ($<1$ m per day) and hence aerobic or anaerobic biological decomposition or precipitation and sorption is to be expected. Drinking water should not be extracted from the ground water within the catchment area. The minimum distance removed may be estimated in the region of 1000 m.

If the wastes are to be deposited on a site with a sealed base, an investigation following the above pattern may be employed to predict whether the organic components of the seepage water can be biologically degraded during the retention time in a sewage plant. Particular attention should be paid to the heavy metal component, which may lead to disruption of the functioning of the sewage plant.

For more detailed assessment of this complicated problem refer to the specialized literature.

### 1.7.17 The importance of site inspections for biological assessment of water resources (cf. also Sections 1.2, 1.4 and 1.5 above)

The results of microbiological analyses can only ever reflect the condition of water at a precise moment in time. The types of germs detected and the relative frequency with which certain of them occur in the sample allow conclusions to be drawn as to the possible nature of contamination. But in many cases where analysis indicates the presence of water pollution, this cannot be explained by the test results themselves; this is also often true with the results of chemical analysis. In order to interpret the findings of laboratory analysis, it is therefore essential to have precise knowledge of the locality where sampling was carried out and the surrounding area. A thorough site inspection is vital, especially when taking samples from a particular water source for the first time or if there is a sudden and serious change in the condition of water which has already been analyzed regularly over a long period.

With ground water recovery systems the following points must be considered:

Design faults and structural damage to the recovery system:

Are structure and system suitably designed and are there any cracks or other weak points through which surface water and rain water, animals or vermin can penetrate?

Is it possible that unauthorized persons can gain access to the recovery plant and pollute the water?

Area surrounding water recovery point and nature of catchment area:

Are the water recovery system and catchment area located in an area where there is a particularly high risk of ground water pollution, e.g. owing to farming, seepage of sewage, nearby refuse tips or centres of population etc.?

Are there other sources of groundwater in the immediate vicinity which could be the cause of pollution (e.g. old wells, test drillings, mine shaft etc.)?

What type of soil is to be found in the catchment area (is the soil permeable and does it lack suitable covering layers)?

Is there a possibility of particularly high infiltration of surface water in the catchment area, e.g. streams, marshy ground after heavy downfall, flood zones?

Do any sewers, water mains or pipelines, roads, etc. pass through the catchment area?

Has building work involving excavation or earth moving taken place in the catchment area, or alternatively any type of dumping?

Obtaining information from local residents and others familiar with the area concerning past conditions:

Have there been refuse tips, building work, sewers or interference with the natural state of the soil in the vicinity of the present catchment area which are no longer visible?

Have there been any other events which could influence the nature of the ground water (this category should include isolated occurrences, e.g. accidents in the catchment area, events such as music festivals and gatherings of campers and the like, natural disaster etc.)?

If there are water towers or other storage facilities in the locality, questions should be asked during the site inspection both about the condition and suitability of the structure (stagnant water in the storage tank owing to inadequate flow, stagnant water in parts of the system through which there is now regular flow, e.g. chambers for fire-fighting reserve, water-level indicators and air chambers in the by-pass system, dead-end pipes, etc.) and also about other possible factors affecting the water.

Can insects, animals or unauthorized persons enter the structure and pollute the water?

Are the overflows and drain-pipes fitted with flaps in working order which prevent small animals from entering the storage tanks where they fall into the water and die?

Is the building equipped with suitably designed ventilation to eliminate the risk of water pollution by air-borne germs and dust? From which direction does the wind mainly come and are there any sources of regularly or sporadically high air-borne germ or dust levels in this direction?

Are there any treatment plants which may cause microbiological contamination of the water (e.g. open plants, sewage irrigation plants or filters colonized by microbes, metering systems with infected chemical solutions)? Are they the correct size for the quantity of water passing through them and how regularly are they maintained?

The site inspection must also establish how the storage tanks are operated by the responsible waterworks:

How long does water remain standing in the individual storage tanks? How often are the latter cleaned and when did this last take place? How are they cleaned and what products are used?

What measurements and checks are made regularly by the waterworks and could these cause contamination of the water, e.g. dirty sounding lines, scoops or other equipment?

If there is a local water distribution network, the site inspection should also establish the size of pipelines used, how and when they were laid and what material they are made of.

Are the pipes laid in trains or in a closed circular layout? Are there high and low points in the pipe trains which could cause so-called water pockets or air pockets to occur?

Are there connecting pipelines in which water stagnates for long periods and becomes colonized with germs (e.g. pipes which supply gardens or connection branches ready fitted for future pipeline extensions)?

How frequently are the pipelines flushed through and, when this is done, is there any recognizable contamination (turbidity, brownness, black coloration of the water plus hydrogen sulphide smell, etc.)?

Are there any users attached to the supply network whose water consumption has sudden peaks causing temporarily high rates of flow in the system as a whole?

Does water in the pipe trains always flow in the same direction or can reverse flow occur, e.g. with pressure and gravity pipelines?

Is water from various recovery plants mixed within the supply network?

Is it possible that private water recovery plants feed into the public supply network with which they are connected either directly or indirectly (it should always be a requirement that private water supply systems are clearly separated from the public supply network)?

Are the connecting pipelines between the individual areas supplied and the public network fitted with valves to prevent water being sucked back if a pressure drop occurs in the network?

Have any large-scale repairs been made to the distribution system or any new pipelines laid, and what steps were taken by the waterworks to prevent any pollution of the existing system (intermittent chlorination, flushing, microbiological tests on the new pipelines, etc.)?

When performing site inspection and water sampling, the water recovery systems should be inspected for biological indications of possible pollution.

Have the roots of nearby trees and plants grown into the fabric of the water recovery systems creating so-called "fox-tails"? (Where these "foxtails" occur in test diggings they hinder the flow of ground water into the spring-intercepting structure by altering the cross-section of the pipelines. The roots provide a path for microorganisms, small animals and also polluted surface water to enter the recovered water).

Do the test diggings, spring-intercepting structures or other water recovery systems bring with them undissolved particles, such as grains of sand, clay or silt, which are deposited in the systems themselves or in the storage tanks providing a breeding ground for microbes and other micro-organisms or bringing about anaerobic conditions and decomposition?

Do well worms (Tubifex species), well shrimps (Gammarus) and water hog lice (Aselus) occur in the water recovery systems? The presence of such animal organisms in drinking water systems is not only unappetizing but is also an indication that harmful influences may be present. Although these creatures are not themselves pathogenic and as far as is known cause no pollution of the water worth mentioning, it is nevertheless possible that other hygienically undesirable influences are finding their way into the water by the same route. Thus it is essential to discover and cut off their means of access.

It must also be established whether springs and test diggings which are near the surface have a constant rate of water yield. If they show a marked increase in yield immediately after precipitation, or even deliver turbid water, this means that they are not adequately protected against hygienically harmful influences and must be regarded as at risk.

Site inspection is especially important with surface water sources (streams, ponds, drainage ditches and artificial ground water sources in building foundations, gravel pits, etc.). Here plant and animal colonization provides a good means of assessing whether anthropogenic influences are present; for example, accumulations of foam on the water surface are generally a visible sign that waste water containing surface-active agents is entering the system. (Foam can occur naturally, especially during spring in the proximity of coniferous forests as a result of saponins leached from the fallen needles. But unlike surfactant foams, this natural sort gradually turns brown with age).

Deposits of brown iron oxide flakes on plants and stones in the water indicate a water source with high iron content flowing into the surface water. Often the ferric hydroxide only separates out once the iron-containing, oxygen-deficient water mixes with the surface water which has a high oxygen level. This means that in flowing water iron deposits on the river bed sometimes do not start until far below the actual point of affluence.

Colonization of the water by the grey, shaggy filaments of the bacterium Spherotilus natans indicates that the water is serving as a receiving body for a sewage disposal plant and that it is being contaminated by household sewage. The occurrence of Spherotilus natans in this type of water is so frequent and characteristic that the bacterium used to be known as "sewage fungus". Although this description is inaccurate from a microbiological point of view, since Spherotilus natans is in fact a filamentous bacterium, it is still no less important today as an indicator of the presence of sewage and so the hygiene of any stream where it is discovered must be regarded as suspect.

On the other hand, deposits of lime on the underwater parts of plants or sinter deposits on stones in the bank area of the stream bed are not symptomatic of a lack of hygiene. This type of phenomenon frequently occurs in springs which have their origin in chalky ground if the so-called lime-carbonic acid balance of relatively hard water is disturbed. Such disturbances can be caused either by the water becoming heated during its passage or by mosses and higher plants growing in the water and drawing off carbon dioxide.

As a rule sedimentation in rapidly flowing mountain streams is minimal, whereas in slowly flowing, lowland streams sediments are usually to be observed on the stream bed. In non-polluted flowing water these sediments are mainly composed of fine sand and clay particles, are not black in colour and do not have a putrid odour. If either of the latter symptoms is recognizable, this is always an indication that the water is polluted with organic substances which are contained in the sediments and are being decomposed by microbes under anaerobic conditions.

The nature of the sediment is an even clearer indicator of contamination in stagnant water than in flowing water and so should always be examined when mak-ing an assessment of long-term water quality. When making site inspections of either stagnant or slowly flowing waters, it is particularly important to observe whether any gas bubbles occur and, if so, whether the gas rising is digester gas.

In addition to the general observations to be made during site inspection, as described above, a more specialized study of the plants and animals in an area of surface water can yield vital information about water quality. Such studies of fauna and flora and microbe life (colonization of the plankton and seston by microorganisms) lead to the classification of water quality according to saprobic systems, but this goes beyond the scope of a simple site inspection and often requires the use of particular equipment and sampling techniques as well as demanding specialist knowledge of the relevant fauna and flora.

When it comes to analyzing the results of a site inspection and producing a report, it has always proved advantageous to supplement these with photographs of the various items dealt with. These not only make the written report more easily comprehensible but also provide a long-term record of developments and changes, so facilitating the accurate assessment of water resources and drinking water supply systems.

# 2 Theoretical Introduction to Selected Methods of Water Analysis (Classic and Instrumental Methods)

## 2.1 Concentration processes such as evaporation, distillation, precipitation, coprecipitation, adsorption, ion exchange and extraction

### General introduction

Depending on their type, substances dissolved in water can occur in a broad range of concentration according to the type of water and its genesis or origin.

There are natural waters which are almost like distilled water, and strongly saline waters such as seawater or so-called brines; the latter can comprise up to more than 100 g/l dissolved salts. The level of dissolved mineral substances for drinking water can be estimated as ranging up to approximately 1,500 mg/l with an optimum range between approximately 100 and 1,000 mg/l.

Waste waters can contain large amounts of organic substances, but also of industrial wastes and hazardous trace elements. The water analyst then has to decide whether he can examine a water sample for one or more substances according to selected methods directly, or whether he must select a concentration process for the substance concerned, and at the same time the disturbing substances in the water sample concerned must be depleted. The customary determination methods used in water analysis, e.g. physico-chemical measurements such as pH value, oxidation-reduction potential, conductivity, and also the detection of carbon dioxide or hydrogen carbonate ions, calcium, magnesium, chloride or sulphate do not in general require pretreatment of the water, although in this case, for example for highly saline waters (brines), dilution may be necessary rather than a concentration process.

It is often necessary in water analysis to remove the water itself by means of evaporation or distillation, or to separate off certain substances from disturbing constituents by means of precipitation, adsorption or extraction.

### 2.1.1 Evaporation

With this method, a measured or weighed water sample is heated in a suitable evaporation vessel until the water is almost completely evaporated and only those substances which were previously dissolved in the water and are not volatile at the selected temperature remain in the dish. The vessel can be a weighed platinum, glass or porcelain dish, and the sample is heated over a boiling water bath, a heated sand bath, an electric hot plate or an air bath. The evaporation residue can then be dried to constant weight and weighed. Various drying temperatures are used, e.g. 105 °C, 180 °C or 260 °C. If the water contains a great deal of dissolved sodium chloride, the high drying temperature is to be used if constant weight is required. This is so as to be certain that the so-called decrepitation water is expelled from the forming NaCl crystals. The dry residue can then be used for analysis or for further separation processes. Spillages are of course to be avoided, especially when the evaporation process is nearing its end. It should also be noted that during evaporation the compounds which are more difficult to dissolve separate out first and the readily soluble compounds in the water such as sodium chloride only later. This means that either the entire dry residue must be used for further analysis or, if aliquot parts are to be used, the residue must be homogenized thoroughly beforehand.

### 2.1.2 Distillation

In this case the water is gently distilled off in suitable distillation apparatus leaving in the flask either a highly concentrated solution or, if distilled to dryness, a salt mass. Distillation to separate off the water itself is carried out in glass or quartz apparatus. It is possible, if distillation is conducted appropriately, to recover highly volatile substances (such as organic substances) as the distillate is cooled (e.g. in a cold trap). By means of distillation under acid or alkaline conditions, with or without the use of additional water vapour or reducing or oxidizing agents, certain groups of substances can be separated off in such a way that they can be analyzed in the distillate with no disturbing factors. Examples of this procedure would be the isolation from acid solution of phenol-type substances volatile in water vapour, or the isolation of trivalent arsenic from total arsenic if a reducing agent for As V is used, and the isolation of cyanides, fluoride, ammonium and organic nitrogen compounds. It is also possible to isolate carbon dioxide, either bonded for example as hydrogen carbonate or carbonate, or in the free state, physically dissolved in the water, as well as hydrogen sulphide and its compounds, boric acid or mercury. In general, any organic substances can be isolated which are volatile under the selected distillation conditions and which are recoverable by condensation of the water vapour.

### 2.1.3 Precipitation

Precipitation was very frequently used in the past as a method of determining certain substances in water by direct gravimetric analysis. Even today, for example for sulphate determination, precipitation from an acidified water sample with barium chloride solution, yielding low-solubility barium sulphate, is not only used for concentrating the sulphate, but also for direct gravimetric analysis (see Section 3.2).

The precipitation of chloride and, if present, dissolved bromide and iodide as the total of dissolved halide ions by silver salts as low-solubility silver halides is also not only used for concentrating traces of bromide and iodide for example, but also for gravimetric analysis, as a method of calibration. Bromide and iodide should then be determined separately (Section 3.2) and subtracted from the total of the precipitated halides. Reliable results are then obtained even at high concentrations.

### 2.1.4 Coprecipitation

This method is closely related to precipitation, but makes use of the fact that when certain substances are precipitated other substances are coprecipitated. In radiochemistry in particular, coprecipitation with isotopic or non-isotopic carriers is still commonly practised today. In classical water analysis coprecipitation is often used to concentrate trace elements. In this way, for example, arsenic can be more or less completely coprecipitated and isolated with iron (III) hydroxide. With manganese in a manganese (III) or manganese (IV) compound, thallium can be coprecipitated and concentrated. Sections 3.2 to 3.5 and Section 3.7 deal with cases where precipitation and coprecipitation are still in general use in water analysis. Section 3.3 in particular covers the concentration of rubidium and caesium together with potassium as tetraphenylborates.

### 2.1.5 Adsorption and ion exchange

If appropriate adsorbents are chosen, e.g. activated charcoal, aluminium oxide, cellulose etc., it is possible to isolate certain substances in water which are otherwise difficult to concentrate. Activated charcoal, for example, is particularly suitable for isolating many organic substances in water. Aluminium oxide is used to isolate and concentrate fluoride, arsenic, phosphate and uranium from aqueous solutions which have been set to approx. pH 6 with carbonic acid. Certain types of cellulose are used for the adsorptive concentration of traces of heavy metals.

Adsorption is carried out either with the so-called batch method or by filtration through a column. It is possible to conduct fractional desorption in many cases, for example by using acids or organic solvents. Certain substances in water can also be concentrated by means of ion exchange resins. By using an appropriate ion exchanger, e.g. cation or anion exchangers of certain types, it is possible not only to separate cations and anions from water samples, but also to differentiate various groups of substances. Elutions with suitable solvents are possible, and the eluates can then be examined separately for the isolated substances (Chapter 3).

Ion exchange is also a good method of determining the total dissolved mineral substances in a water (see Section 3.1).

### 2.1.6 Extraction

Liquid-liquid extraction has become a very important method of concentrating trace substances, in particular metal traces, and of removing disturbing substances. For extraction, certain substances in water can be transformed directly into an organic solvent which is not miscible with water, e.g. traces of oily or fatty substances with hexane, chloroform or trichlorotrifluoroethane. The inorganic compounds are often transformed into an extractable compound and then isolated by extraction. This method is fundamental, and has become highly important in trace analysis. One only has to think of complexing by means of dithizone, diethyl dithiocarbamate etc. using a suitable extracting agent such as chloroform or diethyl ether.

Methods of concentration by extraction are described in Chapters 3 and 4.

Traces of uranium can be extracted from a water sample in the presence of high concentrations of aluminium nitrate with a suitable organic solvent and concentrated for fluorometric determination, for example.

### 2.1.7 Volumetric analysis

In this procedure a substance dissolved in the water is reacted with a titrant of known concentration. The end point of the reaction is indicated either by an indicator changing colour, by a precipitate forming or electrometrically. Volumetric analysis is simple to perform, and, with visual recognition of the endpoint of the reaction, can also be carried out in a laboratory which does not have the latest equipment. It is possible, for example, to carry out the quantitative determination of chloride, hydrogen carbonate or carbonate, or of calcium or magnesium, as well as organic compounds such as surfactants by means of so-called two-phase titration.

Today, volumetric determination of dissolved substances in water can be automated. With appropriate equipment, it is possible not only to direct the reaction quantitatively, but also to evaluate the endpoint of the reaction electrometrically and print it out, for example, or produce a titration diagram. The experience of the authors has shown the volumetric procedures described in Chapters 3 and 4 to be of particular value in practical water analysis in the laboratory.

## 2.2 Electrochemical processes of analysis

### 2.2.1 Introduction

A number of analytical methods are based on electrochemical properties in solutions. If, for example, two metallic conductors are immersed in an electrolyte solution, current can flow when an electric potential is applied. If two different metals are present in the electrolytic cell, an electric potential can be tapped. Its force is dependent on the type of electrode materials and on the composition of the solution, on the gap between the electrodes and on the electrode surface.

The separations and displacements of charges at interfaces can be determined with the aid of electrodes or even produced or changed by applying a current. The molecules in the dissolved substances are partially polarized as a result of anisotropy at the interface. Behind this interface, the solvated ions are enriched or depleted, causing the formation of an induced charge on the electrode surface. This leads to the establishment of an equilibrium dependent on the type of electrode and electrolyte. The following types of electrodes are distinguished, according to the composition of the system and the factors determining the potential difference between electrode and electrolyte:

Electrodes of the first type:

A metal electrode is here immersed in a salt solution of the same metal. The potential difference depends on the activity of the cation in the solution, e.g. $Zn/Zn^{2+}$.

Electrodes of the second type:

These comprise metal electrodes coated with a thin layer of sparingly soluble salt of this metal. The potential difference depends on the activity of the anion of this salt in the solution, e.g. $Ag/AgCl/Cl^-$.

Electrodes of the third type:

The sparingly soluble salt layer also contains a second cation which forms together with the joint anion a sparingly soluble compound with a larger solubility product than the electrode-metal compound. The potential difference is dependent on the activity of the second cation in the solution, e.g. $Ag/Ag_2S/Cu/Cu^{2+}$.

Redox electrodes:

The electrode here consists of a predominantly inert metal and the solution contains no ions of this metal. The potential difference depends on the normal redox potentials of the redox system in the solution, e.g. $Fe^{2+}/Fe^{3+}$. Sub-systems of such a corresponding redox pair are joined together by the processes of oxidation (electron donation) and reduction (electrode acceptance).

Ion-sensitive electrodes:

Metal electrodes invested with different materials (semiconductors, glass, ion exchangers) are immersed in a solution containing the ions to be determined. These ions distribute themselves between electrode and solution phase. The potential difference is selectively dependent on the activity of the ions to be determined. Example: Fluoride electrode.

The electrode potential can be changed by applying a voltage. As a result of this change (polarization), the potential of the electrode deviates from the value obtained using the Nernst equation. A concentration polarization results if enrichment or depletion of ions takes place in the vicinity of the electrode. This produces an electromotive force opposite to the voltage applied. The chemical polarization is based on the fact that during the passage of current on the electrode surface, substances are produced preventing further passage of current. Here too, a polarization voltage arises opposite to the voltage applied. Current cannot start to flow again until the voltage applied attains the same level as this decomposition voltage.

The following electrochemical processes of analysis are discussed below:

a) Coulometry
This process of analysis is based on Faraday's laws, i.e. the relationship of equivalence between total electric charge and chemical reaction.

b) Potentiometry
This process is based on the relationship between the concentration of the ion located in the solution and the electromotive force of an electrochemical cell in which this ion is one of the components.

c) Voltametry
Measuring with this process uses cells in which the one electrode acts as a non-polarizable reference electrode and the other as a polarizable inert electrode. The change in current is recorded against the change in voltage applied. A special case in voltametry is polarography in which, for example, a dropping mercury electrode is used as polarizable electrode. In amperometry, a further special field of voltametry, two polarizable electrodes are used.

## 2.2.2 Coulometry

If DC voltage is applied to two electrodes immersed in an electrolyte solution, the positive ions migrate to the cathode and the negative ions to the anode under the influence of the electric field. Processes of exchange take place on the electrode surfaces. This reaction between the dissolved substance and the electric charge is called primary reaction. It may be accompanied by secondary reactions.

According to Faraday, the amount of any substance primarily dissolved or deposited on an electrode is proportional to the total electric charge passed. The electric charge Q should here be understood as the total electric charge passed in time t.

$$Q = \int_0^t I \, dt \tag{1}$$

The electric charge is given in Coulomb. (1 Coulomb = 1 A · s).

Faraday's laws state that:

1) The amount of any substance dissolved at or deposited on an electrode is proportional to the electric charge necessary for this reaction

$$m \sim I \cdot t \tag{2}$$

2) The amounts of different substances dissolved or deposited by the passage of the same electric charge are in the ratio of their equivalent weights.

$$\text{Equivalent weight} = \frac{\text{Atomic weight}}{\text{Oxidation number}} \tag{3}$$

The electric charge 96487 Coulomb (Faraday constant F) is required to separate an equivalent gram of a substance. If G is the weight of the converted substance in grams, n the quantity of electrons involved, and M the molecular weight of the substance, then the following applies:

$$C = \frac{M \cdot Q}{n \cdot F} = A \frac{i \cdot t}{F} \tag{4}$$

There are essentially two different coulometric processes, namely potentiostatic and galvanostatic coulometry. The former functions with constant, controlled electrode potential, whereas the galvanostatic method - also called coulometric titration - functions with constant current strength and uncontrolled potential. Fig. 13 shows the basic circuit diagram for potentiostatic coulometry.

The potential for working electrode $E_1$ against the reference half-cell $E_3$ is controlled by regulating the voltage applied to cell $E_1E_2$ so that the deflection of the galvanometer G remains constant. The substance to be determined must be dissolved or deposited quantitatively and with 100 % current yield on the electrode. The substance must diffuse on the electrode surface, which means that the current required for electrolysis must not exceed the current required for a diffusion.

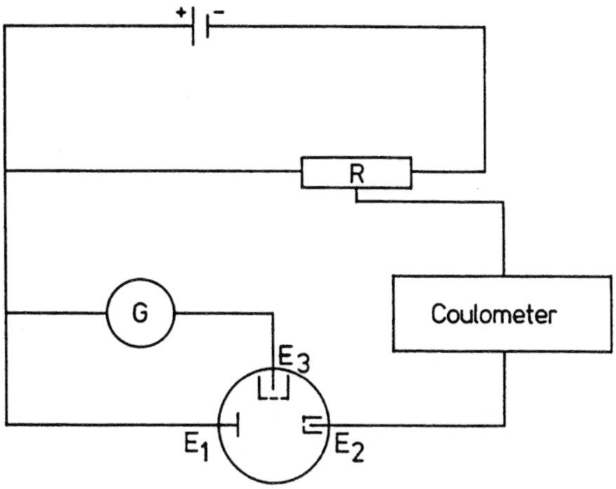


Fig. 13. Coulometry with U = const.

As the concentration of the substance to be determined falls as a result of dissolution or depositing on the electrode, the electrolytic current also approaches zero asymptotically if the electrolytic process is conducted with a potential less than the half-wave potential of the substance in question. The electric charge consumed is measured with a coulometer or determined by integrating Equation (1). The advantage of this process is in the selectivity of the electrode processes, e.g. a determining process can be conducted for two metals with similar depositing behaviour.

Fig. 14 shows the basic circuit diagram for coulometric titration. As the potential of the working electrode $E_1$ is not controlled, the experimental conditions must be selected so that no side reactions can occur.

For this reason, a substance whose electrically produced reaction products react quantitatively with the substance to be analyzed is added to the

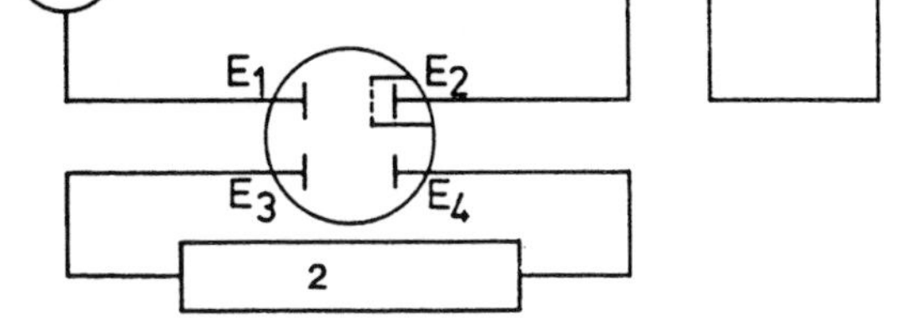

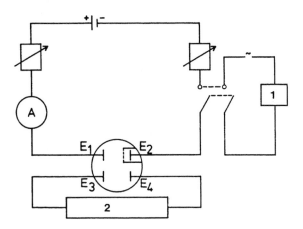


Fig. 14. Coulometry with I = const. and electric end-point determination; 1) = clock; 2) = indication

electrolyte solution. This intermediate reagent is produced at constant current. The time during which the current flows is measured to determine the electric charge.

If the end-point is determined electrically, two separate circuits are necessary. The "generator circuit" with electrodes $E_1$ and $E_2$ serves to produce the titration agent, the "indicator circuit" with electrodes $E_3$ and $E_4$ is used for end-point determination.

The advantages of coulometric titration are as follows:

- exact proportioning in the microgram range
- production of very pure reagents
- production of non-stable reagents
- no standard solutions are necessary
- no determination of titre is necessary
- high speed

In principle, all processes used in volumetric analysis can be used for indicating the end-point. Essentially, however, coulometry is restricted to electrical processes as these allow a high degree of automation. Any of the usual potentiometric indicator systems can be used for end-point determination, e.g. a Pt-electrode and a calomel electrode. The choice of the indicator system depends on the reaction occurring. If the indicator potential measured is plotted against the period of electrolysis, the usual potentiometric titration curves can be produced. The information required can be gained from the potential jumps.

In amperometric end-point determination, the strengths of the diffusion current are measured at prescribed potentials in the indicator electrodes. In this case, the indicator current strength is proportional to the concentration of ions involved. For this purpose, a potential of 100 - 300 mV is applied to two electrodes (double Pt-plate). The indicator current strength is plotted against the period of electrolysis. According to the type of reactions involved, different curves can be observed (Fig. 15):

a) The current is low and almost constant as far as the equivalent point and then rises linearly. - Example: $As^{3+}$ + $Br_2$

b) The current passes through a maximum, falls to the equivalent point and then rises linearly. - Example: $I^-$ + $Br_2$

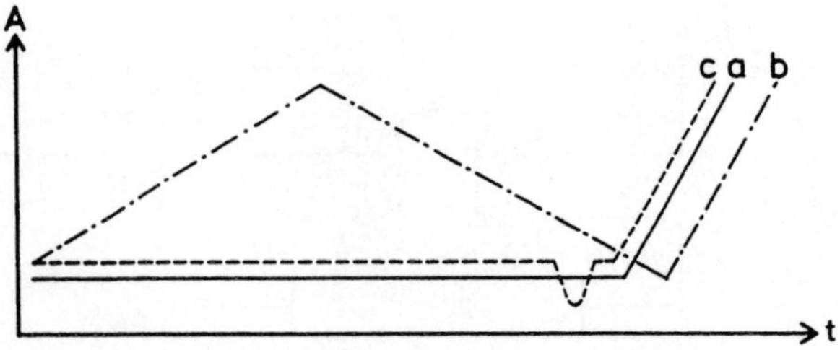


**Fig. 15.** Indicator current A as function of time t

c) The curve is similar to that described under a), but shows a dip shortly before the equivalent point. - Example: titrations with bromine or chlorine with high generator current strength.

A special case in amperometric indication is the "Dead-Stop Method". Current can only flow at the indicator electrodes if both oxidizable and reducible substances are present. The potential difference may be very low (<100 mV). The equivalent point is characterized by the fact that the indicator current is zero, as no reversible ion pairs are present here (e.g. coulometric titration of $Fe^{2+}$ with $Ce^{4+}$).

### 2.2.3 Potentiometry

The relationship between the concentration of an ion in solution and the e.m.f. of the cell in which this ion is located can be expressed under ideal conditions by the Nernst equation.

$$E = E_0 + k \cdot \ln c$$

where E represents the potential in the cell, $E_0$ a constant at certain temperatures, c the concentration of the ion and $k = R \cdot T/n \cdot F$ (R = general gas constant, T = absolute temperature, n = quantity of electrons produced in the semireaction, F = Faraday constant.).

The above-mentioned equation only applies to infinite dilution, whereas the following applies to real solutions:

$$E = E_0 + k \cdot \ln a$$

i.e. the activity a (calculated from the concentration and the coefficients of activity f using the formula $a = c \cdot f$) is used for practical calculation.

If two electrodes are immersed in a solution whose ions can react with the electrode, and the circuit is closed, the potential existing at the electrodes can be measured. This is characteristic of the system electrode/solution/ electrode and dependent on the temperature, pressure and composition of the system. In two electrodes consisting of metals A and B, the following reactions can occur:

$$A \rightleftharpoons A^+ + e^-$$

$$B \rightleftharpoons B^+ + e^-$$

where the total reaction $A + B^+ \rightleftharpoons A^+ + B$, i.e. A is oxidized at the first electrode and produces $A^+$, and $B^+$ is reduced at the second electrode and precipitates metal B. There is therefore a charge transfer between electrode and solution. If one of the reactions is predominant, a potential difference is created between electrode and solution as a result of the charge transfer. The size of the potential difference is above all dependent on the nature of the semireaction, and further on the activity of all substances participating in the reaction. This relationship is expressed by the Nernst equation. $E_o$ represents here the characteristic constants for the semireaction and expresses the potential difference between electrode and solution in equilibrium status (standard potential). If all substances affected in the semireaction have an activity of 1 mol/l, then $E = E_o$.

If the given constants are entered in the Nernst equation, the natural logarithm then converted to the Briggs logarithm, and the activities finally replaced by the concentrations, the result is the Nernst equation in a form that can be used for practical quantitative calculations.

$$E = E_o + \frac{0.0591}{n} \cdot \log \frac{\text{oxidation}}{\text{reduction}}$$

In practical terms, it is impossible to measure the potential difference in a semireaction directly. This difficulty is overcome by introducing a second electrode as reference electrode. The "standard hydrogen electrode" whose semireaction is expressed by the formula:

$$2\ H_3O^+ + 2\ e^- \rightleftharpoons H_2 + 2\ H_2O,$$

can be used for this purpose. It consists in principle of a platinum wire covered with platinum sponge and surrounded with flowing gaseous hydrogen, so that a layer of absorbed $H_2$ molecules is formed.

The potentiometrical principle of measurement is, for example, used in measuring pH with a glass electrode or in determining the concentration of ions with the aid of ion-sensitive electrodes.

The glass electrode comprises a thin-walled glass sphere filled with a solution of known, constant pH and immersed in the solution of unknown pH which is to be examined. Two bridge electrodes are immersed in the internal and external solution, e.g. two saturated calomel electrodes. In this case, we have the following cells:

$Hg/Hg_2Cl_2KCl_{tot.}$/internal solution/external solution/$KCl_{tot.}Hg_2Cl_2/Hg$.

The entire measuring set-up can be represented as in the following diagram:

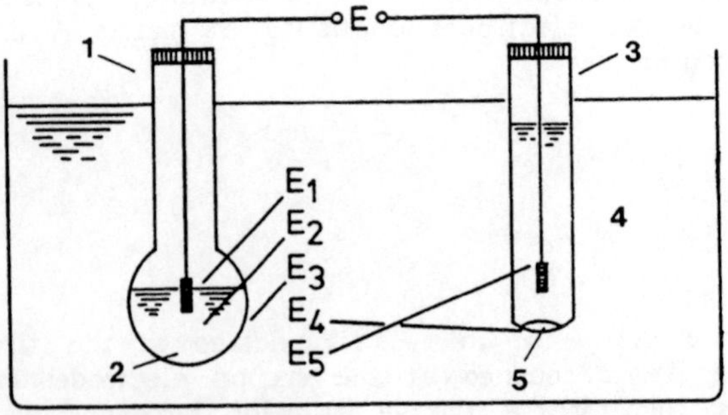


Fig. 16. Schematic diagram of the potentiometric method of measurement; 1) = glass electrode; 2) = Glass membrane; 3) = Reference electrode; 4) = Measuring solution; 5) = Diaphragm

The individual voltages $E_i$ are produced at the following interfaces:

$E_1$ Between the bridge electrode and the solution inside the glass electrode.
$E_2$ Between glass membrane and solution inside the glass electrode.
$E_3$ Between glass membrane and measuring solution.
$E_4$ As diffusion voltage between measuring solution and solution inside the reference electrode.
$E_5$ Between bridge electrode and solution inside the reference electrode.

$E_1$ and $E_5$ can be kept of the same size by setting up the electrodes so that they cancel each other out (symmetrical cell). If they are not of the same size, they are nevertheless independent of the pH of the measuring solution and supply the constant asymmetrical voltage $E_{as}$ as quantity to be added to the measured voltage.

$E_4$ as diffusion voltage is not described by the Nernst equation (Equ. 3). It is considerably dependent on the specific conditions and thus comprises a source of error. However, changes in the diffusion voltage usually only occur very slowly and can be taken into account by calibration. The quantity of $E_4$ seldom exceeds a few mV.

The voltage in the measuring cell is thus essentially determined by $E_2$ and $E_3$. As $E_0$ cannot be measured independently, and only the difference between the two voltages is of interest here, the following equation can be made:

$$E = E_3 - E_2 \qquad \text{(Equ. 4)}$$

Inserting the Nernst equation produces the following result:

$$E = 2.303 \frac{R \cdot T}{n \cdot F} \log a_{H_3O^+ outside} - 2.303 \frac{R \cdot T}{n \cdot F} \log a_{H_3O^+ inside}$$

Rewriting the equation yields the following:

$$E = 2.303 \frac{R \cdot T}{n \cdot F} (pH_{inside} - pH_{outside})$$

or

$$E = 2.2303 \frac{R}{n \cdot F} \cdot T \cdot \Delta pH$$

The above equation shows the temperature-dependence of the measuring sensor, which must be distinguished from the temperature-dependence of the pH in the solution analysed.

Voltage E is also designated as the Nernst voltage for $\Delta pH = 1$. It has the value of 59.16 mV/pH at 25 °C and determines the gradient of the characteristic curve.

In order to keep the pH inside the glass electrode as constant as possible and to make it independent of ageing phenomena (e.g. by exchange of ions with the glass membrane), a buffer solution is used inside the electrode.

In addition to the dependence on temperature described above, the linearity of the characteristic curve also plays a part.

Particularly in the alkaline range from approx. pH 10 upwards, the characteristic curves for normal glass electrodes deviate noticeably from a straight line. This is due to the fact that the glass electrode does not strictly respond specifically to the hydrogen ions only, but is also affected by hydroxide ions.

If the concentration of hydrogen ions is equal to or larger than that of the hydroxide ions, this effect can be ignored. At pH values of over 10, however, the concentration of hydrogen ions is so small that the pH values measured are too low, as the glass membrane only responds to the hydroxide ions. This effect is called the alkali error. Alkali-proof glass electrodes do exist in which the alkali error is low. (For measurement of pH see also Chapter 1.)

If the pH is taken as a measure of the concentration of a substance dissociated in aqueous solution, certain restrictions must be taken into consideration. Clear results are only found in one-component solutions. Furthermore, in cases in which the dissolved substance is not completely dissociated, its activity must be known. The logarithmic relationship also decreases the dependence of the pH on the change in concentration. This can only be seen as an advantage in very large changes of concentration.

An important field of application for potentiometric measurement with the glass electrode is electrometric titration. The e.m.f. in a cell is here used as end-point indicator for a reaction taking place in solution. This has the advantage of allowing one to follow the entire course of titration, e.g. when titrating phosphoric acid with sodium-hydroxide solution (Fig. 17).

This shape of curve, which is characteristic of all such titrations, is a result of the logarithmic relationship between E and the concentration (or rather: activity) of the $H_3O^+$ ions.

### 2.2.3.1 Ion-sensitive electrodes

The method of **measurement with ion-sensitive electrodes** also uses potentiometric principles. Ion-sensitive electrodes are electrochemical half-cells in which potential differences caused by the activity of a specific ion can be determined by measurement. Their design is similar to that of the pH glass electrode; a suitable millivoltmeter and reference

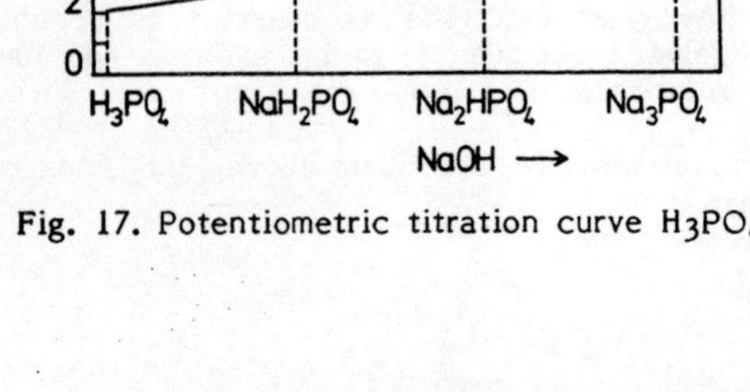

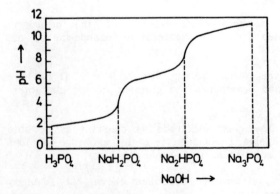


Fig. 17. Potentiometric titration curve $H_3PO_4$ with NaOH

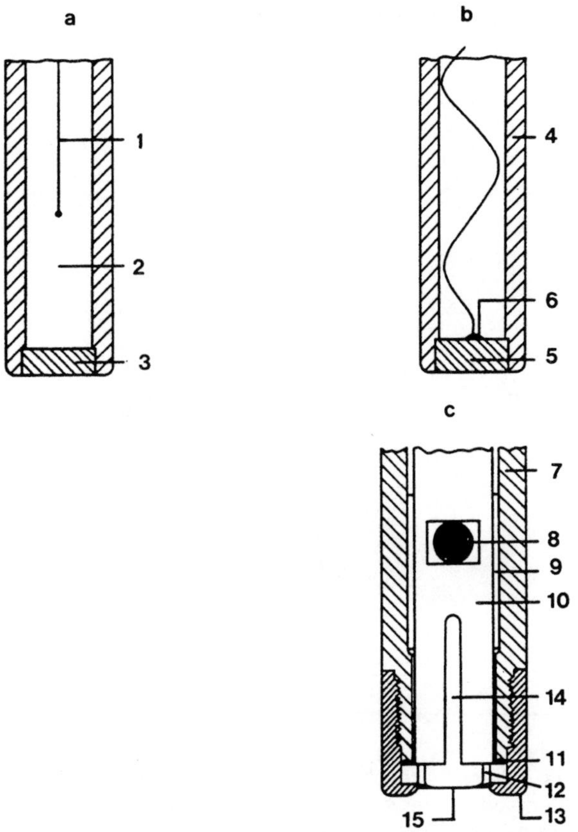


Fig. 18 a, b, c. Different types of ion-sensitive electrodes; a $F^-$ electrode; 1) = Ag/AgCl bridge element; 2) = Electrolyte (NaCl, NaF); 3) = $LaF_3$ monocrystal; b $Ag_2S$ electrode; 4) = Shaft in epoxy resin; 5) = Ion-sensitive solid; 6) = Fixed contact; c Gas-sensitive electrode; 7) = Outside body; 8) = Reference element; 9) = Interior liquid; 10) = Interior body; 11) = O-ring; 12) = Spacer ring; 13) = Sealing cap; 14) = Ion-sensitive solid; 15) = Membrane

electrode are required for measuring the pH. Instead of pH it is also possible to indicate measured values for $pF^-$, $pNH_3$ or $pCa^{2+}$, for example.

There are crystalline and non-crystalline electrodes. The former can be divided into homogeneous (e.g. individual crystals, melted pellets) and heterogeneous (e.g. AgCl in PVC) solid electrodes. The non-crystalline electrodes can also be divided into those supported by porous (e.g. glass filters) or non-porous (e.g. PVC) materials. Liquid ion exchangers and neutral carriers serve as effective electrode substance. Because of the low electrical conductivity of these substances, very thin layers are used, giving them their name of membrane electrodes. Their thicknesses are around 0.1 mm in the case of glass, 1 to 5 mm in the case of organic substances and more than 3 mm for crystals or pellets. The structures of various ion-sensitive electrodes with fixed active phases are depicted in the diagram.

Fields of application for measurement with ion-sensitive electrodes

| | | | |
|---|---|---|---|
| $H^+$ | $Mg^{2+}$ | $CrO_4^{2-}$ | $BF_4^-$ |
| $Li^+$ | $Ca^{2+}$ | $Mn^{2+}$ | $Tl^+$ |
| $Na^+$ | $Sr^{2+}$ | $Fe^{2+}$ | $CO_2$, $CN^-$,$SCN^-$ |
| $K^+$ | $Ba^{2+}$ | $Ni^{2+}$ | $Pb^{2+}$ |
| $Rb^+$ | $La^{3+}$ | $PdCl_4^{2-}$ | $NH_3$, $NH_4^+$, $NO_3^-$ |
| $Cs^+$ | | $Cu^{2+}$ | $PO_4^{3-}$, $HPO_4^{2-}$ |
| | | $Ag^+$ | $SO_2$, $S^{2-}$, $SO_3^{2-}$ |
| | | | $F^-$ |
| | | $Zn^{2+}$ | $Cl^-$ |
| | | $Cd^{2+}$ | $Br^-$ |
| | | $Hg^{2+}$ | $I^-$ |
| | | | Oxalate, Benzoate |

As there are by now approximately 30 different ion-sensitive types of electrode, these and the reactions produced on them cannot be treated in detail. Reference should be made to more detailed works. The Table shows fields of application for direct measurement of various ions and neutral components. Those parameters which are particularly suitable for such measurements in water analysis are underlined. (See also Section 3.2)

### 2.2.4 Polarography

In polarographic analysis, current-voltage curves are recorded as they are formed on a polarizable micro-electrode when the diffusion of the ions in solution at the electrode is the determining step in the electro-chemical reaction.

A dropping Hg-electrode usually serves as polarizable micro-electrode. Either an Hg bottom electrode, a calomel electrode or another electrode of the second type is used as non-polarizable reference electrode.

A variable voltage is applied to both electrodes in the solution to be analyzed to record the current-voltage curve. The solution to be analyzed is electrolyzed. However, this takes place to such a low degree that the composition of the solution remains practically unchanged. The basic circuit diagram for polarography is shown in Fig. 19.

If ions ("depolarizers") which can be oxidized or reduced in a specific voltage interval are present in the solution to be analyzed, the current also changes with changes in voltage. The change in current is dependent on

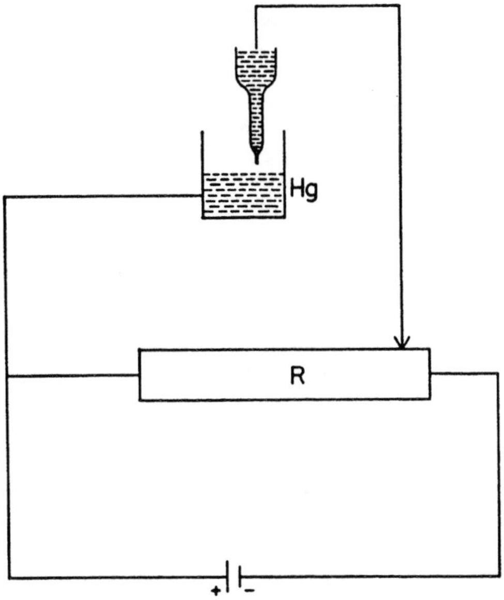


Fig. 19. Basic diagram of a polarograph

the speed of diffusion of the ions on the mercury drop. Working with the dropping mercury electrode has the advantage that a new and thus clean electrode is constantly available. In the ideal situation, the curve shown in Fig. 20 is produced.

The curve is divided into three sections. In field A, the voltage applied is not sufficient to reduce the depolarizer. The low current nevertheless flowing is referred to as residual or basic current. The charging of the Hermholtz double layer at the mercury/solution interface is above all responsible for the generation of this current. This layer acts as a condenser with constantly increasing capacity.

The strength of current rises constantly at the breakpoint between A and B because reduction of the depolarizer commences. At C, the current reaches a

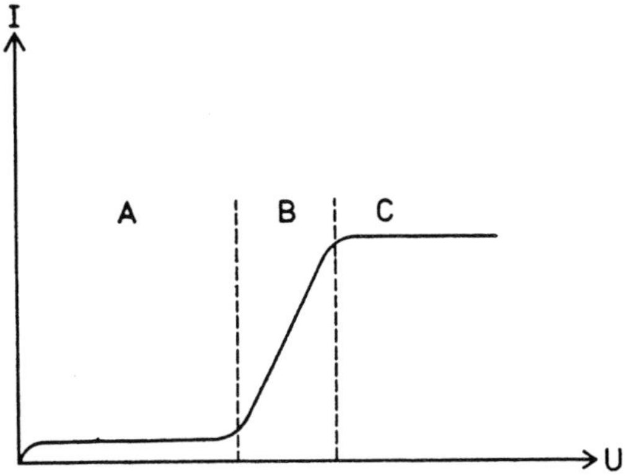


Fig. 20. Current voltage curve

limit value caused by local depletion of depolarizer in the immediate vicinity of the electrode. Those ions reaching the electrode from the solution are immediately reduced, and the state of diffusion is ideally only dependent on the concentration slope solution/electrode. This means that the limit value for the current (limit or diffusion current) is directly proportional to the concentration of the reduced substance.

In order to avoid the situation that current transport in the solution has to be taken over by the ions in the substance to be analyzed, a highly concentrated "supporting electrolyte" is added. This supporting electrolyte must comprise sparingly reducible electrolytes (e.g. KCl, $NH_4Cl/NH_4OH$). As soon as voltage is applied, the supporting electrolyte ions migrate to the anode or cathode. The sparingly reducible supporting electrolyte ions are nevertheless not discharged and surround the cathode. The positive charging for the most part neutralizes the negative field around the cathode so that no gradient of electric field is present at this location. This means that only those depolarizers can be discharged which reach the cathode by diffusion. If several reducible ions are present in the solution, a characteristic step will form for each ion with further changes in voltage, so that several components can be determined in one process with the aid of polarography.

In contrast to the system of DC polarography described here, AC polarography uses a constant AC voltage superimposed on the variable DC voltage. The differences in the current-voltage curves for the two processes are shown clearly in Fig. 21.

In AC polarography, peaks instead of steps are recorded, facilitating the evaluation of the polarogram. The strength of the diffusion current $i_d$ is given by the Ilkovic equation:

$$i_d = 0.732 \cdot n \cdot F \cdot c \cdot D^{1/2} m^{2/3} t^{1/6}$$

where:

$i_d$ = Diffusion current in µA
n = Number of electron transfers at the cathode
F = Faraday constant
c = Concentration of the depolarizer
d = Coefficient of diffusion for the depolarizer

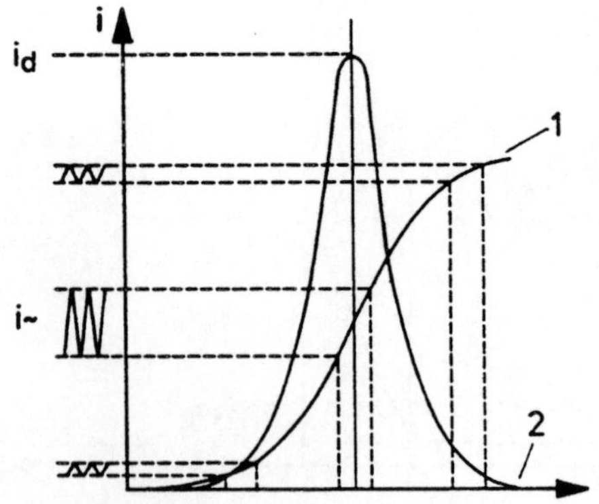


Fig. 21. Current-voltage curves for DC and AC polarography; 1) = DC polarogram; 2) = AC polarogram

m = Mass Hg/sec.
t = Dropping time for the Hg

The numerical factor results from calculating the surface of the Hg drop. The equation for the strength of current as a function of voltage can be derived with the aid of the Nernst equation, provided that the reaction occurs reversibly at the dropping electrode.

Following Nernst, for process

$$A + n \cdot e^- \rightleftharpoons A^{n-} \qquad (1)$$

the following is true

$$E = E_o + \frac{R \cdot T}{n \cdot F} \cdot \ln \frac{°[A]}{°[A^{n-}]} \qquad (2)$$

where °[A] represents the depolarizer concentration on the electrode surface.

In order to generate the equation for the polarographic level, it is necessary to express the concentration in terms of the diffusion current i, since i is determined by the ions diffusing per unit time at the electrode (Fick's law). Furthermore, i is proportional to the decline in concentration:

$$i = k \cdot [A] - °[A] \qquad (3)$$

The decline in concentration arises from the reduction from A to $A^{n-}$. As A disappears from the solution in the same ratio as $A^{n-}$ is formed, the following is true, provided that the diffusion speeds of A and $A^{n-}$ are equal:

$$A = °[A] + °[A^{n-}] \qquad (4)$$

Setting Equ. (4) in Equ. (3) gives

$$i = k(°[A] + °[A^{n-}] - °[A]) = k\ °[A^{n-}] \qquad (5)$$

where

$$k = 0.732 \cdot n \cdot F \cdot c \cdot D^{1/2} m^{2/3} t^{1/6}$$

It is clear from Equ. (3) that i reaches maximum (limit current $i_d$) when a maximum concentration decline is present, i. e. when °[A] = 0

This gives

$$i_d = k \cdot [A] \qquad (6)$$

Deriving from Equations (4), (5) and (6) for the concentrations required:

$$°[A] = [A] - \frac{i}{k} = \frac{i_d}{k} - \frac{i}{k} \quad \text{and} \quad °[A^{n-}] - \frac{i}{k} \qquad (7)$$

Inserting Equ. (7) in Equ. (1), the function required can be arrived at as follows

$$E = E_0 + \frac{RT}{n \cdot F} \cdot \ln \frac{i_d - i}{i} \quad \text{or} \qquad (8)$$

$$i = \frac{i_d}{e^{(E-E_0)nF/RT} + 1}$$

For potentials E which are considerably more positive or negative than $E_0$, Equ. (8) supplies the basic or limit current for intermediate values of the function i = f (U).

In the case of equilibrium, i.e. $E = E_0$, it follows from Equ. (8) that

$$i = \frac{i_d}{2}$$

i.e. if the strength of current achieves precisely half the value of the limit current, the so-called half-wave potential, the potential here applied is identical with the standard potential of the redox system.

The redox system involved can thus be determined qualitatively from the position of this half-wave potential.

In polarographic processes, two main methods are used for determining quantity:

1) Direct comparison:

   It is important in this method that the concentrations in the solution to be analyzed and the reference solution are approximately the same. The solution to be analyzed and the reference solution are polarogrammed consecutively under identical experimental conditions. The content of the solution to be analyzed is calculated using the formula:

$$c_x = c_s \cdot \frac{i_x}{i_s}$$

2) Standard addition:

   The polarogram for the solution to be analyzed is plotted. A solution of the element to be determined is added in a known concentration and a fresh polarogram plotted. The added quantity of standard should be large enough for the diffusion current to be approximately twice as large in the second determining process as in the first polarogram. The unknown concentration is determined from the rise in diffusion current.

$$c_x = \frac{- v c_s h}{h V - H (V+v)}$$

   where:

   $c_x$ = Concentration in the solution to be analyzed
   $c_s$ = Concentration in the standard solution
   $i_x$ = Diffusion current in the solution to be analyzed
   $i_s$ = Diffusion current in the standard solution
   h = Height of the peaks in the solution to be analyzed
   H = Height of the peaks in the solution to be analyzed + standard
   V = Volume of the solution to be analyzed
   v = Volume of the added standard

## 2.3 Spectrophotometry (or photometry)

In this physicochemical method of investigation, light of a certain wavelength is beamed into the solution of a substance. The light is absorbed either entirely or partially by the molecules of the substance.

As a result of the absorption of the light, the intensity of the light beam directed into the solution ($I_0$) is reduced as it passes through the sample. The residual intensity still present on exit (I) is measured with a suitable measuring instrument and compared with the intensity of the irradiated light $I_0$. The amount of light which is absorbed during the passage through the sample solution depends on the structure of the absorbing molecules, their concentration in the solution, and the path length which the light beam traverses in the medium of the solution. This forms the basis for the two principal applications of spectrophotometry.

1. The degree of light absorption in certain spectral regions permits conclusions to be drawn as to the chemical structure of the substances.

2. If the chemical structure of the substances is known, the concentration in the solution concerned may be deduced. The following distinctions are made, according to the wavelength region in which the absorption spectrum is measured.

   a) Spectra in the ultraviolet range (UV spectra), 200 - 400 nm.
   b) Spectra in the visible range (visible spectra), 400 - 800 nm.
   c) Spectra in the infrared range (IR spectra), 0.8 - 50 µm or 50 - 500 µm (far infrared).

If radiation of gradually increasing wavelength (i.e. decreasing energy) is shone through the solution and in so doing the absorbed proportion of light intensity constantly determined, sections of the spectrum are passed through in which the energetic resistance is too large to stimulate an electron transfer (which is responsible for the absorption of light). The sample absorbs little light at these wavelengths. If, however, light of a wavelength is irradiated which possesses exactly the amount of energy required to stimulate an electron transfer, the molecule absorbs very actively. If the wavelength of the irradiated light is again increased, the "suitable" level for optimum stimulation is left behind and the absorption becomes weaker again.

The absorption curve of the substance is obtained by plotting a graph with the wavelength $\lambda$ of the irradiated light on the abscissa and the portion of light intensity absorbed by the sample, i.e. light absorption, on the ordinate. (In practical analysis the usual measuring instruments - spectrophotometers - indicate either the transparency (transmission) or the extinction E). The graph shows the absorption spectrum. The wavelengths of greatest absorption of light, i.e. with greatest energy uptake by the substance, are designated the absorption maxima ($\lambda_{max}$). They correspond to the amount of energy required to just stimulate electron transfers. The wavelengths of maximum absorption have a special role to play in the field of analytical chemistry.

In spectrophotometry in the narrower sense the monochromatic light beam of a discrete wavelength is filtered from a polychromatic light beam in an accessory unit (monochromator: prism or grating). In addition, monochromatic light may also be produced and used for analytical measurements by taking metal vapour lamps (e.g. Hg or Cd vapour lamps) in the place of an incandescent lamp and monochromators. The metal vapour lamps emit light of certain discrete wavelengths which are characteristic of the metal concerned. If an appropriate light filter which specifically allows one of these discrete wavelengths to pass is positioned between the light source and the sample, monochromatic light is also produced. The intensity of the light after it has passed through the cuvette can be measured, for example in a photoelectric cell or by a photomultiplier.

The original photo-optical or visual method is still common in water analysis where the colour intensity of a coloured compound of unknown concentration is compared with the colour intensity of solutions of the same compound of known concentration. Since the intensity of colour is proportional to the concentration of the compound, the concentration of the unknown sample can be ascertained by colour comparison - provided that completely analogous test conditions are adhered to (e.g. equal path length).

Colour comparators are frequently used in practice, the coloured sample solution being compared with standardized colour disks.

### 2.3.1 Measurable variables of light absorption

Three measurable variables are available for characterizing the degree of light absorption, i.e. the light intensity absorbed by a "dissolved" sample substance. These are:

a) transmission T
b) absorption A
c) extinction E

Let $I_o$ be the intensity of light shone into a substance sample and I be the light intensity still present after passage through the sample. Transmission, absorption and extinction may then be defined as follows.

**Transmission** T indicates what fraction of the irradiated light intensity (in %) emerges from the sample:

$$T = \frac{I}{I_o} \cdot 100\ (\%)$$

If the sample solution absorbs no light ($I = I_o$), T = 100 %. If, on the other hand, all the light is absorbed (I = 0), T = 0 %.

**Absorption** A indicates what fraction of the irradiated light intensity (in %) is absorbed by the sample solution:

$$A = \frac{I_o - I}{I_o} \cdot 100\ (\%)$$

A = 0 % if the sample solution absorbs no light ($I = I_0$), and A = 100 % if all the light is absorbed ($I = 0$).

**Extinction** E is the most frequently used measurable variable for light absorption. It is the decimal logarithm of the ratio of the intensity of the irradiated light ($I_0$) to the intensity of the light beam leaving the sample solution (I):

$$E = \log \frac{I_0}{I}$$

The extinction is obtained from Lambert-Beer's law. It is zero if the sample absorbs no light ($I = I_0$; $E = \log 1 = 0$). If light absorption by the sample is absolute ($I = 0$), the extinction is infinite. In practical spectrophotometry, the extinctions measured are between 0 and 1, and also more rarely between 1 and 2. As a logarithmic value, the extinction has no unit of measurement.

For the measurement of an absorption spectrum, the measurable variables A and E should be plotted on the ordinate in ascending order, whereas T should be plotted in descending order. The following two equations show the relationship:

$$A = 100 - T \quad \text{and} \quad E = \log \frac{100}{T}$$

## 2.3.2 The Lambert-Beer Law

In order to measure the absorption of light in a spectrophotometer, the substance is fed in solution into a measuring vessel (cuvette) with a path length d, and the light intensity of a monochromatic beam of light is measured before entering ($I_0$) and after leaving the cuvette (I). If path length d or concentration c of the solution is changed, the absorption of light is also changed. The dependence is described by the Lambert-Beer law. The extinction E is used as a measure of light absorption. The Lambert-Beer law is a combination of Lambert's and Beer's law.

**Lambert's law.** The number of molecules struck by the light beam on its way through a solution is entirely dependent on the path length of the cuvette, given constant concentration. If the path length is doubled, the light beam strikes twice as many molecules on its way through the cuvette, which then also absorbs twice the amount of light energy. The extinction E, as a measure of light absorption, must therefore be proportional to the irradiated path length d (Lambert's law):

$E \sim d$ (d = path length in cm) becomes

$$E = k_1 \cdot d$$

where:

$k_1$ = (substance-specific) proportionality factor

**Beer's law.** Given constant path length d, the number of molecules struck by the light beam is entirely dependent on the molar concentration of the solution. Doubling the concentration causes a doubling of the number of molecules struck and hence a doubling of light absorption. Given constant path length, therefore, the extinction E must be proportional to the molar concentration c of the solution (Beer's law):

E c (c = molar concentration in mol $\cdot$ $litre^{-1}$ becomes:

$$E = k_2 \cdot c$$

where:

$k_2$ = (substance-specific) proportionality factor

**Lambert-Beer law.** If both path length d and molar concentration s of the solution are changed, Lambert's law and Beer's law must be combined to produce the Lambert-Beer law:

$$E = (k_1, k_2) \cdot c \cdot d$$

The proportionality factors $k_1$ and $k_2$ become a new proportionality factor $\varepsilon$

$$E = \varepsilon \cdot c \cdot d$$

where:

E = Extinction (dimensionless)
$\varepsilon$ = Proportionality factor = molar extinction coefficient ($l \cdot mol^{-1} \cdot cm^{-1}$).
c = Molar concentration ($mol \cdot litre^{-1}$)
d = Path length (cm)

The Lambert-Beer law states that light absorption (extinction E) is proportional to molar concentration c and path length d. Plotting extinction against concentration given constant path length (or against path length given constant concentration) therefore produces a straight line (see below).

The proportionality factor $\varepsilon$ of the Lambert-Beer law is designated as the molar extinction coefficient ("molar extinction", for short). $\varepsilon$ is equivalent to the extinction of a 1-molar solution ($c = 1\ mol \cdot litre^{-1}$) over a path length of 1 cm (d = 1):

$$\boxed{\varepsilon = E\,\frac{1 mol}{1 cm}}$$

Plotting a graph of $\varepsilon$ against wavelength $\lambda$ produces the absorption curve of a substance in the same way as plotting A or E against wavelength $\lambda$.

Each substance has a different molar extinction coefficient $\varepsilon$ for each wavelength of its absorption spectrum. Generally, the $\varepsilon$ value quoted is that determined for the absorption maximum $\lambda_{max.}$, namely $\varepsilon_{max.}$.

This value, $\varepsilon_{max.}$, represents a characteristic substance constant which, depending on the structure of the substance, may lie between about 20 and 200 000 $l \cdot mol^{-1} \cdot cm^{-1}$, and which is frequently given and plotted as log $\varepsilon$ so as to avoid excessively high numbers.

Applications of the Lambert-Beer law

The Lambert-Beer law is the fundamental law of absorption spectrophotometry. It is applied in the following ways.

1) By measuring the extinction E of a substance in solution of known concentration c, the **molar extinction coefficient** $\varepsilon$ may be determined:

$$\varepsilon = \frac{E}{c \cdot d} \quad (l \cdot mol^{-1} \cdot cm^{-1})$$

$\varepsilon$, as a substance constant, may be used for testing the identity of substances and determining the structure of organic molecules.

2) If is known, the **concentration of a solution** may be calculated from the Lambert-Beer law by measuring the extinction E of the substance in solution:

$$c = \frac{E}{\varepsilon \cdot d} \quad (mol \cdot l^{-1})$$

This formula forms the basis of the application of the Lambert-Beer law in quantitative spectrophotometric analysis. It should be noted that this law applies strictly only to monochromatic light and, as a limiting law, strictly only to dilute solutions, at constant temperature.

## 2.3.3 Design and mode of operation of an absorption spectrophotometer

In order to be suitable for measuring the absorption spectrum of a substance, a spectrophotometer must be in a position to:

a) produce light of a certain wavelength (monochromatic light); the wavelength should be continuously variable (at the monochromator)

b) measure the intensity $I_0$ of the light shone into the substance solution and the intensity I of the light emerging from the substance solution at the receiver

c) display the absorption A or transmission T or extinction E, and register the entire absorption spectrum of the substance (indicating instrument, plotter).

The light source, for example an incandescent lamp, provides polychromatic light. A bundle of rays is collimated through an aperture and strikes the monochromator. With the aid of a prism or a grating, the monochromator sorts a single wavelength (monochromatic light) from the mixture of many wavelengths of the polychromatic light.

The intensity $I_0$ is assigned to the monochromatic light of a certain wavelength selected in this way. This intensity is recorded by directing the light beam to the receiver (e.g. a photoelectric cell), but in general only after passing the light beam through a "reference cuvette" filled with pure

The diagram shows the design of an absorption spectrophotometer.

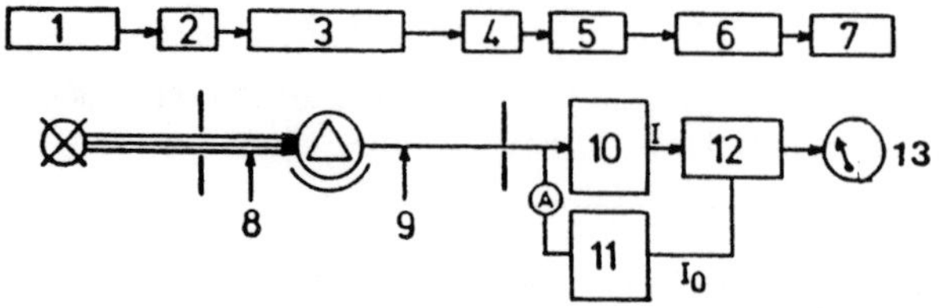


Fig. 22. 1) = Light source; 2) = Aperture; 3) = Monochromator; 4) = Aperture; 5) = Cuvette; 6) = Receiver; 7) = Display; 8) = Polychromatic light; 9) = Monochromatic light; 10) = Substance; 11) = Solvent; 12) = Receiver; I); $I_o$); A); 13) = ATE

solvent (beam path A in the diagram). This precludes errors which may arise due to the individual absorption of the solvent, the reflection of light from the cuvette and dispersion in the cuvette. In addition, the 0/100 % setting of the device is made.

Subsequently, the same light beam passes through the cuvette containing the dissolved sample. The intensity of the light emerging from this cuvette is also measured in the receiver. The (automatic) indicating instrument compares I with $I_o$ and allows the values to be read off as A, T or E.

For identification tests or in order to record the absorption curves of compounds these measurements are repeated, slowly changing the wavelengths by turning the monochromator.

(As the light passes through the prism of the monochromator the various wavelengths are refracted to varying extents; as the prism is turned, light of a different wavelength passes through the exit aperture of the monochromator.)

In order to determine the concentration of any known substance with a known molar extinction, it is sufficient to set the appropriate wavelength and measure I at a certain constant path length.

### 2.3.4 Photoelectric photometers

In the case of simpler photoelectric photometers which can only be used for the quantitative determination of dissolved substances at selected wavelengths (generally sufficient in water analysis), the monochromator is replaced by a spectral filter which only allows light of a certain limited wavelength range to pass. By appropriate selection of a metal vapour lamp, therefore, the same purpose is achieved as with a monochromator.

The light beam emitted by the incandescent lamp or a metal vapour lamp (e.g. Hg or Cd vapour lamp) is "sorted" by the spectral filter in such a way that it is virtually monochromatic. As such it passes through the cuvette containing the solution of the substance to be determined (after exchanging the sample cuvette for a blank-test cuvette containing pure solvent). After leaving the cuvette the residual light intensity I of the beam is converted to an electric current in the photoelectric cell

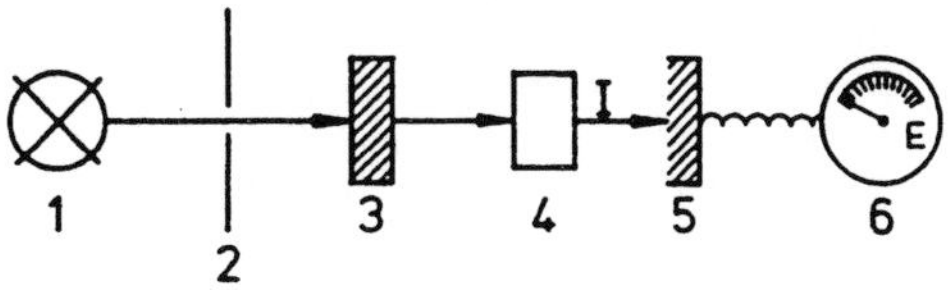


Fig. 23. Design of a photoelectric photometer; 1) = Incandescent lamp; 2) = Aperture; 3) = Spectral filter; 4) = Cuvette; 5) = Photoelectric cell; 6) = Indicating instrument

("photoelectric" indication). This current is passed on to an indicating instrument (galvanometer) fitted with a scale allowing the extinction E to be read off directly. The zero point (E = O) is set using the measured value of the blank-test cuvette (pure solvent, plus - where appropriate - all reagents used).

The selection of the spectral filter is dependent on the colour of the solution of the substance to be determined. This solution always absorbs the colour which is **complementary** to its own. The spectral filter must therefore allow light of this complementary colour to pass. For example, a solution is coloured orange because it absorbs blue, the colour complementary to orange, at approx. 450 nm. In this case, then, a spectral filter must be used which provides blue light in the spectral range around 450 nm. It must be remembered that, in contrast to a monochromator, a spectral filter does not only allow light of a single wavelength to pass. The filter has a "main wavelength range", alongside which, however, neighbouring wavelengths also pass with decreasing intensity. Metal vapour lamps in combination with suitable filters, on the other hand, supply monochromatic light at the respective discrete wavelengths of the metal vapour.

### 2.3.5 Evaluation

**(Spectrophotometric determination of content)**

Extinction $E_x$ of the solution for analysis.
The concentration of the substance to be determined can be derived from the value $E_x$ using one of the following methods:

1) the Lambert-Beer law
2) with the aid of a reference solution
3) from a calibration curve

1) **Application of the Lambert-Beer law**

$E = \varepsilon \cdot c \cdot d$ $\qquad\qquad$ $\varepsilon = E\,\frac{1\text{mol}}{1\text{cm}}$ (in mol · litre$^{-1}$)

the concentration of the substance to be determined is derived from the measured extinction $E_x$ (in mol · $l^{-1}$) as

$$E_x = \varepsilon \cdot c_x \cdot d \text{ , i.e.}$$

$$c_x = \frac{E_x}{\varepsilon \cdot d} \text{ [mol} \cdot \text{litre}^{-1}]$$

2) In order to determine the content with the aid of a reference solution of known concentration $c_V$, the extinction of the reference solution $E_V$ and that of the sample solution $E_X$ are measured with the same path length d. The Lambert-Beer law applies to both solutions.

$$E_V = \varepsilon \cdot c_V \cdot d \qquad \text{and} \qquad E_X = \varepsilon \cdot c_X \cdot d$$

The ratio of the extinction values of the two solutions is the same as that of their concentrations

$$\frac{E_X}{E_V} = \frac{c_X}{c_V}$$ ; converting, we obtain $c_X$:

$$\boxed{c_X = c_V \frac{E_X}{E_V}}$$

3) **Interpretation with the aid of a calibration curve.**

In order to plot the calibration curve, several solutions of known concentration are prepared of the substance to be determined and their extinction measured. These are then plotted on a graph as a function of concentration (Fig. 24). The measured extinction $E_X$ provides the required concentration $c_X$ of the solution of unknown concentration.

The calibration curve provides information as to whether the Lambert-Beer law applies to the required concentration range, i.e. whether extinction E and concentration c are directly proportional. If this is the case, the calibration curve is a straight line. Deviations from the Lambert-Beer law in certain concentration ranges are indicated by the curved course of the calibration curve.

**Determination of the calibration constant**

In all physical methods of measurement used in water analysis it proves advantageous, wherever a **linear** function exists between concentration c and measured value M within the intended measuring range, to dispense with the use of a plotted calibration curve, thereby avoiding unnecessary errors when reading from the calibration lines.

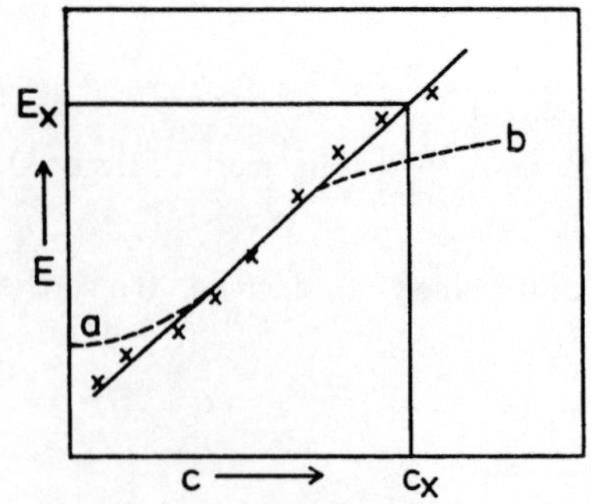


Fig. 24. Calibration curve for the quantitative determination of a substance. a, b: deviations from the Lambert-Beer law.

In such a case, the procedure is as follows: the known concentration c given by measurement of a number n of calibration solutions is obtained from each of the measured values M, the corresponding quotient calculated

$$F = \frac{c}{M}$$

and from this the arithmetic mean determined

$$\bar{F} = \frac{\Sigma F}{n}$$

The value $\bar{F}$ represents the calibration constant.

According to the relation $M\bar{F} = c$, the concentration value of the sample corresponding to each measured value can now be determined. If the original sample is used in diluted form, multiply by the dilution factor to obtain the concentration of the component to be determined in the original water sample.

On the other hand, it can be seen from the manner of deviation of the F values from $\bar{F}$ whether a linear relation between measured value and concentration really exists; this would, for example, not be the case if the F values significantly increase or decrease from n = 1 to n = 10.

The following example is intended to clarify the problem.

10 calibration measurements were carried out with solutions which contained the substance to be determined in concentrations between 0.1 and 5.0 mg/l in the sample for measurement. The table was prepared from the measured values obtained.

From the table it may be calculated that:

$$\Sigma F = 83.989$$

$$n = 10 \qquad \bar{F} = \frac{\Sigma F}{n} = 8.3989$$

| n | c | M | $F = \frac{c}{M}$ |
|---|---|---|---|
| 1 | 0.10 | 0.0120 | 8.333 |
| 2 | 0.30 | 0.0362 | 8.287 |
| 3 | 0.50 | 0.0601 | 8.319 |
| 4 | 0.60 | 0.0722 | 8.310 |
| 5 | 0.80 | 0.0974 | 8.213 |
| 6 | 1.00 | 0.1233 | 8.110 |
| 7 | 2.00 | 0.2346 | 8.525 |
| 8 | 3.00 | 0.3365 | 8.915 |
| 9 | 4.00 | 0.4792 | 8.347 |
| 10 | 5.00 | 0.5794 | 8.630 |

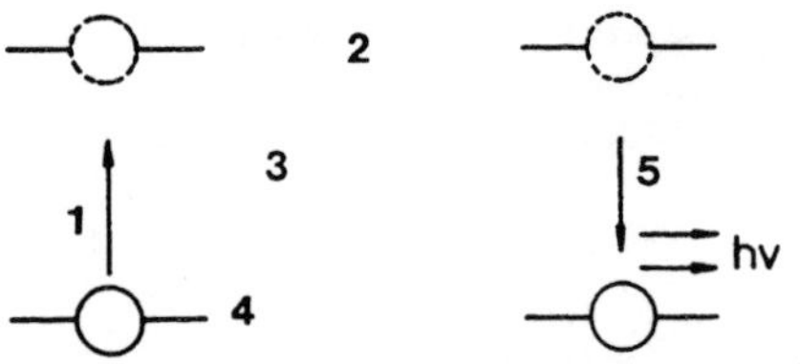


Fig. 25. 1) = Excitation; 2) = Excited state; 3) = Therm. energy; 4) = Atom in ground state; 5) = Emission

## 2.4 Flame-emission spectrophotometry (FES)

In flame-emission spectrophotometry, also frequently referred to for short as "flame photometry", the sample under investigation is converted to atomic vapour by applying thermal energy (a flame). As energy continues to be added, a number of the atoms are changed to an excited state, in which, however, they remain for only a short time ($10^{-7}$ to $10^{-4}$ sec.).

Following this period, the excited atoms, in which the outer electrons had occupied more highly excited energy levels (orbitals), return to the ground state. In this process, the energy difference $\Delta E_\lambda = h\nu$ between the excited level and the electron ground state is released. The atomic line radiation $\Delta E_\lambda$ emitted is characteristic for each element. If the flame burns evenly and the substance is fed into the flame at a constant rate over the period of the measurement, the intensity $I_\lambda$ of the spectral lines observed provides a measure of the concentration of the substances.

The principle of FES is based on the quantitative measurement of the emission radiation of the intensity $I_\lambda$ . The diagram shows the process of atomic emission.

### 2.4.1 Design of the instrument

The following is a simplified schematic diagram of the arrangement of the equipment for FES.

The usual atom reservoir in FES is the flame. The sample enters the flame, where atomization takes place, in the form of a solution via the burner.

The degree of atomization of the sample in the flame and hence the sensitivity, detection limit and magnitude of potential physical and chemical interferences depend on a number of operational parameters of the flame. For example, the degree of atomization increases as flame temperature rises. Parallel to this, detection sensitivity is improved and the influence of chemical interferences is reduced.

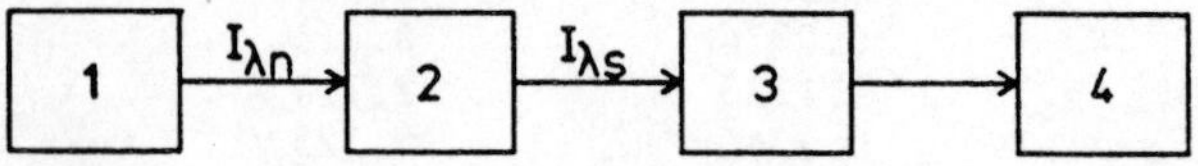


Fig. 26. 1) = Atom reservoir; $I_{\lambda n}$) = Intensity of the polychromatic flame emission; 2) = Wavelength sector; $I_{\lambda s}$) = Intensity of atomic emission of the selected wavelength; 3) = Detector; 4) = Indicating instrument

A **hydrogen/oxygen flame** has proved its worth in applications in FES, particularly when determining alkali metals. With a flame temperature of 2660 °C it is admittedly somewhat cooler than an acetylene/oxygen flame (3130 °C), but the characteristic radiation of the hydrogen/oxygen flame is about 10 times lower than that of the acetylene/oxygen flame.

The predominant burner type in FES is the **turbulent burner**, also known as the direct-atomizing burner.

Generally, the cylindrical turbulent burner consists of two concentric jets around a stainless-steel capillary tube. Hydrogen flows through the outer jet, and oxygen through the inner jet. The two gases are mixed above the rim of the jets. The sample solution is drawn up from the sample container situated beneath the burner by the suction caused by the gas flow at the upper end of the capillary and is sprayed directly into the flame. Since the mixing process occurs above the rim of the jets the system is safe from backfiring and the burner remains cool, making, for example, easy adjustment possible.

Two parts of the flame may be distinguished: an inner, luminescent cone (primary zone), and the outer, less luminescent surrounding area (secondary diffusion zone). The highest temperature is generally found directly above the inner cone. This region is therefore suitable for the measurement, thanks also to the fact that the background radiation is at its weakest here.

In FES, grating or quartz-prism monochromators are commonly used as **wavelength selectors** for the polychromatic line radiation coming from the atom reservoir (flame).

In simpler devices, so-called filter photometers, the monochromator is replaced by filters which have the task of allowing the desired spectral line to pass as easily as possible, but on the other hand to retain the lines of other elements to a large extent. An improvement in line selection can be achieved by arranging several different filters in series. However, one is always dependent on a limited number of filters. In contrast, a monochromator has the advantage of being adjustable to any element detectable in the flame simply by adjusting the prism. It is even possible to select the most favourable line amongst the various lines of an element.

The transmission region of a monochromator can be continuously adjusted by varying the aperture width (adjusting the entrance and exit apertures). This makes it possible to reduce interference, but on the other hand involves a reduction in sensitivity in the case of narrow aperture widths. In order to measure such low radiation power, it is necessary to use a sensitive radiation detector with high-quality electronics.

The exit radiation is measured in a secondary electron multiplier (SEM) used as a detector as the photons hit the photocathode. The latter usually consists of alkali-metal alloys, and is of varying sensitivity, depending on the spectral range of the incident photon radiation. The electrons which are primarily liberated are amplified directly by a dynode cascade by a factor of up to $10^{10}$. The anode collector voltages may reach 2500 V. The degree of amplification is limited by so-called thermal noise.

The current signal proportional to the incident light intensity I produced in the detector or the SEM is fed to an **indicating instrument**. The measured value can be read against time (quantitative measurement) or against wavelength (qualitative measurement). The instruments have current-proportional digital displays giving a readout of absorption or, if a logarithmic converter is connected in series, of extinction.

In FES, the relative detection limit is defined as the concentration for which the intensity corresponds to twice the noise level. The absolute detection limit thus refers to the masses of this concentration. The optimum measuring range generally lies between 15 and 100 times this value.

It should be noted that while sensitivity and selectivity do indeed increase as spectral band width decreases (aperture width of the monochromator), at the same time the total intensity recorded at the detector decreases, worsening the signal-to-noise ratio.

### 2.4.2 Evaluation

Since it is difficult to determine all the values contained in the basic equation for the evaluation of the intensity measurement, the simpler relative method is preferred in FES. This is based on **calibrations** which are carried out with known element concentrations with the same operational parameters as for the measurements on the sample. The content of the sample of the element to be determined is then given by comparing the intensity or extinction of the calibration solution and the sample solution. The aim should always be for the matrices of the two solutions to agree.

A further calibration and evaluation method is the **addition method.** Here, known quantities of the element to be determined are added to the sample solution. Treated and untreated sample solutions are then measured. The measured intensities or extinctions are plotted on a graph against the element concentration with the addition, and the resulting calibration curve is extended to the intersection with the abscissa. The value at the intersection point with the abscissa represents the concentration of the untreated sample solution. The particular advantage of this method lies in the fact that if the matrix of the sample is unknown the same matrix is retained for the calibration. This method should not be confused with the **"internal standard" method,** where a known, constant quantity of an element which was not originally contained in the sample is added to the sample and calibration solutions.

Improved accuracy of evaluation in FES is given by using the so-called **"pincer method"**, which is recommended for example for rubidium and caesium determination. In this method, two calibration solutions as close as possible to the solution for analysis are taken from the series of calibration solutions; the concentration of one of the two calibration solutions should be below and that of the other calibration solution above the concentration of the solution for analysis. Under these conditions (optimum approximation = pincer), determination can be carried out to an accuracy which deviates from the true value to an ever decreasing extent as the difference in concentration between the two calibration solutions decreases and as the calibration curve runs ever straighter.

Flame-photometric measurements are particularly subject to the influence of the following factors:

1. Surface tension, viscosity and density of the solution under investigation

2. Hindrance of evaporation due to accompanying substances which form new, thermally stable compounds in the flame with compounds of the element to be determined

3. Mutual influence of excitation of the excitable elements present in the sample

4. Cross sensitivity, i.e. raising the measured value by the radiation emitted by accompanying elements which reaches the photoelectric detector as a result of incomplete optical separation.

In principle, the simplest way to eliminate these disturbing factors is to adjust the composition of the calibration solutions with regard to all accompanying substances to match the composition of the sample solution under investigation. This procedure presumes at least approximate knowledge of the composition of the sample which in most cases is not available in practice, not to mention the considerable calibration work in each individual case.

Reference is made here to a method perfected by W. Schuhknecht and H. Schinkel (Fresenius' Z. Anal. Chem. 194, 161 - 183 (1963) for the elimination of such interference factors in a mode of operation universally valid for alkali determination. It also forms the basis for the methods of determination of lithium, sodium, potassium, rubidium and caesium presented in this book (Section 3.3).

In the case of alkali determination, interference factor 2 plays no significant role due to the relatively high volatility of all alkali compounds. The cross sensitivity, interference factor 4, of most of the elements potentially present in the sample is so much smaller than the intensity of the alkali lines that it can likewise be ignored. This also applies to the mutual cross sensitivity of potassium, sodium and lithium. Alkaline earths present in the solution form an exception. The interfering alkaline-earth cross sensitivity can be suppressed by adding aluminium nitrate to calibration and measurement solutions as a "physical buffer". Since the relatively high aluminium salt content required here also predominantly determines the physical properties of all solutions (surface tension, viscosity, density), the addition of aluminium nitrate also dispenses with interference factor 1.

The use of caesium chloride as a "spectroscopic buffer" helps towards eliminating interference factor 3. It can be seen that caesium, an element with very low ionization energy (3.89 eV), eliminates the mutual influencing of excitation of potassium, sodium and lithium.

## 2.5 Emission spectrum analysis

Spectral analysis is used for qualitative and quantitative analysis. The analysis is carried out by producing, observing and measuring the electromagnetic spectra of the substances under investigation with the aid of appropriate spectrometers.

In the case of emission spectrum analysis the substance under investigation is vaporized under an electric discharge occurring between two electrodes, and the atoms are excited to produce radiation by supplying electrical or thermal energy. This radiation is produced in the following way: the additional energy raises the electrons of the outermost electron shell of the atoms to a trajectory with a higher energy level (orbital); during the transition to the ground state the energy previously taken up is released again in the form of radiation of particular wavelengths. The radiation is then spectrally decomposed in a grating or prism spectrograph. As a rule, the line spectrum produced is recorded on a photographic plate. Qualitative detection of the elements is made on the basis of the location of the spectral lines, and the intensity of the lines serves to determine their quantity. Compared with photographic recording, photoelectric measurement of the spectra has the disadvantage of lower resolving power.

The main light or energy sources which can be used to excite the elements in trace analysis are the electric arc (direct-current arc) and spark discharge. Arc discharge has the advantage of lower detection limits, while spark discharge exhibits better reproducibility with lower detection power.

See also Section 3.4

## 2.6 X-ray Fluorescence Analysis

X-ray fluorescence analysis is used for qualitative analysis of water samples, in particular to determine the presence of heavy elements, and also for semi-quantitative determination, especially of heavy elements such as barium or thallium whose presence has been demonstrated.

In X-ray fluorescence spectrometry, primary X-ray radiation is produced by means of an X-ray tube, exciting the sample located in this beam to emit fluorescence. This secondary radiation is rendered parallel by a collimator and diffracted and reflected by a movable analyzer crystal. The intensity of the radiation is measured in the receiver, e.g. a geiger, proportional or scintillation counter, at a wavelength specific to the element.

In water analysis, a dry residue of the water sample is most commonly used for X-ray fluorescence analysis. The polychromatic radiation emitted by X-ray tubes excites the heavy elements in the dry residue to emit corresponding radiation.

This secondary radiation is diffracted in an analyzer crystal, thereby breaking it down into the various wavelengths. By measuring the angle of deflection it is possible to calculate the wavelength of the diffracted radiation.

Disregarding higher-order reflections only one value of the angle of deflection exists for a given wavelength. The spectra emitted by an element consist of a small number of characteristic lines. If the ionization of the element to be determined takes place in the K shell of the atom, it is mainly the K-$\alpha$ line that is produced. The K-$\alpha$ line is always accompanied by the weaker K-$\beta$ line.

If ionization of the element to be determined takes place in the L shell, then the L spectrum, which has more lines, is observed. Here the L-$\alpha$ and L-$\beta_1$ lines are preferred for analysis. Thus the fluorescing elements can be identified from their emission lines, thereby permitting qualitative analysis of the sample.

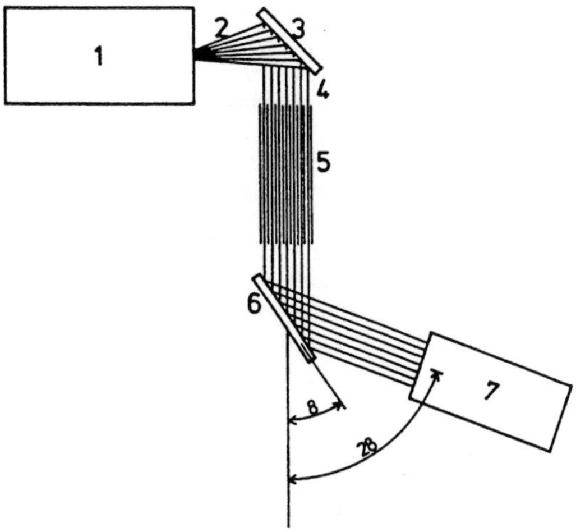


Fig. 27. 1) = X-ray tube; 2) = Primary X-ray beam; 3) = Sample; 4) = Fluorescence radiation; 5) = Collimator; 6) = Analyzer crystal; 7) = Receiver

For quantitative analysis, which is also possible, it is necessary to perform suitable calibrations.

This involves:

a) first identifying the element on the basis of wavelength,
b) measuring the intensity of the radiation emitted, using (for example) a proportional or scintillation counter, and
c) performing the calibration with the aid of calibration samples of appropriate concentration under identical experimental conditions.

Qualitative X-ray fluorescence analysis can form a useful supplement to qualitative emission spectroscopy when analyzing water samples, since the X-ray fluorescence technique is particularly good for identifying the so-called heavy elements in the dry residues.

## 2.7 Atomic Absorption Spectrometry (AAS)

### Theory

The AAS method is based on the fact that atoms in their ground state can absorb light of a particular energy (i.e. frequency). This process is the reverse of the emission of light by atoms excited by being exposed to energy (e.g. thermal energy in flame photometry). In AAS, light of a defined wavelength radiates through the atomizer system (flame or graphite tube cuvette) and is absorbed there by atoms in the ground state. The quantity of absorbed light is proportional to the concentration of non-excited atoms. It is measured as selective resonance in a detector.

Light is emitted at the source of radiation and absorbed by atoms in the atomizer system at exactly defined wavelengths and within strictly limited spectral ranges (half width of hollow cathode discharge lamps acting as source of light, approximately 0.002 nm), whereby each spectrum line is specific for a given element. As every element has its own characteristic emission wavelength (which is identical to the absorption wavelength in the ground state), hollow cathode discharge lamps whose cathode consists of the element to be determined are used as source of radiation. If voltage is applied to the two electrodes in the hollow cathode lamp, a glow discharge forms in the rare gas atmosphere of the lamp. This excites the cathode material, which thus emits its characteristic spectrum

Over a wide range, AAS obeys Beer's law, which describes a linear relationship between extinction and the concentration of the element in question. The concentration required is determined by multiplying the measured extinction by a calibration factor or by reading off with the aid of the calibration curve.

### 2.7.1 Design of the equipment

An atomic absorption spectrophotometer consists of a source of light (hollow cathode discharge lamp) emitting the spectrum of the element to be determined, a monochromator separating from the spectrum the resonance line typical of the element in question and selected for the determining process, and a detector converting the flow of photons into a flow of electrons. An amplifier tuned to the modulation frequency of a rotating-disc shutter can be connected between detector and indicating instrument. This rotating-disc shutter modulates the radiation between source of light and sample in accordance with the amplifier frequency. In this way, any non-modulated radiation, even if emitted spuriously by the flame, can be eliminated.

The flame can be replaced as an atomization device by flame-less "atomizers" (graphite rod, graphite tube cuvette), but special working instructions must be observed when using these (see Fig. 28). The advantage of this kind of modification lies in a considerable improvement in detection sensitivity.

In order to atomize in the flame an element to be determined, the sample is sprayed via a pneumatic sprayer into a mixing chamber where it is mixed intimately with a combustible gas (e.g. acetylene) and an oxidation agent (e.g. air or nitrogen (I) oxide). It then reaches the flame through the burner slit in a laminar burner. As a result of the heat, dissociation takes place in the atoms which absorb the light at a defined wavelength from the hollow cathode lamp.

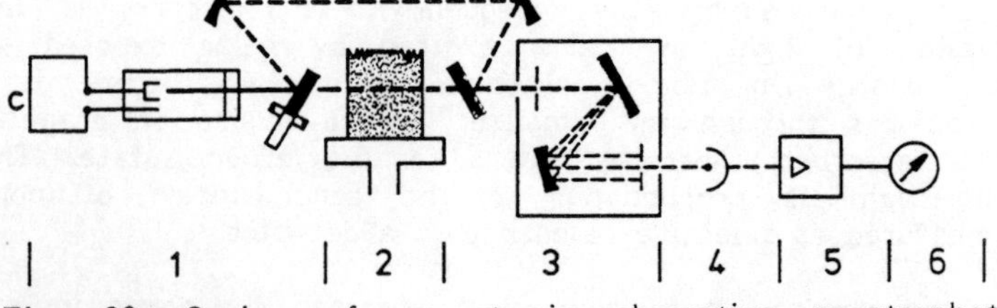


Fig. 28. Design of an atomic absorption spectrophotometer (double-beam alternating light device); 1) = Source of radiation; 2) = Flame; 3) = Monochromator; 4) = Detector; 5) = Amplifier; 6) = Indicating instrument (after B. Welz)

The monochromator used in AAS does not require the high resolving power necessary for other spectrometric processes, because spectral interference is for the most part eliminated by the use of sources of light specific to the element and the selective amplifier. It merely has to eliminate adjacent resonance lines in the element to be determined, not emission lines from accompanying elements. For this purpose, a resolving power of approximately 0.2 nm is generally sufficient. On the other hand, the possibility of working with large slit widths (up to approximately 2 mm) means a high degree of light passage through the monochromator and thus a favourable signal/noise ratio, leading to good identification limits and a high degree of reproducibility.

Disturbances

An important disturbing factor in AAS is "chemical interference". This can occur if the element is present in the flame both in atomic and in molecular form (flame temperature not high enough, secondary reactions with formation of compounds which are difficult to dissociate). In this form the light emitted by the hollow cathode lamp is not absorbed.

In order to eliminate these disturbing effects, which as a rule overlap, attempts must be made to create more favourable conditions for atomization by optimizing the flame and the position of the burner and by increasing the flame temperature (e.g. $N_2O$/acetylene instead of air/acetylene). Individual disturbing factors can be suppressed selectively. Thus, the difficulties caused by phosphate when determining Ca and Mg can be eliminated by adding a lanthanum salt. The addition of calcium similarly eliminates the obstacles presented by silicic acid to the identification of manganese. In the case of barium, the proportion of atoms ionized in the flame - reducing the basic level of the (absorbing) population - is considerable. This problem can be eliminated by adding to sample solution and standard an excess of a more easily ionizable element with a similar or lower ionization potential (e.g. $Na^+$ as NaCl).

Interference due to scattered light is particularly noticeable at short wavelengths. It arises when the flame contains solid particles or droplets of liquid, and also from molecular association of radicals in the flame (O, OH, CN) with the added salts. These disturbances can be eliminated by largely matching the content of foreign matter in both sample and standard, or by making a background correction. The background is measured at a wavelength as close as possible to the analytical line, but at which the element to be measured is not recorded, and this figure is subtracted from the one obtained at the analytical wavelength.

The use of a continuous radiation source can also eliminate background absorption. In this method, radiation from a hollow cathode lamp and radiation from a continuous radiation source are transmitted alternately through the flame. The advantage is that the correction takes place directly on the line. A certain disadvantage in this method is admittedly to be found in the increase in noise.

## 2.7.2 Scope

The table gives a summary of the elements which can be determined with and without flame by means of AAS.

Identification limit µg/l

| Element | Graphite tube furnace (100 µl measuring solution) | Flame |
|---|---|---|
| Ag | 0.005 | 1 |
| Al | 0.01 | 30 |
| As | 0.2 | 20 |
| Au | 0.1 | 6 |
| B | 15 | 1000 |
| Ba | 0.04 | 10 |
| Be | 0.03 | 2 |
| Bi | 0.1 | 20 |
| Ca | 0.05 | 1 |
| Cd | 0.003 | 0.5 |
| Co | 0.02 | 6 |
| Cr | 0.01 | 2 |
| Cu | 0.02 | 1 |
| Fe | 0.02 | 5 |
| Hg | 2 | 200 |
| K | 0.002 | 1 |
| Li | 0.2 | 0.5 |
| Mg | 0.004 | 0.1 |
| Mn | 0.01 | 1 |
| Mo | 0.02 | 30 |
| Na | 0.01 | 0.2 |
| Ni | 0.2 | 4 |
| P | 30 | 50 000 |
| Pb | 0.05 | 10 |
| Pt | 0.2 | 40 |
| Sb | 0.1 | 30 |
| Se | 0.5 | 100 |
| Si | 0.1 | 50 |
| Sn | 0.1 | 20 |
| Te | 0.1 | 20 |
| Ti | 0.5 | 50 |
| Tl | 0.1 | 10 |
| V | 0.2 | 40 |
| Zn | 0.001 | 1 |

Examples of relative identification limits (in µg/l) which can be achieved with the graphite tube furnace and flame techniques. The values for the graphite tûbe furnace technique relate to a sample volume of 100 µl.

The table gives average values which are subject to considerable fluctuations from case to case depending on equipment, composition of the sample and the method selected to optimize the parameters for the instruments.

A number of metals (e.g. Al, Ba and Be) require the use of a (hotter) $N_2O$/acetylene flame to dissociate the molecules. Barium can be sprayed into the flame directly in aqueous solution, whereas for the determination of aluminium and beryllium in the concentrations in which they are found, for example, in water treatment in waterworks, previous enrichment is necessary by chelating and extracting the metal-chelate complexes with an organic solvent.

There are no generally valid working instructions for handling the equipment. Under all circumstances, the instructions supplied with the equipment by the manufacturer are binding. Attention should be paid to good horizontal adjustment to the equipment, to the correct insulation of the hollow cathode lamp and to correct setting of the monochromator to the wavelength of the element in question. Check slit width and lamp current as stated by the manufacturer. After igniting the flame, adjust the flow of oxidizing gas and combustible gas accurately. Adjust burner to maximum absorption and flame stability.

When spraying organic solvents into the flame (after extraction stages, see Sections 3.3. and 3.4), the combustible gas/air ratio should be reduced, as the organic solvent itself acts as combustible gas. If this step is not taken, there is a risk of an interfering luminescence if the flame rises to high above the burner. In order to correctly adjust the ratio of combustible gas to air, commence with the mixture ratio prescribed by the manufacturer for the appropriate analysis, and then reduce the supply of combustible gas step by step, in line with the spray rate of the organic solvent, so that the flame remains at the same height as it was before spraying started. The air/acetylene flame is regulated to give a bluish colour rather than burning brightly. The flow of the acetylene for the laughing gas flame is regulated so that the flame takes on a pink colour. In general, the flow of acetylene for a laughing gas flame must be double the flow for an air/acetylene flame.

### 2.7.3. General aspects of calibration and analysis

After producing suitable series of dilutions from the stock metal solutions, first spray for 2 minutes with demineralized water (4 - 5 ml/min.). The standard solutions are then atomized. A blank sample consisting of demineralized water should be inserted between every two measurements. The readings are plotted as "extinction against concentration" (in µg/l).

In the case of metals which are determined after enrichment by extracting in an organic solvent, all procedural steps must also be applied in the calibrating process. The calibration curves for nearly all elements are practically linear.

The samples for analysis and calibration are treated according to the equipment instructions prescribed by the manufacturer. In order to keep the equipment drift as low as possible, care should be taken that all measurements (blank reading, calibration and sample solutions) are taken speedily one after the other without unnecessary pauses. At least three readings per measurement should be taken to determine the average value. In cases of uncertain values, the "pincer" method should be used. Repeated checking of the zero point is particularly important in the case of samples with low extinction, because a zero point drift can easily occur with increasing soot formation on the burner.

## 2.7.4 AAS measurements using the graphite furnace technique

The sensitivity of the conventional flame techniques is limited in that only 3 to 10 % of the sprayed sample reaches the flame. A feed system consumes approximately 4 ml/min., of which 0.2 ml/min.- reaches the flame. With a gas flow rate of 10 l/min. in an air/acetylene burner, the sample is thus diluted approximately 50 000-fold. If the same solution is dried and the dry residue rapidly heated and atomized, a considerable increase in sensitivity is to be expected. This can be achieved by placing a small volume (2 to 50 µl) in the graphite tube cuvette. The table above shows examples of the relative detection limits of the graphite tube furnace and flame techniques. The graphite tube cuvette is normally 2 to 5 cm long. By applying high currents of low voltage to the ends of the tube, the cuvette is heated. By means of this type of resistance heating, a precise graduation of the heating temperature and thus the selection of optimum temperature conditions for atomization of every single element can be achieved. Atmospheric oxygen should be kept away from the system, which must therefore be flushed with inert gas. The maximum operating temperature is in the region of 3000 °C. The majority of cuvette manufacturers supply tubes coated with pyrolytic graphite. This improves sensitivity and reduces "memory effects" in the case of such carbide-forming elements as molybdenum, silicon, titanium and vanadium. Furthermore, improved resistance to oxidizing mineral acids and longer life at higher temperatures can also be observed.

Nitrogen is frequently used as shielding gas. As a rule, however, argon is used in the case of elements which form nitrides, e.g. barium, molybdenum, titanium and vanadium. In order to reduce condensation of the sample and evaporation of the matrix products at the cooler ends of the cuvette, the shielding gas is usually fed in at the ends and escapes from the sample insertion hole. The flow of gas is usually optimized by the equipment manufacturer.

The sensitivity of this technique depends directly on the volume of the sample. Volumes of between 10 and 20 µl are very frequently used. The use of an automatic feed system is of considerable importance in this graphite tube AAS technique, having a crucial influence on the sample dosing on which precision and accuracy depend. Thus, not only the point in the graphite tube at which a sample is introduced is important, but also how the drop is applied to the tube wall. These factors can only be completely controlled by automation. A degree of reproducibility better than 1 % is achieved in this way.

The majority of commercial appliances allow temperature and time selection for evaporation of the solvent, for the incineration phase and for the atomization phase. It is important for the sample to evaporate completely before the incineration phase commences. If the heating rate is too rapid, the solvent spatters and reproducibility deteriorates appreciably. The temperature for incineration should be selected so that the maximum evaporation of the matrix occurs without reducing the concentration of the element to be determined. In the case of samples with high proportions of volatile substances, mechanical loss can occur if incineration is too rapid. It has proved suitable for evaporation of the solvent to raise the temperature rapidly to near the boiling point of the solvent, to continue heating slowly to boiling point and slightly beyond, and to stay at this temperature approximately 10 to 20 seconds. The rises in temperature are usually followed by an isothermal phase. As rapid a rise in temperature as possible is selected to atomize the element, since the more rapid the heating rate, the greater the density of the atom cloud and hence the greater the sensitivity. If the matrix accompanying the sample is not removed during the atomization phase, changes in sensitivity can be observed after a certain measuring time. This can be overcome by selecting a noticeably increased final temperature after approximately 10 to 20 samples. In complex matrices, e.g. salt waters, a slow heating rate can be selected in order to separate the element to be determined from the matrix.

In the case of non-saline waters, the maximum heating rate is usually selected, as this generally reduces chemical interference. Sensitivity is also increased, particularly with such low-volatile elements as molybdenum, titanium and vanadium. It is of considerable importance to heat the cuvette to a higher temperature than the atomization temperature of the element to be determined.

The signals produced can be evaluated with the aid of the peak-area method or the peak-height method. Suitable integrators are available for this purpose in modern equipment.

Non-specific background absorption occurs to a larger extent in the graphite furnace technique than with flame AAS. For this reason, it is urgently recommended that automatic background compensation be used. Whenever a new method is developed, the background absorption of typical samples should be checked. If such significant absorptions exist, improvement can be achieved by modifying such parameters as incineration temperature or by adding matrix modifiers. Interference phenomena can be further reduced by a variety of methods, e.g. calibration with the aid of the standard addition process, or selective extraction of the element to be determined. Interference should be checked with graphite tube cuvettes of different ages.

### 2.7.5 Example of an AAS measuring system after Perkin-Elmer

Atomic absorption spectrometry (AAS) is one of the most commonly used instrumental techniques of analysis for the quantitative determination of metals and metalloids particularly in water samples, including those from wastewater, refuse seepage water, sludges and wastes. (See preceding Section and Sections 3.3 and 3.4)

The main advantages of AAS are its high specificity and selectivity, the sensitivity being variable over broad ranges depending on the type of atomization selected (flame, graphite, cold vapour or hydride technique).

Information concerning the equipment used, the hollow cathode lamps, the graphite tube, hydride and cold vapour devices can be obtained from the suppliers of this equipment together with further details of fuel gases, safety regulations, fume exhaust facilities and evaluation units. Enquiries are always valuable, amongst other things because the development of this type of apparatus is constantly advancing, as is the entire field of instrumental analysis.

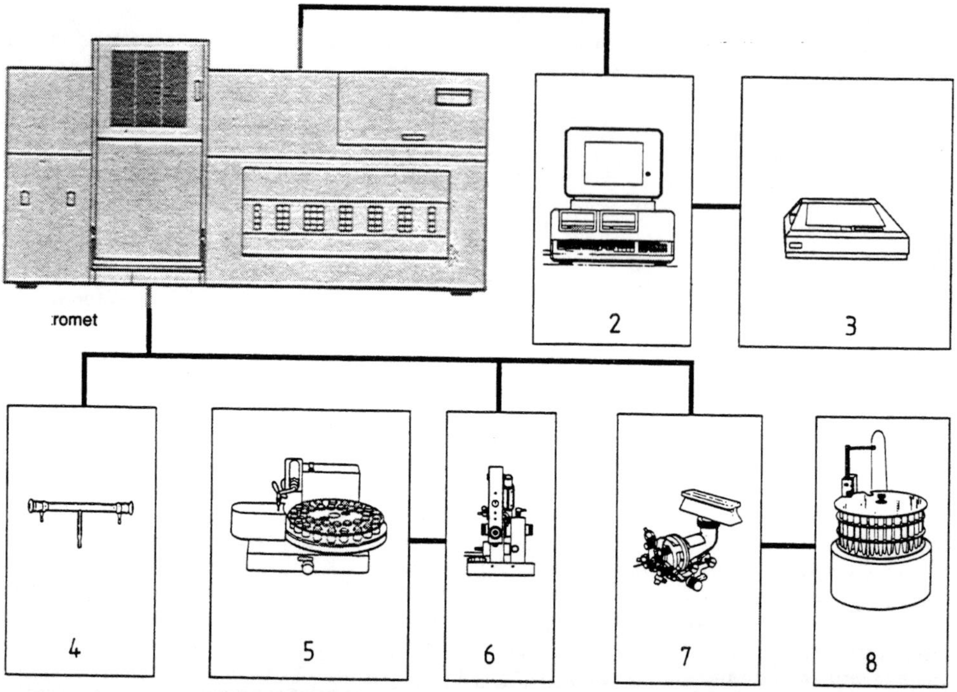


Fig. 29. 1) = Spectrometer; 2) = Data system; 3) = Printer; 4) = Cold-vapour and hydride technique; 5) = Graphite-tube furnace with automated sample feed; 6) = Graphite-tube furnace; 7) = Burner; 8) = Flame unit with automated sample feed (Perkin-Elmer)

## 2.8 Atomic Emission Spectrometry with Inductively Coupled Plasma Excitation (ICP-AES)

### 2.8.1 General

A solid state of aggregation is no longer possible above 6000 °C. At such temperatures, states occur which can be described as "plasma" in the widest sense of the word. Plasma is characterized by the fact that electrically charged particles are created by the breaking up of gas molecules - a process referred to as ionization. A plasma is thus a gas whose atoms or molecules have to a greater or lesser extent broken up into positive and negative charge carriers.

For atoms to ionize, a specific energy, the so-called ionization potential, must first be applied. This energy must furthermore be fed to the atom itself. The ionization potential is expressed in electron-volts (eV). The value for the majority of elements lies between 4 and 25 electron-volts.

In emission spectrometry, plasmas have the task of exciting the atoms to be measured to emission. Plasma is in principle an electrically conductive, fluid system in predominantly gaseous matter. The properties of a plasma are determined by the charge carriers already mentioned and by its quasi-neutrality, as the plasma has an overall electrically neutral effect when observed for a longer period of time.

A plasma is produced by transferring electrical energy to a flow of gas. In the case of inductively coupled plasma, a high frequency generator with an induction coil is used to supply energy for the ionization potential required. The energy is directly proportional to the density and temperature of the plasma. In most cases, easily ionizable argon, and in some cases helium is used as gas. When using a cooler gas, e.g. nitrogen, as cooling jacket, the temperature of the plasma and thus the ionization potential can be increased simultaneously. In this way, ionization potentials in the upper electron-volt range can also be achieved.

It is particularly important for the practical analytic application of plasma that high temperatures speed up thermal ionization. This effect is used as follows in inductively coupled plasma (ICP).

A gas is fed through a system of quartz tubes whose shape facilitates flow and to the end of which a strong current of high frequency is applied. After ionization of the gas, a homogeneous plasma of high thermal intensity is produced. Minutely atomized particles of solution are fed via an added carrier gas into the hot plasma core and atomized or ionized as a result of the residence time of these element particles in the plasma. This causes the elements to emit spectra which can be recorded qualitatively and quantitatively with the usual spectrometric systems (Fig. 30).

In recent years, inductively coupled plasma excitation (ICP) has particularly come to the fore. By means of the high temperature produced in an ICP, which can reach up to 10000 K in the region of the high frequency generator, a very efficient transfer of energy from the plasma to the sample material is achieved, together with relatively long residence time of the aerosol in the system of excitation. Above all, two phenomena are avoided in ICP which occur in conventional atomic absorption spectrometry:

1. Absorption of the radiation emitted within the source of radiation by atoms from the same element. This phenomenon, referred to as self-absorption, has the effect of extending the physical line profile.

2. Absorption by atomic vapour of lower temperature in the peripheral zones of the source of radiation. In this phenomenon, referred to as "self-reversal", radiation at the line centre is lower than that at the flanks. Both in self-absorption and in self-reversal, reductions in sensitivity can occur. It is also advantageous that alkaline-earth elements and boron and silicon can be determined free of any interference on account of the low partial pressure of oxygen in the ICP.

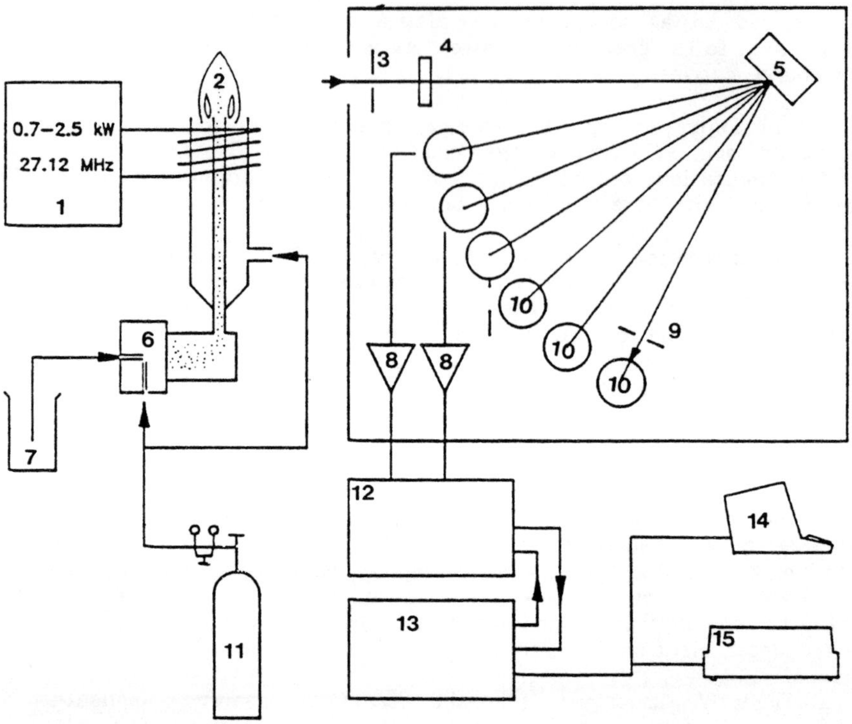


Fig. 30. Schematic diagram of ICP; 1) = HF generator; 2) = plasma 10,000 K; 3) = source slit; 4) = refractor plate; 5) = grating; 6) = atomizer; 7) = sample; 8) = amplifier; 9) = exit slit; 10) = photomultiplier; 11) = argon; 12) = simultaneous measurement system; 13) = computer; 14) = terminal; 15) = printer

Each ICP system consists of a high frequency generator with induction coil, the plasma burner, an atomizer chamber, the atomizer and an appropriate gas supply. Attention must also be paid to further factors necessary for an effective measuring system, such as the actual height of the high frequency output at the induction coil, the use of a gas as cooling jacket, the mechanical flexibility of the burner unit and various technical elements (pump, needle valves, flow meter for gas control).

The conditions for measurement can be selected for routine analysis so that approximately similar conditions of excitation exist for a particular prescribed number of elements. For this reason, ICP is particularly suitable as source of excitation for multi-element analysis of solutions.

### 2.8.2 Equipment

The total configuration of equipment consists of the following modules:

- High frequency generator
- Plasma burner
- Atomizer system

- Spectrometer
- Systems for processing measured values

The high frequency generator supplies the energy for the inductively coupled plasma. As a rule its output lies between 0.7 and 2.5 kilowatt at a normal frequency of 27.12 MHz.

In the majority of cases, argon is used in the plasma burner both as plasma gas and as cooling gas. The burner consists of three quartz tubes inserted one within the other, with the tulip-shaped interior tube initially ensuring a build-up of gas which is only then followed by high acceleration along the interior wall of the outside tube. The cooling gas is fed into the high frequency field via the outside tube. It is required exclusively for the formation of the plasma. The temperature achieved is approximately 10 000 °K. The consumption of cooling gas is 10 to 20 litres per min.

As is also the case in atomic absorption spectrometry, the sample must enter the burner via an atomizer in ICP as well. The efficiency of the atomizer plays a considerable role in producing as finely distributed an aerosol as possible. For example, the Meinhardt atomizer, a concentric pneumatic glass atomizer with fixed capillaries, is often used. The quantity of carrier gas and the efficiency of the atomizer have in the meantime been well matched for ICP analysis. The cross-flow atomizer similarly works on pneumatic principles. The two capillaries used are adjustable and can be adapted to differing plasma conditions. At a carrier gas flow of approximately 1 l/min. 0.5 to 2 ml/min. of sample can be sucked in depending on capillary diameter. The ultrasonic atomizer offers the highest efficiency of all atomizer systems known to date. The sample solution is drawn in by a peristaltic pump and fed via a capillary to an ultrasonic vibrator disc. Its advantage is a considerable increase in intensity compared with pneumatic atomizers, but memory effects as a result of the fine atomization make rinsing periods of several minutes necessary.

Sequence emission spectrometers and multi-element emission spectrometers are used in ICP analysis according to the method of analysis required. In the case of the ICP sequence spectrometer, the constituent elements are measured sequentially, i.e. consecutively. The different wavelength ranges in the spectrum are selected by turning a grating. Depending on which element is selected, a synchronized automatic wavelength selector selects the appropriate wavelength, and this demands a high degree of precision in the movement mechanism of the grating. In this variation of an ICP spectrometer, the lines of analysis can be freely selected, which is advantageous in the case of possible background interference.

In the case of the simultaneous spectrometer, the quantity of a large number of elements can be recorded at the same time within a very short period. Good dispersion, high resolving power, a wide spectral range, and where possible process control coupled with data processing are the requirements for such a system. Limitation to a pre-determined selection of elements is a disadvantage. The information gained in ICP should, like the total programme sequence, be evaluated and/or controlled by coupling to a suitable computer. For example, the following operations can be automated:

- Input of data and user programming by interactivity
- Determination of the spectral window and number of wavelengths to be measured
- Recording the calibration concentration, determining the signal/noise ratio, calculation of mean value and variance.

- Dynamic background measurement at any specified distance from the peak
- Filing of spectra on storage media
- Reporting results of analyses and output of all necessary measuring and result parameters
- Retrieval of stored spectra from the data media
- Control of automatic sample feed equipment

### 2.8.3 Analytical limits

ICP is used to analyze samples of water and waste water, and also solid samples after suitable maceration processes. The detection limits and reproducibility factors (coefficients of variation) are summarized at defined wavelengths for 24 elements in the following table (Hoffmann).

It is clear that the detection limits lie in the same range as for flame AAS. The advantages over flame AAS lie in the possibility of determining

**Reproducibility and detection limits in ICP emission spectrometry**

| Element | Wavelength | Detection limit | Coefficient of variation | |
|---|---|---|---|---|
| | nm | µg/l | for 1 mg/l | for 10 mg/l |
| Ag I | 328.07 | 6 | 0.6 | 0.3 |
| Al I | 308.22 | 20 | 1.5 | 0.9 |
| As I | 193.69 | 30 | 1.2 | 1.0 |
| B I | 249.68 | 5 | 0.6 | 0.5 |
| Ba II | 493.41 | 1 | 0.4 | 0.3 |
| Be II | 313.04 | 0.5 | 0.7 | 0.6 |
| Ca II | 396.85 | 0.5 | 0.5 | 0.4 |
| Cd I | 228.50 | 2 | 0.8 | 0.4 |
| Co II | 228.62 | 3 | 0.8 | 0.7 |
| Cr II | 205.55 | 6 | 0.6 | 0.6 |
| Cu I | 324.75 | 2 | 0.2 | 0.2 |
| Fe II | 259.99 | 3 | 0.6 | 0.5 |
| Mg II | 279.08 | 0.5 | 0.4 | 0.3 |
| Mn II | 257.61 | 1 | 0.7 | 0.4 |
| Mo II | 202.03 | 5 | 0.7 | 0.7 |
| Na I | 589.59 | 20 | 1.0 | 0.5 |
| Ni II | 231.60 | 10 | 0.4 | 0.3 |
| P I | 214.91 | 50 | 1.5 | 0.5 |
| Pb II | 220.35 | 20 | 1.1 | 0.9 |
| Sb I | 206.83 | 30 | | |
| Sn II | 189.99 | 30 | 1.0 | 0.8 |
| Sr II | 407.77 | 0.5 | 0.5 | 0.3 |
| V II | 292.40 | 5 | 0.4 | 0.2 |
| Zn I | 213.86 | 2 | 0.5 | 0.4 |

I) = Atomic line; II) = Ionic line

the elements phosphorus and silicon. Enrichment processes must be carried out before analyzing elements in the lower trace range, such as occur in ground water or drinking water.

## 2.9 Fluorescence Spectrometry

The application of fluorescence spectrometry presupposes that the substances to be analyzed display fluorescence which can be measured, or can be made fluorescent by such measures as producing derivatives. Precisely these processes of producing derivatives are becoming increasingly important, as in many cases they make it possible to take advantage of the benefits of fluorimetry and its low detection limits.

### 2.9.1 Equipment

The four main components of a fluorescence spectrophotometer are as follows:

- Light source for excitation

- Monochromator or filter for the selection of excitation and emission wavelengths

A schematic configuration of a fluorescence spectrophotometer is shown in Fig. 31.

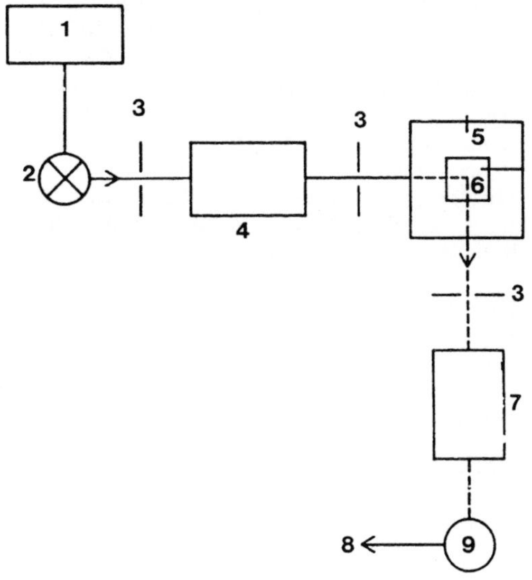


Fig. 31. Schematic diagram of a fluorescence spectrophotometer; 1) = Power supply; 2) = Light source; 3) = Slit; 4) = Excitation monochromator (filter); 5) = Light-absorbing wall; 6) = Measuring cuvette; 7) = Emission monochromator (filter); 8) = Photometer, recorder; 9) = Photomultiplier cell

- Measuring cuvette

- Photomultiplier cell as detector

The sources of excitation frequently employed in spectrophotometers for measuring fluorescence are gas discharge lamps. These have the advantage of a high radiation intensity and of emitting over a wide range of wavelength from approximately 200 nm to above 1000 nm. The most important types are mercury vapour, xenon or mercury-xenon lamps. The lamps most commonly used are high-pressure lamps in which the intensity of excitation must be kept constant. Tungsten lamps are also suitable in the visible range. However, their intensities fall off appreciably in the ultraviolet range.

The wavelengths in fluorescence spectrophotometers are generally selected with the aid of monochromators; line emitters are predominantly used in simple filter fluorimeters, but continuum radiators can also be employed in conjunction with cut-off filters. Fig. 32 clearly shows the basic principle of a monochromator.

The radiation arriving from the fluorescent sample is focussed by condensing lens L onto entrance slit S. The light passing through this entrance slit passes through the collimating lens K and the prism (or grating) D, where it is dispersed into a spectrum. The light emerging from this dispersion unit is focussed by lens F onto exit slit Sa. The exit slit isolates a narrow range of the spectrum. For high standards in the case of mixtures, types of equipment possessing monochromators to diffuse the excitation and emission radiation are necessary.

In contrast to cuvettes for measuring absorption, the standards of measuring cuvette accuracy required in fluorimetry are not so high, as only a comparatively small range of the excitation radiation is recorded. In this context, the effect of flaws in layer thicknesses and parallelism is not as great as when measuring absorption. As a rule, square or round cuvettes are used. In contrast to cuvettes for absorption photometry, cuvettes for fluorimetry are permeable on all sides, as the fluorescent radiation is emitted on all sides. For special tasks (e.g. low cuvette volume in high-pressure liquid chromatography), flow-through cells with very low layer thicknesses are used.

The light signal leaving the cuvette is directed into a detector, which is generally a photomultiplier cell. When light falls on the photo-cathode, electrons can be released which are subsequently attracted and collected by the anode. A potential difference of approx. 100 V is generally maintained between cathode and anode by means of an external voltage supply. The electrical output of a photomultiplier cell is a direct current signal. One disadvantage of such direct current amplifiers is the fact that not only

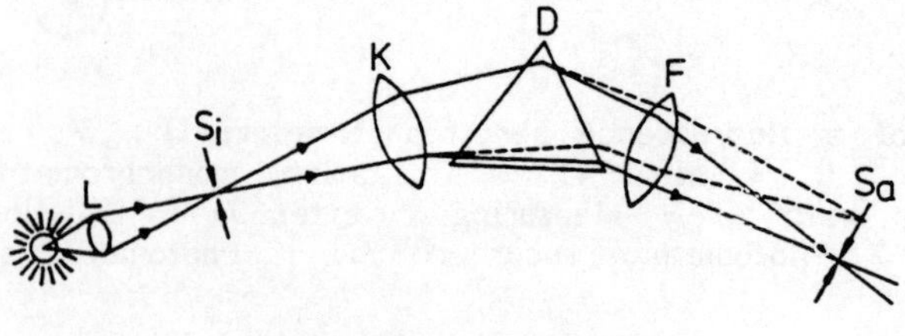


Fig. 32. Schematic diagram of a monochromator

the component of photocurrent due to the emitting sample is amplified, but also undesired components of the photocurrent, e.g. the dark current of the detector.

Compensating apparatus is employed to eliminate such difficulties.

### 2.9.2 Interference

Considerable problems can arise in fluorimetric analysis from the selection of the solvent. Fluorimetric analyses are generally conducted in those solvents whose states of electron excitation are higher in terms of energy than those in the substances to be analyzed. The absorption of excited light is thus hindered by the solvent. In addition, the solvent should be photochemically stable, i.e. during measurement there must be no formation of photochemical species which may under certain circumstances absorb or fluoresce in the spectral range of the sample. As a rule, non-polar solvents, such as n-heptane, cyclohexane or benzene, should be preferred in fluorimetry.

Many organic compounds can be determined in the range of concentration below 1 µg/ml due to their intrinsic fluorescence. A variety of compounds (aromatic hydrocarbons, heterocyclic and conjugated, non-aromatic substances) display high molar absorption coefficients.

In the production of derivatives from functional groups, a fluorescent group of one reagent is transferred onto the substance to be analyzed. A new fluorescent system is thus formed. Such production of derivatives has been described for amino groups and thiol, carbonyl and carboxyl groups. Notable reagents for the production of derivatives are fluorescamine, o-phthaldialdehyde or DANS-Cl (= 5-dimethylaminonaphthalene-1-sulphonyl chloride).

Measurements of the intrinsic fluorescence of substances in thin-layer chromatography can be conducted directly on the thin-layer plate. Polycyclic aromatic hydrocarbons are frequently analyzed in this way. The determination of aflatoxins, a group of highly carcinogenic and toxic substances, is also conducted with the aid of fluorimetry on the thin-layer plate. In addition, numerous pesticides are detected fluorometrically on thin-layer plates, due to their molecular structure. Depending on the type of substance and the nature of the adsorption material, detection limits of between 1 and 20 ng are achieved. Inorganic substances present in water can also be detected by fluorimetric means, e.g. traces of uranium (3.4).

## 2.10 Infrared Spectroscopy

### 2.10.1 General remarks

Analytical significance of the infrared spectrum:

Infrared spectroscopy (IR spectroscopy) is a form of absorption spectroscopy, and is an important and easily applied method of qualitative and quantitative instrumental analysis of the molecule: by means of this method it is possible to establish those functional groups of atoms which are present in the molecule and those which are not present, how they are linked, what form the structure of the molecule takes and the degree of concentration of substance. For the purpose of water analysis, the use of infrared spectroscopy is currently still limited to the testing of concentrations of organic substances, e.g. oils and fats, following extraction. Since the method is constantly being further developed, however, a discussion of the general principles appears useful within the overall framework of water analysis for organic contaminants.

The measured variable is the transmittance of the substance under analysis, as a function of the wave number

$$\frac{1}{\lambda}\ (\mathrm{cm}^{-1})$$

or the wavelength (µm) of the infrared light. The wave number is defined as the number of waves per centimetre of path. The correlation between wave number and wavelength is

$$\frac{1}{\lambda} = X\ \mathrm{cm}^{-1} \Rightarrow \lambda = \frac{10\ 000}{X}\ \mu\mathrm{m}.$$

The sample may be in gaseous, liquid or solid form.

Measurements are made in the infrared range of the spectrum between about 4000 and 400 $cm^{-1}$ or 2.5 and 25 µm. The result of measurement is plotted as a curve and is termed the IR spectrum. The spectral ranges with low transmittance are known as absorption bands. The analytical significance of an IR spectrum is deduced primarily from the position of the bands and band groups within the spectrum and their intensity and form.

Approximately 1 mg of a substance is required to record an IR spectrum. If microscopic techniques are used to prepare the sample, as little as 10 µg of the substance provides a representative IR spectrum. For measuring purposes, this permits infrared spectroscopy and gas chromatography to be combined, and may imply the combination of structure elucidation and substance separation.

Equipment manufacturers now offer Fourier-transform IR spectrometers as standard. These use the multiple-scan technique to analyze about 1 µg of substance and extend the examined spectral range to 10 $cm^{-1}$ or 1 mm wavelength, to the boundary of the microwave range. Fourier-transform IR spectroscopy facilitates the interlinking of gas chromatography and infrared spectroscopy.

The problem as far as water analysis is concerned is the detection limit of IR spectroscopy. This is in the percentage range, or under particularly favourable conditions in the range of mg/kg.

The bands in the IR spectrum result from the ability of the molecule to absorb defined, measurable amounts of energy (energy quanta) from the infrared radiation. The magnitude of these amounts of energy is specific to the molecule and determines the wave number or the wavelength of the band. The absorbed energy excites the atoms in the molecule such that they vibrate. The prerequisite for a molecule to absorb infrared light and thus be caused to vibrate is an electrical dipole moment which changes periodically whenever vibration takes place. Consequently, even molecules which have no permanent electrical dipole moment may also be IR-active.

Atomic vibration parallel to the directed valency is termed stretching vibration. It brings about a periodic change in interatomic spacing. Atomic vibration perpendicular to the directed valency is known as bending vibration. In this case the interatomic spacing remains constant but the valency angles change periodically. Accordingly, a molecule may vibrate in a number of ways (normal vibrations). In IR spectroscopy, however, only a small number of vibrations are activated at any one time.

If stretching and bending vibrations have been localized within the molecule, the characteristic wave numbers of the associated band (group frequencies) indicate the presence of particular functional groups and types of bonds. For example, stretching vibrations of:

| | |
|---|---|
| C-H in aromatic hydrocarbons | 3000 - 3100 $cm^{-1}$ |
| C-O in alcohols, ethers, carbonic acids and esters | 1080 - 1300 $cm^{-1}$ |
| C=O in aldehydes, ketones, carbonic acids and esters | 1690 - 1760 $cm^{-1}$ |
| C-H in alcohols and phenols with hydrogen bridges | 3200 - 3600 $cm^{-1}$ |
| C-N in amines | 1180 - 1360 $cm^{-1}$ |
| C≡N in nitriles | 2210 - 2260 $cm^{-1}$ |

If the vibrations are not localized, but instead the molecule vibrates as a unit, this is known as skeletal vibration. The excitation energy required for skeletal vibration is very closely dependent on the structure of the molecule. The associated bands are typical for the whole molecule. They are to be found in the spectral range between about 1500 and 500 $cm^{-1}$ or 6.5 and 20 µm. This fingerprint range also includes the bands of the stretching vibrations of heavy atoms and metals and the bands of bending vibrations.

It is significant that the characteristic wave numbers of functional groups are different in differing chemical surroundings for the same group of atoms. The vibration sequence, the group frequency and hence the spectral position of the absorption band of a functional group change as a result of interaction (mechanical and electrical coupling) with the rest of the molecule. The greater the difference between the functional groups and the

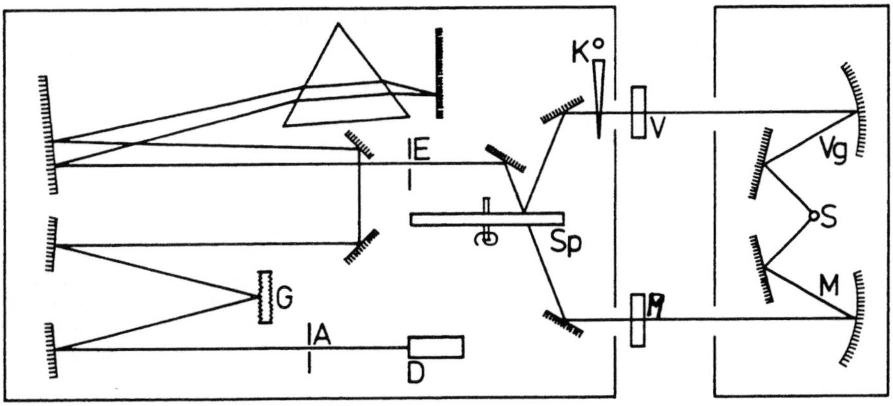


Fig. 33. Schematic structure of a double-monochromator infrared device; S) = radiation source; M) = measuring beam; 2 radiation paths; Vg) = reference beam; Sp) = rotating mirror system; E) = entrance mirror; A) = exit mirror; D) = detector; K) = aperture; the aperture setting is a measure of the absorption of the substance at a given wavelength

rest of the molecule in terms of atomic mass, the type of bond and distribution of electrons, the less marked the change is. This interrelationship is utilized for structure elucidation of the molecule with great success. Hydrogen bridges are a noteworthy form of mechanical and electrical coupling between molecules. They have the effect of shifting the OH group frequencies to lower wave numbers.

Double-beam IR spectrometer (schematic diagram), see Fig. 33.

## 2.10.2 Evaluation of an IR spectrum

When **evaluating an IR spectrum** the first step is to attempt to identify the high intensity bands and band groups. To do this, the practical analyst makes use of a comprehensive collection of IR spectra which is arranged according to families, structural features and functional groups and is equipped with search aids. He should constantly supplement the collection with spectra he has recorded himself. Careful comparison of spectra and familiarity with the possibilities and limitations of the method have been shown to form an essential basis for analytical findings, and this is equally true in IR spectroscopy. The conclusions drawn may be checked against other analytical data. Uncritical application of reference spectra and frequency tables leads to incorrect results.

A number of laboratory rules for use in the experimental evaluation of IR spectra have been established empirically:

The prerequisite for structural elucidation is a chemically pure substance. If its molecular weight is known, this facilitates structural elucidation. The IR spectrum is to be read from left to right.

Atomic oscillators with strong polar bonds provide high-intensity bands. Oscillators which are damped, e.g. by their interaction with hydrogen bridges, produce broad bands.

If the band characteristic of a functional group is missing in the spectral range, it is highly probable that the functional group is absent from the molecule.

If a characteristic spectral range contains the band of a functional group, it is highly probable that a molecule contains this group only if bands of this group also appear in other spectral ranges.

High-intensity bands may almost always be assigned uniquely to functional groups.

In order to be able to determine the substance class at hand, first study the spectral ranges of the C-H stretching vibrations between 2700 and 3100 $cm^{-1}$, the C=C stretching vibrations between 2100 and 2300 $cm^{-1}$, the C≡C stretching vibrations between 1500 and 1690 $cm^{-1}$ and the C-H wagging vibrations, a form of bending vibrations, between 690 and 1000 $cm^{-1}$.

Fig. 34 shows the IR spectrum of an aromatic compound. This is indicated by the bands between 3000 and 3100 $cm^{-1}$ (aromatic C-H stretching vibration), in combination with the high-intensity bands at 1585, 1478 and 1446 $cm^{-1}$ (aromatic C=C stretching vibrations) and with the bands at 695 and 735 $cm^{-1}$ (aromatic C-H wagging vibrations).

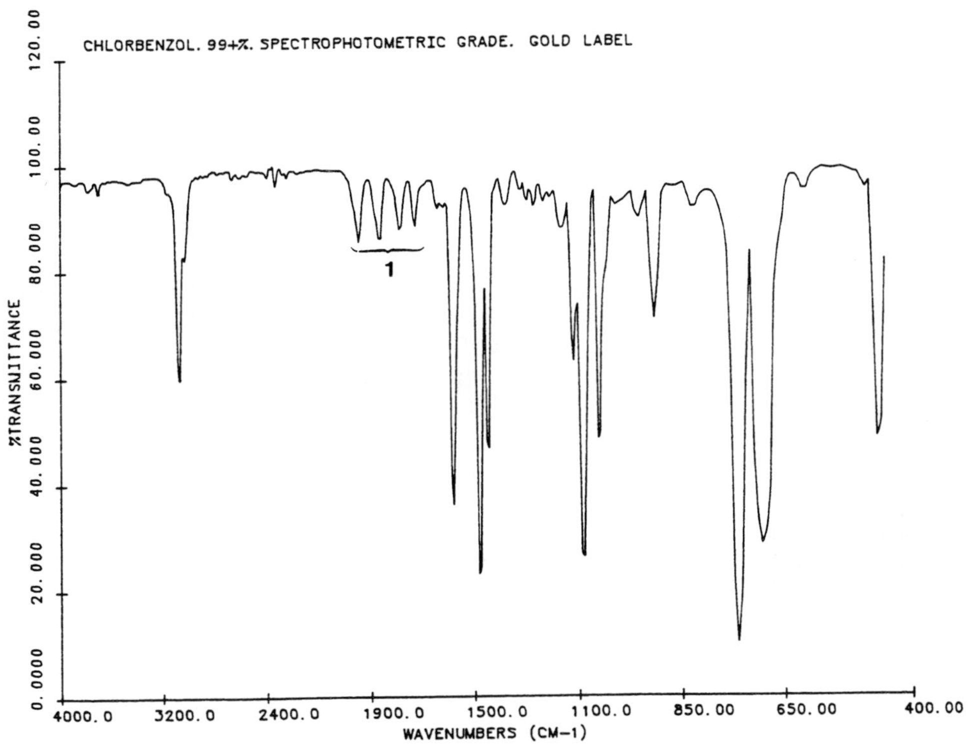


Fig. 34. The infrared spectrum of chlorobenzene. Path length 10 µm; 1) = "Benzene fingers" monosubstituted

The spectral position of the two last-named bands, in conjunction with the previously established facts, points towards a monosubstituted benzene ring.

This interpretation is reinforced by the band shape of the combination vibrations in the "benzene finger" region between 1650 and 2000 $cm^{-1}$. In this case it consists of four main bands; their intensity decreases from left to right. The high-intensity band at 1080 $cm^{-1}$ is the signal of an aromatic C-Cl stretching vibration. It is not possible to determine bands of other functional groups in the spectrum. Accordingly, it can be said that we are here dealing with chlorobenzene.

Fig. 35 shows the IR spectrum of a pure mixture of xylene isomers and is the result of the superimposition of three single spectra.

The aromatic character of the compounds is nevertheless clearly recognizable. The bands at 2920 and 2865 $cm^{-1}$ belong to the C-H stretching vibrations in the $CH_3$ groups. The C-H bending vibrations give rise to the bands at 1375 and 1465 $cm^{-1}$.

The high-intensity, specific analytical bands of the aromatic C-H wagging vibrations at 741 $cm^{-1}$ (o-xylene), 769 $cm^{-1}$ (m-xylene) and 795 $cm^{-1}$ (p-xylene) are of major diagnostic value.

In order to be able to conduct IR spectroscopy and achieve reproducible

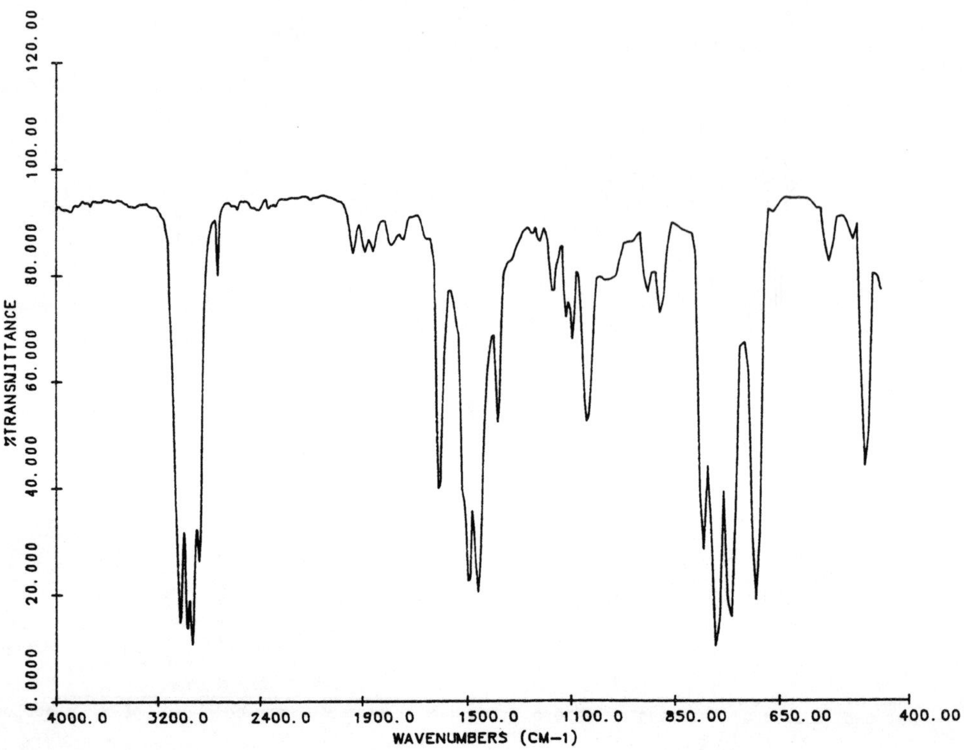


Fig. 35. The infrared spectrum of a pure mixture of xylene isomers. Path length 10 µm; (Mixture of o-, m- and p-xylene)

results, it is necessary to use a double-beam device. The sample to be analyzed is placed in the path of the measuring beam, and standards and known reference samples are placed in the reference beam path of the IR spectrograph. In this way, the spectrum can be compensated and a quantitative result obtained.

The IR spectrum in Fig. 36 also indicates an aromatic compound (D.O. Hummel).

**Evaluation per example in µm:**

Between 3.2 and 3.3 µm, aromatic C-H stretching vibrations; at 6.2, 6.7 and 6.8 µm bands of the aromatic C=C stretching vibrations; at 13.3 and 14.5 µm bands of the aromatic C-H wagging vibrations.

This high-intensity band at 3.0 µm is caused by O-H stretching vibrations and the band at 7.4 µm by O-H bending vibrations. We are in all probability dealing with a phenolic O-H group. The intense band at 8.2 µm supports this assertion. The band indicates a C-C stretching vibration in phenol. The OH and CO bands are key bands for alcohols and phenols. In phenols the C-O bond is attributed more double-bond characteristics than in alcohols. In phenol this brings about a shift of the CO band to lower wavelengths or high wave numbers than in alcohols (C-O stretching vibrations in phenols

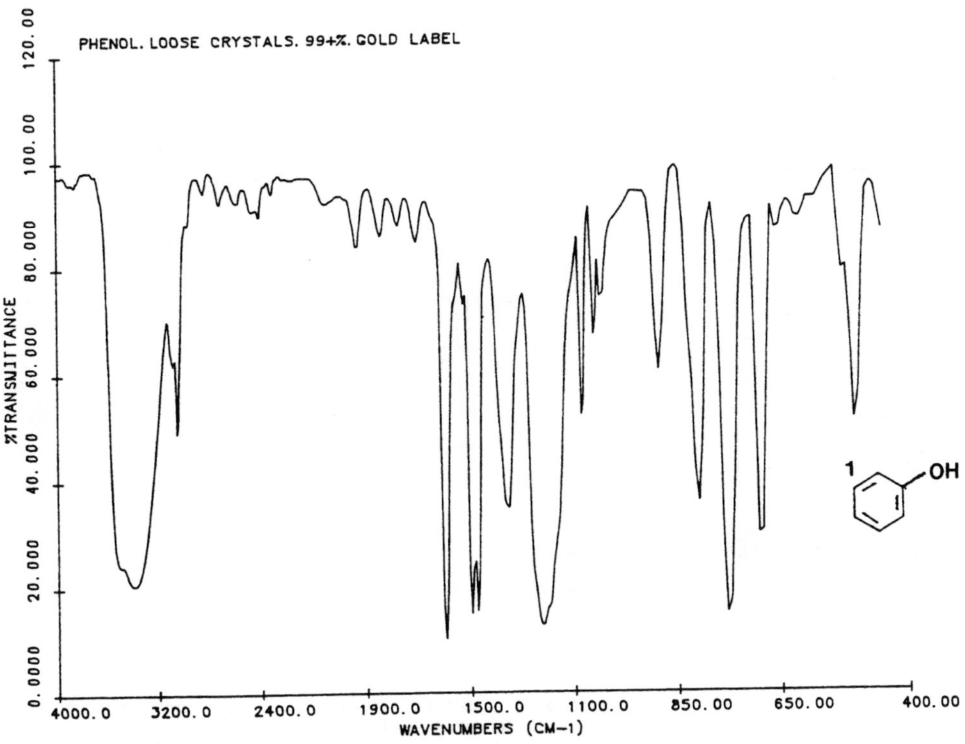


Fig. 36. Infrared spectrum of phenol; 1) = phenol

between 1150 and 1250 $cm^{-1}$, or 8.7 and 8.2 µm, and in alcohols between 1070 and 1170 $cm^{-1}$, or 9.3 and 8.6 µm.)

Since other functional groups are absent, the substance under investigation can be identified as phenol.

The IR spectrum in Fig. 37 does not contain the typical aromatic signals, and therefore originates from a non-aromatic compound. The bands at 2860 and 2940 $cm^{-1}$ indicate C-H stretching vibrations in $CH_2$ and $CH_3$ groups.

At 1375 $cm^{-1}$ one may observe the C-H bending vibrations in $CH_3$ and at 1460 $cm^{-1}$ the C-H bending vibrations in $CH_2$ and $CH_3$. The medium-intensity band at 720 $cm^{-1}$ is characteristic and arises as a result of skeletal vibrations of a methylene chain with generally more than 3 components. The spectrum rules out the possibility of the presence of other functional groups. Accordingly, we are dealing with an aliphatic hydrocarbon. Since the bands of the C=C and the C≡C stretching vibrations are also absent, it is an alkane. After determination of the molecular weight it would be identified as n-hexane.

## 2.10.3 Preparation and handling of the samples

The aim of sample preparation for IR spectroscopy is to obtain thin, plane-parallel layers of the substance under investigation. This is almost always successful; the state of aggregation of the substance plays no decisive

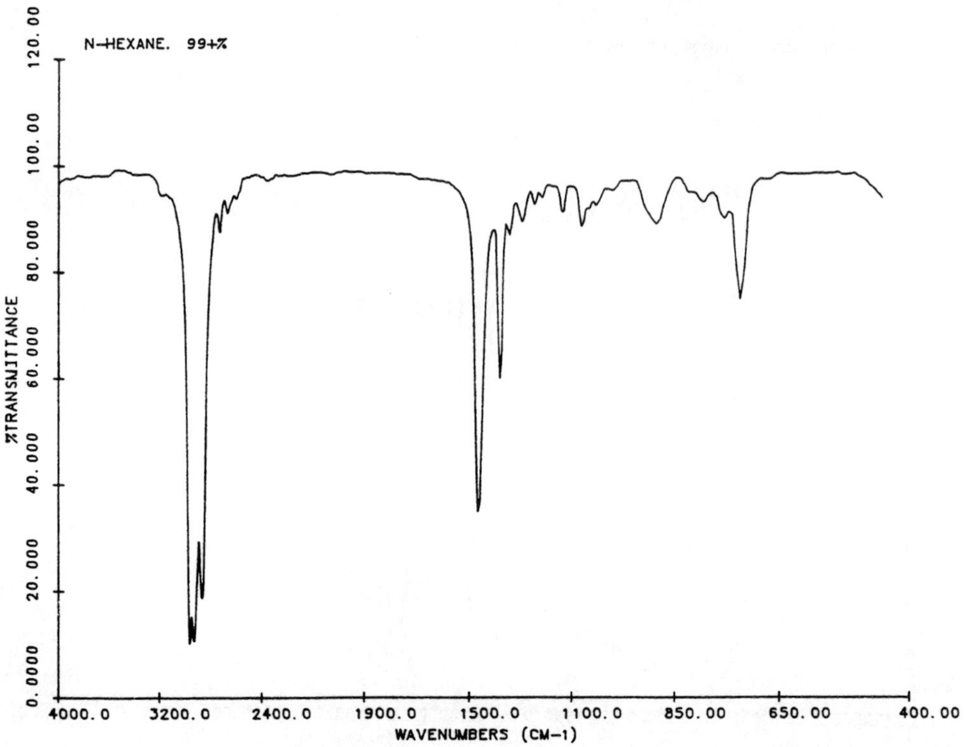


Fig. 37. Infrared spectrum of n-hexane. Path length 10 μm; 1) = n-hexane; 2) = skeletal vibration

part. It is customary to use path lengths between a few μm and 1mm or in special cases somewhat thicker layers. In gas cuvettes light paths up to 40 m long are obtained by multiple reflection.

Of all the techniques of preparation used, a 10- to 20-per cent solution of the substance provides the best IR spectra with the least disturbance. It proves most practical to fill the solution into detachable, sealed liquid cuvettes. The path length is set by means of spacers. The windows consist of plane-parallel NaCl plates (16 μm transmittance limit) or KBr plates (25 μm transmittance limit). In quantitative analysis, the plane-parallelism of the layers and plates must be checked and the path length measured by interferometry with an empty cuvette.

There is no suitable solvent for IR spectroscopy which is sufficiently transparent over the whole range from 4000 to 400 $cm^{-1}$. Between 4000 and 900 $cm^{-1}$ carbon tetrachloride is frequently used, and between 1400 and 400 $cm^{-1}$ carbon disulphide. Below 400 $cm^{-1}$ or 25 μm all the main solvents are sufficiently transparent. If the substance does not dissolve to an adequate extent, solvents such as n-hexane, chloroform or dimethyl formamide may be used. Water should be avoided as a solvent. In cases where this proves impossible, spectroscopy may also be performed with aqueous solutions with the aid of certain manipulative measures. The simplest is to saturate the solution with NaCl or KBr, the same material as the windows. Should this be chemically undesirable, the analysis should be conducted without cuvettes.

According to H. Hausdorff's method the surface energy of the aqueous solution should be reduced by small additions of wetting agents and the solution then stretched as a lamella over a wire ring. A controllable inflow into the lamella and an outflow permit the lamella thickness to be adjusted.

**Note:**

The very low transparency of water between 4000 and 3000 $cm^{-1}$ can be eliminated by using heavy water. The decrease in the transmittance of $H_2O$ and $D_2O$ beneath 1800 $cm^{-1}$ is avoided by appropriate reduction of the light intensity, independent of wavelength, in the reference path.

Good results in the IR spectroscopy of aqueous samples are given by the ATR technique (attenuated total reflection) and a further development of this, the MIR technique (multiple internal reflection); a prerequisite is that the sample space of the IR spectrophotometer is sufficiently large to accommodate the supplementary MIR device. The MIR technique likewise involves measurement of the absorption spectrum of the substance. All of the same molecule-specific and analytical conclusions can therefore be drawn as with an IR spectrum in the transmission of the same substance, even though the two spectra are physically not quite identical.

The advantage of the MIR technique lies in the radically simplified preparation of the samples. Thin layers and cuvettes with infrared-transmitting windows are no longer necessary. Low-intensity bands become measurable, and in addition to aqueous solutions it is also possible to investigate paints or highly viscous samples, for example. The sample under analysis (1) is applied to the upper and lower surfaces of a trapezoidal prism (2), see Fig. 38, and the primary beam of infrared light, PL, is shone into the prism at the angle of incidence specific to the substance (3). After leaving the prism, the beam of infrared light VL enters the IR spectrophotometer. The quality of an MIR spectrum is basically dependent on the refractive index $n_2$ of (2) being everywhere clearly greater than the refractive index $n_1$ of (1) in the analyzed spectral range, and the self-absorption of (2) being low. Only then is undisturbed total reflection, TR, obtained at the boundary surfaces between (2) and (1). On each total reflection the beam of infrared light penetrates the substance under analysis (1) to a depth of approximately 1 µm as a cross-attenuated electromagnetic wave. Each time, the substance impresses its absorption pattern on the beam of infrared light.

When using the MIR technique for the analysis of aqueous solutions, germanium is taken as the prism material (angle of incidence (3) 19° to 20°; infrared-transmitting between 5000 and 500 $cm^{-1}$). It should be noted that the MIR technique yields good results only if used in clean, carefully controlled conditions.

When preparing samples of solid substances, their physical properties must be taken into account. Either sheets of small thickness are used or the substance is partially dissolved, poured as a film onto an infrared-transmitting target and the solvent evaporated. Solid substances are frequently prepared using the KBr moulding technique. Under pressure, potassium bromide flows when cold. Samples are available in the form of KBr mouldings of 13 mm diameter and approximately 0.5 to 1 mm thickness. A useful method of preparation is to grind 200 to 300 mg of dry KBr powder (of spectroscopic quality) as the embedding medium in a vibratory mill,

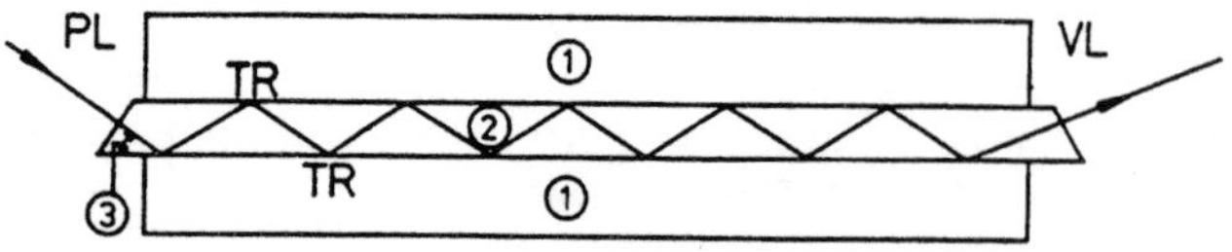


**Fig. 38.** Schematic diagram of the path of the infrared light beam through the sample, using the MIR technique

together with about 1 mg of the substance to be embedded. Substances with low melting points should be ground under liquid nitrogen. The particle size of the substance should not exceed 1 µm; otherwise, the moulding may scatter the infrared light beam to the sides. This is expressed in the IR spectrum as an over-rapid rise in the left-hand slope of bands in the shortwave spectral range (Christiansen effect). The ground mixture is pressed into a mould under a vacuum of 10 - 20 bar for several minutes at a mould pressure of approximately $10^4$ bar. The ideal KBr moulding is clearly transparent. This is achieved when the bonds of the substance are highly polar. Mouldings are generally turbid to a greater or lesser extent. This does not, however, detract from the significance of the IR spectrum. Mouldings which are too turbid may be made clearer by repeating grinding to a finer level and a longer pressing period under higher pressure.

When the KBr moulding technique is used, the potassium bromide, the water it contains or the crystal lattice may react with the embedded substance and alter the structure of the IR spectrum. It must be emphasised that no KBr moulding is stable in the long term.

This problem does not arise with the Mull technique. The finely ground substance is added to a few drops of Nujol, for example, a mixture of long-chain liquid paraffins. The paste thus obtained is then rubbed until it becomes viscous and is pressed out to form a thin film between KBr plates. In this case, too, the scattering of light must be checked. The Mull technique offers rapid preparation of the samples and moderate-quality IR spectra.

The methods of preparing the samples outlined above frequently require compensation of the IR spectrum. If it is necessary to compensate for the self-absorption of the solvent, a reduced layer thickness (path length) should be selected in the reference beam path, the reduction being in inverse proportion to the concentration of the sample. With a little patience it is then possible to eliminate the solvent bands from the IR spectrum of the substance under analysis. However, the result will be satisfactory only if the spectral position, intensity and shape of the absorption bands of the dissolved substance are not altered by solvent effects.

Compensation should not be forced if the transmittance is less than 5 %.

The possibilites of applying IR spectroscopy in water analysis appear, as things stand at present, to be limited. Essentially this means they are confined to investigation of organic contamination, for example by hydrocarbons including aromatics.

## 2.11 Chromatographic methods

### 2.11.1 General

Chromatographic methods, which today provide indispensable assistance not only to analytical but also to preparative chemists in overcoming formerly insoluble problems of separation, may be traced back to around the middle of the last century.

In 1850, F. Runge published his book "Zur Farbenchemie. Musterbilder für Freunde des Schönen und zum Gebrauch für Zeichner, Maler, Verzierer und Zeugdrucker" (On colour chemistry. Examples for lovers of beauty and for use by artists, painters, ornamenters and cloth printers), in which he described for the first time the migration of various dyes on paper using water as an eluent.

"Chromatography", the origins of which are more properly ascribed to the aesthetic and artistic fields, entered the realm of the natural sciences in 1861 when Schönbein, in the course of investigating ozone with iodide, established that different substances are transported by water on paper at different speeds.

The next step was in 1903 when Tswett separated the plant pigments carotene, xanthophyll and chlorophyll in a column; this had not been possible with other, earlier methods.

Not until 25 years later, however, did chromatography come into its own as a result of work by Kuhn, Winterstein and Lederer. Further development then led to "open-column chromatography", thin-layer chromatography, gas chromatography and, as the latest development, high-pressure liquid chromatography.

Chromatographic methods have secured a permanent position among the techniques of investigation used in water analysis. Chromatographic methods are used both for separation and for detection in water analysis. Alongside paper and column chromatography, thin-layer chromatography in particular has gained considerable significance following the work by E. Stahl. With special reference to water analysis, this applies especially to the use of thin-layer chromatography to separate the so-called polycyclic aromatic hydrocarbons (see Section 4.2) or heavy mineral-oil hydrocarbons.

In all chromatographic methods of separation and determination the differing distribution of the substance mixture to be separated between the fixed and liquid phases is important; the use of different mobile phases also results in different separation efficiencies. Depending on the type of chromatographic technique used, the separated substances can be detected with the aid of various detectors.

Definition:

$$R_f = \frac{\text{Length of run of substance}}{\text{Length of run of solvent}}$$

Chromatography is at its simplest when the substances to be separated and determined are coloured, thus making simple, direct detection possible. In addition, detection may be carried out by using certain reagents which form coloured compounds with the separated substances in the same way as with colorimetric methods of analysis. In this context mention must be made of, for example, the separation by thin-layer chromatography of the various phosphorus compounds which may occur in water, i.e. orthophosphate, polyphosphates, condensed phosphates and so on.

### 2.11.2 The main chromatographic techniques used in innumerable variations in water analysis are

Paper chromatography

Thin-layer chromatography (1- and 2-dimensional)

Column chromatography with various column packings

Liquid chromatography (LC)

High-pressure liquid chromatography (HPLC)

Ion chromatography (IC)

Gas chromatography with packed columns and capillary columns, and various stationary and mobile phases and different detectors (see the Section on gas chromatography (2.12) in this chapter)

Even if the instrumental techniques gain ever more ground, some of them in automated form, evaluated and documented by computer, the simple, manual methods such as paper and thin-layer chromatography still have their role to play. One need only think of the separation of the various forms of phosphate such as orthophosphate, pyrophosphate, metaphosphate, polymeric phosphates and condensed phosphates, or the separation of polycyclic aromatic hydrocarbons (specifically with thin-layer chromatography) in water analysis.

Ion chromatography is used to detect not only inorganic anions and cations but also organic group parameters in water samples.

Where instrumental chromatographic techniques have found a place in water analysis they are discussed in this section or the principle is mentioned. Details of individual methods are to be found, for example, in Sections 3.2 (ion chromatography for anions), 3.6 and 4.2 (gas chromatography), and 4.1 and 4.2 (HPLC).

### 2.11.3 General comments on GC-MS techniques in water analysis

Combinations of gas chromatographic techniques of separation with indication by mass spectrometry are used in water analysis for specific purposes, for example to ensure the detection of organic compounds in gas chromatography, and also to clarify their structure. For GC see Section 2.12; schematic diagram of mass-selective detector see Fig. 39 a-d.

The GC-MS technique in water analysis is a typical example of the application of modern technology. Computer-controlled GC-MS coupling is preferred. Small, compact quadrupole mass spectrometers are frequently used for mass spectrometry. The preliminary gas chromatographic separation is preferably conducted with capillary separation columns.

In addition to the quadrupole mass spectrometer, which is used relatively frequently in water analysis, electron impact ionization, chemical ionization, qualitative multiple ion selection and quantitative multiple ion selection (with a certain mass as the internal standard and direct admission of the substances, e.g. for mass calibration) are also used.

The range of sensitivity of the mass spectrometer following preliminary separation by gas chromatography lies well below the µg range and, under favourable conditions, even reaches the nanogram region.

In addition to low-resolution devices (quadrupole or magnetic devices), high-resolution devices are also used, such as double-focusing sector-field devices or maximum-resolution sector field devices such as the triple-quad system, although the latter devices are now hardly encountered in practical water analysis, tending to be used more for research. Mass spectrometry may also be used to measure the isotopic ratios of dissolved gases in water samples, e.g. $^{13}C$-$^{12}C$, $^{15}N$-$^{14}N$, $^{18}O$-$^{16}O$ and to detect inert gases. In order to measure the isotopic ratios, so-called gas devices for the analysis of gaseous samples and thermionic devices for measuring solids are also used directly for mass spectrometry. In many cases, however, the GC-MS analysis must be preceded by a concentration of the substances under investigation in the water sample, for example by liquid-liquid extraction, adsorption processes, ion exchange and so on.

This elegant method of detection using a mass spectrometer as a detector for compounds separated by gas chromatographic means presupposes that an MS can be linked - on line - to a gas chromatograph.

Further conditions are:

a) The use of a carrier gas which has only a low ionization yield under the conditions of the ion source. One such gas with a low ionization yield under these conditions is helium;

b) A gas flow rate as low as possible - of the order of 2 ml/min. - so that the operating pressure of the analyzer can be maintained (capillary column);

and finally

c) The GC and the MS must be pneumatically decoupled: the end of the separation column normally works against atmospheric pressure; the ion source of the MS is in a high vacuum ($10^{-6}$ bar). This pressure gradient is overcome by means of a "restrictor" consisting of a split and an extremely fine hair capillary.

When using the MS as a detector for GC separation, two approaches have proved particularly valuable: firstly, recording mass spectra at intervals of about 1 s - this has the advantage of providing information on the nature of the eluted compound - and secondly, recording specified mass traces (single ion monitoring).

Even though the GC-MS coupling has made possible the detection of certain organic compounds, and also inorganic substances, in water in innumerable cases. One should be extremely careful not to underestimate the possibility of interference occurring in this modern analysis system.

Owing to the multiplicity of individual substances and the resulting combinations of matter, the mixtures of substances which occur in water, particularly surface waters or wastewaters, are so complicated that this analytical technique, particularly when attempting to advance into the ultratrace range, i.e. below µg/l, can only be employed by specialists with appropriate experience and adequate equipment for substance separation and mass-spectrometric detection with corresponding valuation. Thousands of organic compounds have been detected in water samples where the detection limit has been shifted below the ng limit, depending on the ratio of signal to background noise. Nevertheless, with an inexperienced operator and inadequate apparatus, serious mistakes are to be expected, and it must be emphasized once again that evaluation by mass spectrometry also requires verification of the substances detected, and also that in this extreme trace region the relevance of the detected substances should always be judged by a specialist.

It is indispensable when involved with this type of analysis to establish a spectrum library or to link up with and share the use of an existing facility.

Mode of Mass-Separation: a) Quadrupole massspectrometer
b) Magnetic field spectrometer

Mode of Ionization:
a) Electron impact ionization
b) Chemical ionization
c) Fast atomic bombardment

Mode of Detection:
a) Multiple ion selection (qualitative/quantitative)

Diagram of a GC-MS installation as an example:

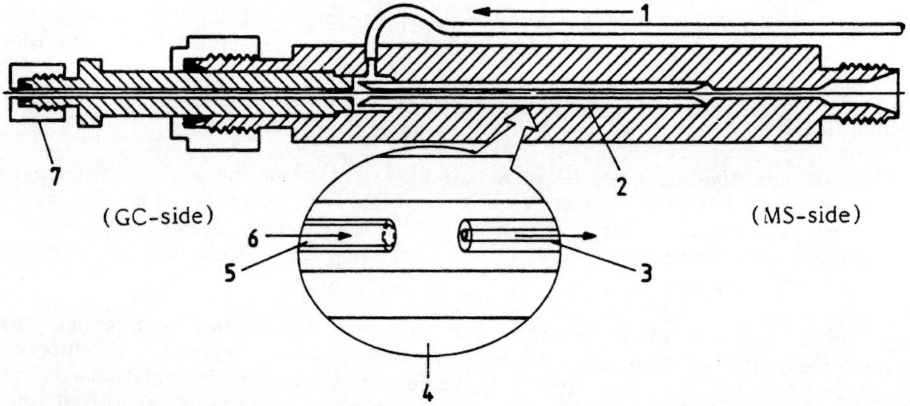


Fig. 39 a. Structure of GC coupling unit; 1) = Helium makeup; 2) = Interchangeable glass insert; 3) = Transfer capillary; 4) = Open coupling; 5) = Capillary column; 6) = Substance flow; 7) = Column connection

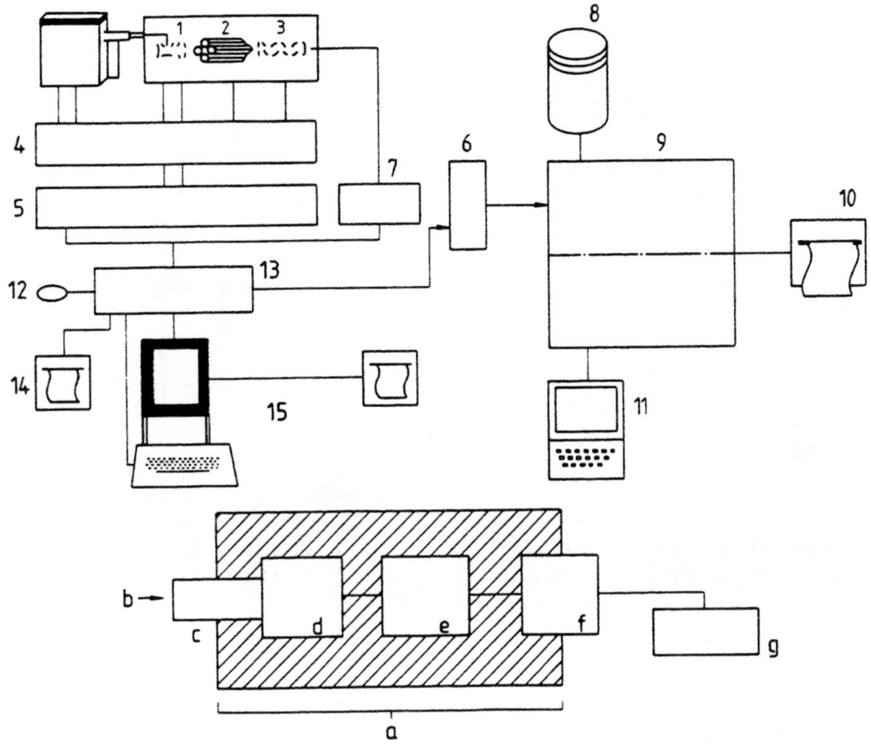


Fig. 39 b. Schematic diagram of a GC-MS system; Example: GC-MS system MAT 44 (Finnigan) with on-line SpectroSystem MAT 44; 1) = Ionization; 2) = Quadrupole; 3) = Photomultiplier; 4) = GC and MS supply; 5) = Interface; 6) = Interface; 7) = Integrating measuring system; 8) = Disk storage; 9) = Memory; 10) = Electrostatic plotter; 11) = Video terminal; 12) = Mass storage; 13) = Processor; 14) = Recorder; 15) = Copier unit; a - g = MS-system

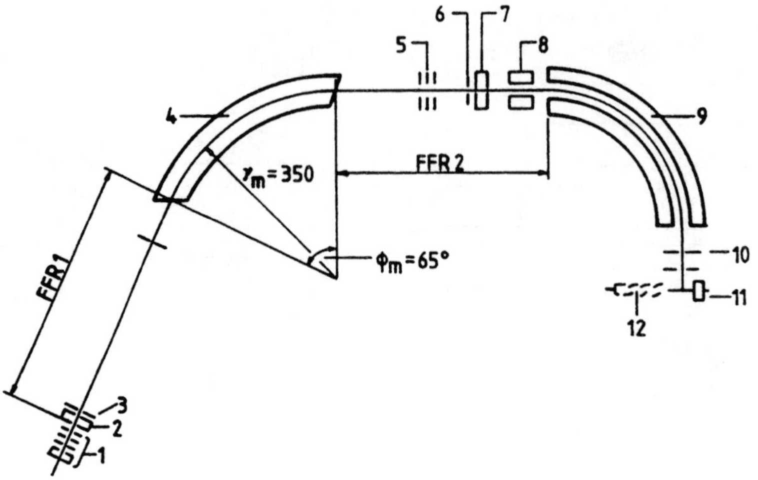


Fig. 39 c. Magnetic sector mass spectrometer; 1) = Ion source; 2) = FFR 1 collision cell; 3) = Source slit; 4) = Magnetic sector; 5) = Focusing element; 6) = Intermediate slit; 7) = FFR 2 collision cell; 8) = Focusing element; 9) = Electric sector; 10) = Collector slit; 11) = Conversion dynode; 12) = Secondary electron multiplyer (SEM)

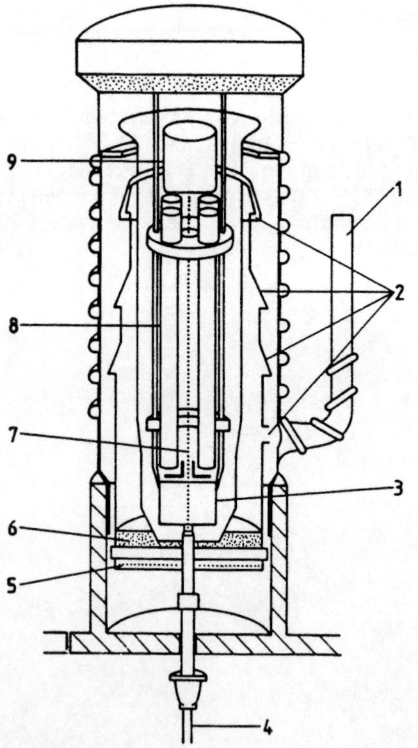


Fig. 39 d. Quadrupole Mass spectrometer or mass-selective detector; 1) = Forevacuum; 2) = Pump stages; 3) = Ion source; 4) = GC column; 5) Heating for oil diffusion pump; 6) = Pump oil; 7) = Ion path; 8) = Quadrupole rods (separation of ions in the electrical field; 9) Secondary emission, electron multiplier (based on Hewlett Packard)

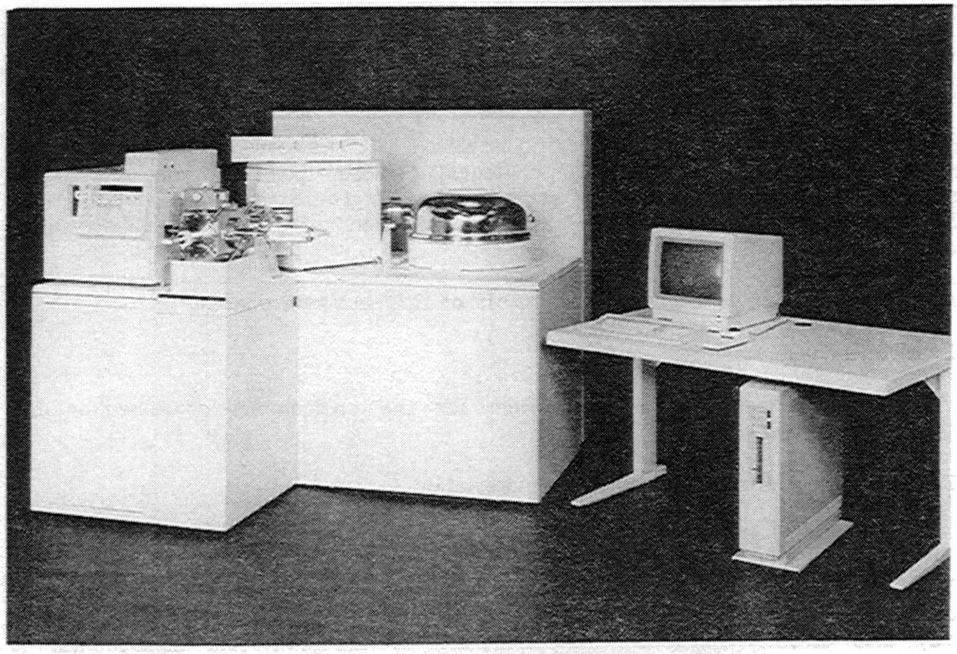

Fig. 39 e. Massspectrometer system MAT 90

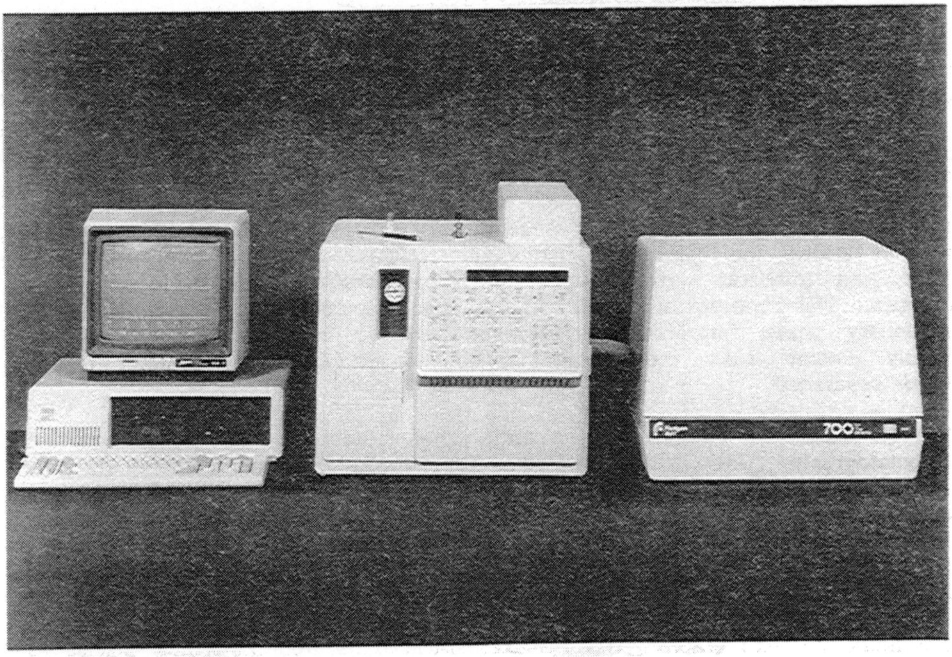


Fig. 39 f. Ion trap detector 700

## 2.12 Introduction to gas chromatography

### 2.12.1 Principles and definition of the method

The term chromatography was introduced by M. Tswett in 1903. It covers various analytical separation techniques, whose primary purpose is to separate multiple-component mixtures of varyingly complicated composition. Separation is usually followed by quantification of the individual sample constituents.

A chromatographic system always consists of at least two phases.

1. Mobile phase (Carrier gas):

   This serves as a transport medium for the components contained in the sample (carrier gas).

2. Stationary phase (Column filling material):

   Separation of the mixture takes place here.

The chromatographic methods are primarily subdivided according to the type of mobile phase used, which in addition to its role as a transport medium may also directly influence the separation of the individual components. A distinction is therefore made between gas-chromatographic separation systems and liquid-chromatographic separation systems. In addition to the state of aggregation of the mobile phase, a further criterion for differentiating chromatographic separation techniques is the distribution mechanism of the separation system.

Gas chromatography (GC) is a physico-chemical separation technique for mixtures of substances which can be evaporated without becoming decomposed up to approximately 400 °C. All gas-chromatographic separations are based on the principle of distributing a substance between two phases. One of the phases consists of a stationary, large-area bed (stationary phase), and the other phase is a carrier gas (He, $N_2$, Ar, $H_2$), which flows past the stationary phase and acts as a transport medium for the individual components of the sample (mobile phase). The character of the stationary phase is such that it allows the carrier-gas molecules to pass with the minimum impediment, but interacts with the various components of the sample to varying degrees. This interaction causes the sample components to be held by the stationary phase for shorter or longer periods, with the result that they finally emerge from the separation column at different times, in other words separated.

If the stationary phase is a solid, the method is known as gas-solid chromatography (GSC). In this case the separating medium and the stationary phase are identical: typical absorbing materials are used, such as silica gel, aluminium oxide, molecular sieve or activated charcoal, to name but a few, whose ability to absorb different molecules to differing degrees on their generally very large surface is utilized to achieve separation. GSC is particularly suitable for the analysis of gases.

Gas-liquid chromatography (GLC) makes use of a liquid stationary phase. The separating medium is made up of two components, a highly porous solid material, the support, whose chemical and sorption properties should be as inert as possible, and the stationary phase proper, which is attached to the surface of the support in the form of a thin film. In GLC, the decisive factor for separation is the different degree of solubility of the sample components in the stationary phase. GLC is the most efficient method of separation which is available to modern analysis, both with regard to the actual separation power and to the field of problems which can be processed with this method.

In addition to GLC and GSC in the outlined form, there are a number of other variants, of which two are of particular importance. Capillary gas chromatography is a type of GLC in which the separating liquid is applied as a thin film to the inner wall of glass or quartz capillaries (internal diameter approximately 0.3 mm, 12 - 150 m long). It is used where maximum separation efficiency is required with the smallest of sample quantities.

The other variant that has to be mentioned here makes use of porous polymers (e.g. Porapak®, Tenax GC®, Chromosorb Century®) as separating media. Separation columns filled with porous polymers are excellently suited to the analysis of complicated mixtures made up of components whose boiling points are not too high and which may also be found in aqueous solution.

### 2.12.2 Structure of gas-chromatographic apparatus

The principal parts of a gas-chromatographic apparatus are shown in Fig. 40 below.

The gas chromatograph shown schematically in Fig. 40 fulfils the necessary conditions for carrying out the intended substance separations, and enables the analyst to control the separation. The system consists of a carrier-gas source (1), e.g. a pressure cylinder with high-purity nitrogen or helium; a pressure-reducing valve and precision adjusting valve ensure that the pressure and/or rate of flow of the carrier gas is set uniformly; this is monitored with a pressure gauge or flow meter. The sections following that are the sample inlet (2), the separation column (3) (incorporated in the GC oven) and the detector (4). All parts (2 - 4) may be thermostatted independently of each other and thus adapted to the respective separation problem.

The substances to be separated are transported through the separation column by the carrier gas. The sample mixture enters a partition process between the gas phase and the stationary phase; for each component of the sample, this process is determined by the coefficient of partition $K_D$, which is substance-specific (a high partition coefficient corresponds to high solubility in the stationary phase). The partition process leads to the formation of discrete bands of substances in the carrier gas, each of

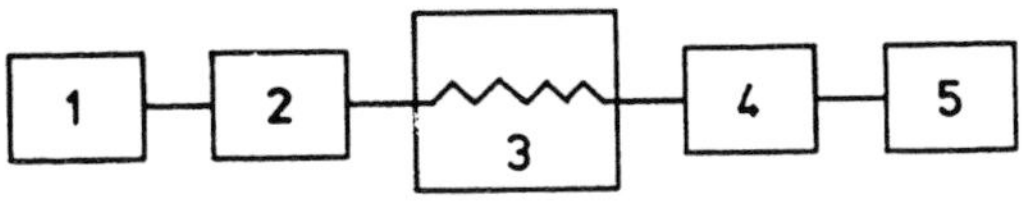


Fig. 40. Gas-chromatographic apparatus; 1) = Carrier-gas supply with pressure and/or flow control; 2) = Injection system; 3) = Column oven with analytical separation column; 4) = Detection unit; 5) = Recording system with data evaluation

which contains one of the sample components. The temperature of the separation column remains constant throughout the analysis. This procedure is known as the isothermal elution technique. The process is allowed to run, at least ideally, until all of the substance bands have left the separation column and have been detected as substance peaks by the detector and recorded.

It must be considered a disadvantage that those components with greater affinity to the stationary phase have a long residence time in the column. The technique of temperature programming may be used in order to reduce the elution time: temperature programming is the technique of increasing the column temperature during the analysis.

The detector of the gas chromatograph supplies an analogue signal which, in the simplest case, is plotted by a self-balancing recorder in the form of a continuous graph (chromatogram) (5). Time is normally taken as the abscissa of the graph and detector output voltage as the ordinate. Fig. 41 shows a chromatogram of this type. The most important information provided by a chromatogram is:

1. Retention time - residence time of the substance in the column. On the chromatogram, this is the distance measured from the injection point to the peak maximum.

2. Peak area - this is the area beneath a substance profile, bordered above by the curve and below by the base line (shaded area in Fig. 41).

Peaks $C_{18}$, $C_{18:1}$ and $C_{18:2}$ in Fig. 41 represent the methyl esters of stearic acid, oleic acid and linoleic acid.

Separation of these substances by other methods is exceedingly difficult or even impossible: the difference in their boiling points is negligible and they differ only in the number of double bonds. With gas chromatography it is possible to separate substances with virtually identical boiling points.

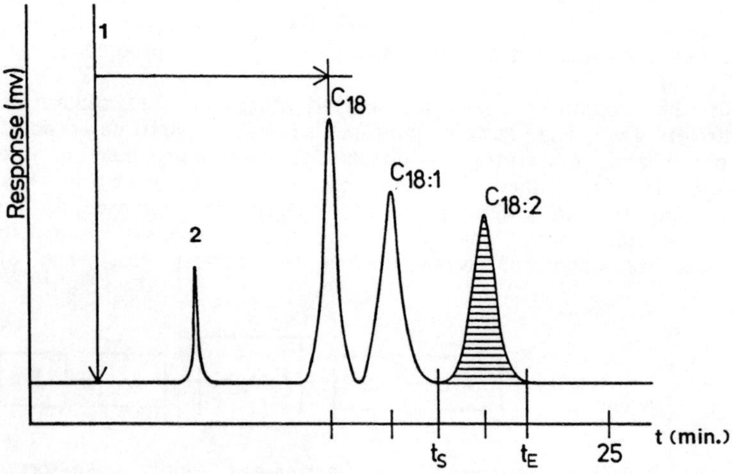


Fig. 41. Relationship of the detector signal (response) to time (refer to text for explanation); 1) = Injection point; 2) = Air

By using selective phases it is possible in this way to achieve a degree of resolution which is unattainable with distillation or any other method.

It is now common practice to supply the detector signal not only to the plotter but also to an electronic integrator which prints out the main information - retention time and peak area - in digital form. Even though the integrator printout contains all the figures necessary to calculate the results of the analysis, the plotted chromatogram remains indispensable for control purposes.

### 2.12.3 Identification of sample components

Gas chromatography is capable of achieving notable results in qualitative analysis, even though it is not, in principle, a qualitative technique. In a particular separation column and under specific operating conditions (column temperature, type of carrier gas, carrier-gas flow) each component of the sample has only one retention time, which is reproduced precisely and can be used to identify the substance causing the peak. It is, however, possible that two substances quite by chance have the same retention time, but a coelution can always be overcome by the use of a more selective stationary phase. By connecting specific detectors and mass spectrometers to the gas chromatograph it is possible to identify substances unambiguously; the results should be confirmed by analysis of reference substances.

### 2.12.4 Determination of sample concentrations

Gas chromatography is excellently suited to quantitative analysis.

The area of each recorded peak is proportional to the concentration of the substance. The area may therefore be used to determine the concentration of each individual substance.

In the chromatogram shown in Fig. 41, the relative areas for the different substances are as follows:

$C_{18}$ : 36.7 (%)
$C_{18:1}$ : 33.0 (%)
$C_{18:2}$ : 30.3 (%)

$$(C_{18} + C_{18:1} + C_{18:2} = 100\ (\%))$$

The actual concentrations are 36.4, 33.2 and 30.4 percent by weight.

The accuracy attainable by gas chromatography depends on the technique used (detector, method of integration and concentration of the sample).

Even with manual injection, quantitative results are normally accurate to 1 - 2 % (RSD)[1]. With the aid of appropriate technology, e.g. electronic integrators or data systems, accuracy may be improved to less than 1 % (RSD).

---

[1] RSD = Relative standard deviation.

### 2.12.5 Sensitivity

The main reason behind the extremely widespread acceptance of gas chromatography is its sensitivity. The simplest version of a thermal-conductivity detector can determine concentrations down to 0.01 % (100 ppm) (depending on the type of carrier gas and the substance to be determined). A flame ionization detection (FID) detects substances whose concentrations are in the ppm to ppb range.

Detectors which are sensitive to specific elements or structures, such as electron capture detectors (ECD) or phosphorus- and nitrogen-selective detectors (TSD) respond even to quantities of substances in the picogram ($10^{-12}$ g) or femtogram ($10^{-15}$ g) ranges.

### 2.12.6 Speed of analysis and automation

The separation of the fatty acid methyl esters as shown in Fig. 41 is completed after approximately 23 minutes. By using gas as the mobile phase, the establishment of equilibrium between the mobile and stationary phases is rapid, thus permitting relatively high carrier-gas flow rates to be used. Standard separations can generally be timed in terms of minutes. Using current technology, however, separations may also frequently be conducted in seconds. If the duration of the analysis is rather long, for whatever reason (e.g. thermal instability of the components or large partition coefficient), or if a large number of samples are to be measured, gas chromatographic analysis can be conducted fully automatically. This has the advantage that the time-consuming measurements described above can be taken overnight. An important constituent of fully automatic measuring apparatus is a data system which undertakes both control and evaluation tasks. Particularly user-friendly data systems permit all the necessary analytical parameters to be stored under one method name on a diskette (magnetic disk). To activate the entire system, the method is then entered into the data system via a disk unit.

Gas chromatographs are simple to operate and to understand. It is generally possible to interpret the data obtained rapidly and directly. In comparison with the value of the results achieved, the costs of a gas chromatograph are low.

### 2.12.7 Theory of the chromatographic process

#### Introduction

When a substance is fed into the separation column via a suitable injection system, it becomes distributed between the stationary and the mobile phases. A short time after injection of the sample, the concentration in the gas phase is very high and that in the stationary phase virtually zero. Some molecules of the substance then become dissolved in the separating liquid and a dynamic equilibrium is established, which means that per unit time the number of molecules desorbed is equal to the number absorbed.

The position of the partition equilibrium is determined by the coefficient of partition $K_D$.

$$K_D = \frac{C_L}{C_M} \tag{2.1}$$

$C_L$ = Concentration of the substance in the stationary phase
$C_M$= Concentration of the substance in the mobile phase (M = Mobile)

$K_D$ is dependent on:

a) the chemical nature of the substance
b) the chemical structure of the phase
c) temperature

It is also necessary that the mobile phase does not go into solution in the stationary phase, nor the stationary phase in the mobile phase. The former possibility can be excluded by using inert carrier gases. The latter eventuality may occur if the vapour pressure of the stationary phase is so high that the separating fluid evaporates. Strictly speaking, the concentration ratio for $K_D$ also only applies if the molecular state of the substance is the same in both the gas phase and in the stationary phase.

If concentrations $C_s$ and $C_m$ are expressed by the quotients of the molar number and the phase volume, we obtain Equation 2.2.

$$K_D = \frac{n_s}{n_m} \cdot \frac{V_m}{V_s} \tag{2.2}$$

The ratio of the mole quantities $n_s/n_m$ is termed the capacity factor k', and the ratio of the volumes $V_m/V_s$ the phase ratio ß.

$$K_D = k' \cdot ß \tag{2.3}$$

The phase ratio ß is a measure of the permeability of a separation column. By way of example, Fig. 42 shows a comparison of a packed column and a capillary column. It has been assumed for the purposes of this comparison that completely non-porous particles (support and phase) have been used for the packed column.

The ß-values of packed columns lie between 5 and 35, and the ß values of capillary columns between 50 and 1500.

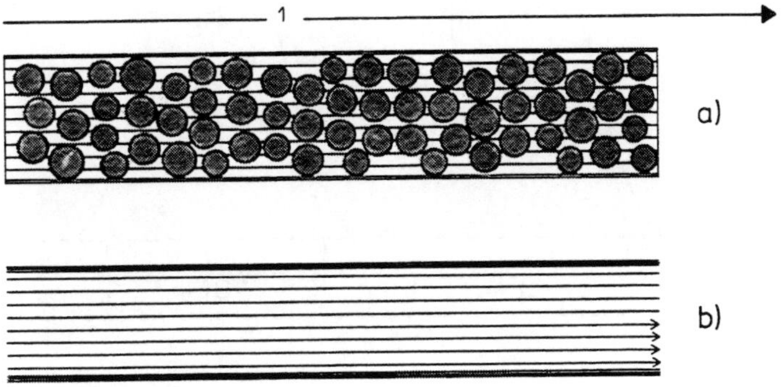


Fig. 42 a and b. Cross-section of a packed column a and a capillary column b (simplified). Only the shaded parts are available to the mobile phase; 1) = Direction of flow

### 2.12.8 The separation process

The efficiency of gas chromatography as a separation method is judged according to the quality of the separation obtained. A measure of the separation quality is resolution, R. Resolution is always calculated for two adjacent peaks. The parameters required for the calculation of R must be taken from the chromatogram (see Fig. 43) or from the printout of a data system.

The defining equations for resolution R are then:

$$R = \frac{2 \cdot \Delta t}{W_B(A) + W_B(B)} \quad (2.4)$$

$$R = 1.177 \cdot \frac{\Delta t}{W_{0.5}(A) + W_{0.5}(B)} \quad (2.5)$$

For good separations the resolution values obtained are greater than 1, and for partial separations the values are less than 1.

The following conclusions may be drawn from Equations 2.4 and 2.5:

Resolution may be increased by increasing $\Delta t$, the difference in retention times, and/or by reducing the peak widths ($W_B$ or $W_{0.5}$). Since the retention times and the peak widths are each influenced by different processes, it is useful to give a simplified explanation of the mechanisms involved. The desired effect, that is the separation of two components, depends on the components displaying different retention characteristics in the column. In gas-liquid chromatography, the chemical nature of the separating medium has the greatest influence on retention. The decisive step towards obtaining a sufficiently large $\Delta t$, therefore, is selection of the "correct" stationary phase.

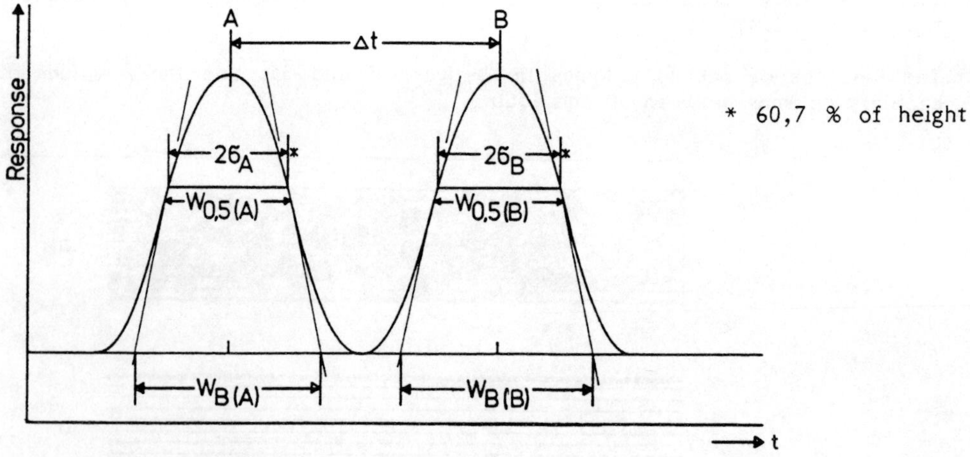


Fig. 43. Chromatogram, illustrating resolution R; $W_B$ = Base-line widths; $W_{0.5}$ = Widths at half height; t = Difference in retention times

The effect which is detrimental to resolution, peak broadening, originates from two undesirable but unavoidable physical processes in the column, which are virtually independent of the chemical nature of the separating medium. It is possible to minimize the peak-broadening processes by technical means, but they can never be fully eliminated.

### 2.12.9 General remarks on the formation of the peak shape

Let us assume that a small quantity (at the most a few μg) of a substance A is injected in vapour form at the top of the separation column. The sample vapour contains a little air, and the detector is of a type which is also capable of detecting air. The separating liquid and column temperature are selected such that for practical purposes the constituents of the air do not interact with the column packing. The peak shape is then formed according to the following pattern.

It should be observed that:

1. The dead time $t_m$ is constant for a specific column and a set carrier-gas flow. It is equal to the time which the carrier-gas molecules themselves or other non-interacting substances (in this case air) require to pass through the separating column. (Therefore $t_m$ = t (mobile phase)).

2. The peak of substance A appears at a time where $t_{ms} > t_m$. Time $t_{ms}$ consists of the time component $t_m$ which the molecules of A have spent in the mobile phase plus the time $t_s$ during which they were dissolved in the stationary phase. This means that $t_s = t_{ms} - t_m$.

   $t_s$ is also known as the adjusted retention time. This is the measured quantity which is actually characteristic for the substance.

3. The peak of substance A is considerably broader than the concentration profile in the injector shortly after injection. The greater the value of $t_{ms}$, the broader and hence flatter the substance peak becomes.

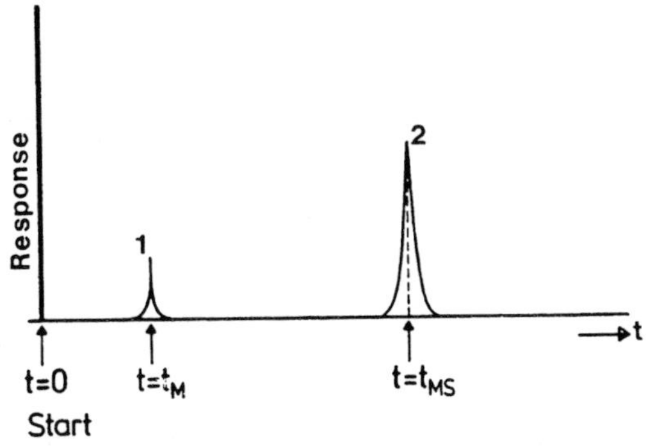


Fig. 44. Chromatogram of a two-component mixture consisting of a substance A with retention time $t_{ms}$ and air with retention time $t_m$; Concentration profile in the injector during dosage: 1) = Air peak in the detector; 2) = Peak of substance A in the detector; $t_M$ = Dead time; $t_{MS}$ = (uncorrected) retention of A

## 2.12.10 The partition isotherms

If a pure liquid is heated, a temperature-dependent vapour pressure is established above its surface. In a two-component system, consisting of two completely miscible liquids A and B, the vapour pressure P is equal to the sum of the partial pressures of the individual components (Dalton). In a gas-chromatographic system, A is equivalent to the substance to be separated, whereas B constitutes the stationary phase. Generally speaking, the working temperature must be selected such that the vapour pressure of the stationary phase is negligible. The vapour pressure $P_i$ above the stationary phase is then produced almost entirely by substance A, and apart from temperature is dependent on the concentration of A in the stationary phase.

The relationship between the concentration of A in the stationary phase ($C_s$) and the concentration of A in the mobile phase ($C_m$) with regard to a particular component is measured at constant temperature. For this reason, the curves generated by plotting $C_s$ against $C_m$ are known as partition isotherms (see Fig. 45).

The shape of the partition isotherms in the chromatographic process has a decisive influence on the shape of the peak. Let us now differentiate between two possible cases:

1. Ideal distribution of substance A between the mobile and stationary phases

   (Curve a in Fig. 45)

   The function displays a linear relationship. The partition coefficient $K_D$ remains constant over a large concentration range.

   Fig. 46 a shows the resultant peak profile. This is symmetrical as regards peak height and is known as a Gaussian curve).

2. Non-ideal distribution of A between the mobile and stationary phases

   a) (Curve b in Fig. 45) (concave isotherm)

   According to curve b in Fig. 45, $C_s$ increases more rapidly than $C_m$. This signifies that the substance tends to be retained more in the mobile phase. The peak shape is asymmetrical (Fig. 46 b).

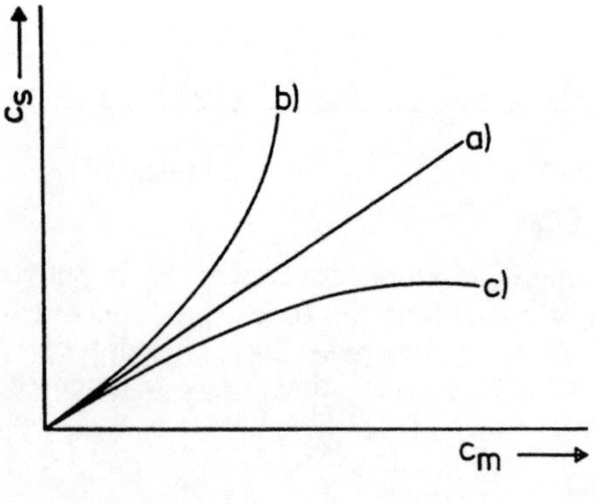


Fig. 45. Relationship of concentrations $C_s$ and $C_m$ (of a component i) in the case of linear a and non-linear b and c partition (T = constant)

b) (Curve c in Fig. 45)

According to curve c, $C_s$ increases less rapidly than $C_m$. In this case, a non-linear, convex isotherm results.

Asymmetry is also the result here, expressed by gradual fading (tailing) of the signals (Fig. 46 c).

In Case a we have an example of linear gas chromatography, and in Cases b and c of non-linear gas chromatography).

## 2.12.11 Causes of peak broadening

### Column efficiency

The shape of the chromatographic peak in gas-liquid chromatography is generally the result of random, diffusion-type processes which occur in the column.

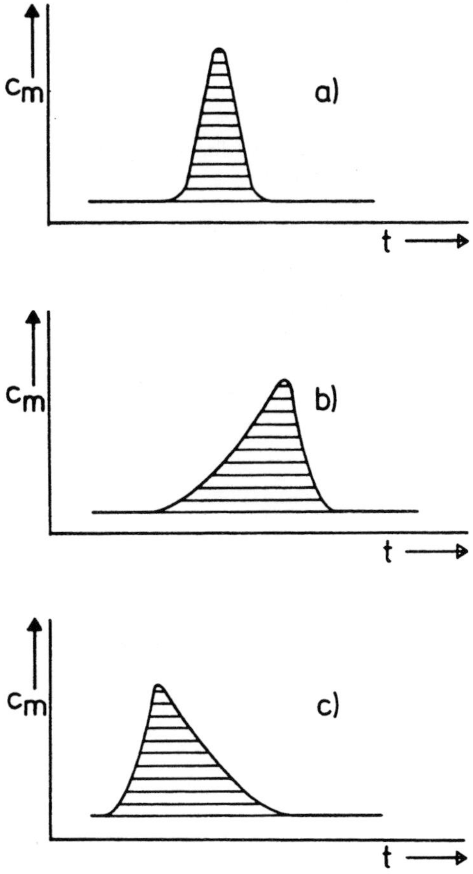


**Fig.** 46 a - c. Relationship of the shape of the peak profile (change of concentration in the carrier gas over time) to the partition isotherm; a = linear; b = concave; c = convex isotherms (Schomburg)

If a diffusible, originally tightly limited system, e.g. a volume of gas, is left to its own resources for a time t, it spreads out by diffusion and is diluted. A Gaussian distribution is always obtained for the characteristic value of the system, e.g. the local concentration. The spreading of the system is described by the standard deviation $\sigma$ which is related to the diffusion coefficient D and the diffusion time t by

$$\sigma = \sqrt{2 \cdot D \cdot t} \tag{2.6}$$

In the separation column there are several simultaneously occurring but by their nature different processes which have the same effect as diffusion.

The phenomenon is called dispersion, the spreading of an originally compact substance peak. In a model described by Giddings (random value model) the various processes can be treated uniformly for calculation and finally combined in an effective diffusion coefficient which describes the overall peak broadening in the column.

Longitudinal diffusion in the mobile phase

As long as sample molecules remain in the mobile phase, i.e. the carrier gas, they will continue to diffuse. Since the concentration of the peak in the direction of the column, at its exit we are interested only in the proportion of diffusion in the direction of the column. In the event of free diffusion the variance produced would be:

$$\sigma^2 = 2 \cdot D \cdot t \tag{2.7}$$

Since diffusion is impeded by the column packing, the actual variance is somewhat smaller:

$$\sigma^2 = 2 \cdot \gamma \cdot D_m \cdot t_m \tag{2.8}$$

$D_m$ = Diffusion coefficient in the mobile phase
$t_m$ = Dead time of the column (equal for all substances)
$\gamma$ = Obstruction factor ($\gamma < 1$)

The dead time is substituted by

$$t_m = L / \bar{v} \tag{2.9}$$

where L is the column length (in cm), and $\bar{v}$ the average linear flow rate (cm · min.$^{-1}$) of the carrier gas. By substituting 2.9 into 2.8, we obtain:

$$\sigma^2 = 2 \cdot \gamma \cdot D_m \cdot \frac{L}{\bar{v}} \tag{2.10}$$

The variance is replaced by a more expressive value which is taken from the theory of distillation columns, the theoretical plate height $H_L$ (the proportion of the plate height which is due to longitudinal diffusion). The relationship between $H_L$ and the variance $\sigma^2$ is:

$$\sigma^2 = H_L \cdot L \tag{2.11}$$

Equating 2.10 and 2.11, we obtain:

$$H_L = 2 \cdot \gamma \cdot D_M \cdot \frac{1}{\bar{v}} \tag{2.12}$$

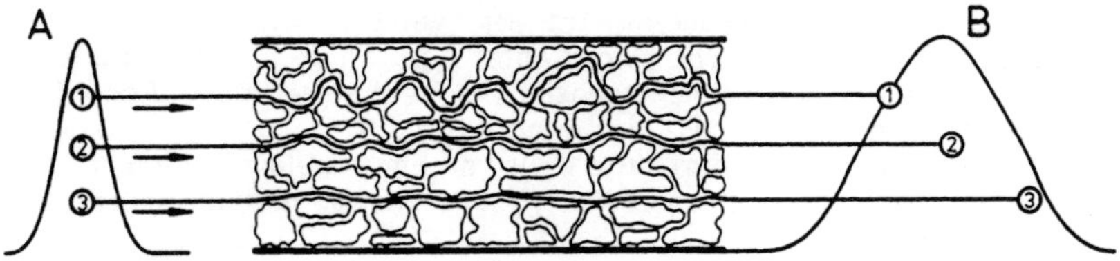


Fig. 47. "Multiple-path effect" (eddy diffusion) as the cause of broadening of a peak A; B) = Resultant substance profile

As a general rule it may be said that low values of $H_L$ lead to narrow peaks.

**The multiple-path effect or eddy diffusion**

The carrier gas "carries" the sample molecules as it flows through the separation column. In so doing, it has to transport the sample molecules through the column packing. Since the packing always displays some form of irregularities, the carrier gas will find a variety of different paths through the column: some virtually unimpeded, others meandering to a greater or lesser extent, and occasionally even "dead ends".

Let us consider a cross section of the carrier gas in which it is divided into individual "flow threads". The majority of these will be of average length. Accordingly, sample molecules of type A which are carried by such flow threads require an average time $t_A$ to pass through the column. Molecules in shorter or longer flow threads travel for correspondingly shorter or longer periods of time. This "multiple-path effect" (eddy diffusion) is another reason why an originally narrow substance peak becomes wider on the way through the column (Fig. 47).

The contribution $H_E$ made by eddy diffusion to the total theoretical plate height H is:

$$H_E = 2 \cdot \lambda \cdot dp \qquad (2.13)$$

$\lambda$ = measure of the irregularity of the packing of the column (measure of the quality of the filling)

dp = diameter of the support particles.

**Table.** Influence of particle size dp on the measure of quality, $\lambda$ .

| Particle size | (mesh) | $\lambda$ |
|---|---|---|
| 200 - 400 | (dp = 0.07 - 0.04 mm) | ~8 |
| 50 - 100 | (dp = 0.3 - 0.15 mm) | ~3 |
| 20 - 40 | (dp = 0.8 - 0.4 mm) | ~1 |

One obvious way of reducing the $H_E$ value would seem to be to use small particles of carrier. The constant $\lambda$, which characterizes the type of packing, is entered in the equation in the same way. According to Klinkenberg and Sjenitzer it is easier to obtain a uniform packing with large particles. This shows that there is a limit below which a smaller particle size brings no improvement. The Table shows the influence of particle size (mesh) on $\lambda$.

A further factor which limits particle size is the pressure drop over the column. This is, of course, increased by small particles. Since with a low pressure drop the flow tends much more towards the laminar, it is desirable to operate the entire column with a flow which supplies optimum separation efficiency.

If it is intended to reduce the $H_E$ value and hence increase separation efficiency, care must be taken to use fine material of uniform grain size, as well as narrow columns, and to keep the size small by even packing. The aim must be to achieve high packing density with as little packing-material fracture as possible.

**Transverse diffusion**

Eddy diffusion, which relates to a variety of flow states, should always be considered in conjunction with transverse diffusion. Transverse diffusion takes place in a radial direction, in other words, in the cross-sectional plane of the column. The flow threads, which play the major role in eddy diffusion, are interlinked by transverse diffusion. Let us consider an extreme case, in which sample molecules enter a stationary flow thread, a "dead end", i.e. cease to flow. These molecules would have to remain permanently in the column if they were not able to jump into a "moving" flow thread by a lateral diffusion step. On the other hand, a molecule which is about to diffuse to a point in the column packing which does not favour transport may be prevented from so doing by being carried further on by the flow. To some extent, then, transverse diffusion and eddy diffusion cancel each other out.

The plate-height component of transverse diffusion is termed $H_Q$; we thus obtain:

$$H_Q = W_i \cdot \frac{dp^2}{D_m} \cdot \bar{v} \qquad (2.14)$$

$W_i$ is a measure of the inequality of the diffusion pathways and of the flow differences bridged by means of the diffusion steps. In simple investigations, $H_Q$ is generally ignored. The plate-height components of eddy diffusion ($H_E$) and of transverse diffusion ($H_Q$) can be combined thus:

$$H_K = \frac{1}{\dfrac{1}{H_Q} + \dfrac{1}{H_E}} \qquad (2.15)$$

**Delayed substance interchange**

The interchange of sample molecules between the mobile and the stationary phases is, of course, a desirable effect, and is what makes chromatographic separations possible. It does, however, also contribute to undesirable peak broadening. The reason for this is that submersion into the separating liquid and returning to the carrier gas take time. If all molecules of a type A were to require exactly the same amount of time for this, no peak broadening would result. But as may easily be imagined, particularly deeply immersed molecules require considerably longer to return to the mobile phase than those which have penetrated only as far as the upper layer of the separating liquid. Relative to the bulk of the molecules, which reach an average submersion depth, the former molecules are eluted later and the latter molecules sooner.

In this way, delayed substance interchange also means that the more frequently the molecules enter the stationary phase the broader an originally narrow substance peak becomes. The plate-height component of the transfer term is:

$$H_T = f \cdot p \cdot (1-p) \cdot df^2 \cdot \bar{v} \cdot \frac{1}{D_S} \qquad (2.16)$$

f = structure dependent shape factor for the stationary phase
df = layer thickness of the separating liquid
p = proportion of sample molecules of type A in the mobile phase, averaged over time
$D_S$ = diffusion coefficient in the stationary phase
$\bar{v}$ = average flow rate (cm/min.) of the mobile phase.

f takes into account the geometric arrangement of the separating fluid on the carrier (spherical, rod-shaped, planar, etc.). $H_T$ is in inverse proportion to $D_S$. Narrow peaks, and hence high resolution, are therefore obtained with separating liquids of as low viscosity as possible. Since the viscosity increases as temperature drops, GLC columns must not be operated at too low temperatures. In the equation for plate height, the layer thickness df of the stationary phase and hence the degree of coating of the column is raised to the power of two and thus exercises a significant influence. Minimum plate height and hence maximum separation efficiency require a minimum degree of coating and layer thickness. There are, however, practical limits to this, in that on the one hand the quantity of stationary phase limits the amount of sample that can be injected, while on the other hand the carrier should be coated as evenly as possible with no gaps.

The product $p \cdot (1-p)$ reaches a maximum at $p = 0.5$. $p = 0.5$ characterizes substances which stay equal periods of time in the mobile and the stationary phases. Considerably larger p values signify extremely short retention times and are not advisable for gas-chromatographic separations. Most separations on packed columns are conducted with retention times $t_{ms}$ equal to between twice and about one hundred times the value of the dead time. It should be considered here that $t_{ms} = 2\ t_m$ is equivalent to a p value of 0.50, i.e. the maximum $H_T$ value, as already mentioned. Accordingly, better resolution is obtained with somewhat lower p values and hence longer retention times.

## 2.12.12 The van Deemter equation

The three components of plate height, $H_E$, $H_L$ and $H_T$, may simply be added together to give a total plate height $H_S$, which describes the total peak broadening of the column:

$$H_S = H_K + H_L + H_T \qquad (2.17)$$

The reason is that the plate height H stands for variances $\sigma^2$ and that variances are additive. By substituting the individual parameters into Equation 2.17 and combining some of the elements, we may simplify as follows:

$$H_S = A + B \cdot \frac{1}{\bar{v}} + C \cdot \bar{v} \qquad (2.18)$$

This is the simplest form of the van Deemter equation. It is quite clear from the graph (Fig. 48) that the total plate height $H_S$ passes through a minimum: the smallest possible plate height $H_{min.}$ and hence minimum peak broadening is obtained at the optimum linear flow rate $\bar{v}_{opt}$. The linear velocities may be replaced at any time by the customarily specified flow rates (cm$^3$/min.) since the two values are proportional to each other.

It is particularly significant that in the range where $\bar{v} < \bar{v}_{opt.}$ the van Deemter curve is much steeper than in the range where $\bar{v} < \bar{v}_{opt}$. Whereas if the carrier gas flow is too low, separation quickly deteriorates until it is no longer useful, $v$ may frequently be increased to a value considerably greater than $v_{opt.}$ while still attaining the required value for resolution. It should be remembered that doubling $v$ results in a halving of analysis time.

In addition to the peak-broadening effects in the separation column, there are also those which may arise in dead volumes and turbulences in column connections, injectors, detectors and valves. It is virtually impossible to quantify these factors. Structure-related effects such as these frequently constitute the difference between good gas chromatographs, in which such disturbances are minimal, and those of inferior quality.

### 2.12.13 The number of theoretical plates

A linear value (e.g. in mm) for plate height $H_S$ may be calculated from Equation 2.18. It should be understood, however, that it is in reality a small volume dV in which a distribution equilibrium of the substance molecules is established between the stationary and the mobile phases.

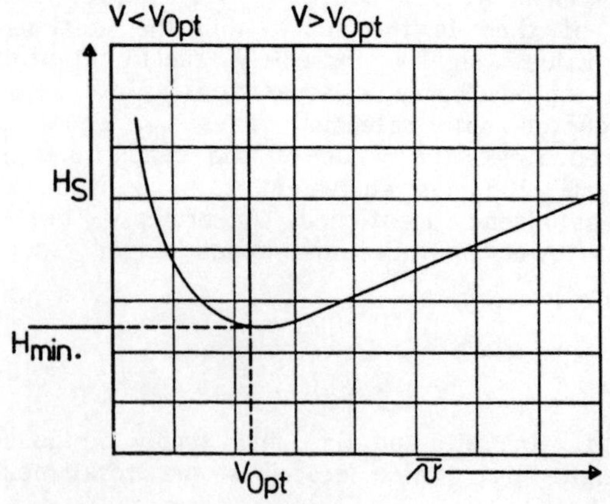


**Fig. 48.** Van Deemter curve: Relationship of plate height $H_S$ to average flow velocity $\bar{v}$ in the separation column.

If the entire column length L is considered as being divided into sections of the length $H_S$ (in the following referred to simply as H), we obtain the theoretical plate number N:

$$N = L/H \tag{2.19}$$

The plate number N may be derived from the chromatogram if the total (uncorrected) retention time $t_{ms}$ is related to the base width $W_B$.

$$N_{th} = 16 \left( \frac{t_{ms}}{W_B} \right)^2 \tag{2.20}$$

In order to determine N it is merely necessary to measure the total uncorrected retention time $t_{ms}$ of the peak and its base width, both, of course, in the same unit of measurement. Since it is frequently not possible to determine the precise position of the inflectional tangents which delimit $W_B$ on the base line, the relationship between $W_{0.5}$ and $W_B$ is often used and N is calculated thus:

$$N = 5.54 \left( \frac{t_{ms}}{W_{0.5}} \right)^2 \tag{2.21}$$

The higher the plate number N in a separation system, the greater is the separation efficiency. In Equations 2.20 and 2.21, the total retention time $t_{ms}$ still includes the dead time, which makes no contribution towards separation.

A somewhat more realistic measure of the separation efficiency of a column is the effective theoretical plate number $N_{EFF}$:

$$N_{EFF} = 5.54 \cdot \left( \frac{t_s}{W_{0.5}} \right)^2 \tag{2.22}$$

$t_s$ = adjusted retention time = hold-up time of the substance in the stationary phase.

### 2.12.14 The efficiency or selectivity of the liquid phase

It has already been mentioned that when using GLC it is possible to separate substances which have a different solubility in the separating liquid. Such substances have different partition coefficients, expressing the distribution between the stationary and mobile phases.

The solubility of sample molecules A in the liquid phase is determined by the strength of the interaction between the molecules concerned. The principal types of interaction are listed below in the order of increasing force of interaction.

1. Dispersion forces
2. Induced dipoles
3. Permanent dipoles
4. Hydrogen bridges

As a rule of thumb it may be said that particularly effective interaction is possible, thus yielding relatively high solubility, if the chemical nature of the sample molecules (dissolved substance) and the liquid phase (solvent) is as similar as possible.

The capacity factor k' and hence the retention time of a sample component A depend primarily on its partition coefficient $K_D$. If a different substance, B, is to be separated from A, $K_D(A)$ must be different to $K_D(B)$. If $K_D(A) < K_D(B)$, A emerges after a shorter retention time than B. The most practical measure of the ability of column to react to differences between A and B with different retention times is its selectivity $\alpha$ or relative retention.

$$\alpha = \frac{t_s(B)}{t_s(A)} \qquad (2.23)$$

Since the adjusted retention times are proportional to the partition coefficients, $\alpha$ is also the ratio of the partition coefficients of the substance pairing under investigation. Symmetrical, Gaussian peaks may only be expected if the partition coefficient is constant over a broad concentration range (see Section 2.12.10).

The selectivity of the separation column obviously determines the distance between the two substances to be separated in the chromatogram. Besides physical separation efficiency, which is expressed in terms of plate numbers, N, selectivity is therefore the second value which determines the attainable resolution. As a final consideration when calculating the resolution, allowance must be made for the fact that during separation peaks which elute later are increasingly far apart. This is accounted for by including the capacity factor k'.

### 2.12.15 The resolution equation

If all the parameters described above are combined, namely plate number, capacity factor and selectivity, we obtain finally a general formula for the resolution R:

$$R = \frac{\sqrt{N}}{4} \cdot \frac{\alpha - 1}{\alpha} \cdot \frac{k'}{k' + 1} \qquad (2.24)$$

The aim of every gas-chromatographic analysis is to separate the individual substances from each other. A measure of the quality of the separation is the resolution R, already defined in Section 2.12.7. According to Equation 2.24, the resolution is equal to the product of the capacity term $(k'/k'+ 1)$, the selectivity term $(\alpha - 1)/\alpha)$ and the column-efficiency term $\sqrt{N}/4$.

In order to improve the resolution R, each term may be optimized.

#### Optimization of the capacity term

According to Equation 2.3, the partition constant $K_D$ is equal to the product of k' and ß. Resolving in terms of k' and substituting $V_m/V_s$ for ß, we obtain:

$$k' = K_D \cdot \frac{V_s}{V_m} \qquad (2.25)$$

Since $K_D$ is a constant for a particular substance/stationary phase pairing (where T is constant), k' can be increased by increasing $V_s$ or reducing $V_m$. This means that k' is a measure of the quantity of stationary phase in the column.

The quantity of stationary phase cannot be increased at will; it is highly dependent, for example, on the nature of the carrier. H. Rotzsche provides an excellent summary.

The maximum value which the capacity term may assume is one. In practice, the k' values should be between 2 and 5, since otherwise the retention times are unacceptable long.

**Optimization of the selectivity term**

The selectivity $\alpha$ of a separating liquid must always be seen in conjunction with the sample molecules to be separated. A separating liquid is considered to be selective if it engages in intensive intermolecular interaction with the molecules of the substance.

According to the definition of $\alpha$ (see Equation 2.23), the function $((\alpha - 1)/\alpha)$ is meaningful for values between 0 and +1 only. ( $\alpha$ in Equation 2.23 cannot be less than 1, and where $\alpha = 1$, $((\alpha - 1)/\alpha) = 0$.

The selectivity term, and hence the choice of the "correct" stationary phase, has a major influence on the resolution R.

**Optimization of the efficiency term**

Optimization of the efficiency term is best illustrated by an example.

How long must a column be in order to obtain a resolution of 1.5?

The theoretical plate number of a 3-meter column is calculated according to Equation 2.20 and the data taken from Fig. 49:

$$N_{th} = 16 \cdot \left(\frac{17}{1}\right)^2 = 4624 \qquad (2.26)$$

The theoretical plate number $N^*_{th}$ required to achieve a resolution of 1.5 is calculated from Equation 2.24 as:

$$N^*_{th} = 16 \cdot R^2 \cdot \left(\frac{\alpha}{\alpha - 1}\right)^2 \cdot \left(\frac{k'_B + 1}{k'_B}\right) \qquad (2.27)$$

($k'_B$ = capacity factor of peak B)

$t_{s(A)}$ = 13 min.; $t_{s(B)}$ = 16 min.; $t_m$ = 1 min.

Total uncorrected retention times may, of course, also be used for the calculation of R.

$W_{B(A)}$ = 0.818 min.; $W_{B(B)}$ = 1 min.

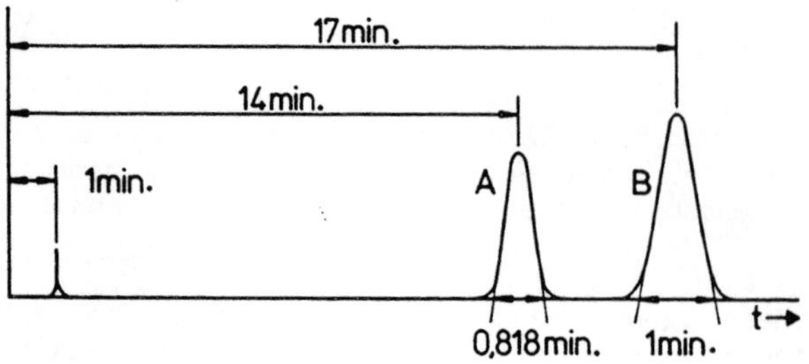


Fig. 49. Separation of two substances on a packed column; Optimization of column length

The calculated resolution of 3.3 is much too large and the analysis time of 17 min. is unacceptable long. Even a resolution of R = 1.5 signifies a "base-line separation" (99.7 % separation). A shorter column with fewer theoretical plates would suffice.

The values for $\alpha$ and $k'_B$ are obtained from Fig. 49.

$$\alpha = \frac{17 - 1}{14 - 1} = 1.231 \; ; \; k'_B = \frac{17 - 1}{1} = 16$$

Substituting these values in Equation 2.27, we obtain:

$$N^*_{th} = 16 \cdot (1.5)^2 \cdot \left(\frac{1.231}{0.231}\right)^2 \cdot \left(\frac{17}{16}\right)^2 \quad (2.28)$$

$$N^*_{th} = 1154 \quad (2.29)$$

Since the plate height $H_{th}$ remains constant, the column length is calculated as:

$$L = L' \cdot \frac{N^*_{th}}{N_{th}} \quad (2.30)$$

$$L = 3 \cdot \frac{1154}{4624}$$

$$L = 0.75 \quad \text{(m)}$$

It can be seen, that the desired separation is achieved with a much shorter column. Since separation is complete on the 3-m column, a higher rate of carrier-gas flow may be used to shorten the analysis time (see Equation 2.18).

### 2.12.16 General remarks on GC detectors

Thanks to their sensitivity or specificity, the following types of detector have proven valuable for performing gas-chromatographic analyses of trace amounts of organic substances in water:

A) Flame ionization detector (FID) (2.12.16.1)
B) Electron capture detector (ECD) (2.12.16.2)
C) Thermionic specific detector, alkali TSD (AFID) (2.12.16.3)
D) Flame-photometric detector (FPD) (2.12.16.4)
E) Thermal conductivity detector (2.12.16.5 and 3.6.1)
F) Microcoulometric detector (2.12.16.6)
G) Mass-selective detector (2.12.16.7 and 2.11.3)

It should be understood that there are two types of detectors: those which produce unequal signals from equal concentrations of different substances in the mobile phase are **specific**. If this only applies to substances incorporating a particular element, a particular functional group or with some other outstanding property, the detector is **selective**.

### 2.12.16.1 Flame ionization detector (FID)

#### A) Flame ionization detector (FID)

The FID is used to measure the electrical conductivity of a flame in an electrical field. The flame is produced with hydrogen as the fuel gas, which is mixed with the carrier gas ahead of the burner nozzle. The air required for combustion is supplied to the flame separately. Positive fragments of the organic substances and electrons are produced in the flame as a result of ionization. The electrons are accepted directly by the anode. The positive ions are discharged at the cathode.

When operating the FID, care must be taken that the flows of $H_2$, $N_2$ and air are set at a constant rate precisely according to the specifications of the equipment manufacturer. The gases must be dry and of high purity. The FID is outstanding for its very high sensitivity and wide range of linear response. Careful calibrations are required for each component under analysis. The FID is capable of taking direct measurements only of substances with CH bonds. Inert gases, $N_2$, $O_2$, $H_2$, CO and $CO_2$ are not recorded.

There is a voltage (300 V) between the flame tip and a cylindrical collecting electrode. The carrier gas from the column is mixed with hydrogen and burned via the nozzle. If an organic component is eluted from the column, charged particles result which reduce the resistance between the electrodes; the current which then flows is proportional to the quantity of eluted organic material.

### 2.12.16.2 Electron capture detector (ECD)

#### B) Electron capture detector (ECD)

Virtually all modern electron capture detectors operate according to the pulse-modulated constant-current technique (PMCC). The ionization source is a $^{63}Ni$ foil ($\beta^-$-emitter), which is attached to the inside of the detector cell. When carrier gas (highly pure nitrogen or a mixture of $Ar/CH_4$) flows through the cell, the gas becomes ionized. The slow electrons formed flow to the anode, thus generating a current.

$$TG + \beta^- \longrightarrow TG^- + e^- \text{ (primary, slow)}$$

(TG = carrier gas)

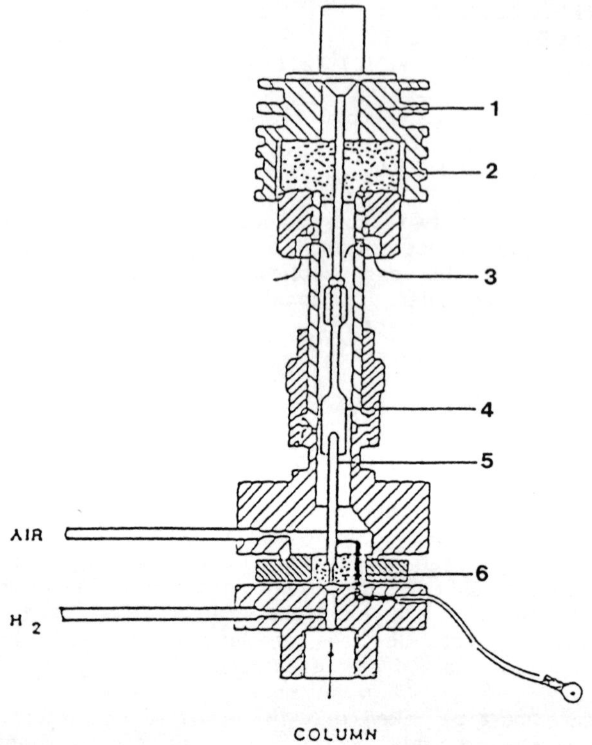


Fig. 50 a. Flame ionization detector (FID); 1) = Cooling fins; 2) = Insulator; 3) = Outlet; 4) = Ring electrode; 5) = Flame; 6) = Ceramic insul.; (Based on Siemens, Sichromat)

If negative pulses with a frequency of $f_o$ are applied to the foil, the electrons are able to reach the collector only during a negative pulse period. In this way, the pulse frequency $f_o$ determines the residual current of the ECD. If an electron-absorbing substance S enters the detector cell, the electron concentration falls. This loss is compensated by an increase in pulse frequency to $f_s$, so that the residual current remains constant. The difference $f = f_s - f_o$ is proportional to the quantity of S and is fed to a recording instrument.

The ECD is dependent on concentration. It must be ensured, therefore, that the volumetric flow rate of the carrier gas is constant. Since sensitivity is a function of detector temperature, the temperature should be kept constant, ± 0.3 - 0.5 °C.

It is advisable before and after each quantitative analysis to calibrate the detector with a known amount of the substance to be detected.

This type of detector is used for the detection and determination of very small quantities of substances with high electron affinities. In the trace range, the ECD is employed amongst other things for the analysis of traces of pesticides, nitro compounds, ozonides, chemical poison gases, pharmaceuticals, carcinogens, metabolites and metallo-organic compounds.

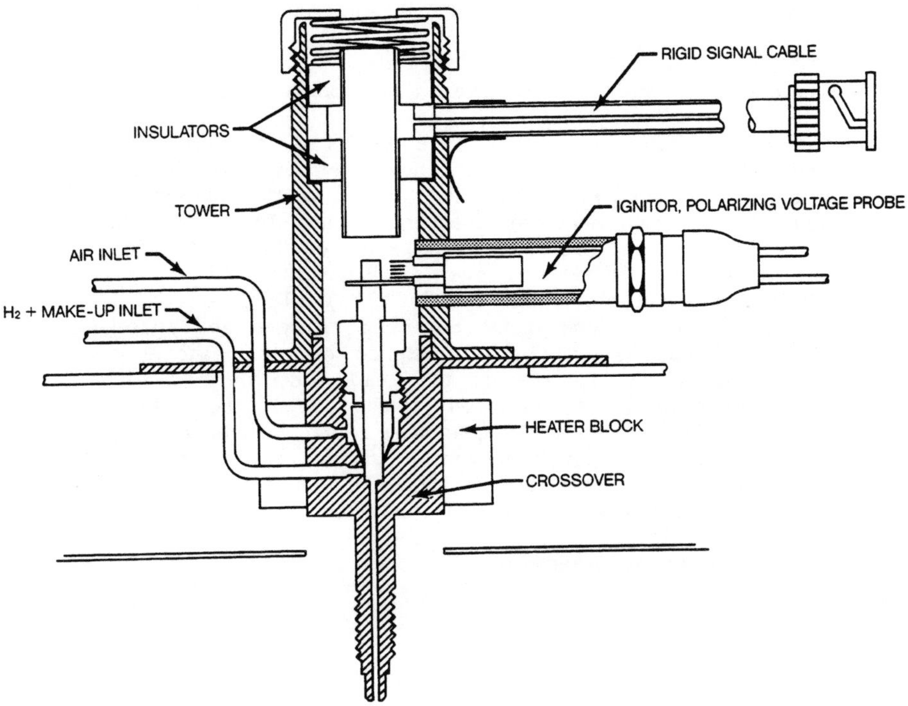


**Fig. 50 b.** Cross-sectional view of flame ionization detector

The carrier gas ($H_2$, $N_2$ or Ar: $CH_4$ - 95 : 5) flows through the detector cell and is ionized by the $\beta^-$ radiation of a radio-active source ($^3H$ or $^{63}Ni$). The "slow" electrons formed migrate to the anode (which usually has a potential of 90 V); as a result, a residual current of approximately $10^{-8}$ A flows. If an organic substance is eluted which is capable of "capturing" electrons, the current flow is reduced.

Modern ECDs achieve a constant current flow with the aid of pulsating voltages at the electrodes. The frequency of the voltage pulses is directly proportional to the concentration of "electron-capturing" molecules.

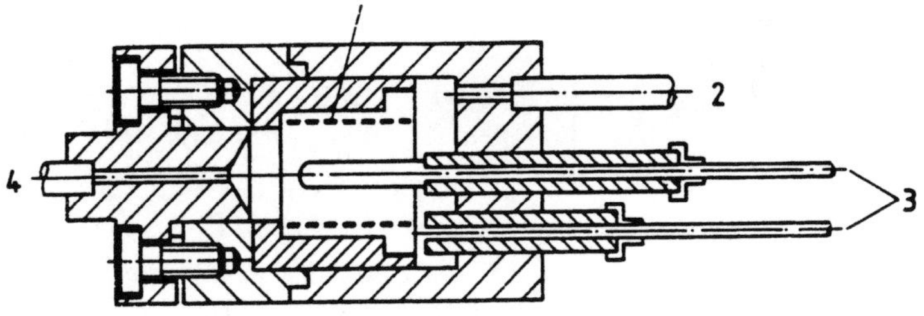


**Fig. 51.** Electron capture detector (ECD); 1) = Radioactive foil; 2) = Gas out; 3) = Electrodes; 4) = Gas in

### 2.12.16.3 Thermionic detector (TSB) and alkali flame ionization detector (AFID)

C) Thermionic detector (TSD), alkali flame ionization detector (AFID)

The alkali flame ionization detector is a variant of the FID; it is available in various versions. They may be specific and selective for organic compounds which contain nitrogen and/or phosphorus, depending on model, and are therefore frequently used in trace analysis of pesticides. The desired selectivity is achieved by holding a bead of alkali salt (containing Rb or Cs) over an FID flame (AFID) (Fig. 52 b) or by heating the bead electrically (Fig. 52 a) (TSD, approx. 500 - 850 °C). The ionization of compounds containing nitrogen and phosphorus, in particular, is made easier in the presence of alkali atoms in the ionization zone close to the surface of the bead. According to recent theory, the alkali atoms serve as electron carriers from the heated salt beads to the compounds. Larger quantities of halogen-containing substances (particularly halogenated solvents) should be avoided, since otherwise there may be irreversible evaporation of alkali halogenides, resulting very quickly in the destruction of the detector.

N- and P-selective detectors

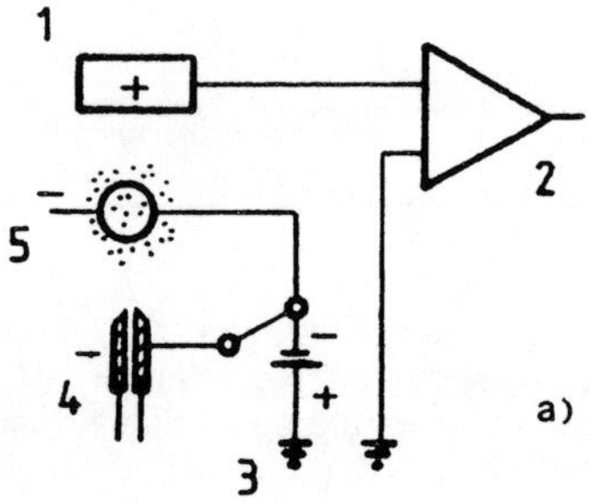


a) TSD with electrical heated bead

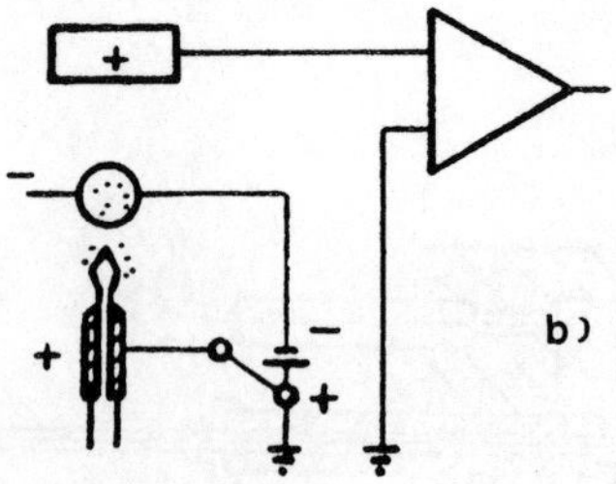


b) AFID

Fig. 52 a and b. a 1) = Collecting electrode; 2) = Amplifier; 3) = Power supply unit; 4) = Nozzle; 5) = Alkali source;

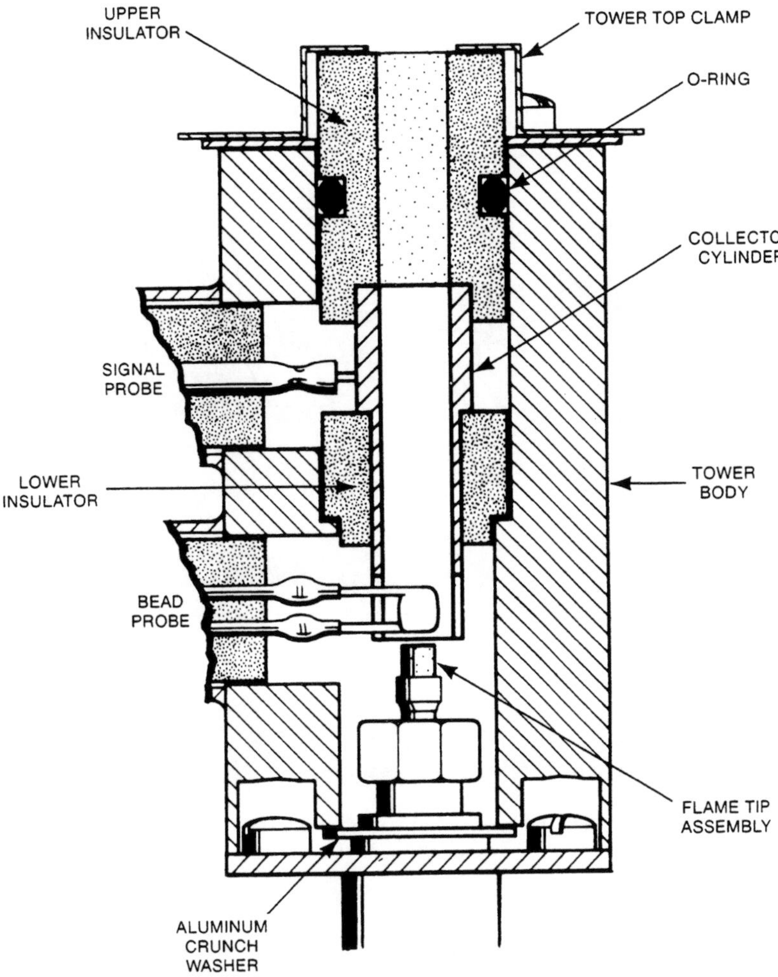


Fig. 52 c. TSD N/P selective

2.12.16.4 Flame-photometric detector (FPD)

D) Flame-photometric detector (FPD)

This type of detector is frequently used in pesticide-residue analysis because of its high specificity and selectivity and its high sensitivity to sulphur and phosphorus compounds. In the arrangement devised by Brody and Cheney the radiation from a hydrogen-air flame is directed to a photomultiplier through an interference filter. Measurements are taken at a wavelength of 526 nm for the determination of phosphorus, and at 394 nm for sulphur. Disturbances due to other products in the flame are largely avoided due to the fact that only the upper part of the flame is in the light beam (Fig. 53). In the presence of a hydrocarbon matrix a double flame FPD is essential.

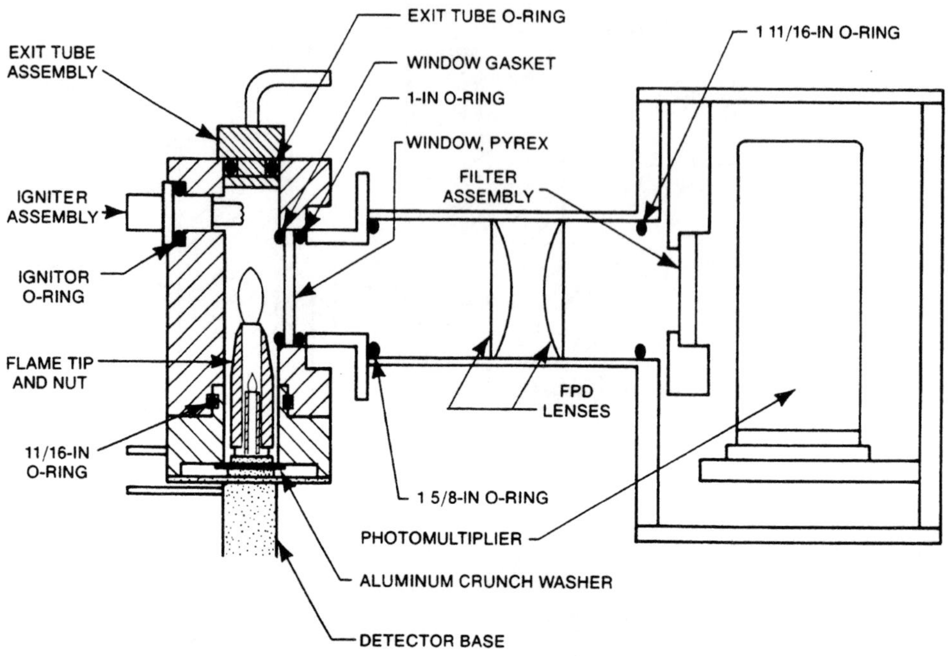


Fig. 53. Flame-photometric detector (FPD)

2.12.16.5 Thermal conductivity detector

E) Thermal conductivity detector (see Section 3.6.1)

2.12.16.6 Microcoulometric detector

F) Microcoulometric detector

The eluted halogen-organic compounds are pyrolized; the resultant hydrogen halide is quantified selectively by microcoulometric titration.

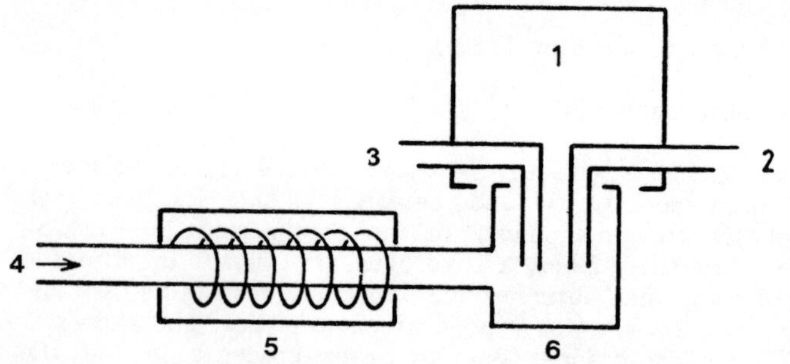


Fig. 54. 1) = Microcoulometer; 2) = Electrolysis circuit; 3) = Measuring circuit; 4) = Column; 5) = Pyrolysis oven; 6) = Titration Cell

### 2.12.16.7 Mass-selective detector

G) Mass spectrometer or mass-selective detector (see 2.11.3)

Note:

It is of advantage to use gas chromatographs with two detectors of different types in conjunction with a two-channel plotter. When operating two such detectors simultaneously, such as FID and ECD, it is possible to make assertions regarding the nature of components separated by gas chromatography: while the FID records all organic compounds containing CH, the ECD indicates only those compounds containing halogens.

## 2.12.17 Theoretical considerations regarding detectors

### 2.12.17.1 General remarks

A GC detector has the task of translating the material signal leaving the separation column into an electrical signal, avoiding distortion at all times. In so doing it should "ignore" the carrier gas, in other words in the absence of substance it should produce a zero signal. The output voltage of the detector is displayed on a strip chart recorder as a chromatogram (Fig. 55).

In some detectors, the electrical output signal is proportional to the mass flow (in g · $sec^{-1}$) and in others the output signal is proportional to the concentration (in g · $ml^{-1}$) of the components. The main representative of the first type is the thermal conductivity detector, and of the second type the flame ionization detector. The most obvious difference between the two categories of detector is in their differing reaction to changes in carrier-gas flow rate.

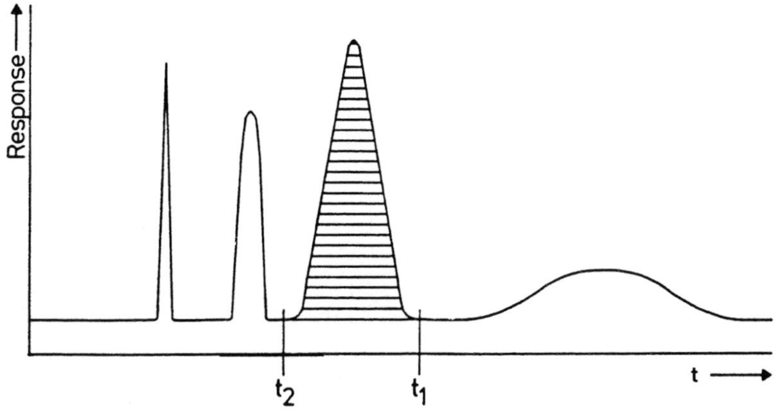


Fig. 55. Chromatogram of a differential detector. The shaded area is proportional to the total mass of the substance emerging from the column in the time $t_2$ - $t_1$

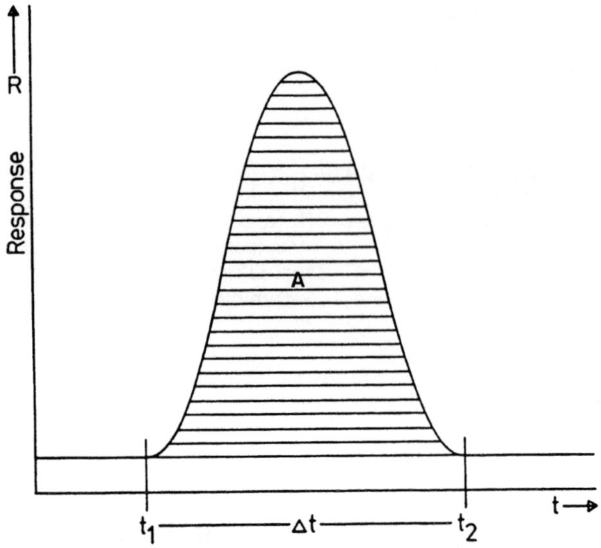


Fig. 56. Detector signal S as a function of time

In a concentration-dependent detector the level of the detector signal is given as follows:

$$S = K \cdot C \qquad (1)$$

S = Detector output signal (mV)
K = Apparatus constant ($mV \cdot g^{-1} \cdot ml$)
C = Concentration of sample substances in the detector ($g \cdot ml^{-1}$)

If the detector signal S is plotted against time t, we obtain the function shown in Fig. 56.

The shaded area A is defined as:

$$A = \int_{t1}^{t2} S \cdot dt \qquad (2)$$

If we substitute the value for S from Equation 1 into 2, we obtain:

$$A = \int_{t1}^{t2} (K \cdot C \cdot dt) = K \cdot \int_{t1}^{t2} C \cdot dt \qquad (3)$$

The limits of integration, $t_1$ and $t_2$, correspond to the times at which peak detection begins and ends. Depending on the degree of accuracy required for the analysis, for quantitative measurement of the peak area $t_2 - t_1 = \Delta t$ should have a value between 6 $\sigma$ and 8 $\sigma$, for symmetrical peaks. ($\sigma$ is the standard deviation or half inflection-point width of the Gaussian peak.)

The actual quantitative measured value, which should be proportional to the quantity of sample substance, is the integral of the detector-signal level (mV) over time (sec.), i.e. the peak area A with the dimension (mV · sec.).

For the sake of simplification let us assume that the substance peak emerging from the column is rectangular (Fig. 57).

During time $\Delta t = t_2 - t_1$ the concentration of sample substance in the detector is constant.

$$C = \frac{M}{V \cdot \Delta t} \tag{4}$$

M = Mass of the sample component causing the peak (g)
V = Average flow rate of the carrier gas (ml · $sec.^{-1}$)
$V \cdot \Delta t$ = Volume of carrier gas contained in the peak (ml)

As a result of 1, during the same time period the detector supplies a constant output signal S, which is not dependent on flow rate. The peak area A, however, here instead of the integral in 2 simply the product $S \cdot \Delta t$, is clearly directly proportional to the time $\Delta t$ during which concentration C is kept constant in the detector. If in the centre of the rectangular peak (Fig. 57) the carrier-gas flow is stopped, the detector would issue signal S for an indefinite time; A would therefore be of indefinite size.

If the above relationships are combined, we obtain:

$$A = S \cdot \Delta t = K \cdot C \cdot \Delta t = K \cdot \frac{M \cdot \Delta t}{V \cdot \Delta t}$$

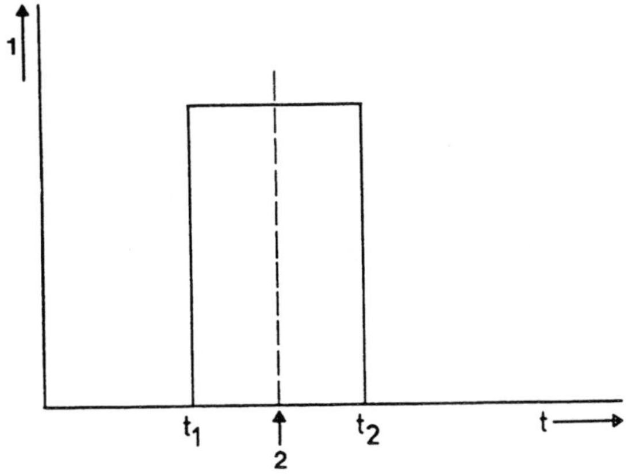


Fig. 57. Idealized relationship of substance concentration to time t (refer to text for details); 1) = Substance concentration; 2) = Retention time

$$A = \frac{K \cdot M}{V} \qquad (mV \cdot sec.) \tag{5}$$

For concentration-dependent detectors the implications of this, therefore, are:

a) The peak area is proportional to the quantity of sample.
b) The peak area is in inverse proportion to the carrier-gas flow.

In order to obtain comparable results, a thermal conductivity detector must be operated with a constant flow rate.

The mass flow sensitive detector (e.g. FID) first translates its input "quantity of substance" into the variable "electric charge". M (g) of substance generates the charge:

$$Q = \dot{M} \cdot q \tag{6}$$

($q$ = charge per g of substance in ($A \cdot sec. \cdot g^{-1}$))

A flow of substance from the separation column of $M(g\ sec.^{-1})$ produces a current:

$$I = \dot{M} \cdot q\ (A) \tag{7}$$

which generates the voltage $U\ (V) = I \cdot R$ via load resistance R. The signal level is then:

$$S = \dot{M} \cdot q \cdot R\ (V) \tag{8}$$

However, with a constant sample concentration in the carrier gas, i.e. within the rectangular peak, the following applies:

$$\dot{M} = C \cdot \dot{V}\ (g \cdot sec.^{-1}) \tag{9}$$

Under these conditions, the signal level is proportional to the carrier-gas flow V.

In this case, too, the peak area A is given by $S \cdot \Delta t$:

$$A = \dot{M} \cdot q \cdot R \cdot \Delta t\ (V \cdot sec.) \tag{10}$$

In the case of real peaks, $\dot{M}$ is not constant, meaning that the following must be used:

$$A = \int_{t_1}^{t_2} q \cdot R \cdot \dot{M} \cdot dt \tag{11}$$

Evaluation of the integral yields, as for Equation 10:

$$A = q \cdot R \cdot \dot{M} \tag{12}$$

from which we may derive for mass flow sensitive detectors that:

a) The peak area is proportional to the sample mass.
b) The peak area is not dependent on the carrier-gas flow.

### 2.12.17.2 Characteristic parameters of detectors

In principle, any property of a substance which can be converted into an electrical signal can be utilized in the design of a gas-chromatographic detector. There are indeed a great many different detectors, but only few types have found broad acceptance.

In general, a distinction is drawn between differential and integral detectors. Differential detectors constantly issue an output signal which corresponds to the quantity of substance emerging from the separation column at that moment.

Integral detectors successively add all input signals together and issue a corresponding summated signal at the output. Only differential detectors play a major role in gas chromatography.

A further distinction is drawn between destructive and non-destructive detectors. The first type alters or destroys the sample components during detection (e.g. thermal conductivity detectors), the second type does this not (e.g. TCD).

The following parameters characterize a detector:

1. Selectivity
2. Linear range and linearity
3. Noise level and detection limit
4. Sensitivity
5. Measuring volume (cell volume)
6. Time constant

Depending on the purpose of the detector, particular emphasis will be placed on one or other of the characteristic parameters, which will then be optimized accordingly.

#### Selectivity

It is generally desirable to have a detector which detects all substances well. Such a device is impossible to achieve, since each detector is based on the measurement of a particular substance property and different substances have correspondingly differing substance properties. There are, however, detectors which are suitable for very broad substance ranges. Flame ionization detectors and thermal conductivity detectors, for example, are virtually universal. Even with these types, though, different properties from substance to substance play a part to the extent that substance-specific correction factors must be used for the correlation between substance mass and peak area.

Highly selective detectors have been developed specifically for the detection of more or less closely delimited substance classes. These are of exceptional value in the analysis of complicated mixtures.

#### Linear range and linearity

The proportionality between the sample input of the detector and the electrical output voltage should extend over as great a range as possible. Let us consider the relationship of the signal to concentration in a concentration-sensitive detector (Fig. 58).

Two characteristic quantities are defined as follows:
the linear range (LR):

$$LR = \frac{C_{max.}}{C_{min.}} \tag{13}$$

and linearity M:

$$M = \frac{\log S_{max.} - \log S_{min.}}{\log C_{max.} - \log C_{min.}} \tag{14}$$

where:

$S_{max.}$ = Signal level at $C_{max.}$
$S_{min.}$ = Signal level at $C_{min.}$

(see Fig. 58)

The linearity value of a detector should be as close to one as possible. Formally speaking, the linearity indicates the slope of the S ⟶ C function within the linear range.

In many detectors the linear range extends over several orders of magnitude. For this reason, the calibration curve is usually shown on a logarithmic scale. Fig. 58 shows a calibration curve for an FID.

From Fig. 58 it can be seen that the detector signal is proportional to mass flow $\dot{M}$ over seven powers of ten. The linear range is accordingly $10^7$.

If the logarithm is taken of Equation (8), we obtain:

$$\log S = \log (q \cdot R) + \log \dot{M} \tag{15}$$

The linear section in Fig. 58 should, therefore, have the gradient 1. Values encountered in practice tend more to be between 0.95 and 0.99, depending on the design of the FID.

The total linear range of a detector, also known as the dynamic range, cannot be mapped with recording instruments such as recorders, plotters or

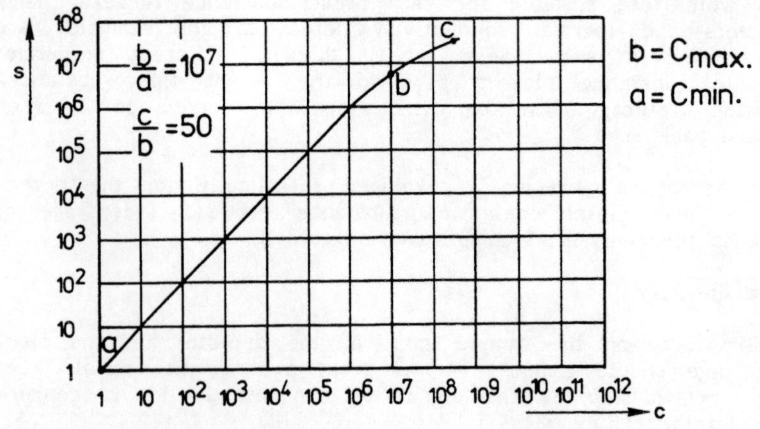


Fig. 58. Linearity curve of a flame ionization detector

magnetic tapes. For this reason, signal-attenuation stages are connected downstream of the detector. The resistance networks are almost much less linear than the detector itself.

## Noise level and detection limit

All detectors also produce an output signal when only the carrier gas is passing through them. When the amplifier is adjusted to a low sensitivity this signal is not visible and the connected recording instrument shows a smooth base line running parallel to the edge of the paper. With high amplification, however, in the least favourable case three effects may be observed:

a) The base line runs away monotonically in a positive or negative direction. This phenomen is known as drift. Drift can almost always be traced to some kind of operator error.

b) The base line appears in a wavy form, thus simulating low peaks. The spacing of these apparent peaks is measured in minutes. "Long-term noise" of this type is also traceable to incorrect operation of the gas chromatograph.

c) The base line is modulated by "short-term noise" and appears in the form of a band (Fig. 59). The interval of noise oscillations is in the range of seconds or less. This noise originates partly from the detector and partly from the amplifier. It can never be completely eliminated and can only be minimized. A genuine detector signal can only be recognized as such when it clearly stands out from the noise band.

If the average noise voltage (peak-to-peak) is designated N, a signal is still considered to be detectable if its level is at least 2 · N. That sample concentration or sample mass which produces a signal level of 2 N is known as the detection limit of the detector system (LLD = lower limit of detection or MDQ = minimum detectable quantity).

## Sensitivity

The slope of the detector calibration curve (Fig. 58) is known as sensitivity. In the case of concentration dependent detectors, sensitivity has the dimension [A · ml · $g^{-1}$], and for mass flow sensitivity detectors [A · sec. · $g^{-1}$]. If the linear range sensitivity is constant, and outside this range it changes to a greater or smaller degree. If the detector is greatly overloaded, it enters the saturation area. In this operating range its sensitivity is zero.

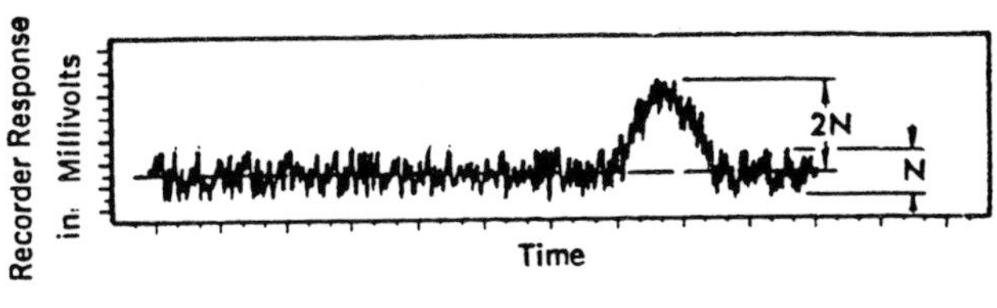


Fig. 59. Relationship of signal level to time; N) = Noise level; 2N) = Minimum detectable quantity

### Measuring volume

The actual measuring volume of the detector should be as small as possible. This requirement is all the more vital the narrower the chromatographic peaks are. If the detector is to reproduce the substance peak as faithfully as possible, it must scan the peak differentially. This means, however, that the detector volume must be small in relation to the carrier-gas volume in which the entire peak is contained. Let us assume that a very narrow peak has a base width of 1 sec. = $4 \cdot \sigma$. The peak elutes almost completely, therefore, within 2 seconds. At a carrier-gas flow of 3 ml/min it may be calculated that the peak is contained in 0.1 ml of carrier gas. If the detector volume were 1 ml, for example, several narrow peaks of this type, eluting in direct succession, could be accommodated in the detector simultaneously. In this way, the effect of separation would be completely cancelled out.

### Time constant

The purpose of any detector is to convert emerging substance peaks into a useful electrical signal without delay and without any distortion of shape. To achieve this it is necessary that the detector be capable of converting the substance peak into the signal considerably more rapidly than changes in the peak occur. The detector must have a small time constant. If one considers that very narrow capillary peaks may well have a base width of only 100 msec., it is clear that the time constant of the detector must be in the region of a few msec. Virtually all detectors in current use fulfil this condition. It must nevertheless be taken into account that a detector can only be operated in conjunction with the associated amplification electronics. In some devices the amplifiers are overdamped so as to give an appearance of producing particularly low noise levels. This may result in such great distortion of the peaks that quantitative measurements are no longer possible.

## 2.13 Gas-chromatographic headspace analysis (cf. also Chapter 4)

### 2.13.1 General remarks

In water, volatile organic substances frequently occur only in the trace range. However, the human senses of smell and taste respond to the minutest traces of such organic substances, with the result that such traces may make drinking water unpalatable. On the other hand, volatile organic substances such as fuels or organic solvents which are only very slightly soluble in water may be emulsified by surface-active substances, with the result that more marked concentrations of such volatile organic products may also occur in water. This applies in particular to wastewater from refineries or from chemical plants, or to the wastewater from petrol (gasoline) separators at filling stations.

It has also been observed that traces of organic solvents may be transferred into the water from drinking-water containers which have been freshly lined, rendering the water undrinkable.

If volatile organic substances are present in a water sample, and the water is kept in a closed thermostated system, a balance of vapour pressure is established between the liquid aqueous phase and the gas phase. The establishment of the vapour pressure equilibrium can be accelerated by raising the temperature or by shaking out. When the vapour pressure equilibrium is established, gas samples are taken from the headspace of the flask and the gas samples are analyzed by gas chromatography. If the presence of certain organic substances is likely in view of the retention times, identification can be carried out using the admixture method under otherwise identical test conditions.

### 2.13.2 Area of application

Gas chromatographic headspace analysis is a trace chemistry technique. Volatile organic substances can be detected even in the picogram range by this method.

Since gas chromatographic analysis, by its very principle, is not a qualitative but a quantitative method of analysis, the signals obtained must be attributed to particular substances. Experience shows that the analyst will recognize contamination by heating oil or petrol by their characteristic chromatograms. If assignment is impossible, a qualitative statement regarding the substance cannot be made; instead, it can only be established that substances are detectable in the gas chromatogram, without being able to describe their nature.

It is possible to conduct calibration according to the admixture method in conjunction with this technique if the organic impurities can be recognized on the basis of the gas chromatogram. Equal quantities of water are measured into similarly shaped glass vessels so that the same size of gas space relative to the water volume is present in all calibration samples. Increasing quantities of the suspected substances are then measured into the water in correspondingly low concentrations, and headspace equilibrium is established, as in the analysis of the sample. In the case of fuel contamination, for example, using the admixture method with fuels of various origin not only enables the fuel itself to be identified, but also allows its origin to be established at least with some degree of probability. Great care should be taken that the external conditions, such as flask volume, water volume and gas-space volume, as well as temperature and the time taken to establish headspace equilibrium, are always kept comparable.

An initial indication as to whether the technique of gas chromatographic headspace analysis is applicable is given by checking the smell of the water; this may also provide information concerning the nature of the volatile organic substances.

It should be noted that the gases naturally occurring in solution in a water, such as carbon dioxide, nitrogen, oxygen or methane, are also concentrated in the gas space and, like methane, may be measured in addition.

Completely odourless glass flasks which have been carefully cleaned should be used for taking the sample.

Glass flasks should be chosen which provide approximately the same remaining gas space when filled with 1 litre of water. The flasks should be heated for 24 hours at 105 °C in a drying oven before taking the sample, and when cool provided with a ground-glass stopper fitted with a septum to allow samples to be removed from the gas space with a gas syringe.

Blank tests without the addition of water but using the same septum and otherwise treating the flask in the same way should be carried out at all events in order to discover any potential influences. (For details of the method see section 4.2)

**Common injection systems for capillary columns**

- Split injector after Grob
- Headspace
  - simple headspace analysis
  - dynamic headspace injection (purge and trap)
- Cool column injection

**Types of columns**

- Packed columns: Glass or metal columns filled with a solid support; this support is coated with a stationary phase (generally a liquid with a high boiling point).
- Capillary columns: Usually glass or quartz capillaries 30 to 100 m long (internal diameter: 0.25 to 0.5 mm) in which the stationary phase is applied as a thin film (thickness 0.5 to 0.1 µm) or is in the form of a "bonded phase" (fused silica).

## 2.14 High-performance liquid chromatography (HPLC)*

### 2.14.1 Basic principles

The following section aims to provide a brief description of the principles of HPLC and its most important parameters in water analysis.

In contrast with classical column chromatography, HPLC uses columns with considerably smaller internal diameters (2 to 5 mm) and considerably smaller particle sizes in the stationary phase (3 to 10 µm).

This results in a build-up of considerably higher back pressure (up to 400 bar) as the solvent passes through the separation column. Today the method achieves separation efficiencies and detection limits which previously could not be expected of anything but the most sophisticated gas chromatographic methods.

A pump (2) delivers a certain quantity of solvent through the system from a reservoir (1) at a very even rate (even against differing back pressure). (Fig. 59 b)

---

* H. G. Deckert, K. Jung, Millipore-Waters, 6236 Eschborn, FRG.

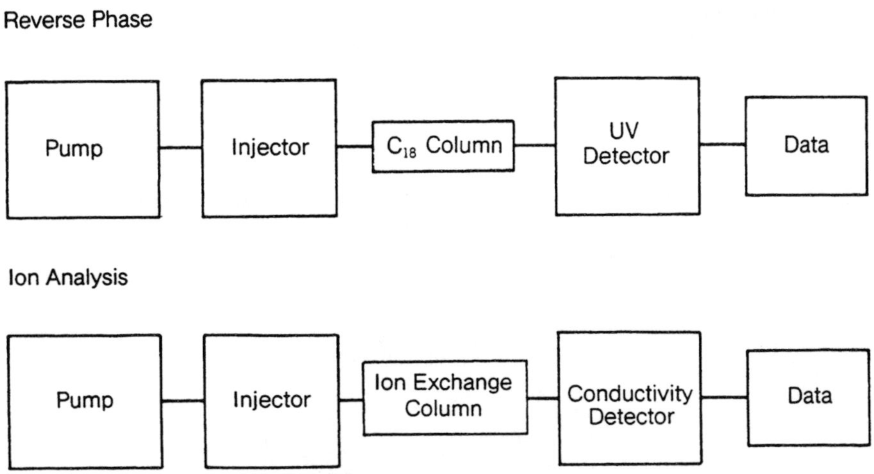


**Fig. 59 a.** Schematics of typical liquid chromatographic systems for organic and inorganic compounds

A special injection system (3) permits a certain quantity of sample (liquid or dissolved) to be introduced precisely into the pressurized overall system. Separation of the substance mixture takes place in the column (4).

Finally, in a detector system connected directly to the column outlet, the individual substances are indicated and the resulting detector signal (5) is transmitted to a recorder, electronic integrator or a data system (6) for qualitative or quantitative evaluation.

In the next section, the parameters which describe or influence chromatographic separation are discussed.

### 2.14.2 Retention

In HPLC, the individual substances reach the column exit, i.e. the detector, at different times, one after the other. This means that with a constant flow of solvent the substances differ in their retention time on the stationary phase, so that the total retention time $t_R$ comprises the retention time on the stationary phase $t'_R$ and in the mobile phase $t_0$.

$$t_R = t_0 + t'_R$$

If a substance failed to interact with the stationary phase at all, it would leave the column after $t_R = t_0$, the "dead time".

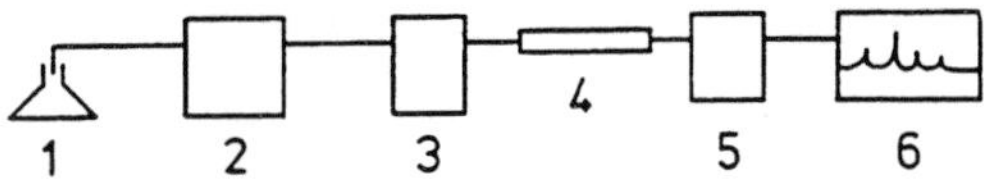


**Fig. 59 b.** Principle of the apparatus layout of an HPLC system 1) = Reservoir; 2) = Pump; 3) = Injector; 4) = Column; 5) = Detector; 6) = Integrator

The retention time is a necessary parameter but not sufficient to identify a substance. Under the same external conditions (temperature, solvent composition, flow, stationary phase), different substances may have the same retention times.

### Capacity factor

$t_R$ is dependent both on column length and on flow rate. In the literature, therefore, the capacity factor, k', of a substance is used to describe its chromatographic behaviour. k' is the ratio of the retention probability of a substance in the stationary to the mobile phase

$$k' = \frac{n\ \mathrm{stat}}{n\ \mathrm{mob}}$$

or

$$k' = \frac{t'_R}{t_o} = \frac{t_R - t_o}{t_o}$$

k' values between 1 and 5 are generally considered to be optimum. In the case of lower values, interaction with the stationary phase is too low, and if $k' > 5$ the analysis time is too long.

### Selectivity factor

The ratio of the capacity factors of two substances, 1 and 2, is designated the selectivity factor $\alpha$.

$$\frac{k'_1}{k'_2} = \alpha\ 1{,}2$$

It follows that $\alpha$ is a measure of the separation of two substances in a defined chromatographic system. If $\alpha = 1$, no separation takes place.

Selectivity factors between 1.3 and 1.5 are optimal.

### Resolution

The resolution of two peaks is designated as the ratio of the distance between the peak maxima $t_{R2} - t_{R1}$ to the average of their base widths

$$\frac{W_1 + W_2}{2}$$

(see also Fig. 60)

$$R = \frac{2\,(t_{R2} - t_{R1})}{W_1 + W_2}$$

The following equation relates R with $\alpha$, k' and N:

$$R_{1,2} = \frac{1}{4} \cdot (\alpha - 1) \cdot \frac{k'}{1 + k'} \cdot \sqrt{N}$$

where $k'_1 \approx k'_2$

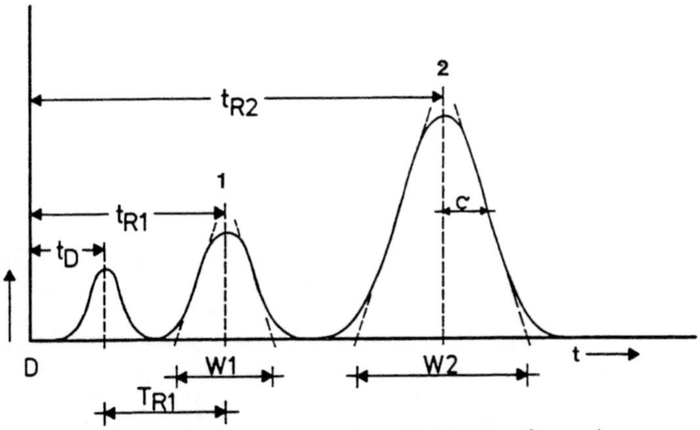


Fig. 60. The most important parameters for characterization of a chromatographic separation

It is worth noting that doubling the number of plates improves resolution by a factor of 1.4.

From a practical point of view, this means that if resolution is unsatisfactory:

First of all an attempt should be made to alter the mobile phase, i.e. the solvent is altered such that its polarity is in fact retained, but the type of interaction possible with the substances to be separated is changed. In practice, these may be for example two solvent components which have the same polarity but only one of which forms hydrogen bonds and the other not.

If no success is obtained in this way, the stationary phase should be changed. This therefore means a change from "reversed-phase" chromatography to "normal-phase", or vice versa.

### 2.14.3 Separation mechanisms

The following section deals briefly with the most important separation mechanisms in HPLC.

#### Adsorption chromatography

Most packing materials used in HPLC are based on silica gel, which may differ in particle size, surface area and pore size. (Typical particle sizes lie between 5 µm and 10 µm. Pore size is 600 to 1250 nm and surface area 100 to 300 $m^2/g$).

Pure, i.e. unmodified, silica gel is highly polar and the separation mechanism is based on the interaction of the substances to be separated with the free Si-OH groups of the stationary phase. In practice, this means that polar substances are retarded to the greatest extent.

#### Reversed-phase chromatography

Silica gel is a highly reactive material to which various functional groups may be bonded relatively easily. If, for example, silica gel is caused to

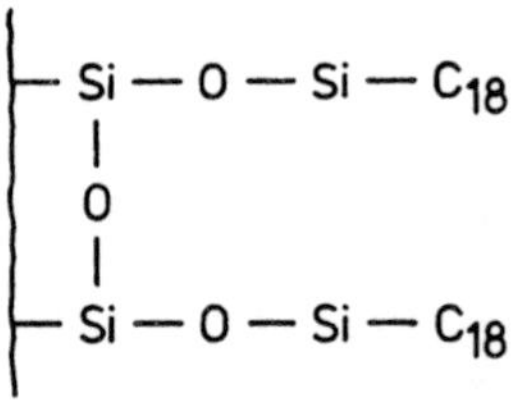

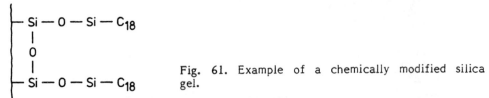

Fig. 61. Example of a chemically modified silica gel.

react with R-Si-Cl (R = $C_8H_{17}$, $C_{18}H_{37}$), the alkyl chains are bonded to the Si-OH groups and HCl is given off. (Fig. 61)

In this process, a highly non-polar silica gel is formed. In normal-phase chromatography, the eluent is less polar than the stationary phase. If the mobile phase is more polar than the stationary phase one speaks of reversed-phase chromatography.

In reversed-phase chromatography, non-polar substances are retarded to a greater extent. Typical solvents are aqueous buffers, methanol, acetonitrile, tetrahydrofuran, and mixtures of buffer and these organic components.

Today, the majority of separations in HPLC are conducted on reversed-phase materials. These materials can be balanced more rapidly, require a solvent with aqueous organic components and are more versatile with regard to their areas of application.

**Ion-exchange chromatography** (see also 2.15)

The stationary phase contains bonded cation- or anion-exchange groups (e.g. $NR_3^+$ or $SO_3^{2-}$ groups) which interact with molecules of opposite charge.

For details of further specialized techniques (ion-pair chromatography, gel permeation), refer to the supporting literature.

## 2.15 Ion chromatography

(Introduction)[1]

The analysis of organic anions, such as organic acids or also phenols, sugars and alcohols, as well as that of organic cations such as amines and amino acids, has long been a constituent part of high-pressure liquid chromatography.

In the case of inorganic ions, however, the situation is different. Well into the seventies, analysts were still dependent on wet analysis or other non-chromatographic techniques. The conductivity of inorganic ions in aqueous solutions suggested itself as an aid to detection.

However, the problem of detection remained unsolved until the eighties. It was, of course, necessary to measure minute changes in conductivity in the presence of very high background conductivity of the appropriate eluents. By chemically suppressing the background conductivity of the eluent, Small, Stevens and Baumann succeeded in getting round the detection problem for

---

[1] H. G. Deckert, Millipore-Waters GmbH, Eschborn, FRG.

the first time in 1975. They described their development of the so-called suppressor technique under the title "Novel Ion Exchange Chromatographic Method".

As early as 1976, then, it was possible to obtain the first analyzers (offered as chloride and sulphate analyzers).

Further development since 1979 has been in two directions. Whereas on the one hand attempts have been made to reduce the disadvantages of the suppressor technique by developments to the suppressor and columns, as a parallel development initial tests are being made using ion chromatography without a suppressor, but with separator columns on a silicate base and a conductivity detector.

In 1982 it proved possible to construct a self-generating hollow-fibre-membrane suppressor.

Fig. 62 is a schematic representation of the structure of an ion chromatograph using the suppressor technique:

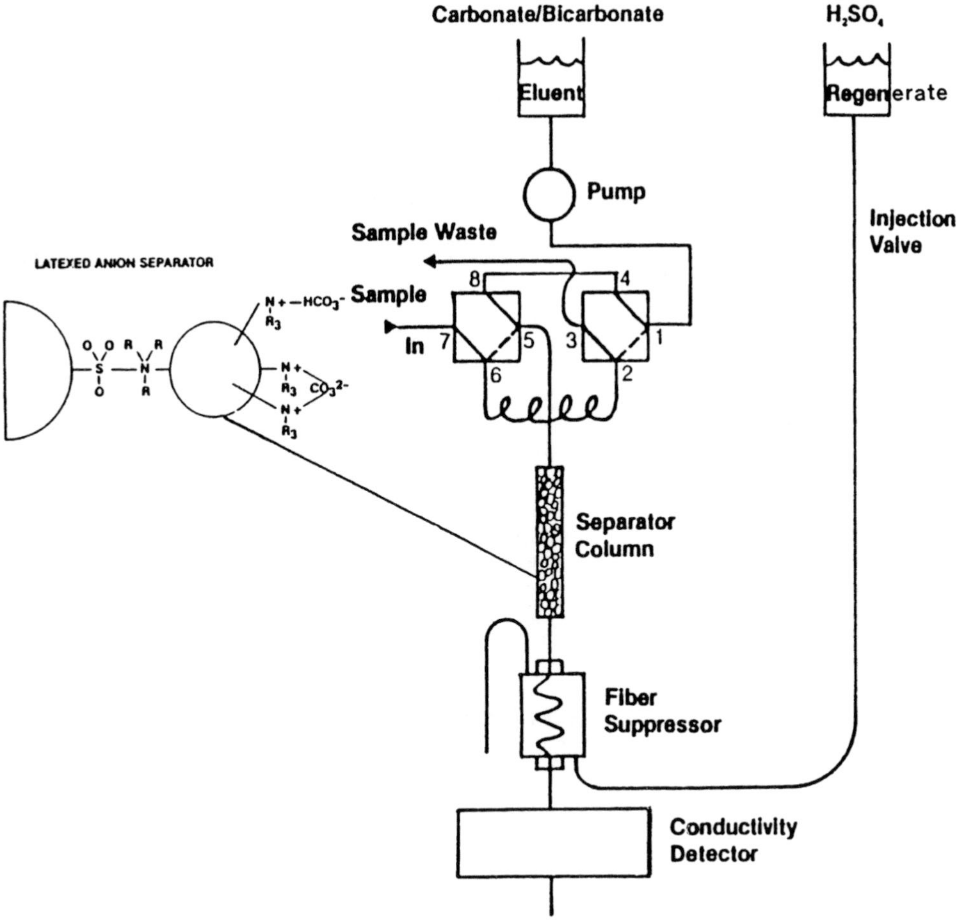


Fig. 62. Structure of a suppressor ion chromatograph (ion chromatography using fiber suppressor)

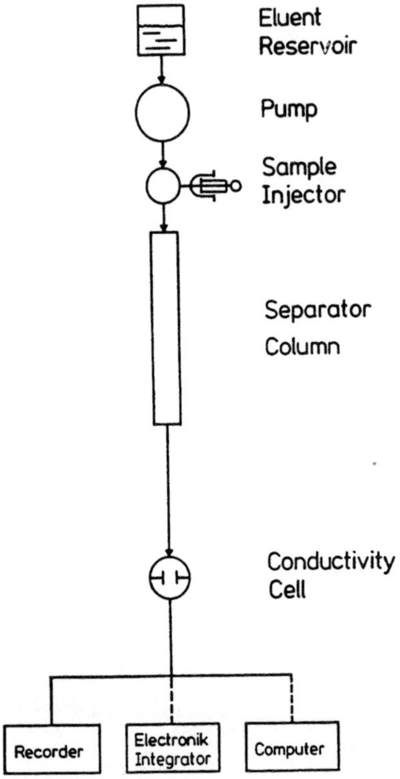


Fig. 63. Structure of an ion chromatograph without suppressor

In 1984, advances in electronics made it possible to develop a highly sensitive thermostated conductivity detector for the single-column technique, and finally help the suppressorless IC technique to come to fruition. Fig. 63 is a schematic representation of the structure of an ion chromatograph without suppressor.

### 2.15.1 Ion chromatographs with suppressor

Ion chromatography with a suppressor has achieved the widest distribution in laboratories. The first analytical separator columns for ion chromatography consisted of ion exchangers with a plastic base. The suppressor which is required for the suppressor technique has two functions. On the one hand it removes the ions of the mobile phase, and on the other it reinforces the ionic character of the sample ions and hence boosts their detection limit.

Today's suppressor technique uses anion columns with exchange material on polystyrene diphenylbenzene copolymers, the surfaces of which are sulphonated and subsequently agglomerated with amine latex (see Fig. 64).

The cation columns consist of sulphonated polystyrene diphenylbenzene (see Fig. 65).

As a result of their relatively high capacity, these columns require strong eluents with high intrinsic conductivity. The problem arising from this is to detect ions in the presence of very high background conductivity, in

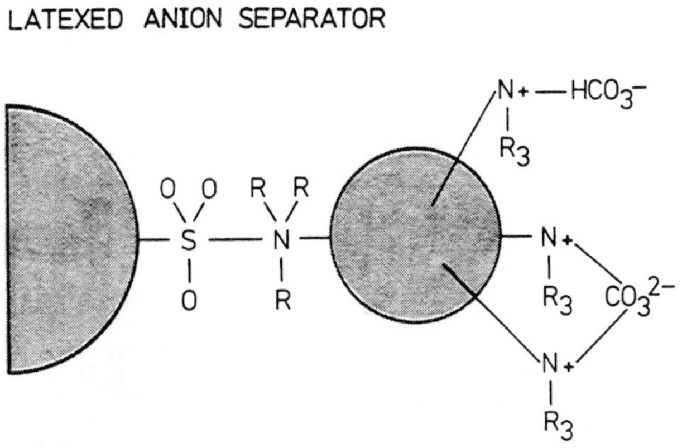


Fig. 64. Anion column for the suppressor technique

other words a similar problem to the measurement of small changes in absorption in UV spectroscopy with a highly UV-absorbent eluent.

The first solution to this problem, as already mentioned, was the development of the suppressor. In the suppressor, the sodium ions in the anion eluent (e.g. sodium carbonate/hydrogen-carbonate) are replaced by hydrogen.

In the case of cation eluents the chloride is replaced by OH ions. This reduces the background conductivity of the eluent from approximately 800 µS to about 25 µS (Fig. 66).

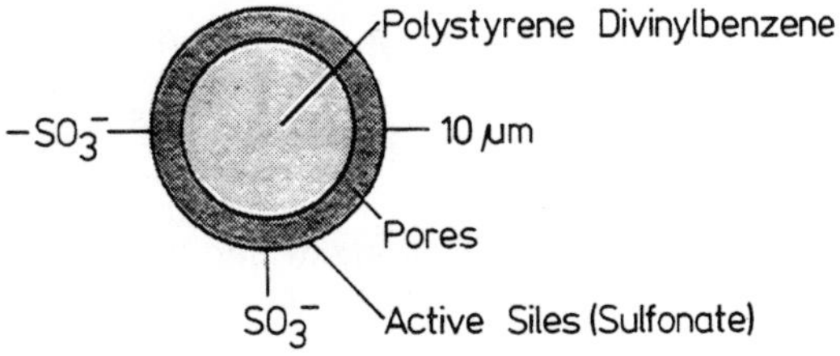


Fig. 65. Cation element

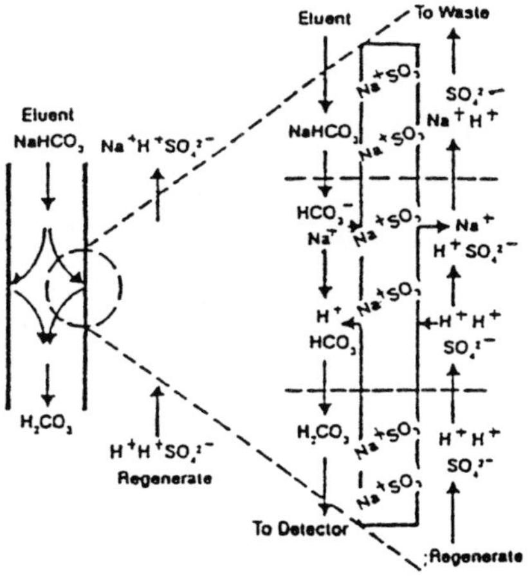


Fig. 66. Schematic of Fibre Suppressor

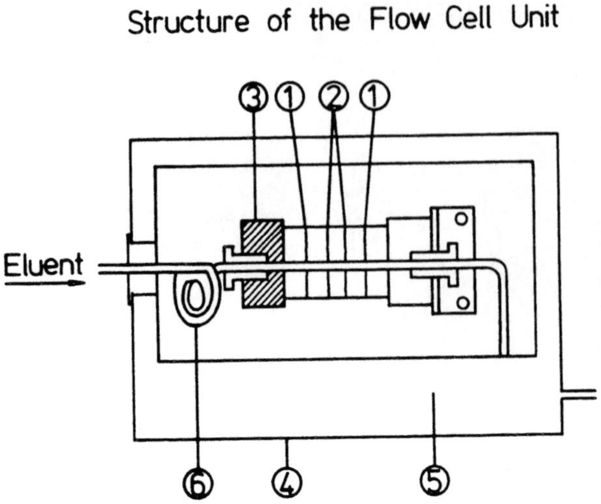


Fig. 67. Thermometric conductivity detector with 5 electrodes; 1) = reference electrode; 2) = detection electrode; 3) = guard electrode; 4) = cell unit; 5) = cell oven; 6) = 2 meters of 0.25 mm i.D. stainless steel tubing

The eluents which can be used for this technique are limited by the suppressor used. Accordingly, changes in the necessary resolving properties make it necessary to change the separation columns.

### 2.15.2 Ion chromatography without suppressor

In ion chromatography without a suppressor the conductivity of the eluent is no longer suppressed chemically but rather by electronic means. Fig. 67 is a schematic representation of the structure of a modern conductivity detector.

As a result of the electronic compensation of the background conductivity, the full dynamic range of a detector of this type is retained. Today, ion-

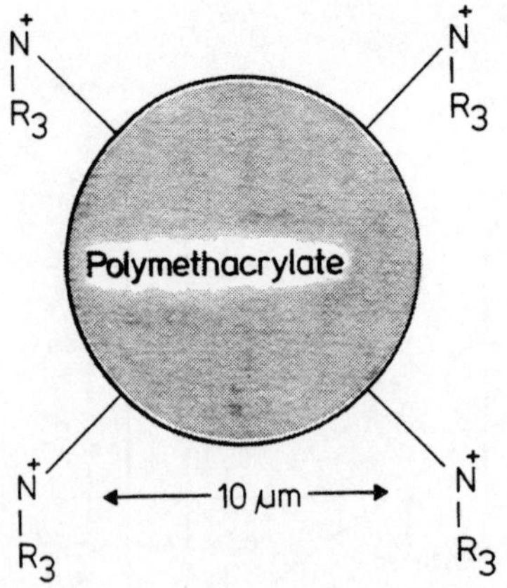


Fig. 68. Anion Column (Waters)

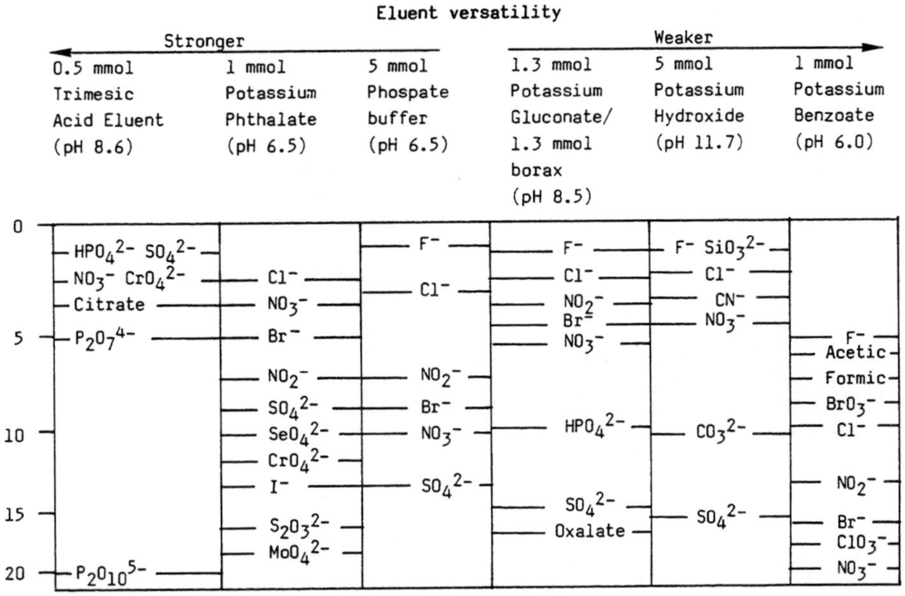


**Fig. 69.** Overview of the eluents most commonly used for ion chromatography without suppressor

exchange materials on a polymethacrylate base, amongst others, are used for this technique, since their use permits correspondingly weaker eluents as a result of reduced capacity (Fig. 68). In addition, in contrast to the silicate-base columns used previously, these columns are stable over the entire pH range.

Very precise temperature control of the detector is necessary for trace analysis since conductivity is heavily dependent on temperature.

## 2.15.3 Anion separation

In the case of ion chromatography without a suppressor the resolving property of the column can be adapted to the particular analysis in question by a simple change of eluent. Fig. 69 shows a schematic representation of the possibilities for anion separation offered by a single-column system of this type.

## 2.15.4 Examples

The following chromatograms illustrate the possibility of mastering different separation problems with one column and different eluents.

The demands made by environmental analysis are very varied. Whereas on the one hand when investigating drinking water the anions are to be detected in the lower ppm range, investigations of the acid-forming anions in rainwater sometimes reveal notably high concentrations.

**Chromatogram of Anions with Potassium Benzoate**

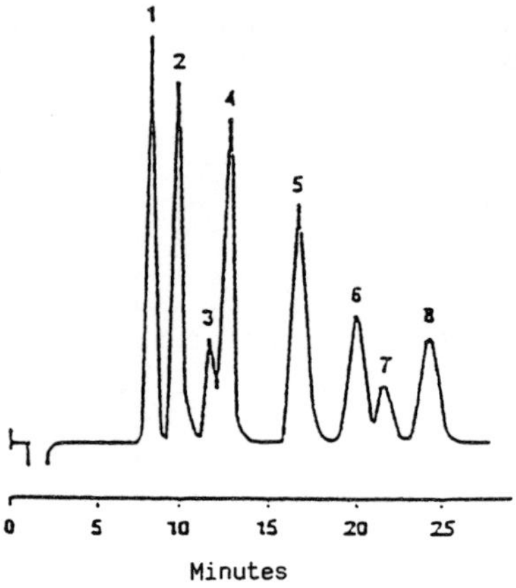


| | |
|---|---|
| Column: | IC-PAK A |
| Detector: | Waters™ 430 Conductivity Detector |
| Eluent: | 1 mmol, potassium benzoate |
| Sensitivity: | 4 µS/s |
| Flow rate: | 1.2 ml/min. |
| Sample volume: | 100 µl |

| | |
|---|---|
| 1. Acetic acid | 50 ppm |
| 2. Formic | 20 ppm |
| 3. $BrO_3^-$ | 20 ppm |
| 4. $Cl^-$ | 10 ppm |
| 5. $NO_2^-$ | 10 ppm |
| 6. $Br^-$ | 10 ppm |
| 7. $ClO_3^-$ | 20 ppm |
| 8. $NO_3^-$ | 10 ppm |

Fig. 70. Anion separation using potassium benzoate eluent

The greatest demands are made of the system at differences in concentration of 1000 and more. In these border regions, columns with a higher exchange capacity prove to be advantageous.

Other detectors are possible in ion chromatography, in addition to the measurement of conductivity. UV detection enables anions absorbent in the UV range (e.g. $NO_2^-$, $NO_3^-$) to be detected specifically, and also makes it possible to work with a UV-absorbent eluent. In the process, the background absorption is reduced by the non-absorbing ions. With this indirect UV absorption, therefore, negative peaks are obtained.

**Chromatogram of Anions with Potassium Phthalate**

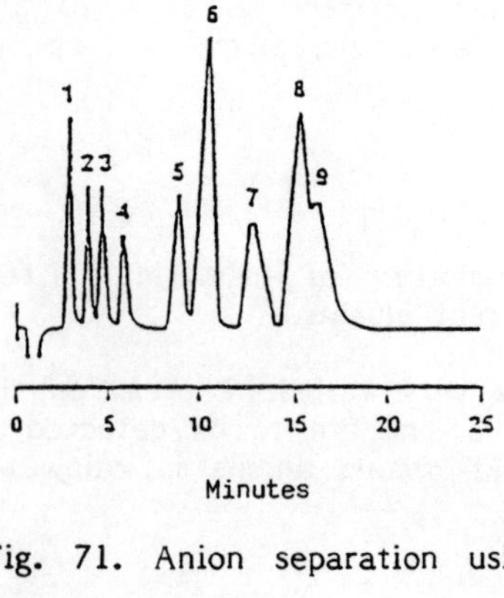


| | |
|---|---|
| Column: | IC-PAK A |
| Detector: | Waters™ 430 Conductivity Detector |
| Eluent: | 1 mmol, potassium phthalate |
| Sensitivity: | 5 µS/s |
| Flow rate: | 1.2 ml/min. |
| Sample volume: | 100 µl |

| | |
|---|---|
| 1. $Cl^-$ | 5 ppm |
| 2. $NO_2^-$ | 5 ppm |
| 3. $Br^-$ | 5 ppm |
| 4. $NO_3^-$ | 10 ppm |
| 5. $SO_4^{-2}$ | 50 ppm |
| 6. $SeO_4^{2-}$ | 50 ppm |
| 7. $CrO_4^{-2}$ | 50 ppm |
| 8. $S_2O_3^{-2}$ | 50 ppm |
| 9. $MoO_4^{-2}$ | 50 ppm |

Fig. 71. Anion separation using potassium phthalate eluent

Anion Capability

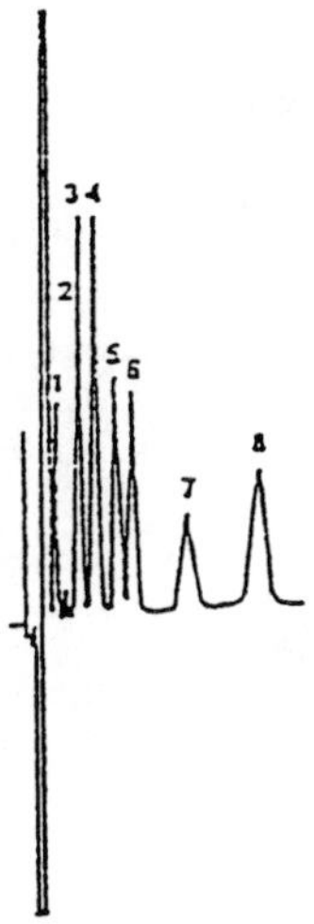

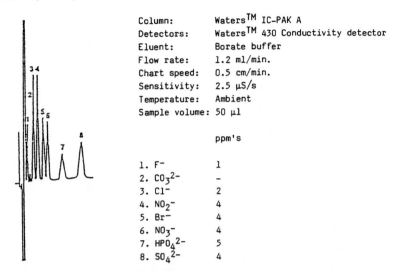

| | |
|---|---|
| Column: | Waters™ IC-PAK A |
| Detectors: | Waters™ 430 Conductivity detector |
| Eluent: | Borate buffer |
| Flow rate: | 1.2 ml/min. |
| Chart speed: | 0.5 cm/min. |
| Sensitivity: | 2.5 µS/s |
| Temperature: | Ambient |
| Sample volume: | 50 µl |

| | ppm's |
|---|---|
| 1. $F^-$ | 1 |
| 2. $CO_3^{2-}$ | - |
| 3. $Cl^-$ | 2 |
| 4. $NO_2^-$ | 4 |
| 5. $Br^-$ | 4 |
| 6. $NO_3^-$ | 4 |
| 7. $HPO_4^{2-}$ | 5 |
| 8. $SO_4^{2-}$ | 4 |

Fig. 72. Anion separation using borate gluconate eluent; anion capability

Another important detector for ion chromatography is the electrochemical detector (ECD). This makes it possible to detect anions such as cyanide, nitrite, sulphide, bromide, iodide and sulphite with maximum sensitivity. Thanks to its specific detection capabilities it is also possible to identify these ions without interference alongside high concentrations of other ions which may also display similar retention values (Fig. 75).

In water analysis it is also important to identify the cations as well as the anions. Fig. 76 shows the determination of the cations in the analyzed rainwater.

Chromatogram of Anions with Potassium Hydroxide

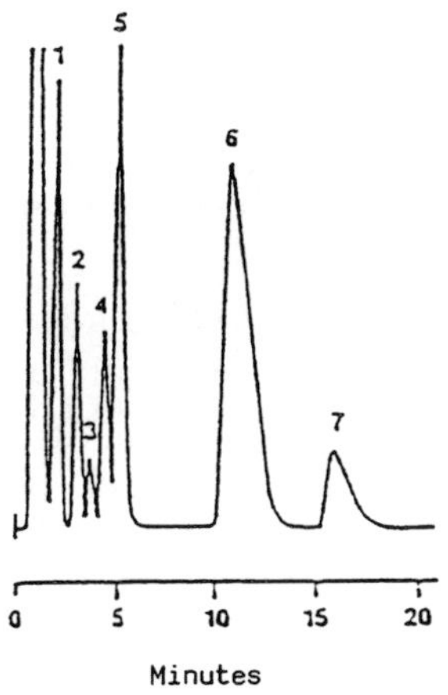

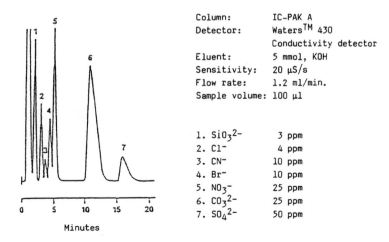

| | |
|---|---|
| Column: | IC-PAK A |
| Detector: | Waters™ 430 Conductivity detector |
| Eluent: | 5 mmol, KOH |
| Sensitivity: | 20 µS/s |
| Flow rate: | 1.2 ml/min. |
| Sample volume: | 100 µl |

| | |
|---|---|
| 1. $SiO_3^{2-}$ | 3 ppm |
| 2. $Cl^-$ | 4 ppm |
| 3. $CN^-$ | 10 ppm |
| 4. $Br^-$ | 10 ppm |
| 5. $NO_3^-$ | 25 ppm |
| 6. $CO_3^{2-}$ | 25 ppm |
| 7. $SO_4^{2-}$ | 50 ppm |

Fig. 73. Anion separation using KOH eluent

**Anions in Rain**

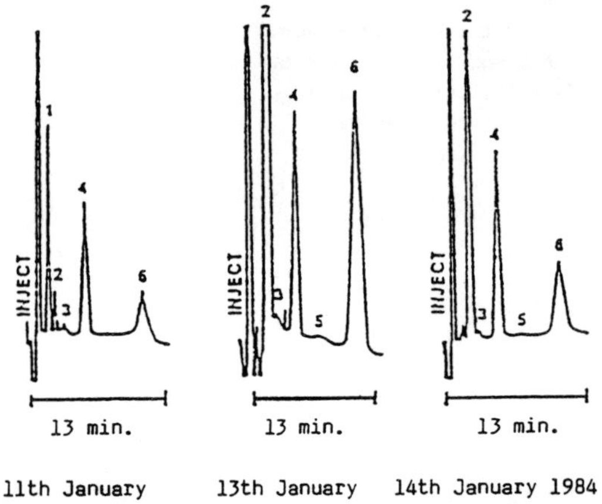

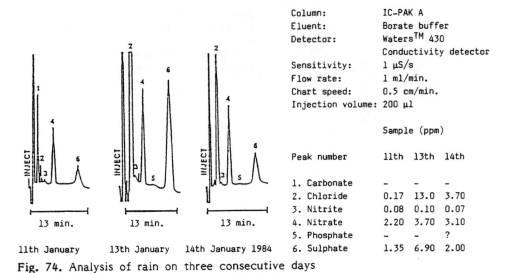

| | |
|---|---|
| Column: | IC-PAK A |
| Eluent: | Borate buffer |
| Detector: | Waters$^{TM}$ 430 Conductivity detector |
| Sensitivity: | 1 µS/s |
| Flow rate: | 1 ml/min. |
| Chart speed: | 0.5 cm/min. |
| Injection volume: | 200 µl |

| Peak number | Sample (ppm) 11th | 13th | 14th |
|---|---|---|---|
| 1. Carbonate | - | - | - |
| 2. Chloride | 0.17 | 13.0 | 3.70 |
| 3. Nitrite | 0.08 | 0.10 | 0.07 |
| 4. Nitrate | 2.20 | 3.70 | 3.10 |
| 5. Phosphate | - | - | ? |
| 6. Sulphate | 1.35 | 6.90 | 2.00 |

Fig. 74. Analysis of rain on three consecutive days

The cation analysis made it possible to confirm the greatly increased concentration of chloride ions on 13th January, 1984. The alkaline-earth metals can be identified alongside the alkali ions, using the same column.

In addition, a further interesting application by contrast with AAS and ICP, is to determine metal ions in different stages of oxidation in a single run.

Both in anion and in cation analysis, automatic concentration makes it possible to go down as far as the ppt range, thus providing a linear range from below 1 ppb (µg/kg) to approx. 400 ppm (mg/kg).

**Iodide Analysis**

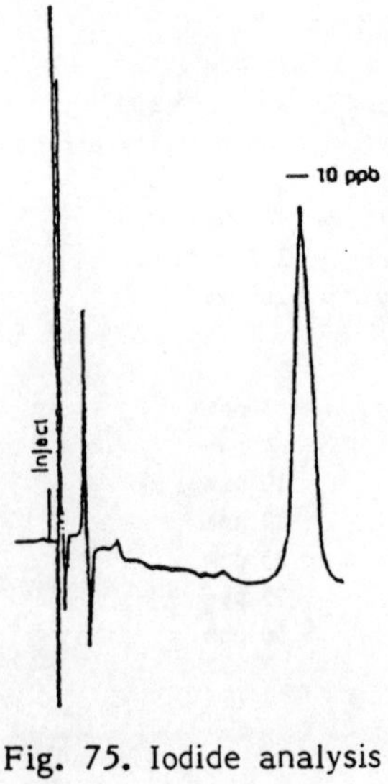

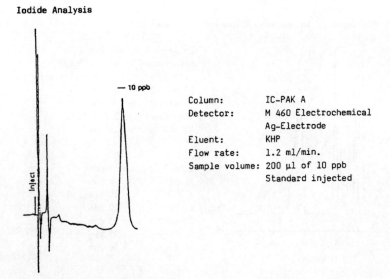

| | |
|---|---|
| Column: | IC-PAK A |
| Detector: | M 460 Electrochemical Ag-Electrode |
| Eluent: | KHP |
| Flow rate: | 1.2 ml/min. |
| Sample volume: | 200 µl of 10 ppb Standard injected |

Fig. 75. Iodide analysis using an electrochemical detector

Cations in Rain

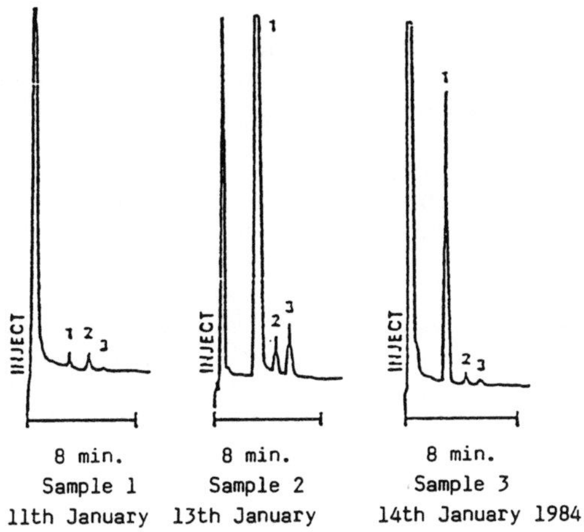


| | |
|---|---|
| Column: | IC-PAK C |
| Mobile phase: | 2 mmol $HNO_3$ |
| Detector: | Waters$^{TM}$ 430 Conductivity Detector |
| Sensitivity: | 20 µS/s |
| Flow rate: | 1.2 ml/min. |
| Chart speed: | 0.5 cm/min. |
| Injection volume: | 100 µl |

| | Sample (ppm) | | |
|---|---|---|---|
| Peak number | 11th | 13th | 14th |
| 1. Sodium | 0.13 | 18.5 | 2.43 |
| 2. Ammonium | 0.16 | 0.38 | 0.12 |
| 3. Potassium | 0.05 | 1.25 | 0.16 |

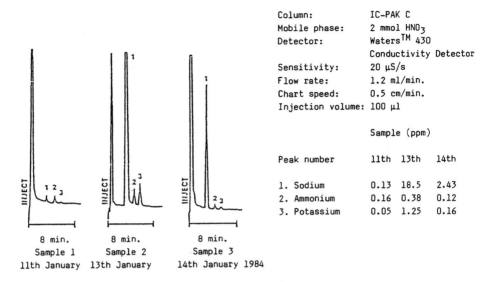

Fig. 76. Analysis of rain on three consecutive days

## 2.16 Radiochemical analysis of water samples (Introduction)

### 2.16.1 General information

Water can contain radionuclides which may be of natural origin, e.g. potassium 40 or uranium and thorium and the members of their decay series. On the other hand, man-made radionuclides may be present, particularly in surface waters. These are created by atomic fission and enter the water via fall-out, or as a result of reactions in nuclear reactors, or even from the waste water from institutions working with radionuclides.

As man has no sensory organ for emitters of radioactivity, and as these generally occur in extremely low concentrations and can be enriched from the water via the food chain to a level conceivably harmful to man and beast, highly sensitive methods for identification and measurement are necessary to analyze radionuclides. (For methods see Section 3.7)

Natural or synthetic radionuclides can occur in the dissolved state in water samples or be linked adsorptively to particles and exist as corpuscular radiators or as emitters of electro-magnetic radiation. In practice, a distinction must be made between emitters of $\alpha$-, $\beta^-$ and $\gamma$-rays.

### 2.16.2 Types of radiation

In the case of $\alpha$-decay, dipositively charged particles (helium nuclei) of mass "4" are ejected from the parent nuclide. This parent nuclide thus suffers the loss of 4 mass units and 2 positive charges, so that a daughter nuclide is created located two places to the left of the parent nuclide in the periodic system. The corpuscular radiation in $\alpha$-decay is characterized by a considerable energy of several million electron-volts and, linked to this, a high level of ionization in the irradiated matter. On the other hand, it is also characterized by only a low range of the $\alpha$-particles on account of the relatively high mass. $\alpha$-particles are absorbed by such thin objects as sheets of paper. The two natural decay series in uranium and thorium result in solid radionuclides which are soluble in water, and in the case of radon, in a gaseous radioactive intermediate product with the character of a rare gas. The decay chains in the uranium series, for example, lead via radium 226 to the radioactive rare gas radon 222 which has a half-life of 3.8 days. The products of radon 222 decay are polonium, lead and bismuth which through multiple $\alpha$- and $\beta^-$-decay are converted into the inactive lead 206, partly via excited states (thallium 210 and thallium 206) with the emission of $\alpha$-radiation but also $\beta^-$-and $\gamma$-rays. Similar conversions take place in the thorium series with radon 220 as a gaseous radionuclide with a half-life of only seconds.

$\alpha$-radiation can occur in water samples, particularly in the case of uranium and thorium decay. Radium 226 and radon 222 are separated by suitable processes and their $\alpha$-radiation is measured.

The second type of corpuscular radiation in radioactive decay is $\beta^-$-radiation. In this case, negatively charged electrons are ejected from the radionuclide. The atomic mass number of the radionuclide does not change.

The emission of $\beta^-$-radiation creates a daughter nuclide of the same mass as the parent isotope and located immediately to the right of it in the periodic system of elements, owing to the loss of a negative charge.

A $\beta^-$-radiating nuclide existing naturally in water is potassium 40. Potassium, which exists dissolved in small quantities in practically every kind of natural water, is a mixture of isotopes with a proportion of approximately 0.012 % of the $\beta^-$-radiating radionuclide potassium 40.

$\beta^-$-emitters have a relatively low range. They are nevertheless important for the organism if they can build up in the body and cause parts of the body to be exposed to local radiation over a lengthy period of time.

Strontium 90 may here be mentioned as example, an artificial radionuclide with a half-life of some 28 years, which can be stored in the bone as $\beta^-$-emitter and, in spite of the low range of $\beta^-$-radiation, can lead for example to abnormal conditions in centres producing blood cells.

$\gamma$-radiation, a no-mass electromagnetic wave-type radiation, does not involve any change in the atomic number of the element or of the atomic mass number. The $\gamma$-rays (photons or $\gamma$-quanta) have a considerable range, representing the excess energy emitted from an excited core on reversion to its ground state.

$\gamma$-emission occurs in the natural decay series of uranium and thorium. $\gamma$-radiating nuclides occur in nuclear fission, and isotopes which emit $\gamma$-rays, such as cobalt 60, are produced and used for particular technical purposes.

### 2.16.3 Radionuclides in water

Depending on the origin of the water, the following radionuclides may be important:

Carbon 14, phosphorus 32, sulphur 35, cobalt 60, strontium 89 or strontium 90, yttrium 90, ruthenium 106, silver 110, iodine 131, caesium 137, barium 137, barium 140, cerium 144,

and also the heavy radionuclides uranium, thorium, the actinoids and their decay products, and potassium 40.

### 2.16.4 Units of measurement

The unit of measurement for radioactivity in the international system of measurement is the bequerel, defined as 1 decay $\cdot$ $s^{-1}$, corresponding to 60 decays/min. The previous unit, Curie (Ci), is also still in use to some extent today:

$$1 \text{ Ci} = 3.7 \cdot 10^{10} \text{ disintegrations} \cdot s^{-1}$$

Figures for radioactivity are given in

$$\text{mCi} = 10^{-3} \text{ Ci}, \ \mu\text{Ci} = 10^{-6} \text{ Ci}, \ \text{nCi} = 10^{-9} \text{ Ci}, \ \text{pCi} = 10^{-12} \text{ Ci}.$$

Thus, for example, nCi is converted into Bq by multiplying the value of nCi by 37.

### 2.16.5 Radioactive decay

Radio-chemical decay takes place according to statistical laws. It cannot be influenced by secondary measures.

A radionuclide decays in a reaction of the first order, whereby the radioactive decay can be described by an exponential function. Of the nuclei ($N_0$) existing at time t = 0, only N nuclei remain after time t, where the difference has decayed in this time. The decay constant $\lambda$ in the dimension $t^{-1}$ describes the probability of decay. This probability is characteristic for every nuclide. The following can therefore be formulated as the decay law:

$$N = N_0 \cdot e^{-\lambda t}$$

The decay rate (activity A) results as

$$A = \lambda N$$

As criterion for the rate of decay of a radionuclide, the half-life is defined as that time in which half of the radioactive nuclei originally present have decayed

$$\frac{t}{2} = \frac{\ln 2}{\lambda}$$

If several radionuclides are decaying or if a nuclide displays several decay probabilities, the total decay probability is the sum of the individual decay probabilities.

So-called decay energy is released in every radioactive decay caused by $\alpha$- or $\beta^-$-emission and in the emission of $\gamma$-quanta.

By measuring the kinetic energy of the emitted $\alpha$- or $\beta^-$-particle, this decay energy can be recorded; the energy from $\gamma$-quanta can also be recorded by diffraction on gratings or by measuring a photoelectric effect. The decay energy is measured in electron-volts (eV), one eV being defined as the energy absorbed by a singly charged particle when passing through a voltage of 1 volt.

One electron-volt per decayed particle corresponds to 23 kcal/mol, which is approximately equivalent to 96.5 kJ/mol.

Knowledge of the half-life of a radionuclide is important for evaluation in water analysis and also for the choice of analytical method.

Very short-lived nuclides will generally have decayed before they can be examined and therefore scarcely cause noticeable changes in the biotope water. Longer-lived radionuclides with half-lives of days or even many years may themselves or via their radioactive decay products be enriched in the foodstuff chain, and may even directly endanger man and beast via water in cases of incorporation.

The analysis and quantitative determination of radioactive substances in water is therefore of special importance. The difficulties in radiochemical water analysis are characterized by the fact that radioactive nuclides exist predominantly in a dissolved state in unweighable quantities, and must therefore be enriched and separated from disturbing emitters before measurement, in spite of modern, highly sensitive measuring equipment.

## 2.16.6 Enrichment of radionuclides

In principle, the following methods are common in water radiochemistry as methods of enrichment before producing and measuring a preparation.

**Evaporation of the water sample**

**Gravimetric determing processes:**
- Precipitation with isotopic carrier
- Precipitation with non-isotopic carrier
- Precipitation with hold-back carrier
- Separation of the precipitations by filtration or centrifugation

**Extraction:**
- Liquid-liquid extraction
- Liquid-solid extraction
- Liquid-gaseous extraction
- (Separation by means of separatory funnel or by distillation columns or perforators)

**Chromatographic methods of separation:**
- Paper chromatography
- Thin-layer chromatography
- Column chromatography (e.g. also high-pressure liquid chromatography)
- Gas chromatography

Electrophoresis

Electrolysis

Electrodeposition

Ion exchanger

Adsorption (e.g. low-temperature adsorption of radioactive gases on activated carbon)

Distillation

Sublimation

Separation of radioactive gases, e.g. by means of extraction into the gas phase by shaking.

Separation by changing the valency of the ions

### 2.16.7 Measuring preparations

According to the operation used for separation, the measuring preparations produced for actual measurement should take account of radio-chemical measuring technology. The following parameters should be observed: e.g. half width, self-absorption, absorption in the vapour phase, wall and distribution effects, activity yield, efficiency of measuring equipment, calibration of the instruments, zero effect. Mixtures of radionuclides exist and a variety of emitters may also be present. $\beta^-$-emitters are frequently in the majority.

Inactive dissolved substances in the water predominate and affect the analysis. The aim should be a simple check procedure to determine whether there is any possibility of danger from the water, in particular to man and animals, as a result of incorporation of the dissolved radionuclides. (see also Section 3.7)

### 2.16.8 Total determination

Totalling determining processes of this type include for example measurements of "gross $\beta^-$-activity", where the total $\beta^-$-activity including the activity of natural potassium 40 is recorded, and of "net $\beta^-$-activity", after allowing for the potassium 40 content. Examinations for $\alpha$- and $\gamma$-emitters are only made as part of the routine check in specific cases where their presence is suspected. Danger to the human and animal body from incorporation of radionuclides is considerable. For this reason, the limit for the level of radionuclides in water is generally set very low. Taking into consideration a daily water consumption of approximately 2 to 3 litres, the proposed value for the tolerable level of $\beta^-$-radioactivity in a water is 1 Bq/l. If the measured value is higher, a check should be made by determining the quantity of potassium and subsequent conversion to see whether this guide value of 1 Bq/l is exceeded even when the natural $\beta^-$-activity of potassium 40 is taken into consideration. In such cases, the water is not safe for regular human consumption, and nuclide separations must be used to determine which radionuclides cause the increased

$\beta^-$-activity. A similar approach should be taken if a high level of $\alpha$-activity or a measurable $\gamma$-activity are determined. The guide value for $\alpha$-activity in water is 0.1 Bq/l, not counting any radon present.

As a result of the large number of known radionuclides, the work of water analysis is particularly difficult if it is found that guide values have been exceeded, and tests are usually made for specific compounds which can be expected in view of the history of the water.

In this context, it should be pointed out that on account of the low measuring rates in the water (1 Bq/l corresponds to one decay per second!) strict care must be taken that the laboratories and measuring rooms are kept absolutely clean for radio-chemical treatment and measurement of water samples. Contamination by radionuclides, e.g. by calibrating emitters, can lead to high zero-effects, making it impossible to carry out measurements in this range. Particular care should therefore be taken in a radio-chemical water laboratory, over and above the cleanliness and care necessary in principle in any analytical laboratory. The same applies to the measuring equipment and detectors used.

The zero-rates should be re-determined at regular intervals and appropriate calibrations with different calibrating emitters are necessary after analytical separation work.

### 2.16.9 Safety regulations

Although work in a water analysis laboratory does not involve the more strongly radioactive preparations, and indeed only very low activities are to be expected, the usual safety regulations should be strictly observed. Smoking, eating and drinking are prohibited. The measuring preparations should be stored under lock and key.

Generally, the low natural radioactivity arising from the uranium and thorium decay series, from potassium 40 and from cosmic radiation will be measured in the water and also in the air samples. In this context, attention should be paid in water analysis to the radioactivity caused by the natural $\beta^-$-emitter potassium 40 and by the $\alpha$-emitters radium 226 or the radioactive rare gas radon 222. Carbon 14 and tritium due to cosmic radiation must be taken into consideration as $\beta^-$-emitters. In the case of man-made radionuclides, particular attention should be paid to the influence on the water of fall-out from the air in the case of nuclear fission experiments or accidents in nuclear reactors. In such cases, increased or even harmful levels of radioactivity can occur as a result of fission products.

Personnel should be monitored with dosemeters to be carried in the pocket or worn on the body as badges. Monitoring units for surveillance are also to be recommended in radio-chemical water laboratories, particularly when using radioactive calibrating preparations.

The waste water from a radio-chemical laboratory should as a matter of principle be collected separately from other waste water, and, as far as possible, also collected separately according to specific activity and type of emitter or half-life. Radioactively contaminated waste waters should be treated. For monitoring purposes, 1 to 2 litres of the waste water samples should be evaporated and the activity of the residue measured.

The **exhaust air** from a radio-chemical laboratory should be filtered to extract dust. In general, however, no additional exhaust air treatment is necessary in a water laboratory measuring radioactivity in water samples. The exhaust air is monitored for radio-active aerosols by filtration of a specific quantity of air via membrane filters, and by measuring the deposit on the membrane filter.

### 2.16.10 Radiation protection

When working with radio-isotopes, strict attention should be paid to the radiation protection regulations in force at the time.

As regards external radiation, the penetrating $\gamma$-rays are particularly active and correspondingly dangerous for organisms. The high-energy $\alpha$-rays and the $\beta^-$-emitters have a particularly damaging effect on life in cases of incorporation, in spite of their lower range.

Definitions for measured values in radiation protection:

rad: If energy of 100 erg is absorbed in 1 g of a substance, irrespective of the type of radioactive emitter, this dose is designated 1 rad.

rep: 1 rep is understood as the radiation dose which $1.6 \cdot 10^{12}$ pairs of ions produce in 1 g of tissue or in 1 g of water, or which corresponds to an energy absorption of 91 erg/g.

rem: This value expresses the relative biological effectiveness (RBE) of the radioactive emitter.

The following definition is true:

$$\text{Dose in rem} = \text{dose in rep} \cdot \text{RBE}$$

| | | | |
|---|---|---|---|
| The RBE factor is for X-, $\gamma$- and $\beta^-$-radiation | | | 1 |
| Proton radiation | | approx. | 5 |
| -radiation | | | 10 - 20 |
| Neutron radiation | a) fast neutrons | | 10 |
| | b) slow neutrons | | 5 |

### 2.16.11 Detectors and measuring equipment

Suitable detectors and radiation measuring equipment are required to be able to detect and measure radioactivity. Depending on the type of activity to be measured, the following are used as detectors:

Ionization chambers
Geiger counters (self-quenched) for $\alpha$, $\beta^-$ and $\gamma$-radiation
Proportional counter - self-quenched counter
Methane flow counters
Scintillation counters
Neutron counters and
Detectors for particular purposes, e.g. gas-filled counters, or counters for very low activities
Counter combinations with anticoincidence circuit (measuring counter surrounded by a ring of protective counters which lower the zero effect).

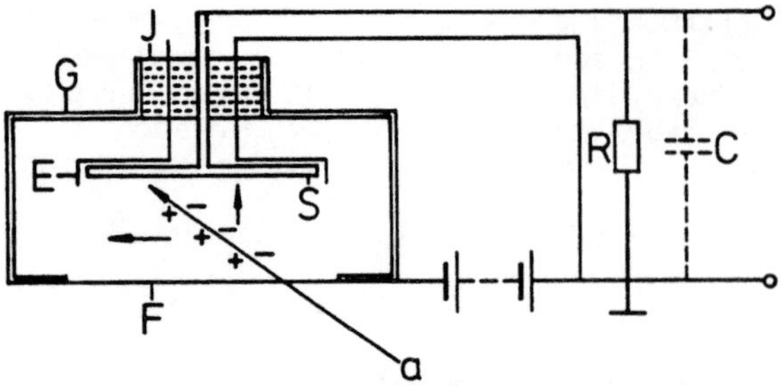


Fig. 77. Basic circuit diagram for a parallel plate ionization chamber; F) = Window; I) = Insulator; R) = Anode AC resistance; C) = Chamber capacity; S) = Collecting electrodes; E) = Protective ring; G) = Housing

Large-area counters with anticoincidence circuit and appropriate lead screening, combined with a suitable measuring and evaluation unit, are frequently selected as detectors in water analysis.

In the **self-quenched counters**, the pulse height is independent of primary ionization. Particles are only counted by such a detector, whereas **proportional counters** can also measure the energy of the particles. At constant counter voltage, the size of the pulse emitted in proportional counters is proportional to the quantity of the ions primarily produced in the counter and thus also to the energy of the appropriate particle. The resolving power of counters lies in the range of $10^4$ to $10^5$ pulse/s. **Ionization chambers**, in which the ionization capacity of charged particles is used to detect them, are often used in water analysis to detect gaseous emitters, e.g. radon.

An ionization chamber functions in the field of saturation, with the saturation voltage applied siphoning off all charge carriers created by ionization of charged particles along their path. In this phase, a linear relationship exists between pulse height and saturation voltage.

A voltage pulse is created at the anode AC resistance R. The quantity of ions formed in the ionization chambers is dependent on the type of particles, their energy and the design of the chamber. The pulse height for a given type of particle is a measure of the energy of the incident radiation; the quantity of pulses is proportional to the number of irradiated particles.

Where the quantity of incident particles is sufficiently large a fall in DC voltage occurs at R, depending on the chamber capacity C, which is proportional both to the number of particles and to their energy. The number of particles can be measured at constant particle energy or, conversely, the particle energy may be measured with constant particle numbers.

If the two figures differ, the chamber flow provides a measure of the dose, which is proportional to particle energy and number.

A **counter** consists of a metal cylinder as the cathode and a thin central wire as the anode. The counter is fitted with an inlet window, e.g. one made of a thin mica insulator, and filled with gases or gas mixtures. Depending on the level of the voltage applied, a counter can be operated in

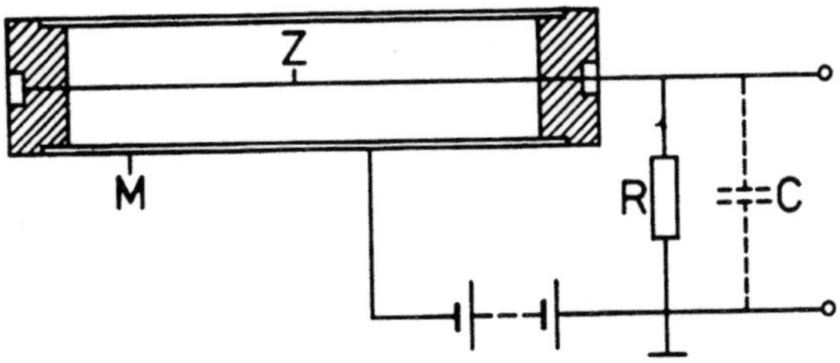


Fig. 78. Basic circuit diagram for metal-sheathed counter; M) = Metal sheath; Z) = Counter wire; R) = Anode AC resistance; C) = Counter capacity

the proportional region, where a primary avalanche is formed, or in the geiger region, in which every incident particle triggers a discharge. The difference between an ionization chamber and a proportional counter lies in the fact that in the ionization chamber the ions formed by the incident particle itself reach the electrodes, whereas in the proportional counter the number of particles is increased by a primary avalanche. $\alpha$-emitters and $\beta^-$-particles can be separated in the proportional region by applying different counter voltages.

In order to prevent continuous discharging, counters are filled with an extinguishing gas. This prevents formation of photons which would cause continuous discharging. If organic substances e.g. alcohol, are used as extinguishing gas, these are gradually consumed. Counters with halogens as extinguishing gases have a longer life, as a certain recombination of the dissociated halogen molecules takes place here. Halogen counters have no proportional region.

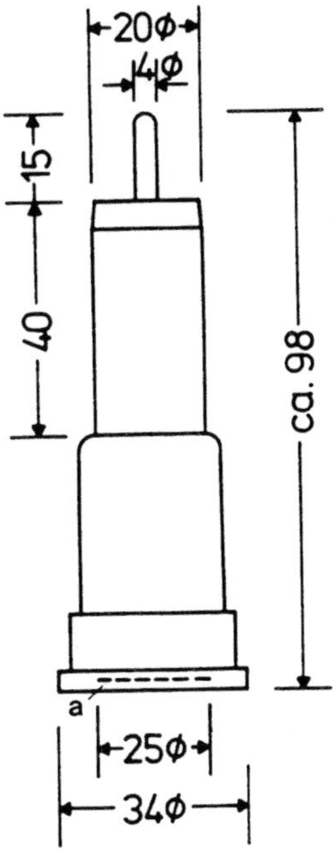


Fig. 79. Geiger counter (Dimensions in mm); a) = Mica window

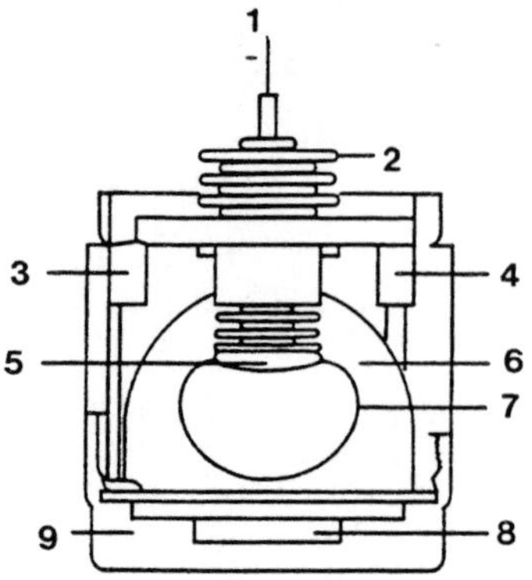


Fig. 80. Methane flow counter; 1) = High-voltage input; 2) = Insulator; 3) = Gas outlet; 4) = Gas inlet; 5) = Brass plate; 6) = Counting chamber; 7) = Counting loop; 8) = Sample; 9) = Screw cap (after G. Tralan, Roentgen-Blätter 12/1960)

A self-quenched counter can no longer register particles during the discharge period, and a pulse can not be registered until the so-called dead time and recovery time have elapsed. The resolving power of a self-quenching counter is thus determined by the dead time and recovery time. It fluctuates between 10 and 30 µs. A counter which is operated in the proportional region has a resolving power of 1 to 10 µs. Several discharges can be measured simultaneously and independently of one another.

A self-quenching counter is operated in the plateau region. The pulses registered rise initially with increasing voltage and remain almost constant in the plateau region. Depending on the technical measuring equipment, the plateau slope can be estimated at approximately 1 to 2 % per 100 volt. This means that a change in voltage in the plateau region causes no noticeable change in the measured results.

The plateau in a counter should be checked frequently, e.g. weekly. A standard preparation giving some 100 or 1000 pulses/min. is used. If the plateau slope rises to above 10 % per 100 volt, the counter should be discarded.

Apart from checking the plateau, the total counter unit should be calibrated with a calibrating preparation of similar emitter composition to that contained in the water. The measuring conditions, including preparation techniques, the geometry of the measuring equipment and screening and registering should be kept comparatively constant from the point of calibration to measuring the water sample.

The **zero effect**, defined as the registration of signals in the counting unit without preparation, is in the range of 10 to 20 pulses/min. for Geiger counters, and in the range of 1 pulse/min. for Geiger counters in the form of end-window counters with anticoincidence circuit. The zero effect can be further reduced by suitable screening and appropriate equipment layout.

The **efficiency** of Geiger counters is approximately 10 %, and for those with anticoincidence circuit it can be estimated at 20 %.

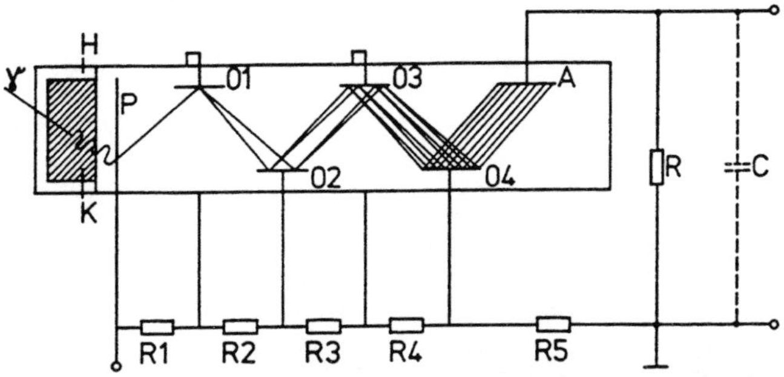


Fig. 81. Basic circuit diagram of a scintillation counter; K) = Scintillation crystal; H) = Lightproof shell; P) = Photo-electric cathode; D) = Dynodes; R) = Dynode resistances; A) = Anode; R) = Anode AC resistance; C) = Circuit capacitance

**Methane flow counters** have an efficiency of around 50 %, on account of the positive geometry of the measuring unit. The zero effect in flow counters is in the region of 20 pulses/min., but this counter equipment can nevertheless compete successfully with an anticoincidence circuit as a result of its high efficiency.

The efficiency is determined not only by the geometry of the measuring unit, the measuring conditions and the detector used, but also by the type of radiation from the radionuclide and its decay energy. When examining water samples, the nuclide mixture is usually of unknown composition, so

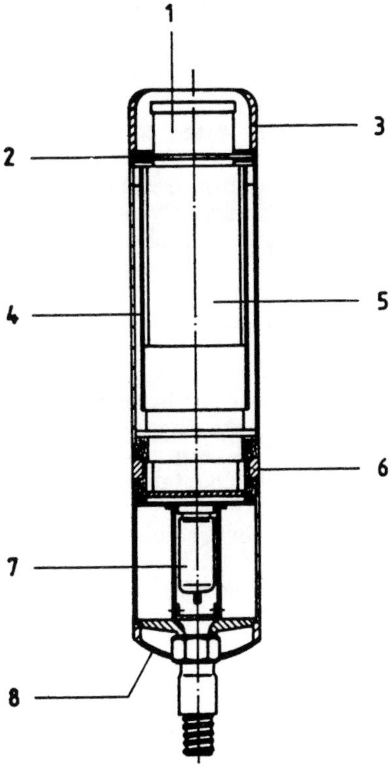


Fig. 82. 1) = NaJ crystal; 2) = Rubber mounting; 3 and 8) = Protective cap; 4) = Mu-metal screen; 5) = Photo-electric multiplier Dumont 6292; 6) = Thermal insulation; 7) = Cathode follower

that an efficiency figure quoted in relation to a particular radionuclide cannot be taken as the efficiency of the measuring unit in water analysis. It should be noted that with low-energy ß⁻-emitters the efficiency falls rapidly. In water analysis, a radionuclide with a maximum ß⁻-energy of a little above 1 MeV is taken as standard.

The so-called average ß⁻-energy of fission nuclides arising as decay products in nuclear fission also lies in the same range.

The natural isotope mixture of potassium, with a proportion of 0.0119 % of the ß⁻-emitter potassium 40, is frequently used as a calibrating preparation for water analyses.

Scintillation counters for ß⁻ and $\alpha$-emitters are hardly ever used for water analysis, as the sensitivity of the measuring unit is generally not sufficient. Scintillation counters for the detection of $\gamma$-radiation can only be used in water analysis if $\gamma$-radiating nuclides, e.g. from nuclear fission reactions, are present with corresponding activities in the water. This is generally not the case in natural water resources.

Scintillation counters use different scintillators for different types of radiation, e.g. thallium-activated sodium iodide crystals for gamma-radiation, anthracene for ß⁻-radiation and silver-activated zinc sulphite screens to measure $\alpha$-radiation. Neutron can also be detected with scintillation detectors using special crystals (e.g. europium-activated lithium iodide crystals).

The radiation quanta or the particles to be measured produce weak flashes in the scintillation crystals which fall onto the photo-electric cathode of a secondary electron-multiplier. Electrons are picked out here by being siphoned off by an electrical field and intensified stage by stage in dynodes connected in series. The intensified flow of electrons arrives at the anode and produces a corresponding pulse.

Scintillation detectors are largely used for $\gamma$-emitters, and with a following $\gamma$-spectrometer (single-channel or multi-channel instrument) it is possible to analyze the energy of the radioactive emitters. A scintillation detector is thus particularly suitable for detecting and measuring $\gamma$-emitters, whereas for $\alpha$ and ß⁻-radiation a methane flow counter is more suitable. The preparation to be measured can be placed directly in the measuring chamber here, and absorption losses, e.g. through windows, do not occur. The error geometry is also particularly advantageous in a methane flow counter.

Cross-section through a scintillation measuring head, comprising a NaI (Tl) crystal as scintillator in which the energy of the radiation particle is converted into light, and the photo-electric multiplier in which the light is converted into an electrical pulse. The cathode follower transmits the pulse to the counter.

Detectors for radioactive radiation are operated in conjunction with radiation meters. The latter can count the particles per unit of time, determine mean values and measure the energy of the particles. The following basic measuring sequence is usual:

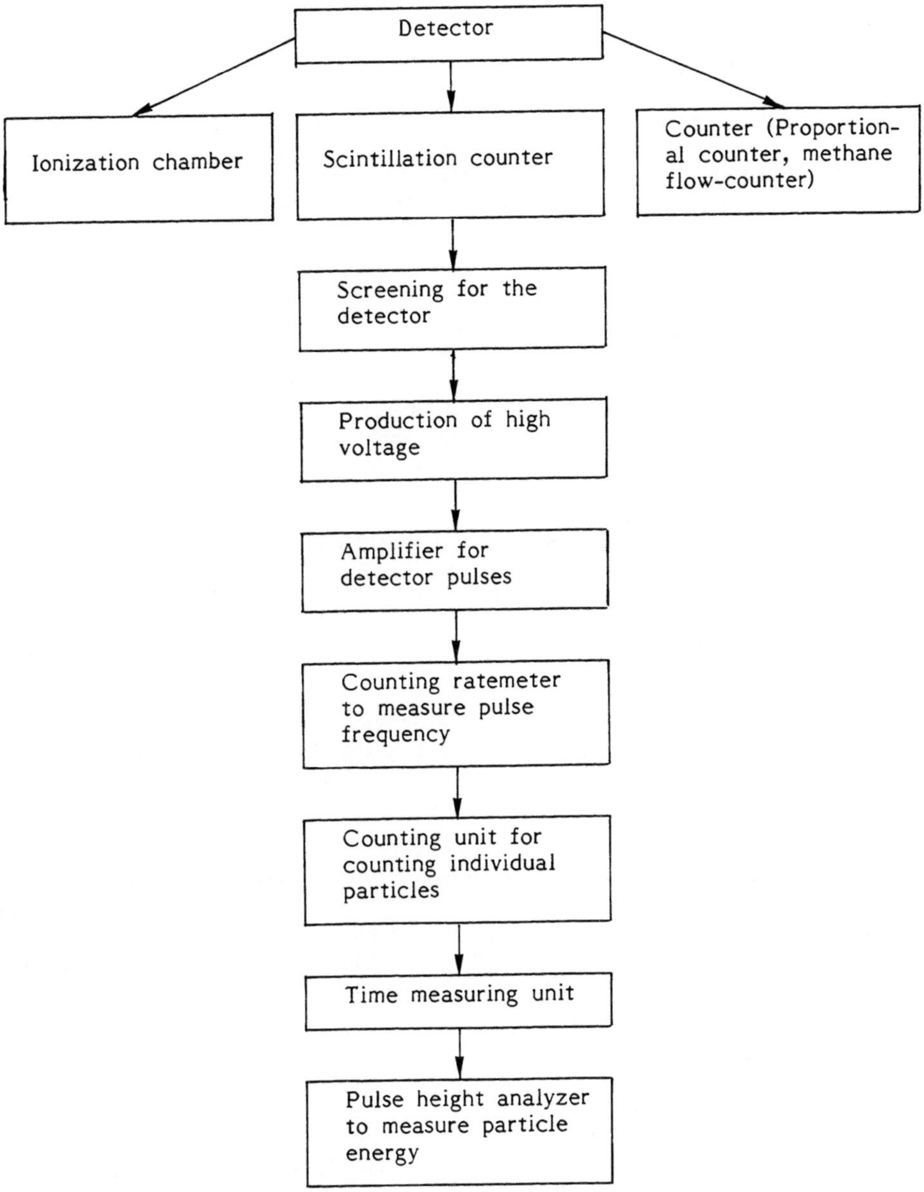


Attention should be paid to the appropriate measurement data for the detector and the adjustment of equipment in every radio-chemical measuring process. The zero effect should always be determined and kept as low as possible.

It is not possible within the scope of this collection of methods to deal with the theoretical processes in the interaction of radioactive emitters with matter (e.g. ionization, photo-electric effect, Compton effect, pair creation), the construction and operation of detectors, the structure and circuitry of the measuring equipment, or the electrical evaluation of the signals received. The reader should consult the specialized radiochemical literature.

## 2.17 Enzymatic Analysis

Enzymes are biological catalysts which the living cell uses to metabolize materials. Enzymes have an effect specific to substance, and allow reactions to occur with very small quantities of this substance. Now that enzymes have been successfully isolated from the living cell, and have even been produced synthetically, enzymatic reactions can also be employed in analytical chemistry. In particular, methods of enzymatic analysis are applied in clinical chemistry, food analysis and biochemistry, and also in the field of water analysis.

Enzymatic analyses have the advantage of being specific for a transformation of a particular material, so that as a rule no separation is necessary, even in the event of small concentrations of this substance existing together with larger quantities of accompanying substances, and the determining process can be conducted directly on the mixture of materials. Handling is generally simple and can even be performed by trainee employees. In addition, these analyses usually take only a short time to perform and are generally economical.

The following general points apply to enzymatic analyses:

1. The enzyme must act specifically, and must supply analytically determinable reaction products. (So far, it has only been possible to achieve this if several enzyme reactions are allowed to take place in succession, or if work is conducted in parallel with complementary enzymes).

2. The enzyme must effect the complete transformation of the material, and the reaction must take place within a practicable period of time.

3. The enzyme must be sufficiently active and have sufficient affinity with the substrate.

Purified enzymes together with operating instructions for conducting enzymatic analyses are offered on the market by specialist firms (e.g. Boehringer Mannheim GmbH, FRG). Care should be taken in storing enzymes that the prescribed storage conditions are strictly adhered to. It should always be remembered that enzymes are often sensitive to heat and light and can possibly be damaged if they are incorrectly stored. The influence of such chemicals as oxygen, ozone, chlorine, acid vapours, water vapour, etc., should be avoided. Enzymes are also frequently subject to natural ageing, which causes a loss of activity. It is therefore absolutely essential to observe the prescribed periods of storage.

In water analysis, enzymatic techniques are today mainly employed in the quantitative determination of sugars, organic acids, alcohols, and organic nitrogen compounds such as urea. The number of standardized enzymatic methods is increasing constantly.

The use of enzymatic techniques in the field of water analysis is described in Section 4.2, taking as an example the enzymatic determination of urea, e.g. in swimming pool water.

# 3 Inorganic Parameters

## 3.1 Total parameters

### 3.1.1 Turbidity measurements (see also Section 1.7.5)

In practical water analysis, turbidity measurements on a visual basis are made for orientation purposes immediately after sampling (Chapter 1).

However, turbidity measurements are also important when untreated water is purified to obtain drinking water. If the untreated water available for purification is surface water, for example, then flocculation using suitable chemicals and flocculation aids such as aluminium sulphate, ferrous sulphate or ferric sulphate is important, and where necessary an organic flocculation aid should be used and the pH adjusted. One very simple method of checking the effectiveness of flocculation on the laboratory scale is the "jar test". This is a laboratory technique in which the flocculation tests are performed in 1-litre or 1.5-litre beakers on a multiple stirrer unit with 4 - 6 heads. Modifications and new developments in equipment for recording flocculation and filtration parameters have been described by a variety of authors.

In addition to pH and p- and m-values (see Chapter 1 and Section 3.2), turbidity measurements are also used in the appraisal of water quality. Instruments which make use of the absorption of a beam of light passing through a turbid solution have become widespread in practice.

Scattered light measurements are also used in turbidity analysis. The instruments may take the form of hand-held devices for discontinuous opera-

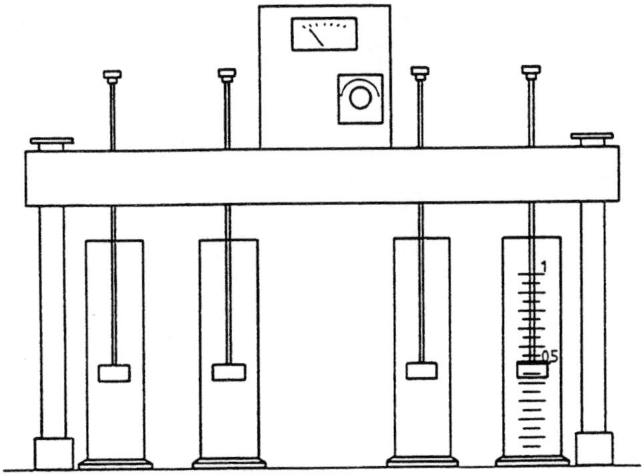


**Fig. 83.** Multiple stirrer for the jar test

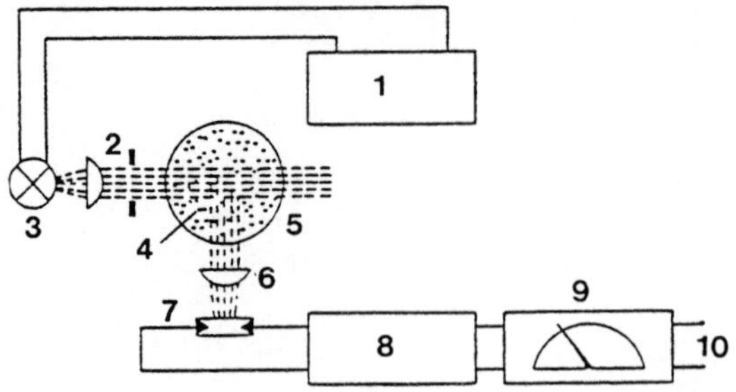


Fig. 84. Schematic diagram of laboratory turbidity photometer LTP 3; 1) = Stabilizer for lamp; 2) = Optical system; 3) = Lamp; 4) = Particles causing turbidity; 5) = Cuvette; 6) = Optical system; 7) = Measuring cell; 8) = Compensating amplifier; 9) = Data display, recording and control unit; 10) = Data output

tion, or continuous flow units which take measurements in flow-through cuvettes, or even with the water in free fall in the case of highly turbid water samples (e.g instruments made by the companies Dr. Lange, Düsseldorf, FRG, or Hach, Loveland, Colorado, USA).

The draft versions of ISO norms (1983) describe both simple techniques for on-the-spot turbidity measurement and optical turbidity meters. The diagram below illustrates the principle of a turbidity photometer for laboratory use (Dr. Lange, model LTP 3).

The following schematic diagram shows a different 90° scattered light photometer by the same company, working on the double-beam principle.

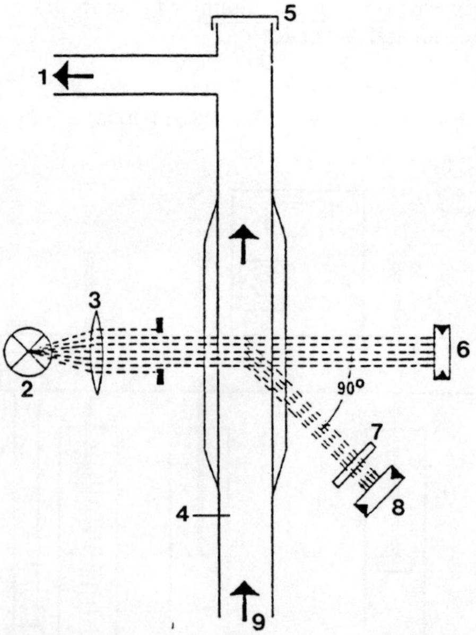


Fig. 85. Precision turbidity photometer TK 3; 1) = Outlet; 2) = Lamp; 3) = Optical system; 4) = Tubular cuvette; 5) = Cleaning access cap; 6) = Reference cell; 7) = Interference filter; 8) = Measuring cell; 9) = Inlet

Formazine ($C_2H_4N_2$) is used as standard. It is prepared in accordance with ISO as follows:

Solution a:
Dissolve 10 g of hexamethylene tetramine ($C_6H_{12}N_4$) in distilled water and dilute to 100 ml.

Solution b:
Dissolve 1 g of hydrazine sulphate ($N_2H_6SO_4$) in distilled water and dilute to 100 ml. To prepare the standard solution, mix 5 ml of solution a with 5 ml of solution b, keep the mixture at 25 $\pm$ 3° C for 24 hours and then dilute to 100 ml with distilled water.

The turbidity of this solution is approximately 400 FNU (formazine nephelometer units). The solution can be kept for 4 weeks in the dark at 25 $\pm$ 3 °C. For calibration purposes the standard formazine solution is diluted to bring it within the measuring range of the water sample to be tested.

It is recommended that measurements be taken at a wavelength of 620 nm, with a bandwidth of 10 nm. The results are given in FNU and rounded to 0.1 FNU, or in the case of turbidity values greater than 10, to 1 FNU.

### 3.1.2 Density

The ratio of mass "m" to volume "V" in a substance is described as its density "d"

$$d = \frac{m}{V}$$

In water analysis this is given in g/ml.

If the density of waste water samples is to be measured, this is carried out with either an aerometer or a pycnometer.

#### Pycnometric process

#### Equipment

Pycnometer with a content of 200 - 250 ml with a flask neck narrowed to 5 - 6 mm clear opening, with etched mm-graduation of 50 mm length

Precision scales, thermostat

#### Method

When taking samples on site, the pycnometer, which should be completely dry on the inside, is filled with the water sample, avoiding gas losses, until the level of the liquid comes halfway up the graduated pycnometer neck. At temperatures below +10 °C, correspondingly less must be filled; at temperatures around +30 °C and above, correspondingly more must be filled, because the warmer water contracts and the colder water expands when the temperature is equalized to +20 °C or at +25 °C. The filled flask should immediately be sealed tight with a rubber stopper and stored for safe transport.

In the laboratory, the temperature of the filled pycnometer is equalized in a thermostat for 1 h to +20 °C or +25 °C ($\pm$ 0.2 °C). The water level is then read off accurately on the scale of the pycnometer neck; care should be taken that no air bubbles are present in the liquid and no droplets of water on the wall of the vessel neck.

After the water level has been noted, the pycnometer is weighed accurately to 1 mg, then emptied, cleaned and filled with distilled water at +20 °C or +25 °C to the water level previously recorded. The vessel is again sealed with the rubber stopper and weighed with the same care, emptied, dried and then weighed at +20 °C or +25 °C in an empty condition (with stopper).

## Calculating density

Calculation should be according to the following formula:

$$d_{20\ °C} = \frac{m_1 - m_3 \cdot 0.9982}{m_2 - m_3}\ (g/ml)$$

where:

$m_1$ = Weight in g of the pycnometer with the water to be analyzed
$m_2$ = Weight in g of the pycnometer filled with distilled water
$m_3$ = Weight in g of the empty, dried pycnometer and its stopper

0.9982 = Density of water at +20 °C (0.9970 at +25 °C)

For densities below 1, the density is given to 4 decimal places; for densities over 1, to 3 decimal places.

Density of water at different temperatures

| t °C | d | t °C | d | t °C | d | t °C | d |
|---|---|---|---|---|---|---|---|
| 0.0 | 0.999840 | 4.5 | 0.999970 | 9.0 | 0.999780 | 13.5 | 0.999312 |
| 0.5 | 0.999871 | 5.0 | 0.999964 | 9.5 | 0.999741 | 14.0 | 0.999243 |
| 1.0 | 0.999899 | 5.5 | 0.999954 | 10.0 | 0.999699 | 14.5 | 0.999172 |
| 1.5 | 0.999921 | 6.0 | 0.999940 | 10.5 | 0.999653 | 15.0 | 0.999099 |
| 2.0 | 0.999940 | 6.5 | 0.999928 | 11.0 | 0.999604 | 15.5 | 0.999022 |
| 2.5 | 0.999954 | 7.0 | 0.999901 | 11.5 | 0.999552 | 16.0 | 0.998942 |
| 3.0 | 0.999964 | 7.5 | 0.999876 | 12.0 | 0.999497 | 16.5 | 0.998879 |
| 3.5 | 0.999970 | 8.0 | 0.999848 | 12.5 | 0.999439 | 17.0 | 0.998773 |
| 4.0 | 0.999972 | 8.5 | 0.999816 | 13.0 | 0.999376 | 17.5 | 0.998685 |
| Continued | | | | | | | |
| 18.0 | 0.998595 | 22.5 | 0.997654 | 27.0 | 0.996531 | | |
| 18.5 | 0.998500 | 23.0 | 0.997537 | 27.5 | 0.996372 | | |
| 19.0 | 0.998403 | 23.5 | 0.997417 | 28.0 | 0.996231 | | |
| 19.5 | 0.998304 | 24.0 | 0.997295 | 28.5 | 0.996088 | | |
| 20.0 | 0.998203 | 24.5 | 0.997170 | 29.0 | 0.995943 | | |
| 20.5 | 0.998098 | 25.0 | 0.997043 | 29.5 | 0.995795 | | |
| 21.0 | 0.997991 | 25.5 | 0.996913 | 30.0 | 0.995645 | | |
| 21.5 | 0.997881 | 26.0 | 0.996782 | | | | |
| 22.0 | 0.997769 | 26.5 | 0.996648 | | | | |

**Example**

$d_{20\,°C}$ (determined pycnometrically) = 1.006 g/ml.

If density is determined at ambient temperature, this must be recalculated for density $d_{20\,°C}$ or $d_{25\,°C}$. In this case, instead of the value 0.9982 (density of the water at 20 °C), the value of factor "F" should be inserted which can be found in the following Table for temperatures from 0 to +30 °C.

**Note:**

The density can also be determined using a "density balance", by measuring the hydrostatic buoyancy of a displacer of known volume.

Measurement using the bending vibration method is also possible.

Although density measurements are comparatively rare in water analysis, they should be performed with great accuracy when they are needed, e.g. when measuring stratification in a body of water.

### 3.1.3 Total determination of dissolved and undissolved substances

Water may contain mineral substances, organic compounds and gases in solution as well as undissolved (suspended) substances. Dissolved substances can be separated from undissolved substances by filtration. For total determination of dissolved and undissolved substances, the following are distinguished:

**Total residue (mg/l)**

The sum of all non-volatile dissolved and undissolved substances weighed after a measured or weighed unfiltered water sample has been evaporated under defined conditions (105 °C, 180 °C or 260 °C) and dried to constant weight.

**Evaporation residue (mg/l)**

The quantity of non-volatile, dissolved substances in a water, determinable under set conditions (type of filter, pore size, evaporation temperature e.g. 105 °C, 180 °C or 260 °C).

**Ignition residue (mg/l)**

Substances contained in water which are weighed after the evaporation residue has been kept glowing at 400 - 450 °C for one hour.

**Loss on ignition (mg/l)**

Substances contained in water which are calculated from the difference between the evaporation residue and the ignition residue.

**Undissolved substances (mg/l or ml/l)**

Suspended, deposited and/or floating substances which are settleable or may be filtered off under defined conditions.

Dissolved gases, volatile substances and substances which decompose during evaporation and drying to form volatile components are not detected by these methods of total determination.

It should be noted that the evaporation residue does not necessarily correspond to the sum of dissolved mineral substances in a water if hydrogen carbonate ions are dissolved. During drying, these are converted into carbonate with the release of $H_2O$ and $CO_2$:

$$2HCO_3^- = CO_3^{2-} + H_2O + CO_2$$

The theoretically calculated loss in weight amounts to 50.4 % of the hydrogen carbonate content.

**Equipment**

Analytical balance

Platinum, quartz or porcelain dish, 250 ml

Suction bottle, water-jet pump

Drying oven 105 - 110 °C, 180 °C or 260 °C

Muffle furnace, 400 - 450 °C

Porcelain or platinum crucible

Ammonium nitrate solution, 1 %

Imhoff cone, 1 litre (see drawing in Chapter 1)

Paper filters, quantitative (medium hard, medium-sized pores):
Before use, the filters should be washed with distilled water and dried and weighed under analysis conditions.

Membrane filter, pore size approximately 0.45 µm

**Method**

**Determination**

After shaking well, take 100 ml (for example) of the unfiltered water sample and evaporate it over the water bath in a previously dried and weighed platinum, quartz or porcelain dish. Dry the residue for two hours at 105 °C, 180 °C or 260 °C in the drying oven, and after cooling (in the desiccator), weigh accurate to 0.1 mg. Continue drying at the same temperature for 30 minutes and weigh again after cooling. If the result of the second weighing does not deviate from the first by more than $\pm$ 10 %, the result can be considered to be constant. Otherwise continue drying for a further 30 minutes.

If less than 20 mg residue remains to be weighed, the determination should be repeated using a larger volume of water.

### Evaporation residue

#### Determination

The procedure for determining the evaporation residue is the same as for the determination of the total residue, with the exception that a **filtered** water sample is used. Take a sufficiently large volume of the filtered sample to provide an evaporation residue of at least 20 mg but not more than 1000 mg.

The results are calculated in mg/l; in addition to the drying temperature, the type of filter used, e.g. membrane filter 0.45 µm, and the filtration time after sampling must also be stated. If iron or manganese which was originally in solution and passed through the filter has separated out, then this should be stated too.

### Ignition residue and ignition loss

#### Determination

After being weighed, the dish with the evaporation residue is heated (ignited) at 450 °C for 1 hour. If the ignited residue has a brown or black coloration (due to organic substances), it should - after cooling - be moistened with a few drops of ammonium nitrate solution, dried carefully and heated again at 400 - 450 °C for 10 minutes. After cooling in the desiccator, weigh the residue to an accuracy of 0.1 mg.

#### Calculation

Total residue (105 °C, 180 °C or 260 °C) in mg/l resp. ignition residue 450 °C (mg/l) =

$$\frac{\text{Amount weighed in mg} \cdot 1000}{\text{Sample volume in ml}}$$

The difference in weight between the evaporation residue and the ignition residue is the loss on ignition.

### 3.1.3.1 Undissolved substances

Amongst the undissolved substances, a distinction may be made between "filterable substances" and "settleable substances". The "filterable substances" are those undissolved substances in the water which can be filtered off using a paper (or membrane) filter. The filterable substances can also be calculated from the difference between the total residue and the evaporation residue.

"Settleable substances" (suspended, deposited or floating substances) are principally determined using the volumetric method (see drawing in Chapter 1).

#### Determination of "filterable substances"

Filter 1 litre of a thoroughly mixed water sample. The paper or membrane filters should be washed with distilled water beforehand and dried at e.g. 105 °C for 2 hours, cooled in the desiccator and weighed. Rinse the filtered residue with about 10 ml of distilled water and vacuum-dry. Lift the paper or membrane filter carefully off the filter base, predry on a watch glass at approximately 50 °C and then dry for 2 hours at 105 °C in the drying oven. Allow to cool in the desiccator and weigh to an accuracy of 0.1 mg. The weighing should be done quickly so as to minimize the uptake of moisture from the air. To avoid secondary separation from the water sample filtration should take place soon after the sample has been taken.

**Determination of "settleable substances"**

In order to prevent errors due to subsequent flocculation, as soon as possible after taking the sample pour 1 litre of the thoroughly mixed water sample into an Imhoff cone and leave to stand for 2 hours away from direct sunlight. 10 minutes before taking the reading, rotate the vessel suddenly around the vertical axis so as to sediment the particles adhered to the walls. Finally, take readings of the volume of the settled substances and, if appropriate, of the substances floating on the surface. Give the result in ml/l. (See drawing in Chapter 1).

### 3.1.3.2 Cumulative determination of dissolved substances with cation exchangers

**Principle**

As the water flows through a column with cation exchangers which have been converted into the hydrogen form by means of hydrochloric acid, all the cations are removed from the water and replaced by H-ions (K. E. Quentin 1955):

$$\underset{\text{H-exchanger}}{H\,[RSO_3]} + NaCl \text{ dissolved in water} \longrightarrow \underset{\text{exchanger (loaded with Na)}}{Na\,[RSO_3]} + HCl$$

The corresponding free acids are therefore present in the column effluent. Hydrogen carbonate and carbonate ions are no longer present in the effluent since free carbon dioxide has formed during the reaction with the H exchanger:

$$\text{H-exchanger} + Ca(HCO_3)_2 \text{ dissolved in water} \longrightarrow \text{exchanger (loaded with Ca)} + 2CO_2 + 2\,H_2O.$$

If 100 ml of water is subjected to this ion exchange process, the value yielded by titration of the effluent with 0.1-m NaOH provides the sum of the anions less the mmol value for $HCO_3^-$. If the latter value, taken from hydrogen carbonate determination, is added, this yields the total mmol of the anions and hence also the cations. After the ion exchange process, the undissociated substances (e.g. $H_2SiO_3$) are contained in the effluent in unchanged form and have no effect on determination. If, before the exchange took place, the $HCO_3^-$ was identified by titration with 0.1-m HCl to pH 4.4, it may be that the following reaction has taken place:

$$Ca(HCO_3)_2 + 2\,HCl \longrightarrow CaCl_2 + 2\,CO_2 + 2\,H_2O.$$

If a sample titrated in this way is subjected to cation exchange, subsequent titration of the effluent yields the total mmol value of anions or cations including $HCO_3^-$. This is because $CaCl_2$ has reacted with the H-ions to form free hydrochloric acid. This method of determination should be used in preference to determination without previous $HCO_3^-$ titration. The strongly acid effluent from the cation exchanger should normally by titrated

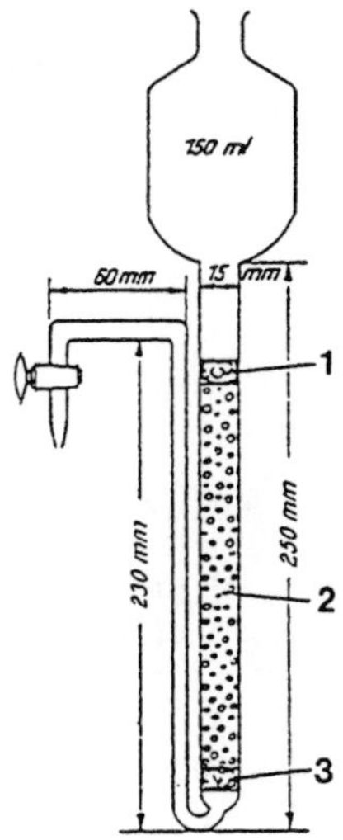


Fig. 86. Equipment: Exchanger column; 1) = Glass wool; 2) = Ion exchanger; 3) = Glass wool;

with 0.1-m NaOH. However, since it is easier to carry out acid titration to pH 4.4, it is advisable to add a measured excess of 0.1-m NaOH to the effluent and then proceed with titration in similar fashion to hydrogen carbonate determination.

Either the entire exchanger effluent volume including the wash water or an aliquot part of the acid effluent may be used for determination. The method described below covers only the main possible alternatives.

Cation exchanger, strongly acid, with sulphonic groups as active groups (e.g. ion exchangers I Merck, Dowex 50, Amberlite IR 120).

Method

Preparation of the exchanger and the column filling:

The cation exchanger filling of the column shown in Fig. 86 provides a useful capacity of approximately 60 - 80 mmol. First treat the exchanger medium with 12 % hydrochloric acid for approximately 2 hours, decanting and adding acid repeatedly, rinse, and rewash with redistilled water. Adding water, wash the exchanger into the column. During filling, open the tap to an extent that ensures even deposition without the formation of air bubbles. Whenever working with the exchanger column care should be taken that the liquid level is always above the resin bed. Next wash the column through with redistilled water until the effluent shows a neutral reaction or the chloride reaction with silver nitrate is negative. The column is then ready for use. When ion exchange is completed, regenerate with approximately 100 ml of the 12 % hydrochloric acid and wash to neutral.

Procedure

Taking the cation or anion mmol sum as a basis, titrate a quantity of water corresponding to approximately 30 - 40 mmol with 0.1-m HCl to pH 4.4. Introduce the titrated solution gradually into the prepared exchanger column and open the tap to set the effluent through-flow rate to approximately 8 - 10 ml per minute. Place an appropriate beaker beneath the outflow. When the water has flowed through, rinse with redistilled water until the effluent shows a neutral reaction. If the volume of effluent plus wash water is very large, concentrate to approximately 100 ml. Then add a measured quantity of 0.1-m NaOH (usually 50 ml), check for an alkaline reaction of the solution and titrate with 0.1-m HCl again to pH 4.4.

Calculation

mmol total of the cations or anions in 1 litre of water

$$= \frac{(a - b) \cdot 100}{\text{ml of water sample}}$$

a = ml of 0.1-m NaOH added after ion exchange
b = ml of 0.1-m HCl consumed in back-titration

### 3.1.4 Sulphide sulphur ($H_2S$, $HS^-$, $S^{2-}$) (see also Chapter 1 and Section 3.6)

#### 3.1.4.1 Iodometric determination of sulphide sulphur

General remarks

The iodometric determination of sulphide sulphur can be carried out in the field at the sampling point. The sulphide sulphur is fixed as zinc sulphide. Oxidize with acidified iodine solution to elemental sulphur. Measure the iodine consumption with sodium thiosulphate solution.

$$ZnS + I_2 + 2\ H^+ \longrightarrow S + 2\ HI + Zn^{2+}$$

The iodometric method is not specifically intended for the detection of divalent sulphur. Iodometric titration basically determines the reducing power of a water sample. The result of photometric determination in the laboratory is therefore also required in order to assess the sulphide sulphur content of a water sample.

The results of the iodometric method are inaccurate with sulphide sulphur concentrations of less than 1 mg/l.

Equipment

Erlenmeyer flask with ground glass stopper, 250 ml

Burette, volume 50 ml, 0.1 ml scale

Zinc acetate solution, 2 %, slightly acetic

0.01 m potassium iodate solution:
Treat 100 ml of 0.1 m potassium iodate solution with 60 ml phosphoric acid (1.70 g/ml) and dilute with water to 1 litre.

Potassium iodide, reagent purity, cryst.

Starch solution:
1 % aqueous solution of "soluble starch"

Sulphide reference solution:
See "Spectrophotometric analysis of sulphide sulphur"

Sodium thiosulphate solution, 0.01 m

Method

**Determination**

Directly after taking the sample, measure a volume of water (up to approximately 200 ml; it should contain at least 0.2 mg sulphide sulphur) into the Erlenmeyer flask, which should contain 10 ml of zinc acetate solution (sufficient for binding 28 mg of S). Leave the sample to stand until adjusted to room temperature, then add a few crystals of potassium iodide, and when these are dissolved swirl constantly and add 10 or 20 ml of 0.01 m potassium iodate/phosphoric acid solution according to the quantity of sulphur expected. Add 5 ml of starch solution and titrate with 0.01 m sodium thiosulphate solution until permanently discoloured.

Titre setting

Carry out determination of the titre of the 0.01 m sodium thiosulphate solution simultaneously. Titrate similarly against the potassium iodate/phosphoric acid solution using distilled water instead of the water sample.

Calculation

$$\frac{A - (B \cdot F)}{V} \cdot 160 = \text{mg/l of } S^{2-}$$

A = Quantity of 0.01 m potassium iodate/phosphoric acid solution in ml
B = Quantity of 0.01 m sodium thiosulphate solution consumed in ml
F = Titre of the standard solution
V = Volume of water sample used in ml

## 3.1.4.2 Spectrophotometric analysis of sulphide sulphur as methylene blue

General remarks

Hydrogen sulphide and dimethyl-p-phenylenediamine form a sulphurous intermediate compound (intermediate stage), which changes into leucomethylene blue. The leucomethylene blue is oxidized by iron (III) ions to methylene blue. The methylene blue is measured photometrically at 670 nm.

The reactions described above may be formulated as follows (after K. E. Quentin and F. Pachmayr):

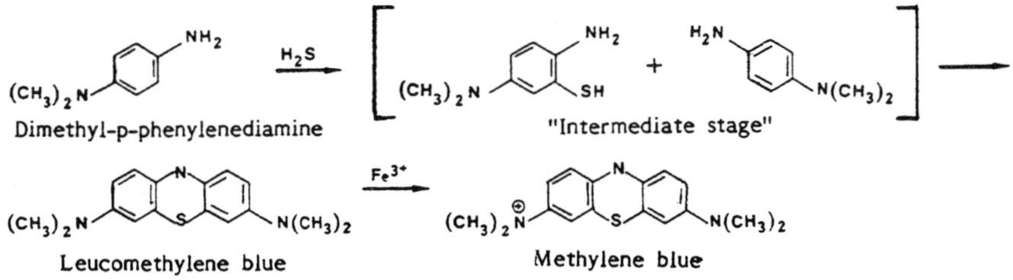

This method can be used for direct determination of sulphide sulphur in water (modification A) as well as after distillation (modification B). In the case of modification A, sulphides and hydrogen sulphide are fixed with zinc acetate as ZnS on sampling. In the laboratory, the acid in the amine reagent liberates hydrogen sulphide, which reacts with dimethyl-p-phenylenediamine in the manner described.

In the case of modification B, the sulphide sulphur fixed as zinc sulphide is distilled in distillation apparatus as hydrogen sulphide after adding phosphoric acid, and collected in a receiver containing zinc acetate solution. Further treatment is as for modification A.

The method is suitable for the determination of sulphide sulphur in a concentration range between 0.005 and 5 mg/l. Up to 70 µg the extinctions obey the Lambert-Beer law. Between 10 and 60 µg of sulphide the average error is $\pm$ 1 %.

Quantities of iodide greater than 1 mg in the aliquot of the water sample used for the investigation have an interfering effect. However, such concentrations of iodide are only rarely obtained in exceptional cases.

Nitrite ions in quantities less than 10 µg in the solution under investigation have no interfering effect; interference due to greater quantities can be eliminated by adding 3 drops of a 5 % urea solution.

Iron (II) ions up to 25 mg/l and sulphite ions up to 10 mg/l do not interfere with the determinations.

If notable quantities of hydrogen carbonate ions or carbonate ions are present, adding the strongly acid dimethyl-p-phenylenediamine reagent solution causes $CO_2$ to form, which may carry off $H_2S$ as it escapes. For this reason, the reagent should be introduced as a layer beneath the sample, and the flask then closed and shaken. When the flask is opened, only $CO_2$ escapes since $H_2S$ has been bonded by the amine.

Equipment

Spectrophotometer (670 nm) or photometer

Cuvettes, 1, 2 and 5 cm

Volumetric flasks with ground glass stopper, 100 ml and 250 ml

Quentin's distillation apparatus (modification B only):
See illustration. A 100 ml volumetric flask serves as the receiver with a hole corresponding to a similar hole in the ground section of the inlet tube

Steel cylinder containing nitrogen

Nitrogen purification solution:
2 % potassium permanganate solution, containing 5 g mercury (II) chloride in 100 ml.

Zinc acetate solution, 2 %, slightly acetic

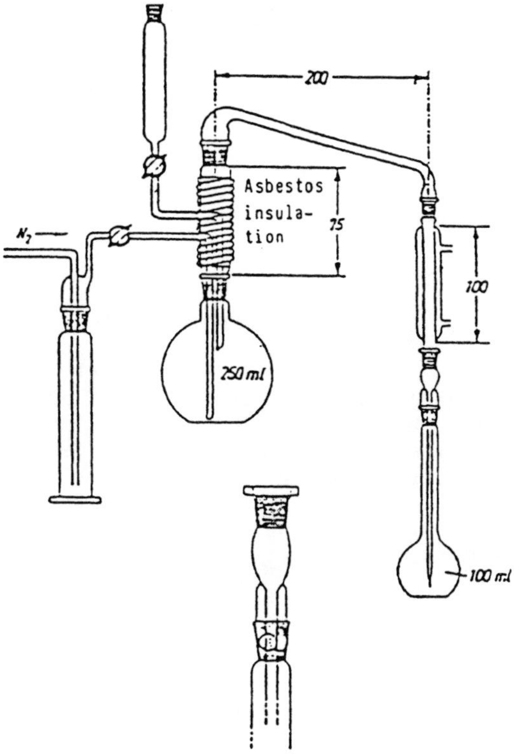


Fig. 87. Quentin's distillation apparatus for $H_2S$

Amine reagent solution:

Suspend 2 g dimethyl-p-phenylenediamine-hydrochloride in 200 ml of dist. water in a 1-litre volumetric flask. Carefully add 200 ml of conc. sulphuric acid (1.84 g/ml) and make up to the mark when cooled to 20 °C.

Iron (III) reagent solution:

Add 10 ml of conc. sulphuric acid (1.84 g/ml) to a solution of 50 g iron (III) ammonium sulphate, $NH_4Fe(SO_4)_2 \cdot 12H_2O$ in distilled water in a 500 ml volumetric flask and dilute to the mark with dist. water.

Phosphoric acid (1.70 g/ml)

Sulphide reference solution:

Add 500 ml of freshly prepared gelatin solution (300 mg/l) to 250 ml of the 2 % zinc acetate solution. Place an amber glass flask containing this solution beneath the distillation apparatus as the receiver. The outlet tube of the condenser should be well immersed into this solution (if necessary lengthen the tube). Direct a moderate flow of nitrogen into the distillation flask, the flow having previously passed through the wash bottle containing nitrogen purification solution. In the distillation flask, bring 150 ml boric acid solution (6 g $H_3BO_3$, reagent purity, dissolved in 150 ml of warm water) to the boil. Prepare sodium sulphide solution by dissolving a washed sodium sulphide crystal in a little water, approx. 40 mg $Na_2S \cdot 9\ H_2O$ cryst., reagent purity, corresponding to approx. 5 mg S, and introduce the solution via the dropping funnel during heating. When the boric acid solution is boiling and water begins to distill over, interrupt the boiling briefly and slowly add the sodium sulphide solution drop by drop. The released hydrogen sulphide distills off within 15 minutes and is bonded as ZnS in the receiver. Put the contents of the receiver in a 1 litre volumetric flask, dilute with water to the mark and determine the sulphide content of the solution iodometrically. (Use boiled out water for all solutions. Store in amber glass flasks.)

## Method

### Calibration (modifications A and B)

Taking the result of the iodometric sulphide determination in the sulphide reference solution as a basis, put varying quantities of this solution, containing between 10 and 70 µg sulphide, into 100 ml volumetric flasks and add water such that the liquid does not quite reach the neck of the flask. Then continue as described under "Modification A determination" until methylene blue coloration appears. Prepare the calibration curve in this way. Up to 70 µg sulphide, this is a straight line. Take measurements at 670 nm.

**Note:**

Dark blue solutions with more than 70 µg sulphide content cannot be measured even after dilution. The calibration curve prepared does not apply to dilutions of this type.

### Determination (modification A)

Place 5 ml of zinc acetate solution in each of a number of 100 ml volumetric flasks (according to the equation $Zn^{2+} + S^{2-} \rightarrow ZnS$, this quantity is sufficient to bind 14 mg of sulphide).

Depending on the expected sulphide content of the water, up to 75 ml of the test water may be added to these flasks at the sampling point. Then add distilled water until the liquid almost reaches the bottom of the neck of the flask. Run in 10 ml of amine reagent solution into the liquid at the flask neck in such a way as to form a lower layer. Close the flask and shake briefly (in waters containing hydrogen carbonate or carbonate, $CO_2$ escapes), and immediately add 0.5 ml of iron (III) reagent solution. This should bring the total liquid level up to the neck of the flask. Shake again vigorously, leave to stand for 10 minutes at room temperature, dilute to the mark with distilled water and mix up. Take a colorimetric measurement at 670 nm against a simultaneously prepared blank test containing distilled water instead of the water sample under test.

**Note:**

All interferences mentioned and also interference due to contaminants and substances causing turbidity or due to excessive concentrations of mineral substances (see "Scope of application") can be eliminated by using the distillation method (modification B).

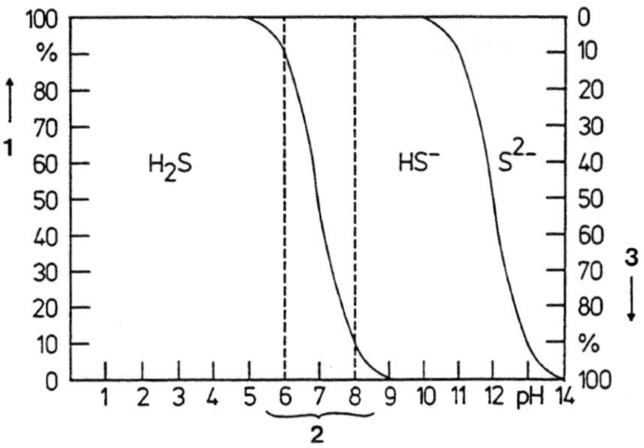


**Fig. 88.** Dissociation $H_2S$ at 18 °C, pH range 1 - 14; 1) = Total sulphur; 2) = pH range of natural waters; 3) = Total sulphur

Determination (modification B)

Introduce 5 ml of zinc acetate solution (sufficient to bind 14 mg sulphide) into each of a number of 250 ml volumetric flasks (distillation flasks). Add up to 150 ml of the test water, depending on the expected sulphide sulphur content of the water, to these flasks at the sampling point. In the laboratory, remove the ground glass stopper and connect the flask to the distillation apparatus. Put 10 ml of zinc acetate solution in the receiving flask. The ground inlet tube with hole should dip into the zinc acetate solution. Turn the tube to match up the holes on the inlet tube and the volumetric flask.

Bring to the boil, applying a moderate flow of nitrogen. When the solution is boiling and water distilling over, interrupt the boiling process and add 5 ml of phosphoric acid (1.70 g/ml) through the dropping funnel. $H_2S$ is then transferred to the receiver with the water vapour where it is again bonded to zinc.

When the volume of liquid in the receiver reaches about 70 ml (10 ml of zinc acetate solution + 60 ml of distillate), stop the distillation process.

Remove the volumetric flask including the inlet tube from the apparatus. Add 10 ml of amine reagent solution through the inlet tube, shake briefly and immediately add 2 ml of iron (III) reagent solution and shake again. When shaking the sample solution, turn the holes in the ground sections away from each other and plug the flask and inlet tube with a ground glass stopper (see illustration). Do not remove the inlet tube until after adding the iron solution and shaking, then rinse it with water. Following this, leave the solution to stand at room temperature for 10 minutes, dilute with distilled water to the mark, mix and measure the extinction of the colour solution against the simultaneously prepared blank test.

Calculation

Read off the sulphide sulphur content of the sample from the calibration curve on the basis of the measured extinction, and taking account of the volume of sample solution used for determination, recalculate the result in terms of 1 litre of the water sample.

The quantitative distribution of divalent sulphur amongst the individual sulphur compounds is dependent on the pH of the water. Fig. 88 shows the dissociation of hydrogen sulphide at 18 °C in the pH range 1 - 14.

### 3.1.5 Water "hardness" (see also Chapter 1, and Sections 3.2 and 3.3)

General remarks

The term water hardness is obsolete and has no satisfactory definition. Nevertheless, it continues to be used. The so-called hardness of a water is due to the ions of magnesium, calcium and strontium dissolved in the water. The sum of these ions is designated total hardness, and today is often determined on site by complexometry.

So-called carbonate hardness, however, if this is to be measured, corresponds to the hydrogen carbonate ions which can be allocated to calcium, magnesium and strontium. The proportion of so-called carbonate hardness in °d (for the various units, see conversion table below) which can be attributed to the hydrogen carbonate ions can be calculated by multiplying the m value of a water by 2.8. There are cases where the carbonate hardness is higher than the so-called total hardness. If this is so, alkali hydrogen carbonates are present dissolved in the water and the carbonate hardness should then be taken as total hardness. The so-called non-carbonate hardness is ascertained purely by calculation, by subtracting the carbonate hardness from the total hardness. The non-carbonate hardness is usually understood to be the hardness produced by the equivalent proportions of chloride and sulphate to magnesium, calcium or strontium.

The conversion table shows the relationships between the various degrees of hardness in use internationally and the measured values in mmol/l or meq/l.

**Conversion table.** (alkaline-earth ions Mg, Ca and Sr)

| | Alkaline-earth ions mmol/l | Alkaline-earth ions meq/l | German degrees of hardness °d | American degr. of hardness ppm $CaCO_3$ | British degr. of hardness °e | French degr.of hardness °f |
|---|---|---|---|---|---|---|
| 1 mmol/l Alkaline-earth ions | 1.00 | 2.00 | 5.60 | 100.0 | 7.02 | 10.0 |
| 1 meq/l Alkaline-earth ions | 0.50 | 1.00 | 2.80 | 50.0 | 3.51 | 5.0 |
| 1 German degree | 0.18 | 0.357 | 1.00 | 17.8 | 1.25 | 1.78 |
| 1 ppm $CaCO_3$ (USA) | 0.01 | 0.020 | 0.056 | 1.0 | 0.0702 | 0.100 |
| 1 British degree | 0.14 | 0.285 | 0.798 | 14.3 | 1.00 | 1.43 |
| 1 French degree | 0.10 | 0.200 | 0.560 | 10.0 | 0.702 | 1.00 |

## 3.2 Anions

### 3.2.1 Fluoride

General remarks

Almost all natural waters contain fluoride ions. The content varies on average between 0.01 and 1.5 mg/l. Fluoride concentrations in natural subterranean waters or waters from particular geological formations, however, may be considerably higher: up to 10 or 15 mg/l. Wastewaters containing fluoride and hydrogen fluoride stem primarily from fluorine factories, pickling houses, electroplating plants, the glass industry and from ore mining and processing.

Particular importance is attached to fluoride as a means of preventing caries. To this end, sufficient fluoride may be added to the drinking water to obtain a fluoride concentration of about 1.0 mg/l, which is considered to be the optimum. Continuous and precise monitoring of the fluoride content of such drinking water is necessary because the natural fluoride content, which must also be taken into account, is variable, and persistent overdosage may have a detrimental effect on health. For this reason, the WHO has set a limit value of 1.5 mg/l for fluoride in drinking water. The 1986 Drinking Water Ordinance in the Federal Republic of Germany adopted this limit value for a life-long intake of approximately 2 litres of water daily.

The substances most commonly used for the fluoridation of drinking water for the purpose of preventing caries are hexafluorosilicic acid, sodium fluoride or sodium hexafluorosilicate. However, opinions differ in many countries as to the advantages, effectiveness or even harmfulness of drinking water fluoridation with the result that to date, no uniform statutory rulings have been achieved. The drinking water fluoridation stations operating in Europe are mainly test plants and their area of influence is regionally limited. In the USA artificial fluoridation is more widespread. Both sides of the question, forced medication by means of treated drinking water and the consequences of dental caries, must be considered. Natural mineral water containing fluoride appears to be a conceivable alternative, e.g. in bottled form. Partial defluoridation is possible by filtration of the water over $Al_2O_3$ after acidification with $CO_2$.

Three methods are described below:

3.2.1.1 Spectrophotometric determination

3.2.1.2 Determination using an ion-selective electrode (if necessary, separation and concentration by steam distillation)

3.2.1.3 Ion chromatography (3.2.11)

#### 3.2.1.1 Spectrophotometric determination with lanthanum alizarin complexone, directly or following steam acid distillation

General remarks

After concentration, if required, by steam acid distillation, the fluoride taken up in 0.1 m sodium hydroxide solution is determined photometrically at 600 - 620 nm as a blue-coloured alizarin complexone/lanthanum fluoride complex. The colour reaction on which this method is based is founded on the fact that the orange-coloured solution of alizarin complexone (alizarin 3-methylamine-N, N-diacetic acid dihydrate) forms a red chelate with lanthanum (III) ions; when small quantities of fluoride are added, this red chelate yields the blue-coloured ternary complex, whose depth of colour is proportional to the concentration of fluoride. In the range from 0 to 30 µg of fluoride the calibration curve forms a straight line. The addition of acetone to the reagent solution increases the sensitivity of the method.

If the fluoride concentration is relatively high, approximately 0.5 mg of $F^-$/l or more, photometric determination can be carried out directly in the water sample (approx. 10 - 25 ml) without recourse to steam distillation. This applies in particular to routine determination in drinking water. Interferences due to other substances are to be expected only rarely in these lightly mineralized waters. It is necessary to conduct preliminary tests in order to be certain of the reliability of the results.

The method may also be applied to waters with lower concentrations by alkaline concentration of a larger volume of water and driving over the fluoride from the evaporation residue by steam acid distillation. The detection limit is approximately 0.005 mg/l.

Interfering chloride ions are bonded by the addition of silver sulphate to the distillation acid. The volume of distillation acid specified in the procedure contains a quantity of silver sulphate which is capable of bonding 135 mg of chloride ions. Should more than 135 mg of $Cl^-$ be present in the water under analysis, either a correspondingly smaller volume of water should be used or more silver sulphate added. The chloride content of the water should in any case always be known. Nitrite ions up to 2 mg do not interfere with the colour reaction.

If organic substances are present, e.g. in wastewaters, the alkaline evaporation residue must be incinerated in a platinum dish. The incineration temperature must not exceed 450 °C. Slight interference is caused by sulphate ions passing over into the distillate. This must be eliminated by adding 0.1-m NaOH to all the blank and calibration solutions and to the distillates, and then neutralizing with 0.05 m $H_2SO_4$. The 15 mg of $SO_4^{2-}$ which are then always present in the measuring flask level out the influence of the distilled sulphate traces. Up to 5 mg of aluminium ions cause no interference.

Equipment

Spectrophotometer

Cuvettes, 2 cm

Measuring flasks, 250 ml and 1 litre

Platinum dish

Alizarin complexone solution, 0.004 m:
Dissolve 1.6854 g of alizarin 3-methylamine-N,N-diacetic acid dihydrate (mol. weight 421.36) in 20 ml of 0.5 m sodium acetate solution (41 g/l sodium acetate, anhydrous), heating gently, and make up to the mark in a 1-litre measuring flask with redist. water at 20 °C.

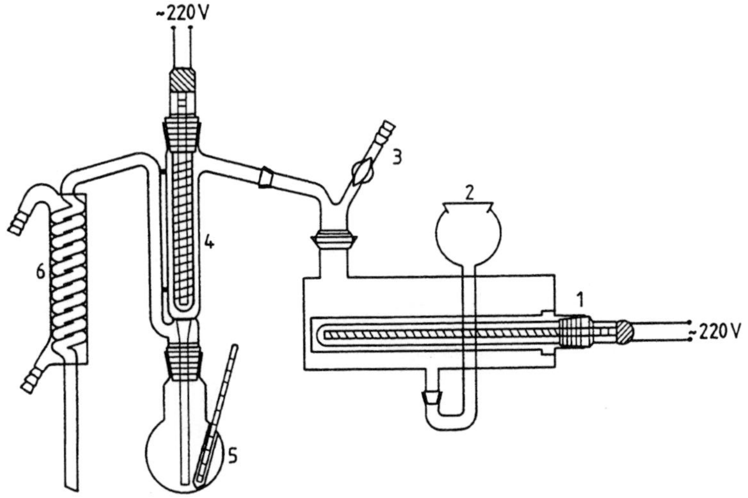


Fig. 89. Distillation apparatus after SEEL; 1) = Steam boiler with rod immersion heater; 2) = Filling funnel; 3) = Safety cock; 4) = Steam superheater (rod immersion heater); 5) = 150-ml sample flask with thermometer pocket (thermometer 0 - 150 °C); 6) = coil condenser (approx. 30 cm, jacket length approx. 15 cm)

Lanthanum nitrate solution, 0.004 m:
Dissolve 1.7322 g of lanthanum nitrate ($La(NO_3)_3 \cdot 6\ H_2O$) in redist. water in a 1-litre measuring flask and make up to the mark at 20 °C.

Acetone, reagent purity

Acetate buffer solution (pH 4.4):
Mix 630 ml of 0.5 m acetic acid (approx. 29 ml of glacial acetic acid + 971 ml of redist. water) with 370 ml of 0.5 m sodium acetate solution.

Lanthanum alizarin complexone solution:
Put 80 ml of acetate buffer solution, 500 ml of acetone, 50 ml of lanthanum nitrate solution and 50 ml of alizarin complexone solution in a 1-litre measuring flask and make up to the mark with redist. water at 20 °C. The pH should be between 5.6 and 5.8; check with a pH meter. The solution may be kept for 8 weeks in a refrigerator.

Sodium hydroxide solution, 0.1 m and 1 m

Sulphuric acid, (1.64 g/ml) and conc. (1.84 g/ml)

Silver sulphate, $Ag_2SO_4$

p-nitrophenol indicator solution (pH transition from yellow/alkaline to colourless/acid at approx. pH 6):
Dissolve 0.2 g of p-nitrophenol in 100 ml of dist. water.

Sulphuric acid for distillation:
Use a 1-litre round-bottom flask with a thermometer pocket at the side (thermometer up to 200 °C) and a ground-glass attachment. The ground-

glass attachment has an inlet pipe for steam which reaches almost to the base of the flask. In the ground-glass stopper there is a discharge pipe, which is in turn connected by a ground joint with the coil condenser to draw off the distillate. The flask should contain conc. sulphuric acid. After connecting the steam boiler and the condenser, heat the sulphuric acid to 120 - 130 °C. Introduce steam and regulate the temperature to approx. 150 °C. Distill off and discard at least 500 ml. Raise the temperature to 171 °C and terminate steam distillation. If these conditions are complied with the sulphuric acid remaining in the flask is of the original concentration again.

Heat 10 g of silver sulphate and 10 ml of sulphuric acid (1.84 g/ml) in a porcelain dish, and continue heating for a further ten minutes after the appearance of $SO_3$ vapour. When cool, transfer the solution to a beaker and mix with the 500 ml of sulphuric acid which remains in the flask following steam distillation, and heat until the solution is clear, stirring constantly. Allow to cool. Store the sulphuric acid for distillation in a dark storage bottle.

Fluoride reference solution:

Dissolve 221 mg of sodium fluoride of reagent purity in redist. water and make up to the mark in a 1-litre measuring flask at 20 °C. Dilute 10 ml of this solution to the mark in a 1-litre measuring flask with redist. water at 20 °C. 1 ml of this solution contains 0.001 mg of $F^-$ ions.

Method

Preparation of the water sample

Add 1 m sodium hydroxide solution (until the alkaline reaction occurs) to 250 ml of the water sample and evaporate in a platinum dish. The chloride content of the sample must be known so that a sufficient quantity of silver sulphate to bond the chloride can be added with the sulphuric acid for distillation. In the case of waters which are loaded with organic substances the alkaline evaporation residue should be incinerated in the platinum dish; the incineration temperature must not exceed 450 °C.

Distillation

Transfer the evaporation residue (if appropriate, following incineration) to the distillation apparatus by small amounts with a maximum of 15 ml of redistilled water. The quantity of water may be increased if it does not suffice. However, the solution must then be concentrated to 15 ml again in the sample flask.

Add 30 ml of sulphuric acid for distillation with sufficient silver content to bond 135 mg of chloride ions (if necessary add additional silver sulphate), and connect the sample flask to the distillation apparatus. Introduce redistilled water into the steam boiler via the feed funnel until the water level is approximately 2 cm above the rod immersion heater. Beneath the condenser outlet position a 250-ml measuring flask containing 5 ml of 0.1 m sodium hydroxide solution and 1 ml of p-nitrophenol indicator solution in order to collect the distillate.

Switch on the rod immersion heater of the steam boiler and, when the steam thus generated reaches the steam superheater, switch on the rod immersion heater in the latter as well. The distillation temperature is between 120 and 130 °C.

When approximately 200 ml of distillate has been collected, neutralize the excess of 0.1 m sodium hydroxide in the receiving measuring flask with 0.05 m sulphuric acid until the colour changes from yellow to colourless. Make up to the mark with redistilled water at 20 °C.

Determination

Transfer an aliquot part of the neutralized distillate - a maximum of 50 ml, but only such a quantity that no more than 30 µg of fluoride ions are contained in the aliquot - to a 100-ml measuring flask, add 50-ml of lanthanum alizarin complexone solution and make up to the mark with redistilled water. It is important when adding the lanthanum alizarin complexone solution that the same drop rate is set and the same pipette is used as for the parallel calibration process. Allow the solution to stand for a further 15 minutes at room temperature. Its pH value is now between 4.8 and 5.0. Measure the extinction at 600 - 620 nm against water.

Calibration

In order to prepare the calibration curve parallel to the determination of the sample, transfer 10, 20 and 30 ml of fluoride reference solution (with 0.01, 0.02 and 0.03 mg of fluoride) and 5 ml of 0.1 m sodium hydroxide solution to 100-ml measuring flasks, add 1 ml of p-nitrophenol indicator solution and titrate with 0.05 m sulphuric acid to the neutral point. In the same way as described under "Determination", add 50 ml of lanthanum alizarin complexone solution by pipette, make up to the mark with redistilled water (20 °C), and after 15 minutes measure at 600 - 620 nm against redistilled water. Prepare and treat a blank test in the same way without the addition of fluoride.

It is absolutely essential that a calibration curve is prepared and a blank test taken into account for each fluoride determination.

Calculation

Read off the appropriate fluoride content from the calibration curve on the basis of the measured extinction, take into account the quantity of water used and the blank test, and convert for 1 litre.

### 3.2.1.2 Determination of the fluoride ion with an ion-selective (ion-sensitive) electrode

General remarks

The quantitative determination of a number of dissolved substances in water using ion-selective electrodes and a corresponding measuring system has found widespread acceptance. This method has proved particularly valuable for the quantitative determination of fluoride ions in water samples in the range above approximately 0.1 mg/l.

In this method the voltage developed between the measuring electrode and the reference electrode is measured. Under certain conditions this measured value is proportional to the logarithm of the fluoride concentration. Attention must be paid to the water temperature, the ionic strength of the dissolved substances and the pH value. These parameters should be standardized in the calibration of the method in such a way that interference in the course of the actual measurement can be largely ruled out.

Fluoride complexes are rendered available for measurement by the addition of a special buffer (TISAB buffer - Total Ionic Strength Adjustment Buffer). If stronger complexing substances such as boron compounds or organic substances are present in waste waters, it is advisable to carry out distillation according to method 1 before the actual measurement.

Equipment

In addition to the standard laboratory equipment, a measuring instrument is required to measure the voltage, with divisions of 1 mV.

Fluoride-selective electrode cell (single-probe measuring cell recommended)

pH measuring device with glass electrode as single-probe measuring cell

Recording apparatus, if required

Thermostat

Stirring device

Hydrochloric acid, 1 m

Sodium hydroxide solution, 1 m

TISAB buffer solution:
Dissolve 300 g of sodium citrate ($C_6H_5Na_3O_7 \cdot 2H_2O$) in approx. 600 to 800 ml of water. Subsequently dissolve 22 g of 1,2-cyclohexylene dinitrilotetra-acetic acid ($C_{14}H_{22}N_2O_8 \cdot H_2O$) and 60 g of sodium chloride in this solution. When the solution is clear, make up to 1000 ml. The pH of this buffer solution is 5.8.

Fluoride calibration solutions:
Use sodium fluoride (NaF) which has been dried at 120 °C for 2 hours. Dissolve 2.210 g and dilute to 1000 ml. Prepare fluoride calibration solutions of varying concentration by diluting this stock solution, e.g. dilute 10 ml of fluoride stock solution to 1000 ml with water, fluoride content 10 mg/l. Prepare further fluoride calibration solutions with fluoride contents of 5, 1, 0.5 and 0.2 mg/l in the same way.

Method

The water sample should be analyzed no later than 3 days after sampling, and it is advisable to filter the sample through an appropriate filter before analysis. If filtration is carried out a note to this effect should be made with the results.

Condition the ion-sensitive electrode by placing it at least 1 hour before starting measurement in a calibration solution which may contain, for example, 0.5 mg of fluoride ions/l. Following this period, rinse the electrode and use for measurement.

Check the calibration curve daily with fluoride calibration solutions of appropriate concentration. This is important in particular due to the fact that both the origin and the gradient of the calibration curves of the ion-selective electrodes may change.

For the measurement itself, put 25 ml of TISAB buffer solution in the receiver and add 25 ml of the sample under analysis by pipette, after filtration if necessary. The pH must be 5.8 ± 0.2. Should it deviate from this value adjust with hydrochloric acid or sodium hydroxide solution. Following this preparation set the solution for measurement to the measuring temperature, e.g. 20 ± 0.5 °C or 25 ± 0.5 °C (the same measuring temperature as was used for preparing the calibration curve). Immerse the ion-selective electrode cell or the single-probe measuring cell into the solution and stir for 5 minutes at a speed of approx. 200 rpm. If after this period the measured value does not change by more than 0.5 mV, turn off the stirrer and read off the measured value after about 10 to 20 seconds.

As a control, especially in the case of low fluoride concentrations, appropriate fluoride calibration solutions of known concentration may be added to a second or third water sample together with the buffer solution.

Calibration

For calibration purposes take 25-ml portions of TISAB buffer solution in the same way and to each of perhaps 5 receivers add calibration solutions with fluoride concentrations between 0.2 and 10 mg/l by pipette. Rinse the electrode between measurements. Adapt the concentration ranges of the calibration solutions approximately to the concentration of fluoride expected in the sample under analysis. Carry out the calibration itself at least twice and repeat daily. Plot the measured values thus established on semi-logarithmic paper. Use the logarithmic abscissa for the fluoride concentration of the calibration solutions in mg/l and the ordinate with decimal divisions for the corresponding measured values in mV.

Read off the concentration of fluoride ions in the water sample under analysis from the calibration curve on the basis of the measured value in mV. Take into account the volume of water used or any dilution made and convert for 1 litre. The measured values should generally be rounded off to one decimal place.

### 3.2.1.3 Ion chromatography determination of fluoride ions (see Section 3.2.11)

## 3.2.2 Chloride

General remarks

3.2.2.1 Gravimetric analysis

3.2.2.2 Volumetric determination with electrometric indication

Modification A: chloride content up to 1000 mg/l
Modification B: chloride content over 1000 mg/l

3.2.2.3 Volumetric determination with visual indication (potassium chromate as indicator)

3.2.2.4 Ion chromatography (3.2.11)

In the course of the methods (see Sections 3.2.2.1 - 3.2.2.3) bromide ions and iodide ions are also detected. If these latter ions are determined separately, for example in mineral waters and brines, the chloride content can be calculated by subtracting these values.

### 3.2.2.1 Gravimetric determination

#### General remarks

In a weak nitric acid solution chloride ions react with silver ions. Sparingly soluble silver chloride is formed, and if bromide ions and iodide ions are present, so too are silver bromide and silver iodide. The silver halides separate out and are deposited, and are then filtered, dried and weighed, and the content of chloride ions is calculated, if necessary after correction for the separately determined bromide ions and iodide ions. This method is particularly suitable for waters with high concentrations of chloride ions, and as a calibration or reference method.

Any sulphite ions or sulphide ions present are oxidized with hydrogen peroxide (about 3 to 5 %).

#### Equipment

Analytical balance

Drying oven (130 °C)

Glass filter crucible

Nitric acid, (1.070 g/ml), chloride-free

Silver nitrate solution: approx. 1 m

Rinsing liquid: distilled water acidified with nitric acid.

#### Method

Acidify a measured volume of water between 100 ml and 1000 ml with nitric acid and heat to between about 60 and 80 °C. Slowly conduct the precipitation, stirring gently, with the approx. 1 m silver nitrate solution. The silver chloride precipitate which now forms settles quickly. The supernatant liquid becomes clear. The completeness of the precipitation should be tested by adding about 0.1 ml of silver nitrate solution.

Keep the precipitate over night in the dark (on exposure to light part of the silver is reduced, with the result that the precipitate may assume a blue-black colour). After the settling period filter the supernatant solution through the dried and weighed glass filter crucible. Using the rinsing liquid, repeatedly make the precipitate into a paste in a beaker and then transfer quantitatively to the filter crucible. Rinse this precipitate of silver chloride until the rinsing liquid shows no further visible reaction with hydrochloric acid or sodium chloride solution. Then dry the filter crucible and contents in the drying oven at 130 °C and weigh when cool. Repeat this operation until constant weight is reached.

1 mg of the weighed silver chloride precipitate corresponds to 0.2474 mg of chloride ions. Calculate the concentration of chloride ions in mg/l of water, taking into account the measured quantity of water and the weighed

quantity of silver chloride. If a correction is made for the separately determined bromide ions and iodide ions this should be stated. In the case of concentrations of chloride ions above 100 mg/l the results should be rounded off to 1 mg/l. A note should be made of the fact that the gravimetric method was used.

### 3.2.2.2 Volumetric determination with silver nitrate (electrometric indication)

(This internal method has proved valuable as a routine technique for analysing waters with high chloride contents and is therefore described here.)

#### General remarks

A) The water sample is titrated directly with 0.1 m or 0.01 m silver nitrate solution at a pH of between 4.5 and 8.5 (modification A: up to about 1000 mg/l $Cl^-$).

B) A water sample with more than 1000 mg/l $Cl^-$ and a pH of between 4.5 and 8.5 is treated with a weighed quantity of 0.1 m silver nitrate solution. This quantity should be less than the amount required for the quantitative precipitation of the chloride ions. The low concentration of chloride ions still present in solution is subsequently titrated with 0.01 m silver nitrate solution.

For electrometric indication a measuring cell consisting of an indicator electrode and a reference electrode linked with a sensitive galvanometer is used (see illustration under "Equipment"). Due to the higher chloride ion concentration in the sample solution the potential of the indicator electrode differs from that of the reference electrode. This potential difference is measured by the galvanometer. During titration the potential of the indicator electrode approaches that of the reference electrode, and at the end point of the titration they are equal. Consequently, the needle of the galvanometer passes through zero at the equivalence point.

Sulphite ions and sulphide ions are eliminated by adding dilute hydrogen peroxide solution drop by drop while cooling the solution.

#### Equipment

Measuring cell with silver plate as indicator electrode and a silver/silver chloride reference electrode (Schneider):

Burettes, volume 50 ml, 0.1 ml scale

Silver nitrate solution, 0.1 m

Silver nitrate solution, 0.01 m:
Prepare by dilution of 0.1 m silver nitrate solution

Sodium chloride solution, 0.1 m

Sodium chloride solution, 0.01 m

Hydrogen peroxide solution, approx. 3 - 5 %

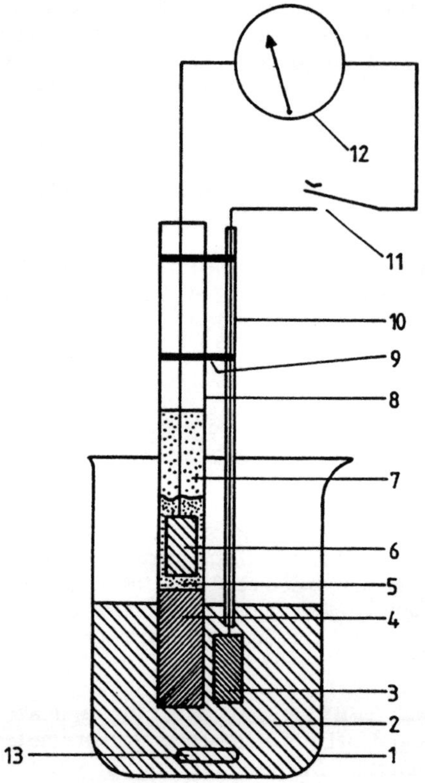


Fig. 90. 1) = Beaker, 400 ml; 2) = Solution for analysis; 3) = Silver indicator electrode, 20 x 10 x 1 mm; 4) = Diaphragm packed with quantitative filter paper; 5) = Silver chloride suspension in redistilled water; 6) = Silver reference electrode, dimensions as for 3; 7) = Redistilled water; 8) = Glass tube, diameter 15 mm; 9) = Connection of the two electrodes to form a measuring cell; 10) = Glass tube, diameter 6 mm; 11) = Switch to close the measuring current circuit; 12) = Sensitive galvanometer as zero instrument; 13) = Magnetic stirrer

Nitric acid, (1.070 g/ml) chloride-free

Sodium-hydroxide solution, 1 m chloride-free

(The two latter items if required for adjustment of the pH range 4.5 - 8.5)

Method

If the chloride content of the water sample is unknown, the approximate concentration should be established initially by pre-titration with 0.1 m silver nitrate solution. If it does not exceed 1000 mg/l, titrate directly according to modification A. If it is above 1000 mg/l, proceed according to modification B.

Determination (modification A)

Into the beaker in the titration apparatus pipette an exact volume of the water sample which corresponds to a consumption of about 20 ml of 0.1 m silver nitrate solution or a consumption of 10 - 50 ml of 0.01 m silver nitrate solution. Introduce the measuring cell and switch on the magnetic stirrer and galvanometer. Subsequently, titrate until the galvanometer indicator passes through zero.

Determination (modification B)

Measure off a volume of the water sample which corresponds to a consumption of between 10 and 100 ml of 0.1 m silver nitrate solution and make up to about 200 ml with distilled water. Precipitate silver chloride with a weighed quantity of 0.1 m silver nitrate solution, stirring slowly in cool conditions. The quantity of 0.1 m silver nitrate solution used is selected such that about 1 or at most 2 ml less silver nitrate solution is consumed than during pre-titration. Now transfer the pre-precipitated solution to the titration apparatus, introduce the measuring cell, and switch on the magnetic stirrer and galvanometer. Subsequently titrate with 0.01 m silver nitrate solution until the galvanometer indicator passes through zero.

Since the density of 0.01 m silver nitrate solution is close to 1 g/ml and this solution was obtained by diluting 0.1 m silver nitrate solution, the quantity of 0.01 silver nitrate solution consumed, in ml, should be divided by 10 and added to the weight of 0.1 m silver nitrate solution. This sum provides the consumption of 0.1 m silver nitrate solution for the measured volume of the water sample.

Titre setting

The titre setting of the 0.1 m and 0.01 m silver nitrate solutions is made in the same way as described under "Determination (modification A)" using 0.1 m and 0.01 m sodium chloride solutions.

As a result of weighing the standard solution and the use of corresponding calibration the deviation between multiple determinations is below an absolute level of 3 mg/l chloride ions, even in the case of very high concentrations of chloride ions.

Calculation (modifications A and B)

When using 0.1 m silver nitrate solution:

$$\frac{V_1 \cdot F \cdot 3.5453 \cdot 1000}{E} = \text{mg/l } Cl^-,$$

when using 0.01 m silver nitrate solution:

$$\frac{V_2 \cdot F \cdot 0.35453 \cdot 1000}{E} = \text{mg/l } Cl^-$$

$V_1$ = Consumption of 0.1 m silver nitrate solution, in the case of modification A in ml, and in the case of modification B in g
$V_2$ = Consumption of 0.01 m silver nitrate solution in ml
F = Titre of the standard solutions
E = Volume of water sample used for determination.

### 3.2.2.3 Volumetric determination with silver nitrate and potassium chromate (visual indication)

General remarks

If all chloride ions are bonded as AgCl, silver ions added in excess, together with chromate ions, form reddish-brown silver chromate ($Ag_2CrO_4$).

The method is suitable for the determination of chloride ions in concentrations between 10 and 150 mg/l. Waters whose chloride-ion concentrations lie outside this range can also be determined according to this method by appropriate dilution or concentration. Its applicability to waste water investigation is dependent on the type of waste water involved.

Coloured or turbid waters should be shaken out with chloride-free, freshly precipitated aluminium hydroxide or chloride free washed activated charcoal, and filtered. Organic substances are destroyed in a weakly alkaline solution with $KMnO_4$, heating gently. Excess $KMnO_4$ is eliminated with $H_2O_2$, the hydrated manganese oxides are filtered off and washed, chloride-free.

Interfering $SO_3^{2-}$can be eliminated by $H_2O_2$ in neutral solution, and $S^{2-}$, $S_2O_3^{2-}$, $CN^-$ and $SCN^-$ can be eliminated by $H_2O_2$ in alkaline solution. Iron ions can be precipitated by shaking with 1 g of chloride-free zinc oxide and can then be filtered off. Phosphate interferes in concentrations greater than 25 mg/l. It is coprecipitated with iron.

Equipment

Burette, volume 50 ml, 0.1 ml scale

Silver nitrate solution, 0.05 m and 0.02 m

Sodium chloride solution, 0.05 m and 0.02 m

Potassium chromate solution:
Dissolve 7.5 g of $K_2CrO_4$ in a sufficient quantity of dist. water and add silver nitrate solution until a reddish-brown precipitate begins to form. Allow to settle for 2 hours, filter the solution and replenish the filtrate with dist. water.

Method

If the chloride content of the water sample is unknown this should first be determined by pre-titration with silver nitrate standard solution of appropriate normality. Take 100 ml of the filtered water sample with a chloride content of 1 - 15 mg, or a smaller volume diluted to 100 ml with distilled water. The pH should lie between 7 and 9 (acid- or alkaline-reacting waters should first be neutralized with chloride-free reagents). Add 1 ml of potassium chromate solution.

According to the chloride concentration, titrate with 0.05 m or 0.02 m silver nitrate solution until the colour changes from greenish-yellow to reddish-brown. In order to assist determination of the end point of the titration, the analyst may add a drop of sodium chloride solution after completion of the titration of the sample. If the yellow-green colour reappears the end point of the titration was reached.

The quantity of silver nitrate which is required to produce a recognizable change of colour should be determined by means of a blank test with 100 ml of distilled water and 1 ml of potassium chromate solution. It should then be deducted from the result of titration.

Titre setting

The titre setting of the 0.05 m and 0.02 m silver nitrate solutions is made according to the procedure described, using 0.05 m or 0.02 m sodium chloride solution respectively. The method used should correspond to the technique described for the analysis in all ways. If work is conducted carefully the determination error of the method can be limited to 1 - 3 mg/l $Cl^-$.

Calculation

When using 0.02 m silver nitrate solution:

$$\frac{V_1 \cdot F \cdot 0.7091 \cdot 1000}{E} = \text{mg/l } Cl^-,$$

When using 0.05 m silver nitrate solution:

$$\frac{V_2 \cdot F \cdot 1.7727 \cdot 1000}{E} = \text{mg/l } Cl^-,$$

$V_1$ = Consumption of 0.02 m silver nitrate solution in ml
$V_2$ = Consumption of 0.05 m silver nitrate solution in ml
F = Titre of the standard solutions
E = Volume of water sample used for determination.

#### 3.2.2.4 Ion chromatography determination (see Section 3.2.11)

### 3.2.3 Bromide and iodide

General remarks

Natural waters contain only very low concentrations of bromide and iodide ions. Higher concentrations may occur in mineral waters, natural salt brines and in certain industrial wastewaters; in most cases they are accompanied by a very large excess of chloride ions.

#### 3.2.3.1 Consecutive iodometric determination of bromide and iodide in one solution

The following method of consecutive determination of $Br^-$ and $I^-$ by iodometry, based on a method developed by P. Höfer, has proven effective.

#### 3.2.3.2 Ion chromatography (see Section 3.2.11)

Analysis by means of ion chromatography has gained significance in recent years. (3.2.11)

#### 3.2.3.1 Consecutive iodometric determination of bromide and iodide in one solution

Bromide and iodide ions can be determined successively in one water sample using a method developed by P. Höfer: hypochlorite ions are used to oxidize bromide ions to bromate ions and iodide ions to iodate ions:

$$Br^- + 3\ ClO^- = BrO_3^- + 3\ Cl^-$$

$$I^- + 3\ ClO^- = IO_3^- + 3\ Cl^-$$

The excess hypochlorite ions are destroyed by formic acid. Subsequently, iodide ions added at pH 3 to 4 are oxidized to iodine by the iodate ions resulting from the above reaction. For each iodide ion originally contained in the sample six equivalents of iodine are liberated. These are titrated with sodium thiosulphate solution:

$$IO_3^- + 6\ I^- + 6\ H^+ = 3\ I_2 + I^- + 3\ H_2O$$

$$3\ I_2 + 6\ S_2O_3^{2-} = 6\ I^- + 3\ S_4O_6^{2-}$$

In this solution, following the addition of hydrochloric acid and adjustment of the pH to < 1, the bromide ions are then detected analogously:

$$BrO_3^- + 6\ I^- + 6\ H^+ = 3\ I_2 + Br^- + 3\ H_2O$$

$$3\ I_2 + 6\ S_2O_3^{2-} = 6\ I^- + 3\ S_4O_6^{2-}$$

Direct determination of bromide ions is possible in a range between 0.01 and 15 mg/l and determination of iodide ions in a range between 0.005 and 5 mg/l of the water sample. If lower concentrations of these ions are to be determined, a larger quantity of water must be concentrated by evaporation under alkaline conditions.

Nitrite ions may be oxidized to nitrate ions in acid solution (pH < 1), before analysis, with 1 ml of approximately 0.4 m sodium hypochlorite solution. The solution should then be neutralized (pH 6 to 8) with reagent-purity sodium hydroxide solution or with reagent-purity solid calcium carbonate.

Should the concentration of bromide and/or iodide be less than 0.01 mg/l, the solution should be concentrated to 100 ml under alkaline conditions. The solution thus obtained may either be processed further directly, in which case precipitates of alkaline earth carbonates should be filtered off, or the evaporation residue can be extracted three times with absolute ethanol. In the latter case, the bromide and iodide ions pass into the ethanol extract. This should be boiled down under alkaline conditions and the residue, after gentle ignition (not more than 400 °C), picked up with approximately 100 ml of distilled water. Acidify the solution obtained (pH < 1) and oxidize with 1 ml of 0.4 m sodium hypochlorite solution. Finally neutralize the solution (pH 6 to 8) and continue according to the procedure described under "Determination".

A maximum of 1.5 mg of bromide ions and 0.5 mg of iodide ions should be contained in the volume of water which is used for analysis in this method. It is important to adhere to the quantities and concentrations of reaction solutions specified in the procedure and to the specified waiting times. Blank determinations and calibrations should be carried out together with each analysis using the same quantities of reagents under precisely the same test conditions.

## Equipment

Sodium thiosulphate solution, 0.005 m

Sodium hypochlorite solution, approx. 0.4 m:
Either use a commercially available solution, in which case the concentration should be determined iodometrically, or prepare the solution from sodium hydroxide solution by introducing chlorine

Formic acid, 2 m

Potassium iodide solution: 16.6 g/100 ml KI

Hydrochloric acid, reagent purity, approx. 10 % (1.05 g/ml)

Sodium hydroxide solution, 1 m

Starch solution:
1 % aqueous solution of "soluble starch"

Buffer solution:
Dissolve
100 g of monosodium dihydrogen phosphate dihydrate,
100 g of disodium monohydrogen phosphate dodecahydrate,
100 g of sodium diphosphate decahydrate and
300 g of sodium chloride, all of reagent purity, in 1400 ml of water.

The pH of this buffer solution is approx. 5.8. If 6 ml of this solution is diluted to 100 ml, a pH of approx. 6.8 is established.

Bromide reference solution:
Dissolve 1.4894 KBr of reagent purity (dried at 105 °C) with dist. water in a measuring flask to 1000 ml. 1 ml of this solution contains 1 mg of bromide ions. If required, a dilution of 1 : 10 may be prepared. 1 ml of the dilute bromide reference solution contains 0.1 mg of bromide ions.

Iodide reference solution:
Dissolve 0.1308 g of reagent-purity KI (dried at 105 °C) with dist. water in a measuring flask to 1000 ml. 1 ml of this solution contains 0.1 mg of iodide ions.

### Method

Establish the quantity of water which is appropriate for analysis by means of a preliminary test with 100 ml of the water sample.

### Determination

Use the volume of the water sample established in the preliminary test, pretreated as required, (pH between 6 and 8), make up to 100 ml and add 10 ml of buffer solution. A pH between 6.5 and 7 is thus established. Turbidity or precipitates of alkaline earth phosphates which may arise after the addition of the buffer solution are of no significance for the further course of the analysis.

Add 4 ml of sodium hypochlorite solution and heat to about 90 °C within 5 to 10 minutes. Swirl the solution frequently while heating, but avoid boiling.

Treat the sample while still hot with 10 ml of 2 m formic acid, drop by drop. In so doing, the excess hypochlorite ions are decomposed and any precipitated alkaline earth phosphates are dissolved again (a pH value of 3 to 4 is established). Allow to cool to room temperature (approx. 20 °C) and then treat the solution with 1 ml of potassium iodide solution. Leave to stand for about 5 minutes and then add 1 ml of starch solution. If the sample contains iodide ions a blue coloration appears. If this is the case, titrate with 0.005 m sodium thiosulphate solution until the blue coloration disappears. The decoloration should persist for at least 1 minute. The quantity of sodium thiosulphate consumed corresponds to the content of iodide ions in the volume of sample used.

In order to determine the content of bromide ions, treat the titrated sample with 10 ml of hydrochloric acid (1.05 g/ml). If bromide ions are present, a blue coloration appears once more. Following a waiting time of about 5 minutes again titrate with 0.005 m sodium thiosulphate solution until the blue coloration disappears. The decoloration must again persist for at least 1 minute. The consumption of sodium thiosulphate solution in the course of the titration corresponds to the content of bromide ions in the volume of sample used.

Conduct a blank test in the same way using 100 ml of distilled water instead of the sample. Any consumption of sodium thiosulphate solution should be deducted from that established in the determination of the sample.

Titre adjustment

The precise titre of the 0.005 m sodium thiosulphate solution must be established for each analysis, if possible simultaneously. To achieve this, take known volumes of bromide and iodide reference solutions instead of the sample, make up to 100 ml, and treat as for the analysis. Calculate the factor of the sodium thiosulphate solution from the results of titration.

Calculation

1 ml of 0.005 m sodium thiosulphate solution is equivalent to 0.10575 mg of iodide ions or 0.06667 mg of bromide ions. The concentration per litre of water may be calculated according to the following formulae:

$$\frac{V \cdot F \cdot 0.10575 \cdot 1000}{E} = \text{mg/l I}^-,$$

$$\frac{V \cdot F \cdot 0.06667 \cdot 1000}{E} = \text{mg/l Br}^-,$$

V = Consumption of 0.005 m sodium thiosulphate solution in ml,
F = Titre of standard solution,
E = Volume of water sample used for analysis.

### 3.2.3.2 Ion chromatography (see Section 3.2.11)

## 3.2.4 Nitrite

### 3.2.4.1 Spectrophotometric analysis with sulphanilic acid and 1-naphthylamine

General remarks

In acid solution, nitrite ions and aromatic amines together form diazonium salts, which for their part couple with aromatic amines to form intensively coloured azo compounds. The nitrite determination described here is based on the diazotization of sulphanilic acid (p-aminobenzene sulphonic acid) with nitrite to form p-diazobenzene sulphonic acid. This acid couples with 1-naphthylamine, forming red-violet p-benzene sulphonic acid azonaphthylamine.

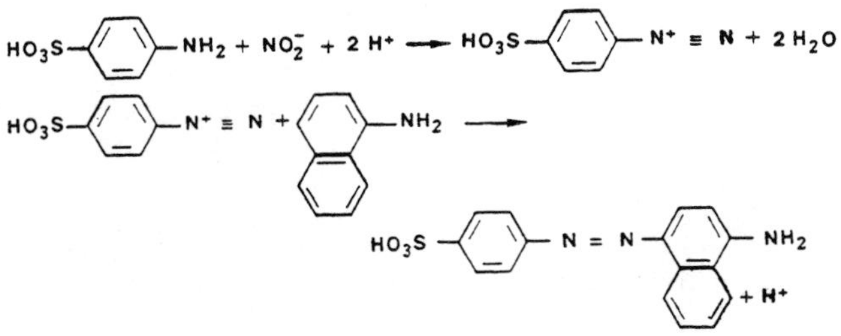

The method is suitable for the determination of nitrite ions in water in concentrations between 0.005 and 0.6 mg/l. If the nitrite concentration is greater than 0.6 mg/l, the solution must be diluted accordingly before analysis since the Lambert-Beer law no longer applies in this range.

In order to prevent secondary formation of nitrite ions by microbial or chemical redox reactions from ammonium ions or nitrate ions, nitrite determination should if possible be carried out within a few hours or at the latest within one day of sampling. Sterilized flasks should be taken for filling. The sample should be cooled to at least +10 °C until the analysis is carried out.

It is possible to compensate for slight turbidity or self-coloration in the sample in the spectrophotometer by measuring the extinction in comparison with the water sample. In the case of high turbidity which cannot be removed by filtration or of strong self-coloration, especially in the presence of colloidally dissolved organic substances and sulphides, add 5.0 ml of aluminium sulphate solution and 5.0 ml of 1 m sodium hydroxide solution per 100 ml of water sample and shake. When the hydroxide precipitate has settled, filter through glass wool or cotton wool; discard the first parts of the filtrate. Take the dilution into account in evaluation by multiplying the result by 1.1.

Colloidal organic substances, humic acids and free chlorine, however, may also be removed by shaking the sample with 1 - 2 g nitrite-free activated charcoal and filtering after 5 minutes reaction time. Before treating with activated charcoal, the pH of the sample must be set to more than 8.5 to prevent adsorption of the nitrite ions by the activated charcoal.

Interference due to copper ions, which accelerate the breakdown of the diazonium salt, or due to other heavy metal ions is largely prevented by the aluminium sulphate/NaOH precipitation. In this case, also, dilution of the sample may be advisable because of the high sensitivity of the colour reaction to nitrite ions.

**Equipment**

Spectrophotometer 530 nm or photometer

Cuvettes, 1, 2 and 5 cm

Volumetric flasks, 100 ml

Glacial acetic acid, reagent purity (1.05 g/ml)

Sulphanilic acid solution:
Heat 1 g sulphanilic acid with 50 ml glacial acetic acid and 50 ml dist. water and dissolve by diluting with 200 ml hot dist. water. Store the solution in a brown flask in a cool place.

1-naphthylamine solution:
Dissolve 0.2 g 1-naphthylamine in 50 ml glacial acetic acid and 100 ml dist. water, then dilute with a further 150 ml dist. water.

Aluminium sulphate solution:
Dissolve 120 g $Al_2(SO_4) \cdot 18\ H_2O$ (superpure) in nitrite-free dist. water and make up to 1 litre. Before use, shake up any sediment formed.

Sodium hydroxide solution, 1 m

Activated charcoal, nitrite-free

Nitrite stock solution:
Dry 0.150 g of reagent purity sodium nitrite at 105 °C for 1 hour before weighing in, dissolve in nitrite-free water, and after adding 1 ml chloroform make up to 1 litre with nitrite-free water. Standardize the solution against 0.02 m potassium permanganate solution and check the titre weekly. 1 ml = 0.1 mg $NO_2^-$.

Nitrite reference solution I:
Dilute 10 ml nitrite stock solution to 100 ml with nitrite-free water. 1 ml of this solution contains 0.01 mg $NO_2^-$. Prepare a fresh batch for each application.

Nitrite reference solution II:
Dilute 10 ml nitrite stock solution to 1 litre with nitrite-free water. 1 ml of this solution contains 0.001 mg $NO_2^-$. Prepare a fresh batch for each application.

## Method

### Calibration

Pipette increasing volumes of nitrite reference solutions I and II with $NO_2^-$ concentrations between 0.005 and 0.05 mg into 100 ml volumetric flasks and treat according to the method specified under "Determination". Prepare the calibration curve from the extinction values.

### Determination

Carry out a preliminary test in order to establish the volume of the water sample required for a 0.0005 to 0.05 mg nitrite content. To do this, pipette 10 - 50 ml of the water sample into a 100 ml volumetric flask and add 4 ml of a freshly prepared mixture of equal parts of sulphanilic acid solution and 1-naphthylamine solution. Make up to the mark at 20 °C, mix, and leave the solution to stand for 2 hours in darkness. After this reaction time, measure the extinction at 530 nm.

Using the result of this preliminary test as a basis, pipette an appropriate volume of the water sample with no more than 0.05 mg $NO_2^-$ into a 100 ml volumetric flask. The volume used must not exceed 95 ml. Add 4 ml of the mixture of equal parts of sulphanilic acid solution and 1-naphthylamine solution to this volume, dilute the solution to the mark with distilled water and mix well. Leave to stand for 2 hours in darkness, and measure the extinction at 530 nm.

Carry out a blank test with distilled water in the same way.

Calculation

Using the measured extinction, read the appropriate nitrite content from the calibration curve and convert the results for the volume of the water sample used to 1 litre.

### 3.2.4.2 Ion chromatography (see Section 3.2.11)

### 3.2.5 Nitrate

### 3.2.5.1 Spectrophotometric analysis with sodium salicylate

General remarks

Nitrate ions and sodium salicylate together form yellow sodium nitrosalicylate, which can be determined spectrophotometrically at 420 nm.

Nitrate concentrations between 0.2 and 40 mg/l may be determined. If concentrations are lower than 0.5 mg/l, however, the method is inaccurate. The water sample should be cooled to + 4 °C and examined within 1 to 2 days at the latest.

Equipment

Spectrophotometer

Cuvettes, path length 0.5 to 5 cm

Glass or porcelain dishes, 50 ml

Sand bath or drying oven

Sodium salicylate solution:
Dissolve 0.5 g of reagent purity sodium salicylate in distilled water to 100 ml. The solution must be freshly prepared daily.
Sulphuric acid, reagent purity (1.84 g/ml)

Sodium hydroxide solution:
Dissolve 400 g of reagent purity sodium hydroxide and 16 g of reagent purity potassium sodium tartrate in dist. water, cooling the mixture, and make up to 1 litre. Store the solution in a plastic flask.

Nitrate stock solution:
Dissolve 0.137 g sodium nitrate of reagent purity in distilled water, add 1 ml of chloroform, and make up to 1 litre. 1 ml of this solution contains 0.1 mg of $NO_3^-$.

Nitrate reference solution:
Dilute nitrate stock solution with distilled water in the ratio 1 : 10. 1 ml of this solution contains 0.01 mg of $NO_3^-$.

Method

Calibration

Transfer aliquots of the nitrate reference solution with between 0.01 and 0.5 mg $NO_3^-$ to a glass or porcelain dish and treat according to the method given under "Determination". Either prepare a calibration curve or calculate the calibration factor from the extinction values obtained.

Determination

The nitrate content of the water sample to be investigated should be between 0.01 and 0.5 mg; this may be determined by a preliminary test. Add 2 ml of sodium salicylate solution to the water sample in a glass or porcelain dish and evaporate over a water or sand bath or in a drying oven at 100 °C. Dry the residue at 100 °C for 2 hours and then cool in a desiccator. Add 2 ml of sulphuric acid (1.84 g/ml), allow the solution to stand for 10 minutes, add 15 ml of distilled water and 15 ml of sodium hydroxide solution, cooling constantly, and transfer the solution to a 50 ml volumetric flask. Make up to the mark with distilled water at 20 °C; shake, and measure the solution within 10 minutes at 420 nm. Prepare a blank test with distilled water in similar manner and take into consideration for the calculation. Even greater accuracy can be achieved by mixing a known quantity of nitrate ions with a further water sample and carrying out determination analogously. It should be noted that the reagents must be added by means of pipettes and the specified times and temperatures strictly adhered to.

Calculation

Using the measured extinction as a basis, read the appropriate nitrate content from the calibration curve or calculate it using the calibration factor, taking the blank reading into account. Recalculate the results to 1 litre to take account of the volume of the water sample used.

### 3.2.5.2 Nitrate with 2,6-dimethylphenol

Photometric determination of nitrate with 2,6-dimethylphenol is conducted in accordance with the German Standard Methods and/or in accordance with DIN 38405, Part 9. In a sulphuric or phosphoric acid solution, nitrate ions react to this reagent within approximately 5 minutes to produce yellowish-coloured 4-nitro-2,6-dimethylphenol.

As this method is admittedly very suitable for drinking water and groundwater, but not for more highly polluted waters such as effluent. A second method, the classic reductive distillation process, is described in Section 3.2.5.3, while Section 3.2.5.4 refers to the ion chromatography technique described in Section 3.2.11. Waters in which the chloride concentration is approximately 10 times as large as the nitrate concentration should be distilled before determining nitrate. Any disturbance due to nitrite ions is eliminated by means of amidosulphuric acid.

### Equipment

Photometer

Cuvettes, Erlenmeyer flasks, transfer pipettes, measuring flasks

Nitrate calibrating solution:
163.1 mg potassium nitrate (dried at 105 °C) is diluted in distilled water to 1 litre. 1 ml corresponds to 0.1 mg nitrate ions ($NO_3^-$).

Acid mixture:
1 part by volume of sulphuric acid (1.84 g/ml) is mixed with 1 part by volume of phosphoric acid (1.71 g/ml).

Amidosulphuric acid ($H_2NSO_3H$)

Reagent solution:
1.2 g of 2,6-dimethylphenol is dissolved in 1 litre acetic acid free from water. This solution should be prepared fresh weekly.

### Analysis

5 ml of the water to be analyzed is pipetted into a 100 ml Erlenmeyer flask. 40 ml of the acid mixture is added and, in the event of disturbing quantities of nitrite (over 0.2 mg/l) being present, approximately 50 mg amidosulphuric acid is also added. After 10 minutes, 5 ml of the reagent solution is added, and the result measured photometrically at 324 nm after a reaction period of a further 10 minutes. The water to be analyzed is employed as reference solution. A blank reading is measured simultaneously; 5 ml distilled water is used instead of the water sample. The calibrating curve is formed by taking different quantities of the calibrating nitrate solution, filling them with distilled water to 5 ml in each case, and then analyzing them in the same way as the sample to be tested.

Where necessary, the nitrate calibrating solutions should be diluted in order to comply with the measuring range of the sample to be analyzed. Taking into consideration the quantity of water employed and the calibrating curve, the results are given in mg $NO_3$/l water, to no more than one decimal place.

## 3.2.5.3 Determination of nitrate after reductive distillation

### General remarks

Nitrate ions are reduced to ammonia by Devarda's alloy (= alloy comprising 50 % Cu, 45 % Al, 5 % Zn) in alkaline solution by means of nascent hydrogen:

$$NO_3^- + 8H^+ \rightarrow NH_3 + 2\ H_2O + OH^-$$

The ammonia is distilled and determined by acidimetric or spectrophotometric means in the distillate.

The process is suitable for determining nitrate in turbid or coloured waters or waste waters and in samples with a relatively high chlorine content (salt waters). In the case of concentrations of $NO_3$ in excess of

5 mg/l, final determination is by titration; in the case of low concentrations, determination is by the spectrophotometric process with indophenol.

Ammonium ions are a disturbing factor and are distilled off from the soda-alkaline solution before adding Devarda's alloy. Nitrite ions are detected together with the nitrate ions. They must be determined separately and allowed for in calculating the findings. Analysis must take place in ammonia-free atmosphere; if filtration is necessary, care must be taken to make use of nitrogen-free filter paper. Nevertheless, blank determining processes are absolutely essential.

### Equipment

Distillation apparatus:
1 litre Erlenmeyer flask with distillation head and Kjeldahl connecting bulb, descending condenser with extended tip

Devarda's alloy in powder form

30 % sodium hydroxide solution of reagent purity

Boric acid solution:
40 g $H_3BO_3$ of reagent purity is dissolved to 1 litre in water free of ammonium ( 1 ml of this solution absorbs approximately 2 mg $NH_4^+$)

Nitrate calibrating solution:
1.631 g $KNO_3$ is dissolved to 1 litre in distilled water. 1 ml corresponds to 1 mg nitrate ions.

### Calibration

Increasing quantities of the nitrate reference solution are distilled according to the working instructions described below. The ammonium ions are determined titrimetrically or photometrically in aliquot parts of the distillate (see Ammonium, Section 3.3.5)

### Analysis

250 ml of the water sample, or a smaller portion filled up to 250 ml with distilled water, is transferred to the distillation flask, and 30 ml of 30 % sodium hydroxide solution added. This is reduced to approximately half the volume by distilling, in order to remove the ammonium ions. The distillation solution is then allowed to cool and a new receiving flask provided. Next, 1 g Devarda's alloy is added to the distillation flask, this is immediately sealed and the end of the condenser submerged in the absorption solution (50 ml boric acid solution). After the Devarda's alloy has completely dissolved at room temperature, 100 ml is distilled over into the boric acid solution. This distillate (100 ml plus 50 ml boric acid solution) is transferred into a 200 ml measuring flask and filled up to the mark with water free of ammonium at 20 °C.

The ammonium ions are now determined either acidimetrically or photometrically in aliquot parts (see Ammonium, Section 3.3.5).

If nitrite ions are present, these are determined separately, converted to nitrate ions by multiplying by 1.3478, and then subtracted from the nitrate content determined. The nitrate ions contained in the water sample to be analyzed are calculated, taking into consideration the original volume of water used and the appropriate calibrations, and given in mg/l to a maximum of one decimal place.

### 3.2.5.4 Ion chromatography (see Section 3.2.11)

## 3.2.6 Sulphite

### General remarks

One tried-and-tested technique is the volumetric method (iodometric titration) described in Section 3.2.6.1, involving the use of a known quantity of iodine and back titration of the iodine not consumed.

Where the sulphite concentration is higher (over 50 mg/l) and where considerable quantities of organic substances are present, the gravimetric method (Section 3.2.6.2) should be used after distillation.

For low concentrations - and in the presence of sulphide and thiosulphate - the polarographic method (Section 3.2.6.3) can be used.

### 3.2.6.1 Iodometric determination of sulphite

#### General remarks

In the iodometric determination method, iodine is used to oxidize the sulphite ions to sulphate ions:

$$SO_3^{2-} + H_2O + I_2 \longrightarrow 2\ H^+ + 2\ I^- + SO_4^{2-}$$

The quantity of elemental iodine consumed in this reaction is equivalent to the sulphite ion content of the water sample under investigation. Any excess iodine is titrated back using sodium thiosulphate solution.

The method is suitable for determining sulphite ions in concentrations ranging from 0.5 to 50 mg/l. With higher concentrations the water sample must be diluted. However, in such cases it is also possible to use the gravimetric method (see Section 3.2.6.2).

All substances capable of being oxidized by iodine may have a disturbing influence. The concentrations of nitrite ions, sulphide ions and free hydrogen sulphide must be determined separately and taken into account when calculating the result. Where substantial quantities of organic substances are present ($KMnO_4$ consumption greater than 60 mg), gravimetric determination as sulphate following distillation and oxidation is to be preferred (Section 3.2.6.2).

#### Equipment

Glass-stoppered bottles:
Capacity 200 - 300 ml; contents to be measured to the nearest 0.1 ml

Phosphoric acid, 1.15 g/ml:
Prepared by diluting 12.5 ml of phosphoric acid, 1.70 g/ml, with 100 ml distilled water

Iodine solution, 0.05 m and 0.005 m

Sodium thiosulphate solution, 0.1 m and 0.01 m

Zinc iodide starch solution

Method

Determination

The glass-stoppered bottle of known capacity is filled to overflowing with the water to be analyzed. Then, disregarding any overflow of water, a pipette reaching to the base of the bottle is inserted and used to add 10 - 30 ml of 0.05 m iodine solution, depending on the expected sulphite ion concentration, and 3 ml of phosphoric acid (1.15 g/ml). The bottle is then closed, avoiding any inclusion of air, and shaken. After 10 minutes the solution is quantitatively transferred to an Erlenmeyer flask and the excess iodine titrated with 0.1 m sodium thiosulphate solution, adding zinc iodide starch solution towards the end of titration. If the consumption of the 0.05 m iodine solution is less than 1 ml, the determination must be repeated using 0.005 m iodine solution and back titrating with 0.01 m sodium thiosulphate solution.

Calculation

1 ml of 0.05 m iodine solution corresponds to 4.003 mg $SO_3^{2-}$.

1 ml of 0.005 m iodine solution corresponds to 0.400 mg $SO_3^{2-}$.

The calculation of the result must take account of the content of the glass-stoppered bottle less the quantities of iodine solution and phosphoric acid added.

If the water sample also contains nitrite ions and a separate nitrite determination is made, 1.7 mg $SO_3^{2-}$ per litre must be deducted from the result of the sulphite determination for every 1 mg of $NO_2^-$ per litre.

Proceed similarly in the case of sulphide, deducting 2.5 mg $SO_3^{2-}$ per litre from the result of the sulphite determination for every 1 mg of $S^{2-}$ per litre.

### 3.2.6.2 Gravimetric determination as barium sulphate after distillation

General

The sulphite ions are distilled over as sulphur dioxide from a solution containing phosphoric acid into a receiver containing a solution of iodine and potassium iodide, where they are oxidized to sulphate ions:

$$SO_3^{2-} + I_2 + H_2O \longrightarrow SO_4^{2-} + 2\,H^+ + 2\,I^-$$

$$SO_4^{2-} + Ba^{2+} \longrightarrow BaSO_4$$

The sulphate ions are precipitated out of the weakly acid solution as barium sulphate by adding barium chloride solution, and then weighed.

The method is suitable for determining sulphite ions in concentrations in excess of 20 - 30 mg/l and where considerable quantities of disturbing organic substances are present.

The distillation process largely eliminates any disturbing factors.

**Equipment**

Distillation apparatus with long condenser end

Phosphoric acid (1.15 g/ml):
See Section 3.2.6.1

Iodine/potassium iodide solution:
Dissolve 1.5 g potassium iodide and 10 g iodine in 1000 ml distilled water.

Barium chloride solution:
Dissolve 2.5 g $BaCl_2 \cdot 2\ H_2O$ in distilled water and dilute to 1000 ml.

Hydrochloric acid (1.05 g/ml):
Dilute 75 ml hydrochloric acid (1.19 g/ml) with 200 ml distilled water.

Carbon dioxide gas

Porcelain crucible

Paper filters

**Determination**

Depending on the expected concentration of sulphite ions indicated by a preliminary test, 300 to 500 ml of the water sample is distilled after addition of 20 ml phosphoric acid (1.15 g/ml). For this purpose the entire distillation apparatus must be filled with carbon dioxide before the sample is introduced, so as to prevent oxidation of the sulphur dioxide by atmospheric oxygen. In the receiver the end of the condenser dips into the iodine/potassium iodide solution, which must be present to excess.

Once distillation is completed, 1 ml of hydrochloric acid (1.05 g/ml) is added to the iodine/potassium iodide solution in the receiver and the sulphate formed is precipitated out at boiling heat with an excess of hot barium chloride solution (the barium chloride solution is added in one go). After standing overnight it is filtered through a fine-grained filter and washed out with hot distilled water. The filter with the barium sulphate is dried and then incinerated in the crucible, ignited at 800 °C until it reaches constant weight, and the weight recorded.

**Calculation**

The factor for converting from barium sulphate to $SO_3^{2-}$ is 0.3430. The sulphite concentration of the sample can thus be calculated by means of the formula:

$$\frac{g \cdot 343}{v} = G$$

where:

g = weighed-out quantity in mg
v = volume of water used in ml
G = $SO_3^{2-}$ concentration in water sample in mg/l

### 3.2.6.3 Polarographic determination of sulphite (including sulphide and thiosulphate)

Equipment

Polarograph with polarography stand

Sodium hydroxide solution, 2 mol/l (2m)

Acetic acid, 2 mol/l (2m)

Sulphite standard:
1.5742 g $Na_2SO_3$ diluted to 1 litre with distilled water. The water used must be oxygen-free.

Sulphide standard:
Dissolve 7.4901 g $Na_2S \cdot 9\ H_2O$ in distilled water, add 200 ml sodium hydroxide solution and dilute to 1 litre.

Thiosulphate standard:
2.2133 g $Na_2S_2O_3 \cdot 5\ H_2O$ diluted to 1 litre with distilled water.

Analysis

A) If sulphide is not present:

Deaerate 19 ml of distilled water and 1 ml of sodium hydroxide solution with nitrogen for 5 min. in the polarograph vessel. Depending on the sulphite concentration, add 1 - 10 ml of sample, mix with nitrogen, and polarograph.

B) If sulphide is present or if the mixture contains all three components:

If only $SO_3^{2-}$ and $S_2O_3^{2-}$ are present, mix 20 ml distilled water with 1 ml sodium hydroxide solution and 2 ml acetic acid, deaerate for 5 min. with $N_2$, add 1 - 10 ml of the sample, and polarograph.

If sulphide also has to be determined, first perform the analysis described under A). Then add 2 ml acetic acid to the solution, mix for 3 min. with nitrogen, and polarograph again.

Working conditions

Method: Differential pulse polarography

$U_{start}$: - 0.2 V for sulphide
0 V for sulphite and thiosulphate

$\Delta U$ : - 2 V

Determination of concentration: standard addition method

The following reduction peaks are obtained:

- 0.72 V for sulphide
- 0.14 to - 0.18 V for thiosulphate
- 0.58 V for sulphite.

## 3.2.7 Sulphate

### 3.2.7.1 Gravimetric Determination as Barium Sulphate

In order to eliminate any disturbing silicates, the water sample is evaporated to dryness after acidification with 5 ml hydrochloric acid (1.125 g/ml). The residue is heated to boiling point, after adding 5 ml of the same hydrochloric acid, diluted with 50 ml distilled water and the mixture filtered while hot. The filter residue is rinsed out with hydrochloric acid diluted 1 : 50.

In the event of sulphites being present, the water sample is not acidified until the quantity of 0.1 m or 0.01 m iodine solution detected in determining sulphite (see sulphite) has been added. The value equivalent to the sulphite content must then be subtracted from the sulphate content.

Equipment

Analytical balance

Quartz dish

Platinum or porcelain crucible

Porcelain filtering crucible A 1 or round filter, diameter 5 cm

Hydrochloric acid (1.125 g/ml)

Silver nitrate solution, approximately 0.1 m

Barium chloride solution:
10 g $BaCl_2 \cdot 2 H_2O$ of reagent purity is dissolved in 90 ml distilled water. 1 ml of this solution precipitates approximately 40 mg $SO_4^{2-}$.

Method

The volume of the water sample to be employed for gravimetric determination of sulphate must be measured so that it contains between 10 and 500 mg sulphate ions. Where necessary, a larger volume of water must be reduced to 400 ml, and/or in the case of lower concentrations of sulphate ions to 100 ml. Hydrochloric acid is added to this volume until any sediment possibly present or any salt precipitate (calcium sulphate) is redissolved. In principle, the entire contents of a sampling should be used for an analysis. Care should be taken that the hydrochloric acid employed for acidification and/or for dissolving residue and salt precipitates does not comprise more than 1 ml per 100 ml of the reduced solution. Any precipitates not dissolving (e.g silicic acid) or other substances remaining undissolved are filtered off before precipitation of sulphate.

The solution prepared for precipitation is heated to boiling point and an excess of hot barium chloride solution added at boiling heat (the barium chloride solution is poured in all at once). After completion of precipitation, the sample is allowed to continue to boil for a further 30 minutes and then covered with a watch glass overnight. On the next day, the barium sulphate precipitate is filtered off, either through a paper filter (possibly with filtering slime), or through a filtering crucible A 1 which has been baked at 800 °C and weighed. The precipitate is washed with hot water until a negative chloride reaction is detected in the filtrate. If a paper filter was used, this is ashed in a platinum or porcelain crucible which has been baked and weighed. The crucible is baked for some 15 minutes at 800 °C and weighed after cooling. When using a filtering crucible, this is dried with the precipitate for 2 hours at 130 °C and then also baked in a muffle furnace for 15 minutes at 800 °C and weighed after cooling. 1 mg barium sulphate corresponds to 0.4115 mg $SO_4^{2-}$. The sulphate content of the water analyzed is calculated from the quantity of barium sulphate weighed, according to the following formula:

$$\text{mg/l } SO_4^{2-} = \frac{\text{mg barium sulphate} \cdot 411.5}{\text{Quantity of water used in ml}}$$

### 3.2.7.2 Nephelometric Determination of Sulphate

**General remarks**

The rapid method described below for determining sulphate in water samples follows a modification by W. Regnet and P. Udluft.

The sulphate dissolved in a water sample is precipitated as $BaSO_4$ and the resulting turbidity determined after 45 minutes with a spectrophotometer at 490 nm. A barium chloride seeding solution containing barium sulphate crystals is employed for this process of precipitation. In each case, a calibrating curve is also produced at the same time as the determining process. The measuring range for sulphate lies between 1 and 45 mg/l when using 1 cm cuvettes. In the case of higher concentrations of sulphate, the water samples should be appropriately diluted beforehand with distilled water:

**Equipment**

Spectrophotometer with 1 cm cuvettes, 490 nm

Test tubes

pH paper

Distilled water

Sodium sulphate stock solution:
Sodium sulphate is dried at 200 to 300 °C; 7.39 g is dissolved in 1 litre of distilled water; corresponds to 5 g sulphate ions/l.

Calibrating solution:
10 ml of the stock sulphate solution is diluted to 1 litre with distilled water; corresponds to 50 mg sulphate ions/l.

Barium chloride seeding solution:

2 ml of the calibrating solution is diluted with 18 ml distilled water and acidified with several drops of 2 m HCl. Approximately 500 mg $BaCl_2$ of reagent purity is added to this solution.

Analysis

9 ml of the water sample to be analyzed is placed in a test tube and acidified with 2 m HCl until the pH is less than 3.

Different quantities of the calibrating solution between 1 and 9 ml are similarly pipetted into test tubes. In the case of quantities below 9 ml, these are topped up to 9 ml in each case with distilled water, and all solutions are acidified with 2 m hydrochloric acid. Then, 1 ml of the barium chloride seeding solution is added to each of the calibrating solutions and the solutions to be analyzed, shaken, and after the solutions have been allowed to stand for a minimum period of 30 minutes and a maximum period of 45 minutes their extinctions are measured against distilled water at 490 nm, using 1 cm or 4 cm cuvettes.

If the water displays any self-colour or turbidity, measurements should be made against the coloured or filtered water sample. This method of determination is easily reproducible, and, with appropriate variations, allows determination of sulphate in concentrations ranging from 1 mg/l when using a 1 cm cuvette, and from 0.2 mg/l when using a 4 cm cuvette, up to 50 mg sulphate/l. The process has proved in practice to be a good, reliable and fast method of measuring sulphate.

#### 3.2.7.3 Ion chromatography (see Section 3.2.11)

### 3.2.8 Phosphate

General remarks

Waters and wastewaters may contain orthophosphates, condensed phosphates (poly- and metaphosphates) and organic phosphorus compounds. Under natural conditions free from human influence, concentrations of phosphorus compounds, calculated as hydrogen phosphate ions, are generally less than 0.1 mg/l. Higher concentrations in ground water and surface waters (rivers and lakes) are very frequently due to the effects of civilization. They play a major part in the eutrophication of standing water.

The pH of the water determines whether the dissolved orthophosphate is present as $PO_4^{3-}$, $HPO_4^{2-}$, $H_2PO_4^-$ or $H_3PO_4$, as illustrated by Fig. 91.

The water samples may be taken in plastic bottles or glass receptacles, but on no account must these have been treated with cleaning agents containing phosphate (not even if they have been thoroughly rinsed!).

The ion $PO_4^{3-}$ does not occur in natural waters. In the pH range of groundwater and surface waters, it is almost always mixtures of $HPO_4^{2-}$ and $H_2PO_4^-$ that are encountered. In conventional water analysis, therefore, the results of the determination of orthophosphate should be calculated in terms of hydrogen phosphate ions. When assessing the nutrient situation in a water or wastewater, it is desirable to differentiate and determine separately the various groups of phosphorus compounds as well as the total phosphate. The following methods are described below:

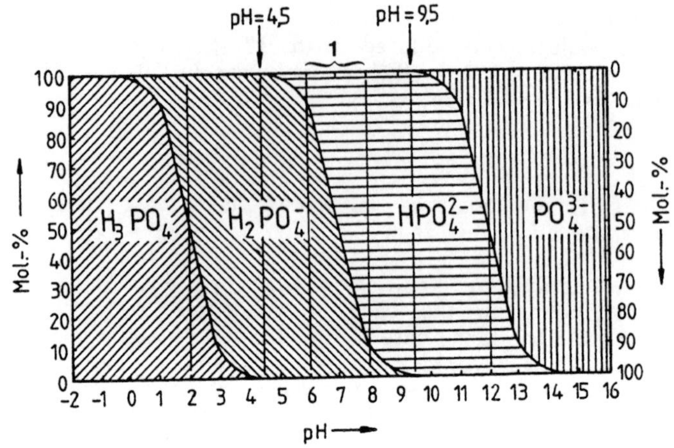


Fig. 91. Ion concentrations as a function of pH in a phosphoric acid/phosphate solution (after Hollemann-Wiberg); 1) = Buffer effect

3.2.8.1 Orthophosphate (by spectrophotometry)

3.2.8.2 Sum of orthophosphate and hydrolyzable phosphorus compounds

3.2.8.3 Total phosphate

3.2.8.4 Phosphate determination using ion chromatography (see Section 3.2.11)

3.2.8.5 ICP-AES (see Section 3.3.12)

### 3.2.8.1 Orthophosphate (calculated as hydrogen phosphate)

#### General remarks

Hydrolyzable phosphates and organic phosphorus compounds are not detected in the course of the determination of hydrogen phosphate ions under the conditions described here (no acid hydrolysis). In order to ensure complete differentiation of the individual phosphate and phosphorus compounds in the water sample, determination of the orthophosphate must be conducted at the latest 3 hours after taking the sample. Longer storage times may lead to shifts in the ratios between the individual forms of bonds as a result of precipitation or gradual hydrolysis.

#### Spectrophotometric determination as a molybdophosphate complex (phosphorus molybdenum blue)

Hydrogen phosphate ions react with ammonium molybdate to form ammonium molybdophosphate:

$$(NH_4)_3\ PO_4 \cdot 12\ MoO_3 \quad \text{or}$$

$$(NH_4)_3[P(Mo_{12}O_{40})] \quad \text{or}$$

$$(NH_4)_3\ [P(Mo_3O_{10})_4]$$

(The three formulae are the various forms of notation which are used for ammonium molybdophosphate in chemical literature).

The ammonium molybdophosphate is extracted with a benzene/isobutyl alcohol mixture and reduced in the organic phase with acid tin (II) chloride solution to intensely blue-coloured phosphorus molybdenum blue:

$$[P(Mo_3O_{10})_4]^{3-} + 11\ H^+ + 4\ Sn^{2+} \longrightarrow$$

$$(MoO_2 \cdot 4\ MoO_3)_2 \cdot H_3PO_4 + 2\ MoO_2 + 4\ Sn^{4+} + 4\ H_2O$$

This compound is measured photometrically at 625 nm. Its blue colour can probably be traced to the mixed oxide formation of the tetravalent and hexavalent molybdenum (Holleman-Wiberg).

The method is suitable for the determination of concentrations of hydrogen phosphate ions between 0.01 and 1.0 mg/l in water. The procedure is described in two variants. Variant A is suitable for the determination of 0.2 - 1.0 mg/l and Variant B for the determination of less than 0.2 mg/l of hydrogen phosphate ions. If a sufficiently sensitive photometer is used, together with 5-cm cuvettes, as little as 0.005 mg/l of $HPO_4^{2-}$ can be detected. If concentrations are higher than 1 mg/l the water sample should be diluted.

Extraction with the benzene/isobutyl alcohol mixture eliminates most of the disturbing factors which might otherwise affect spectrophotometric phosphate determination.

**Equipment**

Spectrophotometer

Cuvettes, path length 1 cm and 5 cm

Pipette with suction apparatus, volume 25 ml

Separating funnels, volume approx. 250 ml and 500 ml

Measuring flask, 50 ml

Benzene/isobutyl alcohol mixture:
1 + 1 (v/v)

Ammonium molybdate solution:
Dissolve 10 g of reagent-purity $(NH_4)_6Mo_7O_{24} \cdot 4\ H_2O$ in 100 ml of dist. water. Mix this solution with reagent-purity sulphuric acid (1.22 g/ml) in the ratio 1 : 1.

Methanolic sulphuric acid:
Dissolve 20 ml of reagent-purity sulphuric acid (1.84 g/ml) in 980 ml of methanol.

Tin (II) chloride solution, conc.:
Dissolve 10 g of reagent-purity $SnCl_2 \cdot 2\ H_2O$ in 25 ml of hydrochloric acid (1.15 g/ml). The solution may be kept for a maximum of 3 weeks.

Tin (II) chloride solution, dilute:
Add 0.5 ml of conc. tin (II) chloride solution to 100 ml of 0.5 m

sulphuric acid by pipette. This solution should be prepared freshly before each determination.

Hydrogen phosphate reference solution I:
Dissolve 0.1855 g of disodium hydrogen phosphate ($Na_2HPO_4 \cdot 2\ H_2O$) reagent purity in dist. water and make up to 1 litre. 1 ml of this solution contains 0.1 mg of hydrogen phosphate ions.

Hydrogen phosphate reference solution II:
Dilute hydrogen phosphate reference solution I with dist. water in the ratio 1 : 10. 1 ml of this solution contains 0.01 mg of hydrogen phosphate ions. It must be prepared freshly each time the calibration curve is checked.

### Method

The specified procedure must be precisely followed in all aspects. This applies in particular to the photometric measurement. The extinction is highly time-dependent; consequently, the samples should be allowed to stand for a predetermined period no shorter than 10 minutes and no longer than 30 minutes, and this period should be the same for each measurement. Only in this way can reproducible values be obtained.

### Calibration

Take volumes of hydrogen phosphate reference solution II containing increasing quantities of $HPO_4^{2-}$, between 0.01 and 0.2 mg, make up to 200 ml with distilled water and subject to the analysis process described under "Determination, Variant B". In the case of water samples with more than 0.2 mg/l of $HPO_4^{2-}$, take appropriate quantities of hydrogen phosphate reference solution II with increasing proportions of hydrogen phosphate ions, make up to 40 ml with distilled water and continue as described under "Determination, Variant A".

### Determination, variant A

Transfer 40 ml of the water sample (or of the diluted sample if the hydrogen phosphate content of the water sample is greater than 1 mg/l), 40 ml of distilled water (as a blank test) and 40 ml of each of the calibration solutions into 250-ml separating funnels, add 50 ml of benzene/isobutyl alcohol mixture and 25 ml of ammonium molybdate solution, and shake vigorously for 15 seconds. Each sample or calibration solution must be shaken immediately after the addition of the ammonium molybdate solution.

After separation of the layers, allow the aqueous layer to run off, withdraw 25 ml of the organic phase using a pipette and suction apparatus, and transfer to a 50-ml measuring flask. Add to this, one after the other, mixing well each time, 15 ml of methanolic sulphuric acid, 1 ml of dilute tin (II) chloride solution, and methanolic sulphuric acid once again to make up to the mark. After 10 to 30 minutes carry out photometric measurement against the reagent blank test at 625 nm (or using an appropriate filter).

### Determination, variant B

Measure off 200 ml of the water sample, 200 ml of distilled water (as a blank test) and 200 ml of each of the calibration solutions into 500 ml separating funnels, add 17 ml of isobutyl alcohol to each (in order to saturate the water with isobutyl alcohol) and shake briefly. Now add 50 ml of benzene/isobutyl alcohol mixture and subsequently 50 ml of ammonium molybdate solution and shake vigorously for 60 seconds. Each sample and calibration batch should be shaken immediately after the addition of the ammonium molybdate solution. Continue with the procedure as described under "Determination, Variant A", second paragraph. Conduct the photometric measurement at as great a path length as possible (e.g. 100-mm cuvettes).

Calculation

Read off the hydrogen phosphate content from the calibration curve on the basis of the measured extinction, take into account the volume of water used and convert to 1 litre. It should be specified whether the analysis was carried out on the unfiltered or the filtered water sample.

If the result is to be given in mg of $PO_4^{3-}$/l, mg of $P_2O_5$/l or mg of P/l, when preparing a phosphorus balance or for some other reason, the following factors are used for conversion:

$$\text{mg } PO_4^{3-}/l = \text{mg } HPO_4^{2-}/l \cdot 0.990$$

$$\text{mg } P_2O_5/l = \text{mg } HPO_4^{2-}/l \cdot 0.7394$$

$$\text{mg } P/l = \text{mg } HPO_4^{2-}/l \cdot 0.3227$$

Note:

A method of phosphate determination described in the German Standard Methods and in DIN 38405 similarly uses molybdate ions, in the presence of antimony ions, to form a complex which is reduced to phosphorus molybdenum blue by ascorbic acid and can be measured photometrically. Experience with this method has so far been good.

### 3.2.8.2 Sum of orthophosphate and hydrolyzable phosphorus compounds

General remarks

Separate quantitative determination of orthophosphate, hydrolyzable phosphates and organic phosphorus compounds is not possible with the means offered by conventional water analysis if these substances are present together in dissolved form, since in the course of hydrolysis not only orthophosphates but also organic phosphate compounds are partially codetected.

If, however, one can be certain that organic phosphates are not present it is possible to determine the inorganic phosphates by hydrolysis with subsequent determination as hydrogen phosphate.

Two groups of hydrolyzable phosphates may be distinguished: the (linear) polyphosphates with chain-like anions and the (cyclic) metaphosphates with ring-shaped anions.

When dealing with problems of water treatment and corrosion protection, the determination of hydrolyzable phosphates is particularly important. As cathodic inhibitors they possess the property of preventing corrosion of the piping in water supply systems by aggressive water. This is due to the formation of a thin protective layer of calcium iron phosphate on the inner walls of the pipes. As a result of their further property of masking the hardening constituents of the water in complex compounds they prevent the precipitation of calcium salts and excessive sedimentation of hardening constituents in the pipe system and in water tanks, heating coils and hot-water boilers.

## Spectrophotometric determination after hydrolysis to hydrogen phosphate ions

The hydrolyzable phosphates are converted to hydrogen phosphate by acid hydrolysis and the hydrogen phosphate is determined by spectrophotometry as a molybdophosphate complex. Since the orthophosphate which was originally present is codetermined in the course of this method, the content of hydrogen phosphate ions determined without hydrolysis must be subtracted from the result of the procedure with hydrolysis.

The method is suitable for the determination of between 0.01 and 1.0 mg/l of hydrogen phosphate ions (sum of hydrogen phosphate ions, both originally present in the water sample and arising by hydrolysis). In the event of higher concentrations the sample must be diluted. If 50-mm cuvettes are used concentrations down to 0.005 mg of $HPO_4^{2-}$/l can be determined.

Should it be required that (linear) polyphosphates and (cyclic) metaphosphates are differentiated or the hydrolytic breakdown of phosphates of higher molecular weight should be observed, separation by thin-layer chromatography is possible.

Any organic phosphorus compounds present in the water sample are partially codetermined in the course of hydrolysis. It is, therefore, impossible to provide absolutely reliable quantitative results with regard to inorganic hydrolyzable phosphates.

### Equipment

In addition to the chemicals listed under "Equipment" in the section "Hydrogen phosphate ions", the following reagents are required:

Phenolphthalein solution:
Dissolve 5 g of phenolphthalein in 95% ethanol. Add 500 ml of dist. water and treat the solution with 0.02 m sodium hydroxide solution until a weak red coloration appears.

Acid mixture:
Slowly add 300 ml of reagent-purity conc. sulphuric acid to 600 ml of dist. water. When cool add 4 ml of reagent-purity conc. nitric acid and then dilute with dist. water to 1000 ml.

Sodium hydroxide solution, 1 m:
Dissolve 40 g of reagent-purity sodium hydroxide in dist. water and make up to 1000 ml.

Method

Determination

Determination should be carried out as soon as possible after taking the sample. Otherwise, reactions may occur which distort the original content of hydrolyzable phosphates.

Take 100 ml of the sample, or a volume which has been diluted to 100 ml with distilled water (conduct a preliminary test), and treat with one drop of phenolphthalein solution. If a red coloration appears add acid mixture until the colour disappears and a further 1 ml of acid mixture in excess.

Heat the solution prepared in this way to boiling temperature for at least 90 minutes. The volume of the solution should be maintained between 25 and 50 ml. This can be achieved by mounting a cold finger and occasionally replenishing with distilled water.

When cool, add sodium hydroxide solution until a weak pink coloration appears. Then make up to 100 ml with distilled water. To determine the content of hydrogen phosphate ions in the solution thus obtained, proceed as described in the section "Hydrogen phosphate ions".

Calculation

The result should be given as the content of hydrogen phosphate ions determined after hydrolysis of the phosphates, converted accordingly for 1 litre of the water sample. This result must then be reduced by the content of hydrogen phosphate ions which has been established directly and without hydrolysis in the determination carried out in parallel.

### 3.2.8.3 Total phosphorus

General remarks

The determination of all groups of phosphates and phosphorus compounds is useful for the nutrient balance of a water since not only orthophosphate (hydrogen phosphate) but also hydrolyzable phosphates and organic phosphorus compounds can be utilized by organisms.

It is also appropriate to determine the total phosphorus content if reliable differentiation of individual phosphate species is no longer possible as a result of difficult conditions when taking or transporting the sample or because of overlong storage of the water sample before analysis is begun.

If determination of hydrogen phosphate and of total phosphorus is conducted in parallel, the difference between the two results provides a measure of the sum of the hydrolyzable phosphates and the organic phosphorus compounds.

**Spectrophotometric determination following decomposition by acid and total hydrolysis to hydrogen phosphate ions**

Organic phosphorus compounds are converted to hydrogen phosphate ions by wet-ashing with sulphuric acid and hydrogen peroxide. The hydrolyzable phosphates are also quantitatively included in this decomposition. Determination is then continued as a molybdophosphate complex by spectrophotometry, as described in the section "Hydrogen phosphate ions".

If the content exceeds 1 mg/l, calculated as $HPO_4^{2-}$, the water sample must be diluted.

Since it is possible that certain types of glass may release phosphate into the sample solution under the prevailing decomposition conditions, increased decomposition blank results must be expected. It is therefore necessary to determine a number of decomposition blank test values in parallel with every analysis series. In the case of low phosphorus contents it is advisable to conduct the decomposition in quartz flasks.

### Equipment

In addition to the chemicals listed under "Equipment" in the section "Hydrogen phosphate ions", the following reagents are required:

Sulphuric acid:
Dilute 500 g of conc. sulphuric acid (1.84 g/ml) to 1 litre in dist. water.

Ammonium hydroxide solution:
Mix ammonium hydroxide (0.907 g/ml) with dist. water in the ratio 1 : 1.

Sodium hydroxide solution:
Dissolve 200 g of reagent-purity NaOH in dist. water and make up to 1000 ml.

p-nitrophenol solution:
Dissolve 0.2 g of reagent-purity p-nitrophenol in 100 ml of dist. water.

Hydrogen peroxide, approx. 30 %.

Potassium permanganate solution, approx. 0.1 m

### Method

Conduct calibration, determination and calculation as described in the section "Hydrogen phosphate ions". Decomposition by acid (wet-ashing) should be conducted either in an unfiltered sample which has been homogenized by shaking or in a filtered water sample. This pretreatment should be stated with the results.

### Decomposition by acid

Measure an appropriate volume (preliminary test) of the water sample into a narrow-mouth Erlenmeyer flask, treat with 4 ml of sulphuric acid and 1 ml of hydrogen peroxide (30 %) and heat to between 160 and 180 °C over a sand bath under a fume hood.

Following evaporation of the water (appearance of white sulphuric acid vapours) heat for at least a further 2 hours. Once again add 1 ml of hydrogen peroxide and heat to between 160 and 180 °C for a further 2 hours. Subsequently add 50 ml of distilled water for the hydrolysis of condensed phosphates which are present or have formed and leave to stand on the hot sand bath for a further 1 hour.

When the solution is cool, add 1 - 2 drops of p-nitrophenol solution and treat with ammonium hydroxide (1 + 1) or sodium hydroxide solution until the colour changes to yellow. In order to avoid interference due to hydroxide precipitation, immediately add one drop of sulphuric acid (decoloration of the indicator) and cool to at least 25 °C.

In order to destroy $H_2O_2$ residues, treat with drops of 0.1 m potassium permanganate solution in excess. Make up to 100 ml and continue as described in the section "Hydrogen phosphate ions".

### 3.2.8.4 **Phosphate determination using ion chromatography** (see Section 3.2.11)

### 3.2.8.5 ICP-AES (see Section 3.3.12)

## 3.2.9 **Carbonic acid, hydrogen carbonate and carbonate** (see also Chapter 1 and Section 3.6)

The concentration of dissolved carbon dioxide, hydrogen carbonate ions and carbonate ions is particularly important as far as the properties and evaluation of a water are concerned, and hence for water analysis in general. In aqueous solution, the chemical species

$$CO_2 \quad H_2CO_3 \quad HCO_3^- \quad CO_3^{2-}$$

are linked together by a system of equilibrium relations which result from the dissociation equilibrium of the carbonic acid and the ion product of the water (pH value).

The sum of $CO_2$, $H_2CO_3$, $HCO_3^-$ and $CO_3^{2-}$ is designated **"total carbon dioxide"**. It represents the content of inorganic carbon in the water. The content of **"free carbon dioxide"** consists of the quantity of $CO_2$ dissolved in the water plus the portion of the $CO_2$ which reacts with water to form $H_2CO_3$. However, the portion of the undissociated carbonic acid - $H_2CO_3$ - can virtually be ignored, since less than 1 % of the dissolved carbon dioxide reacts with water to $H_2CO_3$ (which then dissociates into the corresponding ions). The "free carbon dioxide" is then approximately equal to the "dissolved carbon dioxide".

According to the definition given here, the total carbon dioxide content is compiled from the sum of "free carbon dioxide", "hydrogen carbonate carbon dioxide" and "carbonate carbon dioxide" ($S_c$, calculated as $CO_2$).

$$S_c = c_{CO_2} + c_{HCO_3^-} + c_{CO_3^{2-}}$$

Given that all the forms of carbon dioxide listed here are linked by equilibrium relationships via the pH value, the concentration of these compounds in a water can be calculated if $S_c$ is determined and the pH is measured (for further details, see under "Total carbon dioxide").

The method preferred in practice for determining the content of "free carbon dioxide", hydrogen carbonate ions and carbonate ions, and thereby the total carbon dioxide content, is to titrate the water sample acidimetrically. Carbonic acid is a dibasic acid; it dissociates in two stages. The interrelationship which forms the basis of the acidimetric titration

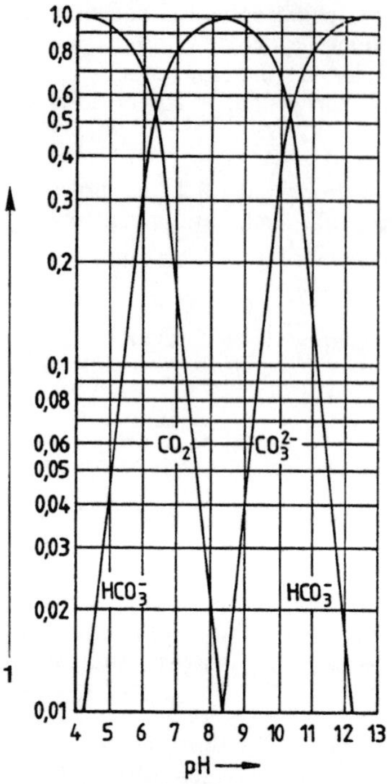


Fig. 92. Proportions of $CO_2$, $HCO_3^-$ and $CO_3^{2-}$ at 25 °C as a function of the pH of the solution (according to K. Haberer); 1) = Proportions of the forms of carbon dioxide in total carbon dioxide

can be ascertained relatively easily from Figs. 92 and 93; detailed reference to the mathematical derivation given by L. Fresenius and O. Fuchs is not necessary here.

It can be seen from Fig. 92 that at pH 6.4 the proportions of hydrogen carbonate ions and free dissolved carbon dioxide, dependent on the pH value, are equal (the point of intersection of the curves); carbonate ions are not present in the water at pH 6.4. At pH 10.3, the proportions of hydrogen carbonate ions and carbonate ions present in the solution are equal, whereas the proportion of free dissolved carbon dioxide approaches zero from pH 8.35 onwards.

Fig. 93 shows clearly that in the vicinity of these two pH values (approximately between 5.0 and 7.8 and between 9.0 and 11.6) there are buffer zones in which only a gradual change in pH is produced by a relatively large addition of the acid or base. Accordingly, sudden changes in pH occur between pH 4.0 and 4.8 and between 8.0 and 8.6.

If it is assumed that the carbonic acid is the only weak acid present in the water sample (this is not always the case, but is very frequently so) and that, correspondingly, its anions are the only weak bases, the water sample is titrated to these two pH jumps as the equivalence points for carbonate ions (pH 8.3; cf. Fig. 93) and hydrogen carbonate ions (pH 4.3). The results achieved are designated in accordance with the indicators originally used: **"p value"** (indicator phenolphthalein at pH 8.3) and **"m value"** (indicator methyl orange pH 4.3). Whether an acid or a base is used for titration to the change in colour depends on the pH of the water sample.

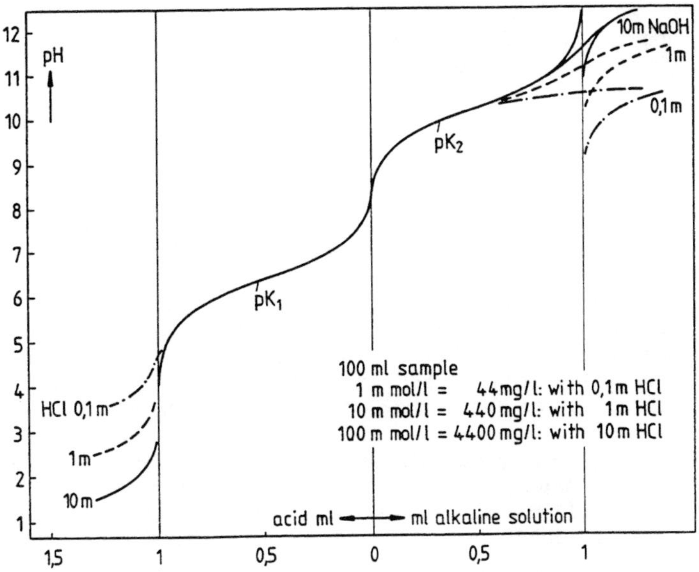


Fig. 93. Titration of hydrogen carbonate solutions (according to A. Grohmann)

### 3.2.9.1 Acid consumption (alkalinity)

#### General remarks

Under certain conditions, there is a stoichiometric relationship between the p and m values of a water and the concentration of hydrogen carbonate ions, carbonate ions and hydroxyl ions. For this reason, the p and m values may be used to determine the concentration of these ions indirectly and also to calculate the so-called carbonate hardness.

Although these conditions apply in many cases, particularly with natural waters, it remains in the first instance an arbitrary assumption. Strictly speaking, therefore, it is more correct in the first instance to specify the results of titration achievable using this method (the determination of the p and m values) without comment as acid consumption. If a base has been used for titration, the result is given as **base consumption** (p and m values then have a negative sign). In more recent publications, there has been much discussion about the detection, calculation and significance of the various carbon dioxide compounds in water. Experience shows that in many water laboratories in a large number of countries around the world, the classical methods tend still to be in use today; consequently, these methods of detecting the carbon dioxide compounds are dealt with in some detail below.

Note also the remarks in Chapter 1 on local investigations, which are repeated in part here.

The acid consumption (alkalinity) of a water is defined as the quantity in mmol/l of a strong acid which is consumed during titration to achieve certain pH values or until certain indicators change colour. If titration is carried out electrometrically up to pH 8.3 or if phenolphthalein is used as an indicator, the resulting acid consumption is the p value. If titration is carried out electrometrically to pH 4.3 or if methyl orange or a mixed indicator is used, the resulting acid consumption is the m value.

### 3.2.9.2 Acid consumption and base consumption (p and m value)

The p value (indicator phenolphthalein, pH 8.3) and the m value (indicator methyl orange, pH 4.3) are primarily measures of **acid consumption** (alkalinity) or **base consumption** (acidity); a positive sign indicates acid consumption and a negative sign base consumption. Acid and base consumption are characteristics which depend on the substances dissolved in a water. For this reason, the result should initially be given as a p or m value (in mmol/l) without further interpretation - as has already been mentioned. Only then should investigations be made as to what conclusions may be drawn with regard to the components of the water.

Since, however, carbonic acid is by far the most frequently occurring weak acid in natural waters, in **practical water analysis** it is assumed that the results of the p and m value determination refer to the anions of the carbonic acid if certain conditions are fulfilled.

The pH limit values used, by definition, in the determination of the p and m values approximate (only) roughly to the limit values of the phase diagram of the carbonic acid salts (cf. Fig. 92). Nevertheless, the proportional concentration of hydroxyl ions, carbonate ions and hydrogen carbonate ions may be estimated in this way; this is generally sufficient for water analysis in practice.

A positive p value can only be obtained if the pH of the water sample is above 8.3. If this is the case, initially the hydroxide is neutralized:

$$OH^- + H^+ \longrightarrow H_2O \qquad (1)$$

As the pH falls, the reaction of the carbonate ions with the acid follows:

$$CO_3^{2-} + H^+ \longrightarrow HCO_3^- \qquad (2)$$

Titration with electrometrical display at pH 8.3 detects the sum of $OH^-$ + $CO_3^{2-}$, if both ions are present in the water.

If titration is continued to pH 4.3, below pH 8.3 the hydrogen carbonate ions are also detected:

$$HCO_3^- + H^+ \longrightarrow H_2O + CO_2 \qquad (3)$$

It must be borne in mind, however, that not only the hydrogen carbonate ions present in the water originally are detected, but also those which were formed from $CO_3^{2-}$ during titration to the end point pH 8.3 according to equation (2). During the second titration, then, the carbonate ion is (indirectly) neutralized a second time, i.e. at the second equivalence point.

If other acid-consuming substances are contained in the water under test apart from carbon dioxide compounds, the results of the calculation will be incorrect. But even if no other such substances are present, the error will only be less than 10 % if the p or m values (consumption of an HCl standard solution) are greater than 1 mmol/l.

### 3.2.9.3 Acidimetric determination of carbonate ions and hydrogen carbonate ions (p and m values)

#### General remarks

Carbonic acid is a dibasic acid. Depending on the pH of the solution of its salts, either just hydrogen carbonate ions are present or a mixture of hydrogen carbonate ions and carbonate ions (see Fig. 92).

Since there are few specific test reactions for both anions, their concentration is calculated from the acid consumption under certain ion conditions. In practice, the p and m values are used for this.

For most practical purposes, it is sufficient to assume that pH 8.3 (p value) represents the limit for the presence (and hence the equivalence point) of carbonate ions, and pH 4.3 (m value) the limit for the presence (and hence the equivalence point) of hydrogen carbonate ions.

The following qualifications apply: if the pH of the water sample is between 4.3 and 8.3, the acid consumption (in mmol/l) at the methyl orange end point (pH 4.3; m value) corresponds to the hydrogen carbonate ion content. Carbonate ions cannot be present in a water sample of this type.

If the pH of the water sample is between 8.3 and 9.5, the acid consumption at pH 4.3 (m value) minus twice the acid consumption at pH 8.3 (p value) provides the hydrogen carbonate content (the p value must be doubled because after $CO_3^{2-} + H^+ \rightarrow HCO_3^-$ only 1/2 $CO_3^{2-}$ has been detected).

The carbonate content of a water sample can be determined from the p value (i.e. = 2 p in mmol/l) only if the pH of the sample is 8.3 or higher (up to 9.5).

If the pH of the water sample is above 9.5, m - 2p + $c_{OH^-}$ holds for the hydrogen carbonate content and 2p - $c_{OH^-}$ for the carbonate content.

If the water sample is turbid or coloured, the acid consumption can only be determined electrometrically.

The analysis of waters containing iron and those which contain calcium hydrogen carbonate (in particular mineral springs) should be conducted at the sampling point. Otherwise the $HCO_3^-$ content of waters containing iron may be altered by the precipitation of iron hydroxide. A similar error occurs when calcium carbonate precipitates in calcium hydrogen carbonate water during transport or storage. Waters with a pH of 6 and below, standing in open containers, may lose free $CO_2$ and therefore also $HCO_3^-$, since $CO_2$ and $HCO_3^-$ are linked by the following equilibrium ratio:

$$\frac{c_{HCO_3^-} \cdot c_{H^+}}{c_{CO_2} + c_{H_2CO_3}}$$

#### Equipment

A) Electrometrical end point indication:

Burette, 10 ml, 0.02 ml divisions

pH meter (with temperature correction)

Measuring cell: glass and calomel electrodes (single-probe measuring cell)

Titration equipment, thermostatted, with magnetic stirrer (or electrometric titration equipment with end point preselection)

Hydrochloric acid, 0.1 m

B) End point indication with colour indicators:

Burette, 25 ml, 0.05 ml divisions

Hydrochloric acid, 0.1 m

Phenolphthalein indicator:
Dissolve 1 g phenolphthalein in 100 ml ethanol.

Methyl orange indicator:
Dissolve 0.5 g methyl orange in 100 ml distilled water

Mixed indicator, pH 4.4 - 4.3, according to Cooper and Mortimer:
Carefully mix and pulverize 20 mg methyl red and 100 mg bromocresol green in an agate mortar. Make a paste with a little 96 % ethanol and dissolve with a total of 100 ml of 96 % ethanol. Store in a brown dropping bottle.

## Method

**A) Determination of the p value (electrometric)**

Pipette 100 ml of the sample into the titration vessel. Insert the single-probe measuring cell and read off the pH. If this is above 8.3, titrate with 0.1 m hydrochloric acid until pH 8.3 is reached. The method employed is to mix with the magnetic stirrer for 5 seconds after each addition and then take the reading with the liquid at rest. The quantity, in ml, of 0.1 m hydrochloric acid consumed provides the p value in mmol/l. If the pH was below 8.3 before titration began, p = 0 (the water sample cannot contain carbonate ions).

**B) Determination of the p value (visual)**

Add 2 drops of phenolphthalein indicator to 100 ml of the sample. If a red coloration of the solution occurs (pH above 8.3), titrate with 0.1 m hydrochloric acid until the solution is colourless. The quantity, in ml, of 0.1 m hydrochloric acid consumed provides the p value in mmol/l. If no red coloration of the water sample occurs on addition of the indicator, p = 0 (see above).

**A2) Determination of the m value (electrometric)**

Titrate the sample which has been titrated to pH 8.3 (p value) with 0.1 m hydrochloric acid in similar fashion to pH 4.3. The total ml of 0.1 m hydrochloric acid consumed for both titration stages (to pH 8.3 and then further to pH 4.3) provides the m value in mmol/l.

## B2) Determination of the m value (visual)

Take the sample titrated visually to the colour change (phenolphthalein pH 8.3), add 2 drops of methyl orange indicator (or 3 drops of mixed indicator) and titrate with 0.1 m hydrochloric acid until the colour changes from yellow to brownish yellow (or from bluish green to red). The total ml of 0.1 m hydrochloric acid consumed during both titration stages (to pH 8.3 and then to pH 4.3) provides the m value in mmol/l.

When using the mixed indicator, it should be noted that a grey coloration occurs shortly before the end point is reached; this then turns to red as the next 1 - 2 drops of the acid are added.

### Calculation

The concentration of hydroxyl ions, carbonate ions and hydrogen carbonate ions can be calculated from the quantity of 0.1 m hydrochloric acid consumed (p ml = p value, m ml = m value). On this point, however, refer to the qualifications made above.

| Acid consumption (0.1-m HCl) | $OH^-$ Ion mmol/l | $CO_3^{2-}$ Ion 1/2 mmol/l | $HCO_3^-$ Ion mmol/l |
|---|---|---|---|
| p = 0 ; m > 0 | 0 | 0 | m |
| 2 p < m | 0 | 2 p | m - 2 p |
| 2 p = m | 0 | 2 p | 0 |
| 2 p > m > p | 2 p - m | 2(m - p) | 0 |
| p = m | p | 0 | 0 |

1 mmol hydroxyl ions = 17.01 mg $OH^-$

1 mmol carbonate ions (1/2) = 30.01 mg $CO_3^{2-}$

1 mmol hydrogen carbonate ions = 61.02 mg $HCO_3^-$

If using 100 ml of a water sample,
1 ml 0.1 m HCl = 17.01 mg/l $OH^-$
= 30.01 mg/l $CO_3^{2-}$
= 61.02 mg/l $HCO_3^-$.

Taking the pH of the water sample into account, the above calculation can be simplified as follows:

| pH of water sample | Hydrogen carbonate mmol/l) | Carbonate (mmol/l) 1/2 |
|---|---|---|
| 4.5 - 8.2 | m | 0 |
| 8.2 - 8.6 | m - 2 p | not calculable |
| 8.6 - 9.5 | m - 2 p | 2 p |
| above 9.5 | m - 2 p + $c_{OH^-}$ | 2 p - $c_{OH^-}$ |

### 3.2.9.4 Base consumption (acidity)

General remarks

The base consumption of a water is defined as that quantity of sodium hydroxide solution in mmol/l which is consumed during titration up to certain pH values or to the transition points of certain indicators. If titration is carried out electrometrically to pH 4.3 or methyl orange or a mixed indicator are used, the base consumption given is the negative m value. If titration is carried out electrometrically to pH 8.3 or a phenolphthalein indicator is used, base consumption is the negative p value.

The base consumption of natural waters is principally caused by the carbon dioxide dissolved in the water. It may also be brought about, however, by humic acids and other weak organic acids. In such instances, the pH values are not below pH 4.3. Mineral acids, on the other hand, may give rise to pH values below 4.3.

Negative m or p values

A) Determination of the negative m value (electrometric)

Pipette 100 ml of the sample into the titration vessel, which is equipped with a magnetic stirring rod. Insert the single-probe measuring cell and read off the pH. If it is below pH 4.3, titrate with 0.1 m sodium hydroxide solution until pH 4.3 is reached. Mix with the magnetic stirrer for 5 seconds after each addition; take the readings when the liquid is at rest. The quantity, in ml, of 0.1 m sodium hydroxide solution consumed indicates the negative m value in mmol/l. If the pH value was at or above 4.3 before titration began, the negative m value $-m = 0$.

B) Determination of the negative m value (visual)

Add 5 drops of methyl orange indicator (or mixed indicator) to 100 ml of the sample. If a red (violet) coloration occurs, titrate with 0.1 m sodium hydroxide solution until the colour turns to brownish yellow (green). The quantity, in ml, of 0.1 m sodium hydroxide solution consumed in the course of this process indicates the negative m value in mmol/l.

If a yellow coloration occurs instead of a reddish (or violet) coloration as the methyl orange (or mixed) indicator is added, then $-m = 0$.

A2) Determination of the negative p value (electrometric)

100 ml of the sample is prepared as for the determination of the negative m value. If the single-probe measuring cell indicates a pH below 8.3, titrate with 0.1 m sodium hydroxide solution until pH 8.3 is reached. Mix for 5 seconds with the magnetic stirrer after each addition; take the readings with the liquid at rest. The quantity, in ml, of 0.1 m sodium hydroxide solution consumed indicates the negative p value in mmol/l. If the pH was at 8.3 or above, the negative p value $-p = 0$.

B2) Determination of the negative p value (visual)

Add 5 drops of phenolphthalein indicator to 100 ml of the sample. If the water does not assume a reddish coloration, titrate with 0.1 m sodium hydroxide solution until the first permanent weak pink coloration appears. The quantity, in ml, of 0.1 m sodium hydroxide solution consumed in the course of this process indicates the negative p value in mmol/l.

If a reddish coloration appears as the phenolphthalein indicator is added, -p = 0.

### Calculation

The results are given in mmol/l.

Negative m value (-m) mmol/l $= a \cdot F$

Negative p value (-p) mmol/l $= b \cdot F$

a = Consumption of 0.1 m sodium hydroxide solution in ml for a 100 ml sample during titration to pH 4.3
b = Consumption of 0.1 m sodium hydroxide solution in ml for a 100 ml sample during titration to pH 8.3
F = Factor of the 0.1 m sodium hydroxide solution

## 3.2.10 Total carbon dioxide (see also 3.6, carbon dioxide)

### General remarks

The total carbon dioxide of a water consists of the so-called free dissolved carbon dioxide (3.6) and the carbon dioxide contained in hydrogen carbonates and carbonates (3.2.9):

$$CO_2\ (\text{total}) = CO_2 + H_2CO_3 + HCO_3^- + CO_3^{2-}\quad (\text{calculated as } CO_2)$$

Expressed as concentrations (c) and in simplified, condensed form:

$$S_C = c_{CO_2} + c_{HCO_3^-} + c_{CO_3^{2-}}\quad (S_C \text{ calculated as } CO_2)$$

The total carbon dioxide of a water ($S_C$) is also designated "the sum of the dissolved carbon dioxide and its compounds". The term "inorganic carbon ($S_{CO_2}$)" is also common.

Since all forms of carbon dioxide (free carbon dioxide, hydrogen carbonate ions and carbonate ions) are linked together by equilibrium relations (dissociation equilibrium of the carbonic acid, ion product of the water) via the pH, it is possible, if $S_C$ is known, to calculate the content of the individual compounds or ions given the pH of the water in question (see Section 3.2.9).

The determination of total carbon dioxide plays such an important role in $CO_2$ analysis of water because $S_C$ represents the only clearly determinable value; whenever individual components are determined, there is the danger that equilibria are shifted or other, e.g. organic anions are detected. ($HCO_3^-$ can also be detected by HPLC). Besides this, $S_C$ determination is also required in many investigations and calculations of the cycle of materials in water.

Quantitative determination



### 3.2.10.1 Calculation of $S_c$ (diss. $CO_2$) from the p and m value

In many cases, the acids and bases contained in a water other than the sum of the dissolved carbon dioxide $S_c$ are negligible. Assuming this to be the case, the acid consumption and the base consumption of the water is caused purely by the carbon dioxide and its anions. Under these conditions, $S_c$ can be calculated from the difference "m value - p value".

If the solution is a pure solution of carbon dioxide and its anions without foreign buffer substances, the following applies (A. Grohmann):

$$\text{m value} = 2\, c_{CO_3^{2-}} + c_{HCO_3^-} + c_{OH^-} - c_{H_3O^+}$$

$$\text{p value} = c_{CO_3^{2-}} - c_{CO_2} + c_{OH^-} - c_{H_3O^+}$$

In the pH range between 4.5 and 9.5 the following applies:

$$\text{m value} = 2\, c_{CO_3^{2-}} + c_{HCO_3^-}$$

$$\text{p value} = c_{CO_3^{2-}} - c_{CO_2}.$$

Hence it follows that:

Therefore, $S_c = m - p$

The method may be applied if the $S_c$ value is greater than 0.5 mmol/l.

Method

The p and m values of the water are to be determined according to the procedure described in Section 3.2.9.3, "Acidimetric determination of carbonate ions and hydrogen carbonate ions (p and m values)".

$S_c$ is calculated according to the equation

$$m - p = S_c$$

in which

m = m value of the water in mmol/l
p = p value of the water in mmol/l
$S_c$ = Total carbon dioxide content (= sum of the concentration of free, dissolved carbon dioxide, carbonate ions and hydrogen carbonate ions) in mmol/l, calculated as $CO_2$.

### 3.2.10.2 Volumetric determination after distillation

#### General remarks

On taking the sample, convert the free, dissolved carbon dioxide and hydrogen carbonates into carbonates by adding sodium hydroxide. In closed apparatus in the laboratory, add acid to the weakly alkaline sample to liberate carbon dioxide which is driven by a flow of inert gas ($CO_2$-free air) into a receiver containing sodium hydroxide solution. $CO_2$ is absorbed faster and more completely if n-butanol is added to the solution.

First neutralize the excess sodium hydroxide solution by adding acid until the transition point of the phenolphthalein/naphtholphthalein mixed indicator is reached (pH 8.6) as carbonate is converted into hydrogen carbonate. After adding the bromocresol green/methyl red mixed indicator (pH 4.5), titrate the hydrogen carbonates with 0.01 m sulphuric acid.

The method is suitable for determining low to medium concentrations (20 - 360 mg/l) of total carbon dioxide. For a sample volume of 25 ml, this corresponds to a concentration of 0.5 - 9 mg/25 ml. Within the range 2 - 7 mg $CO_2$/25 ml, standard deviation is max. ± 2 %.

Sulphides and $H_2S$ have an interfering effect. Before the acid is added, they should be bonded as CuS by introducing 0.5 ml of 10 % $CuSO_4$ solution.

#### Equipment

Apparatus for the conversion, absorption and titration of $CO_2$ from water samples (Fig. 94), consisting of:

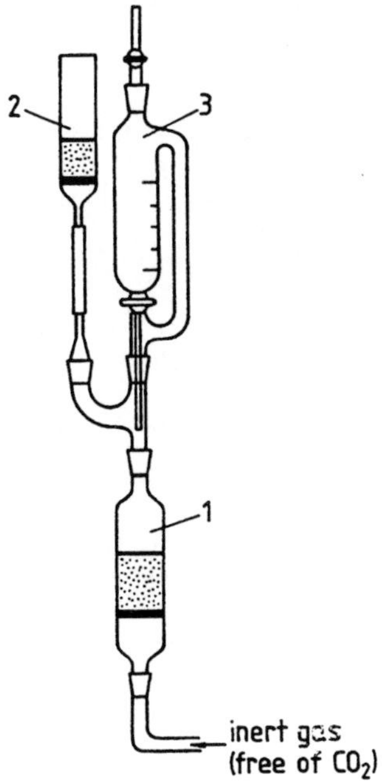


Fig. 94. 1) = Sample container; 2) = Absorption and titration vessel; 3) = Dropping funnel with pressure compensation

Sample container (1) with sintered plate,

U-tube with ground joints,

10 ml dropping funnel with pressure compensation (3),

frit as absorption and titration vessel (2).

All ground parts should be greased with silicone grease; connect all parts of equipment firmly. (from "Ausgewählte Methoden der Wasseruntersuchung")

Flowmeter

Wash bottles with 30 % potassium hydroxide solution

Water, carbon dioxide-free:
Boil distilled water in a flask for a few minutes. Immediately afterwards, close the flask with a plug which has a soda lime tube and a siphon. Use this $CO_2$-free water to prepare all reagent solutions.

EDTA:
Disodium salt of ethylene diamine tetraacetic acid, reagent-purity.

NaOH:
reagent-purity: lozenges, for conservation of the samples. The $CO_2$ content must be determined as described under "Procedure".

Sodium hydroxide solution, 1 m

Sodium hydroxide solution, 0.05 m:
Dilute 25 ml of 1 m sodium hydroxide solution to 500 ml with $CO_2$-free water.

Sulphuric acid, approximately 2 m:
Add 110 ml of reagent-purity sulphuric acid (1.84 g/ml) to 900 ml distilled water, stirring and cooling all the time.

Sulphuric acid, 0.05 m

Sulphuric acid, 0.01 m:
Dilute 200 ml of 0.05 m sulphuric acid to 1 litre with distilled water

n-butanol, reagent-purity

Phenolphthalein/naphtholphthalein mixed indicator:
A. Dissolve 0.25 g phenolphthalein in 100 ml 96 % ethanol
B. Dissolve 0.10 g naphtholphthalein in 100 ml 96 % ethanol
Mix 20 ml of solution A with 50 ml of solution B.

Bromocresol green/methyl red mixed indicator:
A. Dissolve 0.2 g bromocresol green in 100 ml 96 % ethanol
B. Dissolve 0.015 g methyl red in 50 ml 96 % ethanol
Mix solution A and B and store in a dark glass flask.

Method

Sampling

Fill a ground-stopper glass flask of approximately 100 ml to overflowing at the sampling point and immediately add two NaOH lozenges, 3 drops of phenolphthalein/naphtholphthalein mixed indicator and 0.5 g EDTA. Stopper the flask, leaving no bubbles, and shake until the sample is clear.

The samples must display an obvious alkaline reaction; if not, add a further NaOH lozenge.

During transport and storage of the vessels, care should be taken that the water samples are not heated above the original temperature (risk of breakage). The analysis must be carried out within 5 days.

Determination

Introduce 10 ml of approximately 2 m sulphuric acid into the dropping funnel (3) of the apparatus. Set the air flow, which is directed into the sample container (1) from below after passing through two wash bottles with 30 % potassium hydroxide solution, to approximately 60 bubbles/min. (approx 3 l/h).

Following these preparations, introduce 25 ml of the conserved sample (or some other suitable volume with a $CO_2$ content between 2 and 7 mg) into the sample container and immediately connect the latter to the dropping funnel and the absorption vessel (2). Shut off the dropping funnel and allow the air flow to flow through the apparatus for approximately 3 minutes.

Introduce 10 ml of 0.05 m sodium hydroxide solution and 3 drops of n-butanol into the absorption vessel. With the air flow bubbling evenly through the sodium hydroxide solution, add the approximately 2 m sulphuric acid to the sample drop by drop from the dropping funnel. After 30 minutes, the liberated $CO_2$ is completely expelled and absorbed in the receiver.

Continue to allow the air to flow during the subsequent titration in order to ensure that the titrant and titration solution are thoroughly mixed; however, reduce the air flow to 30 bubbles/min beforehand and add a further 10 drops of n-butanol to inhibit foaming. After adding 10 drops of phenolphthalein/naphtholphthalein mixed indicator (end point pH 8.6), add 2 - 3 ml of 0.05 m sulphuric acid drop by drop to convert the major portion of the alkali surplus (the indicator must not change colour!). Then convert the remainder of the carbonate into hydrogen carbonate by adding 0.01 m sulphuric acid, slowly and drop by drop, until the indicator changes from red-purple to pure blue. Subsequently add 5 drops of bromocresol green/methyl red mixed indicator (end point pH 4.5), and titrate with 0.01 m sulphuric acid until the transition occurs from blue through green to yellowish red.

The consumption of 0.01 m sulphuric acid between the two end points (pH 8.6 and 4.5) corresponds to the total carbon dioxide content of the sample. Determine a blank value each day in similar fashion using $CO_2$-free water.

After titration is completed, remove the solution from the absorption vessel and rinse the latter several times with distilled water. In the same way, rinse the entire apparatus thoroughly with distilled water.

In order to determine the average carbonate content of the NaOH lozenges, dissolve 20 lozenges in 100 ml of $CO_2$-free distilled water and treat with 10 drops of phenolphthalein/naphtholphthalein mixed indicator.

First neutralize the sample with the approximately 2 m sulphuric acid until the mixed indicator changes to blue (approximately 10 - 12 ml), reverse the change to red-purple by the addition of 0.05 m sodium hydroxide solution, and add 0.01 m sulphuric acid drop by drop until the colour changes to blue again. Subsequently, add 5 drops of bromocresol green/methyl red mixed indicator and titrate with 0.01 m sulphuric acid, as described above. Calculate the $CO_2$ content in mg per NaOH lozenge.

Calculation

The results may be given in mg/l or in mmol/l and designated total $CO_2$ (total carbon dioxide). 1 ml of 0.01 m sulphuric acid corresponds to 0.44 mg of $CO_2$.

$$\text{Total } CO_2 \text{ mg/l} = \frac{(a - b) \cdot F \cdot 440.1}{V_1} - \frac{z \cdot c \cdot 1000}{V_2}$$

a = Consumption of 0.01 m sulphuric acid between the end points (pH 8.6 to pH 4.5) for the sample
b = Consumption of 0.01 m sulphuric acid in ml for the blank test
F = Factor of the 0.01 m sulphuric acid
$V_1$ = Undiluted sample volume used in ml
c = $CO_2$ content in mg per NaOH lozenge (for conservation)
z = Number of lozenges used
$V_2$ = Volume of the conserved sample taken

$$1 \text{ mmol } CO_2 = 44.01 \text{ mg } CO_2$$

$$1 \text{ mg } CO_2 = 0.02272 \text{ mmol } CO_2$$

### 3.2.10.3 Gravimetric determination after distillation

General remarks

On sampling, free dissolved carbon dioxide and its compounds are fixed as carbonate by filling the water samples into flasks which are charged with sodium hydroxide solution or calcium oxide. In closed apparatus in the laboratory, add acid to the weakly alkaline sample, liberating carbon dioxide, bind the carbon dioxide on soda-asbestos and weigh.

The method is particularly suited to the determination of $CO_2$ in waters rich in carbon dioxide whose $CO_2$ content (total carbon dioxide) is 1000 mg/l or higher.

Equipment

Flat-bottomed flasks with ground joint, ground glass stopper and safety groove in the stopper to secure for transport, volume 250 ml and 500 ml.

Steel cylinder of nitrogen for the introduction of $N_2$ as carrier gas.

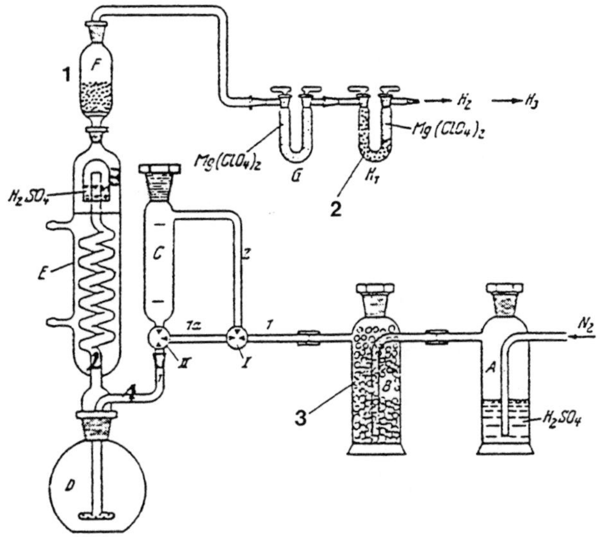


Fig. 95. Distillation apparatus after R. Fresenius and F. Neumüller, modified by K. E. Quentin and L. Feiler; 1) = $CuSO_4$ on pumice stone; 2) = Soda-asbestos; 3) = Soda lime

Structure of the apparatus

A: Wash bottle containing sulphuric acid (1.84 g/ml); it serves as a bubble counter and dries the carrier gas.

B: Absorption tower containing soda lime and with a cotton-wool plug at the tower exit; serves to extract $CO_2$ from the carrier gas.

C: Acid vessel with two engraved marks between which there is 50 ml of hydrochloric acid (1.10 g/ml) with methyl red additive; in addition, glass pipe extensions and 3-way cocks (I, II).

D: Reaction vessel: 250 or 500 ml graduated flask with ground glass stopper.

E: Condenser, with (1) gas inlet tube which ends above the base of D in a 2.5 cm T-piece with holes to allow the gas to enter; (2) cooling section with a small number (5 - 6) of sharp turns; (3) gas-washing cap at the upper end of the condenser, closed off from the cooling water section at the bottom. The distilling pipe ends here in an internal closed bubble cap with a ring of holes near the base. It is filled 1/3 full of conc. sulphuric acid (1.84 g/ml) (approximately 20 ml, acid level above the holes, but below the end of the distilling pipe). This is for drying and purification of the carbon dioxide.

F: Absorption vessel, filled half full of $CuSO_4$ on pumice stone (for binding $H_2S$).

G: Drying tube with $Mg(ClO_4)_2$ or granular $P_2O_5$.

H: Weighed absorption tubes with soda-asbestos and $Mg(ClO_4)_2$ or granular $P_2O_5$ (to bind the water released on absorption of $CO_2$). $H_1$ and $H_2$ serve to absorb the $CO_2$, $H_3$ safeguards the end of the apparatus.

Sodium hydroxide solution, carbonate-free:
1 part NaOH, reagent purity, + 1 part distilled water, treated with 1 spatula-tip of barium chloride to bind $CO_2$; store airtight and with a soda lime protective tube.

Calcium oxide, reagent purity,; before use ignite for 1 h at 1000 °C

Soda-asbestos, millet-seed-size for elementary analysis.

Soda lime, granular, with indicator, reagent purity

Magnesium perchlorate, reagent purity, for drying purposes or

Phosphorus pentoxide, granulated

Pumice stone, boiled out with sodium hydroxide solution and ignited, reagent purity

Copper sulphate, reagent purity

Copper sulphate pumice stone:
Grind pumice stone to pea size, sieve out the fine particles and thoroughly soak with saturated copper sulphate solution which contains a few drops of dilute sulphuric acid. After pouring off the surplus copper sulphate solution, dry the mass in a porcelain dish on an electric hot plate, stirring constantly, and heat in the drying oven at 180 °C until the mass appears white. Store in a firmly sealed plastic reagent bottle.

Hydrochloric acid, reagent purity, (1.10 g/ml)

Sulphuric acid, conc, reagent purity, (1.84 g/ml)

Methyl red indicator solution

## Method

### Sampling

Before sampling, blow purified nitrogen through 6 flat-bottomed flasks in the laboratory (four for $CO_2$ determination and two for the blank reading) and introduce approximately 10 ml of carbonate-free sodium hydroxide solution (use a siphon, in order to prevent the carrying over of barium chloride deposits) or 3 g calcium oxide into the flasks. Lightly grease the ground joints with alkali-resistant silicone grease and close carefully.

At the sampling point, feed measured volumes of the water sample into the flasks charged with sodium hydroxide solution or CaO, carefully avoiding gas losses (cf. Chapter 1).

The procedure should be to pipette between 50 ml and 150 ml into the 250 ml flasks and between 300 and 350 ml of the water sample into the 500 ml flasks, depending on the quantity of $CO_2$ expected. Mix slowly and continuously with a circular motion of the flask, as the water flows in. The flasks are later to be connected directly to the distillation apparatus (D in the illustration). For this reason, they must be filled in such a way that 50 ml of hydrochloric acid may still flow in from the acid vessel C and there is also sufficient headspace for the decoction process.

Sampling may also be undertaken with the aid of weighed quantities. Run approximately 50 - 150 ml into the 250 ml flasks and approximately 300 - 350 ml into the 500 ml flask at the sampling point, and otherwise proceed as described above. Then weigh the filled flasks on returning to the laboratory. The quantity of water to be investigated is given by the difference in weight before and after the sample is taken.

Determination

Assemble the distillation apparatus. Fill the acid vessel C to the upper mark with hydrochloric acid and 10 drops of methyl red indicator solution, charge the bubble cap at the upper end of the condenser with sulphuric acid and the absorption vessel F with copper sulphate pumice. The copper sulphate pumice filling in absorption vessel F serves to bind $H_2S$. Fill the absorption tubes $H_1$, $H_2$ and $H_3$ with soda-asbestos and magnesium perchlorate in the way shown in the illustration. Care should be taken that the coarse-grained soda-asbestos filling is as loose as possible. This prevents the tubes from choking up during the reaction with $CO_2$ with the release of water, which in turn is bound in this tube by $Mg(ClO_4)_2$, impeding the gas through-flow.

The magnesium perchlorate tube G must be prepared separately. After being filled, it should be connected for approximately one hour to a carbon-dioxide-generating device and saturated with this gas. Then expel the excess carbon dioxide by connecting to the nitrogen cylinder downstream of the absorption tower B for one hour. A magnesium perchlorate tube prepared in this way may be used for $CO_2$ determination many times without renewed filling and pretreatment.

Connect reaction vessel D to the apparatus. A means of heating D must be provided.

Connect wash bottle A to the nitrogen cylinder and direct the nitrogen flow straight through the apparatus via I and II at a rate of approximately 2 - 3 bubbles per second. After 15 minutes, turn off the nitrogen flow and simultaneously switch the three-way cock I to acid vessel C. Close off the four absorption tubes G, $H_1$, $H_2$ and $H_3$ and weigh tubes $H_1$ and $H_2$. Before weighing, equalize the pressure by opening the tubes briefly, closing them again and then rubbing with a leather (weights $A_1$ and $A_2$ in mg). Reconnect all the tubes to the apparatus and open the taps.

Turn on the nitrogen flow at 2 - 3 bubbles per second, and simultaneously switch the three-way cock II to open the connection between C and D (blocking connector 1a). The $N_2$ overpressure created in the upper section of C forces the 50 ml of hydrochloric acid (with methyl red additive) into reaction vessel D. The overpressure prevents the carbon dioxide forming in D from escaping through the acid in C. As soon as the acid level reaches the lower mark on C, switch three-way cocks I and II simultaneously to direct the gas flow straight into D. Carbon dioxide evolves in the cool of reaction vessel D. Apply the nitrogen flow for 15 minutes to drive the carbon dioxide via E - F - G into the absorption tubes $H_1$ and $H_2$. A white discoloration of the soda-asbestos indicates a reaction with the carbon dioxide.

After 15 minutes, begin heating reaction vessel D and keep the solution gently boiling for approximately 50 minutes, maintaining the controlled flow of nitrogen. The steam condensation should end at about the second or third turn of the condenser so that no liquid reaches the gas-washing jar containing sulphuric acid in the condenser head.

Allow the reaction vessel to cool, leaving the nitrogen flow turned on for approximately 10 minutes. Remove $H_1$ and $H_2$, seal immediately and weigh in the same way as before (weights $B_1$ and $B_2$ in mg). The sum of the differences in weight is the content of total carbon dioxide in the water sample.

It is possible to check the perfect functioning of the apparatus by determining the $CO_2$ content of a sample quantity of reagent-purity $NaHCO_3$ which is first weighed and then dissolved in water.

Calculation

$$\text{mg/l total carbon dioxide } (S_c) = \frac{(B_1 - A_1) + (B_2 - A_2) \cdot 1000}{\text{Quantity of water used, in ml}}$$

3.2.11 Ion Chromatography of seven anions

General remarks

As an example of the compilation and representation of specifications for water analysis, the chairman of the Working Group (German Standardized Methods or DIN) has made the following draft available. (DIN 38405, D 19, draft manuscript)

Determination of seven anions (fluoride, chloride, nitrite, orthophosphate, bromide, nitrate and sulphate) in (unpolluted) waters using ion chromatography.

Range of application

The method is basically suited to the determination of fluoride, chloride, nitrite, orthophosphate, bromide, nitrate and sulphate in unpolluted water (such as drinking water, rainwater, groundwater and surface water) in the following masses per unit volume:

| | | | | |
|---|---|---|---|---|
| Fluoride | : 0.1 | - | 10 | mg/l |
| Chloride | : 0.1 | - | 50 | mg/l |
| Nitrite | : 0.05 | - | 20 | mg/l |
| Phosphate | : 0.1 | - | 20 | mg/l |
| Bromide | : 0.05 | - | 20 | mg/l |
| Nitrate | : 0.1 | - | 50 | mg/l |
| Sulphate | : 0.1 | - | 100 | mg/l |

In individual cases, the range of application can be altered by varying the working conditions (e.g. sample volume, detectors, separation columns etc.).

General information

In surface water, groundwater, drinking water and rainwater the anions chloride, nitrite, orthophosphate, bromide, nitrate and sulphate occur in widely differing concentrations. The concentration range varies from a few µg/l to g/l. For further details of ion chromatography see Chapter 2.

**Table.** Cross sensitivity of the anions

| Ratio measured ion/ interfering ion | | Max. tolerable interfering ion concentration in mg/l | |
|---|---|---|---|
| $F^-/Cl^-$ | 1 : 500 | 400 | $Cl^-$ |
| $Cl^-/NO_2^-$ | 1 : 50 | 5 | $NO_2^-$ |
| $Cl^-/NO_3^-$ | 1 : 500 | 500 | $NO_3^-$ |
| $Cl^-/SO_4^{2-}$ | 1 : 500 | 500 | $SO_4^{2-}$ |
| $NO_2^-/Cl^-$ | 1 : 250 | 100 | $Cl^-$ |
| $NO_2^-/PO_4^{3-}$ | 1 : 50 | 20 | $PO_4^{3-}$ |
| $NO_2^-/NO_3^-$ | 1 : 500 | 500 | $NO_3^-$ |
| $NO_2^-/SO_4^{2-}$ | 1 : 500 | 500 | $SO_4^{2-}$ |
| $PO_4^{3-}/Cl^-$ | 1 : 500 | 500 | $Cl^-$ |
| $PO_4^{3-}/NO_3^-$ | 1 : 500 | 400 | $NO_3^-$ |
| $PO_4^{3-}/Br^-$ | 1 : 100 | 100 | $Br^-$ |
| $PO_4^{3-}/NO_2^-$ | 1 : 100 | 100 | $NO_2^-$ |
| $PO_4^{3-}/SO_4^{2-}$ | 1 : 500 | 500 | $SO_4^{2-}$ |
| $Br^-/Cl^-$ | 1 : 500 | 500 | $Cl^-$ |
| $Br^-/PO_4^{3-}$ | 1 : 100 | 100 | $PO_4^{3-}$ |
| $Br^-/NO_3^-$ | 1 : 50 | 100 | $NO_3^-$ |
| $Br^-/SO_4^{2-}$ | 1 : 500 | 500 | $SO_4^{2-}$ |
| $NO_3^-/Cl^-$ | 1 : 500 | 500 | $Cl^-$ |
| $NO_3^-/Br^-$ | 1 : 100 | 100 | $Br^-$ |
| $NO_3^-/SO_4^{2-}$ | 1 : 500 | 500 | $SO_4^{2-}$ |
| $SO_4^{2-}/Cl^-$ | 1 : 500 | 500 | $Cl^-$ |
| $SO_4^{2-}/NO_3^-$ | 1 : 500 | 400 | $NO_3^-$ |

Basis of the method

The ions are separated by liquid chromatography with the aid of a column. An anion exchanger of low capacity generally serves as the stationary phase. Detection may be made with various physical techniques. Conductivity detectors are customarily used.

Interference factors

Ion chromatographic anion determination is generally a relatively trouble-free technique. Nevertheless, mention must be made of a number of potential sources of interference:

Relatively high concentrations of organic acids, such as malonic acid, maleic acid and malic acid, may influence the determination of inorganic anions.

Solid substances and organic substances contained in water (such as mineral oils, detergents and humic acids) reduce the service life of the separation column and should therefore be separated off before the analysis.

Fluoride determination may be disturbed by the presence of monocarboxylic acid and carbonates.

In the determination of $F^-$, $Cl^-$, $NO_2^-$, $PO_4^{3-}$, $Br^-$, $NO_3^-$ and $SO_4^{2-}$, cross sensitivity (inadequate separation) may occur if differences in concentration are great. The following ratios of concentration have been tested (Table). If a sample volume of 50 µl is used, no interference arises.

Where the anions bromide and phosphate are not listed they have no interfering effect in the defined range of application.

In buffered eluents (e.g. carbonate/hydrogen carbonate) the determination is not influenced by the pH of the sample (pH 2 - 9).

This information applies only if the quality requirements of the separation column are fulfilled and if the electrical conductivity of the samples at 25 °C does not exceed 1000 µS/cm. In real samples, the peak resolution (R) should be no worse than 1.3.

## Equipment

Ion chromatography system (see also Chapter 2) which fulfils the requirements of quality. It generally consists of the following components (Fig. 96):

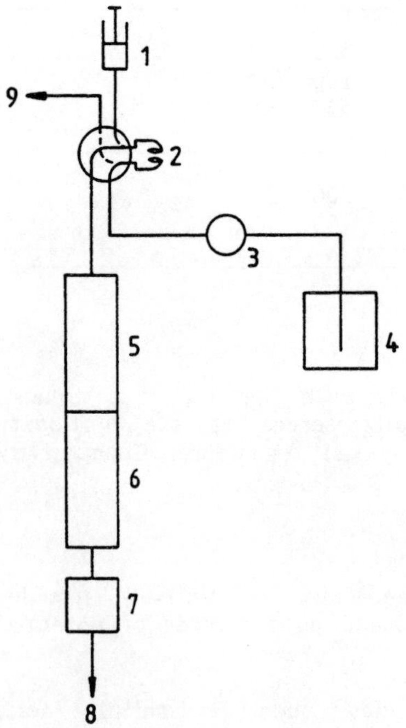


Fig. 96. Diagrammatic representation of an ion-chromatograph; 1) = Sample injector; 2) = 0.3-ml injection system; 3) = Pump; 4) = Eluent; 5) = Precolumn; 6) = Separation column; 7) = Detector (e.g. conductivity detector; 8) = Recording unit; 9) = Outflow

- Eluent reservoir
- Pump
- Sample applicator system (e.g. with sample link, 50 µl)
- Precolumn
- Separation column (in some cases with suppressor) with appropriate separation efficiency
- Detector (e.g. conductivity detector)
- Recording unit (e.g. integrator with plotter)
- Drying oven
- Desiccator
- Measuring flasks, 100, 1000 and 5000 ml e.g. DIN 12664-MSA
- Measuring flasks, plastic, for low concentrations
- Measuring pipettes, 1, 2, 3, 5 and 10 ml, e.g. DIN 12697 MPAS 1 (or microlitre pipette)

Only chemicals of "reagent purity" are used. The electrical conductivity of the water must be < 0.1 µS/cm and the water must contain no particles > 0.45 µm.

Sodium hydrogen carbonate ($NaHCO_3$)
Sodium carbonate ($Na_2CO_3$)
Sodium fluoride (NaF)
Sodium chloride (NaCl)
Sodium nitrite ($NaNO_2$)
Potassium hydrogen phosphate ($KH_2PO_4$)
Sodium bromide (NaBr)
Sodium nitrate ($NaNO_3$)
Sodium sulphate ($Na_2SO_4$)

**Eluents**

Different eluents are used, depending on the separation column and detector. Degasified water must be used for their preparation. During operation, renewed absorption of gas is to be avoided (helium superposition; collapsing flasks). In order to prevent the growth of bacteria and algae, the eluent should be stored in a dark place and renewed every 2 to 3 days.

**Eluents for IC using the suppressor technique**

If the suppressor technique is employed, sodium hydroxide and salt solutions of weakly dissociated acids, such as sodium carbonate/sodium hydrogen carbonate, sodium hydrogen carbonate, or sodium tetraborate are used.

**Sodium carbonate/sodium hydrogen carbonate concentrate**

The addition of the following eluent concentrate has proved effective for preliminary treatment of the samples and for preparation of the eluent:

Weigh 25.44 g of sodium carbonate and 25.2 g of sodium hydrogen carbonate into a measuring flask, volume 1000 ml, and make up to the mark with water.

The solution contains 0.24 mol/l of sodium carbonate/l and 0.3 mol/l of sodium hydrogen carbonate/l and may be stored in cool conditions for about 1 year.

**Sodium carbonate/sodium hydrogen carbonate eluent**

The following eluent has proven effective for the determination of $F^-$, $Cl^-$, $NO_2$, $PO_4^{3-}$, $Br^-$, $NO_3^-$ and $SO_4^{2-}$ in one operation:

- Measure 50 ml of the concentrate into a 5000-ml measuring flask and make up to the mark with water.

The solution contains 0.0024 mol of sodium carbonate/l and 0.003 mol of sodium hydrogen carbonate/l.

**Eluents for IC without suppressor**

For IC techniques without a suppressor facility weakly dissociated salt solutions such as potassium hydrogen phthalate, sodium borate/sodium gluconate and sodium benzoate are used. The concentrations of the salts generally lie between 0.0005 and 0.01 mol/l. Concentrates and eluent solutions are prepared in a similar fashion. The pH of the eluent must be adjusted after dilution of the concentrate.

The exact composition of the eluent should be taken from the specifications of the column manufacturer.

**Stock solutions**

Stock solutions with a concentration of 1000 mg/l ($F^-$, $Cl^-$, $NO_2^-$, $PO_4^{3-}$, $Br^-$, $NO_3^-$, $SO_4^{2-}$):

Pretreat amounts of the substances according to the Table and place each into a 1000-ml measuring flask, and dissolve in a little water; make up to the mark with water.

Commercial stock solutions of appropriate concentration may also be used.

**Standard solutions**

If required, standard solutions may be prepared from the stock solutions with various combinations of anions and different concentrations. The lower

**Table.** Initial weights and pretreatment for the stock solutions

| | | | Pretreatment[a] by drying | |
|---|---|---|---|---|
| Anions | Substances | Weight (g/l) | Duration (h) | Temp. (°C) |
| Fluoride | NaF | 2.2100 | 1 | 105 |
| Chloride | NaCl | 1.6484 | 2 | 150 |
| Nitrite | $NaNO_2$ | 1.4998 | 1 | 150 |
| Phosphate | $KH_2PO_4$ | 1.4330 | 1 | 105 |
| Bromide | NaBr | 1.2877 | 6 | 150 |
| Nitrate | $NaNO_3$ | 1.3707 | 24 | 105 |
| Sulphate | $Na_2SO_4$ | 1.4790 | 1 | 105 |

[a] After drying, cool the substances in the desiccator

the anion concentration selected, the greater the risk of concentration changes due to interaction with the vessel material. The storage of fluoride and chloride standard solutions in teflon (PTFE) or polyethylene containers has proved successful. Experience has shown that better storage of nitrate standard solutions is possible in borosilicate flasks.

The same containers should always be used for the same anions and anion concentrations.

Mixed standard solution I

$$ß\ (F^-,\ NO_2^-,\ PO_4^{3-},\ Br^-) = 10\ mg/l$$

$$ß\ (Cl^-,\ NO_3^-,\ SO_4^{2-}) = 100\ mg/l$$

Pipette the volumes specified in the next Table into a 100-ml measuring flask, and make up to the mark with water.

The solution can be stored for about 1 - 2 days in cool conditions.

Mixed standard solution II

$$ß\ (F^-,\ NO_2^-,\ PO_4^{3-},\ Br^-) = 1\ mg/l$$

$$ß\ (Cl^-,\ NO_3^-,\ SO_4^{2-}) = 10\ mg/l$$

Pipette 10 ml of mixed anion standard solution I into a 100-ml measuring flask and make up to the mark with water.

The solution is stable for only 1 - 2 days, even if cooled.

Mixed standard solution III

$$ß\ (F^-,\ NO_2^-,\ PO_4^{3-},\ Br^-) = 0.1\ mg/l$$

$$ß\ (Cl^-,\ NO_3^-,\ SO_4^{2-}) = 1.0\ mg/l$$

Pipette 1 ml of mixed anion standard solution I into a 100-ml measuring flask and make up to the mark with water.

The solution should be prepared freshly each day.

**Table.** Volumes of stock solution required to prepare mixed standard solution I

| Anions | Stock solution (ml) | Anion concentration (mg/l) |
|---|---|---|
| $F^-$ | 1 | 10 |
| $Cl^-$ | 10 | 100 |
| $NO_2^-$ | 1 | 10 |
| $PO_4^{3-}$ | 1 | 10 |
| $Br^-$ | 1 | 10 |
| $NO_3^-$ | 10 | 100 |
| $SO_4^{2-}$ | 10 | 100 |

**Table:** Concentrations of the reference solutions[a]

| Anions | Concentrations of reference solutions (mg/l) | |
|---|---|---|
| $F^-$ | 0.1; 0.3; 0.5; 0.7; 0.9 | |
| $NO_2^-$ | -- " -- | Working range |
| $PO_4^{3-}$ | -- " -- | 0.1 - 1.0 mg/l |
| $Br^-$ | -- " -- | |
| $Cl^-$ | 1; 3; 5; 7; 9; | |
| $NO_3^-$ | -- " -- | Working range |
| $SO_4^{2-}$ | -- " -- | 1.0 - 10 mg/l |

[a] The concentration of the reference solution is reduced by the addition of 1 ml of eluent concentrate. The discrepancy is compensated, however, by equivalent treatment of the sample.

### Anion reference solutions

Depending on the anion concentrations expected, prepare at least 5 (preferably 10) reference solutions from the stock solution or standard solutions I and II; the reference solutions should be distributed as evenly as possible over the expected range of operation.

For example, for ranges of 0.1 to 1.0 mg/l for $F^-$, $NO_2^-$, $PO_4^{3-}$, $Br^-$ and 1 to 10 mg/l for $Cl^-$, $NO_3^-$, $SO_4^{2-}$, proceed as follows:

Pipette 1, 3, 5, 7 and 9 ml of mixed anion standard solution I into separate 100-ml measuring flasks, make up to the mark with water and add 1 ml of eluent concentrate.

The concentrations of the reference solutions are shown in the Table.

The reference solutions must be prepared freshly for each day of measurement.

### Blank test solution

Fill a 100-ml measuring flask to the mark with water, and add 1 ml of an eluent concentrate by pipette.

### Separation column quality requirements

The principal item in the technique of ion chromatography is the separation column. Its separation efficiency is influenced by various limiting conditions, such as column material and eluents. Within the framework of the standard, only those separation columns may be used which display a baseline-resolved separation of all components following injection of a standard solution with all seven anions ($F^-$, $Cl^-$, $NO_2^-$, $PO_4^{3-}$, $Br^-$, $NO_3^-$ and $SO_4^{2-}$) in a concentration of 1 mg/l (Fig. 97).

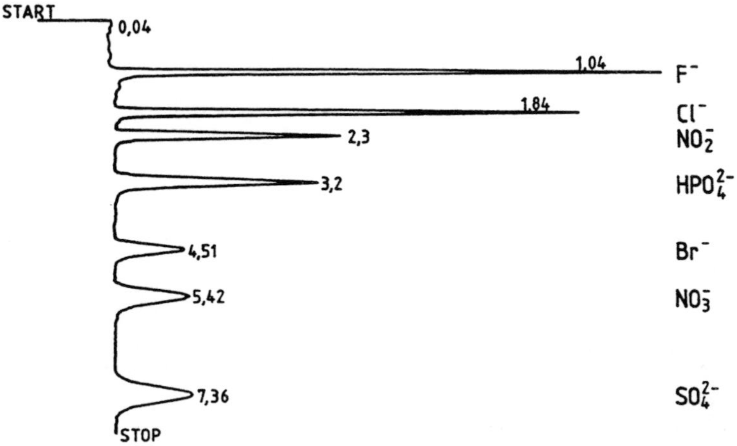


Fig. 97. Example of a chromatogram of a column which conforms to the standard[a]

Where not all the anions shown in Fig. 97 are to be determined, the requirement applies analogously to the ions which are to be determined. The peak resolution should not fall below R = 1.3 (Fig. 98).

## Pretreatment of the sample

On arrival in the laboratory, the sample should be filtered through a membrane filter (pore size 0.45 µm) in order to prevent adsorption of the anions on the solid matter and to prevent bacterial reactions.

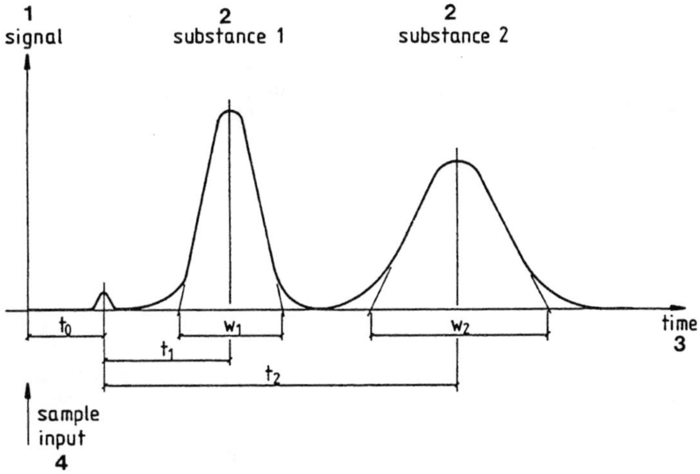


Fig. 98. Idealized representation of chromatographic separation; 1) = Signal; 2) = Component; 3) = Time; 4) = Sample introduction

---

[a] Elution sequences and retention times may vary, depending on the type of column and the composition of the eluent

The sample can be stabilized by cooling until the determination is carried out.

$$R = \frac{2\,(t_2 - t_1)}{W_2 + W_1}$$

$R$ = Peak resolution
$t_1$ = Retention time of 1st peak
$t_2$ = Retention time of 2nd peak
$W_1$ = Peak width of 1st peak at base
$W_2$ = Peak width of 2nd peak at base

In order to prevent precipitation reactions on the separation column, the water sample should be treated with the eluent concentrate, factor 100, before analysis, i.e. 1 part concentrate + 100 parts sample. Dilution effects are eliminated by equivalent treatment of the calibration solution.

If necessary, the sample under analysis must be diluted with water and treated with eluent concentrate.

Directly before injection of the sample, the sample must once again be filtered through a membrane filter (pore size 0.45 µm) in order to remove any solid matter present.

If the sample also contains dissolved organic components, such as humic substances, it is advisable to use a precolumn. This serves to protect the analytical separation column. A basic distinction is made between pre-columns which contain the same resin material as the analytical separation column and those with a macroporous polymer.

Reference solutions must receive the same pretreatment as sample solutions.

Method

The ion chromatograph should be put into operation according to the instructions of the equipment manufacturer. Operational readiness is achieved when the base line is stable. Calibration is carried out as described in the following section.

The pretreated sample is injected into the chromatograph.

Calibration

The anions are identified by means of a comparison of the retention times of the sample with those of the reference solutions. It should be noted that the retention times may be dependent on concentration and matrix. The magnitude of the signal (peak) is proportional to the concentration of the anion.

For measurement, a linear reference function must be prepared, as follows:

Prepare reference solutions and blank test solution
Chromatograph the reference solutions and the blank test solution
Determine the line of regression for the series of measurements thus obtained.

For the ion i which is to be determined the following general equation (reference function) applies:

$$y_i = b_i \cdot \beta_i + a_i \qquad (1)$$

$y_i$ = Signal value (measured value) in peak height or units of peak area (e.g. counts)
$b_i$ = Gradient of the reference function (counts · litre · $mg^{-1}$)
$\beta_i$ = Concentration of the ion (mg · $litre^{-1}$)
$a_i$ = Ordinate intercept of the reference function (calculated blank reading) (counts)

Checks should be made from time to time for significant deviations in gradient $b_i$ and ordinate intercept $a_i$.

**Measurement according to the standard calibration method**

After construction of the reference function, the pretreated sample can be measured. Should the ion concentration of the sample under analysis exceed the range of validity of the reference function, the sample solution must be diluted. It may be necessary to prepare a new reference function for the lower concentration range.

After each sample series, but not later than every 10 to 20 measurements, the validity of the reference function should be checked using two reference solutions of different concentrations, and recalibration should be carried out, if necessary.

**Evaluation**

The concentrations of the anions should be ascertained from the reference function (Equation 1) on the basis of the peak areas or peak heights. The following equation applies:

$$\beta_i = \frac{y_i - a_i}{b_i}$$

All dilution operations should be taken into account in the calculation.

**Presentation of the results**

Values are given rounded off to 0.1 mg/l, but no more than two significant places.

Example: Chloride ($Cl^-$) 45 mg/l
Sulphate ($SO_4^{2-}$) 120 mg/l

**Analytical report**

The report should refer to this method and contain the following details:

a) Precise identity of the water sample
b) Statement of the results
c) Description of any pretreatment of the sample
d) Any deviation from this method and indication of any circumstances which may have influenced the result.

## 3.3 Cations

### 3.3.1 Lithium

**General remarks**

The determination of lithium in water is carried out by AAS (methods 3.3.1.2 and 3.3.1.3), or by flame photometry (method 3.3.1.1) - still in use today - according to the "universal method" proposed by Schuhknecht/Schinkel as described in the section on "Sodium". Since a concentration process is necessary in the case of concentrations of lithium ions below 1 mg/l, a number of departures from the method given for sodium are necessary. Direct determination of traces of lithium is also possible using "stable isotope dilution and field desorption mass spectroscopy" (Schulten et al. 1979).

#### 3.3.1.1 Determination by flame photometry

See the section on "Sodium". The characteristic spectral line for lithium lies at a wavelength of 671 nm.

In the case of lithium levels around or below 1 mg/l, concentration is necessary.

Since calcium ions in concentrations greater than 1000 mg/l have the effect of raising emissions, ethanol extraction must be used for waters of this type (investigation of the sample following concentration).

**Equipment**

See the section on "Sodium" 3.3.2. The following items are required in addition:

Barium chloride solution:
Dissolve 10 g barium chloride, $BaCl_2 \cdot 2\ H_2O$, reagent purity, in 100 ml redistilled water. 1 mg of this solution precipitates approx. 40 mg of sulphate ions.

Ethanol, approx. 95 %, reagent purity, Jena or Pyrex glass flasks, 1 litre, for sample taking and storing the water samples until analysis is made.

Polyethylene vessels are not to be recommended, since LiCl is occasionally used as a catalyst in the production of polyethylene. The polyethylene produced in this way may be contaminated by traces of lithium.

**Method**

**Calibration**

Refer to the section on "Sodium" (3.3.2).

It is advisable to check the calibration curve frequently. The admixture method has proved particularly valuable for this purpose. Add known quantities of lithium to additional off-takes of the water samples under investigation and then subject these solutions to the normal process of analysis.

A) **Investigation of the sample without concentration** for waters with more than 1 mg/l lithium ions. Transfer an exactly measured quantity of the water sample, between 10 and 75 ml, depending on the lithium content, with a pipette to a 100 ml measuring flask, acidify with 1 ml of hydrochloric acid and treat with 5 ml of buffer solution (caesium chloride/aluminium nitrate). Dilute to the mark with redistilled water at 20 °C, shake thoroughly and then continue as described under "Determination" in the section on "Sodium" (3.3.2).

B) **Investigation of the sample with concentration** (for waters with 1 mg/l lithium ions and below). One method is to evaporate a fairly large water sample (at least 500 ml) following acidification with hydrochloric acid to a volume of 50 - 70 ml and then to continue with the investigation as described under A).

If this simple evaporation process is not possible, because it leads to the precipitation of salts, or if the content of calcium ions in the water sample is greater than 1000 mg/l, the procedure is as follows. Acidify between 1 and 5 litres of the water sample with hydrochloric acid (1 ml in excess following neutralization), evaporate to about 500 ml, keep at the boil and treat with a sufficiently large volume of barium chloride solution drop by drop to precipitate the sulphate ions. After at least 2 hours, filter off the barium sulphate precipitate and wash out with water containing hydrochloric acid. Evaporate the hydrochloric filtrate over a water bath until a damp salt mass remains.

When cool, triturate the damp hydrochloric salt mass five times, each time with 50 ml of ethanol (96 %) (glass pestle) and filter off the ethanol from the undissolved residue. The combined ethanol extracts contain all the lithium ions (since lithium chloride is soluble in ethanol). If, following evaporation of the ethanol extract, a quantity of salt remains which it is anticipated is not soluble in 50 ml of dist. water repeat the extraction of this salt mass with ethanol after thorough moistening with hydrochloric acid.

Dissolve the salt mass which finally remains following evaporation of the ethanol in 50 ml of dist. water and transfer to a 100-ml measuring flask. Add 1 ml of hydrochloric acid and 5 ml of buffer solution (caesium chloride/aluminium nitrate) and dilute to the mark with distilled water at 20 °C. Mix thoroughly and proceed as described under "Sodium" ("Determination by flame photometry").

**Calculation** See under "Sodium" (3.3.2.1)

Concentrations of lithium ions below 0.1 mg/l are rounded off to 0.01 mg/l; if concentrations are higher, round off to 0.1 mg/l.

Atomic weight of lithium: 6.941

### 3.3.1.2 Direct determination of lithium by means of atomic-absorption analysis

**Equipment parameters**

| | |
|---|---|
| 1. Hollow-cathode lithium lamp | |
| 2. Wavelength | 670.8 nm |
| 3. Lamp current | 5 mA |
| 4. Aperture width | 1 nm |

| | |
|---|---|
| 5. Fuel gas | Acetylene |
| 6. Oxidant | Air |
| 7. Flame type | Oxidizing |
| 8. Background compensation | No |
| Optimum measuring range | 1 - 5 mg/l |
| Determination limit | 0.05 mg/l |
| Calibration | 0.2 - 5 mg/l |
| Acid concentration | 10 ml hydrochloric acid (1.17 g/ml) per 1000 ml sample solution |

**Remarks**

It is advisable to add 2 ml potassium chloride solution per 50 ml solution for measurement.

Potassium chloride solution:
Dissolve 10 g potassium chloride (KCl) in water and make up to 100 ml.

Concentration by evaporation is possible, if concentrations are low.

### 3.3.1.3 Lithium determination with the graphite tube technique (Furnace method)

**Equipment parameters**

| | |
|---|---|
| 1. Hollow-cathode lithium lamp | |
| 2 Wavelength | 670.8 nm |
| 3 Lamp current | 5 mA |
| 4 Aperture width | 0.5 nm |
| 5. Background compensation | No |
| 6. Carrier gas | Argon |
| 7. Drying | 25 s at 140 °C |
| 8. Incineration | 30 s at 500 °C |
| 9. Atomization | 3 s at 2400 °C |
| 10. Gas stop | Yes |
| Optimum measuring range | 5 - 20 µg/l with 10 µl sample |
| Determination limit | 0.5 µg/l |
| Calibration | Addition method |
| Acid concentration | 10 ml nitric acid (1.40 g/ml) per 1000 ml sample solution |
| Matrix modification | 10 µl 1 % ammonium sulphate solution |
| Ammonium sulphate solution $(NH_4)_2SO_4$ | Dissolve 1 g $(NH_4)_2SO_4$ in water and make up to 100 ml |

**Remarks**

Stabilize the water sample immediately after sampling with 10 ml nitric acid (1.40 g/ml) per 1000 ml sample solution.

## 3.3.2 Sodium

**General remarks**

The determination of sodium in water is carried out either by AAS or flame photometry. These methods are sensitive and fast, and yield sufficiently accurate and reliable results with relatively little effort. In contrast, the gravimetric methods of cumulative determination of $Na^+ + K^+$ as chlorides or sulphates, subsequently subtracting the separately determined potassium values, are hardly used today, in particular because of the difficult and tedious working methods. (Handbuch der Lebensmittelchemie, Volume 8, Part 2, Springer-Verlag 1969). For ICP-AES see Section 3.3.12.

### 3.3.2.1 Flame photometry

Alkali compounds possess the capability to colour a flame by the emission of light rays as they are vaporized in the flame. Each element emits a characteristic spectrum. Over a certain period of time a constant supply of the sample solution is sprayed into the flame, which must burn evenly and non-luminously. The 589 nm spectral line characteristic of sodium is separated out of the light emitted in this way with the aid of a filter, grating or prism, and its intensity is measured with a photometer.

The use of a buffer solution containing caesium chloride as a spectroscopic buffer and aluminium nitrate as a physical buffer largely eliminates the potential interferences otherwise caused by alkali metals in flame photometry. Since it is easily ionizable, the caesium chloride has the effect of almost totally suppressing the ionization of the K, Na and Li atoms which are also present and exert a mutual influence on excitation. To this must be added the spectroscopic buffer action of caesium which has a smoothing effect on operational fluctuations of the burner.

The interfering effect of cross sensitivity which occurs in the presence of alkaline earths, particularly calcium, can be suppressed by adding aluminium nitrate. Sparingly volatile alkaline-earth-aluminium compounds are formed which no longer influence the intensity of sodium emission. (See Chapter 2, Flame Emission).

Depending on the quality of the instrument available, sodium can be determined at concentrations of < 0.1 up to > 100 mg/l. In addition to parallel determinations and three to five readings per individual determination so as to compensate for fluctuations in the readings of the instrument display, the possibility of further interference and the ubiquitousness of sodium make it necessary to carry out a blank test, the result of which should be taken into account in the evaluation.

#### Equipment

Flame photometer with appropriate filter or with monochromator

Pre-chamber atomizer or burner for direct atomization

Air-acetylene or air-propane with pre-chamber atomizer

Hydrogen-air or hydrogen-oxygen with direct atomization

Buffer solution:
Dissolve 50 g reagent-purity caesium chloride (CsCl) and 250 g reagent-purity aluminium nitrate ($Al(NO_3)_3 \cdot 9 H_2O$) in redistilled water to 1 litre.

Alkali standard solution (for the simultaneous determination of $Li^+$, $Na^+$ and $K^+$):
Dissolve 0.6109 g lithium chloride, 0.2542 g sodium chloride and 0.1907 g

potassium chloride (all spectrographically pure) in a 1-litre measuring flask with redistilled water, add 100 ml reagent-purity conc. hydrochloric acid and make up to the 1-litre mark. This solution contains 100 mg each of sodium, potassium and lithium ions per litre.

Alkali guide solution (10 mg each of $Li^+$, $Na^+$, $K^+$ as well as 5 g CsCl and 25 g aluminium nitrate in 1 litre):

Mix 100 ml alkali standard solution with 100 ml buffer solution in a 1-litre measuring flask and dilute to the mark with redistilled water at 20 °C. It is advisable to prepare 2 alkali guide solutions in parallel so as to recognise dilution errors. These alkali guide solutions are used separately to set the 100 point of the instrument, whereas redistilled water is used for the zero point.

Hydrochloric acid, reagent purity (1.19 g/ml)

### Calibration

In order to record the calibration curve, use calibrated pipettes to transfer volumes of alkali standard solution between 1.00 and 10.00 ml into 100-ml measuring flasks. Add 1 ml of reagent-purity hydrochloric acid (exact measurement of the acid is not especially critical) and 5 ml of buffer solution, fill to the mark with redistilled water and mix well. The calibration solutions may contain for example 0.1, 0.5, 1, 2, 5 and 10 mg/l each of lithium, sodium and potassium and 50 ml buffer solution per litre. It is advisable to check the calibration curve against a second dilution series using alkali standard solution prepared separately and also setting the zero and hundred points separately.

In both calibration series it may be necessary to compensate for the flame background. Carry out all individual measurements and each setting of the instruments 3 to 5 times and use the average values as the readings for plotting the calibration curve.

### Determination

Establish the approximate content of sodium ions in the water sample in a preliminary test. Then select the volume of water sample for the actual determination, or the dilution or concentration, as the case may be, such that ideally 0.1 to 10 mg/l sodium ions is present for determination. Treat these volumes with 1 ml of analytical grade hydrochloric acid. Add 50 ml of buffer solution by pipette, make up to 1 litre with redistilled water at 20 °C and mix well. It is of course possible to proceed analogously to 100 ml using proportionally smaller quantities. It is advisable to take further volumes of the water sample separately in the same way (at least five, preferably 10) and carry out determination in measurement series. Correspondingly, carry out a blank test with reagent-purity hydrochloric acid and buffer solution, likewise diluted with redistilled water.

Set the 100 point at 589 nm with the aid of the alkali guide solutions. Check the zero point with redistilled water and compensate for the flame background. It is advisable to carry out the setting of the hundred point, the compensation of the flame background, the calibration and the measurement of the solutions under investigation several times. This is the only way of compensating for unavoidable instrument fluctuations and also reading inaccuracies.

Calculation

Take the average value of the individual measurements, deduct the blank reading and read off the content of sodium ions in the sample from the calibration curve.

The content G in mg/l is given by the formula

$$\frac{a \cdot 1000}{b} = G \text{ (mg Na}^+\text{/l)}$$

a = reading from the calibration curve, $Na^+$ content in mg
b = volume of the analyzed water sample in ml

Round off contents up to 100 mg/l to 0.1 mg/l, and higher contents to 1 mg/l.

Atomic weight of sodium: 22.9898

### 3.3.2.2 Direct determination of sodium by means of atomic-absorption analysis

Equipment parameters

| | | |
|---|---|---|
| 1. | Hollow-cathode sodium lamp | |
| 2. | Wavelength | 589.0 nm |
| 3. | Lamp current | 5 mA |
| 4. | Aperture width | 0.5 nm |
| 5. | Fuel gas | Acetylene |
| 6. | Oxidant | Air |
| 7. | Flame type | Oxidizing |
| 8. | Background compensation | No |

| | |
|---|---|
| Optimum measuring range | 0.2 - 1 mg/l |
| Determination limit | 0.05 mg/l |
| Calibration | 0.1 - 1 mg/l |
| Acid concentration | 10 ml hydrochloric acid (1.17 g/ml) per 1000 ml sample solution |

Remarks

Add 2 ml caesium chloride - aluminium nitrate solution per 50 ml sample solution to reduce interference.

Caesium chloride - aluminium nitrate solution:
Dissolve 50 g caesium chloride (CsCl) and 250 g aluminium nitrate ($Al(NO_3)_3 \cdot 9H_2O$) in water, add 10 ml hydrochloric acid (1.17 g/ml) and make up to 1 litre with water.

Concentration by evaporation is possible if concentrations are low.

### 3.3.2.3 ICP-AES (see Section 3.3.12)

## 3.3.3 Potassium

### General remarks

When determining the radioactivity of water samples the quantitative determination of the concentration of potassium ions is important since the natural isotopic mixture of potassium contains the radioisotope potassium-40, which is a natural $\beta^-$ -emitter and also emits gamma rays. (The proportion of potassium-40 in the natural potassium isotopic mixture is constant at 0.0119 %).

The determination of potassium ions in water is carried out by AAS (method 3.3.3.2) or by flame photometry (method 3.3.3.1). (See Sodium, Section 3.3.2). For ICP-AES (method 3.3.3.3) see Section 3.3.12.

The gravimetric methods of analysis with perchlorate or tetraphenyl borosodium, frequently used in the past, are rarely used today.

### 3.3.3.1 Determination by flame photometry

The instructions given in the section on sodium determination by flame photometry (3.3.2.1) should be applied analogously.

Wherever "sodium ions" are mentioned, the term must of course be replaced by "potassium ions". Measurement and calibration is carried out at the wavelength of potassium, namely 768 nm.

Since the concentrations of potassium ions are generally lower than those of sodium ions, and since the calibration curve approaches an ideal straight line, the results are more accurate than those of sodium determination. For this reason, twin determination from two different measured volumes is sufficient.

Atomic weight of potassium: 39.102

### 3.3.3.2 Direct determination of potassium by means of atomic-absorption analysis

### Equipment parameters

| | |
|---|---|
| 1. Hollow-cathode potassium lamp | |
| 2. Wavelength | 766.5 nm |
| 3. Lamp current | 5 mA |
| 4. Aperture width | 1 nm |
| 5. Fuel gas | Acetylene |
| 6. Oxidant | Air |
| 7. Flame type | Oxidizing |
| 8. Background compensation | No |

| | |
|---|---|
| Optimum measuring range | 0.2 - 1 mg/l |
| Determination limit | 0.05 mg/l |
| Calibration | 0.1 - 1 mg/l |
| Acid concentration | 10 ml hydrochloric acid (1.17 g/ml) per 1000 ml sample solution |

**Remarks**

Add 2 ml caesium chloride - aluminium nitrate solution per 50 ml sample solution to reduce interference.

Caesium chloride - aluminium nitrate solution:
Dissolve 50 g caesium chloride (CsCl) and 250 g aluminium nitrate ($Al(NO_3)_3 \cdot 9H_2O$) in water, add 10 ml hydrochloric acid (1.17 g/ml) and make up to 1 litre with water.

Concentration by evaporation is possible if concentrations are low.

### 3.3.3.3 ICP-AES (see Section 3.3.12)

### 3.3.4 Rubidium and caesium

**General remarks**

In natural waters rubidium and caesium ions occur at the most in traces. Either flame photometry (method 3.3.4.1) or AAS (method 3.3.4.2) following concentration by coprecipitation with potassium as tetraphenylborate is recommended for the quantitative determination of rubidium and caesium ions in water or for their semi-quantitative estimation (in the case of caesium, AAS-method 3.3.4.4 and method with concentration 3.3.4.5).

Potassium, rubidium and caesium ions together with tetraphenylborate ions form a precipitate which does not dissolve readily in water. According to W. Geilmann and W. Gebauhr the solubility products are:

K : $2.25 \cdot 10^{-8}$
Rb: $2.0 \cdot 10^{-9}$
Cs: $8.4 \cdot 10^{-10}$

For concentration, the rubidium and caesium ions are precipitated in the form of tetraphenylborates together with potassium ions, which serve as carriers. The lower solubility of the rubidium and caesium tetraphenylborates (see the solubility products) compared with the potassium salt means that under the selected precipitation conditions the rubidium and caesium yields are virtually quantitative.

### 3.3.4.1 Determination by flame photometry

The precipitate is dissolved in acetone and used for determination by flame photometry (see also detection by atomic absorption, Section 3.3.4.2). The tetraphenylborate anion does not interfere with flame emission (I. Rubeska).

A reference solution is used to compensate for the influence of the potassium. It consists of a precipitate of a quantity of potassium, prepared in the same way, corresponding to the content of potassium ions in the quantity of water used for the rubidium and caesium determination. This generally entails no particular difficulty since a determination of potassium precedes the determination of Rb and Cs in the course of water analysis. Experience shows that the potassium content in water hardly ever exceeds 300 mg/l. Consequently, the method has been adapted to a constant potassium content of 200 mg/l, which may be achieved by adding potassium chloride if the content of the natural water sample is lower.

Calibration is carried out with solutions of equal potassium content (200 mg/l), adding known quantities of rubidium and caesium. Following the instructions of the instrument manufacturer the hundred point of the flame photometer is set using the potassium solution containing the maximum concentrations of rubidium and caesium. Pure acetone is taken to set the zero point. As a result of the potassium content of the calibration solutions, the calibration curves do not pass through the origin of the coordinates. Instead, they begin at a value corresponding to the emission of the line K 768.2 nm at the wavelengths of Rb 795 nm and Cs 852 nm respectively (I. Rubeska).

The calibration curves for rubidium and caesium prepared according to this method display a relatively marked curvature. For this reason, following the first determination and using the Rb and Cs values obtained as a basis, the determination should be repeated with two new volumes of the sample, and so on. Using the pincer method the final values obtained are then considerably more reliable.

The considerable influence of the potassium content on rubidium and caesium emission is not eliminated in the method described, (the calibration curves do not pass through the zero point), but since it is kept at a constant level in all preparations it has little interfering effect.

Interferences occur if the acetone used for the solution contains more than 1 % water. A water content greater than 1 % has the effect of overvaluing rubidium readings and undervaluing caesium readings.

If the Rb : Cs ratio does not exceed 1 : 5 or 5 : 1 the everpresent mutual influence of the two elements on each other can be ignored. Outside these limits separate calibration solutions should be used for rubidium and caesium, but in all other respects the directions given remain the same.

It should be noted that when spraying acetone solutions only direct atomizers can be used without danger. During measurement, keeping the operating speed as constant as possible is the only way of excluding the influence of varying quantities of evaporating acetone.

## Equipment

Flame spectrometer

Burner for direct atomization

Fuel gas: hydrogen

Oxidation gas: oxygen

Glass filter crucible

Rubidium chloride, reagent purity, or preferably spectrally pure

Caesium chloride, reagent purity, or preferably spectrally pure

Potassium chloride, reagent-purity, or preferably spectrally pure

Sodium tetraphenylborate solution, 0.1 m:
Dissolve 3.4 g reagent purity sodium tetraphenylborate, $NaB(C_6H_5)_4$, in 100 ml of dist. water. In order to clarify the slightly turbid solution treat with 0.5 g reagent-purity alkali-free aluminium hydroxide, and

shake thoroughly. Leave to stand for 10 minutes, filter through a quantitative filter and use the clear filtrate.

Wash solution:
Dilute the 0.1-m sodium tetraphenylborate solution with dist. water 1 + 20 and acidify with acetic acid to pH 4 - 5.

Acetone, reagent-purity, maximum water content 1 %.

$Rb^+/Cs^+$ standard solution:
Dissolve 1.4148 g of RbCl + 1.2668 g of CsCl in redistilled water, add one drop of reagent purity conc. hydrochloric acid, and make up to 1 litre at 20 °C (corresponds to 1 mg/ml each of $Rb^+$ and $Cs^+$).

Dilute $Rb^+/Cs^+$ standard solution:
Dilute the standard solution in the ratio 1 : 10 with redistilled water (corresponds to 0.1 mg/ml each of $Rb^+$ and $Cs^+$).

## Method

### Calibration

In each of six 500-ml beakers dissolve 381 mg of potassium chloride in 100 ml of dist. water, such that 200 mg of potassium ions are contained in each solution. In ascending order, add 2.0 ml, 5.0 ml, 10.0 ml, 20.0 ml and 40.0 ml of the dilute $Rb^+/Cs^+$ standard solution (corresponding to 0.2 mg, 0.5 mg, 1.0 mg, 2.0 mg and 4.0 mg of rubidium and caesium ions) to five of these solutions. The sixth solution, without any Rb/Cs added, serves to determine the pure potassium calibration point (which marks the intersection of the calibration curve with the ordinate).

Acidify all six solutions with acetic acid to pH 4 - 5, heat to about 40 - 50 °C and stir in 75 ml of the sodium tetraphenylborate solution to each solution. Leave to stand for 10 minutes and filter by suction through a glass filter crucible. Carefully and quantitatively wash out the precipitates in the beakers into the filter crucible with the wash solution. A maximum of 100 ml of the wash solution may be used per precipitation for rinsing the beakers and washing out the precipitates into the filter crucibles (danger of Rb and Cs losses).

Dry the glass filter crucibles with the precipitates at 120 °C for 1 hour. The following procedure has proved effective for the subsequent solution of the precipitates in acetone. First mechanically remove the bulk of the dry precipitates from the filter crucibles with the aid of a spatula and transfer it carefully to 100-ml measuring flasks via small dry funnels. Dissolve the small amounts of precipitate remaining in the filter crucibles with acetone and transfer the acetone solutions through the funnels to the appropriate measuring flasks. Make up to the mark with acetone at 20 °C.

During subsequent flame spectrometry, first carry out the correction of the background and the setting of the zero point with pure acetone. The hundred point is set using the solution containing 4 mg each of $Rb^+$ and $Cs^+$ in addition to 200 mg of $K^+$. Finally take the measurements for the other calibration solutions and the pure potassium solution.

Measurement of the rubidium content is made at 795 nm. Subsequently take the same solutions in the same way for measurement of the caesium content at 852 nm. Repeat each measurement, including compensation of the background and setting the hundred point, at least four times. Take the average value as the reading.

**Determination**

Measure a quantity of the sample containing exactly 200 mg of $K^+$ on the basis of the potassium content of the sample previously determined by flame photometry. If the potassium content of the water is so low that 10 litres of water contain less than 200 mg of $K^+$, measure off 10 litres of the water sample with known potassium content and add the missing proportion of potassium ions to make up to 200 mg of $K^+$ by adding an appropriate weighed quantity of potassium chloride.

Acidify the sample with acetic acid to pH 4 - 5 and reduce the volume of the sample, which varies according to the original potassium content, to about 200 ml by evaporation. When the solution has been brought to a temperature of 40 - 50 °C, stir in 75 ml of sodium tetraphenylborate solution and after 10 minutes filter the precipitate by suction (glass filter crucible).

Continue the further treatment of the sample according to the directions described for calibration solutions under "Calibration". Measurement of the rubidium content is made at 795 nm, and measurement of the caesium content at 852 nm.

In order to obtain more reliable figures, the first determination of the content of rubidium and caesium ions should be followed by a second measurement of the water sample using the same procedure. In order to apply the **pincer method** two new calibration solutions are prepared at the same time, in which the concentrations of rubidium and caesium are about 0.1 - 0.3 mg above and below the measured value of the first analyzed sample respectively. This second determination makes more precise approximation possible.

**Calculation**

Refer to the section on "Sodium".

Atomic weight of rubidium: 85.4678
Atomic weight of caesium: 132.9055

### 3.3.4.2 Determination of rubidium and caesium in water samples by means of AAS

**Principle**

The elements rubidium and caesium are precipitated with sodium tetraphenylborate solution together with potassium, dissolved in a mixture of organic solvents (acetone and methyl isobutyl ketone) and subsequently determined by means of atomic absorption flame analysis.

**Equipment**

Apparatus for atomic absorption flame analysis

**Reagents**

Acetone

Methyl isobutyl ketone

Potassium-chloride solution (20 mg/l potassium)

4.5 % sodium tetraphenylborate solution

Glass fibre filters made of borosilicate glass, substance weight approximately 70 g/m$^2$, diameter: 5 cm, fibre diameter: 0.5 - 1.5 µm, pore size: 0.3 - 1 µm

Method

Depending on the rubidium and caesium content to be expected, up to 1 litre of the water sample acidified with hydrochloric acid is concentrated to approximately 200 ml by evaporation. After cooling, the total concentration of potassium in the samples is raised to some 50 mg by means of the potassium solution. The sample to be analyzed is subsequently heated to 40 - 50 °C, 20 ml of a 4.5 % sodium tetraphenylborate solution added and stirred, and rubidium and caesium precipitated jointly with potassium.

The precipitate is isolated on a glass fibre filter, washed out with distilled water, subsequently released from the glass fibre filter with 10 ml acetone, and raised to a volume of 50 ml with 40 ml methyl isobutyl ketone. A blank reading and calibrating solution, each containing between 10 and 200 µg rubidium and caesium, are treated in the same way.

Final determination is by means of atomic absorption flame analysis.

Wavelength for rubidium 780.0 nm, for caesium 852.1 nm.

See also Sections 3.3.4.3 and 3.3.4.5

### 3.3.4.3 Direct determination of rubidium by means of atomic-absorption analysis

Equipment parameters

| | |
|---|---|
| 1. Hollow-cathode rubidium lamp | |
| 2. Wavelength | 780.0 nm |
| 3. Lamp current | 5 mA |
| 4. Aperture width | 0.2 nm |
| 5. Fuel gas | Acetylene |
| 6. Oxidant | Air |
| 7. Flame type | Oxidizing |
| 8. Background compensation | No |

| | |
|---|---|
| Optimum measuring range | 1 - 5 mg/l |
| Determination limit | 0.1 mg/l |
| Calibration | 0.2 - 5 mg/l |
| Acid concentration | 10 ml hydrochloric acid (1.17 g/ml) per 1000 ml sample solution |

Remarks

It is advisable to add 2 ml potassium chloride solution per 50 ml sample solution.

Potassium chloride solution (KCl):
Dissolve 10 g potassium chloride (KCl) in water and make up to 100 ml.

Concentration by evaporation is possible if concentrations are low.

It may prove advantageous to perform concentration by coprecipitation with potassium as a sparingly soluble tetraphenylborate compound.

### 3.3.4.4 Direct determination of caesium by means of atomic-absorption analysis

Equipment parameters

| | |
|---|---|
| 1. Hollow-cathode caesium lamp | |
| 2. Wavelength | 852.1 nm |
| 3. Lamp current strength | 10 mA |
| 4. Aperture width | 1.0 nm |
| 5. Fuel gas | Acetylene |
| 6. Oxidant | Air |
| 7. Flame type | Oxidizing |
| 8. Background compensation | No |

| | |
|---|---|
| Optimum measuring range | 1 - 5 mg/l |
| Determination limit | 0.1 mg/l |
| Calibration | 0.2 - 5 mg/l |
| Acid concentration | 10 ml hydrochloric acid (1.17 g/ml) per 1000 ml solution for measurement |

Remarks

It is advisable to add 2 ml of potassium chloride solution per 50 ml of solution for measurement.

Potassium chloride solution:
Dissolve 10 g potassium chloride (KCl) in water and make up to 100 ml.

It may prove advantageous to coprecipitate $Cs^+$ together with $K^+$ and $Rb^+$, using sodium tetraphenylborate (see Section 3.3.4.5).

### 3.3.4.5 Basis of the method of concentration

Using sodium tetraphenylborate solution, caesium is precipitated together with rubidium and potassium, separated off, dissolved in a mixture of acetone and methyl isobutyl ketone and determined by atomic absorption analysis.

Reagents

Hydrochloric acid (1.17 g/ml)

Potassium chloride solution:
Dissolve 4 g potassium chloride (KCl) in water and make up to 100 ml.

Sodium tetraphenyl borate solution:
Dissolve 45 g sodium tetraphenyl borate $Na(B(C_6H_5)_4)$ in water and make up to 1000 ml.

Methyl isobutyl ketone (MIBK)

Acetone

Glass-fibre filter of borosilicate glass without binding agent, 5 cm diameter, pore size 0.3 - 1 µm.

Porcelain suction filter or glass filtration device with glass filter plate and removable top part (the filter is held between glass plate and top part by spring pressure).

Procedure

Acidify the necessary quantity of water with hydrochloric acid (5 ml HCl (1.17 g/ml) per 1000 ml of water sample) and concentrate to 150 - 200 ml by evaporation.

If necessary, add potassium chloride solution to bring potassium content to 50 mg.

Heat to 50 °C and treat with 20 ml sodium tetraphenyl borate solution.

Leave to stand for about 30 minutes.

Use filtration unit to isolate the precipitate on a glass-fibre filter.

Dissolve the isolated precipitate in a mixture of four parts by volume of MIBK and one part by volume of acetone and dilute with this mixture to a defined volume.

Treat calibration solutions and blank tests in the same way as samples for investigation.

Final determination by means of atomic-absorption analysis.

Equipment parameters as for direct determination.

### 3.3.5 Ammonium (ammonia)

General remarks

Ammonium ions may occur naturally in so-called reduced subterranean waters but may on the other hand need to be evaluated as an indication of secondary influencing of the water. A decision - either natural and harmless or secondary and dubious - can only be made if a number of factors are taken into account (geological conditions, overall analysis, redox potential, presence of $Fe^{2+}$, $CO_2$, $H_2S$, biological parameters etc).

In the methods of determination commonly used in water analysis it is the sum of ammonium ions and free ammonia that is determined. The ratio of the two components is dependent on the pH and temperature of the water and also on its salt content. If details of individual concentrations are required, they may be estimated on the basis of the following curve published by H. Woker (Int. Verh. Limnol. 10 575 (1948)).

#### 3.3.5.1 Photometric determination of $NH_4^+$ as indophenol

General remarks

In the presence of a catalyst, ammonium ions react with substances containing phenol by means of an oxidizing agent, forming a blue-coloured indophenol

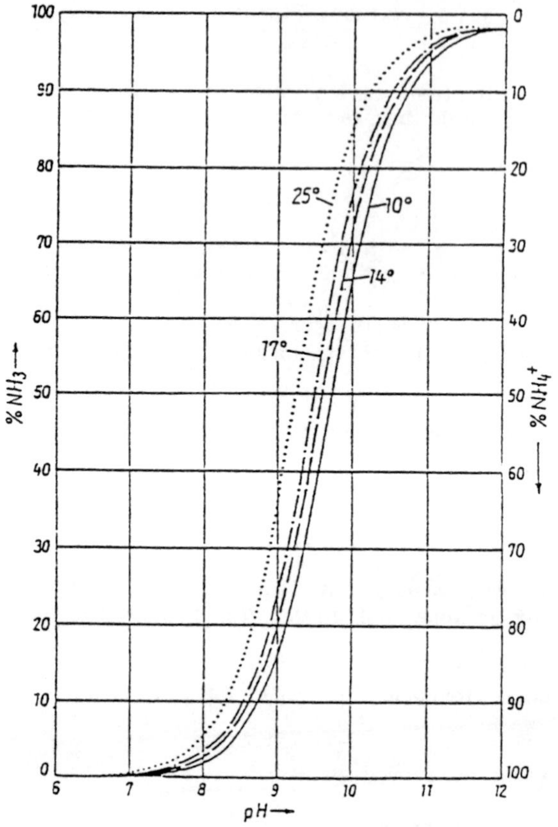


Fig. 99. Proportion of free ammonia in the total of ammonium ions and ammonia as a function of temperature and pH according to Woker. (Values refer to an ionic concentration of 0.01 mol/l, corresponding to a salt content of approx. 500 mg/l).

compound. The sodium salt of dichloroisocyanuric acid serves as the oxidizing agent and sodium salicylate as the phenolic component. The reaction is catalyzed by disodium pentacyanonitrosylferrate (sodium nitroprusside).

For the taking and conservation of water samples see Chapter 1.

Equipment

Photometer

Cuvettes, path length 10 to 50 mm

Citrate solution:
20 % trisodium citrate · 2 $H_2O$, reagent purity and 2 % sodium hydroxide, reagent purity.

Disodium pentacyanonitrosylferrate-sodium salicylate solution:
0.2 g of $Na_2$ $(Fe(CN)_5NO)$ · 2 $H_2O$ (0.2 %) and 17 g of reagent purity

sodium salicylate (17 %) in dist. water. The solution will keep for a limited period only.

Dichloroisocyanuric solution:
0.58 % (sodium salt of dichloroisocyanuric acid, $C_3N_3Cl_2NaO_3$). The solution must be freshly prepared daily.

Oxidation reagent:
Mix 100 ml of citrate solution and 25 ml of dichloroisocyanuric solution before use.

Ammonium stock solution:
Dissolve 2.9655 g of reagent-purity ammonium chloride (dried in a desiccator over silica gel) in ammonium-free dist. water and make up to 1 litre (1 ml = 1 mg of $NH_4^+$)

Ammonium reference solution I:
Before use, dilute 100 ml of the ammonium stock solution to 1000 ml with ammonium-free dist. water (1 ml = 0.1 mg of $NH_4^+$)

Ammonium reference solution II:
Before use, dilute 100 ml of ammonium reference solution I to 1000 ml with ammonium-free water (1 ml = 0.01 mg of $NH_4^+$).

Preparation of ammonium-free water:
Distill 1 litre of dist. water after adding 3 ml of alkaline potassium permanganate solution (dissolve 10 g of potassium permanganate and 50 g of NaOH to 1 litre in dist. water) and discard the first 200 ml of the distillate.

Alternatively, pass distilled water through a strongly acid cation-exchanger column ($H^+$-type)

## Method

### Calibration

Measure into 100-ml beakers 0 - 0.10 - 0.20 - 0.40 - 1.0 - 2.0 - 4.0 - 8.0 ...... 20.0 ml of freshly prepared ammonium reference solution II with 0 - 0.001 - 0.002 - 0.004 - 0.01 - 0.02 - 0.04 - 0.08 ....... 0.2 mg of $NH_4^+$, make up to 50 ml (mark beaker appropriately) with ammonium-free dist. water and continue treatment as below. If need be, the range of the calibration curve can be changed by preparing another dilution series accordingly.

### Determination

Treat 50 ml of the water sample and the calibration samples one after the other with 2 ml of the disodium pentacyanonitrosylferrate/sodium salicylate solution and 2 ml of oxidation reagent. Mix thoroughly after each addition.

After 90 minutes reaction time at room temperature in $NH_3$-free atmosphere measure the solution photometrically at a wavelength of 690 nm in cuvettes of appropriate path length.

Perform a blank test in the same way, using 50 ml of ammonium-free dist. water instead of the water sample.

## Calculation

Take the average value of several individual measurements, deduct the blank reading, read off the content of ammonium ions from the calibration curve and recalculate for 1 litre of water.

For concentrations up to 0.50 mg/l $NH_4^+$ round off the figures to 0.01 mg/l, and for concentrations above 0.5 mg/l $NH_4^+$ round off to 0.02 mg/l.

### 3.3.5.2 Acidimetric determination of $NH_4^+$ or $NH_3$ after distillation

#### General remarks

Ammonium ions and dissolved ammonia are distilled off with water vapour as $NH_3$ at pH 7.4 and collected in a receiver containing boric acid. The primary ammonium salt of boric acid thus formed undergoes hydrolysis until equilibrium is established according to

$$(NH_4)\,H_2BO_3 + H_2O \longrightarrow H_3BO_3 + NH_4^+ + OH^-$$

with the result that titration can be carried out directly with sulphuric acid.

It is also possible to collect the $NH_3$ in a known quantity of sulphuric acid and back-titrate the excess acid with NaOH standard solution. In both cases an acid/base titration is carried out. The boric-acid method offers the advantage that only the factor of the sulphuric acid is required, whereas in the case of back-titration of the excess sulphuric acid it is also necessary to determine the factor of the sodium hydroxide solution.

**Note:**

Laboratory air etc., chemicals and the water used to produce the water vapour must be ammonium-free. If precipitates appear after the addition of phosphate buffer solution, e.g. in the case of calcium concentrations greater than 250 mg/l, the quantity of buffer per 100 ml of water must be doubled, the pH value subsequently checked and if necessary adjusted to pH 7.4 with $H_2SO_4$ or NaOH.

In order to remove free chlorine, add 1 ml of sodium thiosulphate solution per mg of chlorine before distillation.

Urea and other acid amides cause interference by splitting off ammonia during distillation from solutions which are too strongly alkaline. Distillation is carried out at pH 7.4 in order to minimize this error. Nevertheless, strict separation of inorganic and organic nitrogen compounds is not possible since the splitting off of $NH_3$ from amines cannot be completely avoided.

The method is suitable for the investigation of waters with more than 5 mg of $NH_4^+$/l, or of heavily polluted waters such as waste waters, refuse seepage waters and so on, and of waters in which accompanying substances prevent direct photometric determination. Determination of the ammonium ions in the distillate can be carried out by photometry according to 3.3.5.1 if the concentration range of 2 mg/l $NH_4^+$ is not exceeded. If the concentrations are higher, the ammonium in the distillate should be determined by acid/base titration.

### Equipment

Distillation apparatus with receiver; volume of the distillation flask approx. 500 ml

Phosphate buffer solution, pH 7.4:
Dissolve 14.3 g of reagent-purity $KH_2PO_4$ and 68.8 g of reagent purity $K_2HPO_4$ in ammonium-free water and make up to 1 litre.

Boric acid solution:
Dissolve 40 g of reagent-purity $H_3BO_3$ in ammonium-free water and make up to 1 litre (1 ml absorbs about 2 mg of $NH_4^+$)

Sulphuric acid, 0.025 m

Methyl red solution:
Dissolve 0.1 g methyl red (sodium salt) in 100 ml of 96 % ethanol

Sodium thiosulphate solution:
Before use, dissolve 0.1 g of reagent-purity $Na_2S_2O_3 \cdot 5\ H_2O$ in ammonium-free water and make up to 100 ml (1 ml corresponds to about 1 mg of chlorine)

Preparation of ammonium-free water: see under 3.3.5.1

Ammonium stock and reference solutions: see under 3.3.5.1

### Procedure

Depending on the ammonium concentration of the water, transfer up to 200 ml of sample water with a maximum of 20 mg of $NH_4^+$ to the distillation flask. If necessary pretreat the water (e.g. with sodium thiosulphate solution). If the volume is smaller, dilute to about 200 ml with ammonium-free water.

Treat the sample with 20 ml of phosphate buffer solution. If a precipitate forms (calcium phosphates), treat the sample with a further 20 ml of phosphate buffer solution and then set to pH 7.4 with $H_2SO_4$ or NaOH.

With water cooling distill at least 100 ml into the receiver containing 50 ml of boric acid solution. Care should be taken that the tip of the condenser is immersed in the solution. After completion of the distillation, rinse the tip of the condenser with distilled water.

The residue in the distillation flask may be used to determine the organically bonded nitrogen.

Carry out a blank test treating 200 ml of ammonium-free water instead of sample water in the same way.

### A) Photometric determination

Transfer the distillate to a 200-ml measuring flask and make up to the mark with ammonium-free water. Analyze aliquot parts of this solution according to the method described under 3.3.5.1 "Photometric determination as indophenol".

Calculation

As described in Section 3.3.5.1

B) Titration

Treat the distillate with five drops of methyl red solution and titrate with 0.025 m $H_2SO_4$ from yellow to red. Boil up the solution briefly so as to expel $CO_2$, cool off to room temperature and titrate again to the same colour.

The corresponding distillate of the blank test serves as the control sample.

Calculation for titration method (B)

1 ml of 0.025 m $H_2SO_4$ corresponds to 0.902 mg of $NH_4^+$.

The content of ammonium ions in the water sample is therefore calculated according to the formula

$$\frac{(a - b) \cdot F \cdot 902}{V} = G$$

a = Consumption of 0.025 m $H_2SO_4$ in ml for the distillate of the sample
b = Consumption of 0.025 $H_2SO_4$ in ml for the distillate of the blank test
F = Factor of the 0.025 m $H_2SO_4$
V = Undiluted volume of sample used for distillation, in ml
G = Content of ammonium ions in the water sample, in mg/l

Round off the results to 0.1 mg/l $NH_4^+$.

Conversion factors

| | mg $NH_4^+$ | mg $NH_3$ | mg N | µmol N |
|---|---|---|---|---|
| 1 mg $NH_4^+$ corresponds to | 1 | 0.94 | 0.78 | 55.44 |
| 1 mg $NH_3$ corresponds to | 1.06 | 1 | 0.82 | 58.72 |
| 1 mg N corresponds to | 1.29 | 1.22 | 1 | 71.39 |
| 1 µmol N corresponds to | 0.018 | 0.017 | 0.014 | 1 |

## 3.3.6 Magnesium

General remarks

### 3.3.6.1 AAS with flame

The determination of magnesium by atomic-absorption spectrometry is described schematically in method (1). Determination is carried out by direct atomization of the aqueous sample solution doped with lanthanum ions in a laughing gas-acetylene flame.

### 3.3.6.2 Complexometric (chelatometric) analysis

Calcium and magnesium ions are determined side by side in a solution by chelatometric titration, with for example voltammetric end-point indication. First the sum of $Ca^{2+}$ + $Mg^{2+}$ is determined with $Na_2EDTA$, and then in a second aliquot $Ca^{2+}$ alone is determined with $Na_2EGTA$. The magnesium content is calculated from the difference (see 3.3.7 Calcium, and 3.1 Hardness).

3.3.6.3. ICP-AES (see 3.3.12)

### 3.3.6.1 Direct determination of magnesium by means of atomic-absorption analysis

**Equipment parameters**

| | |
|---|---|
| 1. Hollow-cathode magnesium lamp | |
| 2. Wavelength | 285.2 nm |
| 3. Lamp current | 3.5 mA |
| 4. Aperture width | 0.5 nm |
| 5. Fuel gas | Acetylene |
| 6. Oxidant | Laughing gas |
| 7. Flame type | Reducing |
| 8. Background compensation | Yes |

| | |
|---|---|
| Optimum measuring range | 0.5 - 2 mg/l |
| Determination limit | 0.05 mg/l |
| Calibration | 0.2 - 2 mg/l |
| Acid concentration | 10 ml hydrochloric acid (1.17 g/ml) per 1000 ml solution for measurement |

**Remarks**

Add 5 ml lanthanum nitrate solution per 50 ml solution for measurement to reduce matrix disturbances.

Lanthanum nitrate solution:
Dissolve 50 g lanthanum nitrate ($La(NO_3)_3 \cdot 6\ H_2O$) in water and make up to 1000 ml.

Concentration by evaporation is possible.

### 3.3.6.2 Complexometric analysis (see Calcium, Section 3.3.7)

### 3.3.6.3 ICP-AES (see Section 3.3.12)

## 3.3.7 Calcium

**General remarks**

### 3.3.7.1 AAS

The method of atomic-absorption spectrophotometry described involves atomization in a laughing gas/acetylene flame and measurement at 422 nm (see method outline of 3.3.7.1).

### 3.3.7.2 Complexometric (chelatometric) determination

Calcium ions are determined together with magnesium ions in a solution by chelatometric titration, with for example voltammetric indication. For this purpose, first the sum ($Ca^{2+}$ + $Mg^{2+}$) is titrated with $Na_2EDTA$, and then $Ca^{2+}$ is titrated alone with $Na_2EGTA$ under working conditions in which $Mg^{2+}$ is not determined. The $Mg^{2+}$ content is given by the difference.

### 3.3.7.3 ICP-AES (see Section 3.3.12)

### 3.3.7.1 Direct determination of calcium by means of atomic-absorption analysis

**Equipment parameters**

| | |
|---|---|
| 1. Hollow-cathode calcium lamp | |
| 2. Wavelength | 422.7 nm |
| 3. Lamp current | 3.5 mA |
| 4. Aperture width | 0.5 nm |
| 5. Fuel gas | Acetylene |
| 6. Oxidant | Laughing gas |
| 7. Flame type | Reducing |
| 8. Background compensation | Yes |

| | |
|---|---|
| Optimum measuring range | 0.5 - 2 mg/l |
| Determination limit | 0.05 mg/l |
| Calibration | 0.2 - 2 mg/l |
| Acid concentration | 10 ml hydrochloric acid per 1000 ml solution for measurement |

**Remarks**

Add 5 ml lanthanum nitrate solution per 50 ml solution for measurement to reduce matrix disturbances.

Lanthanum nitrate solution:
Dissolve 50 g lanthanum nitrate ($La(NO_3)_3 \cdot 6\ H_2O$) in water and make up to 1000 ml

Concentration by evaporation is possible.

### 3.3.7.2 Chelatometic determination of calcium ions and magnesium ions (Fresenius, Schneider, Thielecke)

**General remarks**

Volumetric analysis with $Na_2EGTA$ ($Ca^{2+}$) and $Na_2EDTA$ ($Ca^{2+}$ + $Mg^{2+}$) by electrometric (voltammetric) indication.

Calcium ions are complexed by ($Na_2$-EGTA) at pH 10.5 (triethanolamine/ethanolamine) as the calcium chelate is formed, and the sum of calcium and magnesium ions is complexed by ($Na_2$-EDTA) as analogue chelates are formed.

Since the complex-formation constants of the metal chelates of calcium, strontium and barium are so different in comparison with magnesium, titration can be carried out without previously separating off the magnesium. End-point indication is based on the anodic oxidation of the mercury at an amalgamated silver electrode. The presence of ethanolamine promotes the oxidation.

Until the end-point is reached, the electrode reaction

$$Hg + HY^{3-} \longrightarrow HgY^{2-} + H^+ + 2\ e$$

is largely inhibited. As soon as the complexing solution starts to be in excess, the anodic dissociation of the mercury by depolarization begins.

The method can be applied to any waters whose interference ions can be masked by the addition of citrate/tartrate solution and triethanolamine solution, such as iron, manganese, copper and aluminium. Notable concentrations of other heavy-metal ions must be separated off beforehand.

Barium and strontium ions are titrated with EGTA and EDTA together with calcium, or calcium and magnesium. The magnesium value is obtained by taking the difference between the results of the two titrations. When determining calcium, strontium must be determined separately, e.g. by flame spectrometry, and deducted from the calcium value after conversion.

**Equipment**

Device for carrying out voltammetric titrations with adjustable current strength.

Titration stand, in which the motor-ram burette is synchronized with the measuring device and recording instrument, permitting the potential curve to be plotted during titration.

Silver rod electrode

Graphite electrode

Amalgamation of the silver rod electrode:
After degreasing, dip the electrode into nitric acid (1 : 1; v/v) for a few seconds, rinse thoroughly with dist. water, dry with filter paper and immerse approx. 5 cm into a thick-walled tube filled with reagent-purity mercury for a few minutes. When the silver rod is taken out, a drop of mercury forms at its lower end which should be removed by knocking briefly against the inner wall of the test tube.

EGTA solution, 0.1 m:
Gently heat 38.035 g of ethylene glycol-bis-(2-aminoethyl) N,N,N',N'-tetraacetic acid in 100 ml of 1 m sodium hydroxide solution; when cool, add dist. water and make up to 1 litre in the measuring flask.

EGTA solution, 0.01 m:
Prepare by diluting 0.1 m EGTA solution by a factor of 10.

EDTA solution, 0.1 m:
Dissolve 37.225 g of the disodium salt of ethylendinitrilotetraacetic acid · $H_2O$ (Dinatriumdihydrogenethylendiamintetraacetic acid) in dist. water and make up to 1 litre in the measuring flask.

EDTA solution, 0.01 m:
Prepare by diluting 0.1 m EDTA solution by a factor of 10.

Calcium solution, 0.1 m:
Dry 10 g of reagent-purity calcium carbonate, make into a suspension with a little dist. water and dissolve with 60 % perchloric acid of reagent purity using a little in excess, briefly boil off the carbon dioxide, and when cool make up to 1 litre with dist. water in the measuring flask.

Calcium solution, 0.01 m:
Prepare by diluting 0.1 m calcium solution by a factor of 10.

Magnesium solution, 0.1 m:
Dry 8.4321 g of reagent-purity magnesium carbonate, make into a paste with a little dist. water and dissolve with 60 % perchloric acid of reagent purity with a little in excess. Boil off the carbon dioxide, and when cool make up the solution to 1 litre with dist. water in the measuring flask.

Magnesium solution, 0.01 m:
Prepare by diluting 0.1 m magnesium solution by a factor of 10.

Masking solution:
Dissolve 5 g of reagent-purity ammonium citrate and 5 g of reagent-purity sodium potassium tartrate each in about 100 ml of dist. water, pour together in a 500-ml measuring flask and make up to the mark with dist. water. Store in a cool place.

Ethanolamine solution:
Dilute 30 g of ethanolamine to 500 ml with dist. water. Store in a cool place.

Triethanolamine solution:
Dilute 15 g of triethanolamine to the mark in a 1 litre measuring flask with dist. water. Store in a cool place.

Hydrofluoric acid, 40 %, reagent purity

Perchloric acid (1), 60 %, reagent purity

Perchloric acid, (2):
Dissolve approx. 330 g of 60 % perchloric acid with dist. water to 1 litre.

Nitric acid, 1 : 1 (v/v)

Mercury, reagent purity

Store all solutions in plastic flasks.

**Method**

Measure 50 ml of the water sample into each of two 250-ml beakers (if concentration < 10 mg/l $Ca^{2+}$ use correspondingly more; evaporate to 50 ml) and acidify with perchloric acid (2) (pH 2 - 3). Boil until the carbon dioxide is expelled, and when cool treat with 10 ml of masking solution. Leave the samples to stand for 10 minutes, then add 1 ml of triethanolamine solution and adjust to pH 10.5 with ethanolamine solution (thereby preventing any major pH change during titration due to liberated hydronium ions). Titration is carried out with either 0.01 m or 0.1 m EGTA or EDTA solution, depending on the calcium content (or in the case of the second sample calcium and magnesium content). Titration must begin immediately after adjusting the pH value.

Connect the amalgamated silver rod electrode, i.e. the indicator electrode, as the anode, and the graphite electrode, i.e. the counter-electrode, as the cathode. Set a constant current of 1 μA for calcium determination with EGTA. Arrange the two electrodes such that they are at the same height above the base of the beaker and about 0.5 cm apart; immersion depth should be 3 to 4 cm. Stir during titration. Immerse the fine tip of the burette, not quite touching the base of the beaker, at the greatest possible distance from the electrodes. Before beginning titration and immersing the tip of the burette, eliminate instrument play by allowing a few drops of titration solution to flow out. Provisionally set the paper feed of the recording instrument to a fairly fast speed and the voltage range to optimum sensitivity. About 1 ml before the expected end point titrate at a reduced rate. The end of the titration is shown by a sharp bend to the left in the curve.

To determine the sum of calcium and magnesium in the second sample, adjust to pH 10.5 and a constant current of 20 μA and titrate in the way described above, but using EDTA solution, until the curve bends sharply. Evaluation is carried out graphically by extending straight lines from both curve branches. The intersection of the two lines makes it possible to read off the ml directly from the paper strip.

For each unknown sample, it is expedient to carry out an orientation analysis beforehand and subsequently two further determinations. The orientation titration should be conducted using the faster paper feed rate and, initially, the following titrations as well, until about 1 ml before the expected end point the speed should be reduced.

Calculate the average consumption from the two sets of results. Carry out the titre setting of the solutions in the usual way.

### Calculation

mg of $Ca^{2+}$/l:

Consumption (ml) of 0.01 m EGTA solution · titre of 0.01 m EGTA solution · 0.4008 · 1000 divided by volume in ml.

mg of $Mg^{2+}$/l:

(Consumption (ml) of 0.01 m EDTA solution · titre of 0.01 m EDTA solution minus consumption (ml) of 0.01 m EGTA solution · titre of 0.01 m EGTA solution) · 0.24312 · 1000 divided by volume in ml.

Give the results rounded off to 0.1 mg/l.

**Note:**

Instead of the chelatometric method with electrometric (voltammetric) end-point indication as described above, complexometric titration with $Na_2$-EDTA may also be carried out using colour indicators for end-point indication. In this case, first the sum of $Ca^{2+}$ and $Mg^{2+}$ is determined with $Na_2$-EDTA against eriochrome black T and then (second sample) $Ca^{2+}$ alone.

$Na_2$-EDTA is again used for titration, but with 2-hydroxy-1-(2-hydroxy-4-sulpho-1-naphthylazo) naphthalene-3-carbonic acid (calcon carbonic acid) as an indicator.

See also Section 3.1 Hardness

### 3.3.7.3 ICP-AES (see Section 3.3.12)

## 3.3.8 Strontium

**General remarks**

In its chemical behaviour, strontium is very similar to calcium and they are generally detected together. Natural waters are low in strontium (less than 1 mg/l). Higher concentrations - up to several mg/l - may occur in subterranean waters. Sea water contains an average of 6 - 10 mg $Sr^{2+}$/l.

In the human body strontium is assimilated in the same way as calcium and is incorporated together with calcium as a component of the skeletal material of the bones. These characteristics of strontium have no harmful effect on the organism but may lead to health problems if instead of the naturally occurring non-radioactive isotopic mixture the long-lived isotope strontium-90 ($\beta^-$emitter, half-life 28 years), which is formed during nuclear fission of uranium, enters the human body, for example via radio-actively contaminated water. (See also Section 3.7, Radioactivity).

The AAS technique (3.3.8.1) described below is recommended for quantitative determination of strontium. If an AAS device is not available, the flame photometry method (3.3.8.2) may also be used. For ICP-AES (3.3.8.3) see Section 3.3.12.

### 3.3.8.1 Direct determination by means of atomic-absorption analysis

**Equipment parameters**

| | |
|---|---|
| 1. Hollow-cathode strontium lamp | |
| 2. Wavelength | 460.7 nm |
| 3. Lamp current | 5 mA |
| 4. Aperture width | 0.5 nm |
| 5. Fuel gas | Acetylene |
| 6. Oxidant | Laughing gas |
| 7. Flame type | Oxidizing |
| 8. Background compensation | No |
| Optimum measuring range | 1 - 5 mg/l |
| Determination limit | 0.05 mg/l |
| Calibration | 0.2 - 5 mg/l |
| Acid concentration | 10 ml hydrochloric acid (1.17 g/ml) per 1000 ml solution for measurement |

**Remarks**

Add 5 ml lanthanum nitrate solution per 50 ml solution for measurement to reduce matrix disturbances.

Lanthanum nitrate solution:
Dissolve 50 g lanthanum nitrate ($La(NO_3)_3 \cdot 6\ H_2O$) in water and make up to 1000 ml.

Concentration by evaporation is possible.

### 3.3.8.2 Determination by flame photometry

**General remarks**

When sprayed into a flame, strontium compounds colour the flame red. The spectral line which is characteristic for strontium is separated from the emitted light at 460 nm with the aid of a filter, prism or grating and its intensity is measured with a photometer.

Waters containing strontium mostly have a high calcium content as well. Consequently, the effect of calcium, intensifying the emission of strontium must be compensated for. In practical water analysis calcium is always determined in conjunction with strontium determination. For this purpose, in a separate preparation a quantity of calcium carbonate is weighed in corresponding to the $Ca^{2+}$ content of the quantity of water used for the analysis. This sample is dissolved in HCl and water and made up to the volume of the solution for measurement (100 ml). The emission of this solution is determined under the same measuring conditions which applied to the sample for analysis and subtracted from the measured value of the analysis.

Since strontium losses cannot be excluded in the course of the concentration process, it is advisable to treat a further weighed sample of water with a known quantity of strontium and to process and the measure this sample in the same way. This admixture control makes it possible to recognise losses of strontium due to the process of analysis or measurement errors. It is important that reagents of the greatest possible purity are used.

The method of concentration described for barium (3.3.9.2), for example, may be recommended; it may be followed up to the point where the carbonates are dissolved in HCl.

**Equipment**

Flame photometer with pre-chamber atomizer and a spectral filter or interference filter or

Flame spectrometer with direct atomization

Fuel gases in steel bottles:
An acetylene/air flame is to be recommended in the case of pre-chamber atomization and a hydrogen/oxygen flame in the case of direct atomization.

Semimicro balance: reading accuracy 0.01 mg

Platinum crucible with lid

Porcelain and glass dishes for evaporation, 1000 and 500 ml

Measuring flasks, 100 ml

Measuring pipette, 10 ml, calibrated

Sodium carbonate, reagent purity

Ammonium chloride, reagent purity

Hydrochloric acid, reagent purity (1.19 g/ml)

Ammonium hydroxide solution:
Carbonate-free, approx. 27 % (0.90 g/ml)

Strontium standard solution:
Dissolve 0.1685 g of strontium carbonate ($SrCO_3$, anhydrous), spectrally pure, in water, adding 10 ml of reagent-purity conc. hydrochloric acid and make up to 1 litre. This solution contains 0.1 mg/ml Sr.

Calcium carbonate, spectrally pure

**Calibration**

In order to prepare the calibration curve, take increasing quantities of strontium standard solution and dilute to the mark with water in 100-ml measuring flasks (eg 0.1, 0.2, 0.5, 1.0 and perhaps also 1.5, 2.0 mg/100 ml SR). The zero point is set using water, and the hundred point of the flame spectrometer is set with the strontium standard solution. The measured values for the calibration curve result in a straight line.

**Determination**

Measurements are taken of the solution of Ca-carbonates and Sr-carbonates in HCl. The influence of calcium must be taken into account. To achieve this, a quantity of calcium carbonate corresponding to the (known) calcium content of the sample solution is dissolved in hydrochloric acid and water and made up to 100 ml. The emission of this solution (at 460 nm) determined under the same measuring conditions as for the sample for analysis should then be subtracted from the measured value of the analysis. The difference provides the net emission for strontium.

A second water sample, treated with a known quantity of Sr and prepared in the same way as the actual sample for analysis, should also be measured.

**Calculation**

The measured value of the pure calcium solution is subtracted from the overall measured value for Sr + Ca emission. The difference (net emission) equals the test value for strontium. This should be corrected according to the value obtained by the admixture method. The strontium content is taken from the calibration curve and recalculated for 1 litre of water.

**3.3.8.3 ICP - AES (see Section 3.3.12)**

Atomic weight of strontium: 87.62

## 3.3.9 Barium

### 3.3.9.1 Direct determination of barium by means of atomic-absorption analysis

Equipment parameters

| | |
|---|---|
| 1. Hollow-cathode barium lamp | |
| 2. Wavelength | 553.6 nm |
| 3. Lamp current | 10 mA |
| 4. Aperture width | 0.5 nm |
| 5. Fuel gas | Acetylene |
| 6. Oxidant | Laughing gas |
| 7. Flame type | Rich in fuel gas |
| 8. Background compensation | No |

| | |
|---|---|
| Optimum measuring range | 1 - 10 mg/l |
| Determination limit | 0.2 mg/l |
| Calibration | 0.5 - 10 mg/l |
| Acid concentration | 10 ml hydrochloric acid (1.17 g/ml) per 1000 ml solution for measurement |

Remarks

Add 2 ml potassium chloride solution and 5 ml lanthanum nitrate solution to reduce matrix disturbances and improve sensitivity.

Potassium chloride solution:
Dissolve 100 g potassium chloride (KCl) in water and make up to 1 litre.

Lanthanum nitrate solution:
Dissolve 50 g lanthanum nitrate ($La(NO_3)_3 \cdot 6H_2O$) in water and make up to 1 litre.

The direct determination of barium is exceptionally susceptible to interference, particularly from calcium ions. For this reason, concentration is not advisable. Despite the length of time involved, the method of enrichment described in the following may be recommended.

### 3.3.9.2 Method of enrichment of barium

Basis

Barium ions are precipitated together with calcium by means of sodium carbonate, the carbonates dissolved in hydrochloric acid, and subsequently the barium ions are precipitated with chromate ions from acetate-buffered solution together with lead ions as carriers. The chromates are separated off, dissolved, and subsequently measured by AAS.

Reagents

| | |
|---|---|
| Ammonium hydroxide solution | (0.91 g/ml) |
| Hydrochloric acid | (1.17 g/ml) |
| Methyl orange | |

Sodium carbonate solution (saturated)

Calcium chloride solution:
Dissolve 5.0 g of $CaCl_2$ in water and make up to 1000 ml

Tartaric acid ($C_4H_6O_6$)

Lead nitrate solution:
Dissolve 2.5 g of $Pb(NO_3)_2$ in water and make up to 1000 ml

Sodium acetate ($NaCH_3COO$)

Potassium chromate solution (saturated)

Potassium chloride solution:
Dissolve 100 g of KCl in water and make up to 1 litre

Lanthanum nitrate solution:
Dissolve 50 g $La(NO_3)_3 \cdot 6H_2O$ in water and make up to 1000 ml

Membrane filter, pore size 0.45 µm

**Method**

Treat up to 2 litres of water with 10 ml hydrochloric acid (1.17 g/ml) and evaporate to a volume of about 100 - 200 ml.

Treat low-calcium waters with 25 ml of the calcium chloride solution, and add 100 mg tartaric acid.

Neutralize against methyl orange with the saturated sodium carbonate solution and add 5 ml of the latter in excess.

Leave to stand at 5 - 10 °C for 24 hours.

Filter through paper filter (group 2), rinse with cold water, and dissolve residue with a few ml of hot hydrochloric acid 1 + 1.

Set solution to about pH 3 with ammonium hydroxide, treat with 5 ml lead nitrate solution and add 5 g sodium acetate (max. total volume approx. 80 ml).

Add 1 ml saturated potassium dichromate solution and leave to stand at about 5 °C for 48 hours.

Separate precipitate through membrane filter (pore size 0.45 µm).

Treat filtrate with 5 ml lead nitrate solution and again leave to stand at approx. 5 °C for 24 hours (postprecipitation).

Dissolve the residue separated on the membrane filter with 5 ml hot hydrochloric acid 1 + 1, treat with 2 ml potassium chloride solution and 5 ml lanthanum nitrate solution and make up to the mark in 50-ml measuring flask.

Proceed in similar manner with the postprecipitation

Prepare 3 calibration readings and a blank reading in the same way.

Final determination: flame-absorption spectroscopy

Equipment parameters as described for direct determination

Decomposition with nitric acid - hydrogen peroxide is required for organically loaded samples.

Final determination by X-ray fluorescence techniques is also possible, making use of the Ba-L$\alpha$ line (vacuum, 20 mA, 60 kV).

### 3.3.9.3 Barium determination with the graphite tube technique (Furnace method)

**Equipment parameters**

| | |
|---|---|
| 1. Hollow-cathode barium lamp | |
| 2. Wavelength | 553.6 nm |
| 3. Lamp current | 10 mA |
| 4. Aperture width | 0.5 nm |
| 5. Background compensation | No |
| 6. Carrier gas | Argon |
| 7. Drying | 30 s at 125 °C |
| 8. Incineration | 30 s at 1000 °C |
| 9. Atomisation | 3 s at 2700 °C |
| 10. Gas stop | No, reduce carrier gas flow to 0.5 l/min. |

| | |
|---|---|
| Optimum measuring range | 50 - 200 µg/l with 10 µl sample |
| Determination limit | 5 µg/l |
| Calibration | Addition method |
| Acid concentration | 10 ml nitric acid (1.40 g/ml) per 1000 ml sample solution |

**Remarks**

Immediately after sampling, stabilize the water sample with 10 ml nitric acid (1.40 g/ml) per 1000 ml sample solution.

### 3.3.9.4 ICP - AES (see Section 3.3.12)

Atomic weight of barium: 137.34

## 3.3.10 Iron

**General remarks**

Ionogenic or complexed, divalent or trivalent iron may occur dissolved, collodially dissolved or dispersed in water. It occurs in ionic form if the water is free from oxygen or its pH is below about 3. At pH values higher than 3 or 4, iron (II) ions are converted to sparingly soluble hydrated oxide, and at pH values greater than 8, iron (II) ions are partly converted to insoluble iron (II) hydroxide. In waters containing oxygen, iron (II) ions are oxidized to iron (III) ions, which in turn form insoluble ferric hydroxide at pH values above 3.

For these reasons, surface waters (containing oxygen) are relatively low in iron. Considering the prevailing hydrogen-ion concentrations, trivalent ionogenic iron can hardly be expected, and the water-soluble compounds of the divalent iron which can be brought to the surface waters from underground are quickly converted to sparingly soluble iron (III) hydroxides under the influence of oxygen from the air and are hence separated from the water.

In ground water, in the absence of dissolved oxygen and if fairly high concentrations of carbonic acid are present, the situation is different from that of surface waters. Here relatively high iron concentrations (up to several mg/l) may be found. In most cases iron is present in the water in the soluble Fe -(II)- hydrogen carbonate form. In addition, iron also occurs in ground water in conjunction with organic substances, in particular with humic acids with which it forms stable complexes, partly in dissolved and partly in collodial form, which are also fairly resistant to atmospheric oxygen.

For the reasons described, subterranean waters may contain considerably greater quantities of dissolved iron as long as they do not come into contact with atmospheric oxygen. Even at a concentration of > 0.5 mg/l, dissolved iron gives drinking water a clearly metallic taste. Waters with high iron content are undesirable from an economic point of view (rust stains on household washing, the same in laundries, detrimental effect on taste and colour in dairies and breweries, and so on).

For this reason, iron concentrations below 0.1 mg/l are suitable for central drinking water supply plants. 0.2 to 0.3 mg/l Fe can be tolerated. However, concentrations of this order may be sufficient to propagate the development of iron bacteria (Leptothrix, Gallionella, Crenothrix). These promote and accelerate the conversion of divalent to trivalent iron which, as a hydroxide precipitate together with the gelatinous bacteria colonies (and in conjunction with manganese bacteria), may lead to the accumulation of mud and sedimentation.

### 3.3.10.1 Direct determination of iron by means of atomic-absorption analysis

**Equipment parameters**

| | |
|---|---|
| 1. Hollow-cathode iron lamp | |
| 2. Wavelength | 248.3 nm |
| 3. Lamp current | 5 mA |
| 4. Aperture width | 0.5 nm |
| 5. Fuel gas | Acetylene |
| 6. Oxidant | Air |
| 7. Flame type | Oxidizing |
| 8. Background compensation | Yes |

| | |
|---|---|
| Optimum measuring range | 0.5 - 5 mg/l |
| Determination limit | 0.1 mg/l |
| Calibration | 0.2 - 5 mg/l |
| Acid concentration | 10 ml hydrochloric acid (1.17 g/ml) per 1000 ml solution for measurement |

Remarks

Concentration by evaporation is possible, if concentrations are low. A blank reading is necessary since the danger of contamination is very high.

Decomposition with nitric acid/hydrogen peroxide is required if the samples contain organic loads.

### 3.3.10.2 Iron determination with the graphite tube technique (Furnace method)

Equipment parameters

| | |
|---|---|
| 1. Hollow-cathode iron lamp | |
| 2. Wavelength | 371.99 nm |
| 3. Lamp current | 5 mA |
| 4. Aperture width | 0.5 nm |
| 5. Background compensation | Yes |
| 6. Carrier gas | Argon |
| 7. Drying | 40 s at 180 °C |
| 8. Incineration | 45 s at 500 °C |
| 9. Atomization | 3 s at 2300 °C |
| 10. Gas stop | Yes |

| | |
|---|---|
| Optimum measuring range | 10 - 50 µg/l with 10 µg/l sample |
| Determination limit | 2 µg/l |
| Calibration | Addition method |
| Acid concentration | 10 ml nitric acid (1.40 g/ml) per 1000 ml sample solution |

Remarks

Stabilize the water sample immediately after sampling with 10 ml nitric acid (1.40 g/ml) per 1000 ml sample solution.

### 3.3.10.3 Spectrophotometric determination of iron (II) ions with 2,2'-bipyridyl (Fresenius-Schneider)

General remarks

With iron (II) ions 2,2'-bipyridyl forms a stable red-coloured complex which is not susceptible to oxidation and is not affected by iron (III) even in excess. It is therefore possible to differentiate between divalent and trivalent iron. The reagent is used to fix the ions at the sampling point.

The red-coloured iron (II)-2,2'-bipyridyl complex maintains its colour for up to 14 days under the measuring conditions selected. Spectrophotometric measurement is carried out at $\nu_{max.}$ = 522 nm ($\varepsilon_{522\ nm}$ = 7 844 l/mol · cm).

The detection limit for iron (II) ions is 0.02 mg/l. The extinction range E = 0.100 to 1.100 corresponds to iron (II) concentrations of 0.1 to 3 mg/l, if 50-mm cuvettes are used for measurement. Between 0 and 1 mg/100 ml $Fe^{2+}$ the calibration curve is a straight line.

## Equipment

Photometer

Cuvettes, path length 1 cm and 5 cm

Measuring flasks, 100 ml

2,2'-bipyridyl solution 0.1 %:
Dissolve 0.1 g of reagent-purity 2,2'-bipyridyl in 100 ml of dist. water.

Iron (II) reference solution I:
Dissolve 0.7022 g of ammonium iron (II) sulphate ($(NH_4)_2Fe(SO_4)_2 \cdot 6H_2O$) in dist. water and make up to 1 litre. 1 litre of this solution contains 0.1 mg $Fe^{2+}$.

Iron (II) reference solution II:
Dilute 100 ml of reference solution (I) to 1 litre with dist. water. 1 ml of this solution contains 0.01 mg $Fe^{2+}$.

(Both reference solutions are unstable. They must be freshly prepared before use each time.)

## Method

### Sampling

Measure 20 ml of the 2,2'-bipyridyl solution into a 100-ml measuring flask. At the sampling point, pipette 10, 20, 50 or 75 ml of the water sample rapidly and without shaking into the measuring flask. The quantity taken depends on the estimated $Fe^{2+}$ content. The red compound forms immediately. Turbid water samples should be filtered quickly before being transferred.

### Calibration

Take increasing quantities between 0 and 1.0 mg of $Fe^{2+}$ of freshly prepared iron (II) reference solutions I or II and transfer to 100-ml measuring flasks containing 20 ml of 2,2'-bypyridyl solution. Dilute to the mark with dist. water and measure at 522 nm.

### Determination

Dilute to the mark with dist. water at 20 °C and measure the extinction at 522 nm in 1-cm or 5-cm cuvettes. The pH of the solution for measurement should be between 3 and 9. If in exceptional cases the value is outside this range, a little solid sodium acetate should be used as a buffer (see under "Method" in Section 3.3.10.4 "Spectrophotometric determination of total iron with 2,2'-bipyridyl").

### Calculation

Using either the calibration curve or a calibration constant, work out the content of iron (II) ions in mg from the extinction values as measured, and convert to mg $Fe^{2+}$/l taking into account the quantity of water transferred to the measuring flask at the sampling point.

### 3.3.10.4 Spectrophotometric determination of total iron with 2,2'-bipyridyl (Fresenius-Schneider)

#### General remarks

In order to be able to determine the total iron content of a water spectrophotometrically with 2,2'-bipyridyl it is necessary to reduce the trivalent iron to iron (II) ions. Ascorbic acid is used as a reducing agent.

If undissolved or colloidal iron is also to be determined, the water is first acidified with hydrochloric acid. As it is heated, undissolved iron (III) compounds and those in colloidal solution are converted to the dissolved ionic form. In this form they are then reduced by ascorbic acid. In order to detect all the forms of iron, a separate sample should be filtered on the spot and the residue taken back to the laboratory to be decomposed, for example with HCl and $HNO_3$, evaporated and analyzed. The filtrate should be analyzed separately. It is advisable to use the entire weighed contents of a flask for the analysis. Any undissolved residues adhering to the walls are dissolved by adding hydrochloric acid. If insoluble residues still remain, eg silicic acid, they should be filtered off and decomposed before determination is carried out.

#### Equipment

See Section 3.3.10.3

In addition:

Hydrochloric acid, reagent purity (1.19 g/ml)

Sodium hydroxide solution, approx. 5%

Sodium acetate, $CH_3COONa \cdot 3\ H_2O$, reagent purity

Iron (III) reference solution I:
Dissolve 0.1430 g of reagent-purity iron (III) oxide in about 25 ml of hydrochloric acid (1.19 g/ml) at about 60 °C, dilute the solution with water, transfer to a 1-litre measuring flask and make up to the mark at 20 °C. 1 ml of this solution contains 0.1 mg iron (III) ions.

Iron (III) reference solution II:
Dilute 100 ml of reference solution I with dist. water to 1 litre. 1 ml of this solution contains 0.01 mg iron (III) ions.

#### Method

#### Determination

Measure or weigh the water sample, acidify distinctly with hydrochloric acid, and transfer to a beaker. Dissolve any undissolved residues in the sample container with hydrochloric acid (1.19 g/ml) and likewise transfer to the beaker. Leave the water sample in the beaker to evaporate for about 30 minutes at 90 - 95 °C. This procedure should also be followed even if no visible iron precipitates are present, since oxidized iron may be present in colloidal solution (since there is always a risk that traces of iron are introduced in the laboratory, a blank determination must always be carried out in parallel).

When cool, transfer the solutions to measuring flasks of appropriate size and make up to the mark with distilled water.

Transfer aliquot parts of these solutions and the blank test as well as the volumes of iron (III) reference solutions for the preparation of the calibration curve to 100-ml measuring flasks and basify with 5 % sodium hydroxide solution to pH 2 or max. 3. Add 300 - 500 mg of ascorbic acid to reduce the iron (III), treat with 20 ml of 2,2'-bipyridyl solution, buffer with solid sodium acetate (about 2 - 5 g) to a pH of about 5 to 6 and make up to the mark at 20 °C. Measure the extinction in 1-cm or 5-cm cuvettes at 522 nm.

Calculation

Read off the iron content from the calibration curve on the basis of the extinctions measured, take into account the dilution factor, the cuvette length and the initial volume and calculate the content of total iron in 1 litre of the water sample. A calibration constant may also be used for calculation instead of a calibration curve.

A blank test must always be carried out and taken into account.

### 3.3.10.5 Spectrophotometric determination of total iron with thioglycolic acid

General remarks

With iron (II)/(III), thioglycolic acid forms a red-purple colour complex which is measured at $\lambda_{max.} = 530$ nm ($\varepsilon_{530\ nm} = 3\ 765$ l (mol · cm)). The technique is one of the less sensitive methods of determining iron in water analysis but is simple and reliable.

It is suitable for the determination of total iron in concentrations greater than 0.3 mg/l.

Equipment

Photometer

Cuvettes, path length 1 cm and 5 cm

Measuring flasks, volume 50 ml

Thioglycolic acid, reagent purity, approx. 80 %

Ammonia solution (0.91 g/ml)

Method

Determination

Treat 45 ml of the sample solution with 0.5 ml of thioglycolic acid in a 50-ml measuring flask. Alkalize the solution by adding ammonia solution drop by drop; a red-purple coloration appears. Dilute to the mark with distilled water and measure the extinction of the solution at 530 nm.

Calibration and calculation are carried out in the same way as described under 3.3.10.4 "Spectrophotometric determination of total iron with 2,2'-bipyridyl".

### 3.3.10.6 ICP-AES (see Section 3.3.12)

Atomic weight of iron: 55.847

## 3.3.11 Manganese

General remarks

3.3.11.1 and 3.3.11.2

The determination of manganese by atomic-absorption spectrometry is a sensitive method which is virtually free from interference. Either flame AAS (3.3.11.1) or graphite tube AAS (3.3.11.2) may be used.

3.3.11.3 Photometric technique

After oxidation of the lower valency stages to manganese (VII), manganese can be determined in water photometrically as the permanganate ion. Oxidation is carried out in the presence of silver ions either using periodate ("Standard Methods for the Examination of Water and Wastewater" 1975) or using peroxodisulphate ("ausgewählte Methoden der Wasseruntersuchung", 1973, and "Deutsche Einheitsverfahren zur Wasser-, Abwasser- und Schlamm-Untersuchung", 1972).

3.3.11.4 ICP-AES (see Section 3.3.12)

### 3.3.11.1 Direct determination of manganese by means of atomic-absorption analysis

Equipment parameters

| | |
|---|---|
| 1. Hollow-cathode manganese lamp | |
| 2. Wavelength | 279.5 nm |
| 3. Lamp current | 5 mA |
| 4. Aperture width | 0.2 nm |
| 5. Fuel gas | Acetylene |
| 6. Oxidant | Air |
| 7. Flame type | Oxidizing |
| 8. Background compensation | Yes |

| | |
|---|---|
| Optimum measuring range | 1 - 2 mg/l |
| Determination limit | 0.2 mg/l |
| Calibration | 0.1 - 2 mg/l |
| Acid concentration | 10 ml hydrochloric acid (1,17 g/ml) per 1000 ml sample solution |

Remarks

Concentration by evaporation is possible if concentrations are low.

If samples have organic loads, decomposition is necessary using nitric acid-hydrogen peroxide.

### 3.3.11.2 Manganese determination with the graphite tube technique (Furnace method)

**Equipment parameters**

| | | |
|---|---|---|
| 1. | Hollow-cathode manganese lamp | |
| 2. | Wavelength | 279.5 nm |
| 3. | Lamp current | 5 mA |
| 4. | Aperture width | 0.5 nm |
| 5. | Background compensation | Yes |
| 6. | Carrier gas | Argon |
| 7. | Drying | 25 s at 180 °C |
| 8. | Incineration | 20 s at 500 °C |
| 9. | Atomization | 3 s at 2400 °C |
| 10. | Gas stop | Yes |

| | |
|---|---|
| Optimum measuring range | 1 - 5 µg/l with 10 µl sample |
| Determination limit | 0.05 µg/l |
| Calibration | Addition method |
| Acid concentration | 10 ml nitric acid (1.40 g/ml) per 1000 ml sample solution |
| Matrix modification | 10 µl 0.1 % ammonium nitrate solution |
| Ammonium nitrate solution | Dissolve 0.1 g $NH_4NO_3$ in water and make up to 100 ml. |

**Remarks**

Immediately after sampling, stabilize the water sample with 10 ml nitric acid (1.40 g/ml) per 1000 ml sample solution.

### 3.3.11.3 Spectrophotometric determination as permanganate following oxidation by peroxodisulphate

**General remarks**

Manganese ions are quantitatively oxidized to permanganate ions in acid solution by peroxodisulphate in the presence of silver ions acting catalytically:

$$2\ Mn^{2+} + 5\ S_2O_8^{2-} + 8\ H_2O \xrightarrow{Ag^+} 2\ MnO_4^- + 10\ SO_4^{2-} + 16\ H^+$$

The coloration of the permanganate ions, taken as a measure of the total manganese content of the water sample, is evaluated photometrically at 525 nm (absorption maxima at 528 and 548 nm). Beer's law applies in the concentration range from 0.6 to 25 mg/l.

The method is directly applicable to drinking water, ground water and surface water, and to waste waters following mineralization. Concentrations between 0.05 and 5 mg/l can be determined without dilution or concentration.

Chloride ions, organic substances and more than 5 mg/l iron ions have an interfering effect.

Up to 300 mg/l, chloride ions can be converted to mercury (II) chloride by adding 0.1 m mercury (II) nitrate solution. In contrast to silver chloride, this remains in solution and causes no additional turbidity.

Organic substance up to a $KMnO_4$ consumption of 60 mg/l can be destroyed by boiling for between 5 and 10 minutes with 10 ml of nitric acid (1.2 g/ml).

In the presence of higher chloride concentrations than 300 mg/l and/or organic substances with a $KMnO_4$ consumption greater than 60 mg/l, mineralize with conc. $H_2SO_4$ and conc. $HNO_3$. During this process, $Cl^-$ will be volatilized.

The interfering yellow coloration caused by iron (III) ions is eliminated by adding 1 ml of conc. phosphoric acid before oxidation with ammonium peroxodisulphate.

Coloration caused by other substances or turbidity of the water sample can be eliminated for measurement purposes by carrying out a second extinction measurement after destroying the permanganate coloration by means of nitrite or sodium azide.

**Equipment**

Photometer

Cuvettes, path length 5 cm

Measuring flasks, volume 100 ml

Phosphoric acid, conc., reagent purity (1.70 g/ml), manganese-free

Sulphuric acid, conc., reagent purity (1.84 g/ml), manganese-free

Nitric acid, conc., reagent purity (1.40 g/ml), manganese-free

Nitric acid (1.2 g/ml):
Add 200 ml of $HNO_3$ (1.40 g/ml), to 300 ml dist. water and mix.

Mercury (II) nitrate solution:
Moisten 17.13 g of $Hg(NO_3)_2 \cdot H_2O$ of reagent purity with 2 ml of conc. $HNO_3$, dissolve in a little dist. water, and dilute to 200 ml. 1 ml of the solution masks 17.75 mg of chloride.

Silver nitrate solution, 0.02 m.

Ammonium peroxodisulphate solution:
10 % (the solution should be freshly prepared daily.)

Sodium nitrate solution:
5 % (the solution is stable for a few days only.)

Manganese stock solution:
Dissolve 658.8 mg of manganese (II) perchlorate, $Mn(ClO_4)_2 \cdot 6\ H_2O$, dried to constant weight over silica gel in an evacuated desiccator in dist. water and make up to 1 litre. 1 ml of this solution contains 0.1 mg $Mn^{2+}$.

Manganese reference solution:
Make up 10 ml of the manganese stock solution to 100 ml with dist. water. 1 ml of this solution contains 0.01 mg $Mn^{2+}$.

Sodium azide ($NaN_3$):
Instead of sodium nitrite solution.

## Method

### Calibration

Measure off increasing quantities of the manganese reference solution containing between 0.05 and 1 mg of manganese, make up to 100 ml and treat according to the method given under "Determination" (add 5 ml of $HNO_3$ (1.2 g/ml) and 0.5 ml of 0.02 m silver nitrate solution, boil etc.).

### Determination

Take 100 ml of the water sample with a $Mn^{2+}$ content between 0.05 and 1 mg. If necessary, dilute a smaller quantity of sample to 100 ml or concentrate a larger quantity of sample to 100 ml. Acidify with 5 ml of nitric acid (1.2 g/ml) and treat with a volume of $Hg(NO_3)_2$ solution equivalent to the chloride content of the sample. In addition, an excess of 2 ml of $Hg(NO_3)_2$ solution is required[1].

If chloride concentrations above 300 mg/l and/or organic substances with a $KMnO_4$ consumption above 60 mg/l are present, first evaporate the 100-ml water sample (see above) with 1 ml each of conc. sulphuric acid and conc. nitric acid under a fume hood until white sulphur trioxide vapour appears. In the case of a brown coloration, dilute with dist. water and repeatedly evaporate with 1 ml of nitric acid until the mixture is colourless or no longer changes. Pick up with 5 ml of nitric acid (1.2 g/ml) and 100 ml of dist. water.

After masking the chloride or eliminating it, add 0.5 ml of 0.02 m silver nitrate solution and boil the solution for about 5 minutes adding boiled-out boiling chips so as to decrease the volume of liquid and simultaneously to destroy reducing substances, especially organic substances. Treat with 10 ml of freshly prepared 10 % ammonium peroxodisulphate solution, bring to the boil again for 5 minutes, cool immediately under running water and transfer to a 100-ml measuring flask with a glass stopper for photometric evaluation. Rinse with freshly boiled-out dist. water, dilute to the mark and mix thoroughly.

Measure the sample at 525 nm against dist. water within 30 minutes ($E_1$). Treat a blank solution in the same way.

Some water samples are inherently coloured or turbid. In such cases, particularly if the manganese concentration is low, it is advisable to decolour the solution with 1 - 2 drops of 5 % sodium nitrite solution after measuring the extinction in the cuvette and subsequently to measure the extinction once again ($E_2$). The net extinction E, to be used in the calculation, is given by $E_1 - E_2$.

An alternative procedure is to take the remaining solution not used for measurement, treat with about 50 mg of sodium azide in a 100-ml measuring flask, leave to stand for 10 minutes and measure extinction $E_2$.

---

[1] For the detoxification of residual Hg solutions according to W. Fresenius and W. Schneider see "Note".

Calculation

Read off the manganese content from the calibration curve (C) on the basis of the measured extinction value (or if appropriate the net extinction $E_1 - E_2 = E$). Using the formula

$$\frac{C \cdot 100}{V} = G \text{ (mg/l } Mn^{2+})$$

convert to the total manganese content, given as mg/l manganese (II) ions. In the formula

G = manganese ion concentration in mg/l $Mn^{2+}$
C = manganese from the calibration curve
V = undiluted volume of sample used, in ml.

### 3.3.11.4 ICP-AES (see Section 3.3.12)

Atomic weight of manganese: 54.9380

Note:

Detoxification of residual mercury solutions by precipitation with sodium thiosulphate and sodium hydroxide solution

Procedure

Collect the reaction solutions containing mercury ions which remain following completion of the manganese determination. Per 500 ml of solution treat with 30 g of sodium thiosulphate and then with 300 ml of 30 % sodium hydroxide solution. In this process the mercury is removed from the solution as sulphide. $Hg^{2+}$ ions are no longer detectable in the filtrate of the supernatant colourless liquid (flameless atomic-absorption spectrophotometry; detection limit 0.001 mg/l).

## 3.3.12 Atomic-emission spectrometry with inductively coupled plasma (ICP-AES, see also Chapter 2)

General remarks

Determination of 24 elements (Ag, Al, B, Ba, Ca, Cd, Co, Cr, Cu, Fe, K, Mg, Mn, Mo, Na, Ni, P, Pb, Sb, Sr, Ti, V, Zn and Zr) by atomic emission spectrometry with inductively coupled plasma (ICP-AES). The following draft was made available to us by the chairman of the Working Group (German Standardized Methods/DIN) as an example of the compilation and presentation of specifications for water analysis (draft manuscript, February 1987).

Area of Application

The technique is fundamentally suited to the determination of Ag, Al, B, Ba, Ca, Cd, Co, Cr, Cu, Fe, K, Mg, Mn, Mo, Na, Ni, P, Pb, Sb, Sr, Ti, V, Zn and Zr in water (e.g. wastewater and surface water) and in sludge (following appropriate decomposition processes) at or above the concentrations listed in Table a. Both the dissolved concentrations and the total concentrations of the above mentioned elements in the water can be determined. In order to determine the dissolved elements, the sample is filtered through a membrane filter (pore size 0.45 µm) before stabilization.

## Basis of the Method

ICP-AES is a technique of measurement used for the detection and determination of elements with the aid of atomic emission. The solution for measurement is atomized and the aerosol is transported into an inductively coupled plasma (ICP) with the aid of a carrier gas. There, the elements are excited such that they emit radiation. This is spectrally dispersed in a spectrometer and the intensities of the emitted element lines are measured by means of detectors (photomultipliers). A quantitative statement is possible by means of calibration with reference solutions, there being a linear relationship between the intensities of the emission lines and the concentrations of the elements over a broad range (usually several powers of ten). The elements may be determined either simultaneously or consecutively.

## Interference

Basically, ICP-AES is a low-interference method. In individual cases, however, the spectral and non-spectral types of interference described below may occur. Of these, line coincidences and interference due to sample feeding are the most significant in practice.

### Description of interference factors

### Spectral interference

#### Line coincidences

This interference arises as a result of overlapping of the spectral lines. Line coincidences become apparent only when a critical concentration ratio between the interfering and analyzed elements is reached. They are dependent on the spectral resolution of the spectrometer. The line coincidences which may occur in the analysis of wastewater have been established in a test. The results of this test are summarized in Table b.

#### Band coincidences

Molecules or radicals (e.g $N_2^+$, OH, NO, CN and NH), which are formed in the plasma from the solvent, the surrounding air, the gases used or incompletely dissociated compounds, (e.g. AlO), may emit bands which coincide with the lines under analysis.

#### Background influence

The spectral background is caused by recombination continua (e.g. Al, Ar) and is dependent, amongst other things, on the matrix and the wavelength.

#### Line reversal

If the concentration of the elements to be determined increases considerably in the outer (cooler) zones of the ICP, the radiation from the substance being analyzed which is coming from the hotter zones is absorbed here. Consequently the measured intensities are lower than they should be for the concentrations of the substances being analyzed.

Spurious radiation

A non-element-specific signal component may be produced by reflections in the spectral apparatus (stray radiation) or by radiation from other orders.

Non-spectral interference

Interference due to physical properties of the solution under analysis

This interference may arise in the sample-feed system as a result of differences in viscosity, surface tension or density between the reference solution and the solution under analysis.

Interference due to deposits

The properties of the sample-feed system may be altered by particles and relatively high concentrations of salts.

Interference due to carryover

Measurement influenced by residues in the sample-feed system (memory effect).

Interference in distribution

Interference due to alterations in the geometric distribution of the mass transportation (axial or radial) of the element to be determined in the radiation source.

Ionization interference

Easily ionizable elements (e.g. alkali metals and alkaline earth metals) may cause a slight displacement of the ionization balance for the element to be determined.

Change in electrical coupling efficiency

In the case of solutions with high total salt concentrations, the coupling efficiency may change from sample to sample.

Methods of eliminating or reducing interference

Spectral interference

If possible, switch to lines of analysis which are not disturbed (Table b).

In order to detect the interfering radiation instrumentally, devices may be used to rotate a quartz refractor plate in the path of rays, to displace the slit or also to rotate the grating. Corrections can be made by measuring the influence of the interfering element (mathematical correction) or by matrix simulation.

Mathematical corrections are based on:

a) Taking the difference between the total signal and the interfering signal: the net intensity ($I_X$) is given by Equation 1 and Fig. 100.

$$I_X = I_{XU} - 2 \cdot \frac{(bI_{U1} + aI_{U2})}{(a + b)} \quad (1)$$

In more complicated cases, further corrections may be necessary.

b) Calculation of the interfering signal in the case of line overlaps: after determining the calibration functions of the interfering elements ($a_1$, $a_2$ ...) and their concentrations ($c_1$, $c_2$ ...), the net intensity $I_X$ can be calculated from Equation 2, see below.

$$I_X = I_{XU} - \Sigma a_i \ c_i \quad (2)$$

In order to eliminate spurious radiation, screens, filters or special photomultipliers may be used.

where:

$I_X$ is the corrected intensity of the line
$I_{XU}$ is the measured intensity of the line
$a_i$ is the gradient of the reference function of the interfering element i in relation to the element line
$c_i$ is the concentration of the interfering element i.

**Non-spectral interference**

Interference due to uneven mass transportation may be avoided by dilution of the solution for analysis, by adjustment of the matrix of the sample solution and reference solution or by humidification of the carrier gas. In addition, the influence of variations in mass transportation can be corrected with the aid of an internal standard.

The influence of carryover can be avoided by sufficient rinsing of the sample-feed device between analyses.

Changes in electrical coupling efficiency are minimized by manual or automatic readjustment to ensure matching of the HF generator and the plasma coil.

**Designation**

Designation of the technique for determining 24 elements:

**Equipment**

Immediately before use, clean the glass and plastic containers and the pipettes with warm, dilute nitric acid and subsequently rinse with water.

ICP atomic-emission spectrometer with background compensation

Atomization system with low-pulsation pump (e.g. peristaltic pump)

Gas supply:
Internal gas: argon
External gas: argon, nitrogen or other gases
Atomizer gas particularly pressure- and flow-stabilized depending on type of atomizer

Bulb pipettes, 1 and 10 ml, e.g. DIN 12 691 VPAS 1 and 10, or millilitre pipettes, 0.1 - 5 ml

Measuring flasks, 100 and 1000 ml, glass (e.g. DIN 12 664 MSA) or plastic (e.g. polypropylene)

Beakers, 250 ml, e.g. DIN 12331 - HF 250

Stoppered plastic containers to store stock and standard solutions, e.g. high-pressure polyethylene or PTFE.

The chemicals used must be of at least "reagent purity", and water should be either redistilled or of an equal degree of purity.

The concentration of the elements under analysis in the water or reagents must be negligible in comparison with the lowest concentration to be determined.

Nitric acid, ($HNO_3$) (1.40 g/ml)

Hydrogen peroxide, ($H_2O_2$) 30 %
(When determining phosphorus, it should be noted that hydrogen peroxide is frequently stabilized with phosphoric acid).

Element stock solutions:
(Ag, Al, B, Ba, Ca, Cd, Co, Cr, Cu, Fe, K, Mg, Mn, Mo, Na, Ni, P, Pb, Sb, Sr, Ti, V, Zn, Zr) = 1000 mg/l each solution.

ICP-AES element solutions with appropriate specifications, which may be used as stock solutions, are available commercially. They should be prepared according to the producer's instructions. Generally, hydrochloric or nitric stock solutions are provided which can be stored for several months. The formulations of element stock solutions are compiled in Table c.

Element standard solutions I:
(Ag, Al, B, Ba, Ca, Cd, Co, Cr, Cu, Fe, K, Mg, Mn, Mo, Na, Ni, P, Pb, Sb, Sr, Ti, V, Zn, Zr) = 100 mg/l each solution.

- Pipette 10 ml of element stock solution into a 100-ml measuring flask.
- Add 1 ml of nitric acid and make up to the mark with water.
- For storage, the solutions should be transferred to polyethylene or PTFE vessels. Silver solutions are stable only in dark flasks.

The solutions may be stored for several months in a cool place.

Element reference solutions:
Prepare multi-element reference solutions which cover the working range from the stock and standard solutions corresponding to the element concentrations expected.
Taking into account the lower limits of the range of application, the following procedure, for example, may be followed:

Multi-element reference solutions I:
(Mn, Mo, Cd, Zn, Ti) = 1 mg/l, (Pb, P) = 10 mg/l.
Transfer 1 ml of element stock solution (Pb, P) and 1 ml each of element standard solutions I (Mn, Mo, Cd, Zn) into a 100-ml measuring flask, by pipette.
Add 1 ml of nitric acid.
Make up to the mark with water.

Multi-element reference solution II:
(Sr, Ba, Cu, Fe, V, Co, Zr) = 1 mg/l, (Al) = 10 mg/l.

Transfer 1 ml of element stock solution (Al) and 1 ml each of element standard solutions I (Sr, Ba, Cu, Fe, V, Co, Zr) into a 100-ml measuring flask by pipette.
Add 1 ml of nitric acid.
Make up to the mark with water.

Multi-element reference solution III:
(Ca, Na, K) = 100 mg/l.
Transfer 10 ml of element stock solution (Ca, Na, K) into a 100-ml measuring flask by pipette.
Add 1 ml of nitric acid.
Make up to the mark with water.

Multi-element reference solution IV:
(Sb, Mg) = 10 mg/l, (B, Ag, Cr, Ni) = 1 mg/l.
Transfer 1 ml of element stock solution (Sb, Mg) and 1 ml each of element standard solutions I (B, Ag, Cr, Ni) into a 100-ml measuring flask by pipette.
Add 1 ml of nitric acid.
Make up to the mark with water.

When preparing multi-element reference solutions with other combinations as well, attention must always be paid to the chemical compatibility of the elements. In order to avoid interference, the decomposing reagents should be added to the reference solutions.

Blank solution:
Measure 1 ml of nitric acid and 100 ml of water into a plastic container.

Zero-value solution:
Transfer 1 ml of nitric acid into a 100-ml measuring flask, by pipette, and make up to the mark with water.

**Sampling**

Basically, when taking samples of water the specifications of DIN 38 402 A 11 to 15 (cf. Chapter 1) must be taken into account. Further to this, proceed as follows:

- Take the water sample in glass, silica glass or plastic containers.
- Add 1 ml of nitric acid per 1000 ml of sample.
- The pH must be < 2; if necessary, increase the addition of acid.

When taking samples from sludge, observe the instructions in Chapter 1.

**Pretreatment of the sample**

In order to determine the total concentrations of the elements in the water, the sample is treated in the following way.

- To 100 ml of homogenized sample add 1 ml of conc. $HNO_3$ and 1 ml of $H_2O_2$
- Concentrate the mixture until a damp residue remains (complete drying may lead to reduced results).
- If decomposition is incomplete, add a little water to the residue and repeat the treatment.
- Pick up the residue with 1 ml of $HNO_3$ and a little water, and make up to 100 ml with water.

Decomposition can be dispensed with if complete detection of the elements is possible even without this pretreatment.

## Procedure

Before conducting the measurement, adjust the equipment-related parameters of the ICP-AES device according to the directions in the manufacturer's operating instructions. Particular attention must be paid to the following points:

- Set spectrometer precisely
- Check and ensure stability of spectrometer, generator and atomization system
- Establish wavelengths for background correction
- Measure each solution at least twice
- After each measurement, rinse the atomization system thoroughly with the zero-value solution.

## Measurement according to the standard calibration method

At the start of the measurement a reference function must be drawn up for each element to be determined. It is the result of two measured points (zero value and calibration concentration). The procedure is as follows:

- Prepare single-element and multi-element reference solutions and zero-value solution
- Measure the intensity of the selected emission lines of the elements in the reference solutions and the zero-value solution
- Determine the reference line for the series of measured values obtained in this way.

The following general equation (3) applies to the element to be determined, i:

$$y_i = b_i \cdot \beta_i + a_i \qquad (3)$$

$y_i$ = Signal level (measured value) (intensity in counts)
$b_i$ = Gradient of the reference function (sensitivity) (counts · litre · $mg^{-1}$)
$\beta_i$ = Mass concentration of the element (mg · $litre^{-1}$)
$a_i$ = Ordinate intercept of the reference function (measured value of zero value solution)

- After drawing up the reference lines, measure samples.
- Regularly check the validity of the reference function with zero-value solution and with the reference solutions.
- Should the element concentrations of the sample solutions exceed the validity range of the reference functions, dilute the sample solution with zero value solution.

In commercial ICP-AES devices, provision is also made for calibration techniques with several reference solutions and other algorithms.

## Measurement with internal standard

By using the internal standard, interferences in the sample-feed system which are caused by differences in viscosity, surface tension or density between the reference solution and the solution under analysis can be corrected. The method is used if matrix simulation is not possible.

If the internal standard is used, both the accuracy and, if measurement is simultaneous, the precision of the technique can be improved.

Elements which are not a constituent of the sample under analysis (e.g. scandium or yttrium) are particularly suited for use as the internal standard. Should it be the case that the element chosen as internal standard is in fact a constituent of the sample under analysis, the concentration of mass added to the sample under analysis should be approximately three orders of magnitude higher than the initial concentration.

The internal standard should possess emission lines which yield high intensities even at a low concentration of mass so that small additions to the solution under analysis suffice (e.g. 1 to 5 mg/l). This prevents uncontrolled influencing of the sample matrix.

The element which is used as the internal standard should display as few lines as possible in the emission spectrum in order to avoid spectral interference. When conducting measurement with the internal standard, proceed as follows:

- Establish the reference function for the elements to be determined and for the internal standard
- Add the internal standard to the solutions under analysis.
- Measure the solution under analysis with the internal standard.
- Measure the blank solution with the internal standard.

## Evaluation

### Evaluation with the standard calibration method

The mass concentrations of the elements are to be determined from the reference function (Equation 1) on the basis of the intensities. The following equation (4) applies:

$$\beta_i = \frac{(y_i - a_i)}{b_i} - \beta_i \, Bl \tag{4}$$

$y_i$ = Signal level of element i (measured value) (intensity in counts)
$b_i$ = Gradient of the reference function (sensitivity) (counts · litre · $mg^{-1}$)
$\beta_i$ = Concentration of element (mg · $litre^{-1}$)
$a_i$ = Ordinate intercept of the reference function (measured value of the zero value solution)
$\beta_i$ Bl = Measured blank reading

All dilution operations must be taken into account in the calculation.

### Evaluation using an internal standard

The correction factor R is determined from the measurements as follows (Equation 5):

$$R = I_{IS}/I^* \tag{5}$$

R = Correction factor
$I_{IS}$ = Intensity of the added internal standard
$I^*$ = Intensity of the internal standard in the sample under analysis

The corrected concentration of an element, $\beta_i$, is calculated according to Equation 6:

$$\beta_i = \frac{(y_i - a_i)\ R}{b_i} - \beta_{iBl} \cdot R_{Bl} \qquad (6)$$

$\beta_i$ = Concentration of element (mg · litre$^{-1}$)
$y_i$ = Signal level of element i (measured value) (intensity in counts)
$b_i$ = Gradient of reference function (sensitivity) (counts · litre · mg$^{-1}$)
$a_i$ = Ordinate intercept of the reference function (measured value of the zero value solution)
$b_{iBl}$ = Measured blank reading
R = Correction factor (Equation 5)

The internal standard is added to the blank solution before measurement, and the measured concentrations of the elements are likewise corrected (Equation 6).

## Presentation of the results

State the results to a maximum of three significant places, but no more than permitted by the lower range of application according to Table a.

| | | |
|---|---|---|
| Example: | Boron (B) | 0.04 mg/l |
| | Phosphorus (P) | 154 mg/l |

## Analysis report

The report should refer to this technique and include the following details:

a) Precise identity of the water sample
b) Statement of the results
c) Description of sample pretreatment, if any
d) Any deviations from this technique and complete information as to any circumstances which may have influenced the result.

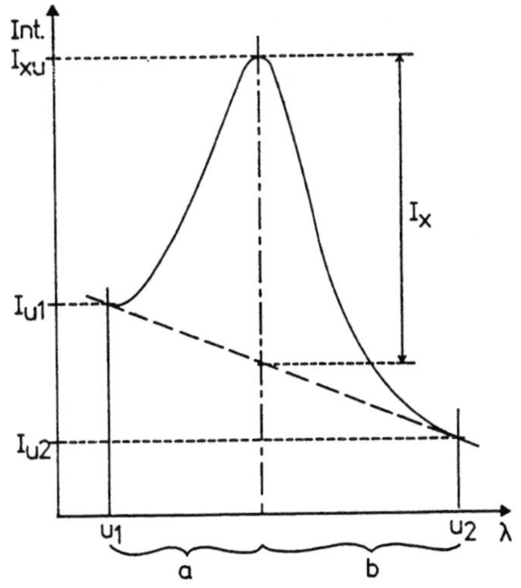


Fig. 100. Spectral interference; (see also equation 1)

**Table a.** Lower limits of application range in ICP-AES

| Element | Line (nm) | Lower limit (mg/l) | Element | Line (nm) | Lower limit (mg/l) |
|---|---|---|---|---|---|
| Ag | 328.068 | 0.01 | Mn | 257.610 | 0.002 |
| | 338.289 | 0.01 | | 293.306 | 0.02 |
| Al | 308.215 | 0.1 | Mo | 202.030 | 0.01 |
| | 396.152 | 0.1 | | 204.598 | 0.02 |
| B | 208.959 | 0.005 | Na | 589.592 | 0.1 |
| | 249.678 | 0.006 | | | |
| | 249.773 | 0.01 | Ni | 231.604 | 0.02 |
| Ba | 233.527 | 0.004 | P | 178.267 | 0.1 |
| | 455.403 | 0.002 | | 213.618 | 0.1 |
| | 493.409 | 0.003 | | 214.914 | 0.1 |
| Ca | 315.887 | 0.1 | Pb | 220.353 | 0.1 |
| | 317.933 | 0.01 | | | |
| | 393.366 | 0.0002 | Sb | 206.833 | 0.1 |
| | | | | 217.581 | 0.1 |
| Cd | 214.438 | 0.01 | | | |
| | 226.502 | 0.01 | Sr | 407.771 | 0.0005 |
| | 228.802 | 0.01 | | 421.552 | 0.01 |
| | | | | 460.733 | 0.1 |
| Co | 228.616 | 0.01 | | | |
| | | | Ti | 334.941 | 0.005 |
| Cr | 205.552 | 0.01 | | 336.121 | 0.01 |
| | 267.716 | 0.01 | | 337.280 | 0.01 |
| | 283.563 | 0.01 | | 368.520 | 0.01 |
| | 284.325 | 0.01 | | | |
| | | | V | 290.882 | 0.01 |
| Cu | 324.754 | 0.01 | | 292.402 | 0.01 |
| | 327.396 | 0.01 | | 310.230 | 0.01 |
| | | | | 311.071 | 0.01 |
| Fe | 259.940 | 0.02 | | | |
| | | | Zn | 206.200 | 0.01 |
| K | 766.490 | 2 | | 213.856 | 0.005 |
| Mg | 279.079 | 0.03 | Zr | 343.823 | 0.01 |
| | 279.553 | 0.0005 | | | |

**Table b.** Potential spectral interference sources when analyzing wastewater with ICP-AES

| Element | Line (nm) | Interfering elements | Element | Line (nm) | Interfering elements |
|---|---|---|---|---|---|
| Ag | 328.068 | | Mn | 257.610 | Fe |
| | 338.289 | | | 293.306 | |
| Al | 308.215 | Mn, V | Mo | 202.030 | |
| | 396.152 | Mo | | 204.598 | |
| B | 208.959 | | Na | 589.592 | |
| | 249.678 | Fe | | | |
| | 249.773 | Fe | Ni | 231.604 | Co |
| Ba | 233.527 | Fe, V | P | 178.267 | |
| | 455.403 | | | 213.618 | Cu, Fe, Mo, Zn |
| | 493.409 | | | 214.914 | Cu |
| Ca | 315.887 | Co | Pb | 220.353 | Al |
| | 317.933 | Fe | | | |
| | 393.366 | | Sb | 206.833 | Cr, Mo |
| | | | | 217.581 | |
| Cd | 214.438 | Fe | | | |
| | 226.502 | Fe | Sr | 407.771 | |
| | 228.802 | As | | 421.552 | |
| | | | | 460.733 | |
| Co | 228.616 | Ti | | | |
| | | | Ti | 334.941 | |
| Cr | 205.552 | Fe, Mo | | 336.121 | |
| | 267.716 | Mn, V | | 337.280 | |
| | 283.563 | Fe, Mo | | 368.520 | |
| | 284.325 | | | | |
| | | | V | 290.882 | Fe, Mo |
| Cu | 324.754 | | | 292.402 | Fe, Mo |
| | 327.396 | | | 310.230 | |
| | | | | 311.071 | Fe, Mn, Ti |
| Fe | 259.940 | | | | |
| | | | Zn | 206.200 | |
| K | 766.490 | | | 213.856 | Cu, Ni |
| Mg | 279.079 | | Zr | 343.823 | |
| | 279.553 | | | | |

**Table c.** Preparation of element stock solutions (1000 mg/l)

| Element | Component | Weight | Solvent |
|---|---|---|---|
| Aluminium | Al | 1000.0 | HCl, 6 mol/l |
| Antimony | Sb | 1000.0 | HCl, 1 mol/l |
| Barium | $BaCl_2$ | 1516.3 | water |
| | $Ba(NO_3)_2$ | 1902.9 | water |
| Boron | B | 1000.0 | $HNO_3$, 4 mol/l |
| | $H_3BO_3$ | 5719.5 | water |
| Cadmium | Cd | 1000.0 | $HNO_3$, 4 mol/l |
| | CdO | 1142.3 | $HNO_3$, 4 mol/l |
| Calcium | $CaCO_3$ | 2497.2 | $HNO_3$, 0.5 mol/l |
| Chromium | Cr | 1000.0 | HCl, 4 mol/l |
| Cobalt | Co | 1000.0 | HCl, 4 mol/l |
| Copper | Cu | 1000.0 | $HNO_3$, 4 mol/l |
| Iron | Fe | 1000.0 | HCl, 4 mol/l |
| | $Fe_2O_3$ | 1429.7 | HCl, 4 mol/l |
| Lead | $Pb(NO_3)_2$ | 1598.5 | water |
| Magnesium | MgO | 1658.1 | $HNO_3$, 1 mol/l |
| Manganese | Mn | 1000.0 | $HNO_3$, 4 mol/l |
| Molybdenum | $(NH_4)_2MoO_4$ | 2043.0 | water |
| Nickel | Ni | 1000.0 | HCl, 4 mol/l |
| Phosphorus | $(NH_4)_2HPO_4$ | 4260.0 | water |
| Potassium | KCl | 1906.7 | water |
| Silver | $AgNO_3$ | 1574.8 | $HNO_3$, 1 mol/l |
| Sodium | NaCl | 2542.1 | water |
| Strontium | $SrCO_3$ | 1684.9 | HCl, 1 mol/l |
| Titanium | Ti | 1000.0 | HCl, 4 mol/l |
| Vanadium | V | 1000.0 | $HNO_3$, 4 mol/l |
| Zinc | Zn | 1000.0 | $HNO_3$, 4 mol/l |
| Zirconium | $ZrOCl_2 \cdot 8H_2O$ | 3532.6 | HCl, 2 mol/l |

## 3.4 Trace substances (inorganic) (For ICP-AES determination of 24 elements see Section 3.3.12)

### 3.4.1 Trace detection of elements in parallel by emission spectrography survey analysis (see also Chapter 2)

#### General remarks

It is possible to detect many elements or their ions in water with relatively little time and effort by analyzing their emission spectra. The line spectrum produced is generally recorded on a photographic plate. Qualitative identification of the elements can be made on the basis of the position of the spectral lines. An element is considered to be detected if a line characteristic of this element has been found and identified with certainty. Identification is made with the assistance of a projector, with which comparison spectra may also be projected, and a table listing the wavelengths of the analytical lines. Generally speaking, the strongest lines are used for this purpose, lying in the ultraviolet (200 to 400 nm) and in the visible spectral range (400 to 800 nm). The iron spectrum may be recommended as a comparison spectrum.

Sufficient homogeneity of the sample which is filled into the hollow of the cathode is important. In order to obtain sufficient quantities of the elements contained in a water sample to permit detection by spectroanalysis, a concentration process is required before the analysis is made. In certain respects, this concentration process must also always be a depletion process, in which elements or ions present in considerable excess are isolated. In this way, their influence on the detection of elements present only in traces is considerably reduced.

#### Concentration process

Trace precipitation and extraction techniques, for example, have been used for multi-element concentration. Trace precipitation (if necessary with carriers) offers the advantage that it involves relatively little work. Since it is possible to use up to 50 litres of a water sample the detection of extreme traces is practicable. Generally, 1 litre of water sample is sufficient for trace concentration by extraction. Extraction methods may be varied in many ways. The techniques may be adapted accordingly from case to case.

#### 3.4.1.1 Concentration by trace precipitation (classical method)

#### Equipment

Quartz or porcelain dish for evaporation, volume at least 1 litre

Beakers of various sizes

Porcelain crucible

Sodium hydrogen carbonate solution, 10 %

Hydrochloric acid, 25 %

Water containing hydrochloric acid, 2 % HCl

Ammonium chloride, reagent purity

Hydrogen sulphide gas (e.g. Kipp's apparatus or $H_2S$ gas bottle)

Ammonium hydroxide solution, approx. 27 %

Method

For concentration by trace precipitation or coprecipitation the entire contents of a flask must be used. In general, a volume of 10 litres is taken initially. Boil down the weakly alkaline water sample (without any sediments which may be contained in the flask, see below). This is the principal solution. In many waters a pH of 7 - 8 is established automatically during evaporation. If necessary, adjust to this pH by adding 10 % sodium hydrogen carbonate solution in order to prevent the volatilization of e.g. mercury or germanium.

Treat the precipitates (sediments) in the sample flask with 25 % HCl and wash into a beaker together with the residue which is insoluble in acid. Evaporate the principal solution (see above) to a volume of about 0.5 - 1 litre and carefully acidify with this acid solution. Filter the acid solution and rinse out the insoluble residue with water containing hydrochloric acid (2 % HCl). If the water contains large amounts of calcium sulphate, the residue must be hot extracted several times with 2 % hydrochloric acid.

Incinerate the insoluble residue collected on the filter. Grind the ignition residue to uniform grain size; this is used for recording the **first spectrum.**

Combine the principal solution with the various weakly hydrochloric extracts of the insoluble residue and treat with about 10 g of ammonium chloride. Subsequently introduce a moderate flow of hydrogen sulphide into the cold solution for about 4 - 6 hours. Leave to stand for a few hours (until the sulphur functioning as trace catcher is completely settled), filter off the insoluble sulphides and rinse with weakly hydrochloric water containing hydrogen sulphide. Following this, rinse a few more times with chloride-free water containing hydrogen sulphide and incinerate at about 400 °C. The residue is used for recording the **second spectrum.**

Use (approx. 27 %) ammonium hydroxide solution to render the filtrate of the hydrogen sulphide precipitation in acid solution ammoniacal. Again pass hydrogen sulphide into the solution for about 1 - 2 hours. Filter off the precipitated sulphides of the ammonium sulphide group and rinse with water containing ammonium sulphide. Incinerate the filter with the residue (400 °C). The ignition residue is used to record the **third spectrum.**

The alkaline earth and alkaline elements are not detected by this method since they are identified elsewhere in the course of the general analysis, e.g. either by AAS or flame emission. Barium appears clearly as early as the first and second spectrum.

A blank test, using the same type and quantity of reagents and the devices and water used for the concentration process, should be carried out at the same time as the analytical process by trace precipitation. Filtration must also be carried out, even if no observable precipitation occurs in the blank test. The filters must also be subjected to the same process. The purity of all reagents must at least be of the grade "reagent purity".

In order to increase detection sensitivity, it is advisable to replenish the cavity in the electrode and arc over again under the same conditions for the same recording.

### 3.4.1.2 Concentration by extraction

(Experience has shown that this relatively old technique provides useful results).

Ammonium pyrrolidine dithiocarbamate (APDC), for example, is suitable for group extraction in water. It is added as a 2 % aqueous solution to the water sample, which should be set to pH 2 (5 ml to 1 litre of water sample). Repeat extraction with about 15 ml of chloroform each time until the chloroform phase is almost colourless (at least three times).

Add a quantity of a 10 % aqueous diammonium citrate solution sufficient to prevent the precipitation of calcium and magnesium ions, set the pH to 8, treat with 5 ml of APDC solution and extract again with chloroform. One extraction generally suffices.

Evaporate the chloroform extracts almost to dryness over a water bath, wash over to a porcelain crucible, bring to dryness on a sand bath and incinerate in a muffle furnace at 350 °C. Proceed with spectrographic analysis as described under "Concentration by trace precipitation".

In order to make the circle of trace elements detectable by group extraction as wide as possible, F. A. Pohl has described a method in which the extractions are performed using a mixture of three complexing reagents: Sodium diethyl dithiocarbamate (DDTC), 8-hydroxiquinoline (oxine) and dithizone.

Method

The extraction of 1 - 5 litres of the water sample is usually sufficient for a spectrographic survey analysis. The water sample must be completely clear. Any sediments present must be dissolved with hydrochloric acid.

Use the entire contents of the transport flask for the extraction. In the case of volumes larger than 1 litre, evaporate to 1 litre. Transfer the water sample prepared in this way, including the solution of the sediment, to a 1.5-litre shaking funnel, set the pH to about 3 by adding hydrochloric acid or ammonium hydroxide solution dropwise (test by spotting onto indicator paper), treat with 2 ml of DDTC solution, shake for 10 seconds, then add 15 ml of oxine solution and shake vigorously for 1 minute.

After separation of the phases, run off the chloroform phase and repeat the same process about three times, again adding DDTC and oxine, until the chloroform phase is virtually colourless after shaking. A slight yellow coloration which recurs if the extraction process is repeated in the same way stems from the oxine which is rearranged to a weakly yellow form on contact with water.

Adjust the solution to pH 5 in the shaking funnel by adding ammonium hydroxide solution a drop at a time and shake out with DDTC and oxine solution in the same way until the chloroform layer is colourless. Collect all the chloroform fractions in a distilling flask.

Add 10 ml of ammonium tartrate solution. In the case of waters which are very rich in minerals: increase this quantity accordingly, so as to prevent hydroxide precipitation at higher pH values. Shake out at pH 7 with 2 ml of DDTC solution, 15 ml of oxine solution and 5 ml of dithizone solution until the chloroform phase remains green. Repeat the same process at pH 9, at which point the concentration is generally completed after a single extraction process.

This method of extractive concentration of trace elements in water for the purpose of subsequent qualitative spectral analysis (survey analysis) using the three complexing reagents, may be used to detect the following elements in concentrations down to 0.5 µg/l or below:

Ag, Al, As, Au, Bi, Cd, Co, Cr, Cu, Fe, Ga, Hf, Hg, In, La, Mn, Mo, Ni, Pb, Pd, Pt, rare earths, Sb, Sc, Sn, Th, Ti, Tl, U, V, Y, Zn and Zr (after F. A. Pohl).

After distilling off the chloroform, transfer the combined chloroform extracts to a quartz dish, add a drop of concentrated nitric acid and incinerate in a muffle furnace at 350 °C.

Insert the thoroughly homogenized residue into the cavity of the electrode for spectral analysis and apply current according to the instructions for use given by the manufacturer of the instrument.

As with all trace analysis, particular attention should be paid to immaculate cleanliness of the laboratory devices, and to the purity of the reagents (at least "reagent purity") and of the solvents, since the risk of introducing other substances is greater than that of losing trace elements (control with blank test).

Pay particular attention to the safety regulations for the use of chloroform. When testing for heavy elements, X-ray fluorescence may also be used for identification. A method which has proved valuable for the detection of Ag, Bi, Cd, Co, Cu, Ni, Pb, Tl and Zn following concentration by extraction is described in DIN 38406, Part 21, September 1980.

### 3.4.2 Aluminium

#### General remarks

Atomic-absorption spectrophotometry can be recommended as a method of determining aluminium. The techniques described here are the direct method using flame-absorption spectrometry (3.4.2.1), where necessary with the method of concentration outlined below, and AAS using the graphite tube technique (3.4.2.2).

Photometric or spectrophotometric methods are also suitable for the determination of small quantities of aluminium in water. In these methods organic reagents form colour lakes with aluminium ions. The absorption of light by these lakes can be measured at the appropriate wavelength (3.4.2.3).

The method with alizarin S and the method with eriochromcyanine R are both included in the "Deutsche Einheitsverfahren zur Wasser-, Abwasser- und Schlamm-Untersuchungen" (E 9, 1972), (German Standard Methods for the

Examination of Water, Wastewater and Sludges). The eriochromcyanine R method is also recommended as a standard method in addition to determination by atomic absorption spectrophotometry in the "Standard Methods for the Examination of Water and Wastewater" of the American Public Health Association, 13th Ed. The authors' experience also indicates that this method is valuable if no AAS device is available.

3.4.2.4 ICP-AES see Section 3.3.12

### 3.4.2.1 Direct determination by means of atomic-absorption analysis

#### Equipment parameters

| | |
|---|---|
| 1. Hollow-cathode Al lamp | |
| 2. Wavelength | 309.3 nm |
| 3. Lamp current | 5 mA |
| 4. Aperture width | 0.5 nm |
| 5. Fuel gas | Acetylene |
| 6. Oxidant | Laughing gas |
| 7. Flame type | Reducing |
| 8. Background compensation | Yes |

| | |
|---|---|
| Optimum measuring range | 2 - 50 mg/l |
| Determination limit | 0.1 mg/l |
| Calibration | 0.2 - 2 mg/l<br>2 - 50 mg/l |
| Acid concentration | 10 ml hydrochloric acid (1.17 g/ml) per 1000 ml solution for measurement |

#### Remarks

Add 5 ml lanthanum nitrate solution per 50 ml solution for measurement to reduce matrix interferences and to improve sensitivity.

Lanthanum nitrate solution:
Dissolve 50 g $La(NO_3)_3 \cdot 6\ H_2O$ in water and make up to 1 litre.

Concentration by evaporation is possible. One extraction method is described below.

#### Reagents

| | |
|---|---|
| Acetic acid | (1.05 g/ml) |
| Hydrochloric acid | (1.17 g/ml) |
| Ammonium hydroxide solution | (0.91 g/ml) |

Oxine solution:
Dissolve 50 g oxine (8-hydroxyquinoline) in 120 ml acetic acid, and make up to 1 litre with water.

Ammonium acetate solution:
285 ml acetic acid + 565 ml water + 285 ml ammonium hydroxide solution, set pH 6.6 (electrometrically).

Chloroform

Methyl red

**Procedure**

Evaporate the required quantity of water with a little hydrochloric acid to approximately 200 ml

Neutralize with ammonium hydroxide with respect to methyl red

Acidify with one drop of hydrochloric acid and transfer to a 250 ml separating funnel

Add 1 ml oxine solution, 5 ml ammonium acetate solution and 10 ml chloroform

Shake for 2 minutes and after phase separation run off the organic phase into a quartz glass

Repeat extraction with 10 ml and 5 ml chloroform

Evaporate the accumulated organic extracts in the quartz glass on a water bath and ignite at 450 °C in the muffle furnace

Allow to cool, pick up with 2.5 ml hydrochloric acid, and heat

Treat with 2.5 ml lanthanum nitrate solution, transfer to 25 ml measuring flasks and fill up to the mark

Carry out final determination as previously described for atomic-absorption analysis

Take a blank reading in the same way

Decomposition with nitric acid/hydrogen peroxide is necessary for organically loaded samples.

### 3.4.2.2 Aluminium determination with the graphite tube technique (Furnace method)

**Equipment parameters**

| | |
|---|---|
| 1. Hollow cathode Al lamp | |
| 2. Wavelength | 309.3 nm |
| 3. Lamp current | 5 mA |
| 4. Aperture width | 0.5 nm |
| 5. Background compensation | Yes |
| 6. Carrier gas | Argon |
| 7. Drying | 25 s at 140 °C |
| 8. Incineration | 15 s at 500 °C |
| 9. Atomization | 3 s at 2400 °C |
| 10. Gas stop | Yes |

| | |
|---|---|
| Optimum measuring range | 5 - 50 µg/l with 10 µl sample |
| Determination limit | 2 µg/l |

| | |
|---|---|
| Calibration | Addition method |
| Acid concentration | 10 ml nitric acid (1.40 g/ml) per 1000 ml sample solution |

Remarks

Immediately after sampling, stabilize the water sample with 10 ml nitric acid (1.40 g/ml) per 1000 ml sample solution.

### 3.4.2.3 Spectrophotometric determination with eriochromcyanine R

General remarks

In acid solution aluminium ions together with eriochromcyanine R form a red-purple colour lake, the colour and intensity of which depend on the pH value of the solution. Optimum colour-lake formation occurs in the region between pH 4 and 6, but in this range the reaction is very slow. The reaction time can be considerably reduced by initially adding the eriochromcyanine R in a medium containing thioglycolic acid (pH 2.5) and subsequently adjusting to about pH 5.6 with the buffer solution. The extinction maximum of the colour complex is at 530 nm.

The method is suitable for the determination of aluminium ions in concentrations between 0.02 and 0.7 mg/l. If concentrations are higher the sample should be diluted accordingly.

In the absence of fluoride and polyphosphates approximately 6 µg of Al per litre can be determined.

Fluoride ions interfere (negative interference) even in low concentrations. In order to eliminate fluoride ions the sample must be evaporated with sulphuric acid (1.84 g/ml), in a platinum or quartz dish.

The influence of polyphosphates is also eliminated in this way. Hydrolysis to orthophosphates must be completed in the course of this process.

If relatively high concentrations of organic substances are present, 5 ml of ammonium peroxodisulphate solution should be added following evaporation and the sample brought to the boil for 50 minutes. Provided the procedure is carefully followed the influence of other interfering factors is only slight in the method as described.

Equipment

Spectrophotometer or filter photometer

Cuvettes, path length 0.5 and 1 cm

Platinum or quartz dish, 500 ml

Sand bath or surface radiant heater

Measuring flasks, volume 1 litre, 250 ml, 100 ml

Bulb pipettes, volume 100 ml, 25 ml, 10 ml, 5 ml, 2 ml, 1 ml

Hydrochloric acid, reagent purity, (1.12 g/ml)

Hydrochloric acid, dilute:
Mix 10 ml of hydrochloric acid, (1.19 g/ml) with 90 ml of dist. water.

Sulphuric acid, reagent purity (1.84 g/ml)

Ascorbic acid solution:
Dissolve 1 g of reagent-purity ascorbic acid, $C_6H_8O_6$, in 100 ml of dist. water.

Thioglycolic acid solution:
Mix 10 ml of thioglycolic acid, $HSCH_2COOH$, approx. 80 %, with 90 ml of dist. water. This solution has a useful life of only one day.

Eriochromcyanine solution:
Dissolve 0.1 g of reagent-purity eriochromcyanine R, $C_{23}H_{15}Na_3O_6S$, in 100 ml of dist. water. The solution has a useful life of only one day.

Buffer solution:
Make up 274 g of reagent-purity ammonium acetate, $CH_3COONH_4$, 109 g of reagent-purity sodium acetate, $CH_3COONa \cdot 3\ H_2O$, and 6 ml of reagent-purity acetic acid, $CH_3COOH$, to 1 litre with dist. water.

Ammonium peroxodisulphate solution:
Dissolve 10 g of reagent-purity $(NH_4)_2S_2O_8$ in 100 ml of dist. water.

Aluminium stock solution:
Dissolve 1000 mg of aluminium, metal strip, superpure, in 20 ml of hydrochloric acid, (1.12 g/ml); make up the solution to 1 litre with dist. water. 1 ml of the solution contains 1 mg of Al.

Aluminium reference solution:
Dilute 10 ml of aluminium stock solution to 1 litre with dist. water (1 ml = 0.01 mg of Al).

## Method

### Calibration

Prepare the calibration curve after taking volumes of the aluminium reference solution which contain 0.01/0.05/0.10/0.20/0.50/0.75/1.0 mg of $Al^{3+}$ per litre according to the procedure described under "Determination".

### Determination

Depending on the concentration of aluminium ions boil down up to 500 ml of the water sample with 2 ml of sulphuric acid in a platinum or quartz dish and evaporate to dryness. Pick up the residue with 1 ml of dilute hydrochloric acid and a few ml of distilled water. Boil the mixture until the residue is completely dissolved.

If relatively high concentrations of organic substances are present, add 5 ml of ammonium peroxodisulphate solution following evaporation and boil the sample for 50 minutes.

Neutralize the hydrochloric solution and transfer to a 100-ml measuring flask. Add 5 ml of ascorbic acid solution and 1 ml of thioglycolic acid solution, dilute to approximately 60 ml with distilled water and mix.

Immediately treat the sample with 10 ml of eriochromcyanine solution, and after 20 minutes with 10 ml of buffer solution. The pH should now be 5.6.

Dilute the solution to the mark with distilled water and measure its extinction after 30 minutes at 530 nm against a blank test which has been treated in the same way, including evaporation with sulphuric acid. The blank test should be conducted with distilled water instead of the water under investigation. Its extinction, measured against distilled water, is 0.6/cm. The colour lake is constant for at least 1 hour.

## Calculation

Read off the aluminium content from the calibration curve on the basis of the measured extinction. Take into account the cuvette path length of the blank test and the initial volume and convert to mg of $Al^{3+}$/l. Instead of reading from the calibration curve a calibration factor may also be used for calculation.

Atomic weight of aluminium: 26.9815

## 3.4.3 Arsenic

### General remarks

Arsenic is geologically widespread and traces of arsenic occur naturally in ground water and surface waters. In some subterranean water geogenic arsenic concentrations even occur in the mg/l range. Anthropogenic arsenic from waste waters, particularly from the metallurgical, chemical or mining industries, may cause high levels of arsenic contamination. Particular attention must also be paid to seepage water from ash heaps and rubble tips.

### Quantitative determination

3.4.3.1 Determination by means of hydride AAS

3.4.3.2 Spectrophotometric determination of total arsenic with silver diethyl dithiocarbamate (Fresenius-Schneider)

3.4.3.3 Iodometric determination of arsenic (III)

3.4.3.4 Separation of arsenic (III) and arsenic (V) (Gorbauch)

### 3.4.3.1 Determination of arsenic using the hydride AAS technique

#### Principle

Using sodium borohydride, arsenic ions are reduced to arsenic hydride, transferred to a heated quartz cuvette with the aid of a current of inert gas, decomposed thermally, and the absorption of the atoms is measured in the beam of an atomic-absorption spectrometer. In the hydride technique, the element which is to be determined is volatilized as a gaseous hydride and separated off from the matrix. Interferences may occur if there is a considerable excess of elements such as antimony, tin, bismuth, mercury, selenium or tellurium, which may also be volatilized using this technique.

Above all, heavy metals such as copper and nickel have a disturbing effect during the hydride formation itself. These interferences may be diminished by adding 300 mg of solid 2-pyridine aldoxime to the solution for measurement.

Since arsenic (III) and arsenic (V) are not equally sensitive when determined by the hydride technique, arsenic (V) should be reduced to arsenic (III) before determination (prereduction).

### Reagents

Arsenic stock solution (1000 mg/l)

Hydrochloric acid (1.17 g/ml)

Sodium borohydride solution:
Dissolve 3 g $NaBH_4$ and 1 g NaOH in water and make up to 100 ml.

Prereduction solution:
Dissolve 5 g ascorbic acid ($C_6H_8O_6$) and 0.5 g potassium iodide (KI) in water and make up to 100 ml.

Prereduction of the water samples and decomposition solutions:
Introduce 20 ml of the water sample or the decomposition solution into the reaction vessel of the hydride system (prereduction may also be carried out in a different vessel), treat with 4 ml hydrochloric acid (1.17 g/ml) and 2 ml prereduction solution and leave to stand for at least 15 minutes.

Treat the blank and calibration solutions in the same way.

### Equipment parameters

Atomic-absorption spectrometer with hydride attachment:

| | |
|---|---|
| 1. Hollow-cathode arsenic lamp | |
| 2. Wavelength | 193.7 nm |
| 3. Lamp current | 5 mA |
| 4. Inert gas | Argon |
| 5. Gas through-flow | 30 l/h |
| 6. Cuvette temperature | 900 °C |
| 7. Aperture width | 1 nm |
| 8. Integration | Peak area |
| 9. Integration time | 25 s |
| 10. Background compensation | No |
| 11. Measuring range | 2.5 - 10 µg/l |
| 12. Determination limit | 0.5 µg/l |

The parameters quoted here for the hydride technique apply to a Varian AAS unit, model 775, in conjunction with a hydride system manufactured by Messrs. Berghoff, Tübingen, FRG.

### Method

Connect the reaction vessel containing the prereduced solution to the hydride system. Allow a constant current of inert gas (argon) to flow through the system at 30 l/h to expel the air. Following this, continuously add 3 % sodium borohydride solution using a peristaltic pump. Arsenic ions are reduced to arsenic hydride and directed by the current of inert gas into a quartz cuvette heated to 800 °C where they are thermally decomposed. Measure the absorption.

Remarks

Since the hydride technique only permits quantitative detection of arsenic which is present ionogenically in solution, decomposition is required for organic or complexed arsenic. Organically loaded water samples must also be decomposed before measurement. Two methods of decomposition are described below.

**A) Decomposition with sulphuric acid/hydrogen peroxide**

**Reagents**

Sulphuric acid (1.84 g/ml)

Hydrogen peroxide (30 %)

Method

Introduce 50 ml of the water sample into a 150-ml beaker, treat with 5 ml sulphuric acid and add 5 ml hydrogen peroxide

Heat in a sand bath until $SO_3$ vapour appears

If the organic load of the water sample is high, repeat the addition of hydrogen peroxide to the decomposition mixture.

When the decomposition mixture is cool, transfer the sample to a measuring flask, 50 ml, and dilute with water to the mark.

Care should be taken that the sample never evaporates to dryness.

**B) Decomposition with magnesium oxide/magnesium nitrate**

**Reagents**

Hydrochloric acid (1.17 g/ml)

Magnesium oxide (MgO)

Magnesium nitrate ($Mg(NO_3)_2 \cdot 6\ H_2O$)

**Procedure**

Introduce 50 ml of the water sample into a 100-ml quartz-glass beaker, neutralize if necessary, and subsequently treat with 1 g magnesium oxide and 1 g magnesium nitrate

Evaporate over a water bath

Carefully cover the evaporation residue with 2 g magnesium nitrate

Dry at 180 °C in a drying oven for 20 minutes

Place in a cold muffle furnace, heat to between 500 and 550 °C and leave at this temperature for 1 hour

Pick up the ignition residue with 10 ml hydrochloric acid, transfer to a 50-ml measuring flask, and dilute with water to the mark

### 3.4.3.2 Spectrophotometric determination of total arsenic with silver diethyl dithiocarbamate

#### General remarks

Arsenic ions are converted to gaseous arsenic hydride (arsine, $AsH_3$) by nascent hydrogen in acid solution. Arsenic hydride reacts with a silver diethyl dithiocarbamate solution in pyridine forming a red compound. The depth of the colour is proportional to the arsenic content. Photometric measurement can either be carried out at 560 nm (green filter) or at 546 nm if a mercury vapour lamp is used.

The method is suitable for the determination of arsenic concentrations between 0.002 and 2 mg/l in water.

Interference due to hydrogen sulphide or substantial quantities of organic substances is eliminated by evaporation with sulphuric acid/nitric acid. (If no evaporation is carried out, a reaction tube containing glass wool soaked with lead acetate must be connected in series to safeguard against interference by hydrogen sulphide.) Oxidizing substances which could inhibit a reduction of arsenic ions to arsine are reduced by tin (II) chloride. Antimony ions interfere, since under the reaction conditions described they are converted to antimony hydride (stibine), and the stibine likewise forms a red colour complex with silver diethyl dithiocarbamate. If this is the case and measurement is carried out at 540 nm, Sb contributes 8 % of the extinction of the arsenic complex (Koch and Koch-Dedic). A correction can therefore be made on the basis of the Sb content, measured separately.

It should be noted that deposits of iron hydroxide serve as trace catchers for arsenic and may concentrate the entire arsenic in the precipitate. For this reason, precipitates must be complètely dissolved before analysis.

The spherical joints must be sealed very carefully with concentrated sulphuric acid. Other sealants are less suitable.

#### Equipment

Arsine generator:
100-ml Erlenmeyer flask with NS 19, absorption attachment with widened connecting piece (to hold lead-acetate glass wool when working without evaporation), spherical joint with clip and absorption vessel (see Fig. 101).

Sulphuric acid, con., reagent purity (arsenic-free), (1.84 g/ml)

Nitric acid, conc., reagent purity, (1.4 g/ml)

Sulphuric acid, dil.:
Dilute 220 ml of conc. sulphuric acid with distilled water to 1 litre.

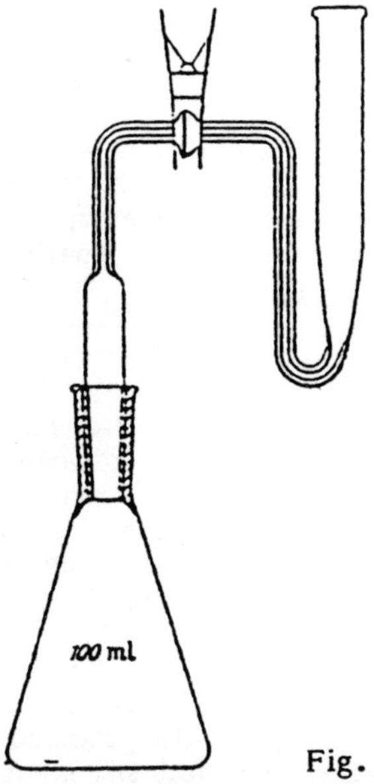


Fig. 101. Arsine generator

Potassium iodide solution:
Dissolve 15 g of reagent-purity potassium iodide in dist. water to 100 ml.

Tin (II) chloride solution:
Dissolve 40 g of reagent-purity $SnCl_2 \cdot 2 H_2O$ in 100 ml of conc. hydrochloric acid.

Coppered granulated zinc:
Copper-plate 100 g of granulated zinc in a 1 % solution of copper sulphate in dist. water until it is uniformly black. Rinse with dist. water and dry.

Lead-acetate glass wool:
Soak glass wool in 10 % lead acetate solution and then dry.

Silver diethyl dithiocarbamate solution:
Dissolve 1 g of reagent-purity silver diethyl dithiocarbamate, $(C_2H_5)_2N \cdot CS_2Ag$ in 200 ml of freshly distilled reagent-purity pyridine. The solution is stable for several weeks in an amber flask.

Arsenic stock solution:
Dissolve 0.1320 g of reagent-purity $As_2O_3$ in 12 ml of 2 m sodium hydroxide solution, neutralize with 1 m sulphuric acid and dilute to 1 litre with dist. water at 20 °C. The solution contains 0.1 mg As/ml, corresponding to 100 mg As/l.

Arsenic reference solution:
Dilute arsenic stock solution in the ratio 1 : 100 with dist. water. This solution contains 0.001 mg As/ml (prepare a fresh solution daily).

## Method

### Calibration

In order to prepare the calibration curve, measure off increasing quantities of arsenic reference solution containing between 0.001 and 0.02 mg of arsenic and transfer to the flask of the arsine generator. Make up the volume of the solution to about 50 ml with distilled water and continue treatment in the manner described under "Determination".

### Determination

Take 50 ml of the water sample either as such or concentrated or diluted, and treat with 10 - 20 ml of conc. sulphuric acid (or 2 - 3 ml of conc. sulphuric acid, 5 ml of conc. nitric acid + 0.5 ml of 30 % $H_2O_2$), reducing the sample in a quartz dish until sulphur trioxide vapour appears. It is advisable to conduct simultaneous parallel determinations and in addition to add a known quantity of arsenic to a measured volume of water and to process this as a blank test in the same way.

When cool, pick up the evaporated solutions with about 20 - 30 ml of distilled water and transfer to the flask of the arsine generator including any undissolved constituents. The volume of the liquid should not amount to more than about 50 ml.

Add 30 ml of diluted sulphuric acid and mix well. Now add 2 ml of potassium iodide solution and 8 drops of tin (II) chloride solution, shake, and leave to stand for 15 minutes at room temperature.

During this time introduce exactly 2 ml of silver diethyl dithiocarbamate solution into the absorption tube. Seal the spherical joints with reagent-purity conc. sulphuric acid. Following the standing period of 15 minutes introduce 6 $\pm$ 0.1 g of the coppered granulated zinc into the flask and immediately seal the arsine generator tightly. For the quantitative course of the reaction leave the equipment to stand for 1 hour at room temperature. Following this period transfer the absorption solutions to 1-cm cuvettes and measure against silver diethyl dithiocarbamate solution at 560 nm (or 546 nm if an Hg lamp is used).

### Calculation

Read off the appropriate arsenic content from the calibration curve on the basis of the measured extinction, take into account the quantity of water used and recalculate for 1 litre of water.

## 3.4.3.3 Iodometric determination of arsenic (III)

### General remarks

Trivalent arsenic may occur in arsenical subterranean waters. It is sensitive to oxidation, which means that determination should be carried out directly at the source. Arsenic III is oxidized to arsenic (V) by iodine in a solution containing sodium hydrogen carbonate.

The method is applicable to arsenic (III) contents greater than about 0.3 mg/l.

Concentrations of iron ions above 1 mg/l may lead to the precipitation of iron hydroxide as the sodium hydrogen carbonate solution is added, carrying down arsenic ions. If this happens, the findings for arsenic will be depressed.

If the water sample contains $H_2S$ or hydrogen sulphide ions they are also detected by iodometric titration. If this is the case it is necessary to determine the sulphide sulphur content separately and deduct this from the previously established total of arsenic (III), $H_2S$ and hydrogen sulphide.

### Equipment

Iodine/potassium iodide solution, 0.005 m:
Dissolve 4 g reagent-purity potassium iodide in 20 ml of dist. water, add 1.3 g of iodine and stir until solution is complete. Dilute with dist. water to 1 litre. Sodium arsenite may be used as the primary standard for setting the pH.

Starch solution:
Mix 1 g of soluble starch with 100 ml of dist. water and heat until clear.

Arsenic stock solution:
See "Spectrophotometric determination of total arsenic with silver diethyl dithiocarbamate"

Sodium hydrogen carbonate solution: cold saturated solution

Hydrochloric acid, 20 %

### Determination

Introduce 50 ml of the cold saturated sodium hydrogen carbonate solution into a 1-litre Erlenmeyer flask and treat with 1 ml of 20 % hydrochloric acid. The $CO_2$ which is generated partially expels the air from the flask.

Measure off 500 ml of the water sample immediately after taking the sample and transfer immediately to the Erlenmeyer flask containing the sodium hydrogen carbonate solution: Add 2 ml of starch solution to the solution and titrate with 0.005 m iodine/potassium iodide solution until a blue coloration begins to appear.

### Titre setting

This should also be performed at the sampling point. Proceed as described under "Determination". Instead of the water under investigation take a quantity of arsenic stock solution corresponding approximately to the arsenic (III) content of the water sample. Dilute this volume with boiled (redistilled) water and titrate. It is advisable to make two titre settings independently of each other.

### Calculation

1 ml of 0.005 m iodine/potassium iodide solution corresponds to 0.375 mg of $As^{3+}$.

Atomic weight of arsenic: 74.9216

### 3.4.3.4 Separating arsenic (III) and arsenic (V)

Pentavalent arsenic compounds occur most frequently but trivalent arsenic ions are also found, particularly in arsenical subterranean waters. If the trivalent arsenic is to be detected separately the water should be sampled in the absence of air since As (III) is very sensitive to oxidation and is converted to the pentavalent oxidation stage merely by the oxygen in the air.

Distill the trivalent arsenic directly from the strongly hydrochloric water sample and carry out determination in the distillate using the AAS hydride technique.

The distillation residue, which contains the pentavalent arsenic, should then be examined in the same way. Compare the results with the total arsenic content established by the same means. (3.4.3.1)

**Note:**

If arsenic is to be removed from the water, this can be achieved by coprecipitating it on iron or manganese hydroxides, or by filtering the water, after acidification with $CO_2$ (pH 6 - 6.5), over activated aluminium oxide. (W. Fresenius and W. Schneider)

### 3.4.4 Antimony

**General remarks**

In natural waters antimony occurs in traces and rarely exceeds 0.01 mg/l.

Antimony is determined either by means of hydride AAS or by spectrophotometry.

**Methods**

3.4.4.1 Hydride AAS

3.4.4.2 Spectrophotometric determination with rhodamine B

Antimony is concentrated by coprecipitation with manganese dioxide hydrate, extracted with methyl isobutyl ketone as antimony triiodide and finally measured at 565 nm as hexachloroantimonate (V) -rhodamine B complex in $CCl_4/C_6H_5Cl$.

3.4.4.3 ICP-AES see Section 3.3.12

### 3.4.4.1 Determination of antimony using the hydride AAS technique

**General remarks**

Antimony ions are reduced to antimony hydride by means of sodium borohydride, transferred to a heated quartz cuvette with the aid of a current of inert gas, decomposed thermally, and the absorption of the atoms is measured in the beam of an atomic-absorption spectrometer. In the hydride technique, the element which is to be determined is volatilized as a gaseous hydride and separated off from the matrix.

Interferences may occur if there is a considerable excess of elements such as tin, arsenic, bismuth, mercury, selenium or tellurium, which may also be volatilized using this technique. Above all heavy metals such as copper and nickel have a disturbing effect during the hydride formation itself. Since antimony (III) and antimony (V) are not equally sensitive when determined by the hydride technique, antimony (V) should be reduced to antimony (III) before determination (prereduction).

**Equipment**

Antimony stock solution (1000 mg/l)

Hydrochloric acid (1.17 g/ml)

Sodium borohydride solution:
Dissolve 3 g $NaBH_4$ and 1 g NaOH in water and make up to 100 ml.

Prereduction solution:
Dissolve 5 g ascorbic acid ($C_6H_8O_6$) and 0.5 g potassium iodide (KI) in water and make up to 100 ml.

Atomic-absorption spectrometer with hydride attachment

| | |
|---|---|
| 1. Hollow-cathode antimony lamp | |
| 2. Wavelength | 217.6 nm |
| 3. Lamp current | 5 mA |
| 4. Aperture width | 0.5 nm |
| 5. Background compensation | Yes |
| 6. Inert gas | Argon |
| 7. Gas through-flow | 30 l/h |
| 8. Cuvette temperature | 900 °C |
| 9. Integration | Peak area |
| 10. Integration time | 25 s |
| 11. Measuring range | 2.5 - 10 µg/l |
| 12. Determination limit | 0.5 µg/l |

The instrument parameters quoted here for the hydride technique apply to a Varian AAS 775 in conjunction with a hydride system manufactured by Messrs. Berghoff, Tübingen, FRG.

**Method**

Prereduction of the water samples and decomposition solutions: Introduce 20 ml of the water sample or the decomposition solution into the reaction vessel of the hydride system (prereduction may also be carried out in a different vessel), and allow to stand for at least 15 minutes with 1 ml prereduction solution and 4 ml hydrochloric acid.

Treat the blank and calibration solutions in the same way.

Connect the reaction vessel containing the prereduced solution to the hydride system.

Allow a constant current of inert gas (argon) to flow through the system at 30 l/h to expel the air.

Following this, continuously add 3 % sodium borohydride solution using a peristaltic pump. Sb ions are reduced to the hydride and directed by the current of inert gas into a quartz cuvette heated to 900 °C where they are thermally decomposed. Measure the absorption.

### Remarks

Since the hydride technique only permits detection of antimony which is present ionogenically in solution, decomposition is required for organic or complexed antimony. Organically loaded water samples must also be decomposed before measurement.

One method of decomposition is described below:

### Reagents

Sulphuric acid (1.84 g/l)

Hydrogen peroxide (30 %)

### Method

Introduce 50 ml of the water sample into a 150-ml beaker, treat with 5 ml sulphuric acid and add 5 ml hydrogen peroxide.

Heat in a sand bath until $SO_3$ vapour appears.

If the organic load of the water sample is high, repeat the addition of hydrogen peroxide to the decomposition mixture.

When the decomposition mixture is cool, transfer the sample to a 50-ml measuring flask, and dilute with water to the mark.

Care should be taken that the sample never evaporates to dryness.

## 3.4.4.2 Spectrophotometric determination with rhodamine B

### General remarks

In hydrochloric solution antimony (V) together with rhodamine B forms a red-purple complex insoluble in water.

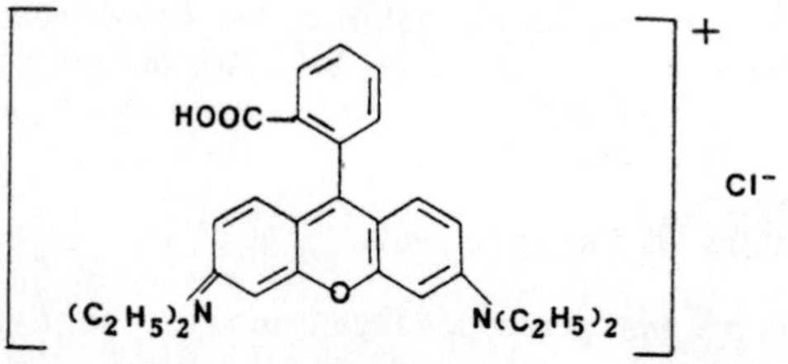


rhodamine B = tetraethyl rhodamine hydrochloride

The complex is extracted with $CCl_4$/chlorobenzene (1 : 4) and determined spectrophotometrically at 565 nm ($\varepsilon_{565}$ = 70 000). Due to the low antimony content in natural waters it is necessary to concentrate the element by coprecipitation with manganese dioxide hydrate. The addition of ethanol to the $KMnO_4$ solution improves the adsorption characteristics of the carrier.

Sb is isolated from manganese, iron and interfering elements by extraction with MIBK from 5 m $H_2SO_4$ with 0.01 m iodide. After back-extraction with 0.4 m hydrochloric acid, excess iodide is removed by oxidation with $H_2O_2$ and extraction of the liberated iodine with chloroform. Any $H_2O_2$ still present is decomposed by addition of a piece of platinum foil, the aqueous solution is concentrated to about 2 ml, and in order to form the hexachloro-antimonate (V) complex 4 m hydrochloric acid and sodium nitrite are added. As the latter is added as a mixing reagent together with rhodamine B the oxidation of the antimony to the pentavalent stage and the formation of the rhodamine B complex with the hexachloroantimonate (V) practically occur simultaneously. This prevents the occurrence of antimony losses due to hydrolysis of the $SbCl_6^-$ which begins immediately after formation of the complex anion. After extraction with $CCl_4$/chlorobenzene (1 : 4) measure photometrically at 565 nm. The total yield of the method is approx 80 %. In the 1-µg range the coefficient of variation is ± 2 %. The calibration curve is linear from 1.5 - 3.0 µg Sb/ml.

Antimony concentrations up to about 0.1 µg/l can be determined using this method.

Waste waters with antimony concentrations above 0.05 mg/l can be analysed directly without previous concentration by oxidatively evaporating a suitable volume of the sample with $H_2SO_4/H_2O_2$ and reducing it to 2 ml. After making up the solution to 30 ml with 4 m hydrochloric acid, extract with $CCl_4/C_6H_5Cl$ and measure photometrically as described under "Method".

In the course of the double extraction, first as antimony triiodide with methyl isobutyl ketone and then as hexachloroantimonate (V)-rhodamine B complex with $CCl_4$/chlorobenzene, manganese, iron and other elements are removed to a sufficient extent to prevent interference with photometric determination.

**Equipment**

Erlenmeyer flasks, 5 litres and 50 ml

Separating funnel, 50 ml

Glass-fibre filter, e.g. Whatman GF/B and GF/A

Centrifuge with centrifuge tubes, approx. 10 ml

Photometer, 565 nm

Cuvettes, 1 cm, 4 cm and/or 5 cm

Platinum foil, coated with platinum black

Hydrochloric acid, conc. (1.19 g/ml)

Hydrochloric acid, 4 m and 0.4 m

Sulphuric acid, conc. (1.84 g/ml)

Potassium permanganate solution, 0.2 m

Ethanol, reagent purity

$SO_2$ solution, saturated in 2 m sulphuric acid

Potassium iodide solution, 5.8 % in dist. water

Hydrogen peroxide, 30 %

Rhodamine B solution: 0.05 g rhodamine B in 100 ml dist. water

Rhodamine B mixing reagent:
Mix 9.2 ml of rhodamine B solution with 0.8 ml of 0.2 m sodium nitrite solution (prepare freshly before use)

Methyl isobutyl ketone, reagent purity

Chloroform, reagent purity

$CCl_4$/chlorobenzene (1 : 4):
Mix one part by volume of carbon tetrachloride with four parts by volume of chlorobenzene.

Antimony stock solution:
Dissolve 100 mg of reagent-purity metallic antimony in 15 ml of a 10 : 1 mixture (v/v) of conc. HCl and conc. $HNO_3$ and dilute with conc. HCl to 100 ml. 1 ml of this solution contains 1 mg of Sb.

Antimony reference solution:
Dilute 1 ml of antimony stock solution with 4 m HCl to 1 litre. 1 ml contains 1 µg of Sb.

**Method**

**Calibration**

In order to prepare the calibration curve, measure off increasing volumes of antimony reference solution containing between 1 and 10 µg or between 0.1 and 5 µg of antimony, make up to the volume of the water sample used for the analysis and treat further as described under "Determination".

**Determination**

Filter e.g. 5.0 litre of water sample through an acid-washed glass-fibre filter (Whatman GF/B or equivalent quality) into a 5-litre flask, add successively 0.7 ml of conc. hydrochloric acid, 11 ml of 0.2 m potassium permanganate solution and 11 ml of ethanol, mix well and leave to stand for 2 - 3 days to allow the manganese dioxide hydrate precipitate to form. Filter off the precipitate through a glass-fibre filter (suction filter), rinse with a saturated $SO_2$ solution in small amounts and transfer to a 50-ml separating funnel. Introduce a further 6 ml of saturated $SO_2$ solution used for rinsing, mix, add 1 ml of potassium iodide solution and 5.8 ml of conc. sulphuric acid, mix and leave to cool.

When cool, shake out the solution vigorously for 2 minutes with 25 ml of methyl isobutyl ketone and after separation discard the aqueous phase. Shake out the organic extract three times, each time with 10 ml of 0.4 m hydrochloric acid. The organic extract contains antimony as triiodide. Subsequently discard the organic phase, combine the aqueous extracts and add about 0.8 ml $H_2O_2$ (30 %) and extract the liberated iodine four times, each time with 10 ml of chloroform. Wait for 10 minutes between each shaking.

Transfer the iodide-free/iodine-free aqueous solution to a 50-ml Erlenmeyer flask, add a small piece of platinum foil coated with platinum black to destroy the excess $H_2O_2$ and evaporate to about 2 ml. Make up the concentrated solution to about 30 ml with 4 m hydrochloric acid, transfer to a shaking funnel, add 1.0 ml of rhodamine B mixing reagent, mix and leave to stand for 25 minutes.

Following this, add 5.0 ml of $CCl_4$/chlorobenzene (1 : 4), shake for 2 minutes, decant the organic phase into a centrifuge tube, centrifuge for a few minutes and then measure the extinction of the clear organic phase in a 5-cm cuvette (in the case of Sb concentrations in the 0.1 - 2.5 µg/l range) or in a 1-cm cuvette (approx. 0.3 - 12 µg/l) at 565 nm against $CCl_4$/chlorobenzene (1 : 4).

Conduct a reagent blank test in the same way. For this purpose take 28 ml of saturated $SO_2$ solution and continue treatment as for the corresponding $SO_2$ solution of the analysis sample in which the manganese dioxide hydrate was dissolved.

**Calculation**

Read off the appropriate antimony content from the calibration curve on the basis of the extinction measured, taking the blank reading into account. Insert the volume of water sample used and recalculate for 1 litre of water.

Atomic weight of antimony: 121.75

## 3.4.5 Beryllium

**General remarks**

Beryllium occurs in water in no more than trace quantities (µg/l). It is therefore necessary to use suitably sensitive direct methods, for example flameless AAS (graphite-tube technique), or to concentrate the beryllium traces, for example by coprecipitating on $Fe^{3+}$.

Three methods and a concentration process are described below:

3.4.5.1 Direct determination of beryllium by AAS

3.4.5.2 Concentration by coprecipitation on $Fe(OH)_3$

3.4.5.3 Determination of beryllium by flameless AAS (graphite tube)

3.4.5.4 Spectrophotometric determination using chromazurol S

### 3.4.5.1 Direct determination by means of atomic-absorption analysis

**Equipment parameters**

1. Hollow-cathode beryllium lamp
2. Wavelength 234.9 nm

| | |
|---|---|
| 3. Lamp current | 5 mA |
| 4. Aperture width | 0.5 nm |
| 5. Fuel gas | Acetylene |
| 6. Oxidant | Laughing gas |
| 7. Flame type | Lean in fuel gas |
| 8. Background compensation | Yes |
| Optimum measuring range | 0.05 - 1 mg/l |
| Determination limit | 0.01 mg/l |
| Calibration | 0.05 - 0.25 mg/l |
| Acid concentration | 10 ml hydrochloric acid (1.17 g/ml) per 1000 ml solution for measurement |

Remarks

Concentration by evaporation is possible if concentrations are low. The method of concentration described below can be recommended.

### 3.4.5.2 Basis of the method of concentration

Beryllium is separated from the matrix by coprecipitation with iron hydroxide.

Reagents

Hydrochloric acid I: 1.17 g/ml

Hydrochloric acid II:
Dilute 100 ml hydrochloric acid I with 100 ml water.

Ammonium hydroxide solution: 0.91 g/ml

Iron (III) chloride solution:
Dissolve 3 g iron (III) chloride in water and make up to 1000 ml.

Method

Slightly acidify sample solution (max. 5 litres) with hydrochloric acid

Add 5 ml iron (III) chloride per litre of sample solution

Shake for 10 minutes (if necessary mechanical shaker)

Allow to settle for 24 hours

Separate precipitate with group II filter

Dissolve with 4 ml hot hydrochloric acid 1 + 1

Transfer to 25-ml volumetric flask and measure by AAS; equipment parameters and calibration are as for direct determination by means of atomic-absorption analysis.

### 3.4.5.3 Beryllium determination with the graphite tube technique (Furnace method)

Equipment parameters

| | |
|---|---|
| 1. Hollow-cathode beryllium lamp | |
| 2. Wavelength | 234.9 nm |
| 3. Lamp current | 5 mA |
| 4. Aperture width | 0.5 nm |
| 5. Background compensation | Yes |
| 6. Carrier gas | Nitrogen |
| 7. Drying | 25 s at 180 °C |
| 8. Incineration | 20 s at 800 °C |
| 9. Atomization | 3 s at 2400 °C |
| 10. Gas stop | Yes |

| | |
|---|---|
| Optimum measuring range | 0.1 - 2 µg/l with 10 µl sample |
| Determination limit | 0.02 µg/l |
| Calibration | Addition method |
| Acid concentration | 10 ml nitric acid (1.40 g/ml) per 1000 ml sample solution |

Remarks

Immediately after sampling, stabilize the water sample with 10 ml nitric acid (1.40 g/ml) per 1000 ml sample solution.

### 3.4.5.4 Spectrophotometric determination using chromazurol S

General remarks

In the presence of pyridinium chloride, beryllium ions form with chromazurol S at pH 6.0 a metal chelate complex which after 15 min. remains stable for at least 5 hours. The absorption maximum of this complex at 575 nm is suitable for photometric determination of beryllium. The molar ratio of beryllium to chromazurol S is 1 : 1.

The technique can be used to determine between 2 and 100 µg of beryllium ions per litre of water. Since such concentrations only occur in very exceptional cases, it is first necessary to concentrate the beryllium. It is usual to start with 5 litres of water sample (see Section 3.4.5.2).

For concentrations of between 0.2 and 10 µg of Be per 100 ml of solution for measurement the reproducibility of the chromazurol S method is in the region of ± 0.2 µg Be.

Equipment

Spectrophotometer

5-cm cuvettes

Polyethylene bottles, approx. 5 litres (for sampling)

Polyethylene beakers, 250 ml and 500 ml

pH paper

White filter strip, 12.5 cm

Measuring cylinder, 100 ml

Iron (III) chloride solution (3 g/l)

Hydrochloric acid I (1.17 g/ml), reagent purity

Hydrochloric acid II:
100 ml hydrochloric acid I and 100 ml dist. $H_2O$

Ammonium hydroxide (0.91 g/ml)

Sodium hydroxide solution, 4 m

Washing liquid containing sodium hydroxide:
Dissolve 10 ml of 4 m sodium hydroxide solution in 1 litre water.

Chromazurol S solution, 0.05 %:
Dissolve 0.50 g chromazurol S ($C_{23}H_{13}Cl_2Na_3O_9S$) in 1 litre water in which 2 g of gum arabic has been dissolved. Leave to stand for several days, filter and store in an amber bottle. The solution keeps for several weeks.

Hydroxylamine hydrochloride solution, 2 % in dist. water

Pyridine/HCl solution:
Slowly add 3.5 ml conc. hydrochloric acid to 21.5 ml pyridine. The solution keeps for at least 3 weeks.

Beryllium stock solution:
Dissolve 0.7658 g beryllium carbonate ($BeCO_3$) (superpure) in dist. water, adding 10 ml hydrochloric acid I, and make up to 1 litre. The solution contains 100 mg Be/l.

Beryllium standard solution:
Make up 10 ml of beryllium stock solution to 1 litre with dist. water. The solution contains 1 µg Be/ml.

**Method**

**Concentration process**

When sampling, fill each 5-litre batch of water into a plastic bottle such that there is still room in each bottle for the hydrochloric acid needed for acidification.

Pour the acidified solution into a beaker with approx. 50 ml of cold 4 m sodium hydroxide solution. Beryllium and aluminium remain in solution as beryllate and aluminate respectively, while iron, titanium and manganese are precipitated. Leave to stand for several hours, filter, collect the filtrate in a polyethylene beaker, and wash the residue with the sodium hydroxide washing liquid. The volume of the filtrate should not exceed 100 ml.

**Determination**

Slightly acidify the filtrate (maximum volume 100 ml) with hydrochloric acid, concentrate by evaporation to about 50 ml, transfer the solution to a 100-ml measuring cylinder, and set up to pH 5.5 - 6.5 with 1 m HCl or 1 m $NH_4OH$. Allow the solution to cool and add 3 ml of the 2 % solution of hydroxylamine hydrochloride followed by 2 ml of pyridine/HCl solution. After mixing, add 1 ml chromazurol S solution by pipette. Then top up to the mark with dist. water at 20 °C, allow to stand for 15 min. and measure at 575 nm in a 5-cm cuvette. The measurement is taken against a blank solution which is prepared in the same way, i.e. put through the entire concentration and measurement procedure, but using dist. water instead of the water sample.

**Calibration**

The calibration curve is plotted for a range from 2 - 30 µg Be/100 ml, using the method described under "Determination". It is not a straight line.

**Calculation**

On the basis of the extinction measured (after deducting the value for the blank solution), read off the Be content from the calibration curve. Taking into account the original volume of the water sample, recalculate for 1 litre water.

Atomic weight of beryllium: 9.0122

## 3.4.6 Lead

**General remarks**

In ground waters natural lead concentrations are almost exclusively in the trace range, less than 0.01 mg/l. Lead compounds in concentrations greater than 0.1 mg/l occurring in surface waters signalize the presence of waste water from the mining, metal and/or chemical industries.

Lead compounds may be found in drinking water where lead piping is used for plumbing purposes. Galvanized water pipes may also release traces of lead since zinc is always accompanied by traces of lead. Similarly, lead stabilizers in polyvinyl chloride pipes may lead to the release of traces of lead into water passing through the pipes.

Lead ions can be determined by means of the following methods:

3.4.6.1 Spectrophotometrically with dithizone (if no AAS device is available)

3.4.6.2 By direct AAS with flame

3.4.6.3 Concentration method for 3.4.6.2

3.4.6.4 Flameless AAS using the graphite-tube technique

3.4.6.5 ICP-AES see Section 3.3.12

### 3.4.6.1 Spectrophotometric determination of lead with dithizone

#### General remarks

In virtually neutral to alkaline solution with dithizone, lead (II) ions form a di-dithizonate,

$$PbX_2 + 2\ HDz \longrightarrow PbDz_2 + 2\ HX$$

which is soluble in organic solvents (e.g $CHCl_3$ and $CCl_4$), and has a carmine red colour. These solutions are stable in darkness; in sunlight, considerable decomposition rapidly sets in.

The $PbDz_2/CCl_4$ solution shows a distinct maximum at 520 nm. However, pure dithizone may also absorb at this point. An error may therefore arise if excess dithizone is incompletely eliminated. Over-thorough washing out of the dithizone, on the other hand, may lead to decomposition of the $PbDz_2$. In order to minimize this source of error the lead dithizonate is treated with hydrochloric acid immediately after removing the excess dithizone. Lead enters the hydrochloric aqueous phase and the HDz equivalent to the lead is measured in the organic phase at $\lambda_{max.}$ 620 nm.

The method is suitable for the determination of lead ions in concentrations between 0.005 and 20 mg/l.

The determination of lead with dithizone is not specific. If the procedure described is followed only bismuth ions and tin (II) ions have an interfering effect. If these are found to be present in marked quantities, for example by spectral analysis, bismuth can be extracted in acid solution at pH 2 before lead with dithizone. Before analysis, tin (II) ions should be volatilized as bromide following oxidation with bromine.

#### Equipment

Spectrophotometer or filter photometer (620 nm)

Cuvettes, path length 1 cm and 5 cm

Measuring flask, 25 ml

Separating funnel, 500 ml

Nitric acid, reagent purity, (1.4 g/ml), tested for freedom from lead with dithizone

Ammonium acetate solution, 10 %:
Make up 10 g reagent-purity ammonium acetate to 100 g with water.

Ammonium hydroxide solution (1 + 5), (about 0.98 g/ml)

Potassium cyanide solution, 10 %:
Dissolve 10 g reagent-purity KCN in 90 ml redist. water. (Observe the precautions).

Potassium cyanide solution, 0.5 %:
Dilute 10 % KCN solution 1 to 20 with redist. water. (Caution)

Hydroxylammonium chloride solution:
Saturated aqueous solution

Carbon tetrachloride, reagent purity

Dithizone solution I:
Dissolve 25 mg dithizone (diphenyl-thiocarbazone, $C_{13}H_{12}N_4S$) in 100 ml reagent-purity tetrachloride. When using reagent-purity dithizone further purification of the dithizone is generally not necessary. If, however, when conducting the blank test the organic extract displays a yellowish-brown coloration, purify as follows: Add 1 ml ammonium hydroxide solution (0.98 g/ml) and 200 ml water to the 25 mg/100 ml dithizone solution. Transfer the dithizone to the aqueous phase by shaking in the separating funnel. Separate off the carbon tetrachloride to form a layer beneath the aqueous phase. Acidify with hydrochloric acid and shake until the dithizone is again present as a green solution in the organic phase. The solution is stable no more than approx. 1 day in an amber flask.

Dithizone solution II:
Dilute dithizone solution I in the ratio 1 to 10 with reagent-purity carbon tetrachloride

Lead stock solution:
Dissolve 0.1599 g lead nitrate, $Pb(NO_3)_2$, reagent purity, in redist. water, treat with 2 ml nitric acid (1.42 g/ml) and make up to 1 litre in the measuring flask at 20 °C. 1 ml of this solution contains 0.1 mg of $Pb^{2+}$.

Lead standard solution:
Make up 10 ml of lead stock solution to the mark in a 1-litre measuring flask at 20 °C with redist. water. 1 ml of this solution contains 0.001 mg of $Pb^{2+}$.

## Method

### Calibration

Take increasing quantities of lead standard solution with lead concentrations between 0.002 and 0.1 mg and analyze as described under "Determination". Either a calibration curve or a calibration factor may be used for evaluation in this range.

### Determination

Use a volume of water which contains between 0.005 and 0.1 mg of lead ions. Weakly acidify with nitric acid in order to oxidize sulphide ions and dissolve precipitates.

If the quantity of water taken exceeds 250 ml, evaporate to this volume. Should an insoluble residue remain following acidification with nitric acid and concentration, filter off the residue and spray off the filter into a beaker. Decoct the residue in the beaker with 10 - 30 ml of 10 % ammonium acetate solution and then filter into the main solution through the same filter. Rinse the filter with hot ammonium acetate solution.

Combine the solutions under analysis in a separating funnel (500 ml) and neutralize (about pH 6 - 7) with ammonium hydroxide solution (0.98 g/ml), testing by spotting onto pH paper. Now add 20 ml of the 10 % potassium cyanide solution (caution!) and 1 ml of the saturated hydroxylammonium chloride solution. The pH should now be between 9 and 10.

Extract the lead with 5-ml portions of dithizone solution II, shaking each time for 1 minute, until the last extracts no longer show a red coloration but rather a colour between light green and yellow. Combine the lead-containing extracts in a second separating funnel. Extract excess dithizone with 5 ml of 0.5 % potassium cyanide solution and immediately carry out phase separation. Then wash the organic phase a second time with 5 ml of 0.5 % potassium cyanide solution and likewise separate immediately.

Next shake the red lead dithizonate solution into a separating funnel with 10 ml of reagent purity hydrochloric acid (1.025 g/ml). In the process the red coloration turns to green. The organic solution contains the green dithizone equivalent to the red lead dithizonate. Separate off this solution and filter through a small filter into a 25-ml measuring flask. (Discard the aqueous hydrochloric phase containing the lead.) Rinse the filter with carbon tetrachloride and make up the solution to the mark in the measuring flask at 20 °C.

Measure the green solution at 620 nm in a 1-cm or 5-cm cuvette. Conduct a blank test analogous to the analysis process, using the same reagents and apparatus.

Calculation

Calculate the lead content on the basis of the measured extinction values with the aid of the calibration curve or the calibration factor. Take the blank reading and the initial volume for determination into account and recalculate for a volume of 1 litre.

Atomic weight of lead: 207.19

### 3.4.6.2 Direct determination by means of atomic-absorption analysis

**Equipment parameters**

| | | |
|---|---|---|
| 1. | Hollow-cathode lead lamp | |
| 2. | Wavelength | 217.0 nm |
| 3. | Lamp current | 5 mA |
| 4. | Aperture width | 1 nm |
| 5. | Fuel gas | Acetylene |
| 6. | Oxidant | Air |
| 7. | Flame type | Oxidizing |
| 8. | Background compensation | Yes |

| | |
|---|---|
| Optimum measuring range | 1 - 10 mg/l |
| Determination limit | 0.1 mg/l |
| Calibration | 0.5 - 10 mg/l |
| Acid concentration | 10 ml hydrochloric acid (1.17 g/ml) per 1000 ml solution for measurement |

**Remarks**

Concentration by evaporation is possible if concentrations are low. However, since in particular a high alkaline-earth content or iron gives rise to interferences, extractive concentration is to be preferred.

The method of extractive concentration described for the element silver with hexamethylene ammonium hexamethylene dithiocarbamate (HMDC) may be used in this case. A further method is described below.

Decomposition with nitric acid - hydrogen peroxide is required for organically loaded water samples.

### 3.4.6.3 Basis of the method of concentration

Lead is converted to its dithizonate in a citrate-buffered medium at pH 9, and the lead dithizonate is extracted with chloroform. The lead dithizonate in the organic phase is reextracted to the aqueous phase by adding a defined quantity of 1 m hydrochloric acid, and subsequently determined in the aqueous phase by atomic-absorption analysis.

**Reagents**

Hydrochloric acid I (1.17 g/ml)

Hydrochloric acid II (1 mol/l)

Ammonium hydroxide (0.91 g/ml)

Diammonium hydrogen citrate solution:
Dissolve 500 g diammonium hydrogen citrate $(NH_4)_2C_6H_5O_7$ in water, set to pH 9.2 with ammonium hydroxide., and dilute to 1000 ml with water. Clear the solution of heavy metals by repeated extraction, adding dithizone solution and chloroform.

Dithizone solution:
Dissolve 50 mg dithizone in 200 ml chloroform

Chloroform ($CHCl_3$)

**Procedure**

Treat the necessary quantity of water with 5 ml of hydrochloric acid I, evaporate to approx. 100 ml and transfer to separating funnel.

Treat with 10 ml diammonium hydrogen citrate solution and set to pH 9.2 with ammonium hydroxide.

Add 10 ml dithizone solution and shake for approx. 1 minute.

Separate off organic phase and repeat extraction twice, each time with 10 ml dithizone solution. The final organic phase should be green. If necessary repeat the extraction and/or increase the addition of dithizone solution.

Treat the collected organic extracts with 10 ml of hydrochloric acid II, shake for approx. 1 minute and separate off aqueous phase. Final determination by atomic-absorption analysis. Equipment parameters are as described for direct determination.

Prepare a blank reading in the same way.

### 3.4.6.4 Lead determination with the graphite tube technique (Furnace method)

**Equipment parameters**

| | |
|---|---|
| 1. Hollow-cathode lead lamp | |
| 2. Wavelength | 217.0 nm |
| 3. Lamp current | 5 mA |
| 4. Aperture width | 1 nm |
| 5. Background compensation | Yes |
| 6. Carrier gas | Argon |
| 7. Drying | 25 s at 220 °C |
| 8. Incineration | 20 s at 500 °C |
| 9. Atomization | 3 s at 2400 °C |
| 10. Gas stop | Yes |

| | |
|---|---|
| Optimum measuring range | 5 - 20 µg/l with 10 µl sample |
| Determination limit | 1 µg/l |
| Calibration | Addition method |
| Acid concentration | 10 ml nitric acid (1.40 g/ml) per 1000 ml sample solution |
| Matrix modification | 10 µl 0.1 % ammonium dihydrogen phosphate solution |
| Ammonium dihydrogen phosphate solution | Dissolve 0.1 g $NH_4H_2PO_4$ in water and make up to 100 ml |

**Remarks**

The use of a so-called platform may be recommended.

Immediately after sampling stabilize the water sample with 10 ml nitric acid (1.40 g/ml) per 1000 ml sample solution.

### 3.4.6.5 ICP-AES see Section 3.3.12

## 3.4.7 Cadmium

**General remarks**

In view of the high toxicity of cadmium it is necessary to be able to determine very small concentrations in water. AAS techniques (Sections 3.4.7.1 and 3.4.7.2) are particularly suitable for determining cadmium in water. Where no AAS facilities are available, it is still possible to use the classical spectrophotometric technique (Section 3.4.7.3) with dithizone and extraction in chloroform.

For ICP-AES (3.4.7.4) see Section 3.3.12

### 3.4.7.1 Direct determination by means of atomic-absorption analysis

Equipment parameters

| | |
|---|---|
| 1. Hollow-cathode cadmium lamp | |
| 2. Wavelength | 228.8 nm |
| 3. Lamp current | 3.5 mA |
| 4. Aperture width | 0.5 nm |
| 5. Fuel gas | Acetylene |
| 6. Oxidant | Air |
| 7. Flame type | Oxidizing |
| 8. Background compensation | Yes |

| | |
|---|---|
| Optimum measuring range | 0.05 - 2 mg/l |
| Determination limit | 0.01 mg/l |
| Calibration | 0.02 - 0.2 mg/l |
| Acid concentration | 10 ml hydrochloric acid (1.17 g/ml) per 1000 ml solution for measurement |

Remarks

High concentrations of alkaline earths have a disturbing effect. If cadmium concentrations are low, concentration by evaporation is possible. The extractive concentration methods described for the elements silver and lead may also be applied to cadmium.

Decomposition with nitric acid/hydrogen peroxide is necessary with organically loaded samples.

### 3.4.7.2 Cadmium determination with the graphite tube technique (Furnace method)

Equipment parameters

| | |
|---|---|
| 1. Hollow-cathode cadmium lamp | |
| 2. Wavelength | 228.8 nm |
| 3. Lamp current | 3.5 mA |
| 4. Aperture width | 0.5 nm |
| 5. Background compensation | Yes |
| 6. Carrier gas | Argon |
| 7. Drying | 25 s at 180 °C |
| 8. Incineration | 20 s at 600 °C |
| 9. Atomization | 3 s at 2400 °C |
| 10. Gas stop | Yes |

| | |
|---|---|
| Optimum measuring range | 0.5 - 2 µg/l with 10 µl sample |
| Determination limit | 0.05 µg/l |
| Calibration | Addition method |
| Acid concentration | 10 ml nitric acid (1.40 g/ml) per 1000 ml sample solution |
| Matrix modification | 10 µl 0.1 % ammonium dihydrogenphosphate solution |
| Ammonium dihydrogenphosphate solution | Dissolve 0.1 g of $NH_4H_2PO_4$ in water and make up to 100 ml |

Remarks

It is advisable to use a so-called platform. Decomposition with sulphuric acid/nitric acid is required for water samples which contain more than 10 mg/l of organically bonded carbon.

### 3.4.7.3 Spectrophotometric determination of cadmium with dithizone

General remarks

In virtually neutral to alkaline solution with dithizone (HDz), cadmium ions form cadmium di-dithizonate, which is soluble in organic solvents and has a pinkish-red colour:

$$CdX_2 + 2\ HDz \rightarrow CdDz_2 + 2\ HX$$

Its solubility is greater in chloroform than in carbon tetrachloride. The $CdDz_2/\ CHCL_3$ solution is more stable than the $CdDz_2/CCl_4$ solution.

The formation of di-dithizonate takes place quickly and quantitatively in the pH range > 7. Alkaline reaction of the solution is therefore of great importance for the selectivity of the method. In the variant described below, interference factors are largely eliminated, but it is still necessary to check for possible influences due to the admixture technique. Blank tests must be performed.

Where the water sample contains substantial quantities of organic substances (permanganate consumption in excess of 30 mg/l), such substances are eliminated by concentrating the sample by evaporation and fuming off with the addition of 5 ml of reagent-purity sulphuric acid (1.84 g/ml) and 5 ml of reagent-purity nitric acid (1.40 g/ml). The residue is picked up with a little redistilled water and treated as described under "Method".

Interference caused by oxidizing substances and by the induced oxidizing effect of manganese (II) at pH < 14 is eliminated by adding hydroxylamine hydrochloride.

The method is suitable for direct determination of cadmium ions in concentrations of between 0.01 and 20 mg/l. If the cadmium concentration is expected to be below 0.01 mg/l, and also in the case of drinking water, correspondingly larger volumes of water sample must be concentrated by evaporation to raise the cadmium concentration sufficiently to bring it within the range of the method.

Equipment

Photometer

Separating funnel, 100 ml

Hydrochloric acid, 2 m

Sodium hydroxide solution, 2 m

Potassium sodium tartrate solution:
Dissolve 250 g of $KNaC_4H_4O_6 \cdot 4\ H_2O$ in 1 litre of redist. water.

Methyl orange solution:
Dissolve 0.5 g methyl orange (sodium salt) in 100 ml dist. water.

Sodium hydroxide/potassium cyanide solution:
Solution a: Dissolve 400 g NaOH and 10 g KCN in 1000 ml redist. water.
Solution b: Dissolve 400 g NaOH and 0.5 g KCN in 1000 ml redist. water.
(WARNING! Potassium cyanide (KCN) is extremely toxic. Never put the pipette to the mouth. Observe safety rules.)

The solutions described so far are stored in polyethylene bottles and will keep for 1 - 2 months.

Hydroxylamine hydrochloride solution:
Dissolve 250 g $NH_2OH \cdot HCl$ in 1000 ml redist. water.

Chloroform, reagent purity

Dithizone solutions:

Dithizone solution a:
Dissolve 50 mg reagent-purity dithizone in 500 ml chloroform. Put in an amber bottle, covered with 0.05 m sulphuric acid and stored in a refrigerator, the solution will keep for approx. 1 week.

Dithizone solution b:
Dilute 50 ml of dithizone solution a to 500 ml with chloroform. The solution is stored as for solution a.

Tartaric acid solution:
Dissolve 20 g of $(CHOH)_2(COOH)_2 \cdot H_2O$ in 1 litre redist. water and store in a refrigerator.

Cadmium stock solution:
Dissolve 100.0 mg of pure cadmium metal in a mixture of 20 ml redist. water and 5 ml of reagent-purity hydrochloric acid (1.19 g/ml), heating all the time. Transfer the solution quantitatively to a 1-litre measuring cylinder and top up to the mark with redist. water. Store the solution in a polyethylene bottle. 1 ml contains 100 µg cadmium.

Cadmium standard solution:
Pipette 10.0 ml of cadmium stock solution into a 1-litre measuring cylinder, add 10 ml of reagent-purity hydrochloric acid (1.19 g/ml) and top up to the mark with redist. water. A fresh batch of this solution must be made from the stock solution every time. 1 ml contains 1 µg cadmium.

Washing liquid (acid):
Thoroughly mix 250 ml hydrochloric acid (1.19 g/ml) with 250 ml dist. water. The solution is used to clean the glass vessels.

**Method**

**Calibration**

Pipette aliquots of cadmium standard solution containing 2.0, 5.0, 8.0, 11.0, 14.0, 17.0 and 20.0 µg $Cd^{2+}$ into separate beakers and proceed exactly as described under "Determination". The readings are plotted as "extinction" against "µg cadmium ions".

### Determination

Pipette into a beaker a suitable volume of the water sample, concentrated to 50 ml by evaporation if necessary and containing between 3 and 20 µg Cd. Any undissolved cadmium compounds are dissolved by adding approx. 5 ml of 2 m HCl and heating the acidified sample for 10 min. If the solution is still turbid, it must be filtered.

Allow the solution to cool, add 5 ml of potassium sodium tartrate solution and one drop of methyl orange solution, and then add 2 m NaOH drop by drop until the colour of the solution changes from red to yellow.

Transfer quantitatively to a separating funnel and add 5 ml of sodium hydroxide/potassium cyanide solution a followed by 2 ml of hydroxylamine hydrochloride solution, mixing after adding each reagent. Then add 15 ml dithizone solution a and shake for 1 min.

Once the layers have separated, run off the (lower) chloroform phase, which contains the cadmium and a small proportion of the interfering accompanying elements, into a second separating funnel containing 25 ml of the tartaric acid solution. Introduce 10 ml of chloroform into the first separating funnel, which still contains the aqueous phase. After shaking for 1 min., run off the chloroform phase into the second separating funnel as well. (This operation should be performed as quickly as possible; none of the aqueous phase must be allowed to pass into the second separating funnel.) Shake the tartaric acid/chloroform phases thoroughly for 2 min.

When the layers have separated out, draw off and discard the chloroform phase. Once again shake the aqueous phase for 1 min. with 5 ml of chloroform, run off the chloroform phase, again without delay, and discard it. To the tartaric acid solution add a further 5 ml of potassium sodium tartrate solution, 5 ml of sodium hydroxide/potassium cyanide solution a, 2 ml of hydroxylamine hydrochloric solution and 15 ml of dithizone solution a, and repeat the operation in the same way.

Now add to the tartaric acid solution one drop of hydroxylamine hydrochloride solution and 15 ml of dithizone solution b. Then add 5 ml of sodium hydroxide/potassium cyanide solution b and immediately shake for 1 min.

After the layers have separated, run off the chloroform phase and filter through a wad of cotton wool into a cuvette, discarding the initial portion of the filtrate. Measure the extinction at a wavelength of 530 nm against a blank test treated in exactly the same way except that distilled water is used instead of the water sample.

### Calculation

Read off the cadmium content of the sample solution from the calibration curve. Taking into account the original volume and any dilutions etc. recalculate in terms of mg $Cd^{2+}$/l. Give cadmium concentrations of less than 1 mg/l to the nearest 0.005 mg/l.

## 3.4.7.4 ICP-AES see Section 3.3.12

Atomic weight of cadmium: 112.40

## 3.4.8 Chromium

### General remarks

Traces of chromium compounds in the trivalent or the hexavalent form may occur in water.

### Methods of determination

3.4.8.1 Direct determination by atomic-absorption flame photometry

3.4.8.2 Total chromium after extraction, using atomic-absorption flame photometry

3.4.8.3 Chromium VI after extraction, using flame AAS or graphite tube

3.4.8.4 Total chromium with flameless AAS (graphite tube)

3.4.8.5 Spectrophotometric methods for use when no AAS facilities are available

3.4.8.6 For ICP-AES see Section 3.3.12

### 3.4.8.1 Direct determination by means of atomic-absorption analysis

Equipment parameters

| | |
|---|---|
| 1. Hollow-cathode chromium lamp | |
| 2. Wavelength | 357.9 nm |
| 3. Lamp current | 3.5 mA |
| 4. Aperture width | 0.2 nm |
| 5. Fuel gas | Acetylene |
| 6. Oxidant | Laughing gas |
| 7. Flame type | Reducing |
| 8. Background compensation | Yes |
| | |
| Optimum measuring range | 0.5 - 5 mg/l |
| Determination limit | 0.1 mg/l |
| Calibration | 0.2 - 5 mg/l |
| Acid concentration | 10 ml hydrochloric acid (1.17 g/ml) per 1000 ml solution for measurement |

Remarks

Add to 50 ml sample 5 ml lanthanum nitrate solution to reduce matrix disturbances.

Lanthanum nitrate solution:
Dissolve 50 g lanthanum nitrate ($La(NO_3)_3 \cdot 6\ H_2O$) in water and make up to 1 litre.

Concentration by evaporation is possible, if concentrations are low. A method of extraction for the determination of total chromium and for the determination of hexavalent chromium is described below.

### 3.4.8.2 Total chromium

Basis of the method

Using a diisopropyl-ketone-xylene mixture, chromium is extracted as an anionic thiocyanate complex with thioethyl methyl ammonium chloride and measured in the extract in an air-acetylene flame or in a graphite tube cuvette.

Reagents

Bromophenol blue solution:
Dissolve 0.1 g bromophenol blue in 2 ml 0.1 m sodium hydroxide solution and dilute to 100 ml with water.

0.1 m Adogen 464 extraction solution:
Dissolve 20.2 g Adogen 464 (trioctyl methyl ammonium chloride) in 150 ml xylene and dilute to 500 ml with diisopropyl ketone.

Ammonium thiocyanate buffer solution:
Dissolve 685 g ammonium thiocyanate and 77 g ammonium acetate in max. 330 ml water, heating gently, add 5 ml bromophenol blue solution and set to pH 5.5 with concentrated acetic acid.

Extract the entire solution 5 times, each time with 40 ml Adogen 464 extraction solution for purification, discard organic phase in each case and make up extracted solution to 1 litre.

Formic acid (HCOOH) (1.22 g/ml)

Chromium stock solution (1000 mg/l)

Preparation of the chromium calibration solution:
Transfer 1 ml of chromium stock solution 1000 mg/l to a dry 100-ml measuring flask, add 50 ml formic acid and make up to the mark with diisopropyl ketone (1 mg Cr/100 ml).

Chromium calibration solution 0.1 - 0.2 - 0.3 mg/l for flame measurement:
Pipette 0.25 - 0.50 - 0.75 ml of the chromium calibration solution, 1 mg Cr/100 ml, or 0.01 mg/ml into dry 25-ml measuring flasks and make up to the mark with Adogen 464-extraction solution.

Method

Add 50 ml ammonium thiocyanate buffer solution to 125 ml of water sample or a sample evaporated to this volume, transfer to separating funnel and shake well.

Leave to stand for one hour.

Add 10 ml Adogen extraction solution and shake for one minute.

Separate off organic phase (centrifuge off any water still present).

Prepare a blank reading in the same way and deduct accordingly.

Final determination is carried out by atomic-absorption analysis, the equipment parameters being as for direct determination.

If concentrations are very low, the organic extract can also be measured using the graphite tube technique.

Decomposition with nitric acid - hydrogen peroxide is required for organically loaded samples.

### 3.4.8.3 Determination of hexavalent chromium

**Basis of the method**

Using ammonium pyrrolidine dithiocarbamate/methyl isobutyl ketone, hexavalent chromium is extracted from a solution containing potassium hydrogen phthalate and determined by atomic absorption.

**Reagents**

Buffer solution:
Dissolve 40 g potassium hydrogen phthalate ($C_8H_5KO_4$) and 4 g sodium hydroxide (NaOH) in 500 ml water.

Ammonium pyrrolidine dithiocarbamate solution:
Dissolve 1.0 g ammonium pyrrolidine dithiocarbamate (APDC) in water and make up to 100 ml.

**Methyl isobutyl ketone (MIBK)**

**Procedure**

Transfer 50 - 20 ml of sample solution to a 100-ml separating funnel.

Treat with 1.0 ml buffer solution and 0.5 ml APDC solution and mix.

Leave to stand for at least 10 minutes.

Add 2 ml MIBK and shake for 10 minutes (mechanical shaker).

Separate off organic phase and determine its chromium content by means of atomic-absorption analysis or the graphite tube technique.

Conduct a blank test and calibrate the method in the same way.

### 3.4.8.4 Chromium determination with the graphite tube technique (Furnace method)

**Equipment parameters**

| | |
|---|---|
| 1. Hollow-cathode chromium lamp | |
| 2. Wavelength | 357.9 nm |
| 3. Lamp current | 3.5 mA |
| 4. Aperture width | 0.5 nm |
| 5. Background compensation | Yes |
| 6. Carrier gas | Argon |
| 7. Drying | 25 s at 140 °C |
| 8. Incineration | 20 s at 800 °C |
| 9. Atomization | 3 s at 2600 °C |
| 10. Gas stop | Yes |

| | |
|---|---|
| Optimum measuring range | 1 - 10 µg/l with 10 µl sample |
| Determination limit | 0.2 µg/l |
| Calibration | Addition method |
| Acid concentration | 10 ml nitric acid (1.40 g/ml) per 1000 ml sample solution |
| Matrix modification: | 10 µl 0.1 % ammonium nitrate solution |
| Ammonium nitrate solution: | Dissolve 0.1 g $NH_4NO_3$ in water and make up to 100 ml |

**Remarks**

Immediately after sampling, stabilize the water sample with 10 ml nitric acid (1.40 g/ml) per 1000 ml sample solution.

### 3.4.8.5 Photometric determination of total chromium using diphenyl carbazide

**General remarks**

In mineral acid solution, hexavalent chromium compounds form with diphenyl carbazide an intensely red-purple coloured complex with an absorption maximum at 540 nm ($\varepsilon$ = 34,200). The measurement is performed in the aqueous phase. The complex forms instantaneously and is stable in sulphuric acid solution. The colour intensity of the solution is proportional to the concentration of chromate ions in the water sample. Trivalent chromium compounds are oxidized to chromate before complex formation and as such are also included in the measurement.

A) The method is suitable for determining total chromium in water in concentrations of approx. 0.01 mg/l and over.

To prevent chromium losses the water sample should be stored in polyethylene containers. Pyrex or normal glass is not suitable. Interference due to self-coloration of the sample can be eliminated by measuring the extinction after a determination run without the addition of diphenyl carbazide.

**Equipment**

Photometer

Cuvettes: path length 1, 2 and 5 cm

Separating funnel, 200 ml

Measuring flasks, 100 ml and 1 litre

Bulb pipettes, 2, 5, 10 and 100 ml

Phosphoric acid:
Dilute 300 ml $H_3PO_4$ (approx. 1.71 g/ml), reagent purity, with 300 ml $H_2O$.

Hydrogen peroxide solution, reagent purity, 30 % by weight $H_2O_2$

Sulphuric acid, conc.

Sulphuric acid, 0.5 m

Sodium chloride solution, 0.1 m

Silver nitrate solution, 0.1 m

Ammonium peroxodisulphate solution, 10 %

Sodium azide solution:
Dissolve 0.5 g $NaN_3$ in 100 ml dist. water.

Diphenyl carbazide solution:
Dissolve 1 g of reagent-purity diphenyl carbazide in acetone. Add one drop of glacial acetic acid and store an an amber bottle.

Chromate stock solution:
Dissolve 0.3740 g of reagent-purity potassium chromate, $K_2CrO_4$, in distilled water in a measuring flask and make up to 1 litre with dist. water. 1 ml corresponds to 0.1 mg Cr.

Chromate reference solution I:
Dilute 100 ml of chromate stock solution to 1 litre with dist. water. 1 ml corresponds to 0.01 mg Cr.

Chromate reference solution II:
Dilute 10 ml of chromate stock solution to 1 litre with dist. water. 1 ml corresponds to 0.001 mg Cr.

### Method

### Calibration

Prepare a suitable series of dilutions containing concentrations of 0.005 to 0.1 mg of chromium (VI) ions, each in 100 ml, and treat these as described under "Determination".

### Determination

Use up to 1000 ml of water sample, the chromium content of which should lie between 0.005 and 0.1 mg. Add 5 ml conc. sulphuric acid to the sample, concentrate by evaporation in a sufficient large beaker and fume off the sulphuric acid. To destroy organic matter add a few drops of $H_2O_2$ while fuming off the sulphuric acid. Where considerable quantities of organic matter are present (dark coloration), repeat the addition of $H_2O_2$ until the organic matter is completely oxidized. (Caution!)

Use 10 ml of 0.5 m sulphuric acid to pick up the residue after fuming off, add 1 ml of $H_3PO_4$ and dilute to approx. 100 ml with dist. water; heat gently to dissolve the salts. Ignore any remaining residue ($SiO_2$, $CaSO_4$).

Allow the sample to cool, then add in succession 5 ml of 0.1 m silver nitrate solution and 10 ml of 1 % ammonium peroxodisulphate solution. Heat to gently boiling for 30 min. The total volume of the sample must not be allowed to drop below about 70 ml. If necessary top up with hot water during boiling.

Allow the sample solution to cool, and add 6 ml of 0.1 m sodium chloride solution and a few drops of $NaN_3$ solution to reduce $MnO_4^-$. The precipitate is filtered off through a glass frit. Rinse out beaker and frit with a little distilled water.

Transfer the filtrate to a 100-ml measuring flask, add 2 ml of diphenyl carbazide solution, top up to the mark with dist. water, and after 5 minutes measure at 540 nm or with a suitable filter photometer.

Conduct a blank test in exactly the same way, but using distilled water instead of the water sample.

### Calculation

On the basis of the extinction measured, read off the appropriate chromium content from the calibration curve. Taking into account the quantity of water used, recalculate for 1 litre of water.

## B) Determination of hexavalent chromium using diphenyl carbazide

### General remarks

By adding certain "trap reagents", which prevent the reduction of hexavalent chromium originally present in the water and the oxidation of any trivalent chromium present to the hexavalent form, it is possible to control the process such that only the original hexavalent chromium is determined. The concentration of trivalent chromium in the water sample can be calculated by subtracting the value thus obtained from the total chromium content (see A "Determination of total chromium using diphenyl carbazide").

The method is suitable for determining chromium (VI) ions in the concentration range 0.1 to 40 mg/l, i.e it is mainly applicable to wastewaters etc.

Interference may be caused not only by all substances which render the water coloured or turbid, but also by sizeable quantities of organic substances and by heavy metals and especially iron (III) ions. Interference is also produced by all substances - compounds and ions - which have an oxidizing or reducing effect, e.g. free chlorine or sulphite ions. The elimination of such interference is dealt with under "Determination".

### Equipment

Apparatus and chemicals as listed under "Determination of total chromium using diphenyl carbazide". Also:

Sulphuric acid (1.18 g/ml):
Carefully add 1 part by volume of reagent-purity sulphuric acid (1.84 g/ml) to 3 parts by volume of dist. water.

Methyl orange solution:
Dissolve 0.1 g of methyl orange in 100 ml of dist. water.

Methanol, reagent purity

Sodium hypochlorite solution:
Approx. 13 % active chlorine

Sodium hydroxide lozenges, reagent purity

Sodium hydroxide solution:
Approx. 2 m

Phosphate solution:
Dilute 100 g of reagent purity disodium hydrogen phosphate $Na_2HPO_4 \cdot 2\ H_2O$ to 1 litre with distilled water.

## Method

### Calibration

Prepare a suitable series of dilutions with concentrations of between 0.1 and 40 mg of chromium (VI) ions per litre and treat the various components as described under "Determination".

### Determination

1) To block the effect of any oxidizing agents in the water (i.e. to stabilize the trivalent chromium compounds), add 25 ml of phosphate solution to the sample and add sodium hydroxide lozenges until an alkaline reaction is observed. After adding 10 ml of methanol, leave to stand overnight.

2) To block the effect of any reducing substances in the water (i.e. to stabilize the hexavalent chromium compounds), make a preliminary test by adding 3 ml of sulphuric acid and five drops of methyl orange solution to 1 litre of the water sample. Then add sodium hypochlorite solution drop by drop until the red coloration produced by the methyl orange disappears.

2a) Treat a second batch by measuring another 1 litre of water sample and adding 10 ml of phosphate solution, approx. 5 sodium hydroxide lozenges and the same quantity of sodium hypochlorite solution as was used for the preliminary test (see 2) above) plus an additional 1 ml. Mix vigorously, add 5 ml of methanol, and leave to stand overnight.

To determine the concentration of chromium (VI) ions, take from the supernatant solution of the water sample treated according to 1) or 2) a volume containing between 0.01 and 0.15 mg of chromate ions.

Pour this solution into a measuring flask, add 2 ml of phosphoric acid and 2 ml of diphenyl carbazide solution, and top up to the mark (e.g. 100 ml) with dist. water.

5 minutes after adding the colour reagent, measure the extinction of the coloured solution in a cuvette of suitable path length at 540 nm or with an appropriate filter photometer.

### Calculation

On the basis of the extinction value, read off the corresponding chromium (VI) content from the calibration curve. Taking into account the quantity of water used, recalculate for 1 litre of water.

### 3.4.8.6 ICP-AES (see Section 3.3.12.)

Atomic weight of chromium: 51.996

## 3.4.9. Germanium

### General remarks

Germanium can be detected by means of emission spectrum analysis on the lines of the qualitative survey analysis described in Section 3.4.1.

In addition, germanium traces may also be detected in conjunction with the determination of molybdenum (cf. also Section 3.4.12.1).

The determination of germanium in water samples is difficult since concentrations are generally only in the µg/l range. However, studies from East Asia attribute favourable properties to germanium with regard to the human organism; this not only gives rise to a new interest in the presence of germanium, particularly in natural mineral waters and spa waters, but also makes it desirable to obtain reliable information on the concentration of germanium in such waters.

The important factor in this instance is not so much the precise, absolute concentration but rather the establishment of the range of concentration of dissolved germanium compounds in water samples.

The following procedure has proved sufficiently reliable.

### 3.4.9.1 Determination of germanium by emission spectrography

Evaporate about one or more litres of the water sample under investigation in a water bath or over a sand bath, almost to dryness. Acidify the residue thus obtained with hydrochloric acid, heat gently and filter the hydrochloric solution through a fine paper filter. Treat the hydrochloric acid filtrate with a mercury (II) chloride solution (corresponding to approximately 20 mg of mercury) and precipitate the aqueous solution containing hydrochloric acid and mercury with hydrogen sulphide. A deposit of mercury sulphide is precipitated which coprecipitates virtually all of the germanium. Isolate this deposit using a membrane filter, remove mechanically from the membrane filter and transfer mechanically to the carbon electrode of an emission spectrograph with arc excitation. Conduct a spectrographic measurement and record the spectra obtained photographically. Process solutions with known quantities of germanium, for example with 0.2, 0.5, 1 and 5 µg, in a parallel procedure and likewise measure spectrographically. The densities of the lines in the spectrum make it possible to allocate the germanium concentration in comparison with the calibration method. The following may be used as germanium lines: 303.91 / 275.46 / 259.24 / 269.14 nm.

A blank test should be conducted in parallel using all the chemicals and including evaporation of an identical quantity of distilled water under conditions exactly equivalent to those of the analysis. The result of the blank test should be taken into account when evaluating the emission-spectrographic analysis.

Experience shows that considerable uncertainty prevails both in the qualitative detection of the presence of germanium traces in water and in particular in the quantitative estimation of the concentration of germanium in water.

Since germanium is claimed to have various beneficial effects on the well-being of the human organism it is important that there should be a reliable and reproducible method of determining germanium in the waters concerned.

### 3.4.9.2 Determination of germanium using graphite-tube AAS technique

General remarks

The germanium is extracted from hydrochloric acid solution (molar concentration 8) with the aid of chloroform and subsequently reextracted by shaking out with water.

Equipment

Hydrochloric acid (1.17 g/ml)

Nitric acid (1.40 g/ml)

Chloroform ($CHCl_3$)

Separating funnel (glass)

AAS unit with graphite tube

Method

Add 100 or 200 ml of hydrochloric acid to 50 or 100 ml respectively of the water to be analyzed. The molar concentration of hydrochloric acid should then be approximately 8. Transfer to a separating funnel, add 25 ml of chloroform, shake for 3 min and separate off the chloroform phase. Extract the aqueous phase again, using 15 ml of chloroform. Shake the combined chloroform extracts with 10 ml of deionized water for 3 min., reextracting the germanium into the aqueous phase. Separate off the aqueous phase, acidify with 50 µl $HNO_3$, and determine the germanium using the graphite-tube technique. (See schematic outline below).

**Instrument parameters for graphite-tube (furnace) method** (schema)

| | |
|---|---|
| 1. Hollow-cathode germanium lamp | |
| 2. Wavelength | 265.1 nm |
| 3. Lamp current | 5 mA |
| 4. Aperture width | 0.5 nm |
| 5. Background compensation | Yes |
| 6. Carrier gas | Argon/nitrogen |
| 7. Drying | 30 s at 90 °C<br>30 s at 120 °C<br>20 s at 180 °C |
| 8. Incineration | 32 s at 800 °C |
| 9. Atomization | 3 s at 2700 °C |
| 10. Gas stop | Yes |

| | |
|---|---|
| Optimum measuring range | 10-20 µg/l with 10 µl of sample |
| Determination limit | 0.5 µg/l |

| | |
|---|---|
| Calibration | Addition method |
| Acid concentration | 5 ml of nitric acid (1.40 g/ml) per 1000 ml sample solution |
| Matrix modification | 10 µl of 0.25 % $Ni(NO_3)_2$ |
| Nickel nitrate solution | Dissolve 0.4 g of $Ni(NO_3)_2 \cdot 6H_2O$ in water and make up to 100 ml |

**Remarks**

Direct determination of germanium is not possible with the furnace method owing to uncontrolled matrix effects. It is necessary to concentrate the germanium by extraction.

Atomic weight of germanium: 72.59

## 3.4.10 Cobalt

**General remarks**

In natural waters, cobalt only occurs in traces (less than 0.005 mg/l). Occasionally, somewhat higher concentrations of cobalt, perhaps a few tenths of a milligram, may be found in subterranean and waste waters.

**Methods of determination**

3.4.10.1 By spectrophotometry as a thiocyanate/tetraphenyl arsonium complex, if no AAS apparatus is available.

3.4.10.2 Atomic absorption method with flame analysis.

3.4.10.3 Flameless AAS using the graphite tube technique.

3.4.10.4 For ICP-AES see Section 3.3.12.

### 3.4.10.1 Spectrophotometric determination as thiocyanate/tetraphenyl arsonium complex

With thiocyanate ions cobalt ions form a complex $(Co(SCN)_4)^{2-}$. When tetraphenyl arsonium chloride $(C_6H_5)_4AsCl$ is added a blue compound is formed which can be directly extracted with chloroform and measured photometrically at 620 nm.

The method can be applied directly in the water sample in a concentration range from 0.1 to 10 mg of cobalt ions/l. In the case of lower concentrations the water sample must be concentrated by evaporation.

A tenfold surplus of nickel ions has no noticeable interfering effect.

**Equipment**

Photometer

Cuvettes, 5 cm

Glass dishes, 1 litre

Separating funnels, 500 ml and 100 ml

Measuring flasks, 25 ml

Hydrochloric acid, conc., reagent purity, (1.19 g/ml)

Ammonium hydroxide solution, (0.91 g/ml)

Potassium thiocyanate solution, 50 %, reagent purity

Tetraphenyl arsonium chloride solution:
Dissolve 2.094 g of reagent-purity $(C_6H_5)_4AsCl$ in dist. water to 100 ml.

Ammonium fluoride solution, 20 % and 1 %, reagent purity

Potassium iodide solution, 10 %, reagent purity

Sodium thiosulphate solution, 10 %, reagent purity

Chloroform, reagent purity

Cobalt stock solution:
Dissolve 1.00 g of metallic cobalt (spectrographically standardized) in 10 ml of reagent-purity nitric acid (1.42 g/ml), adding 15 ml of dist. water. Fume off the solution with 10 ml of reagent-purity sulphuric acid (1.84 g/ml). Dilute with dist. water and make up to 1 litre at 20 °C. 1 ml of this solution contains 1 mg of cobalt.

Cobalt reference solution:
Dilute 1 ml of cobalt stock solution to 100 ml with dist. water. 1 ml of this solution contains 0.01 mg of cobalt.

## Method

### Calibration

In order to prepare the calibration curve take increasing quantities of cobalt reference solution containing between 0.005 and 0.15 mg of cobalt and subject the solutions to the process described under "Determination". In the linear range from 0.005 to 0.15 mg of cobalt a calibration factor may also be used for calculation.

### Determination

The approximate cobalt concentration should be established from the intensity of the cobalt lines in a preliminary qualitative emission spectrum analysis (Section 3.4.1). Take a portion of the water sample measured according to this method, in natural waters generally 5 litres, and acidify with concentrated hydrochloric acid. It is convenient to use the entire contents of one bottle for each batch for analysis. Dissolve any sedimented residues in the transport bottle in hydrochloric acid and add to the main solution. Then boil down, at most until salt precipitation begins.

Following evaporation adjust the pH to 3 - 4 with ammonium hydroxide solution, (0.91 g/ml), transfer to a separating funnel of appropriate size and add 8 ml of potassium thiocyanate solution to form the Co (II) thiocyanate complex, plus 20 ml of 20 % ammonium fluoride solution for the purposes of masking iron. Mix the preparation and add 2 ml of tetraphenyl arsonium chloride solution to form the triple complex and then extract the latter by shaking vigorously (1 minute) with 5 ml of chloroform. Repeat the extraction three more times, each time with 2 ml of tetraphenyl arsonium chloride solution and 5 ml of chloroform, separating the phases after each extraction.

In a 100-ml separating funnel treat the combined chloroform extracts with 10 ml of 1 % ammonium fluoride solution, 2 ml of potassium thiocyanate solution and 1 ml of tetraphenyl arsonium chloride solution. Shake for 1 minute.

Discard the aqueous phase. If, after washing, the organic phase displays a yellowish-green coloration, this indicates the presence of interfering copper (II). If this is the case add 1 ml of potassium iodide solution and shake. During the reduction of $Cu^{2+}$ to $Cu^{+}$ free iodide is formed. This should be removed from the organic phase by shaking with 1 ml of sodium thiosulphate solution.

After separating the phases filter the chloroform solution through a small filter into a 25-ml measuring flask, rinse with 1-ml portions of chloroform and make up to the mark with chloroform (20 °C). Immediately measure the extinction against chloroform in a 5-cm cuvette at 620 nm.

Conduct a blank test following the entire analysis process described above.

Calculation

Read off the concentration of cobalt from the calibration curve on the basis of the measured extinction values, take into account the blank reading and the initial quantity of water used and recalculate for 1 litre of water. It is also possible to work with a calibration factor instead of a calibration curve.

Atomic weight of cobalt: 58.9332

### 3.4.10.2 Direct determination of cobalt by means of atomic-absorption flame analysis

Equipment parameters

| | |
|---|---|
| 1. Hollow-cathode cobalt lamp | |
| 2. Wavelength | 240.7 nm |
| 3. Lamp current | 5 mA |
| 4. Aperture width | 0.5 nm |
| 5. Fuel gas | Acetylene |
| 6. Oxidant | Air |
| 7. Flame type | Oxidizing |
| 8. Background compensation | Yes |

| | |
|---|---|
| Optimum measuring range | 0.5 - 5 mg/l |
| Determination limit | 0.05 mg/l |
| Calibration | 0.1 - 5 mg/l |
| Acid concentration | 10 ml hydrochloric acid (1.17 g/ml) per 1000 ml solution for measurement |

Remarks

Concentration by evaporation is possible if concentrations are low.

The extractive concentration method described for the element silver using hexamethylene ammonium hexamethylene dithiocarbamate (HMDC) may also be used for cobalt.

Decomposition with nitric acid/hydrogen peroxide is required if the samples contain organic loads.

### 3.4.10.3 Cobalt determination with the graphite tube technique (Furnace method)

Equipment parameters

| | |
|---|---|
| 1. Hollow-cathode cobalt lamp | |
| 2. Wavelength | 240.7 nm |
| 3. Lamp current | 5 mA |
| 4. Aperture width | 0.5 nm |
| 5. Background compensation | Yes |
| 6. Carrier gas | Argon |
| 7. Drying | 25 s at 140 °C |
| 8. Incineration | 15 s at 700 °C |
| 9. Atomization | 3 s at 2400 °C |
| 10. Gas stop | Yes |

| | |
|---|---|
| Optimum measuring range | 5 - 20 µg/l with 10 µl sample |
| Determination limit | 0.5 µg/l |
| Calibration | Addition method |
| Acid concentration | 10 ml nitric acid (1.40 g/ml) per 1000 ml sample solution |

Remarks

Immediately after sampling, stabilize the water sample with 10 ml nitric acid (1.40 g/ml) per 1000 ml sample solution.

### 3.4.10.4 ICP-AES see Section 3.3.12

## 3.4.11 Copper

In natural waters, copper ions only occur in very low concentrations. However, corrosive water or water rich in oxygen may put copper from copper plumbing into solution.

Methods of determining copper in water

3.4.11.1 Direct determination by AAS, flame technique

3.4.11.2 Direct determination by flameless AAS, graphite-tube method

3.4.11.3 Spectrophotometric analysis using sodium diethyl dithiocarbamate in cases where no AAS facilities are available.

3.4.11.4 For ICP-AES see Section 3.3.12

### 3.4.11.1 Direct determination by means of atomic-absorption flame analysis

Equipment parameters

| | |
|---|---|
| 1. Hollow-cathode copper lamp | |
| 2. Wavelength | 324.8 nm |
| 3. Lamp current | 3.5 mA |
| 4. Aperture width | 0.5 nm |
| 5. Fuel gas | Acetylene |
| 6. Oxidant | Air |
| 7. Flame type | Oxidizing |
| 8. Background compensation | Yes |
| Optimum measuring range | 0.5 - 5 mg/l |
| Determination limit | 0.05 mg/l |
| Calibration | 0.2 - 5 mg/l |
| Acid concentration | 10 ml hydrochloric acid (1.17 g/ml) per 1000 ml solution for measurement |

Remarks

Concentration by evaporation is possible if concentrations are low.

The extractive concentration method described for the element silver using hexamethylene ammonium hexamethylene dithiocarbamate (HMDC) may also be applied to copper.

Decomposition with nitric acid/hydrogen peroxide is possible with organically loaded samples.

### 3.4.11.2 Copper determination with the graphite-tube technique (Furnace method)

Equipment parameters

| | |
|---|---|
| 1. Hollow-cathode copper lamp | |
| 2. Wavelength | 324.8 nm |
| 3. Lamp current | 3.5 mA |
| 4. Aperture width | 0.5 nm |
| 5. Background compensation | Yes |
| 6. Carrier gas | Argon |
| 7. Drying | 25 s at 140 °C |
| 8. Incineration | 20 s at 600 °C |
| 9. Atomization | 3 s at 2300 °C |
| 10. Gas stop | Yes |
| Optimum measuring range | 5 - 20 µg/l with 10 µl sample |
| Determination limit | 0.5 µg/l |
| Calibration | Addition method |
| Acid concentration | 10 ml nitric acid (1.40 g/ml) per 1000 ml sample solution |
| Matrix modification | 10 µl 0.1 % ammonium nitrate solution |
| Ammonium nitrate solution | Dissolve 0.1 g ammonium nitrate ($NH_4NO_3$) in water and make up to 100 ml |

Remarks

Immediately after sampling, stabilize the water sample with 10 ml nitric acid (1.40 g/ml) per 1000 ml sample solution.

### 3.4.11.3 Spectrophotometric determination using NaDDTC

General remarks

At a pH between 4 and 11, copper (II) ions react with sodium diethyl dithiocarbamate (NaDDTC) to form a brownish-yellow complex

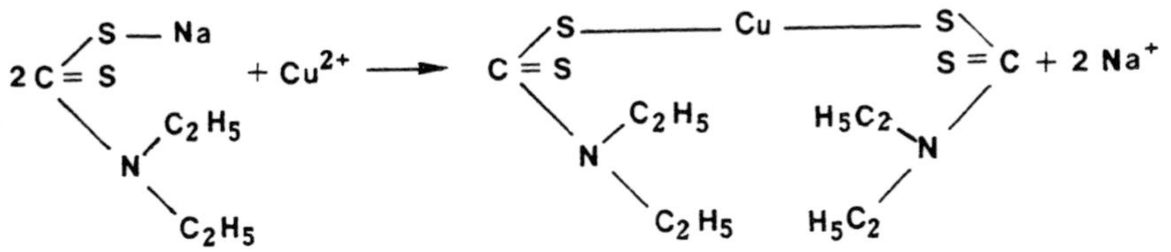

which does not dissolve readily in water and can be extracted with chloroform or carbon tetrachloride. Interference due to other metallic ions can be eliminated by adding $Na_2EDTA$ to the sample under certain pH conditions.

Direct determination of between 0.005 and 0.5 mg $Cu^{2+}$ is possible with 100 ml of water sample.

If cyanide, cyanide complexes or considerable quantities of organic substances are present, the water sample must be mineralized: take for example 100 or 1000 ml of water sample and concentrate by evaporation with 1 ml of conc. $H_2SO_4$ and 1 ml of conc. $HNO_3$ under a fume hood until white sulphur trioxide vapour starts to appear. Should a brown coloration occur, dilute with deionized water and evaporate once again with 1 ml conc. $HNO_3$ until dry. Moisten the residue with 1 ml conc. HCl and again evaporate to dryness. Pick up with deionized water, if necessary, filter, then neutralize and make up to 100 ml with deionized water.

Equipment

Spectrophotometer or photometer, 435 nm

Cuvettes, path length 1 - 5 cm

Water, deionized and copper-free

Sodium diethyl dithiocarbamate solution:
Dissolve 1 g of reagent-purity NaDDTC $(C_2H_5)_2N \cdot CS_2Na \cdot 3\ H_2O$ in water, filter, and make up to 100 ml. The solution keeps for 1 month in a dark glass bottle.

Ammonium hydroxide solution (1 : 1):
Add 100 ml of conc. reagent-purity ammonium hydroxide solution (0.907 g/ml) to 100 ml of water.

Carbon tetrachloride, reagent purity

Ammonium citrate solution, approx. 2 m:
$(NH_4)_3C_6H_5O_7$: 486 g $(NH_4)_3C_6H_5O_7$, reagent purity, per litre

EDTA solution, approx. 0.1 m:
Dilute 37 g of reagent-purity $Na_2EDTA$ with water to 1 litre

Copper stock solution:
Dissolve 0.200 g copper foil or copper wire, reagent purity, in 10 ml of reagent-purity $HNO_3$ (1 : 1). Then add 1 ml of reagent-purity conc. $H_2SO_4$ (1.84 g/ml) and evaporate until white sulphur trioxide vapour appears. Pick up the residue with water and make up to 1 litre. 1 ml contains 0.200 mg $Cu^{2+}$.

Copper standard solution I:
Shortly before use, make up 100 ml of copper stock solution to 1000 ml with water. 1 ml contains 0.020 mg $Cu^{2+}$.

Copper standard solution II:
Shortly before use, make up 100 ml of copper standard solution I to 1000 ml with water. 1 ml contains 0.002 mg $Cu^{2+}$.

## Method

### Calibration

To establish the calibration curves, use copper standard solution II to prepare series containing concentrations of 0.005 - 0.10 mg $Cu^{2+}$/l and 0.10 - 0.50 mg $Cu^{2+}$/l. Then proceed as described under "Determination".

### Determination

Take up to 1 litre of the original or pretreated sample containing between 0.005 and 0.5 mg $Cu^{2+}$ and neutralize against pH paper with NaOH or HCl in a separating funnel.

Add 5 ml of 2 m ammonium citrate solution, 10 ml of 0.1 m EDTA solution, 10 ml of ammonia solution (1 : 1) and 10 ml $CCl_4$. Shake the mixture thoroughly and separate off the organic phase. Keep repeating the extraction until the $CCl_4$ layer remains colourless. Discard the organic extracts.

Then add 10 ml of NaDDTC solution to the aqueous phase. Extract the copper complex that forms by shaking for approx. 5 minutes with 10 ml of $CCl_4$. Separate off the organic phase and filter it through a medium-hard paper filter into a 25-ml measuring flask. Extract the sample again for 1 min. with 10 ml $CCl_4$, and again filter the extract into the measuring cylinder. Fill up to the mark at 20 °C with $CCl_4$, rinsing the filter at the same time.

Handle samples quickly and measure the extinction against $CCl_4$ immediately, as the CuDDTC complex is sensitive to light. If stored in the dark the samples can be kept for 3 h without any change in the extinction values occurring.

Every test series must be accompanied by a blank test in which deionized, copper-free water instead of the sample is subjected to the entire analytical procedure described above.

Calculation

State the results in mg $Cu^{2+}/l$. Read off the value from the calibration curve and recalculate to take account of the volume of water used.

### 3.4.11.4 ICP-AES see Section 3.3.12

Atomic weight of copper: 63.546

## 3.4.12 Molybdenum

General remarks

In natural waters molybdenum occurs only rarely in detectable quantities. A test should nevertheless be carried out for the element, for example by means of emission spectroscopy (Section 3.4.1) to provide a general overview. Some subterranean waters, as well as waste waters, may contain traces of molybdenum. Trace amounts of molybdenum are vital for plants, animals and human beings since it forms a constituent of various enzymes.

3.4.12.1 Spectrophotometric determination (if no AAS equipment is available)

3.4.12.2 Direct determination by means of flame AAS

3.4.12.3 Concentration process for determination by means of flame AAS

3.4.12.4 For ICP-AES see Section 3.3.12

Methods 3.4.12.2 combined with 3.4.12.3 are recommendable.

### 3.4.12.1 Spectrophotometric determination as a molybdenum (V) thiocyanate complex

General remarks

In a mineral-acid solution and in the presence of a reducing agent molybdenum (V) reacts with thiocyanate to form a yellowish-orange Mo(V)SCN complex. The reducing agent used is tin (II) chloride. Extraction of the colour complex, presumably as the oxonium salt of the acid $H_2MoO(SCN)_5$ (Wünsch), allows interfering substances to be separated off.

Assuming an initial quantity of 10 litres of the water sample, which is then evaporated to less than 1 litre and treated in accordance with the directions given, concentrations of molybdenum down to approximately 0.02 mg/l can be determined.

Potential sources of interference are largely eliminated by precipitation with hydrogen sulphide, extraction as cupferronate with methyl isobutyl ketone and final extraction of the Mo(V)SCN complex with diisopropyl ether. All reagents must be of the highest grade of purity. A blank test must be carried out in parallel.

## Equipment

Photometer

Cuvettes, 1 cm and 5 cm

Hydrogen sulphide generator (Kipp's apparatus)

Glass dish, 1 litre

Measuring flask, 50 ml

Quartz or porcelain crucible

Separating funnel, 50 - 100 ml

Sodium hydrogen carbonate solution, saturated in the cold state

Ammonium chloride, reagent purity

Potassium hydrogen sulphate

Hydrochloric acid, 25 %

Sulphuric acid, 20 % and 1.75 m

Ammonium hydroxide solution, 27 %

Cupferron solution, saturated solution in cold water; prepare fresh solution each time.

Methyl isobutyl ketone

Hydrogen peroxide, 30 %

Potassium thiocyanate solution:
Dissolve 5 g of KSCN in 100 ml of $H_2O$

Tin (II) chloride solution:
30 % (dissolved in 5 % hydrochloric acid (not stable))

Iron (III) sulphate solution:
Saturated in the cold state in $H_2O$

Diisopropyl ether

Sodium hydroxide solution, 2 m

Molybdenum stock solution:
Dissolve 0.1500 g of reagent-purity $MoO_3$ in $H_2O$, adding 2 m sodium hydroxide solution; lightly acidify with 20 % sulphuric acid and make up to 1 litre with dist. water. 1 ml contains 0.1 mg of $Mo^{6+}$ ions.

Molybdenum reference solution:
Mix molybdenum stock solution 1 : 10 with dist. water. 1 ml contains 0.01 mg of $Mo^{6+}$ ions.

## Procedure

### Calibration

In order to prepare the calibration curve take increasing quantities of molybdenum reference solution containing between 0.005 and 0.25 mg of molybdenum, dilute with distilled water to approximately equal volume and treat according to the method described under "Determination". This begins with the addition of the 2 m sodium hydroxide solution to the solution containing sulphuric acid after destruction of the organic matter, and continues with the addition of 15 ml of 20 % sulphuric acid and transfer to the separating funnel.

### Determination

In order to determine trace amounts of molybdenum transfer 10 litres of the water sample in small portions to a 1-litre glass dish and boil down in a weak alkaline solution (adding saturated sodium bicarbonate solution). The entire contents of the flask should be used for one sample batch; any residues in the flask should therefore be transferred quantitatively to the glass dish.

Evaporate the weakly alkaline sample to less than 1000 ml, allow to cool, and acidify with 25 % hydrochloric acid until all the soluble substances are dissolved. Now dissolve 50 g of ammonium chloride in the solution. Alkalize with ammonium hydroxide solution and subsequently introduce a moderate flow of hydrogen sulphide.

After about 30 to 60 minutes acidify slightly with hydrochloric acid, without interrupting the flow of hydrogen sulphide. In the ensuing process the sulphides of the elements of the $H_2S$ group are precipitated together with elementary sulphur acting as a carrier. Continue passing hydrogen sulphide into the acid solution for a further hour and then filter off the sulphides. Rinse free of chloride with water containing a little hydrogen sulphide, and carefully incinerate the sulphides in a quartz or porcelain crucible at a maximum of 400 °C. Decompose the residue with a little potassium hydrogen sulphate and, when cool, pick up and dissolve the melt with 1.75 m sulphuric acid. Transfer the solution to a separating funnel, add 0.1 ml of cupferron solution and 0.5 ml of methyl isobutyl ketone for each 1 ml of solution and extract by shaking.

When the phases are separated repeat the extraction of the aqueous phase in the same way, adding cupferron solution and methyl isobutyl ketone. Separate off the aqueous phase, rinse the latter with methyl isobutyl ketone two more times and combine the organic phases for the purpose of further analysis. Discard the aqueous phase, unless it should be required for analysis by emission spectroscopy to detect any tungsten (3.4.1) and/or germanium (3.4.9) which may be present.

Add 5 ml of 20 % sulphuric acid to the organic extract containing the molybdenum and concentrate by evaporation. Add a few drops of hydrogen peroxide (30 %) and heat over a sand bath until the organic matter is completely destroyed. Pick up the remaining sulphuric solution with water, neutralize with 2 m sodium hydroxide solution, add 15 ml of 20 % sulphuric acid and transfer to a separating funnel.

Allow the solution to cool to 20 °C and treat with 3 drops of saturated iron (III) sulphate solution and then 5 ml of potassium thiocyanate solution. Mix, and add 5 ml of tin (II) chloride solution by pipette for reduction. Cool to 20 °C under running water and extract with 25 ml of diisopropyl ether. The diisopropyl ether should previously be saturated with potassium thiocyanate and tin (II) chloride by shaking with 5 ml of each solution.

After shaking for 1 minute, separate the phases, extract the aqueous phase twice more, each time with 10 ml of diisopropyl ether, transfer the organic extracts to a 50-ml measuring flask and make up to the mark with diisopropyl ether at 20 °C. Conduct the measurement in a 1-cm or 5-cm cuvette at 460 nm against diisopropyl ether.

Carry out a blank test in parallel with the entire analysis process. It is also useful to mix known quantities of molybdenum to further sample volumes (admixture method or standard addition).

Calculation

Read off the appropriate molybdenum content from the calibration curve on the basis of the extinction measured for the solution under analysis. Taking account of the blank test and the initial quantity of water used, recalculate for 1 litre of the water sample. Correction may also be required on the basis of the results obtained according to the admixture method.

Atomic weight of molybdenum: 99.94

### 3.4.12.2 Direct determination of molybdenum by means of atomic-absorption flame analysis

Equipment

| | |
|---|---|
| 1. Hollow-cathode molybdenum lamp | |
| 2. Wavelength | 313.3 nm |
| 3. Lamp current | 5 mA |
| 4. Aperture width | 0.5 nm |
| 5. Fuel gas | Acetylene |
| 6. Oxidant | Laughing gas |
| 7. Flame type | Reducing |
| 8. Background compensation | Yes |

| | |
|---|---|
| Optimum measuring range | 1 - 10 mg/l |
| Determination limit | 0.05 mg/l |
| Calibration | 0.2 - 10 mg/l |
| Acid concentration | 10 ml hydrochloric acid (1.17 g/ml) per 1000 ml solution for measurement |

Remarks

It is advisable to add 2 ml of aluminium nitrate solution per 50 ml of solution for measurement.

Aluminium nitrate solution:
Dissolve 100 g aluminium nitrate ($Al(NO_3)_3 \cdot 9\ H_2O$) in water and make up to 100 ml.

In the case of low concentrations, concentration by evaporation is only possible for low-mineral-matter waters (electrical conductivity at 25 °C below 100 µS/cm). The method of extractive concentration described below can be recommended.

### 3.4.12.3 Principle of the method of extractive concentration

Molybdenum ions are extracted with chloroform as molybdenum cupferronate after adding cupferron, the organic phase is evaporated, mineralized and determined by atomic-absorption analysis in the presence of aluminium ions.

**Reagents**

Hydrochloric acid (1.17 g/ml)

Nitric acid (1.40 g/ml)

Sulphuric acid (1.84 g/ml)

Cupferron solution:
Dissolve 6 g cupferron in water and dilute to 100 ml with water

Aluminium nitrate solution:
Dissolve 100 g $Al(NO_3)_3 \cdot 9\ H_2O$ in water and make up to 1000 ml

Chloroform

**Method**

Transfer sample solution (max. 2 litres) to a separating funnel and add 5 ml cupferron.

Extract 3 times, each time with 25 ml chloroform (shake for 2 minutes).

Separate off organic extracts and evaporate the combined organic extracts on a water bath.

Oxidize residue with sulphuric acid/nitric acid.

Fume off the acids.

Pick up the dry residue with 1 ml hydrochloric acid and 0.4 ml aluminium nitrate solution, transfer to a 10-ml measuring flask and fill up to the mark with distilled water.

Prepare a blank reading in the same way.

Final determination by flame AAS as described in 3.4.12.2.

Decomposition with nitric acid - hydrogen peroxide is required for organically loaded water samples.

### 3.4.12.4 ICP-AES see Section 3.3.12

### 3.4.13 Nickel

**General remarks**

As a rule, nickel occurs in natural waters only in traces. Nickel may be contained in waste water from the chemical industry, metal production or ore mining. Low nickel concentrations are also often found in subterranean waters. Nickel is one of the most mobile heavy metals.

**Quantitative determination**

3.4.13.1 Spectrophotometry with diacetyl dioxime
Modification A: without extraction
Modification B: with extraction

3.4.13.2 Direct determination by flame AAS

3.4.13.3 If appropriate, method 3.4.13.2 with concentration

3.4.13.4 Nickel determination by flameless AAS (graphite tube technique)

3.4.13.5 For ICP-AES see Section 3.3.12

#### 3.4.13.1 Spectrophotometric determination with diacetyl dioxime (dimethylglyoxime)

In the presence of a strong oxidizing agent such as bromine or ammonium peroxodisulphate, nickel (II) ions form a wine-red to brown-coloured complex with diacetyl dioxime in alkaline (ammoniacal) solution. This complex contains nickel in a multivalent (presumably tetravalent) form. Its exact nature has however not been finally established:

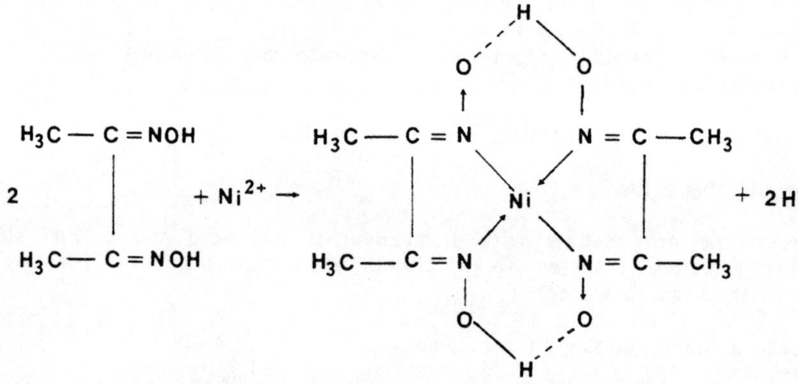

The nickel-diacetyl complex can be measured photometrically in the aqueous solution directly (modification A) or following extraction with chloroform and re-extraction (modification B).

The method in the form of modification A is suitable for ground water and surface water, and for waste waters after appropriate pretreatment. Without dilution or concentration 0.02 - 5 mg/l of nickel (II) can be determined.

Modification B is suitable for the detection of lower concentrations of nickel (II) ions (down to about 0.005 mg/l), in particular in ground waters and subterranean waters. As a basic rule, however, if an AAS device is available methods 3.4.13.2 with 3.4.13.3 or 3.4.13.4 are to be given preference.

In the presence of cyanides the samples should be mineralized due to the complex bonding of $Ni(CN)_4^{2-}$. In this process interfering organic impurities are simultaneously eliminated. Evaporate the samples (50 - 1000 ml) under a fume hood with 2 ml of concentrated sulphuric acid and 2 - 5 ml of concentrated nitric acid until $SO_3$ vapour appears. If the solution is still not clear or colourless, repeat the process after adding 3 ml of concentrated nitric acid. Then carefully evaporate almost to dryness, pick up the residue with 20 - 30 ml of distilled water, neutralize with concentrated ammonia and transfer the solution quantitatively to a 100-ml measuring flask.

Equipment

Photometer

Cuvettes, 1 cm and 5 cm

Bromine water, saturated (only modification A):
Dissolve about 12 ml of bromine in 1 litre dist. water.

Ammonium peroxodisulphate solution (only modification B):
Dissolve 10 g of reagent-purity $(NH_4)_2S_2O_8$ in 90 ml of dist. water.

Diacetyl dioxime solution:
1.2 g of reagent-purity diacetyl dioxime in 100 ml of ethanol or 1.5 g of the sodium salt in 100 ml of dist. water.

Triethanolamine

Chloroform, reagent-purity

Sodium hydroxide solution, 10 m (only modification B)

Ammonium hydroxide solution, conc., (0.91 g/ml)

Nitric acid, conc., (1.4 g/ml) (only modification A)

Hydrochloric acid, 0.5 m (only modification B)

Tartaric acid solution:
Dissolve 10 g of reagent-purity tartaric acid in 90 ml of dist. water (only modification A).

Sodium citrate solution:
Dissolve 20 g of reagent-purity sodium citrate in 80 ml of dist. water.

Hydroxylamine hydrochloric solution:
Dissolve 10 g of hydroxylamine hydrochloride in 90 ml of dist. water.

Phenolphthalein solution (only modification B)

Nickel stock solution:

Dissolve 1.00 g of reagent-purity nickel wire (spectrographically standardized) with 15 ml of $HNO_3$ (1 + 1). Add 5 ml of $H_2SO_4$ (1 + 3) and evaporate until $SO_3$ vapour appears. Subsequently make up to 1 litre with dist. water. 1 ml contains 1 mg of $Ni^{2+}$.

Nickel reference solution I:

Dilute nickel stock solution 1 to 100 with dist. water. 1 ml contains 0.01 mg of $Ni^{2+}$.

Nickel reference solution II:

Dilute reference solution I in the ratio 1 to 10 with dist. water. 1 ml contains 0.001 mg of $Ni^{2+}$.

## Method

### Modification A (without prior extraction)

#### Calibration

In order to prepare the calibration curve take increasing quantities of nickel reference solution (I and/or II) with between 0.02 and 5 mg of nickel, make up to 50 ml with distilled water, and treat as described under "Determination".

#### Determination

Transfer to a 100-ml measuring flask either 50 ml of the water sample or a smaller quantity which has been made up to 50 ml with distilled water (nickel content 0.001 to 0.25 mg). Add 10 ml of bromine water and shake. Add one after the other 8 ml of triethanolamine, 12 ml of concentrated ammonium hydroxide solution and 4 ml of diacetyl dioxime solution and immediately make up to the mark with distilled water. Shake the mixture and without further delay measure the extinction at 460 nm (blue filter) against distilled water. Each individual sample must be treated in the prescribed order without interruption, including the extinction measurement. Conduct a blank test in the same way and take into account accordingly.

#### Calculation

Read off the appropriate nickel content from the calibration curve on the basis of the measured extinction, take into account the quantity of water used and the blank reading, and recalculate for 1 litre of water.

### Modification B (with extraction in chloroform)

#### Calibration

In order to prepare the calibration curve take increasing quantities of nickel reference solution II with between 0.005 and 0.025 mg of nickel and subject the solutions to the process of analysis described under "Determination".

#### Determination

Acidify between 1 and 5 litres of the water sample with concentrated hydrochloric acid and evaporate to a volume of less than 500 ml. Transfer the solution to a separating funnel and add one after the other, shaking constantly, 20 ml of sodium citrate solution, 2 ml of hydroxylamine hydrochloride solution, 2 ml of diacetyl dioxime solution and 1 drop of phenolphthalein solution. Then neutralize with concentrated ammonia (0.91 g/ml) and following the colour change add 3 drops in excess.

Extract the nickel-diacetyl dioxime complex by shaking vigorously (2 minutes) after adding 20 ml of chloroform. Separate off the organic phase and extract the aqueous phase once again with 10 ml of chloroform after adding 1 ml of diacetyl dioxime solution. In the separating funnel wash the combined chloroform phases, containing the nickel, twice with 10 ml of ammonium hydroxide solution (1 + 50) and then shake vigorously for 2 minutes with 15 ml of 0.5 m hydrochloric acid.

After separation of the phases transfer the hydrochloric aqueous solution into which the nickel has been re-extracted to a 25-ml measuring flask. Wash the chloroform twice more, each time with 1 ml of distilled water, adding the washing fractions likewise to the 25-ml measuring flask.

Next treat the hydrochloric solution containing the nickel with 3 ml of diacetyl dioxime solution, 1 ml of 10 m sodium hydroxide solution and 0.3 ml of ammonium peroxodisulphate solution, mixing well each time. Dilute with water to the mark at 20 °C, leave to stand for about 2 hours, and measure in a 5-cm cuvette at 460 nm against distilled water. Carry out a blank test concurrently and take into account accordingly.

Calculation

Read off the appropriate nickel content from the calibration curve on the basis of the measured extinction, take into account the quantity of water used and the blank reading, and recalculate for 1 litre of water.

Atomic weight of nickel: 58.71

### 3.4.13.2 Direct determination by means of atomic-absorption flame analysis

Equipment parameters

| | |
|---|---|
| 1. Hollow-cathode nickel lamp | |
| 2. Wavelength | 232.0 nm |
| 3. Lamp current | 3.5 mA |
| 4. Aperture width | 0.2 nm |
| 5. Fuel gas | Acetylene |
| 6. Oxidant | Air |
| 7. Flame type | Oxidizing |
| 8. Background compensation | Yes |

| | |
|---|---|
| Optimum measuring range | 0.2 - 5 mg/l |
| Determination limit | 0.05 mg/l |
| Calibration | 0.1 - 5 mg/l |
| Acid concentration | 10 ml hydrochloric acid (1.17 g/ml) per 1000 ml solution for measurement |

Remarks

Concentration by evaporation is possible, if concentrations are low.

Decomposition with nitric acid/hydrogen peroxide is required if the samples contain organic loads.

The extractive concentration method described for the element silver using hexamethylene ammonium hexamethylene dithiocarbamate (HMDC) may also be used for nickel.

A further method of extraction is described below.

### 3.4.13.3 Method with concentration

**Basis of the method**

After adding diacetyl dioxime (dimethylglyoxime) solution, nickel ions are extracted from citrate-buffered solution with chloroform, re-extracted with 1 m hydrochloric acid from the organic phase and determined by means of atomic-absorption analysis.

**Reagents**

Hydrochloric acid I (1.17 g/ml)

Hydrochloric acid II, 1 mol/l

Ammonium hydroxide solution I (0.91 g/ml)

Ammonium hydroxide solution II:
Dilute 2 ml ammonium hydroxide solution I with water to 100 ml

Sodium citrate solution:
200 g sodium citrate · 2 $H_2O$/l

Phenolphthalein solution

Diacetyl dioxime solution:
Dissolve 15 g diacetyl dioxime (disodium salt) ($C_4H_6N_2Na_2O_2 \cdot 8\ H_2O$) in water and make up to 1000 ml

Chloroform ($CHCl_3$)

**Method**

Acidify the sample volume with 5 ml hydrochloric acid (1.17 g/ml); if necessary evaporate samples to 200 ml and transfer to a 250 ml separating funnel.

Treat with 10 ml of 20 % sodium citrate solution.

Neutralize with ammonium hydroxide solution I with respect to phenolphthalein and add 3 drops of ammonium hydroxide solution I in excess.

If clouding occurs, acidify again and repeat neutralization.

Add 5 ml diacetyl dioxime solution and 15 ml chloroform and shake well for 2 minutes.

Separate off the chloroform phase and shake the aqueous phase once again for 2 minutes with 10 ml chloroform.

Combine the organic phases and wash with 10 ml of ammonium hydroxide solution II for 30 minutes.

Discard the wash water and treat the chloroform phase with exactly 10 ml of 1 m hydrochloric acid.

Shake vigorously for 2 minutes, separate off the hydrochloric phase and determine nickel in this using atomic-absorption flame analysis.

The equipment parameters are as for direct determination.

### 3.4.13.4 Nickel determination with the graphite tube technique (Furnace method)

Equipment parameters

| | |
|---|---|
| 1. Hollow-cathode nickel lamp | |
| 2. Wavelength | 232.0 nm |
| 3. Lamp current | 3.5 mA |
| 4. Aperture width | 0.5 nm |
| 5. Background compensation | Yes |
| 6. Carrier gas | Argon |
| 7. Drying | 15 s at 140 °C |
| 8. Incineration | 25 s at 700 °C |
| 9. Atomization | 3 s at 2400 °C |
| 10. Gas stop | Yes |
| Optimum measuring range | 5 - 20 μg/l with 10 μl sample |
| Determination limit | 1 μg/l |
| Calibration | Addition method |
| Acid concentration | 10 ml nitric acid (1.40 g/ml) per 1000 ml sample solution |
| Matrix modification | 10 μl 0.1 % ammonium nitrate solution |
| Ammonium nitrate solution | Dissolve 0.1 g $NH_4NO_3$ in water and make up to 100 ml |

Remarks

Immediately after sampling, stabilize the water sample with 10 ml nitric acid (1.40 g/ml) per 1000 ml sample solution.

### 3.4.13.5 ICP-AES see Section 3.3.12

### 3.4.14 Mercury

General remarks

Inorganic mercury compounds (ionogenic) and organic compounds such as methyl mercury, ethyl mercury and so on may occur in water. The latter may be concentrated to a certain extent in organisms (e.g. fish). Although the absolute quantity involved is mostly only very small, in view of their

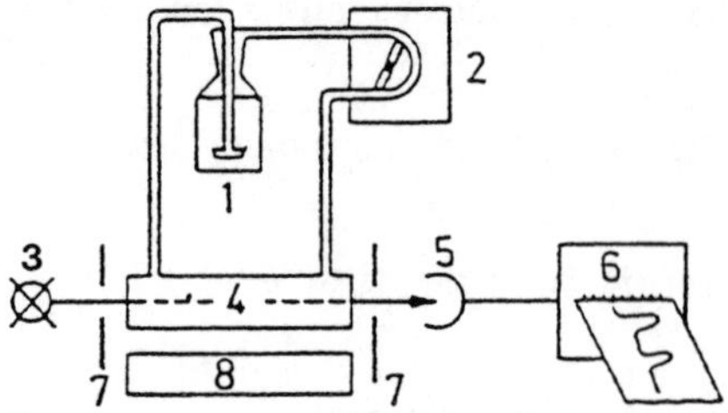


Fig. 102. 1) = Sample; 2) = Flow inducer; 3) = Hollow-cathode mercury lamp; 4) = Cuvette; 5) = Detector; 6) = Plotter; 7) = Diaphragm; 8) Heating element

toxicity these compounds must be taken into account owing to their accumulation in the food chain. It is therefore of interest in analysis to differentiate between inorganic and organically bonded mercury (see the decomposition methods described).

The method of quantitative determination which has gained widest acceptance is the variant of AAS using the cold vapour technique with and without decomposition.

### 3.4.14.1 Mercury AAS with cold vapour method

#### Principle of the method

Mercury ions are reduced to metallic mercury with tin(II) chloride. The metallic mercury is transferred into a quartz cuvette with the aid of a current of inert gas and the absorption of the atoms is measured in the beam of an atomic-absorption spectrometer (cold vapour method).

#### Reagents

Stabilizing reagent:
Treat 500 ml of nitric acid (1.40 g/ml) with 5 g potassium dichromate ($K_2Cr_2O_7$) and make up to 1000 ml with water.

Reduction solution:
Treat 10 g tin (II) chloride ($SnCl_2 \cdot 2\ H_2O$) with 10 ml hydrochloric acid (1.17 g/ml) and make up to 100 ml with water, shaking constantly.

Pretreatment of the water sample on sampling:

Immediately after sampling add 1 ml of the stabilizing reagent per 100 ml of sample volume to the water samples in order to prevent losses by absorption of mercury at the vessel walls.

#### Equipment parameters

Atomic absorption spectrometer with attachment for the reduction and measurement of mercury.

| | |
|---|---|
| 1. Hollow cathode mercury lamp | |
| 2. Wavelength | 253.7 nm |
| 3. Lamp current | 3.5 mA |

| | | |
|---|---|---|
| 4. | Inert gas | Argon or nitrogen |
| 5. | Gas through-flow | 10 l/h |
| 6. | Cuvette temperature | Room temperature |
| 7. | Aperture width | 0.5 nm |
| 8. | Integration | Peak area |
| 9. | Integration time | 25 s |
| 10. | Background compensation | Yes |
| 11. | Measuring range | 1 - 10 µg/l |
| 12. | Determination limit | 0.5 µg/l |

**Procedure**

Transfer 20 ml of the stabilized water sample into the reaction vessel of the system.

Expel the air from the reaction vessel by means of a constant flow of 10 l/h of inert gas (argon or nitrogen).

Begin measurement and at the same time continuously add 10 % tin (II) chloride solution.

Transfer the mercury vapour by means of the carrier gas flow into a quartz cuvette and measure in the beam of the atomic-absorption spectrometer.

Measure blank and calibration solutions in similar fashion.

**Remarks**

Since this technique quantitatively detects only mercury which is present ionogenically in solution, decomposition is required for organically or complex bonded mercury. Organically loaded water samples must also be decomposed before measurement.

Two methods of decomposition are described below:

A) **Pressure decomposition** in a closed Teflon vessel inside a steel cylinder:

Introduce 20 ml of water into the decomposition vessel and treat with 2 ml nitric acid.

Close the lid of the decomposition vessel, insert into the steel cylinder and close with a screw top. Heat the steel cylinder to 180 °C in the heating block or the drying oven and leave at this temperature for 2 hours.

After cooling, open the steel cylinder, take out the decomposition vessel and transfer the decomposition solution quantitatively to measuring flask of 50 ml.

Fill the measuring flasks up to the mark.

Analysis as above (3.4.14.1).

B) **Wet decomposition**

**Reagents**

Nitric acid (1.40 g/ml)

Sulphuric acid (1.84 g/ml)

Potassium permanganate:
Dissolve 50 g potassium permanganate ($KMnO_4$) in water and make up to 1000 ml

Potassium peroxide sulphate solution:
Dissolve 40 g potassium peroxide sulphate ($K_2S_2O_8$) in water and make up to 1000 ml

Hydroxylammonium chloride solution:
Dissolve 10g hydroxylammonium chloride ($NH_3OHCl$) in water and make up to 100 ml.

### Method

Introduce 50 ml of the water sample into a duran glass flask, add 5 ml sulphuric acid (1.84 g/ml), 5 ml nitric acid (1.40 g/ml) and 15 ml potassium permanganate solution.

Leave for 15 minutes at room temperature.

Add 10 ml potassium peroxide sulphate solution.

Cover the duran glass flask and leave for 2 hours at 95 °C in a water bath.

After cooling, add 5 ml hydroxylammonium chloride solution and make up to 100 ml final volume.

Analysis as above (3.4.14.1).

Sensitivity and reproducibility is improved if the mercury vapour formed is concentrated by the formation of amalgams on gold wool. Heating the gold wool causes the mercury to be released suddenly and it may then be measured. This can improve sensitivity by at least a factor of 10.

## 3.4.15 Selenium

### General remarks

In natural waters selenium occurs only in very low concentrations. Sea water contains an average of about 0.005 mg/l. Higher concentrations are rarely found in surface waters. Regionally limited ground waters form an exception: if they occur in an area with rocks and soils containing large amounts of selenium, they may contain between 0.05 and 0.1 mg/l or in exceptional cases even more.

If selenium concentrations which greatly exceed these values occur in surface waters it must be assumed that seleniferous waste water is gaining access. Above all, this may stem from the processing (roasting) of sulphidic ores, the manufacture of sulphuric acid, or from the chemical industry.

**Quantitative determination**

3.4.15.1 Hydride AAS

3.4.15.2 Spectrophotometric determination with o-phenylenediamine

### 3.4.15.1 Determination of selenium using the hydride AAS technique

**Principle**

Using sodium borohydride, selenium ions are reduced to selenium hydride, transferred to a heated quartz cuvette with the aid of a current of inert gas, decomposed thermally, and the absorption of the atoms is measured in the beam of an atomic-absorption spectrometer. In the hydride technique, the element which is to be determined is volatilized as a gaseous hydride and separated off from the matrix. Interferences may occur if there is a considerable excess of elements such as antimony, arsenic, tin, bismuth, mercury, or tellurium, which may also be volatilized using this technique. Above all, heavy metals such as copper and nickel have a disturbing effect during the hydride formation itself. These interferences may be diminished by adding 300 mg of solid 2-pyridine aldoxime to the solution for measurement.

Since the hydride technique only permits quantitative detection of selenium (IV), selenium (VI) must be converted to selenium (IV) by prereduction (boiling in a strongly hydrochloric solution at the reflux).

**Reagents**

Selenium stock solution (1000 mg/l)

Hydrochloric acid (1.17 g/l)

Sodium borohydride solution:
Dissolve 3 g $NaBH_4$ and 1 g NaOH in water and make up to 100 ml.

Prereduction of the the water samples and decomposition solutions:
Treat 100 ml of the sample for investigation with 50 ml hydrochloric acid (1.17 g/ml) and boil for 15 minutes at the reflux. Treat blanks and calibration solutions in the same way.

**Equipment parameters**

**Atomic-absorption spectrometer with hydride attachment:**

| | | |
|---|---|---|
| 1. | Hollow-cathode selenium lamp | |
| 2. | Wavelength | 196.0 nm |
| 3. | Lamp current | 7 mA |
| 4. | Inert gas | Argon |
| 5. | Gas through-flow | 30 l/h |
| 6. | Cuvette temperature | 800 °C |
| 7. | Aperture width | 1 nm |
| 8. | Integration | Peak area |
| 9. | Integration time | 25 s |
| 10. | Background compensation | No |

| | |
|---|---|
| Measuring range | 1 - 5 µg/l |
| Determination limit | 0.5 µg/l |

The instrument parameters quoted here apply to a Varian AAS 775 with a hydride system manufactured by Messrs. Berghoff, Tübingen, FRG.

Method

Transfer 30 ml of the prereduced solution to the hydride system. Allow a constant current of inert gas (argon) to flow through the system at 30 l/h to expel the air. Following this, continuously add 3 % sodium borohydride solution using a peristaltic pump. Selenium ions are reduced to selenium hydride and directed by the current of inert gas into a quartz cuvette heated to 800 °C where they are thermally decomposed. Measure the absorption.

Remarks

Since the hydride technique only permits quantitative detection of selenium which is present ionogenically in solution, decomposition is required for organically or complex-bonded selenium. Organically loaded water samples must also be decomposed before measurement. Two methods of decomposition are described below.

**A) Decomposition with sulphuric acid/hydrogen peroxide:**

Reagents

Sulphuric acid (1.84 g/ml)

Hydrogen peroxide (30 %)

**Method**

Introduce 50 ml of the water sample into a 150-ml beaker, treat with 5 ml sulphuric acid and add 5 ml hydrogen peroxide.

Heat in a sand bath until $SO_3$ vapour appears.

If the organic load of the water sample is high, repeat the addition of hydrogen peroxide to the decomposition mixture.

When the decomposition mixture is cool, transfer the sample to a measuring flask, 50 ml, and dilute with water to the mark.

Care should be taken that the sample never evaporates to dryness.

**B) Decomposition with magnesium oxide/magnesium nitrate:**

**Reagents**

Hydrochloric acid (1.17 g/ml)

Magnesium oxide (MgO)

Magnesium nitrate ($Mg(NO_3)_2 \cdot 6H_2O$)

Procedure

Introduce 50 ml of the water sample into a 100-ml quartz glass beaker, neutralize and subsequently treat with 1 g magnesium oxide and 1 g magnesium nitrate.

Evaporate over a water bath.

Carefully cover the evaporation residue with 2 g magnesium nitrate.

Dry at 180 °C in a drying oven for 20 minutes.

Place in a cold muffle furnace, heat to between 500 and 550 °C and leave at this temperature for 1 hour.

Pick up the ignition residue with 10 ml hydrochloric acid, transfer to a 50-ml measuring flask, and dilute with water to the mark.

### 3.4.15.2 Spectrophotometric determination of selenium with o-phenylenediamine

General remarks

In acid solution selenium (IV) reacts with aromatic o-diamines forming yellow compounds. The intensity of the colour of the compounds is measured directly, or following extraction with toluene, at 334 nm. In the method described here the colour reaction is performed with o-phenylenediamine:

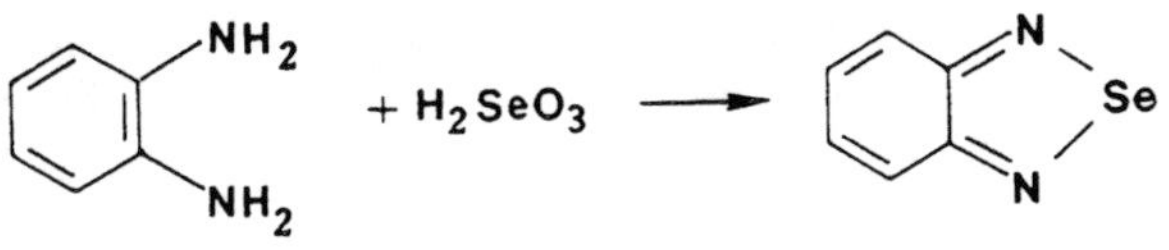


The extinction maximum of 3,4-benzo-1,2,5-selenodiazole lies in the region of 330 - 335 nm ( = 17 750). Beer's law applies in the range from 0.13 - 3 µg Se/ml in toluene.

High selenium contents cannot be expected in natural waters and interference must be anticipated from other substances contained in the water. Consequently, it is advisable to concentrate the selenium and carry out distillation before photometric determination. Alkaline concentration of 1 - 2 litres of water in a vacuum rotation evaporator and subsequent distillation of the concentrate with hydrobromic acid containing bromine leads to concentration and isolation of the selenium without loss (K.-E. Quentin and L. Feiler). This is because in a solution containing sulphuric acid Se (IV) reacts with surplus bromide to $SeBr_4$ which can be distilled off thanks to its volatility.

The $SeBr_4$ is decomposed in aqueous solution (in the distillation receiver) according to $SeBr_4 + 3\ H_2O \longrightarrow H_2SeO_3 + 4\ HBr$ to give selenious acid which reacts with o-phenylenediamine. Excess bromine is bonded with hydroxylammonium chloride. Any Se(VI) which is present is reduced to Se(IV) by HBr. The detection therefore always encompasses total selenium.

The method is suitable for the determination of selenium in natural waters with concentrations between 0.001 and 0.25 mg of Se/l.

Interference due to other substances contained in the water is eliminated by the distillation process.

### Equipment

Vacuum rotation evaporator

Distillation apparatus with distillation flask (250 ml), dropping funnel and connecting piece for introducing $N_2$, condenser and receiver.

pH meter

Separating funnel, 100 ml

Sodium hydroxide solution, 10 %

Hydroxylammonium chloride solution, 2.5 % in water

o-phenylenediamine solution, 0.4 % in water

Toluene, reagent purity

Potassium bromide, reagent purity

Sulphuric acid, conc, (1.84 g/ml)

Bromine water: saturated solution of $Br_2$ in dist. water

Selenium stock solution:
Accurately weigh a piece of selenium (eg. 1 g of superpure metal) and dissolve in 5 ml of conc. nitric acid, heating gently. Carefully evaporate to dryness, pick up with dist. water and dilute to 1 litre. 1 ml then contains approx. 1 mg of selenium.

Selenium reference solution:
Dilute eg 1 ml of selenium stock solution to 1 litre with dist. water. 1 ml of this solution contains 0.001 mg of Se.

### Method

#### Calibration

In order to prepare the calibration curve measure increasing volumes of selenium reference solution with concentrations between 0.001 and 0.25 mg into measuring flasks and make up to 1 litre with dist. water. Subject the solutions to the same process as indicated for the water sample.

#### Determination

After adding 10 ml of 10 % sodium hydroxide solution reduce 1 litre of the water sample to about 10 ml in the vacuum rotation evaporator and then transfer quantitatively to the flask of the distillation apparatus. The receiver of the distillation apparatus should contain 30 ml of 2.5 % hydroxylammonium chloride solution.

Add 0.70 g of reagent-purity KBr and 40 ml of conc. sulphuric acid treated with 8 - 10 drops of saturated bromine water and distill the sample in a nitrogen flow (approx. 25 ml min.) until $SO_3$ vapour appears.

Adjust the seleniferous hydroxylammonium chloride solution to pH 2.4 (measure pH electrometrically) and then add 5 ml of o-phenylenediamine solution. After a reaction time of 150 minutes shake out for 3 minutes with 5 ml of toluene the benzoselenodiazole which has formed. After separation of the phases allow the aqueous layer to run off, remove the toluene solution with a pipette and transfer to cuvettes of appropriate path length. Carry out photometric measurement at 334 nm against pure toluene.

Calculation

Read off the corresponding selenium content from the calibration curve on the basis of the extinction measured, taking the blank reading into account. If a 1-litre water sample was used initially this value corresponds to the selenium content in mg/l. If a larger or smaller volume of sample was taken, recalculate as appropriate.

Atomic weight of selenium: 78.96

### 3.4.16 Silver

General remarks

In drinking water, silver ions occur only as traces. Even if silver ions are added to the water for the purposes of germ inhibition, on the basis of silver's oligodynamic action, concentrations are low (approximately between 0.03 and 0.1 mg/l).

A sensitive method is therefore required. As a general rule atomic-absorption procedures and spectrophotometric methods may be employed.

3.4.16.1 Direct atomization of the aqueous solution (flame absorption), if necessary with concentration by extraction

3.4.16.2 Determination by means of the graphite tube method

3.4.16.3 Spectrophotometric determination with dithizone

3.4.16.4 For ICP-AES see section 3.3.12

In comparison with the spectrophotometric method the atomic-absorption procedures are simpler, faster and virtually free of interference. Where no atomic-absorption spectrophotometer is available, sufficiently accurate and reliable results can be obtained with the dithizone spectrophotometric method.

#### 3.4.16.1 Direct determination of silver by means of atomic-absorption flame analysis

Equipment parameters

| | |
|---|---|
| 1. Hollow-cathode silver lamp | |
| 2. Wavelength | 328.1 nm |
| 3. Lamp current | 3.5 mA |
| 4. Aperture width | 0.5 nm |
| 5. Fuel gas | Acetylene |

| | |
|---|---|
| 6. Oxidant | Air |
| 7. Flame type | Oxidizing |
| 8. Background compensation | Yes |
| Optimum measuring range | 0.2 - 5 mg/l |
| Determination limit | 0.05 mg/l |
| Calibration | 0.2 - 5 mg/l |
| Acid concentration | 10 ml nitric acid (1.40 g/ml) per 1000 ml solution for measurement |

**Remarks**

Concentration by evaporation is possible if concentrations are low.

The method of concentration described below can be recommended.

**Concentration by extraction**

**Basis of the method**

After chelation with hexamethylene ammonium hexamethylene dithiocarbamate (HMDC), silver is extracted from formate-buffered solution with xylene-diisopropyl ketone and the organic extract is determined directly by means of atomic-absorption analysis (3.4.16.1).

**Reagents**

Nitric acid (1), $HNO_3$, (1.40 g/ml)

Nitric acid (2), dilute:
Dilute one part by volume of nitric acid (1) with 9 parts by volume of water.

Hydrogen peroxide, ($H_2O_2$) = 30 %

Formic acid, HCOOH, (1.22 g/ml)

Citric acid monohydrate, $C_6H_8O_7 \cdot H_2O$

Sodium hydroxide, NaOH

m-cresolsulphonphthalein (m-cresol purple)

Xylene (mixture of isomers), $C_8H_{10}$

2,4-dimethyl 3-pentanone - (3), (diisopropyl ketone), $C_7H_{14}O$, distilled, boiling point 124.5 °C

Solvent mixture:
Take 30 ml xylene in a 100-ml measuring flask, and make up to the mark with diisopropyl ketone.

Methanol, $CH_3OH$

Hexamethylene ammonium hexamethylene dithiocarbamate (HMDC), $C_{13}H_{26}H_2S_2$

Formate buffer solution (pH 4):

Dissolve 368 g formic acid and 14 g citric acid monohydrate in about 350 ml of water. Keeping well cooled and stirring constantly, slowly add 243 g sodium hydroxide (caution!). After adding 50 mg of m-cresol purple, dilute the solution to 1 litre. To clear the buffer solution, shake out twice with 50 ml HMDC extraction solution each time (shake for 3 to 5 minutes).

HMDC extraction solution:

Dissolve 1.7 g HMDC in 75 ml of xylene in a dry 250-ml measuring flask, warming gently; dilute the solution with diisopropyl ketone to the mark. The extraction solution should be prepared freshly every day.

HMDC solution in methanol:

Dissolve 5.5 g HMDC in methanol in a 100-ml measuring flask, warming moderately; after cooling to room temperature, make up to the mark with methanol.

Aqueous stock silver solution (1000 mg/l)

Organic standard solution:

Treat 5 ml of the stock solution (1000 mg/l) in a dry 100-ml measuring flask with 50 ml formic acid and 0.2 to 0.5 g citric acid monohydrate, and dilute to the mark with diisopropyl ketone. 1 ml of this solution contains 50 µg silver.

Organic calibration solutions:

Dilute the organic standard solution with HMDC extraction solution in such a way as to produce calibration solutions which correspond to the water samples, taking the desired concentration factor into account.

**Method**

Remove high concentrations of organic substances in the water by boiling down and evaporating, adding 4 ml of nitric acid (1) and 4 ml hydrogen peroxide per 400 ml of sample. (In general, more than 10 mg/l total carbon or more than 50 mg/l chemical oxygen demand counts as a high concentration). Pick up the residue with 40 ml of warm nitric acid (2). Make up the solution to 400 ml with water, and treat further as described below.

Add 20 ml of formate buffer solution to 400 ml of the sample, which should be pretreated if necessary. (The indicator colour must be a clear yellow; if a red coloration appears, add a further 20 ml of formate buffer solution).

Add 2 ml of methanolic HMDC solution, shake, and leave to stand for 5 minutes. Add 20 ml of HMDC extraction solution and shake vigorously for 3 minutes.

After 15 minutes, separate off the organic phase, and centrifuge off any water present.

Take a blank reading in the same way and deduct it.

Zero adjustment of the AAS device should be made with HMDC extraction solution. Use xylene/isopropyl ketone (75 ml xylene/175 ml diisopropyl ketone) to rinse the suction system. It is advisable to use either an adjustable atomizer or the injection method. The organic extract may be kept for 24 hours.

### 3.4.16.2 Silver determination with the graphite tube technique (Furnace method)

Equipment parameters

| | |
|---|---|
| 1. Hollow-cathode silver lamp | |
| 2. Wavelength | 328.1 nm |
| 3. Lamp current | 3.5 mA |
| 4. Aperture width | 0.5 nm |
| 5. Background compensation | Yes |
| 6. Carrier gas | Argon |
| 7. Drying | 25 s at 180 °C |
| 8. Incineration | 22 s at 500 °C |
| 9. Atomization | 3 s at 2000 °C |
| 10. Gas stop | Yes |

| | |
|---|---|
| Optimum measuring range | 1 - 10 µg/l with 10 µl sample |
| Determination limit | 0.1 µg/l |
| Calibration | Addition method |
| Acid concentration | 10 ml nitric acid (1.40 g/ml) per 1000 ml sample solution |

Remarks

Immediately after sampling, stabilize the water sample with 10 ml nitric acid (1.40 g/ml) per 1000 ml sample solution.

### 3.4.16.3 Spectrophotometric determination with dithizone

In acid (to almost neutral) solution silver ions react with dithizone (diphenyl dithiocarbazone), which is coloured green in organic solvents, to form primary silver dithizonate. Silver dithizonate is gold-coloured in organic solvents.

$$Ag^+ + H_2Dz \rightleftharpoons AgHDz + H^+$$

(Dz signifies the radical $C_{13}H_{10}N_4S$ of the dithizone without the two hydrogen atoms which are capable of interchange with metals in the formation of the dithizonates).

The extinction maximum of AgHDz is at 462nm. Since the excess dithizone is removed from the carbon tetrachloride extract by washing twice with 0.01 m $NH_4OH$ solution, the silver dithizonate solution can be measured at the extinction maximum of this compound in the single-colour method. The calibration curves are linear as far as approximately 2.5 µg of Ag/ml of organic solvent. The stoichiometric relationship Ag : Dz was established as 1 : 1 in the fundamental study by Fischer, Leopoldi and von Uslar, as long as the extraction is carried out in acid solution. (In a neutral or weakly alkaline solution a red-purple secondary $Ag_2Dz$ is formed which is only slightly soluble in $CCl_4$).

The method is suitable for the determination of 2 - 300 µg of Ag per litre of undiluted water sample. This range may be extended in either direction by means of dilution or evaporation. In the form specified here (evaporation and twofold extraction) contaminated surface waters and waste waters can be examined in addition to drinking and mineral waters.

Since dithizone does not react specifically with silver ions, the influence of interfering accompanying ions must be eliminated by the choice of appropriate test conditions. If extraction of the silver is carried out in acid solution, metal ions which react with the reagent only in neutral or alkaline solution do not interfere. Apart from bismuth, such metal ions are primarily lead (II) ions and zinc (II) ions. The presence of these substances in water is to be expected. In acid solution they form a dithizonate only if their concentration exceeds that of silver ions by a factor of $10^4$.

Iron (II) ions form a dithizonate under alkaline reaction conditions only. Mercury and copper which are co-extracted in acid solution are separated off by re-extraction with sulphuric KSCN solution, since only the silver passes over to the aqueous phase.

Interference by halide ions is eliminated by treatment with sulphuric acid.

It should always be noted that the photometric dithizone method is susceptible to interference and that consequently the procedure should be closely followed.

### Equipment

Spectrophotometer or filter photometer

Cuvettes, path length 1 cm

Quartz dish, 500 ml

Separating funnels, 500 ml and 250 ml

Measuring flask, 25 ml

Bulb pipettes, 100 ml, 10 ml, 5 ml, 2 ml, 1 ml

Sulphuric acid, (1.84 g/ml), reagent purity

Sulphuric acid, 0.1 m

Ammonium hydroxide solution 0.01 m

Carbon tetrachloride, reagent purity, tested with dithizone for absence of silver

Dithizone solution I:

Dissolve 25 mg of dithizone in 100 ml of reagent-purity carbon tetrachloride. In order to remove traces of Cu add 1 ml of ammonium hydroxide solution (approx. 0.91 g/ml) and 200 ml of dist. water. Transfer the dithizone to the aqueous phase by shaking in the separating funnel. Separate off and discard the organic phase. Add 100 ml of reagent-purity carbon tetrachloride, acidify the aqueous phase with about 20 - 30 ml of hydrochloric acid (1.25 g/ml) and shake until the dithizone is again present as a green solution in the organic phase or until the aqueous phase is colourless. The solution is stable for a one day in an amber flask.

Dithizone solution II:
Dilute dithizone solution I in the ratio 1 : 10 with reagent-purity carbon tetrachloride before use.

Potassium thiocyanate solution:
2 % aqueous solution (store in a dark flask containing 25 ml of dithizone solution).

Silver stock solution:
Dissolve 1.5746 g of reagent-purity silver nitrate in redist. water, add 10 ml of reagent-purity nitric acid (1.42 g/ml) and make up to 1 litre. 1 ml of this solution contains 1 mg of $Ag^+$.

Standard silver solutions:
Ag solution A:
Dilute silver stock solution 1 to 100 with redist. water (0.01 mg of $Ag^+$/ml)
Ag solution B:
Dilute Ag solution A 1 to 10 with redist. water (0.001 mg of $Ag^+$/ml)
Ag solution C:
Dilute Ag solution A 1 to 100 with redist. water (0.0001 mg of $Ag^+$/ml)

The standard silver solutions are to be freshly prepared immediately before use.

## Method

### Calibration

Take volumes of standard silver solution containing increasing quantities of $Ag^+$ between 0.1 and 100 µg and subject each one to the entire process of analysis. Take the measurements at 462 mn. In the range up to 100 µg of $Ag^+$/25 ml the calibration curve is slightly curved.

### Determination

Select a volume of water such that it contains between 1 and 90 µg of $Ag^+$. This volume can be established in a preliminary test with 500 ml of water sample.

Acidify the selected volume in a quartz dish with 5 - 20 ml of sulphuric acid (1.84 g/ml) and heat until fuming. Continue heating until approximately half of the added sulphuric acid is fumed off. When cool dilute with water so as to obtain approximately 2 m $H_2SO_4$, taking the remaining quantity of sulphuric acid into account (for approximately 10 ml of $H_2SO_4$ following evaporation about 80 ml of water is required). Transfer the solution to a separating funnel and rinse the quartz dish with 2 m sulphuric acid.

Extract the approximately 2 m sulphuric acid solution with 5-ml portions of dithizone solution II until the final extract remains a constant green colour (corresponding to complete extraction of Ag). In the presence of considerable quantities of Cu terminate the extraction when the final extract remains red to red-purple, in other words when the extract is pure copper dithizonate, after previously extracting all the silver dithizonate.

Combine the dithizone extracts containing the gold-coloured silver dithizonate and any traces of red-purple copper dithizonate in the separating funnel and rinse with 20 ml of 0.1 m sulphuric acid. Discard the sulphuric wash solution and shake the remaining organic solution three times, for 1 minute each time, with a mixture of 5 ml of potassium thiocyanate solution and 1 ml of 0.1 m sulphuric acid in order to decompose the silver dithizonate. Transfer the three aqueous extracts to another separating funnel, and wash with 10 ml of carbon tetrachloride. After phase separation, transfer the acid KSCN extracts to a quartz dish, treat with 2 ml of sulphuric acid (1.84 g/ml) and heat until fuming.

When half of the sulphuric acid is evaporated, allow to cool and dilute with water such that the solution is about 2 m sulphuric acid (for about 1 ml residual $H_2SO_4$ following evaporation approximately 8 ml of water is required). Again transfer the solution to a separating funnel and rinse the quartz dish with 2 m sulphuric acid.

Now extract the silver a further 2 - 4 times with 5-ml portions of dithizone solution II until the final extract remains green. Combine the dithizone extracts and wash twice with 0.01 m ammonium hydroxide solution. This has the effect of removing the excess dithizone. Filter the carbon tetrachloride solution with the purely gold-coloured silver dithizonate through a small filter into a 25 ml measuring flask and rewash with reagent-purity carbon tetrachloride. Make up to the mark at 20 °C, mix, and measure photometrically at 462 nm.

Subject a blank solution to the same process of analysis as for the water sample, using silver-free redistilled water and all the reagents, and measure photometrically. Subtract the result from the result obtained for the sample.

Calculation

Read off the result from the calibration curve on the basis of the measured extinction value (after deduction of the extinction of the blank solution), take into account the initial volume of the sample solution and recalculate for 1 litre of water.

### 3.4.16.4 ICP-AES (see Section 3.3.12)

Atomic weight of silver: 107.868

### 3.4.17 Thallium

General remarks

Thallium can exist in the trace range (µg/l) in natural waters. It can be detected in the dry matter of water samples by emission spectro-analysis on the basis of an intensely green line at 535 nm.

On account of thallium's toxicity, the element was for example included in the list of 129 so-called "Priority Pollutants" by the US Environmental Protection Agency (EPA).

Previous maceration and enrichment or matrix separation is necessary in a number of analytical methods for determining thallium. The determination limits can be noticeably increased by higher concentrations of matrix elements, particularly in water samples of complex composition. Relatively little is so far known about the causes of systematic errors.

For this reason various methods are described below. Direct flame AAS (3.4.17.1) is only suitable for industrial wastewater, while flameless - graphite-tube - AAS (3.4.17.2) is capable of general application where appropriate equipment is available. Various suggestions are made regarding the preparation and decomposition of water samples and their analysis when working with inversion polarography (3.4.17.3) or X-ray fluorescence analysis (3.4.17.4), assuming such equipment is available.

**A) Chemicals for the thallium determining process**

Saturated $KMnO_4$ solution
50 % $(NH_4)_2H$-citrate solution
0.1 m Disodium salt of ethylenediamine tetraacetic acid dihydrate (EDTA)
0.05 m EDTA solution
Conc. ammonium hydroxide solution
0.1 % sodium diethyldithiocarbamate (DDTC) solution
Tetrachloromethane ($CCl_4$)
Conc. $H_2SO_4$
Conc. $HNO_3$
$HNO_3$ fumans
Nitrogen

Thallium stock solution:

Weigh 1 g of thallium immediately after granulation in order to prevent oxidation in the air, dissolve in 50 ml nitric acid (1 + 3) and then dilute to 1 litre with dist. water in the measuring flask (1 mg/ml).

Thallium standard solution:

By appropriate dilution prepare the thallium stock solution standard dilutions to suit working conditions.

**B) Maceration and extraction for determination by atomic absorption spectrometry.**

Add $H_2SO_4/HNO_3$ to the sample and evaporate to approximately 10 ml. Then neutralize the acid decomposition with ammonium hydroxide solution, taking care that no precipitation occurs. Add 5 ml saturated $KMnO_4$ solution. If manganese dioxide is precipitated, more $KMnO_4$ solution must be added. Then add to the sample 10 ml diammonium hydrogen citrate (50 %) and 10 ml of 0.1 m EDTA solution. Set pH 9 with ammonium hydroxide. Add 10 ml of 0.1 % sodium diethyldithiocarbamate (DDTC) and extract 3 times with tetrachloromethane. Then add a further 50 ml DDTC and extract three more times with $CCl_4$. Evaporate the extracts, adding a number of grains of sodium nitrate. Wet-ash any organic residues in a beaker with conc. $HNO_3$ and subsequently with $HNO_3$ fumans. Pick up the residue with $HNO_3$ (1 m) and make up to 5 or 10 ml.

(This is followed by, for example, flame AAS as described in Section 3.4.17.1)

### 3.4.17.1 Direct determination of thallium by means of atomic-absorption analysis

Equipment parameters

| | |
|---|---|
| 1. Hollow-cathode thallium lamp | |
| 2. Wavelength | 276.8 nm |
| 3. Lamp current | 10 mA |
| 4. Aperture width | 0.5 nm |
| 5. Fuel gas | Acetylene |
| 6. Oxidant | Air |
| 7. Flame type | Oxidizing |
| 8. Background compensation | Yes |

| | |
|---|---|
| Optimum measuring range | 1 - 10 mg/l |
| Determination limit | 0.2 mg/l |
| Calibration | 0.5 - 10 mg/l |
| Acid concentration | 10 ml hydrochloric acid (1.40 g/ml) per 1000 ml solution for measurement |

Remarks

The method of extractive concentration described for the element silver using hexamethylene ammonium hexamethylene dithiocarbamate (HMDC) may also be applied to thallium.

Decomposition with nitric acid - hydrogen peroxide is required for organically loaded samples.

It is also possible to use the method of decomposition and concentration described under B) above.

### 3.4.17.2 Thallium determination with the graphite tube technique (Furnace method)

Equipment parameters

| | |
|---|---|
| 1. Hollow-cathode thallium lamp | |
| 2. Wavelength | 276.8 nm |
| 3. Lamp current | 10 mA |
| 4. Aperture width | 1 nm |
| 5. Background compensation | Yes |
| 6. Carrier gas | Argon |
| 7. Drying | 35 s at 180 °C |
| 8. Incineration | 30 s at 220 °C |
| 9. Atomization | 3 s at 2200 °C |
| 10. Gas stop | Yes |

| | |
|---|---|
| Optimum measuring range | 5 - 50 µg/l with 10 µl sample |
| Determination limit | 2 µg/l |
| Calibration | Addition method |
| Acid concentration | 10 ml nitric acid (1.40 g/ml) per 1000 ml sample solution |
| Matrix modification | 10 µl 1 % sulphuric acid<br>Dilute 1 g sulphuric acid (1.84 g/ml) to 100 ml with water |

**Remarks**

Immediately after sampling, stabilize the water sample with 10 ml nitric acid (1.40 g/ml) per 1000 ml sample solution.

### 3.4.17.3 Determination of thallium by inversion voltammetry

**Maceration and extraction**

To a beaker containing 10 to 250 ml of the water sample add 2 ml conc. $H_2SO_4$ and 2 ml conc. $HNO_3$ to oxidize organic substances, and subsequently evaporate the solution. If higher levels of organic substances are contained (particularly in waste water), adding of $HNO_3$ must be repeated. Once the organic substances are destroyed, evaporate the solution to dry matter. Pick up the dry matter with 50 ml of 1 m HBr solution. Add bromine water to this solution until it remains brown coloured, and subsequently extract 3 times with diethyl ether (15 ml, 10 ml, 10 ml). If the extract displays no brown coloration in the first extraction process, bromine water should again be added.

Evaporate the combined extracts to dry matter in a 50 ml beaker over a waterbath. Oxidize this evaporation residue 1 to 2 times with a few drops of $HNO_3$ fumans and again evaporate to dry matter.

**Measurement**

To determine thallium by inversion voltammetry, pick up the residue from the maceration process in the beaker with 20 ml of 0.05 m EDTA, expose to nitrogen and reduce cathodically on a dropping mercury electrode. The amalgam thus produced is subsequently dissolved anodically. The current-voltage curve is registered with a recording instrument.

In many matrices, surface-active substances can hinder analysis by mechanically inhibiting access of the depolarizer to the electrode or by influencing electrode reduction during measurement. In addition to a fall in peak currents, a shift in peak potentials can frequently also be observed. A breakdown into several peaks is possible here with considerable distortion of the current-voltage curves. These disturbances are however eliminated by maceration.

**Calibration**

Produce a calibration curve by preparing solutions of between 1 µg Tl/l and 100 µg Tl/l. Use the standard-addition process for the analysis topping up the water samples with the known quantities of thallium added for use as an internal standard.

**Calculation**

Evaluate the peaks occurring in inversion voltammetry by comparing the peak heights of the measurement signals received.

### 3.4.17.4 Determination of thallium with X-ray-fluorescence analysis

Concentration

Evaporate 1 to 20 litres of the water sample to dry matter, pick up with hydrochloric acid and precipitate the thallium as sulphide together with the iron sulphide which serves as carrier. Collect the preparation on a membrane filter and analyze spectrometrically with X-ray fluorescence.

Calibration

Draw up a calibration curve by preparing solutions containing between 10 µg Tl/l and 1000 µg Tl/l. Use the standard addition method for the analysis, topping up the water samples with the known quantities of thallium added for use as an internal standard and subjecting them to the same analytical procedure (see also Chapter 2).

Atomic weight of thallium: 204.383

## 3.4.18 Titanium

General remarks

In water, titanium is found only in traces. The determination of titanium in water is carried out by spectrophotometry with chromotropic acid, following concentration (3.4.18.1). ICP-AES Section 3.3.12

### 3.4.18.1 Photometric determination of titanium

The small quantities of titanium in water samples are concentrated by precipitation on iron hydroxide as carrier. Photometric determination is carried out with the aid of chromotropic acid, which forms a red chelate with titanium (IV) at pH 2 ($\varepsilon$ approximately 17,000 at $\lambda_{max.}$ = 470 nm).

Two molecules of chromotropic acid

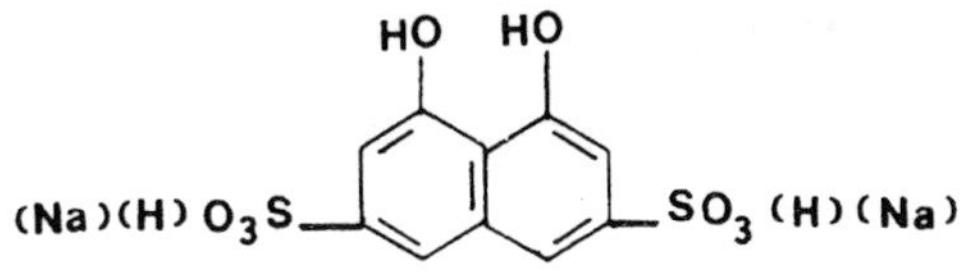


combine with titanium IV to form the complex described by Wuensch:

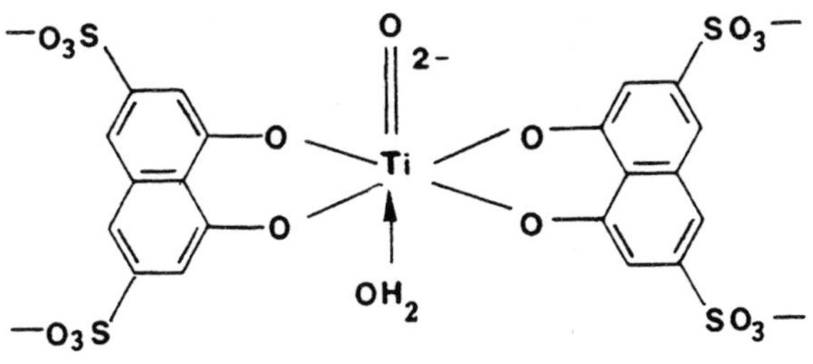


Titanium chelate of chromotropic acid (1,8-dihydroxy - 3,6-disulphonic acid).

The method is suitable for the determination of titanium (IV) concentration down to about 0.02 mg/l in water.

Precipitation with iron hydroxide as a non-isotopic carrier has the effect that virtually no interfering matter is brought into the solution for measurement, as long as the specified analysis procedure is followed. Iron (III), which forms a dark green complex with chromotropic acid, is reduced to iron (II) with ascorbic acid. Under the prevailing conditions the iron (II) remains colourless. Ascorbic acid added in excess guards against reoxidation by atmospheric oxygen, but does not inhibit the formation of the red titanium-chromotropic acid colour complex (W. Koch and H. Ploum).

Titanium (IV) readily undergoes hydrolysis and subsequently no longer reacts with chromotropic acid (G. Wünsch). If this occurs there is not necessarily any visible indication in the form of turbidity or precipitation. The danger of such an occurrence is prevented by decomposition of the potassium hydrogen sulphate melt in a sulphuric acid solution. For the same reason the titanium stock and reference solutions must also be sufficiently acidic (pH less than 1) at all times.

Equipment

Photometer

Cuvettes, path length 1 and 5 cm

Polyethylene flasks, approx. 12 litres (for sample taking)

Platinum crucible

pH meter with single-probe measuring cell

Measuring flask, volume 50 ml

Titanium (IV) oxide

Iron (III) oxide

Ammonium chloride, reagent purity

Potassium hydrogen sulphate, reagent purity

Hydrochloric acid, conc., reagent purity, (1.19 g/ml)

Sulphuric acid, approx. 10 %

Sulphuric acid, approx. 48 %

Ammonium hydroxide solution, approx. 25 %

Washing water contai

Acetate buffer solution (pH 2.6):
Dissolve 100 g of reagent-purity sodium acetate in 120 ml of dist. water. Add 800 ml of reagent-purity acetic acid (1.06 g/ml) and make up to 1 litre with dist. water.

Chromotropic acid solution:
Dissolve 6 g of 1,8-dihydroxy-3,6-disulphonic acid-$Na_2$ (disodium salt of chromotropic acid) in dist. water and make up to 100 litres. This solution is stable for about 1 week.

Titanium stock solution:
Ignite $TiO_2$ in the platinum crucible for 30 min. at 1000 °C. Weigh 166.8 mg of $TiO_2$ and decompose by melting with about 2.5 g of $K_2S_2O_7$ in the platinum crucible. When cool, dissolve the melt in 150 ml of hot 10 %-strength $H_2SO_4$, allow to cool and make up to 1 litre with 10 %-strength $H_2SO_4$ at 20 °C. 1 ml = 0.1 mg of Ti.

Titanium reference solution:
Dilute titanium stock solution 1 to 10 with 10 %-strength $H_2SO_4$ before use. 1 ml = 0.01 mg of Ti.

## Method

### Concentration process

Measure a volume of approximately 10 litres of sample water into the polyethylene sampling bottle at the sampling point, acidify immediately with hydrochloric acid and leave to stand for several hours, shaking occasionally to remove $CO_2$. After this period add 10 - 20 mg of iron (III) oxide dissolved in hydrochloric acid to act as a carrier.

Add about 50 g of solid ammonium chloride to the solution, shaking constantly until dissolved. Now add 25 % ammonium hydroxide solution in order to precipitate the iron as iron (III) hydroxide at pH 9. The titanium contained in the sample (as well as aluminium and beryllium) is coprecipitated with the iron (III) hydroxide.

Leave the sample to stand overnight at about 20 °C to achieve complete sedimentation of the precipitate, filter the supernatant solution through a 12.5-cm paper filter, if necessary with the aid of a siphon, and also transfer the precipitate to the filter as completely as possible. Rinse the flask thoroughly, wash 3 to 5 times with water containing ammonium chloride, the water having been set to pH 8 - 9 with $NH_4OH$, and pour through the filter.

Next rinse out the plastic flask with about 20 ml of concentrated reagent-purity hydrochloric acid, dilute with approximately the same quantity of distilled water and use this solution to dissolve the precipitate on the filter. Rinse the flask and filter several more times with water containing HCl (5 % HCl).

Add about 5 g of solid ammonium chloride to the hydrochloric filtrate and dissolve by swirling. Now repeat the precipitation with 25 % ammonium hydroxide solution and filter the precipitate through the same filter as before. Following reprecipitation incinerate the filter in a platinum crucible and decompose the residue with a little potassium hydrogen sulphate. Dissolve the melt in water, adding five drops of 48 % sulphuric acid, and transfer the solution, whose volume should not exceed 20 ml, to a 100-ml beaker. Rinse the crucible three to five times with 1-ml portions of distilled water.

### Determination

Treat the acid solution in the beaker with 5 ml of ascorbic acid solution in order to reduce the iron (III). Add 2 ml of chromotropic acid solution, mix, and add a further 10 ml of acetate buffer solution.

Adjust the pH to 2 $\pm$ 0.1 by adding 25 % ammonium hydroxide solution, drop by drop. Now transfer the contents of the beaker to a 50-ml measuring flask, rinse the beaker with a few ml of distilled water and make up to the mark with distilled water at 20 °C. The solution is coloured red if titanium is present. Measure the solution in a 5-cm or 1-cm cuvette at 470 nm against distilled water.

Conduct a blank test, repeating the entire procedure with distilled water instead of the sample solution. It is advisable to treat a second water sample with a known quantity of titanium (admixture method) and to analyze this solution in parallel with the solution under analysis and the blank test.

### Calibration

In order to prepare the calibration curve take increasing quantities of titanium reference solution containing between 0.02 and 0.25 mg of titanium and treat as described under "Determination", beginning with the addition of the ascorbic acid solution.

### Calculation

Read off the appropriate titanium content from the calibration curve on the basis of the measured extinction. Take into account the volume of water sample used and the result of the blank test and recalculate for 1 litre of water.

## 3.4.18.2 ICP-AES (see Section 3.3.12)

Atomic weight of titanium: 47.90

## 3.4.19 Uranium (see also Section 3.7)

In nature, uranium occurs predominantly (99.27 %) in the form of the isotope $^{238}_{92}U$ (half-life $4.5 \cdot 10^9$ years; parent element of the radioactive (4n + 2) decay series) it additionally occurs in the form of two other isotopes (0.72 and 0.0056 % respectively), $^{235}_{92}U$ (half-life $7 \cdot 10^8$ years; intermediate in the radioactive (4n + 3) decay series) and $^{234}_{92}U$ (half-life $2.5 \cdot 10^5$ years; intermediate in the (4n + 2) decay series).

The uranium content of natural waters is generally determined by the geological conditions. The most important uranium mineral is uraninite (pitchblende, $xUO_3 \cdot yUO_2$), of which there are sizeable deposits in Zaire, Canada and in Czechoslovakia (Joachimsthal), but also in Colorado (USA), North Africa (phosphate companion), South Africa (gold companion), Australia and the USSR. Natural waters contain approximately 0.0001 - 0.05 mg of uranium/l. In more recent years, it has become necessary to take account of uranium contamination through wastewater from atomic power station reprocessing plants and storage sites for radioactive wastes.

Quantitative determination

Because of their long half-lives, the uranium isotopes which occur naturally have relatively low specific activities (uranium 238 = $3.4 \cdot 10^{-7}$ Ci/g; uranium 235 = $2.1 \cdot 10^{-6}$ Ci/g). It is therefore preferable to determine concentrations by means of physicochemical or chemical methods of analysis.

The fluorimetric method described here is fast, highly sensitive and not particularly expensive.

### 3.4.19.1 Fluorimetric determination of uranium in a sodium fluoride/alkali carbonate melt

General remarks

The weakly nitric-acid solution obtained as a result of concentration of the water sample is extracted with ethyl acetate following the addition of aluminium nitrate. The organic phase containing the uranium is separated off, evaporated in a platinum crucible, and the uranium is melted at 650 °C after adding sodium fluoride/potassium carbonate/sodium carbonate as a fluxing agent. The yellow-green fluorescence of the melt which appears when irradiated with an ultraviolet lamp is compared with that of standard samples prepared under equivalent conditions.

The optimum measuring range lies between 0.05 and 0.7 µg of U per melt sample. This is equivalent to a uranium concentration of 0.05 - 0.7 µg/l if a 1-litre water sample is used as a basis for each melt preparation. In the case of higher uranium contents a smaller quantity of water must be used, if necessary, following previous dilution.

As a result of the extraction with ethyl acetate, the uranium is separated from virtually all interfering elements which occur in natural waters.

Equipment

Platinum dish, approx. 250 ml

Platinum crucible, 25 ml

Measuring flask, 50 ml

Nitric acid (1.40 g/ml), reagent purity

Aluminium nitrate, $Al(NO_3)_3 \cdot 9\ H_2O$, reagent purity

Ethyl acetate, reagent purity

Fluxing agent:
Mix 9 g of sodium fluoride, 45.5 g of sodium carbonate and 45.5 g of potassium carbonate, all of reagent purity, and melt at 650 °C in a platinum dish. Care should be taken that the melt, which should be homo-

genized by thorough swirling, is not heated to a temperature higher than that specified, since otherwise there is a risk of platinum beginning to dissolve out from the dish walls. Reduce the melt obtained to a very fine powder and pass through a sieve. Mix the powder thoroughly once more and store in a tightly closing bottle (or a desiccator) until used.

Uranium stock solution:
Dissolve 0.2109 g of reagent-purity uranyl nitrate, $UO_2(NO_3)_2 \cdot 6\ H_2O$ in dist. water and dilute to 1000 ml. 1 ml of this solution contains 0.1 mg of U.

Uranium reference solution:
Dilute 10 ml or 1 ml of uranium stock solution with dist. water to 1000 ml. 1 ml of this solution contains 1 μg or 0.1 μg respectively of U.

## Method

### Calibration

Prepare standard samples with between 0.05 and 0.7 μg of U by transferring appropriate volumes of uranium reference solution to 50-ml measuring flasks by pipette, diluting with distilled water to about 40 ml, treating with exactly 7.5 ml of nitric acid (1.40 g/ml) and then making up to the mark with dist. water at 20 °C. Continue with further treatment as described under "Determination".

### Determination

Very lightly acidify 1 litre of the water sample with nitric acid and boil down to about 40 ml. A different volume of water may be used, if more appropriate, with a uranium content between 0.1 and 0.7 μg, which should be established in a preliminary test.

If the salt content of the sample is so high that interfering quantities of the salts separate out beforehand, use a correspondingly smaller quantity of water; thanks to the high sensitivity of the method, this generally presents no difficulties.

For every 10 ml of the concentrated sample, add precisely 1.5 ml of nitric acid (1.40 g/ml), or a quantity reduced by the amount used previously to acidify the water sample, if this involved more than 1 - 2 drops, and heat to boiling (do not filter off any precipitate formed).

When cooled to room temperature, add 19 g of aluminium nitrate as a salting-out agent for every 10 ml of solution, and dissolve by heating. When cool, transfer to a separating funnel, add 10 ml of ethyl acetate by pipette, shake for 60 seconds, and leave to stand for about 5 minutes in order to separate the phases.

When the layers are completely separate, withdraw about 8 ml of the organic phase with a measuring pipette and filter through a dry 7-cm filter into a dry test tube. Pipette exactly 5 ml of the filtrate into a 25-ml platinum crucible and evaporate to dryness on a water bath. Add 1.5 g of fluxing agent and melt at 650 °C in an electric furnace for 25 minutes. Mix the melt by swirling several times during this period. It is particularly important to keep strictly to the specified melting temperature and time. It is expedient to prepare the samples under investigation and the reference standards simultaneously in an electrically heated furnace, the temperature of which can be measured by a thermoelement. Conduct a blank test in the same way.

When cool, carefully tap the crucibles to detach the melt pellets, and, without touching the upper surface with the fingers, irradiate with a UV lamp (366 nm) on a matt-black base. Compare the sample and calibration standards visually, or instrumentally, if required, in a reflection fluorimeter (with a photoelectric cell or a secondary electron multiplier; fluorescence line at 555.0 nm). The melt pellets may be kept for about 12 hours.

It should be noted that the intensity of fluorescence is dependent on the melting conditions to a considerable extent. Care should be taken that these are closely reproducible. The specified melting temperature and the melting period must not be exceeded since otherwise platinum dissolves out from the crucibles and may interfere with the uranium fluorescence.

### Calculation

When performing the calculation, take into account the volume of the water sample used and the calibration standards.

Atomic weight of uranium: 238.029

## 3.4.20 Vanadium

### General remarks

In natural waters vanadium is to be found, if at all, in trace amounts only (less than 0.01 mg/l).

In waste waters (for example from ore processing or metallurgical industries) higher concentrations may also be expected.

The recommended method of determination is direct analysis by means of flame absorption spectroscopy (3.4.20.1). Alternatively initial concentration is required (3.4.20.2) or the flameless AAS technique should be used (3.4.20.3).

3.4.20.4 For ICP-AES (see Section 3.3.12)

Qualitative general analysis by emission spectrography (Section 3.4.1) is an aid to deciding which method to use. If the vanadium concentration is expected to be greater than 0.1 mg/l it is also possible to conduct photometric analysis in water using N-benzoyl-N-phenylhydroxylamine (Deutsche Einheitsverfahren zur Wasseruntersuchung etc. (German Standard Methods of Water Analysis), Verlag Chemie (1985), 6940 Weinheim, FRG).

### 3.4.20.1 Direct determination of vanadium by means of atomic absorption analysis

### Equipment parameters

1. Hollow-cathode vanadium lamp
2. Wavelength 318.5 nm

| | |
|---|---|
| 3. Lamp current | 10 mA |
| 4. Aperture width | 0.2 nm |
| 5. Fuel gas | Acetylene |
| 6. Oxidant | Laughing gas |
| 7. Flame type | Reducing |
| 8. Background compensation | Yes |

| | |
|---|---|
| Optimum measuring range | 1 - 10 mg/l |
| Determination limit | 0.1 mg/l |
| Calibration | 0.2 - 10 mg/l |
| Acid concentration | 10 ml nitric acid (1.40 g/ml) per 1000 ml solution for measurement |

**Remarks**

It is advisable to add 2 ml of aluminium nitrate solution per 50 ml of solution for measurement.

Aluminium nitrate solution:
Dissolve 100 g aluminium nitrate ($Al(NO_3)_3 \cdot 9H_2O$) in water and make up to 1000 ml.

In the case of low concentrations, concentration by evaporation is only possible for low-mineral-matter waters (electrical conductivity at 25 °C below 100 µS/cm). The method of extractive concentration described below can be recommended.

### 3.4.20.2 Concentration: Principle of the method of extractive concentration

**General remarks**

Vanadium ions are extracted with chloroform at pH 3.4 as a vanadium oxinate complex, the extracts evaporated, and after oxidation with nitric acid determined by atomic-absorption analysis in the presence of aluminium ions.

**Reagents**

Sulphuric acid ($H_2SO_4$) 5 mol/l

Nitric acid (1.40 g/ml)

Acetic acid (glacial acetic acid) (1.05 g/ml)

Oxine solution:
Dissolve 50 mg oxine in 120 ml glacial acetic acid, dilute to 1000 ml with dist. water.

Sodium carbonate ($Na_2CO_3$), solid

Aluminium nitrate solution:
Dissolve 100 g $Al(NO_3)_3 \cdot 9\ H_2O$ in water and make up to 1000 ml.

Chloroform ($CHCl_3$)

Method

Evaporate the sample amount to approx. 200 ml and transfer to a separating funnel.

Set to pH 3.4 with sulphuric acid or sodium carbonate.

Add 5 ml oxine solution and swirl.

Add 5 ml chloroform and shake for 1 minute, repeating until the last extract is colourless.

Evaporate the collected chloroform extracts in a 50-ml beaker, adding 100 mg $Na_2CO_3$.

Oxidize the organic matter with nitric acid.

Pick up the residue with 1 m $HNO_3$, treat with 1 ml aluminium nitrate solution, transfer to 25-ml measuring flask and make up to the mark with dist. water.

Prepare a blank test in the same way.

Final determination by atomic-absorption analysis (3.4.20.1). Equipment parameters are as described for direct determination.

Decomposition with nitric acid-hydrogen peroxide is required for organically loaded samples.

### 3.4.20.3 Vanadium determination with the graphite tube technique (Furnace method)

Equipment parameters

| | |
|---|---|
| 1. Hollow-cathode vanadium lamp | |
| 2. Wavelength | 318.5 nm |
| 3. Lamp current | 10 mA |
| 4. Aperture width | 0.5 nm |
| 5. Background compensation | Yes |
| 6. Carrier gas | Nitrogen |
| 7. Drying | 25 s at 120 °C |
| 8. Incineration | 20 s at 800 °C |
| 9. Atomisation | 3 s at 2900 °C |
| 10. Gas stop | Yes |

| | |
|---|---|
| Optimum measuring range | 10 - 100 µg/l with 10 µl sample |
| Determination limit | 2 µg/l |
| Calibration | Addition method |
| Acid concentration | 10 ml nitric acid (1.40 g/ml) per 1000 ml sample solution |

Remarks

Immediately after sampling, stabilize the water sample with 10 ml nitric acid (1.40 g/ml) per 1000 ml sample solution.

### 3.4.20.4 ICP-AES (see Section 3.3.12)

## 3.4.21 Zinc

### General remarks

The occurrence of zinc in drinking water is frequently a result of corrosion in galvanized water pipes. Soft water containing carbon dioxide, and also waters rich in chlorides and sulphates, pick up zinc from galvanized pipes particularly easily, gradually destroying the pipes in the process.

### Quantitative determination

3.4.21.1 Photometric/spectrophotometric determination as red zinc dithionate in organic solvent

3.4.21.2 Determination by atomic-absorption spectrometry, with direct atomization of the aqueous solution in an air/acetylene flame

3.4.21.3 Determination of zinc by flameless AAS (graphite-tube technique)

3.4.21.4 For ICP-AES see Section 3.3.12

### 3.4.21.1 Spectrophotometric determination with dithizone

### General remarks

In more or less neutral solution, zinc reacts with dithizone ($C_{13}H_{12}N_4S$) to form zinc dithionate as follows:

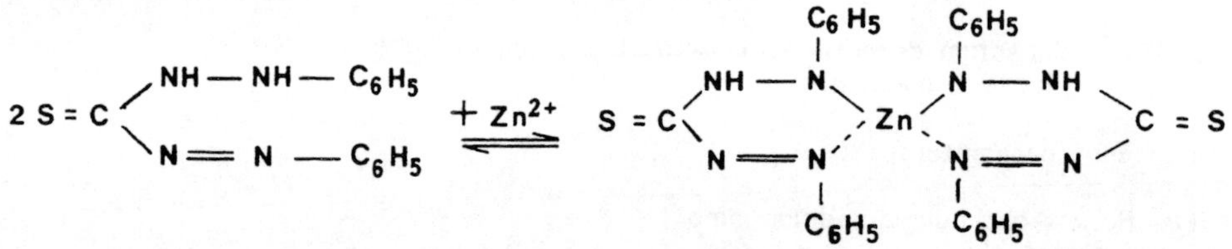

Dithizone + zinc ions = zinc dithionate

Zinc dithionate dissolves in organic solvents, producing a red colour and can be extracted. Although dithizone is not a specific reagent for zinc ions, it is possible, by choosing appropriate conditions for the reaction (pH approximately neutral, addition of masking solution), to apply the method in a largely selective way.

Between 0.05 and 0.50 mg of zinc ions per litre can be determined without dilution or concentration, but the method can also be adapted to varying concentrations. In view of the sensitivity of the reaction (0.00088 - 0.0010 µg Zn · $ml^{-1}$ · $cm^{-1}$ at 538 nm, according to Koch and Koch-Dedic) it is essential to ensure that all reagents and glass apparatus are zinc-free. It is also important to perform the determination in exactly the same way every time and to run a completely identical blank test.

### Equipment

Spectrophotometer (536 nm) or filter photometer

Cuvettes, path length 1 - 5 cm

Separating funnel, approx. 100 ml (grease with silicone grease)

Deionized, zinc-free water

Sulphuric acid, conc., reagent purity (1.84 g/ml)

Nitric acid, conc., zinc-free, reagent purity (1.40 g/ml)

Hydrochloric acid, approx. 1 m:
To avoid high blank values, reagent-purity conc. HCl may be diluted 1 : 1 with dist. water and distilled. Mix 90 ml of the distillate with 400 ml of deionized water.

Masking solution:
10 ml of 2 m $NH_4OH$, reagent purity, + 3 ml of 5% ammonium oxalate solution, reagent purity, + 30 ml of 5% potassium cyanide solution, reagent purity, + 70 ml of 1 m HCl, reagent purity, + 90 ml of 10 % sodium acetate solution, reagent purity, + 240 ml of 50 % sodium thiosulphate solution ($Na_2S_2O_3 \cdot 5\ H_2O$), reagent purity, + 150 ml of 10 % sodium acetate solution, reagent purity, + 400 ml of water. The reagent solutions are to be added in the order listed above. Slight clouding of the solution due to precipitated sulphur is unimportant.

Carbon tetrachloride, reagent purity

Dithizone solution:
Dissolve 25 mg of dithizone in 100 ml of reagent-purity carbon tetrachloride. If reagent-purity dithizone is used there is normally no need for any further purification of the dithizone. However, if a yellowish-brown discoloration of the organic extract occurs during the blank test, the dithizone solution should be purified as follows:

Take the solution of 25 mg dithizone in 100 ml carbon tetrachloride and add 1 ml of ammonium hydroxide solution (0.91 g/ml) and 200 ml of water. Shake in the separating funnel to transfer the dithizone to the aqueous phase. Separate off the carbon tetrachloride layer, and introduce 100 ml of reagent-purity carbon tetrachloride as a layer beneath the aqueous phase.

After acidification with hydrochloric acid, shake until the dithizone is once again present as a green solution in the organic phase.

Dilute dithizone solution: 1 : 10 with carbon tetrachloride.

Zinc standard solution:
Dissolve 1.000 g of reagent-purity zinc in 10 ml of reagent-purity nitric acid (1 : 1, approx. 1.22 g/ml). Add 10 ml of reagent-purity sulphuric acid (1 : 1, approx. 1.52 g/ml), concentrate the solution by evaporation and heat on a sand bath until fuming. Dilute with water, rinse the resulting zinc solution into a 1000-ml measuring flask, make up to the mark at 20 °C and mix. 1 ml of this solution contains 1 mg of zinc.

Dilute zinc standard solution:
Solution A:
Dilute 1 : 100 with water (0.01 mg Zn/ml)

Solution B:
Dilute solution A 1 : 10 with water (0.001 mg Zn/ml).

The dilute zinc standard solution will not keep and must be freshly prepared every time.

## Method

### Preparation of water sample

If the zinc content of the sample is not in the range 0.05-0.50 mg/l, it is necessary to dilute the sample or concentrate it in a quartz dish. If sulphuric acid was used at the sampling point to preserve the sample, evaporate carefully to dry matter in a quartz dish. If neutralization with NaOH or ammonia is necessary, the same operation must also be performed in the blank test because of the possibility of zinc being introduced. Pick up the residue with deionized water, heating gently. Set unpreserved samples to pH 2 - 3 with 1 m hydrochloric acid and take the dilution into account.

### Calibration

To establish the calibration curve, use the zinc standard solution to prepare a series of concentrations in the range 0.05 - 0.50 mg $Zn^{2+}$/l. Treat 10.0 ml of each as described under "Determination". The calibration curve must be checked every few weeks.

### Determination

Neutralize the acid solution in the separating funnel with sodium hydroxide solution to approx. pH 5. Add 20 ml of masking solution and check the pH by spotting on indicator paper. The value should now be between pH 6 and 7. If the pH is outside this range, set it with 0.5 m sulphuric acid or 1 m sodium hydroxide solution. Now extract the zinc with dilute dithizone solution a little at a time. Use 5 ml of dilute dithizone solution each time and shake for 1 min. Separate off the red organic phase and repeat the extraction until the last extract retains its green colour. Collect the combined organic extracts, containing all the zinc and the excess dithizone, in a separating funnel and wash twice with 10 ml of 0.01 m ammonium hydroxide each time. This washing operation must be performed quickly so that no decomposition of the zinc dithionate can occur. After adding the ammonium hydroxide solution, shake for 10 sec and separate the phases immediately. If necessary, rewash the aqueous phases with pure carbon tetrachloride. Filter the red solution of zinc dithionate in carbon tetrachloride through a small dry filter into a 50-ml measuring flask and rinse the filter with carbon tetrachloride. After topping up to the mark with carbon tetrachloride, measure against the solvent at 536 nm in a 1-cm or 5-cm cuvette. Perform a parallel blank test and take it into account.

### Calculation

Read off the zinc concentration from the calibration curve on the basis of the extinction values measured, or calculate it with the aid of a calibration factor, and recalculate for 1 litre of water.

Atomic weight of zinc: 65.37

### 3.4.21.2 Direct determination by means of atomic-absorption analysis

**Equipment parameters**

| | |
|---|---|
| 1. Hollow-cathode zinc lamp | |
| 2. Wavelength | 213.9 nm |
| 3. Lamp current | 5 mA |
| 4. Aperture width | 1 nm |
| 5. Fuel gas | Acetylene |
| 6. Oxidant | Air |
| 7. Flame type | Oxidizing |
| 8. Background compensation | Yes |

| | |
|---|---|
| Optimum measuring range | 0.2 - 2 mg/l |
| Determination limit | 0.01 mg/l |
| Calibration | 0.1 - 2 mg/l |
| Acid concentration | 10 ml hydrochloric acid (1.17 g/ml) per 1000 ml solution for measurement |

**Remarks**

Concentration by evaporation is possible if concentrations are low.

One method of extraction has been described for the element silver. The blank reading presents a major problem, primarily caused by the chemicals used. For this reason, the furnace method should be used if at all possib-le.

Decomposition with nitric acid/hydrogen peroxide is possible with organical-ly loaded samples (take a blank reading).

### 3.4.21.3 Zinc determination with the graphite-tube technique (Furnace method)

**Equipment parameters**

| | |
|---|---|
| 1. Hollow-cathode zinc lamp | |
| 2. Wavelength | 213.9 nm |
| 3. Lamp current | 5 mA |
| 4. Aperture width | 1 nm |
| 5. Background compensation | Yes |
| 6. Carrier gas | Argon |
| 7. Drying | 25 s at 140 °C |
| 8. Incineration | 30 s at 300 °C |
| 9. Atomization | 3 s at 1900 °C |
| 10. Gas stop | Yes |

| | |
|---|---|
| Optimum measuring range | 0.2 - 4 µg/l with 10 µl sample |
| Determination limit | 0.1 µg/l |
| Calibration | Addition method |
| Acid concentration | 10 ml nitric acid (1.40 g/ml) per 1000 ml sample solution |

**Remarks**

Immediately after sampling, stabilize the water sample with 10 ml nitric acid (1.40 g/ml) per 1000 ml sample solution.

### 3.4.21.4 ICP-AES (see Section 3.3.12)

## 3.4.22 Tin (Sn)

The concentrations of tin due to natural resources in ground and surface water are as a rule extremely low. Higher concentrations are only found in waste waters from tin-processing factories.

By contrast, particularly the lower-molecular alkyl derivates of tin are extremely toxic. These latter compounds were therefore included in the "Priority List" of the Environmental Protection Agency (EPA) in the USA and in the "Black List" of the EC Directive for the protection of the Community's waters from contamination. Organic compounds of tin are used to some extent in the production of organic chemicals.

On account of the very low levels of tin in water it is first necessary to concentrate the tin by precipitation. This is done using hydrated manganese (IV) oxide as a trace trap.

### 3.4.22.1 Turbidimetric determination with nitrophenol arsonic acid (if no AAS facilities are available)

**General remarks**

Tin (IV) reacts with nitrophenol arsonic acid (3-nitro-4-hydroxyphenyl arsonic acid-2-nitrophenol arsonic acid-(4)) in sulphuric acid solution to produce a constant turbidity which is proportional to the tin content and can be measured photometrically at 436 nm. For determination purposes the tin must be present in the quadrivalent form.

With the method described below, 5-500 µg Sn/l can be recorded. Concentration can be dispensed with at higher levels of tin.

Precipitation with nitrophenol arsonic acid leads to the separation of tin from many elements, e.g. Cu, Zn, Fe and Pb. According to POHL, up to 5 mg of ammonium ions cause no disturbances, whereas Karsten, Kies and Walraven consider double the quantity of reagent to be necessary when Fe (III) is present. If it is suspected that other elements have also produced turbidity with nitrophenol-arsonic acid, the turbidities in samples and calibrated samples are isolated by means of membrane filters and determined by X-ray fluorescence-spectrometry (Sn-K$\alpha$ line), if such equipment is available.

**Equipment**

Photometer

Cuvettes, path length 4 cm and 5 cm

Porcelain dish, 1000 ml

Measuring flask, 50 ml

Glas-filtering crucible

Nitric acid, 20 %

Conc. sulphuric acid (1.84 g/ml)

Hydrochloric acid 22.5 % (1.12 g/ml)

Conc. hydrochloric acid (1.19 g/ml)

Hydrogen peroxide, 30 %

Sodium peroxide reagent purity

Iron crucible, 10-20 ml

Manganese (II)-sulphate solution:
Dissolve 5 g crystallized manganese sulphate, $MnSO_4 \cdot 4\ H_2O$, to 100 ml in distilled water

Potassium permanganate solution (0.2 m)

Nitrophenol arsonic acid solution:
Dissolve 1 g nitrophenol-arsonic acid in 15 ml methanol and make up with distilled water to 50 ml.

Tin stock solution:
Dissolve 0.500 g tin (99.99 %) in 10 ml hydrochloric acid (1.12 g/ml), add 50 ml conc. hydrochloric acid (1.19 g/ml) and make up to the mark in a 1000 ml measuring flask with distilled water. 1 ml of this solution contains 0.5 mg Sn.

Tin reference solution:
Add 30 ml conc. hydrochloric acid (1.19 g/ml) to 5.00 ml of the tin stock solution and fill to 500 ml with distilled water. 1 ml of this solution contains 0.005 mg Sn.

## Method

### Calibration

To establish the calibration curve, take increasing quantities of the tin reference solution from 5 to 50 µg and make up with distilled water to the volume of the water sample used. Treat as described below for the sample.

### Concentration

Depending on the tin content to be expected, add nitric acid in excess to between 250 and 2000 ml of the water sample and concentrate by evaporation to a volume of approx. 250 ml. For the Blumenthal enrichment process, add 5 ml manganese (II) sulphate solution and 3 ml of 0.2 m potassium permanganate solution and heat to boiling point while stirring. The permanganate coloration disappears and hydrated manganese (IV) oxide is precipitated; the supernatant solution becomes clear.

After adding a further 3 ml 0.2 m potassium permanganate solution and re-heating to boiling point while stirring continuously, the precipitate, which agglutinates well, can be filtered off and washed out with hot water. Although the filtrate is clear, it may still contain traces of tin. For this reason, repeat the hydrated manganese (IV) oxide precipitation with 2 ml manganese (II) sulphate solution and 3 ml 0.2 m potassium permanganate solution. The precipitates are combined by spraying them off the filter into the same vessel and releasing the residue of hydrated manganese (IV) oxide adhering to the precipitates and the glass walls with diluted hydrochloric acid to which a little $H_2O_2$ has been added.

Dilute the combined precipitate with 10 ml hydrochloric acid (1.12 g/ml) and 3 ml conc. sulphuric acid (1.84 g/ml), and add a few drops of $H_2O_2$ (30 %). Heat until the $H_2SO_4$ fumes. After cooling, any salts precipitated are dissolved by heating with 20 ml $H_2O$.

Filter off any insoluble residue remaining (lead(II) sulphate, silicic acid, etc.), wash out, first with hot water acidified with hydrochloric acid and then with pure distilled water, and ash together with the filter. Subsequently, macerate by melting with a little sodium peroxide in a small iron crucible; pick up the melt with warm distilled water. After acidification with hydrochloric acid, combine the solution with the filtrate from the insoluble residue.

**Determination**

Filter the combined solutions through a filtering crucible into a 50-ml measuring flask, add 10 ml nitrophenol arsonic acid solution and make up to the mark with distilled water. After thorough shaking, allow to stand for 2 h and then measure the extinction at 436 nm against distilled water.

A blank test is taken through the entire analytical process and measured in the same way. The blank reading is subtracted from the measured value for the sample.

**Calculation**

On the basis of the extinction measured, read off the corresponding tin content from the calibration curve and recalculate for 1 litre of water, taking into account the quantity of water used.

Atomic weight of tin: 118.69

### 3.4.22.2 Direct determination of tin by means of atomic-absorption analysis

**Equipment parameters**

| | |
|---|---|
| 1. Hollow-cathode tin lamp | |
| 2. Wavelength | 286.3 nm |
| 3. Lamp current | 10 mA |
| 4. Aperture width | 0.5 nm |
| 5. Fuel gas | Acetylene |
| 6. Oxidant | Laughing gas |
| 7. Flame type | Oxidizing |
| 8. Background compensation | Yes |

| | |
|---|---|
| Optimum measuring range | 5 - 300 mg/l |
| Determination limit | 0.5 mg/l |
| Calibration | 2 - 300 mg/l |
| Acid concentration | 100 ml hydrochloric acid (1.17 g/ml) per 1000 ml solution for measurement |

**Remarks**

Concentration by evaporation is possible, if concentrations are low, but the technique described in Section 3.4.22.1 is to be preferred.

Organically loaded waters should be decomposed with hydrochloric acid - nitric acid.

### 3.4.22.3 Determination of tin using the AAS hydride technique

#### General remarks

Tin ions are reduced to tin hydride from a boric-acid-buffered medium by means of sodium borohydride, transferred to a heated quartz cuvette by a current of inert gas, decomposed thermally, and the absorption of the atoms is measured in the beam of an atomic-absorption spectrometer. In the hydride technique, the element which is to be determined is volatilized as a gaseous hydride and in this way separated off from the matrix. Interference may occur if there is a considerable excess of elements such as antimony, arsenic, bismuth, mercury, selenium or tellurium which can also be volatilized with this technique. Above all, heavy metals such as copper and nickel in the solution have a disturbing effect during hydride formation itself. Interference due to phosphoric acid and hydrochloric acid may also be observed. It is therefore vital to check the method by the addition technique.

#### Equipment

Atomic absorption spectrometer with hydride attachment

Stock tin solution (1000 mg/l)

Boric acid ($H_3BO_3$) crystalline

Hydrochloric acid (1.12 mg/l)

Sodium hydroxide solution (NaOH):
Dissolve 5 g NaOH in water and make up to 100 ml.

Sodium borohydride solution:
Dissolve 3 g sodium borohydride ($NaBH_4$) and 1 g sodium hydroxide (NaOH) in water and make up to 100 ml.

#### Method

Transfer 20 ml of the sample solution to the hydride system and adjust to a pH of 0.6 to 1 by adding sodium hydroxide solution or hydrochloric acid. Then add 600 mg of solid, crystalline boric acid as a buffer. Allow a constant current of inert gas (argon) to flow through the system at a rate of 60 l/h in order to expel the air. Following this, continuously add the 3 % sodium borohydride solution using a peristaltic pump. Tin ions are reduced to tin hydride and directed by the inert gas current into a quartz cuvette heated to 850 °C where they are thermally decomposed. Measure the absorption.

#### Equipment parameters

| | |
|---|---|
| 1. Hollow-cathode tin lamp | |
| 2. Wavelength | 286.3 nm |
| 3. Lamp current | 7 mA |
| 4. Aperture width | 0.5 nm |
| 5. Background compensation | Yes |

| | |
|---|---|
| 6. Inert gas | Argon |
| 7. Gas through-flow | 60 l/h |
| 8. Cuvette temperature | 850 °C |
| 9. Integration | Peak area |
| 10. Integration time | 25 s |
| 11. Measuring range | 1 - 7.5 µg/l |
| 12. Determination limit | 0.5 µg/l |

Remarks

The equipment parameters quoted here for the hydride technique apply to a Varian AAS 775 with a hydride attachment manufactured by Messrs. Berghoff, Tübingen, FRG.

Since the hydride technique only permits quantitative detection of tin present ionogenically in solution, organically bonded or complex-bonded tin must be decomposed. Organically loaded water samples should also be decomposed before measurement.

A method of decomposition is described below

Reagents

| | |
|---|---|
| Hydrochloric acid | (1.17 g/ml) |
| Nitric acid | (1.40 g/ml) |

Method

Introduce 50 ml of the water sample into a 150-ml beaker and add 9 ml hydrochloric acid (1.17 g/ml) and 3 ml nitric acid (1.40 g/ml).

Evaporate to dryness in a water bath.

If the organic load is high repeat the addition of nitric acid.

When cool, pick up the evaporation residue with 5 ml hydrochloric acid, transfer to a 50-ml measuring flask and dilute with water to the mark.

Carry out measurements as above under 3.4.22.2 or 3.4.22.3

### 3.4.23 Zirconium (Zr)

If it is necessary to determine zirconium in water, refer to the ICP-AES method described in Section 3.3.12.

## 3.5 Undissociated substances (for $H_2S$ see also Sections 3.1 and 3.6)

### 3.5.1 Boron compounds

General remarks

Groundwaters, surface waters and drinking water usually contain only low concentrations of compounds which include boron. The range tends to be below 0.01 to 0.1 mg of boron per litre rather than above. Should the latter be the case, it is generally due to the influence of wastewater of industrial or commercial origin, or also domestic wastewaters, since boron compounds are frequently present in detergents and thereby find their way into the environment. Nevertheless, subterranean waters, particularly thermal springs, are known which have boron contents of several tens of milligrams per litre.

The boron tolerance of cultivated agricultural plants lies within very tight limits, so consequently the boron content of irrigation water must be checked. Boron concentrations above 1 mg/l (sensitive plants) or 3 mg/l (tolerant plants) render water unusable for agricultural irrigation projects.

The following method may be used for boron determination:

3.5.1.1 Spectrophotometry with 1,1'-dianthrimide (0.01 to 0.2 mg of B/l or following dilution to the mg/l range)

3.5.1.2 Spectrophotometry with azomethine-H

3.5.1.3 Volumetric analysis following distillation (above 0.2 mg of B/l)

3.5.1.4 Following extraction of boron with a solution of 2-ethyl-1,3-hexanediol in methyl isobutyl ketone, boron may also be determined in the organic solution by atomic-absorption spectrophotometry

3.5.1.5 A further proven method of determining boron is that using ICP-AES (see summarized procedure for 24 elements in 3.3.12)

### 3.5.1.1 Spectrophotometric determination with 1,1'-dianthrimide

General remarks

1,1'-dianthrimide =
1,1'-dianthraquinonylamine

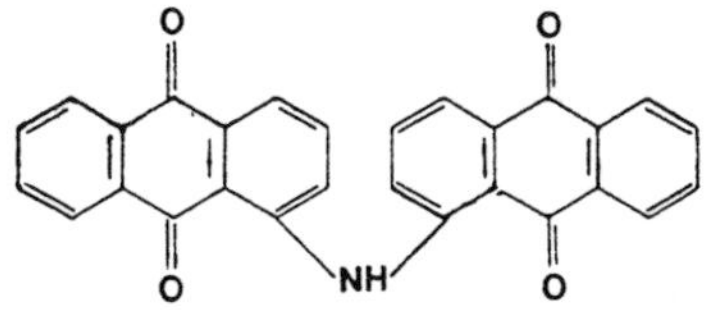


The red 1,1'-dianthrimide dissolves in concentrated sulphuric acid, displaying a dark olive-green colour. If this solution is heated in the presence of small quantities of boron, a blue coloration appears, the intensity of which is dependent on the boron content of the solution. The absorption maximum of the colour is at 630 nm. The extinction maximum is reached when the solution is heated at 70 °C for 5 hours.

The method is suitable for the determination of between 0.01 and 0.2 mg of B/l. In the case of higher boron concentrations appropriate take-offs of the water sample or dilutions should be selected. No more than 0.01 mg and no less than 0.001 mg of boron should be contained in the colour-complex solution to be measured.

The specifications of the procedure (sulphuric acid concentrations, reagent quantities, temperature and heating time of the colour-complex solution) must be strictly adhered to.

Should comparatively large quantities of nitrate and nitrite (more than 10 mg/l) be present, ignite the alkaline evaporation residue (see "Method") at 600 °C for about 30 minutes.

If the water sample contains Ge, Te, Co, Ni, Cu, Cr, V, fluoride, bromide or iodide in concentrations of 1 mg/l or more, boron may be separated from these interfering substances by extraction with 2-ethyl-1,3-hexane-diol in methyl isobutyl ketone and subsequent reextraction with 0.5 m sodium hydroxide solution (see Section 3.5.1.4).

In the presence of fluoride, apparatus made of teflon or some other synthetic material should be used instead of glassware. Acid fluoride solutions may lead to the solution of boron from glass walls.

If the water is coloured or turbid and does not become clear after filtration, the boron may be separated off as boric acid methyl ester (see "Volumetric determination following distillation, Section 3.5.1.3).

Organic substances are eliminated in the course of the analysis by heating for 30 minutes with sulphuric acid containing hydrazine sulphate.

**Equipment**

Spectrophotometer with 1-, 2- and 5-cm cuvettes

Platinum or quartz dish with 100-ml mark on the outer wall

Measuring flasks, 10, 25 and 50 ml

Measuring pipettes, 1 and 5 ml

Separating funnels, 125 ml and 250 ml, glass and if appropriate teflon

Hydrochloric acid, 1+1 and 2 m

Sulphuric acid, conc. (1.84 g/ml)

Sulphuric acid, 1 m

Sulphuric acid containing hydrazine sulphate:
Hydrazine sulphate dissolved in 1 m sulphuric acid, 1 g per litre

Sodium hydroxide solution, 0.5 m
2-ethyl-1,3-hexanediol solution, 20 % solution in methyl isobutyl ketone (MIBK)

Dianthrimide solution:
Dissolve 250 mg of 1,1'-dianthrimide in 500 ml of conc. sulphuric acid, (1.84 g/ml). May be stored in a dark bottle if precautions are taken to prevent moisture attraction.

Boron reference solution:
Dissolve 0.143 g of reagent-purity boric acid ($H_3BO_3$), crystalline, in redist. water and make up to 1 litre. Dilute 100 ml of this solution with redist. water to 1 litre. 1 ml of this solution contains 2.5 µg of boron.

**Method**

**Calibration**

Pipette increasing volumes of boron reference solution:
0.5, 1.0, 1.5, 2.0 and 2.5 ml, equivalent to 1.25, 2.50, 3.75, 5.00 and 6.25 µg of boron, into 50-ml measuring flasks. Make up the solutions which

are less than 2.5 ml to 2.5 ml with sulphuric acid containing hydrazine sulphate. Add 12.5 ml of conc. sulphuric acid (1.84 g/ml) and 5 ml of dianthrimide solution to each measuring flask. Prepare blank tests in the same way but without the addition of boron.

Seal the measuring flasks with a glass stopper and keep at 70 °C for 5 hours in the thermostat. Allow to cool and measure the extinction at 620 nm against the blank solution, for example in 2-cm cuvettes. In the specified range (up to 7.5 µg of B) the calibration curve is a straight line.

Determination

Measure 100 ml of the water sample, or a volume containing between 0.001 and 0.1 mg of B, into a platinum or quartz dish, alkalize weakly and evaporate to dryness. If the sample contains more than 10 mg/l of nitrate ions and nitrite ions, ignite the alkaline evaporation residue at 600 °C for about 30 minutes. No problems arise if extraction is carried out with 2-ethyl-1,3-hexanediol/MIBK and reextraction in 0.5 m sodium hydroxide solution (see Section 3.5.1.4).

Dissolve the evaporation residue with a total of 25 ml of sulphuric acid containing hydrazine sulphate, and transfer to a 50-ml measuring flask. Heat the measuring flask in a boiling water bath for 30 minutes and, when cool, make up to the mark with 1 m sulphuric acid.

Set up seven 50-ml measuring flasks, and into five of these pipette 0.5, 1.0, 1.5, 2.0 and 2.5 ml from the made-up 50-ml sample; add 1 m sulphuric acid to those volumes below 2.5 ml to reach this volume. The two remaining 50-l measuring flasks serve to prepare a blank test and a control with boron reference solution.

To each flask add 12.5 ml of concentrated sulphuric acid (1.84 g/ml) and 5 ml of dianthrimide solution. Seal the flask with glass stoppers, maintain at 70 °C for 5 hours in the thermostat, and when cool measure at 620 nm against the blank solution.

Calculation

Read off the appropriate boron concentration from the calibration curve on the basis of the measured extinction, take into account the quantity of water used and the dilutions, and recalculate for 1 litre of the water sample.

The result may be converted to $HBO_2$ (metaboric acid) by multiplying by the factor 4.049.

### 3.5.1.2 Spectrophotometric determination with azomethine-H

General remarks

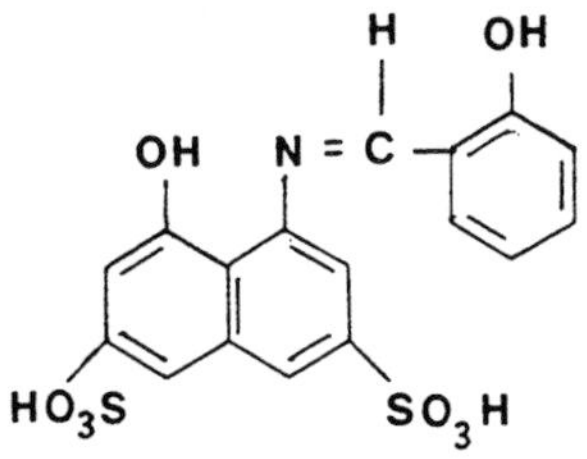


Azomethine-H

In buffered solution, borate ions react with azomethine-H to produce a yellow-coloured compound which when in solution complies with the Lambert-Beer law.

If the water sample under analysis is turbid, filtration must be carried out.

If interference occurs, the boron can either be separated by distillation (see Section 3.5.1.3) and then determined by spectrophotometry, or extracted with a solution of 2-ethyl-1,3-hexanediol in methyl isobutyl ketone (see Section 3.5.1.4). In the latter case, the boron can be determined either in the organic solution by atomic-absorption spectrophotometry, or by spectrophotometry with azomethine-H in the aqueous phase following reextraction with sodium hydroxide solution.

Equipment

Photometer

Cuvettes with 1-, 2- or 5-cm path length

Measuring flasks, 1000 ml and 100 ml, quartz or polyethylene

Polyethylene flasks, 1000 or 500 ml, for sampling

Polyethylene flasks, 100 ml, for performing the analysis

Pipettes

Azomethine-H:
Dissolve 1.0 g of the sodium salt of azomethine-H together with 3.0 g of ascorbic acid in water. Make up to 100 ml in a plastic flask and store in a cool place.
This solution is stable for about 1 week.

Buffer solution:
Dissolve 250 g of ammonium acetate in 250 ml of water, add 80 ml of sulphuric acid, (1.21 g/ml) and 5 ml of phosphoric acid, (1.71 g/ml), and subsequently dissolve in this solution 1 g of citric acid and 1 g of the disodium salt of ethylenediamine tetraacetic acid (EDTA).

Reagent solution:
Mix equal volumes of azomethine-H solution and buffer solution immediately before the analysis and use promptly.

Boron reference solution:
(see Section 3.5.1.1)

Take 10 to 25 ml of the water sample and filter, dilute, or extract and reextract with sodium hydroxide solution as appropriate. Pipette the prepared sample into a dry 100-ml polyethylene flask. Add 10 ml of the reagent solution described above. Concurrently, take portions of boron reference solution of appropriate concentration and treat analogously. In the same way, conduct a completely identical blank test in parallel. Leave all the flasks in the dark for 2 hours and then conduct the measurement at 414 nm.

From the prepared calibration curve, read off the boron concentration, taking the blank reading into account, and recalculate for one litre of water, making adjustment for dilutions where appropriate.

The results may be calculated as boron (atomic weight 10.811) or, by multiplying by the factor 4.049, as $HBO_2$. The results may also be presented as borate ions calculated as boron. Round off the results to 0.01 mg/l, or, if over 1 mg/l, to 0.1 mg/l.

### 3.5.1.3 Volumetric determination following distillation

#### General remarks

With methanol, boric acid forms the readily volatile boric acid methyl ester ($B(OCH_3)_3$), boiling point 68.7 °C. This may be volatilized quantitatively from the acid solution in a methanol vapour flow. The ester which passes over is collected in an alkaline receiver (1 m sodium hydroxide solution) and saponified.

After boiling away the methanol, $CO_2$ is expelled from the acidified solution. It is neutralized, and treated with mannitol, which - as do other polyalcohols - forms a didiol complex with boric acid:

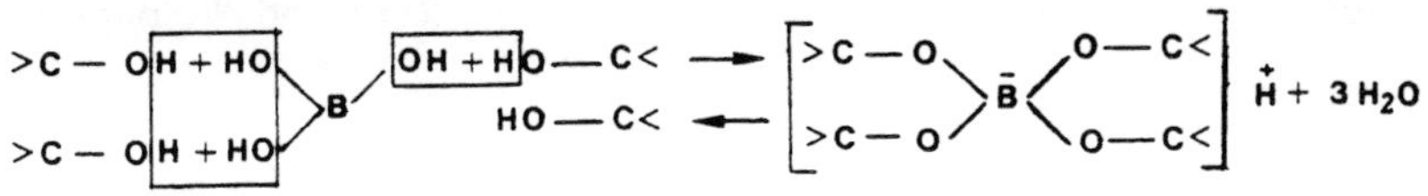

The didiol complex reacts or dissociates in aqueous solutions as a complex acid of the strength of acetic acid and can be titrated as such with sodium hydroxide solution.

This method is particularly suitable for waters with boron contents higher than 0.2 mg/l and those which are coloured or contain non-filterable turbidities.

#### Equipment

Flat-bottomed quartz-glass flask, 250 ml

Device for electrometric pH titration or

Burette, 10 ml, 0.02-ml scale

Soda lime tube

Platinum or quartz dish, 200 ml

Sulphuric acid, conc., (1.84 g/ml)

Methanol, reagent-purity

Pentane, superpure

Hydrochloric acid, 10 %

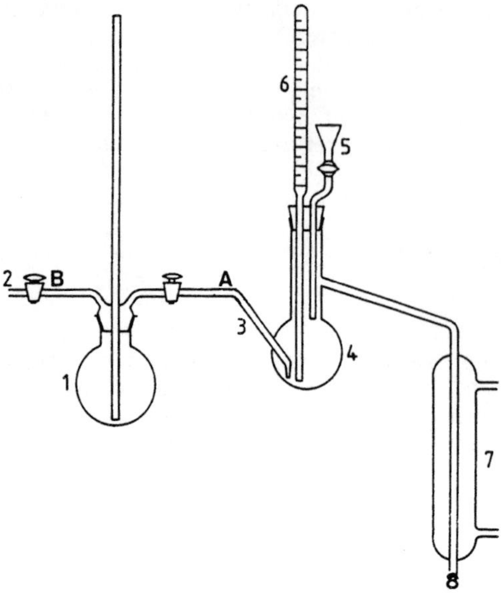


Fig. 103. Methanol distillation apparatus (illustration); 1) = Methanol vaporizer (250-ml flask); 2) = Discharge pipe with stopcock (B); 3) = Inlet pipe with stopcock (A); 4) = Distillation flask (200 ml); 5) = Feed funnel with stopcock and delivery tube; 6) = Thermometer, up to approx. 120 °C; 7) = Liebig condenser, jacket length approx. 20 cm; 8) = Receiver

Apparatus according to K. E. Quentin following H. Roth and W. Beck

Hydrochloric acid, 0.1 m

1-naphtholphthalein solution:
Dissolve 0.2 g in 96 % ethanol. (Only required if visual titration to be carried out.)

D(-)-mannitol

Sodium hydroxide solution, 1 m

Sodium hydroxide solution, 0.01 m

Method

General remarks

In a platinum or quartz dish, evaporate a suitable volume of the water sample to dryness under weakly alkaline reaction conditions. If more than 10 mg/l of nitrate or nitrite is present (see Section 3.5.1.1), ignite the alkaline evaporation residue at 600 °C for 30 minutes.

Transfer the dry residue remaining in the dish to the distillation flask, using a mixture of 4 ml of water and 40 ml of methanol. Insert the ground-glass stopper with thermometer and feed funnel, and carefully add 4 ml of concentrated sulphuric acid via the funnel, cooling the flask. Rinse with 5 ml of methanol, assemble the apparatus, and charge the quartz flask serving as a receiver with 10 ml of 1 m sodium hydroxide solution. The receiver should be marked at 100 ml on the outer wall.

When the vaporizer (1 in the diagram) has been charged with methanol, close stopcock B and open stopcock A, and turn on the heating both for the vaporizer and the distillation flask. When the methanol in flask 1 begins to boil (64.8 °C) and the temperature of the sample in flask 4 has reached about 70 °C, close stopcock B, simultaneously opening stopcock A, and allow the distillation to run until a volume of 100 ml is obtained in the receiver (quartz flask). During the entire distillation (about 15 minutes) the temperature of the sample should not exceed 75 °C, but neither should it fall below 65 °C (boiling point of boric acid methyl ester: 68.7 °C).

Determination

Add 60 ml of distilled water to the 100-ml distillate in the quartz flask. Carefully evaporate the methanol in a boiling water bath, blowing purified air onto it. Neutralize with 10 % hydrochloric acid, lightly acidify with 0.1 m hydrochloric acid, and boil briefly in order to expel carbon dioxide. In so doing, swirl the flask several times in order to swill the walls - if necessary to maintain the slightly acid reaction.

Subsequently seal the flask with a pierced rubber stopper in which a soda lime tube is inserted, and cool under running tap water. In order to prevent the sample solution from absorbing carbon dioxide from the air, cover with a layer of 5 ml of pentane. Adjust the solution electrometrically to pH 8 with 0.01 m sodium hydroxide solution.

If visual titration is to be carried out, add a few drops of 1-naphthol-phthalein solution instead, and set to the final point with the NaOH standard solution.

Now dissolve 2 g of mannitol in the sample solution which has been set to pH 8. A didiol complex forms with boric acid, and as a result of this complex, which dissociates in aqueous solution as a weak acid, the pH falls. Titrate once more with 0.01 m sodium hydroxide solution to pH 8 or the end point of the indicator solution. Then add a further 0.5 g of mannitol. If the indicator colour persists, the end point of the titration was reached; if not, titrate further with sodium hydroxide solution until the colour change occurs again and establish by once more adding mannitol that the equivalent point has definitely been reached.

The total number of ml of 0.01 m sodium hydroxide solution consumed corresponds to the boron content of the sample.

Conduct a blank test in the same way.

Calculation

1 ml of 0.01 m sodium hydroxide solution = 0.1081 mg of boron. Take into account the quantity of water used for the analysis and recalculate the result for 1 litre of water. The result may also be converted to $HBO_2$ (metaboric acid) by multiplying by the factor 4.049.

Atomic weight of boron: 10.811

### 3.5.1.4 Extraction of boron

Transfer a suitable aliquot of the filtered water sample to a separating funnel and adjust to pH 1 with hydrochloric acid 1 + 1. Treat with 20 ml of 20 % 2-ethyl-1,3-hexanediol solution in MIBK and shake for 1 minute; following separation of the phases, drain off the aqueous phase into a second separating funnel and there repeat the extraction with a further 20 ml of the extracting agent. Once again separate off the aqueous phase, and extract the combined organic phases by shaking for 1 minute with 20 ml of 0.5 m sodium hydroxide solution. After separation of these phases, neutralize the aqueous alkaline solution with 1 m sulphuric acid against phenolphthalein, dilute with twice the quantity of distilled water, and boil down in a platinum or quartz dish.

Boron may be determined, for example by AAS, in the evaporated extract.

#### 3.5.1.5 ICP-AES (see Section 3.3.12)

### 3.5.2 Silicic acid

#### 3.5.2.1 Spectrophotometric analysis with ammonium molybdate

General remarks

The cooled water sample is decomposed using hydrofluoric acid/perchloric acid, thereby depolymerizing the entire silicic acid and converting it into a form capable of reaction with ammonium molybdate (W. Fresenius and W. Schneider). After adding boric acid to complex the hydrofluoric acid, yellow-coloured silico-molybdic acid is formed by reaction with ammonium molybdate solution at a pH of 1 to 1.2. The extinction of the silico-molybdic acid is measured at 400 nm.

Experience has shown that after a short familiarization period the method described below is simple to carry out and provides rapid, reliable results. If a distinction is to be made between dissolved and colloidal silicic acid, one water sample should be examined directly and a second sample after membrane filtration.

The method is suitable for determination of between 0.5 and 250 mg of $SiO_2$/l in natural waters. If the content of silicic acid is above 25 mg/l, silicic-acid-free distilled water must be used for dilution.

All solutions used must be stored in plastic flasks so as to ensure that no silicic acid is dissolved out from glass.

Only silicic-acid-free water may be used to prepare the reagent solutions.

Phosphate ions only have an interfering effect at concentrations greater than 1 mg/l. If this is the case, shake the sample solution for 1 minute after adding the ammonium molybdate reagent with 50 ml of a benzene/isobutyl alcohol mixture (1 : 1). After separation of the phases, the organic phase contains virtually the entire phosphomolybdate complex, whereas the silico-molybdate complex remains quantitatively in the aqueous phase. This can then be measured photometrically after a development time of 15 minutes.

Equipment

Spectrophotometer with 1-cm and 5-cm cuvettes

Plastic flasks and beakers, 100 ml

Hydrofluoric/perchloric acid mixture:
45 ml 40 % hydrofluoric acid and 45 ml 70 % perchloric acid together with 10 ml silicic-acid-free water

Ammonium molybdate reagent:
Dilute 50 g reagent-purity ammonium molybdate $(NH_4)_6Mo_7O_{24} \cdot 4\ H_2O$ to 500 ml with silicic-acid-free water. Treat the solution with sodium hydroxide solution until phenolphthalein colour change occurs.

Benzene/isobutyl alcohol mixture, 1 + 1

Boric acid, cryst., reagent purity

Nitric acid (1.22 g/ml):
Mix conc. nitric acid (1.40 g/ml) with silicic-acid-free water 1 + 1

20 % sodium hydroxide solution, reagent purity

Sodium acetate solution, 2 m

Sodium carbonate, reagent-purity, anhydrous

Silicic acid for the reference solution:
Filter precipitated silicic acid, rinse with distilled water containing hydrochloric acid and ignite in the platinum or quartz crucible. Cool in the desiccator. If fine analytical-quality quartz sand is used, it should be extracted with hydrochloric acid, washed with silicic-acid-free water and then ignited in the crucible.

Silicic acid reference solution I:
Melt 100 mg of $SiO_2$ with 2 g of reagent-purity anhydrous sodium carbonate in the platinum (or possibly iron or nickel) crucible. Dissolve the molten mass in silicic-acid-free water in a plastic receptacle and dilute to the mark in a 1 litre volumetric flask at 20 °C. Immediately filter the solution into a plastic flask. 1 ml of this solution contains 0.100 mg of $SiO_2$.

Silicic acid reference solution II:
Dilute 10 ml of silicic acid reference solution I to 100 ml with silicic-acid-free water and likewise transfer immediately to a plastic flask. 1 ml of this solution contains 0.010 mg of $SiO_2$.

Silicic-acid-free water:
In order to remove any last traces of $SiO_2$ from the distilled water to be used for silicic acid determination, run the water through a column with a strongly basic anion exchanger ($OH^-$-type). The column should be used for the discharge piping. Store the silicic-acid-free water in plastic flasks.

## Method

### Calibration

The calibration curve is to be prepared according to the method given under "Determination" for the sample to be analyzed. Instead of an aliquot of the water sample, appropriate portions of silicic acid reference solutions I or II in a 100 ml polyethylene flask are used with 0.025 mg or between 0.25

and 5.0 mg of $SiO_2$. After adding the hydrofluoric acid/perchloric acid mixture and if necessary diluting to a volume of 52 ml, continue the analysis with 20 ml of either solution. The calibration curve at 400 nm is a straight line at least between 0 and 1.0 mg of $SiO_2$/100 ml.

## Determination

Introduce approximately 2 g ± 0.2 g of hydrofluoric acid/perchloric acid mixture into 100 ml plastic flasks. Pipette in 50 ml of the water sample, mix and leave to stand for a few hours. If the plastic flasks are charged with the decomposition acid and weighed beforehand, the water sample may also be filled into the flask directly at the sampling point. The transport time to the laboratory is then used for the acid decomposition of the silicic acid. The quantity of water is calculated by weighing. If dissolved and colloidal silicic acid are to be differentiated, two samples - with and without filtration - should be examined in parallel.

After completion of the acid treatment, take an aliquot in a plastic beaker, add 4 ml of 20 % sodium hydroxide solution with a pipette and then 2 g ± 0.1 g solid boric acid. When the boric acid is dissolved, allow to stand for a further 15 minutes, and add 10 ml of nitric acid (1.22 g/ml) and 10 ml of 2 m sodium acetate solution. The pH should then be in the range of 1 to 1.2.

Next add 10 ml of ammonium molybdate reagent with a pipette and transfer the solution to a 100 ml volumetric flask. Rinse the plastic beaker with small volumes of water; in this way, remnants of deposited or undissolved boric acid are dissolved. Transfer to the volumetric flask.

In the presence of interfering quantities of phosphate ions, add the ammonium molybdate reagent in the way described above. Then transfer the solution to a separating funnel containing 50 ml of benzene/isobutyl alcohol mixture and rinse with small quantities of water. Shake vigorously (2 minutes) to extract the phosphomolybdate complex quantitatively to the organic phase, while the silico-molybdate complex remains in the aqueous phase. After phase separation, which is usually complete after approximately 10 minutes, run the aqueous phase directly into the 100 ml volumetric flask.

Add silicic-acid-free water to the 100 ml volumetric flask to the mark (in the same way as for the standard analysis process and after the removal of phosphate), wait for a total reaction time of 15 minutes (maximum deviation ± 1 minute) after adding the ammonium molybdate reagent and immediately measure the extinction in a 1 cm or 5 cm cuvette at 400 nm.

Take a blank reading concurrently using all the reagents used in the analysis.

## Calculation

Take the appropriate $SiO_2$ content from the calibration curve on the basis of the measured extinction and relate it to 1 litre of the water sample, taking into account the measured quantity of water and the dilution factor.

### 3.5.2.2 Gravimetric determination as silicon dioxide (silica)

#### General remarks

The water sample, containing a little hydrochloric acid, is evaporated. As dehydration takes place, the silicic acid is converted into a sparingly soluble form. When the dry residue is taken up with hot water, most of the deposited silicic acid remains undissolved and is separated off by filtration.

The small quantity of silicic acid (up to 5 %) which has re-entered solution is further dehydrated by repeating evaporation of the filtrate twice, until finally, separation is virtually quantitative. Igniting the combined precipitates yields the quantity of silicon dioxide corresponding to the silicic acid content of the water sample.

The silicic acid still present in the filtrate in solution may also be determined photometrically (see Section 3.5.2.1). The quantity of $SiO_2$ measured should then be added to the result of the gravimetric determination.

The method is suitable for the determination of silicic acid in waters with more than 10 mg of $SiO_2$/l.

After igniting to constant weight, the silicon dioxide generally still contains impurities in the form of salts or complex compounds of multivalent cations (e.g. of iron, manganese and calcium). The weighed ignition residue should therefore be evaporated with sulphuric/hydrofluoric acid, and the fluoridation residue ignited and weighed. The difference between the two weights indicates the silicon dioxide content.

#### Equipment

Analytical balance

Platinum or porcelain dish, 500 ml, with perfectly smooth internal surface

Platinum crucible, 25 ml, with lid

Hydrochloric acid, conc. (1.19 g/ml)

Washing water containing hydrochloric acid:
Mix 95 parts distilled water and 5 parts conc. hydrochloric acid

Sulphuric acid (1 + 3):
Add 1 part conc. sulphuric acid to 3 parts distilled water

Hydrofluoric acid, reagent purity, 38 - 40 %

#### Method

Acidify 1 litre of the water sample with HCl and evaporate to dryness in a platinum or porcelain dish on a water bath.

After evaporation, heat the residue at approximately 130 °C for about 1 hour in a drying oven or over a sand bath. Moisten the cooled contents of the dish with 5 to 10 ml of conc. hydrochloric acid, and following a reaction time of approximately 5 minutes dilute with hot water until all salts are dissolved.

Filter off the silicic acid, which is now largely insoluble, with a quantitative filter and rinse with hot water containing hydrochloric acid. Transfer the filter containing the bulk of the silicic acid to the ignited and weighed platinum crucible.

Evaporate the hydrochloric filtrate again, separate and filter off the remaining silicic acid in the same way. After rinsing, likewise transfer the second filter to the platinum crucible. Finally, evaporate the filtrate and the washing water for a third time to separate off the remaining silicic acid (or determine photometrically the quantities of $SiO_2$ still contained in the latter after decomposition by $HF/HClO_4$ (see Section 3.5.2.1)).

Carefully incinerate the filters in the platinum crucible, ignite to constant weight at 1000 °C and weigh. Soak the crude silicic acid with 0.5 - 1 ml of sulphuric acid (1 + 3) and after adding approximately 2 ml of reagent-purity hydrofluoric acid, fume off over the sand bath. Ignite the fluoridation residue to constant weight at 1000 °C again and weigh.

The difference between the two weights indicates the actual silicon dioxide content. If the fluoridation residue amounts to more than 1 - 2 mg, the determination must be repeated. Care should then be taken to rinse particularly carefully and thoroughly with the water containing hydrochloric acid since it may be assumed that the increased fluoridation residue is traceable to a high calcium sulphate content or to barium sulphate in the water.

Calculation

$$\frac{G_1 - G_2}{V} = \text{mg/l } SiO_2$$

$G_1$ = Weight before fuming off with hydrofluoric acid (in mg)
$G_2$ = Weight after fuming off with hydrofluoric acid (in mg)
V = Volume of the water sample in litres.

1 mg $SiO_2$ corresponds to 1.30 mg $H_2SiO_3$

1 mg $H_2SiO_3$ corresponds to 0.77 mg $SiO_2$

## 3.6 Gaseous Substances

(see also Chapter 1, Sampling and Local Testing e.g. in the case of $H_2S$, and Sections 3.2, Carbon Dioxide Compounds, and 3.7, Radon.)

### 3.6.1 Sampling and gas-chromatographic analysis

General remarks

In virtually all water resources there are gases dissolved in the water. Carbon dioxide is almost always present in more or less considerable quantities (see Chapter 1 and Section 3.2). The content of carbon dioxide is particularly significant in groundwater at greater depths or in subterranean water and may reach concentrations of up to several g/l.

Dissolved nitrogen may also frequently be detected in the latter type of water. Other substances which may be identified in subterranean waters in the form of trace amounts of dissolved gases are hydrogen, methane, carbon monoxide, inert gases (including radon - 222), hydrogen sulphide and hydrocarbons.

Oxygen virtually never occurs in subterranean waters of this type. If it is not possible to identify oxygen at the sampling point and iron is present in the water in dissolved divalent form, together with perhaps hydrogen sulphide and a measurable negative redox potential, a test for secondary influences must always be carried out should it be the case that dissolved oxygen has been determined by means of gas-chromatographic tests in the laboratory. Such influences may arise due to the method of sampling, leaky stopcocks during transport of the sample, or other causes. It is, incidentally always advisable to transport the water and gas samples (which have been taken in gas collecting vessels) weighed down under water in a larger container so as to be able to exclude secondary oxygen access via the air as far as possible. In addition to the above-mentioned dissolved gases, many types of water from deeper strata also contain so-called freely rising gases, which emerge with the water in gaseous form. Frequently, these freely rising gases escape with the subterranean water in the form of small, evenly distributed bubbles, but occasionally also as large bubbles of gas. It may also sometimes be observed that the freely rising gases emerge from the depths with the water in surges. In many cases it is virtually impossible to differentiate between dissolved gases and so-called freely rising gases. It should always be taken into consideration that the freely rising gases may be obtained from a spring or a deep boring with relief of pressure. It is only rarely possible, with the aid of appropriate sampling equipment for subterranean water, to bring water samples to the surface at the pressure applying at depth. It is almost impossible to determine the ratio of dissolved to freely rising gases under the conditions of pressure and temperature which apply to deep sampling. The normal course of events will be that if it is necessary to differentiate between the dissolved and the freely rising gases, the separation is conducted following relief of pressure in a suitable sampling device (see Section 1.4.5).

Following pressure equalization, the gases from the water collect at the top of the gas collecting flask. In the laboratory, the gases are then drawn off via the septum using a gas syringe in order to be analyzed. The volume of the spontaneously collecting gases is measured.

By raising the temperature and applying a vacuum, it is then possible for the dissolved gases to be liberated from the remaining water and analyzed by gas chromatography. If no freely rising gases were observed during sampling, the gas collecting tubes are filled completely with the water sample, transported under water at the temperature of the spring, and then degassed in the laboratory in order to determine the dissolved gases.

The importance of correct transport of the gas or water samples to the analytical laboratory cannot be stressed enough. The samples are best transported under water, and in a quantity of water sufficiently large to ensure that virtually no change in temperature can occur.

The dissolved gases are liberated at 80 - 90 °C in a water-jet vacuum. In order to achieve this, the gas collecting container which was filled during sampling is connected at the top to a further, somewhat smaller, gas collecting tube which is completely filled with sealing liquid (saturated

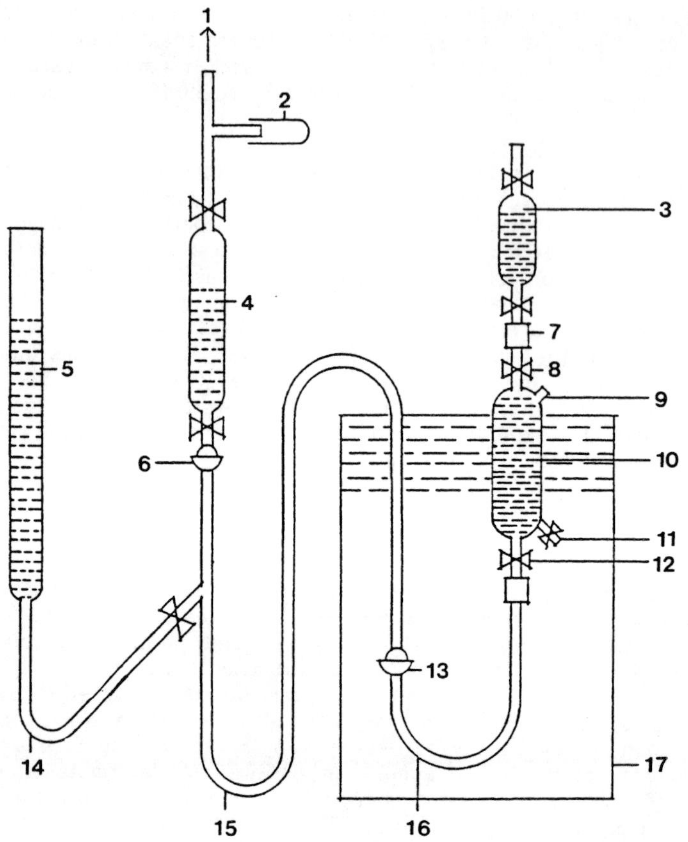


Fig. 104. 1) = Water-jet pump; 2) = Aeration system is briefly closed until sample boils; 3) = Degassed volume; 4) = Sealing liquid; 5) = Level equalization; 6) = Spherical ground-glass joint; 7) = Flexible tube; 8) = Gas-tight ground-glass stopcock; 9) = Adapter with septum; 10) = Gas collecting tube with water sample; 11) = Extension with gas-tight ground-glass stopcock; 12) = Gas-tight ground-glass stopcock; 13) = Spherical ground-glass joint; 14) = Flexible tube connection; 15) = Glass tube, 8 mm; 16) = Glass tube, 8 mm; 17) = Water bath > 80 °C

and weakly acidified sodium sulphate solution). The lower end is connected to a water-jet pump via a glass tube, which is likewise filled with sealing liquid. The sample in the gas collecting tube is immersed as completely as possible into the water bath and heated to approximately 80 - 90 °C. A water-jet vacuum is applied so as to transfer the previously dissolved gases to the upper gas collecting tube, where they may then be measured (including quantitative measurements), when the process is completed. In the course of this procedure, some of any carbon dioxide which may be present is also released. Following termination of the degasification process, the carbon dioxide contained in the degassed portion of the previously dissolved gases is absorbed with 30 % potassium hydroxide solution (KOH). First the volume of the gases which are not absorbed in the

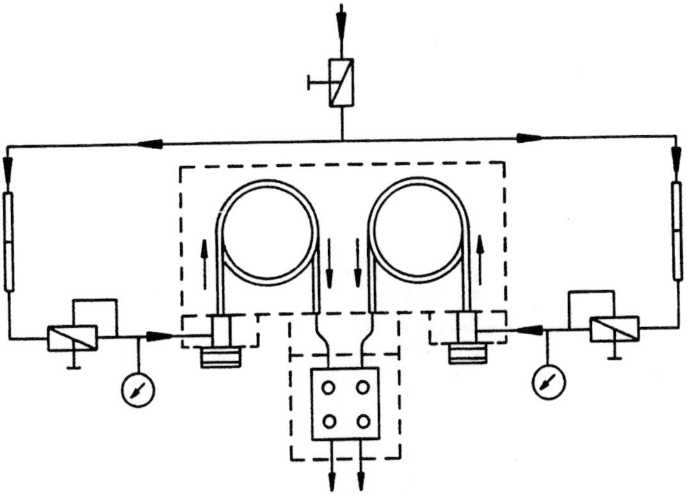

Fig. 105. Compensated gas-chromatographic system with two packed columns

potassium hydroxide solution is measured, and they are then used for gas-chromatographic analysis. In this way it is possible to differentiate approximately between the gases which rise freely with the water and the gases which are soluble in the water under normal pressure and at the exit temperature of the water. This process is particularly important for the analysis of water from deep borings, and also for the assessment of mineral and medicinal waters.

**Gas-chromatographic analysis** (see also Chapter 2 and Section 4.2)

The freely rising gases and the previously dissolved gases, following their liberation as described above, are determined by means of gas chromatography on packed columns using a thermal conductivity detector. It is advisable to use a double-column device (see Fig. 105) in which the two separation columns may be installed in parallel.

The columns are to be filled with 5 Å molecular sieve and Porapak Q. The gases can be separated and determined in a 4-metre-long steel column with 5 Å molecular sieve using argon as the carrier gas. Any argon which may be contained in the sample is detected by using the same separation column with helium as the carrier gas. Carbon dioxide is determined chemically on location and any hydrocarbons which may be present, such as ethane, can be segregated separately in a Porapak Q column and subsequently detected.

Gas dosage is performed using the dosing loop in accordance with Fig. 106.

Attention is once again drawn to the fact that a gas-tight syringe may also be used for dosage of the gas-chromatographic system from the septum of the gas-sampling device. 1 ml of gas is generally used as the sample volume.

A signal generated by the thermal conductivity detector is recorded by a mV plotter, and the peak area, which is proportional to concentration, is determined using an integrator. Test gases of appropriate composition are required for calibration. The test gases are prepared by gravimetric or

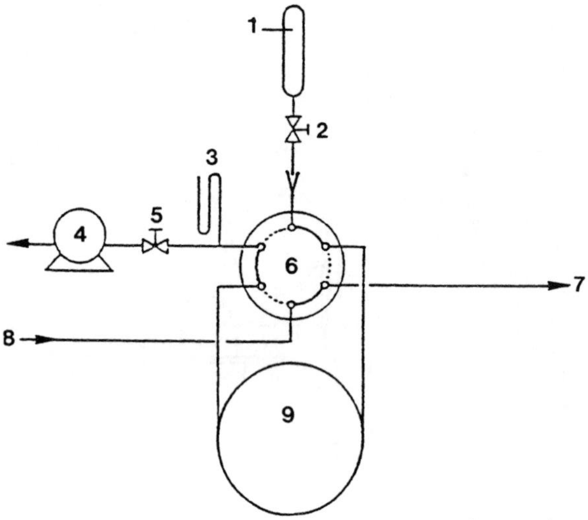


Fig. 106. Feeding of gaseous samples using a variable sample or dosing loop. Also applicable to samples which are at lower than atmospheric pressure. Variation of the dosage quantity is possible both by replacing the dosing loop (different volume), and by admitting more or less sample gas (pressure to be read off at the Hg manometer); 1) = Vacuum-gas ampoule; 2) = Valve 2; 3) = Hg manometer; 4) = Vacuum pump; 5) = Valve 1; 6) = Dosing valve; 7) = to separation column; 8) = Carrier gas; 9) = Dosing loop

volumetric methods. In the gravimetric method, an evacuated gas cylinder is weighed on a precision balance with high load-bearing capacity, the individual gases are fed in and the respective mass proportions determined. Gravimetrically prepared test gases are produced to order by laboratory-gas suppliers. Experience has shown that their composition is precise and reproducible.

Volumetric preparation is somewhat less precise but satisfies the requirements for the analysis of spring gases. A small volume of gas, in accordance with the expected concentration, is fed into a gas collecting tube filled with the main component using the gas-tight syringe. Alternatively, in the case of higher concentrations, the gas is mixed in a gas burette by dosing the individual gases by syringe.

The calculation is conducted according to the principle of the external standard.

$$Rf_i = \frac{\text{Vol. }\%_{iE}}{\text{Peak area}_{iE}}$$

| | | |
|---|---|---|
| $Rf_i$ | = | Response factor for the component to be measured |
| Vol. $\%_{iE}$ | = | Percentage by volume of component i in the test gas |
| Peak $\text{area}_{iE}$ | = | Peak area of component i in the test gas |

From the composition of the dissolved gases thus determined, it is then possible to calculate the content of the individual constituents in ml/l and mg/l of water.

$$\text{Content (ml/l)} = \frac{\text{Vol. }\%_i \cdot V_E}{V_{H_2O} \cdot 100}$$

Vol. $\%_i$ = Percentage by volume of component i in the liberated gas
$V_{H_2O}$ = Volume of degassed water sample (l)
$V_E$ = Volume of liberated gas (ml)

$$\text{Content (mg/l)} = \frac{\text{Molecular weight} \cdot \text{content (ml/l)}}{\text{Molar volume}}$$

The results of the gas-chromatographic analysis procedure are given either in ml/l of freely rising gases, or per litre of the previously dissolved and now liberated gases. The freely rising or previously dissolved gases may also be converted to mg or ml per litre of water. By way of examples, the following pages show chromatograms with a 5 Å molecular sieve column and argon as the carrier gas, a chromatogram on Porapak Q, and the separation of argon and oxygen on a 5 Å molecular sieve with helium as the carrier gas.

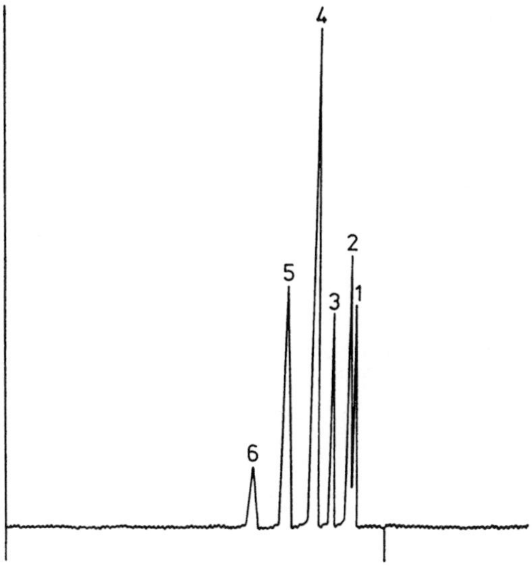


**Fig.** 107. Separation on molecular sieve 5 Å; Carrier gas: argon; 1) = He; 2) = $H_2$; 3) = $O_2$; 4) = $N_2$; 5) = $CH_4$; 6) = CO

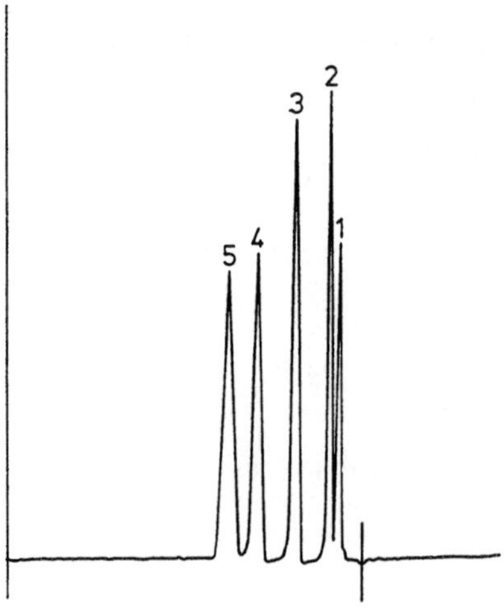


Fig. 108. Separation on Poropak Q; 1) = air; 2) = $CH_4$; 3) = $CO_2$; 4) = $C_2H_4$; 5) = $C_2H_6$

3.6.2 Carbon dioxide (see also Chapter 1 and Section 3.2)

Carbon dioxide dissolved in water (so-called free, dissolved carbonic acid) represents a significant factor in the chemical and biochemical make-up of a water. All natural waters contain dissolved carbon dioxide.

The designation "free, dissolved carbon dioxide" stems from the ratio $CO_2 : H_2CO_3$ = >99 % : <1 % in water. It is "free" in contrast to bonded hydrogen carbonate and carbonate, and "dissolved" differentiates it from the gaseous carbon dioxide frequently occurring in spring waters.

Free, dissolved carbon dioxide is determined acidimetrically by titration with sodium hydroxide solution. The method described by U. Hässelbarth involving direct titration following the reaction $CO_2 + OH^- \rightarrow HCO_3^-$ with electrometrical end-point indication or against phenolphthalein (pH 8.3) has become popular above all in the analysis of waters with $CO_2$ concentrations up to 200 mg/l. In a further method, proposed by R. v. d. Heide

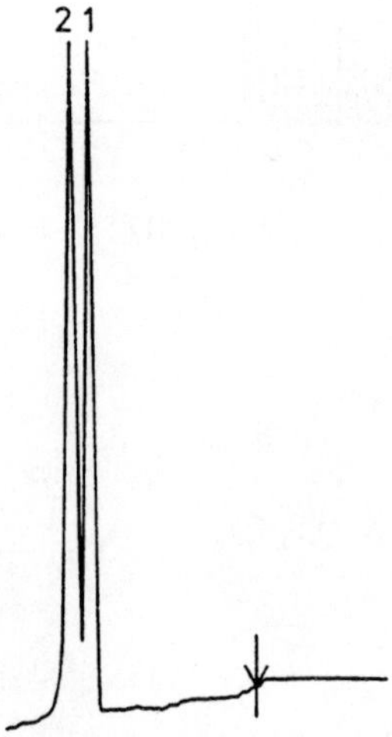


Fig. 109. Separation of argon and oxygen on molecular sieve 5 Å; 1) = Ar; 2) = $O_2$;

and A. Dietl and modified by K. E. Quentin and colleagues, carbon dioxide is fixed with sodium hydroxide solution in the receiver and then the excess solution is back-titrated. In contrast with the above method, this technique has the advantage that it can also be used for higher concentrations of carbon dioxide, for example up to $CO_2$ saturation.

The determination of free, dissolved carbon dioxide in mineral waters containing particularly large quantities of carbonic acid (more than 1 g of $CO_2$/l) can be carried out gravimetrically. The $CO_2$ is fixed with sodium hydroxide solution or calcium oxide at the sampling point, distilled off in the laboratory, bonded to soda-asbestos and weighed. The result obtained is the total carbonic acid content of the water. The content of free, dissolved carbon dioxide can be calculated by subtracting the separately determined hydrogen carbonate content.

**Acidimetric determination of free, dissolved carbon dioxide ("free, dissolved carbonic acid")**

### 3.6.2.1 Direct titration with sodium hydroxide solution

The method is based on the conversion of carbon dioxide with sodium hydroxide solution to sodium hydrogen carbonate:

$$CO_2 + OH^- \qquad HCO_3^-$$

Following complete conversion, the pH of the sodium hydrogen carbonate solution is 8.3. The end point of the titration can be established electrometrically using a glass electrode.

The method is suitable for the determination of $CO_2$ concentrations up to 200 mg/l in natural waters. If concentrations are higher, $CO_2$ losses may occur during titration. Waters of this type should be analyzed using the indirect method (back-titration of excess NaOH with hydrochloric acid).

In waters with sizeable concentrations of $Ca(HCO_3)_2$, interference may be caused by the precipitation of calcium carbonate. This may be prevented by adding 50 % Seignette salt solution which has been set to pH 8.3.

Similarly, interference due to iron (II) ions, which may lead to the precipitation of ferric hydroxides if concentrations are above 3 mg/l, can also be prevented by adding Seignette salt solution.

Transfer of the samples from one container to another should be avoided because of the volatility of dissolved $CO_2$. For this reason, the sample containers should be filled to overflowing from the base upwards at the sampling point and closed for transport, leaving no bubbles. Care should be taken that there is no rise in temperature during transport. A special pump pipette is used to transfer the water sample into the titration vessel.

**Equipment**

Titration flask with ground joint to take the single-probe measuring cell. (See Fig. 110)

Pump pipette: Delivery pipette with twin-bored stopper, glass elbow pipe and hand pump. (See Fig. 111)

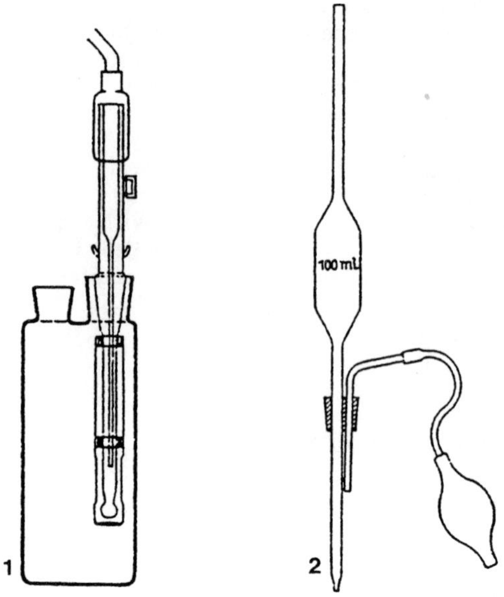


**Figs.** 110 and 111. 1 = Titration flask 2 = Pump pipette (According to U. Hässelbarth)

Magnetic stirrer with thermostat bath, 250 rpm synchronous, with teflon-coated magnetic stirring rods.

Single-probe measuring cell with standard ground joint

pH meter with temperature correction (measuring accuracy ± 0.05 pH)

Burette, volume 10 ml and 0.02-ml scale, with soda lime tube upstream to protect the NaOH solution against access of atmospheric $CO_2$.

Sodium hydroxide solution, 0.02 m

Seignette salt solution:
Dissolve 50 g Seignette salt (potassium sodium tartrate) in distilled water and make up to 1000 ml. Leave to stand for four days and set pH to 8.3 (glass electrode).

**Procedure**

The sample flask should be filled at the sampling point leaving no bubbles. Attach the pump pipette to the neck of the sample flask and carefully force a little more than 100 ml of the water sample into the pipette using the hand pump. Then remove the pipette and set to exactly 100 ml. Introduce the measured quantity into a dry titration flask in which a magnetic stirring rod has previously been placed. The tip of the pipette should reach the base of the titration flask during discharge. Finally, add 1 ml of the Seignette salt solution set to pH 8.3 in order to prevent calcium carbonate and ferric hydroxide precipitation.

Place the titration flask on the magnetic stirrer and add the glass electrode (single-probe measuring cell). Insert the burette outlet into the second pipe connection of the flask.

Titrate with 0.02 m sodium hydroxide solution to pH 8.3. Turn on the magnetic stirrer for 2 - 3 seconds after each addition of the standard solution.

It is advisable to repeat the titration in the following way. Use the same quantity of 0.02 m sodium hydroxide solution as was consumed during the initial titration and add this drop by drop in a single operation. Towards the end, slow the rate of addition to titrate to the equivalence point.

If more than 20 ml of sodium hydroxide solution are consumed, repeat the titration taking 50 ml (instead of 100 ml) of water sample, diluting it to 100 ml with (carbonic-acid-free) redistilled water.

### Calculation

1 ml of 0.02 m sodium hydroxide solution corresponds to 0.88 mg of $CO_2$. If 100 ml of water sample was taken, 1 ml of 0.02 m sodium hydroxide solution indicates 8.8 mg/l of $CO_2$.

### Observations

The accuracy of the quantitative determination of free, dissolved carbon dioxide ("free, dissolved carbonic acid") in natural waters by direct titration with alkaline standard solutions is problematic for a number of reasons. According to J. Kegel, this can be traced to two causes in particular:

1. The method is based on the assumption that a constant pH of 8.3 is assigned to the stoichiometric point of neutrality. In fact, this point is variable and depends on the hydrogen carbonate concentration. According to U. Hässelbarth (1969), the relative systematic error for a $HCO_3^-$ concentration of 10 mmol/l at the end point of the titration amounts to approximately 0.4 %; it diminishes with decreasing $HCO_3^-$ concentrations. As a result, the stoichiometric point of neutrality varies from 8.3 (according to J. Kegel) to the extent of an entire pH unit.

2. In natural waters, analytically determined pH ($pH_a$) and computed values ($pH_r$) fail to agree in the majority of cases. When analyzed, most waters display a pH approximately 0.2 pH units lower than that obtained from the formula. Kegel traces this to the influence of the activity coefficients. In addition, the difference $pH_r - pH_a$ is influenced by the assumption that the acid and base consumption of natural water is exclusively determined by the free and bonded carbonic acid and that this likewise exclusively determines the value of hydrogen activity. Since this assumption does not generally hold in its entirety, additional pH deviations arise which may distort the results of the $CO_2$ determination.

### 3.6.2.2 Alternative: Gas diffusion method

One method of determining free, dissolved carbon dioxide which avoids the errors outlined in 1. and 2. and which, moreover, is specific to $CO_2$, is described by J. Kegel in Fresenius' Z. Anal. Chem. 276, 45 - 50 (1975).

This method is based on accommodating the water sample under test in one section and $CO_2$-free distilled water in a separate section in a gas-tight vessel with a $CO_2$-free atmosphere (air, nitrogen or the like). In this way it is impossible for the two waters to become mixed, but gas exchange may take place without hindrance. Within a certain time, which may be considerably shortened by steady, careful stirring of the vessel, the $CO_2$ from the water sample is transferred to the gaseous atmosphere and from there to the distilled water until a solubility equilibrium of the $CO_2$ is established between the gas phase and the two liquids (Ostwald's solubility coefficient).

Taking as a basis the distribution factor of the $CO_2$ between the gas phase and water at the measured temperature, the $CO_2$ content of the pure $CO_2$ solution can now be determined, without external influence. The result can then be converted to yield the content of free, dissolved carbon dioxide in the water sample taken. Hydrogen carbonate ions and alkaline earth ions in the natural water are certain to be excluded from the distilled water. Consequently, the titration curve is steeper and the setting of the stoichiometric point of neutrality is more accurate and free from the influence of the activity coefficients. The negative influences mentioned in 2. are also eliminated.

This method should be considered for application if a high degree of accuracy is required or if considerable interference is to be expected as a result of the ideal conditions laid down in 2. not being met.

### 3.6.2.3 Back-titration method for the determination of "free, dissolved carbon dioxide"

**General remarks**

The water sample containing carbonic acid is added to excess sodium hydroxide solution, and sodium carbonate is formed. During back-titration with acid to pH 8.3, the hydroxyl ions of the excess sodium hydroxide solution are neutralized according to $OH^- + H^+ \quad H_2O$, and the carbonate formed from carbon dioxide and sodium hydroxide solution is converted to hydrogen carbonate. The two reactions

$$CO_2 + 2\,OH^- \qquad CO_3^{2-} + H_2O$$

$$CO_3^{2-} + H^+ \qquad HCO_3^-$$

indicate that one mole of consumed base is equivalent to one mole of $CO_2$, and accordingly one ml of 1 m NaOH = 44 mg of $CO_2$.

The method is suitable for the determination of free, dissolved carbon dioxide over a very broad concentration range (approximately 100 to 3000 mg/l). It may be used for the analysis of natural waters which contain considerable quantities of calcium and magnesium ions in addition to iron (II) ions. The method is intended primarily as a field method. It allows the determination of $CO_2$ with an average error of +2 to 3 %.

See also Section 1.7.12

Interference due to alkaline-earth ions and heavy metal ions can be eliminated by the addition of a suitable masking agent. According to an investigation by K. E. Quentin, I. Weber and D. Eichelsdörfer, it is possible to obtain good results with an equimolar mixture of potassium sodium tartrate and sodium citrate.

When taking the sample for titration, care must be taken to avoid $CO_2$ losses. For this reason, the sampling containers must be filled to overflowing from the base upwards at the sampling point. Submerge the pipette in order to extract the sample and introduce into the flask containing reagents.

## Equipment

Titration flask, 250 ml

Bulb pipettes, 5 ml, 25 ml and 50 ml

Burette, 25 ml, 0.02-ml scale

Sodium hydroxide solution, 0.25 m

Hydrochloric acid, 0.25 m

Masking reagent:
Dissolve 1 mole of potassium sodium tartrate (282.2 g $C_4H_4KNaO_6 \cdot 4\ H_2O$) and 1 mole of sodium citrate (294.1 g $C_6H_5Na_3O_7 \cdot 2\ H_2O$) in 2 litres of dist. water. Before use, set the solution with lye to the phenolphthalein transition point.

Phenolphthalein solution, 1 % in 96 % ethanol.

## Method

Take a mixture of 25 ml of 0.25 sodium hydroxide solution and 5 ml of masking reagent and add, for example, 50 ml of water sample and 1 ml of phenolphthalein solution. Swirl, and back-titrate with 0.25 hydrochloric acid.

Conduct a blank test analogously using carbonic-acid-free redistilled water instead of the water sample.

The consumption of 0.25 m sodium hydroxide solution in ml is obtained by subtracting the quantity of 0.25 m hydrochloric acid in ml which is consumed during back-titration of the lye in the receiver from the quantity of 0.25 m hydrochloric acid in ml which is consumed during back-titration of the blank test. 1 ml of 0.25 sodium hydroxide solution corresponds to 11 mg of $CO_2$.

## Calculation

$$\text{mg/l } CO_2 = \frac{\text{ml of NaOH consumed} \cdot 11\ 000}{\text{ml of water sample}}$$

## 3.7 Radioactivity measurements in water (see also Chapter 2)

### 3.7.1 General

Standard values for tolerable radioactivity levels in water have been agreed upon between the various international organizations and also established by way of legislation in the various countries. Thus, for example, the tolerable level for the total α-activity is 0.1 Bq/l corresponding to 2.7 pCi/l.

The figure for the total $\beta^-$-activity is 1 Bq/l (corresponding to 27 pCi/l).

These standard values are extremely low and are designed to totally preclude any risk to human beings and animals caused by radionuclides in water. If these total values are exceeded, this does not however mean that the water would be unsuitable for use by humans and for animals, but rather that it is then necessary to determine the radionuclides responsible for the radioactivity in the water. Radium 226 has become an established tracer for α-activity and strontium 90 (under certain circumstances also iodine 129) for $\beta^-$-activity.

General mention should be made of the α-emitters; radium 226, polonium 210, radon (radioactive inert gas) 220 and 222, uranium isotopes and thorium isotopes.

$\beta^-$-emitters: radium 228, lead 210, strontium 89, strontium 90, iodine 129, carbon 14 and tritium.

$\beta^-$ and γ-emitters: iodine 131, caesium 137 with barium 137-m.

#### 3.7.1.1. General guidelines

**α-activity**

If the measured α-activity is less than 0.1 Bq/l or less than 3 pCi/l, no additional analysis is required.

If the α-activity is considerably higher, radon 222 is to be determined and subtracted if it was included in the total α-activity.

If the α-activity is then still clearly in excess of 3 pCi/l, radium 226 is to be determined. If α-activity due to radium 226 is substantiated, the water is not to be used as drinking water for permanent consumption. If, on the other hand, it is demonstrated that the increased α-radioactivity does not stem from radium 226, the water can be used even in the event of an increased α-radioactivity level.

If, however, the measured α-radioactivity is in excess of 0.5 Bq/l (in excess of roughly 15 pCi/l), a nuclide analysis is required in order to establish which radionuclides are causing the increased α-radioactivity. A decision is then to be taken as to whether the water can be used or treated.

**ß⁻-activity**

If the ß⁻-activity of water is less than 1 Bq/l - corresponding to less than 27 pCi/l -, no further analyses are required.

If the ß⁻-activity is between roughly 30 pCi/l and 100 pCi/l, the radioactivity attributable to potassium 40 is to be subtracted.

If the radioactivity is then still greater than approximately 30 pCi/l, a quantitative determination of strontium 90 is to be performed. If the ß⁻ radioactivity was caused by strontium 90 and the radioactivity level exceeds a value of 30 pCi/l, the water is not to be used for permanent human consumption.

If the ß⁻-radioactivity is between 100 and 1000 pCi/l, not only the potassium 40 activity is to be subtracted, but strontium 90 and iodine 129 analyses are also to be performed. If the values determined for strontium 90 and iodine 129 are less than 30 pCi/l and 100 pCi/l respectively, the water can be used. If this is not the case, the water is not suitable for permanent human consumption. It can thus be seen that the activities of radium 226, strontium 90 and iodine 129 are envisaged as limiting radionuclides if the α-and ß⁻-activities of water clearly exceed 0.1 Bq/l and 1 Bq/l respectively.

In addition to the above-mentioned radionuclides, mention should also be made of cobalt 58, cobalt 60, iodine 131, caesium 134, caesium 137 and plutonium 239.

**γ-activity**

If γ-emitting substances are thought to be present in water, then these substances are to be analyzed with a scintillation crystal, for example in the geometrical configuration of a well-type crystal, and using a multichannel pulse height analyzer. Concentration of the radionuclides in the water is usually necessary for this measurement as well if a sensitivity level of, for example, 1 Bq/l is to be attained.

### 3.7.2 α-activity

**General**

The α-activity, which is limited to a maximum of 0.1 Bq/l corresponding to approximately 3 pCi/l, may be caused by the α-emitting radioactive inert gas, radon 222, which is contained in many bodies of water, in particular in groundwater and deep water, and decays with a half-life of 3.82 days. Thus, if the standard value of approximately 3 pCi/l is exceeded, the radon activity would have to be subtracted. However, the normal methods used to determine the so-called total α-activity in water are preceded by an evaporation process, with the result that the radioactive inert gas is no longer present in such a water sample which has been subjected to evaporation.

The determination of the so-called total α-activities thus only covers those α-emitting radionuclides which are not volatilized in the evaporation process and therefore remain in the dry residue of the water.

For measurement of α-radiation, use can be made of all measuring instruments and counters which permit detection of the very low standard value of 0.1 Bq/l - corresponding to approximately 3 pCi/l - with an appropriate degree of efficiency.

Proportional counters and scintillation counters are used as detectors. The proportional counters should preferably be designed without windows and take the form of methane flow counters. $\alpha$-rays generally feature extremely high energy levels of several MeV, but only a very limited range. Water can thus not be measured directly, and instead it is necessary to achieve a concentration of $\alpha$ -emitting nuclides, with particular attention being paid to radium 226, but also to uranium and thorium.

When performing analyses, it is essential to ensure that the counting facility and its geometry, and the concentration operation, effected for example by evaporation of the water, are completely analogous and fully comparable both as regards the analytical sample and as regards calibration values and blank analyses. Generally speaking, use is made of commercial equipment, i.e. equipment available for purchase, for example the above-mentioned proportional counters or scintillation counters as detectors, and in addition the corresponding power supplies, counters and computers.

In our experience, suppliers provide suitable instructions for utilization and checking of the counting facility as a whole. It is therefore urgently recommended that attention be paid to these suppliers' instructions for the measuring instruments, if tests are to be performed by an inexperienced water analyst. Reference is also to be made to the information on low-level measurements in Section 3.7.5 (Maushart).

When taking the sample, it is necessary to establish whether the $\alpha$-measurement is to be taken with the homogenized sample or with the filtered sample. If the sample is filtered on site, the filtrate is to be acidified to a pH of between 2 and 3 using for example nitric acid. Filtration and acid treatment are to be noted in the sampling record.

A uranium standard roughly corresponding to the standard value of 3 pCi/l is added to a further water sample treated in a completely analogous manner.

One µg of uranium corresponds to approximately 0.68 pCi. For calibration purposes, differing quantities of uranium are then to be appropriately prepared from stock solutions by way of dilution and added - in increasing amounts - to the water to be analyzed and also to the calibration solutions.

The amount of water envisaged for analysis, e.g. 1 or 2 litres of the acidified water sample is evaporated in platinum, quartz or glass dishes, in precisely the same manner as the made-up water sample, the calibration samples (at least three) and the blank. The dry residues of these analytical, calibration and blank samples are then measured in completely identical fashion in a suitable measuring facility, e.g. with a proportional or scintillation detector.

A calibration curve is produced from the calibration values and this is then used to obtain the analytical value for the total $\alpha$-activity (taking into account the blank reading). All values are to be expressed in terms of 1 litre and given as Bq/l, or as pCi/l or nCi/l.

If the measured value obtained with this summation analysis exceeds 0.1 Bq/l corresponding to 3 pCi/l, radium 226 is to be determined (see Section 3.7.2.2).

The radiochemical analysis of water samples to establish the total $\alpha$- and $\beta^-$-activity, as well as where applicable $\gamma$-activities, should only be carried out in an analytical laboratory which has not only the appropriate measuring instruments and detectors at its disposal, but also the experience to critically evaluate the analytical data determined.

The information given below on the radiochemical analysis of water samples is to be viewed in this light.

If non-specialist analysts come across radiochemical measured values which exceed the suggested maximum values, it is always advisable to consult an appropriate specialist.

The aim will always be to measure total $\alpha$- and total $\beta^-$-activity (the latter corrected by the potassium 40 content) before going to the trouble of analyzing for radium or radon as $\alpha$-emitters and strontium 90 or iodine 129 as $\beta^-$-emitters. (Guidelines for Drinking Water Quality, WHO, 1984).

In the USA 15 pCi/l is quoted for the total alpha radioactivity, for example, with radium 226 being included in the measured value, but not radon and uranium. A figure of 5 pCi/l applies to radium 226 and radium 228 (see also Chapter 6).

Such overall determinations are difficult to realize for these stated values from the measurement point of view. Thus, ASTM 1943/66 (1977) states that the overall determination can be performed for activities in excess of 0.5 pCi/ml, corresponding to 0.5 nCi/l. With this sensitivity it would scarcely be possible to attain the above-mentioned standard values by way of overall determination. This also applies to alpha spectrometry with prepared water samples. The $\alpha$-activity must therefore be concentrated.

This can be effected for example by evaporating a larger quantity of water. This does however involve the loss of the volatile $\alpha$-emitters such as radon 222. The dry residue of a larger water sample can then be measured, for example using a windowless proportional counter, a scintillation counter or, to advantage, a large-surface counter.

A relatively simple method, which is based on the $\alpha$-radioactivity of the radioactive inert gas radon 222 and enables radon 222 to be determined directly at the source (3.82 days half-life), is described below. After several half-lives, the so-called residual activity is measured analogously and, if the presence of radium 226 is indicated, then radium 226 is determined in a separate analysis which is likewise outlined below.

### 3.7.2.1 Direct measurement of radon 222

#### General

Either this measurement can be made directly at the source or the time and date of sampling must be recorded so that, following measurement in the laboratory, it is possible to calculate back to the time at which the sample was taken.

If the presence of radon 222 in a water sample can be verified in this manner, and if an airtight water sample is examined again after, for example, 5 half-lives and radon 222 radioactivity (residual activity) is

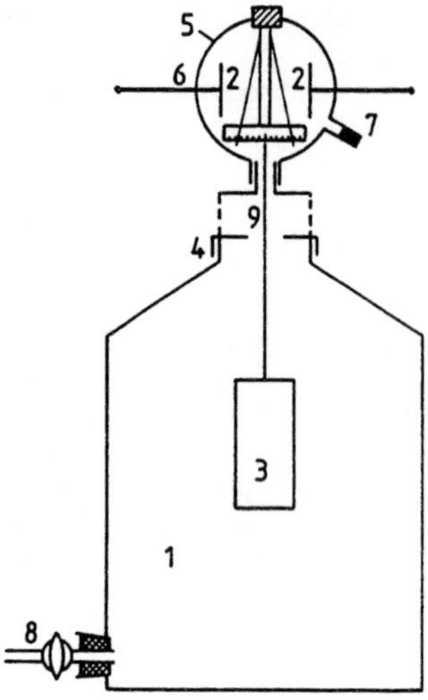


Fig. 112. Schematic view of fontactoscope; 1) = Measurement vessel (5 or 10 litres; made of brass or zinc); 2) = Protective plates; 3) = Dispersion cylinder; 4) = Metal cover; 5) = Electroscope, 6) = Metal ring; 7) = Metal pipe; 8) = Glass stopcock; 9) = Connecting rod

found, then it is possible to use the measured values to establish whether the parent nuclide, radium 226, which constantly produces radon 222, is present. In the latter case, the direct determination of radium 226 is often necessary (see Table 'Radon decay and production of radon from radium').

Radon 222 is shaken out of the water and the discharge of an electroscope is measured in the gas phase following the establishment of equilibrium distribution between the aqueous and the gas phase. This simple, long-practised method is still proving its worth today.

Method

(Radon 222)

Measure the amount of water, e.g. 1 litre, with a graduated cylinder and pour into the zinc or brass vessel. Seal the vessel with a rubber bung and then shake it for 1 minute. This produces the vapour-space equilibrium between the liquid and gas phases. Carefully discharge any gauge pressure and then fit the dispersion cylinder and measuring head in position in the vessel.

Charge the electroscope with a hard-rubber or celluloid charging rod which has previously been rubbed with a woollen cloth. The charging process should be continued until the saturation voltage of 200 volts is exceeded. This can easily be read off from the scale on the electroscope. Charging the electroscope causes the leaves to spread, and these leaves then gradually return to their rest position as the potential decreases. The time sequence during which the leaves return to their rest position is to be observed (stopwatch). A magnifying glass is attached to the measurement section.

Commence measurement roughly 5 minutes after charging, so as not to record the short-lived thoron (radon 220). Measure the discharge of the electroscope once a minute for example at the start and then every 5 minutes.

Read off the position of the leaves to the right and left of the zero point, estimating the tenths of a scale division. Add up the scale divisions to the right and left of the zero point. Continue measurement until the electroscope is discharged.

Empty the measurement vessel immediately after measurement and rinse it in distilled water.

Prior to the actual measurement of the radon 222, effect a blank reading (standard loss) in exactly the same way as the actual analysis (but using distilled water).

Repeatedly effect re-charging from time to time, if the electroscope is discharged too quickly during the measurement of the sample as a result of too high a radon 222 content.

The measurements for the analytical sample and the blank reading can conveniently be entered in the following table:

Radon 222 activity sheet

Designation of water sample:

Measurement with Engler and Sieveking fontactoscope No. ...

a) Standard loss of instrument

Electroscope No.: Measurement vessel:

| Time in minutes | Scale-division reading Left | Right | Total | Voltage reading Volts | Voltage drop Volts | Duration of measurement | Voltage drop per second Volts/sec. |
|---|---|---|---|---|---|---|---|

Voltage drop per second (standard loss) volts/sec. $V_n$ =

b) Voltage drop due to radon 222

Amount of water used W = ml, diluted to litres

Sample taken on: at hours

Distribution coefficient of radon at 20 °C = 0.25

Ratio of air volume to water volume m = 9.3 (example)

Capacity of instrument C = cm

Duane effect: $1 - 0.517 \frac{s}{v} = 0.85$

Factor f for calculating back to time t = 0 minutes from table "Conversion-factors".

Shaken out on (date), (number of) hours after taking sample

| Time t following shaking out minutes | Scale division reading Left Right Total | Voltage reading Volts | Voltage drop Volts | Duration of measurement | Factor f t=0 Voltage drop per second t=0 in volts/sec. |
|---|---|---|---|---|---|

Voltage drop per second calculated for time t = 0 (volts/sec.)

The saturation current is calculated in accordance with the following formula:

$$I = \frac{1000}{W} \cdot \frac{C}{0.3} \cdot \frac{f}{1 - 0.517 \cdot \frac{s}{v}} \cdot \frac{V \ (m + \ )}{m} - V_n \cdot 0{,}36$$

I = nCi/l

### c) Testing for residual activity
(possible radium content)

Radioactivity of water immediately following sampling
$I_o$ = nCi/l

Radioactivity of water following t = days
$I_t$ = nCi/l

The radioactivity A caused by radon 222 is given , for example, in nCi/l and is calculated in accordance with the following formula:

$$A\ (nCi/l) = \frac{1000}{W} \cdot \frac{C}{0.3} \cdot \frac{f}{1 - 0.517\ \frac{S}{V}} \cdot \frac{a(m + \alpha)}{m} - a_n \cdot 0.364$$

where:

A = Radioactivity in nCi/l (1 nCi/l corresponds to 1000 pCi/l or 37 Bq/l)
W = Amount of water used in ml
C = Capacity of instrument measured in cm
f = Conversion factor for calculating back for time t = 0 (table)
S = Internal area of fontactoscope chamber in $cm^2$
V = Volume of chamber in ml
a = Voltage drop observed in volts/sec
$a_n$ = Standard loss of instrument in volts/sec
m = Ratio of water to air in measurement vessel
= Distribution coefficient of radon between water and air (table)
0.346 is the conversion factor for $10^{-3}$ electrostat. units/l to nCi/l

The expression $1 - 0.517\ \frac{S}{V}$ (so-called Duane-Laborde effect)

is a correction factor which takes account of the loss of ionization energy suffered by the $\alpha$-particles of the radon and its disintegration products at the walls of the ionization chamber. This is a function of the size of the measurement chamber.

The correction for the standard loss $a_n$ takes account of the weak natural radioactivity of the air which causes a blank reading when the apparatus voltage drops.

| Distribution coefficient (dependent on temperature) | | Conversion factors (Calculation back for t = 0) | | | |
|---|---|---|---|---|---|
| Temperature °C | Distribution coefficient of radon between water and air | Observation period min. | Factor for calculating back for time t = 0 | Observation period min. | Factor for calculating back for time t = 0 |
| 0 | 0.515 | 0 | 1.000 | 9 | 0.653 |
| 10 | 0.347 | 1 | 0.883 | 10 | 0.647 |
| 20 | 0.252 | 2 | 0.810 | 11 | 0.640 |
| 30 | 0.196 | 3 | 0.761 | 12 | 0.635 |
| 40 | 0.161 | 4 | 0.727 | 13 | 0.632 |
| 50 | 0.140 | 5 | 0.703 | 14 | 0.628 |
| 60 | 0.123 | 6 | 0.686 | 15 | 0.625 |
| 70 | 0.121 | 7 | 0.671 | 180 | 0.460 |
| 80 | 0.119 | 8 | 0.662 | | |

## Residual activity

Water samples for determining so-called residual activity following a specific decay period can be taken in special sampling flasks or blunger flasks. Care is to be taken to ensure that there are no air bubbles in the flask.

Given the simultaneous presence of radium and radon in water, the individual concentrations can be calculated by determining the residual activity. If the measurement is taken immediately after sampling, the result obtained indicates the radon content ($E_{Rn}$). After 30 days and following renewed shaking out (isolation of radon 222), the measurement is repeated. Conclusions can then be drawn about the radium content of the water from the radon 222 then determined. If, as is often the case, the radon measurement is only taken several hours after sampling, the measurement result encompasses the total content of the radon still present and the radon formed from radium ($E_{tot}$.) A second measurement for radium determination after 30 days reveals the radium content ($E_{Ra}$). In accordance with the laws of radioactivity, a certain percentage of the radon will have decayed (B, see Table) after a certain period of time t. The radium forms an equivalent percentage of radon corresponding to the amount of radium (C).

After 7 days (A see Table) the residual activity of radon 222 is approximately 25 % of the initial activity. A share in excess of 25 % indicates the presence of radium 226, the decay of which leads to the constant formation of radon 222.

### 3.7.2.2 Radium 226

For the purpose of this analysis, the radium is concentrated by means of precipitation from a larger amount of water using barium as non-isotopic carrier. Generally speaking, between 10 and 50 litres of water is used for this determination, with the water being measured into a glass balloon flask or an appropriate plastic vessel at the source. Some 500 mg of barium is added to this amount of water as a non-isotopic carrier in the form of barium chloride solution, and acidification is effected with sulphuric acid. The resulting precipitation of barium sulphate is accompanied by precipitation of most of the radium. After several days, the precipitate

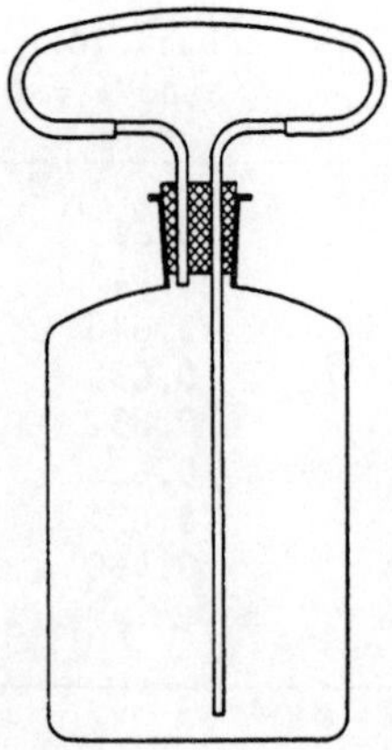

Fig. 113. 1) = Sampling flask for determining residual activity

Radon decay and production of radon from radium

| A | B | C | A | B | C | A | B | C |
|---|---|---|---|---|---|---|---|---|
| 0 | 100 | 0 | 2 days = 48 | 69.60 | 30.40 | 8 days + 12 | 21.43 | 78.57 |
| 0.5 | 99.62 | 0.38 | + 2 | 68.65 | 31.35 | 18 | 20.48 | 79.52 |
| 1 | 99.25 | 0.75 | 4 | 67.50 | 32.50 | 9 days | 19.57 | 80.43 |
| 2 | 98.50 | 1.50 | 6 | 66.51 | 33.49 | + 6 | 18.71 | 81.29 |
| 3 | 97.86 | 2.14 | 8 | 65.61 | 34.39 | 12 | 17.88 | 82.12 |
| 4 | 97.03 | 2.97 | 10 | 64.51 | 35.49 | 18 | 17.09 | 82.91 |
| 5 | 96.29 | 3.71 | 12 | 63.57 | 36.43 | 10 days | 16.33 | 83.67 |
| 6 | 95.57 | 4.43 | 14 | 62.72 | 37.28 | + 6 | 15.61 | 84.39 |
| 7 | 94.85 | 5.15 | 16 | 61.65 | 38.35 | 12 | 14.91 | 85.09 |
| 8 | 94.13 | 5.87 | 18 | 60.75 | 39.25 | 18 | 14.25 | 85.75 |
| 9 | 93.43 | 6.57 | 20 | 59.94 | 40.06 | 11 days | 13.62 | 86.38 |
| 10 | 92.72 | 7.28 | 22 | 58.92 | 41.08 | + 6 | 13.02 | 86.98 |
| 11 | 92.03 | 7.97 | 3 days | 58.06 | 41.94 | 12 | 12.44 | 87.56 |
| 12 | 91.34 | 8.66 | + 2 | 57.29 | 42.71 | 18 | 11.89 | 88.11 |
| 13 | 90.64 | 9.36 | 4 | 56.31 | 43.69 | 12 days | 11.36 | 88.64 |
| 14 | 89.96 | 10.04 | 6 | 55.49 | 44.51 | 12.5 | 10.38 | 89.62 |
| 15 | 89.29 | 10.71 | 8 | 54.76 | 45.24 | 13 | 9.481 | 90.519 |
| 16 | 88.61 | 11.39 | 10 | 53.81 | 46.19 | 13.5 | 8.660 | 91.340 |
| 17 | 87.95 | 12.05 | 12 | 53.03 | 46.97 | 14 | 7.910 | 92.090 |
| 18 | 87.29 | 12.71 | 14 | 52.34 | 47.66 | 14.5 | 7.225 | 92.775 |
| 19 | 86.62 | 13.38 | 16 | 51.42 | 48.58 | 15 | 6.599 | 93.401 |
| 20 | 85.97 | 14.03 | 18 | 50.68 | 49.32 | 15.5 | 6.027 | 93.973 |
| 21 | 85.33 | 14.67 | 20 | 50.02 | 49.98 | 16 | 5.505 | 94.495 |
| 22 | 84.68 | 15.32 | 22 | 49.14 | 50.86 | 16.5 | 5.028 | 94.972 |
| 23 | 84.05 | 15.95 | 4 days | 48.44 | 51.56 | 17 | 4.592 | 95.408 |
| 24 | 83.43 | 16.57 | + 4 | 46.96 | 53.04 | 17.5 | 4.195 | 95.805 |
| 25 | 82.78 | 17.22 | 8 | 45.57 | 54.43 | 18 | 3.831 | 96.169 |
| 26 | 82.16 | 17.84 | 12 | 44.24 | 55.76 | 18.5 | 3.499 | 96.501 |
| 27 | 81.55 | 18.45 | 16 | 42.93 | 57.07 | 19 | 3.196 | 96.804 |
| 28 | 80.93 | 19.07 | 20 | 41.65 | 58.35 | 19.5 | 2.919 | 97.081 |
| 29 | 80.32 | 19.68 | 5 days | 40.41 | 59.59 | 20 | 2.667 | 97.333 |
| 30 | 79.73 | 20.27 | + 4 | 39.21 | 60.79 | 20.5 | 2.436 | 97.564 |
| 31 | 79.11 | 20.89 | 8 | 38.04 | 61.96 | 21 | 2.225 | 97.775 |
| 32 | 78.52 | 21.48 | 12 | 36.91 | 63.09 | 21.5 | 2.032 | 97.968 |
| 33 | 77.93 | 22.07 | 16 | 35.81 | 64.19 | 22 | 1.856 | 98.144 |
| 34 | 77.34 | 22.66 | 20 | 34.75 | 65.25 | 22.5 | 1.695 | 98.305 |
| 35 | 76.76 | 23.24 | 6 days | 33.71 | 66.29 | 23 | 1.548 | 98.452 |
| 36 | 76.20 | 23.80 | + 4 | 32.71 | 67.29 | 23.5 | 1.414 | 98.586 |
| 37 | 75.60 | 24.40 | 8 | 31.74 | 68.26 | 24 | 1.292 | 98.708 |
| 38 | 75.04 | 24.96 | 12 | 30.79 | 69.21 | 25 | 1.078 | 98.922 |
| 39 | 74.48 | 25.52 | 16 | 29.88 | 70.12 | 26 | $8.989 \cdot 10^{-1}$ | 99.101 |
| 40 | 73.91 | 26.09 | 20 | 28.99 | 71.01 | 27 | $7.499 \cdot 10^{-1}$ | 99.250 |
| 41 | 73.36 | 26.64 | 7 days | 28.12 | 71.88 | 28 | $6.256 \cdot 10^{-1}$ | 99.374 |
| 42 | 72.82 | 27.18 | + 4 | 27.29 | 72.71 | 29 | $5.219 \cdot 10^{-1}$ | 99.478 |
| 43 | 72.25 | 27.75 | 8 | 26.48 | 73.52 | 30 | $4.354 \cdot 10^{-1}$ | 99.565 |
| 44 | 71.71 | 28.29 | 12 | 25.69 | 74.31 | 35 | $1.760 \cdot 10^{-1}$ | 99.824 |
| 45 | 71.17 | 28.83 | 16 | 24.92 | 75.08 | 40 | $7.111 \cdot 10^{-2}$ | 99.929 |
| 46 | 70.64 | 29.36 | 20 | 24.18 | 75.82 | 45 | $2.873 \cdot 10^{-2}$ | 99.971 |
| 47 | 70.10 | 29.90 | 8 days | 23.46 | 76.54 | 50 | $1.161 \cdot 10^{-2}$ | 99.988 |
| | | | + 6 | 22.42 | 77.58 | 60 | $1.896 \cdot 10^{-3}$ | 99.998 |

A = time between sampling and measurement in hours/days
B = radon content still present in %
C = radon formed from radium in % of radioactive equilibrium

is separated from the water and a second barium sulphate precipitation is effected in the filtrate. This barium sulphate precipitate is likewise separated and two barium sulphate precipitates are then incinerated together in a platinum crucible and slightly annealed. The sulphate mixture is decomposed with a sodium/potassium carbonate melt and then dissolved in water. The aqueous, alkaline decomposition liquid contains the carbonates of barium and radium in its residue. Following washing with water containing sodium carbonate and after changing the receiver, the carbonates of barium and radium are dissolved in hydrochloric acid, and the radon 222 inert gas which has been formed is driven off by boiling out. The weak hydrochloric acid solution is poured into a washing bottle and the radon is expelled completely by means of air bubbles. The hydrochloric acid solution is hermetically sealed and left to stand for 30 days. During this period, radon 222 is formed as a daughter product of radium 226.

At the same time, and using exactly the same method of working, distilled water is to be mixed with a radium preparation and all the above-mentioned concentration operations and separation procedures are to be performed. Radium 226 is calculated from the radon 222 formed in the measurement and calibration preparations.

The radon formed can be measured, for example, with a Pohl emanometer using a double ionization chamber, or by using electroscopes or other detectors suitable for the $\alpha$-radiation emitted by radon 222.

A further proven method for radon and radium is described below.

### General

In a vibrating-capacitor electrometer the ionization current flows via a high-impedance resistor (up to $10^{12}$ ohms). The resultant drop in d.c. voltage is converted into a corresponding a.c. voltage by means of a capacitor connected in parallel and oscillating periodically at 300 Hz. This a.c. voltage is electronically amplified and measured. The vibrating-capacitor electrometer (e.g. Victoreen Co., Cleveland, USA) features a higher level of sensitivity of the measurement configuration than static electrometers. Further advantages are the ease of operation, the broad measuring range and, above all, the possibility of continuously recording the ionization current. The method of radon/radium measurement elaborated by K. E. Quentin and G. Schretzenmayr (1968) is outlined below.

### Calibration of the apparatus

The apparatus is best calibrated using a radium solution, the activity of which is of the order of magnitude of approximately 10 nCi. The radon - which is in equilibrium - is transferred from the radium solution to the ionization chamber by means of a suitable de-emanation method (degasification). The use of this Ra calibration solution makes it possible to precisely determine the chamber factor K and to simplify the complex calculation. A d.c. voltage of 150 V is used as current source for the chamber voltage.

### Calculation of radon content

The determination is made as described under "calibration of the apparatus". Taking into account the radon decay in the period between sampling and measurement (see Table), calculation is performed using the following formula:

$$nCi/l = \frac{M \cdot K \cdot V \cdot Z \cdot 1000}{S_T \cdot V_4}$$

**Example**

$M = 0.78 \cdot 10^{-12}$A in 100 ml of water ($V_4$) (measuring value)
$K = 4.65$
$V = 1.39$ (Volume correction)
$S_T = 4.6$ (measurement time 30 minutes) (table)
$Z = \frac{100}{67.5} = 1.49$

(Z = Decay factor, the measurement was taken 52 hours after sampling. The radon decay can be seen from the table; in this case only 67.5 % of the original radon content is still present).

$$nCi/l = \frac{0.78 \cdot 4.65 \cdot 1.39 \cdot 1.49 \cdot 1000}{4.6 \cdot 100} = 16.33$$

**$S_T$ values as allowance for the increase in current due to daughter products**

| Measurement times in minutes | $S_T$-value |
|---|---|
| 2 | 3.31 |
| 3 | 3.51 |
| 4 | 3.67 |
| 5 | 3.80 |
| 6 | 3.90 |
| 7 | 3.98 |
| 8 | 4.04 |
| 9 | 4.10 |
| 10 | 4.14 |
| 12 | 4.22 |
| 14 | 4.28 |
| 16 | 4.33 |
| 18 | 4.37 |
| 20 | 4.41 |
| 30 | 4.60 |
| 40 | 4.80 |
| 50 | 5.00 |
| 60 | 5.19 |
| 75 | 5.41 |
| 90 | 5.58 |
| 105 | 5.73 |
| 120 | 5.83 |
| 150 | 5.94 |
| 180 | 6.00 |
| 210 | 6.01 |
| 240 | 6.01 |

**Example**

t = 52 hours, i.e. radon formed in time t as percentage of radium content = 32.5 %, radon still present = 67.5 %;

$E_{tot.}$ = 41.95 nCi/l;

$E_{Ra}$ = 0.69 nCi/l;

$E_{Ra'} = 0.69 \cdot \frac{32.5}{100} = 0.224$ nCi/l (radon formed in 52 hours)

$E_{tot} = E_{tot.} - E_{Ra'} = 41.95 - 0.224 = 41.726$ nCi/l;

$$E_{Rn} = E_t \cdot \frac{100}{67.5} = \frac{41.726 \cdot 100}{67.5} = 61.8 \text{ nCi/l}$$

With the normal equipment configuration of the vibrating-capacitor electrometer, radium can only be determined if the amount of radon formed is in the nanocurie range. Smaller concentrations require special methods.

The chamber factor K is calculated in accordance with the following formula:

$$K = \frac{\text{nCi} \cdot S_T}{M \cdot V}$$

where:

nCi = Radon activity of calibration solution ($V_4$)

$S_T$ = Correction value for increase in current during measurement period caused by short-lived daughter products of radon ($^{218}Po$, $^{214}Pb$, $^{214}Po$).
This correction value is taken from the $S_T$ table. The measurement period is the time span between the commencement of the transfer process and the taking of the reading + 5 min. As, in the first few minutes of the recommended transfer period of 15 minutes, the chamber does not contain all the activity, an empirical correction of 5 minutes is added. A measurement period of roughly 30 minutes following commencement of de-emanation (degasification) has proved advantageous.

M = Measured value, read off from the indicating instrument in picoamperes ($10^{-12}$A), minus the blank reading of the apparatus.

V = Volume correction for the air volume in the de-emanation vessel (degasification) ($V_1$), in the tubes and pipes ($V_2$) and in the ionization chamber ($V_3$), and for the solubility of the radon (see Table 'Distribution coefficient') at the measurement temperature in the amount of water used ( $\cdot V_4$). The correction factor takes account of the partial amount of the activity effective in the ionization chamber, i.e. the ratio of the volume of the ionization chamber to the total volume.

The volume correction is calculated as follows:

$$V = \frac{V_1 + V_2 + V_3 + ( \cdot V_4)}{V_3}$$

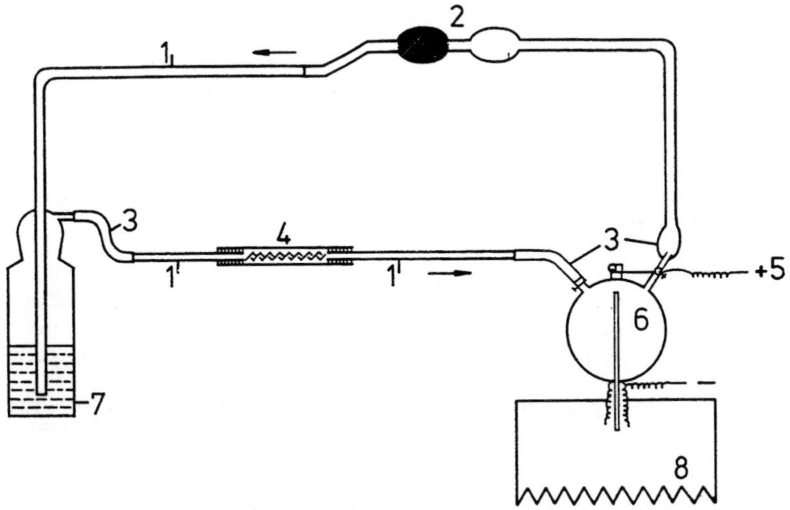


Fig. 114. Degasification apparatus for water samples containing radon (schematic); 1) = Glass tube; 2) = Rubber blower (or diaphragm pump); 3) = Flexible tubing; 4) = Drying tube (with CaCl2); 5) = Lead for chamber voltage; 6)= Ionization chamber; 7) = De-emanation vessel with water containing Rn (300 ml washing bottle); 8) = Vibrating-capacitor electrometer

### 3.7.2.3 De-emanation of water sample (isolation of radon 222)

Suitable methods of transferring the radon 222 into the ionization chamber in an approximately quantitative manner are the vacuum and the pumping method. With the first method the gas is drawn into a large-volume, evacuated ionization chamber from a glass vessel filled with the water sample containing radon. This is done by attaching one connection of the glass vessel to the evacuated chamber, whilst the other connection - reaching down to the bottom of the glass vessel - supplies compressed air via an air compensator. When the connections are opened, the compressed air is drawn through the blunger vessel and the radon is thus transferred to the ionization chamber. To ensure complete removal of the radon, the air must flow through the blunger vessel in finely distributed form.

With the pumping method, a glass bottle (washing bottle or 1 litre flask with washing-bottle head) is connected - with the inclusion of a rubber blower or a diaphragm pump and a short drying tube with $CaCl_2$ - to the ionization chamber in such a manner that the dry gas flow is routed into the chamber. For de-emanation (degasification) purposes, the air is pumped through the bottle for roughly 15 minutes. Experience has shown that extending this period does not produce better results.

### 3.7.3 Measurement of $\beta^-$-activity of water samples

#### General

There is no universal concentration process for measuring the radioactivity of water. The evaporation method has proven suitable for measuring $\beta^-$-radioactivity. Detecting the radioactivity of volatile radionuclides is deliberately dispensed with in this process (for example, only 10 - 20 % of iodine 131 is recorded).

If applicable, special measurement methods are to be used here, for example concentration precipitation using silver salts (3.2.2.1), with the silver chloride serving as non-isotopic carrier for the iodide and roughly 80 % of the radioactive I-131 being precipitated as well. The iodine in this precipitate can be determined for example by means of gamma spectrometry ( $\gamma$ 0.36 MeV).

The calibration of the measurement set-up is performed with natural potassium containing 0.0119 % of the natural $\beta^-$-emitter, potassium 40. If the composition of the nuclide mixture (dissolved in water) to be analyzed is known, a preparation with a similar content can be used for calibration.

When employing evaporation, the most favourable recovery rate for $\beta^-$-emitters in water, namely in the over 90 % range, is obtained if the water sample is acidified after the sample has been taken.

If the water contains undissolved substances, either the total activity of the dry residue including the undissolved substances is to be established, or the water is to be filtered prior to the evaporation process. In the latter case, it is possible to make separate statements about the amount of $\beta^-$-emitters dissolved in the water or adsorbed on solids. Experience has shown that roughly 5 - 10 % of the $\beta^-$-activity is adsorbed on the solids.

Great significance is attached to the separate determination of the radioactive substances actually dissolved in the water or bonded to solids, or which accumulate in micro-organisms, especially bearing in mind the question of accumulation in the food chain by way of fish.

The person taking the sample must therefore know whether the water is to be filtered for on-site measurement of the $\beta^-$-activity in the event of turbidity. If filtration is to be effected, the residues on the filter are to be washed out 2 - 3 times with a few millilitres of distilled water. The washing water is to be added to the filtrate. The measured-out and, if applicable, filtered water samples are to be acidified on site with sulphuric acid or nitric acid until a pH of between 1 and 2 is attained.

The water sample prepared in the laboratory for evaporation is to be evaporated, taking case to avoid boiling. Investigations have shown that if the water sample boils, radionuclide losses of between 20 and 30 % may occur. The most favourable method of evaporating the water is in platinum, quartz or glass dishes at a temperature of roughly 90 to 95 °C.

Method

Take a measured-out water sample of between 1 and 2 litres which has been filtered on site if applicable and to which nitric or sulphuric acid has been added until a highly acidic reaction takes place (pH 1 - 2), and evaporate the water sample at between 90 and 95 °C to dryness or until a moist salty mass is obtained.

If nitric acid has been used for acidification, dry the dry residue for 3 hours at 105 °C in a drying cabinet.

If sulphuric acid has been used for acidification, heat on a sand bath after evaporation of the water until the sulphuric-acid vapour disappears.

Perform calibration and a blank test in parallel. For this purpose, measure out corresponding amounts of distilled or deionized water as described for the water sample itself and add weighed-out amounts of potassium chloride, e.g. 0.2, 0.5 and 1 g, to the calibration samples. Effect acidification with the same amount and type of acid as used for the actual analysis and the treat these calibration samples in exactly the same way as the water samples to be analyzed. Determine the blank reading from the dry residue of the distilled or deionized water with the same quantity and type of acid as for the analytical sample (mixed and evaporated in the same way etc.).

Following evaporation, transfer the residue of the water sample to be analyzed, the dry residue of the blank test and the dry residue of the calibration samples to counting dishes. If use is made of a large-surface counter, it is possible on the one hand to use a larger quantity of water (up to 5 litres) to increase the sensitivity, or on the other hand to use smaller quantities of saltier water for the analysis.

If the salt content of the water to be analyzed is known, approximately the same amount of salts (generally dissolved calcium chloride and sodium chloride) can be added to the blank reading and to the calibration solutions, since this makes the conditions as regards the weight per unit area of the counting dishes similar to those for the water samples to be analyzed.

Prior to measurement, dry all preparations at 105 °C. Once they have been dried, the preparations can be treated with a thin protective layer of collodion (collodion cotton dissolved in ethyl acetate) to prevent the re-absorption of moisture.

Measure the calibration values, the analytical samples and the blank reading/background of the counter.

On account of the statistical nature of radioactive decay it is advisable to perform each measurement three times.

The measurement time can be extended to 60 minutes. Here again analogous conditions must be created as regards blank reading, calibration values and measurement sample. Compliance with the same measurement geometry is also to be ensured.

## Calculation

The blank reading and thus also the background counting rate of the background are subtracted from the measurement rates for the 3 calibration samples and for the analytical sample.

By definition, 28 $\beta^-$-decays per second or 1 680 $\beta^-$-decays per minute are taken as a basis (28 Bq/g of potassium) for 1 g of the natural isotopic mixture of potassium 40.

The amount of potassium 40 is calculated and the theoretical decay rate determined on the basis of the weighed amount of potassium salt. The efficiency of the counting set-up is calculated from the ratio of the theoretical decay rate to the measured pulse rates of the calibration samples. The mean value of the degree of efficiency is to be determined from the three individual measurements.

To calculate the gross $\beta^-$-activity of the water, the measured rate - minus the background rate - is corrected to take account of the degree of efficiency of the counting set-up as determined by calibration.

1 g of natural potassium/l corresponds to $0.76 \cdot 10^{-3}$ µCi/l, corresponding to 28 Bq/g

1 mg of natural potassium/l corresponds to $0.76 \cdot 10^{-6}$ µCi/l, corresponding to 0.028 Bq/mg

Gross $\beta^-$-activity can thus be calculated in accordance with following:

$$A_1\ (\mu Ci/l) = \frac{M \cdot 0.76 \cdot 10^{-3}}{K}$$

$$A_2\ (\mu Ci/ml) = \frac{M \cdot 0.76 \cdot 10^{-6}}{K}$$

M = Pulse rate of measured value of water sample, in terms of 1 litre of water taking into account the background.

K = Mean pulse rate of the calibration taking account of the background, referenced to 1 g of potassium/l or 1 mg K/l.

Net $\beta^-$-activity

Quantitative determination of the potassium content of a water sample can be effected by way of flame emission, atomic absorption or precipitation as potassium tetraphenylborate (see Section 3.3.3).

The net $\beta^-$-activity of the water is determined by subtracting the natural potassium 40 activity (0.0119 %) corresponding to the determined potassium content of the water.

The measured values for gross $\beta^-$-activity or for net $\beta^-$-activity are given in either Bq/l or pCi/l.

If, following subtraction of the K-40 activity, these measured values are in excess of 30 pCi/l, strontium 90 is to be determined (see also introduction to 3.7).

For measurement purposes, use can be made of all radioactive measurement facilities with suitable detectors, for example a Geiger counter, preferably in the form of a large-surface counter or proportional gas flow counter.

### 3.7.3.1 Analysis of radioactive strontium and barium in water samples

#### General

If determination of the net $\beta^-$-activity reveals that a measured value of 30 pCi/l has been exceeded, a special analysis is to be performed to establish which nuclides have caused the radioactivity in the water. Of the radioactive fission products, strontium 90 represents the greatest danger to the organism. It decays as a $\beta^-$-emitter with a half-life of approximately 28 years. Yttrium 90, which is likewise a $\beta^-$-emitter with a half-life of roughly 65 hours, is produced from strontium 90 in the course of the $\beta^-$ decay.

The relatively short-lived strontium 89, which decays with a half-life of approximately 51 days to form yttrium 89, is also a $\beta^-$-emitter, but it is of less significance on account of its relatively short half-life.

As regards water analysis, it should also be noted that barium isotopes may be present, in particular barium 140 which has a half-life of 12.8 days and is both a $\beta^-$-and a emitter.

The daughter product resulting from the $\beta^-$ decay of barium 140 is lanthanum 140 which, with a half-life of roughly 40 hours, becomes inactive Cer-140. Like barium 140, lanthanum 140 is both a $\beta^-$-and a -emitter.

In view of the fact that when the tolerance limit of 30 pCi/l for an unknown nuclide mixtures is exceeded in the course of water analysis, it is primarily a question of testing for and determining strontium 90 on account of its long half-life and its biological activity on incorporation in the organism. Attention must be paid in the analysis to the disturbance caused by the radioactive alkaline earths, strontium 89 and barium 140, and their daughter products.

Changes in the lime - carbon dioxide balance may cause calcium compounds to be precipitated from the water and strontium 90 to be entrained. For this reason, the water samples are to be acidified with nitric acid immediately after taking the sample.

This also eliminates disturbances which may be caused by the precipitation of iron (III) hydroxide and the co-precipitation of strontium 90.

For calibrating the entire process and the measurement configuration, use is made of distilled water which contains the same quantity of calcium in the form of calcium chloride per litre of water as the water sample to be analyzed and which also has a strontium 90 activity of, for example, 30 or 50 pCi/l. As a double calibration sample, this preparation is subjected to the entire process. The results of the actual water analysis are calculated and expressed in terms of strontium 90 on the basis of the measured values obtained in relation to the doped strontium 90 activity.

**Method**

Measure out 2 litres of the water sample to be analyzed and acidify it with concentrated nitric acid immediately following sampling. The excess of acid should be between 0.5 and 1 ml $HNO_3$.

The undissolved portion is separated from the water sample by means of membrane filtration, washed out and if applicable subjected separately to activity measurement in the test laboratory.

Add known quantities of strontium and barium (roughly 20 - 30 mg of strontium and approximately 10 - 20 mg of barium as chlorides) to the nitric-acid filtrate as inactive carriers. Then neutralize the sample with cold-saturated sodium-carbonate solution and precipitate the alkaline earth carbonates from the hot solution with an excess of approximately 50 ml of cold-saturated sodium-carbonate solution.

To convert any precipitated alkaline earth sulphates, keep the precipitate with the surplus sodium-carbonate solution hot for approximately 10 minutes. After cooling, filter off the precipitate of the alkaline earth carbonates and wash out with distilled water containing sodium carbonate (1 %). Using approximately 5 ml of 20 % nitric acid, dissolve the precipitate, which contains calcium carbonate, strontium carbonate and barium carbonate and possibly also iron hydroxide and manganese hydroxide, and add roughly 140 - 150 ml of 100 % nitric acid to this solution.

This leads to the precipitation of strontium and barium as nitrates, whereas most of the calcium remains in the solution. After roughly 10 minutes draw off the cooled solution via a glass filtering crucible with a fine frit.

Wash the nitrate precipitate of strontium and barium on the frit with 3 - 5 portions of 80 % nitric acid (1 ml in each case) and then dissolve the precipitate in approximately 20 ml of water. Add approx. 5 mg of iron in the form of an iron (III) chloride solution as hold-back carrier for the rare earths (yttrium 90 and lanthanum 140 as daughter products of strontium 90 and barium 140) and then precipitate the iron as iron (III) hydroxide with an approximately 10 % ammonium hydroxide solution (free from carbon dioxide), thus likewise bringing about precipitation of the rare earths. Filter off this precipitate and wash with 3 - 5 portions of distilled water (1 - 2 ml in each case). Add an excess of 10 % $Na_2CO_3$ solution (approximately 3 - 10 ml) to the filtrate and then filter off the strontium and barium which are precipitated as carbonates. Wash the precipitate twice with 1 - 2 ml of washing water containing $Na_2CO_3$ and then dissolve in 5 ml of 20 % nitric acid. Precipitate strontium and barium again as nitrates by adding approximately 140 - 150 ml of 100 % nitric acid and, following a cooling period of 10 minutes, filter again via a glass filtering crucible of corresponding fineness. Wash the nitrates three times with 80 % nitric acid (1 ml in each case) and then dissolve in 20 ml of water. Precipitate the carbonates of strontium and barium again from this solution with 10 % sodium carbonate solution, filter and then dissolve in approximately 6 ml of 20 % hydrochloric acid.

Precipitate the barium from this hydrochloric acid solution as barium chloride with ice cooling. Precipitate with 30 ml of diethyl ether and 90 ml of 36 % hydrochloric acid. Approximately 50 ml of ether-hydrochloric acid is used, the precipitated solution is cooled by placing it in an ice bath and the barium chloride is filtered off via a membrane filter. The precipitate can be used to measure the barium.

Evaporate the filtrate, which contains the strontium compounds and traces of residual barium, to approximately 10 ml, add between 10 and 20 mg of inactive barium as carrier and precipitate again with 50 ml of ether - hydrochloric acid employing ice cooling. The precipitate of barium chloride is filtered off via a membrane filter and can also be used for barium measurement if applicable.

Evaporate the filtrate of the second barium precipitate to roughly 5 ml and then dilute to approximately 25 ml with water. Precipitate strontium as strontium carbonate using 10 % sodium carbonate solution. Centrifuge off the strontium carbonate precipitate, pour off the mother liquor above it, cause the precipitate to swirl up with 10 % $Na_2CO_3$ and centrifuge again.

Separate off, dry and weigh the strontium carbonate. Measure the activity of the preparation using an end-window counter or a methane flow counter. Bear in mind the formation of yttrium 90 as a daughter product of strontium 90. Either perform a fresh hydroxide precipitation for purification purposes or make allowance for the fact that in the first few hours following separation the apparent activity of strontium 90 increases by roughly 1 % per hour.

Using precisely the same method, the calibration values are to prepared with known strontium 90 activities and a blank test is also to be performed.

Generally speaking, the chemical yield for strontium is roughly 80 %.

The strontium 90 content of the water sample examined is to be calculated on the basis of the calibration values with known strontium 90 activities and indicated in pCi/l or in Bq/l.

If the presence of strontium 89 is suspected, or if the tolerance limit for strontium 90 is exceeded, the strontium 89 content must be taken into account. This can be effected for example by separating the yttrium 90 daughter nuclide formed from the strontium 90. With analytical and calibration solutions the yttrium 90 is separated from the strontium by way of precipitation with carbonate-free ammonium hydroxide following the addition of between roughly 5 and 10 mg of iron (III) solution as non-isotopic carrier. The hydroxides are washed out and prepared. The activity of the parent nuclide, strontium 90, and thus also the proportion of strontium 89 in relation to the total measured value can be calculated from the measured yttrium 90 activity.

Notes:

Following measurement of the total activity, a piece of aluminium foil can also be placed on the measurement preparation (mixtures of strontium 89 and strontium 90). The $\beta^-$-radiation of strontium 90 in the preparation is not detected. This method is made feasible by the difference in the maximum $\beta^-$ energies of the two radionuclides (strontium 89: 1.46 MeV, strontium 90: 0.61 MeV). The aluminium foil for covering the preparation should have a substance of roughly 100 mg/cm$^2$.

If strontium 89 is also to be detected, the calibration solutions are to be doped with a mixture of strontium 89 and strontium 90 and the analytical procedure is to be performed accordingly.

The system strontium 90/yttrium 90 and strontium 89 can also be separated by means of paper chromatography or by way of liquid-liquid extraction.

If the precipitates of barium in the form of barium chloride are dissolved in water, the barium can be precipitated as barium carbonate and likewise subjected to activity measurement.

The chemical yield of the isolated strontium or barium component can be calculated on the basis of the amount of inactive carrier material added.

### 3.7.3.2 Rapid method of estimating strontium 90 content in water

General

Haberer and Stürzer describe a rapid method of estimating the strontium 90 content in water using ion exchange papers. Highly acidic, Type SAE-2 cation exchange papers as manufactured by Serva are employed. These papers contain roughly 45 - 50 % of the ion exchanger Amberlite IR-120 in $Na^+$ form. The exchange capacity is given as roughly 0.5 - 1 mmol/filter for 7 cm circular filters (approximately 2 mmol/g). Several exchanger filters can be positioned one above the other in a nutsche. Recommended, for example, is a layer of 6 SAE-2 cation exchange papers with a diameter of 7 cm in a porcelain nutsche for water filtration. Between 100 ml and 1000 ml of water is used for the analysis. The amount of water to be employed is limited by the calcium and magnesium ions in the water. If use is made of 6 filters one above the other in the nutsche, there can be between 100 and 150 mg of calcium in the water used.

Method

Add 5 mg of strontium and 5 mg of barium in the form of strontium chloride and barium chloride dissolved in water to the filtered water as inactive carriers (analyze filtration residue separately for $\beta^-$.-activity). Then filter the water sample prepared in this manner by way of six cation exchange filters positioned one above the other. Wash the filters first with 150 ml of 5 % citric acid solution (pH 3.5) and then with 50 ml of water. Suction-dry the ion exchange filters and then draw through 200 ml of 0.1 mol sodium chromate solution. 200 ml of this 0.1 mol sodium chromate contains 4 ml of glacial acetic acid, 8 ml of 25 % ammonium acetate solution and 2 ml of 0.1 mol strontium nitrate solution. Pour this elution solution for strontium onto the porcelain nutsche and allow the 200 ml to run through in a period of roughly 20 minutes without applying a vacuum. Once the 200 ml of reaction solution have passed through, add a further 50 ml. In this elution process it is mainly the strontium which is selectively dissolved, whereas the barium is retained on the cation exchange filter. Add ammonium carbonate solution to the eluate, heat it to 80 °C and then filter the precipitated strontium carbonate. Wash the strontium carbonate on the filter with water containing ammonium carbonate, dry at 80 °C and measure the $\beta^-$-activity on the filter. A yield of around 80 % can be expected for strontium 90 and strontium 89. Measurement of the strontium following precipitation and filtration as a carbonate must take place as quickly as possible, so as to minimize the disturbance caused by the formation of yttrium activity as a strontium 90 daughter product.

The measurement makes it possible to roughly estimate the sum total of strontium 90 and strontium 89 in the water sample.

If this rapid method reveals the likelihood of disquieting amounts of radioactive strontium in the water, the water is to be re-analyzed using the nitric-acid separation method described.

### 3.7.4 Other radionuclides in water (General analytical information)

Some information on the chemical separation of other fission radionuclides is given below.

**Iodine 131:** Following oxidation using nitrite ions, iodine 131 can be separated as elementary iodine by means of distillation or extraction. The addition of inactive iodine as a carrier makes it possible to precipitate silver iodide as well. Measurement can be effected for example using gamma spectrometry.

**Caesium 137 with barium 137m:** Following the addition of inactive caesium as isotopic carrier or of potassium as non-isotopic carrier, caesium can be precipitated and measured as caesium perchlorate or as caesium phenylborate (3.3.4.5). The formation of barium 137m activity (2.6 minutes) can also be measured (gamma 0.62 MeV).

It is also possible to precipitate the barium with an inactive barium carrier as barium sulphate or barium chloride and to measure the drop in its activity. The separation of caesium and barium is also possible by means of cation exchangers and using fractionated elution. Evaluation on the basis of a $\gamma$ spectrum is another possibility.

**Cobalt 60** can be separated for example as caesium cobalt hexanitrite following the addition of a carrier in the form of a slightly acidic solution or with $\alpha$-nitroso-ß-naphthol. If cobalt 60 is present, the $\gamma$-spectrum shows $\gamma$-lines at 1.17 and 1.33 MeV.

**Sulphur** 35 can be precipitated as barium sulphate (following oxidation to form a sulphate and, where applicable, the addition of a carrier) and purified by means of reprecipitation. Separation and concentration by means of anion exchangers is also possible.

**Phosphorus** 32 is likewise separated as phosphate by means of anion exchangers. Chemical precipitation with inactive phosphate as carrier can be effected as ammonium phosphorus molybdate, as ammonium magnesium phosphate and by co-precipitation with iron (III) hydroxide.

If **carbon** 14 is present, separation can be effected by combustion to form carbon dioxide. Depending on the type of bonding of the carbon, use can also be made of extraction with organic solvents of different polarity to effect separation and concentration.

## 3.7.5 Measurement of radionuclides in water (Low-level method with advanced measurement technology)

(By Rupprecht Maushart, from the Prof. Dr. Berthold Laboratory in 7547 Wildbad, FRG)

3.7.5.1 Radionuclide determination in water

3.7.5.2 Total activity or individual nuclide determination?

3.7.5.3 Criteria governing the selection of the detection method

3.7.5.4 Modern measuring instruments for determining total activity?

3.7.5.5 Detection limit in theory and practice

3.7.5.6 Notes on measurement

### 3.7.5.1 Radionuclide determination in water

#### A) Taking stock of radioactivity

The beginnings of radionuclide determination in water for reasons of radiation protection have their roots in the "zero level measurements" performed in the fifties which were aimed at taking stock of the radioactivity primarily in the vicinity of planned nuclear power stations. On the basis of the above, Kiefer and Maushart in Karlsruhe created as early as 1958 the measurement prerequisites for the method of global determination of radioactivity in water which is still most frequently used today; namely the evaporation of the water sample and the measurement of the residue with a large-surface proportional counter.

In the years that followed the methods were refined and the measurement technique improved. Standardized, generally accepted and practised methods were soon available.

#### B) Radioactive fallout

The rapid developments in this period were not so much the result of the increasing use of radionuclides or the beginnings of nuclear energy, but rather they stemmed from the massive scale of nuclear weapons tests carried out above-ground in the fifties and sixties. The radioactivity liberated by these tests necessitated a wide range of measurements in the biosphere, on the one hand to research into the behaviour and spread of the radioactive fallout, and on the other hand to determine the resultant radiation hazard for human beings. Accordingly, the methods developed concentrated primarily on the detection of fission products.

#### C) Today's new tasks and new measuring instruments

Water analyses are still performed today with the aim of immission monitoring, but the significance of the measurement task has shifted considerably, towards emission measurement, i.e. towards the monitoring of users or producers of radioactive substances. Together with the more stringent requirements and other international recommendations regarding the detection sensitivity of measurement techniques, there has been a marked effect on the methods employed. New measurement instructions meeting these requirements have been available for some years.

However, there have also been pronounced changes in the measurement technique itself. The introduction of microprocessor technology and improvements to detectors, above all in the case of low-level, large-surface counters, has made measurement easier, more precise and more reliable. The state of the art will thus probably not change considerably in the foreseeable future.

Nevertheless, depending on the task involved, decisions on the extent to which a selective analysis of certain nuclides or groups of nuclides is necessary and possible still have to be taken on a case-to-case basis.

### 3.7.5.2 Total activity or individual nuclide determination

For many pollutants there are no binding figures for the maximum concentration in water. By way of contrast, there are precise, internationally recognized limit values (ICRP, 1985) for every single one of some 200 radionuclides. These limit values are derived from the so-called AAA figures (Annual Addition of Activity). These values may differ by up to 6 powers of ten depending on the type of rays and the chemical composition of the radionuclides.

Accordingly, if a definite hazard is to be assessed, i.e. if the exceeding of individual limit values is to be established, there is no further option but to carry out a selective radionuclide analysis when effecting measurement. This analysis consists of a combination of radiochemistry and spectrometry and can be extremely involved.

The more simple method of measuring total activity is thus generally preferred to the determination of individual nuclides. The decision as to whether the activity level is so high that a more in-depth analysis is required is taken on the basis of the results obtained from the total activity measurement (see Section 3.7.1.1).

The term "total" activity measurement is, however, somewhat misleading. There is no method available which can cover all radionuclides equally. Even the total activity measurement displays a greater or lesser degree of selectivity, with the result that the selection of the method inevitably implies a decision as to which radionuclides are to be measured and which are not.

Some idea of the origin of the possible activity in water is thus a prerequisite. Some prime examples would be nuclear medicine therapy units (iodine 131), radionuclide laboratories in the pharmaceutical industry (H3) and nuclear power stations (Cs137, Sr90).

### 3.7.5.3 Criteria governing the selection of the detection method

The choice of a suitable detection method for establishing the total activity of a water sample is determined by three factors of a differing nature.

The required detection limit depends on whether direct measurement of the radiation from the volume of water without prior treatment of the sample is sufficient, or whether concentration of the radionuclides is required. Concentration will always be necessary for detecting the limit value of 0.1 Bq/l. The possibilities offered by and the limitations of direct measurement are discussed elsewhere.

For concentration purposes, the method must be oriented towards the chemical constitution of the water sample and the nature of the radionuclides. Generally speaking, concentration is effected by way of evaporation until the dry residue is obtained. If, however, individual radionuclides, such as iodine isotopes, are present in volatile compounds, specific precipitation reactions are required beforehand. Unusual concentration methods may also be necessary on a case-to-case basis, for example $H^3$ as $H^3H^1O$.

Finally, the measuring instrument with which the activity is detected must be selected with a view to the type of rays and the radiation energy emitted by the radionuclides. If the radionuclides to be detected are not known, this selection process is restrictive from the outset, unless different measurement systems are used consecutively for the same sample.

An overview of types of radiation, measurement methods and detection limits is given in the table in this part.

### 3.7.5.4 Modern measuring instruments for determining total activity?

#### A) Special measuring instruments for the various types of rays

Depending on the type of radiation emitted, at least three different types of measuring instruments will be required in order to be able to carry out total activity measurements for all radionuclides or radionuclide groups:

A **low-level dish measuring station** with large-surface proportional counter for alpha and for virtually all beta emitters, a **scintillation counter** with sodium iodide crystal for gamma emitters and a **liquid scintillation measuring station** for tritium and several other low-energy beta or K-emitters.

The two scintillation measuring stations already represent a transition to spectroscopic nuclide analysis. In conjunction with a multi-channel analyzer, these measuring stations permit at least rough determination of the radiation energy. If, however, spectroscopic nuclide analysis is the main aim of the measurement, then two further measurement configurations must be employed, namely a high-resolution gamma spectroscopy measuring station with GeLi or superpure germanium detector and an alpha spectroscopy measuring station with surface-junction semiconductor detector. The measuring stations for determining total activity are discussed below.

#### B) Low-level measuring station with large-surface counter

The so-called low-level counter for alpha and beta emitters is the most commonly used in water analysis. For determining total activity, such a system must be in a position to measure large sample quantities with the maximum possible sensitivity. A large-surface proportional counter with a very thin window (approx. 0.1 - 0.3 $mg/cm^2$) and a measurement area of roughly 300 $cm^2$ is a suitable detector. It permits the measurement of sample dishes with a diameter of 200 mm in which water samples can also be directly evaporated (Fig. 115).

On the other hand, the effect of ambient radiation on the background of the measurement set-up must be reduced. This is effected by way of anticoincidence systems to eliminate the effect of cosmic radiation, lead shields to attenuate the ambient gamma radiation, and extensive use of electrolytic copper with a particularly low radioactivity level as detector material. Nowadays, beta background rates considerably below 10 pulses per minute are achieved with modern low-level systems employing 200 mm dishes.

A counting-gas supply is necessary on account of the thin detector window. Between 1 and 2 litres of counting gas is required per hour. Non-combustible gas is currently replacing the previously used methane to an increasing extent.

#### C) Single-nuclide analysis/multiple counter

If a selective radiochemical separation of individual radionuclides is performed instead of determination of the total activity, smaller quantities of solids are involved and dishes with a diameter of 30 mm or 60 mm are sufficient. Measurement stations for such dishes have background rates of 0.3 - 1 pulses per minute and achieve correspondingly low detection limits.

In recent years the automatic changers normally used with such systems have become the subject of increasing competition from multi-counter systems in which, for example, 10 dishes can be measured at once (Fig. 116). The throughput times for large numbers of samples can thus be considerably reduced.

A further development is the use of hybrid systems which link the parallel measurement of several samples with an automatic changer for the sample trays.

**D) Gamma-emitter scintillation measuring station**

The second supplementary measurement system is a gamma scintillation measuring station with a sodium iodide (Tl) crystal. The sample to be measured can either be fitted into the well of the crystal as precipitation residue in a test tube, or introduced directly as an untreated water sample in a ring beaker arranged around the crystal (Fig. 117).

This scintillation measuring station is used primarily with beta/gamma and gamma-emitting radionuclides to obtain a rapid overview without sample preparation of the level and - by way of energy spectroscopy - of the type of activity. Direct measurement is however far less sensitive than the concentration method. With beta/gamma emitters, the scintillation measuring station thus only serves to supplement the low-level measurement system described above.

On the other hand, such a measuring station is indispensable for pure gamma emitters.

It is advisable to use a ring-beaker volume of approximately 1.5 litres and a crystal which is not smaller than 7.5 cm x 7.5 cm (3" x 3") in a 100 mm thick lead shield, even if such a set-up is considerably more expensive than the frequently encountered 5 cm x 5 cm crystals with 0.4 litre ring beaker. With medium-energy and high-energy gamma emitters the gain in terms of the detection limit is a factor of approximately three and, provided spectroscopy is used, the energy resolution is better.

**E) Tritium: Liquid-scintillation measuring station**

On account of its low maximum level of radiation energy (only 18 keV), tritium (H3 or T), a pure beta emitter, cannot be detected with the measurement set-ups described above. Moreover, in view of the fact that tritium is normally present in oxidized form as tritiated water (HTO), there is no possibility of concentration by way of evaporation or precipitation on account of its volatility.

Tritium is thus measured directly in the liquid by mixing with an organic scintillator and effecting measurement at the liquid-scintillation measuring station. Two factors may influence the measurement result: attenuation of the fluorescent light ("quench") due to coloration or chemical reactions, and superimposition by other radionuclides in the sample. Both effects can sometimes, but not always, be offset by way of measurement techniques. The only possibility then left open is to purify the water sample, for example by means of distillation.

For tritium measurements there is no practical alternative to the liquid-scintillation measuring station. On the other hand, this apparatus can also be used to measure all other beta emitters, as well as $\gamma$-emitters such as I 125.

However, on account of the small sample quantity which can be used (5 - 10 ml), the detection limits obtained are poorer by a factor of 100 than those achieved with concentration.

F) **Evaluation electronics**

Nowadays, data evaluation systems feature a high level of user-friendliness thanks to the use of microprocessors and advanced electronics.

Details are given by Maushart. The most important features of modern evaluation electronics are the automatic conversion of the results into absolute or specific activity units (Bq and Bq/g or Bq/l) and the independent resolution of gamma spectra into the individual nuclide components. This is made possible by prior storage of calibration and background data which are stored in so-called files or lists for the most widely varying measuring conditions.

Furthermore, simultaneous, separate alpha/beta measurement with low-level measurement systems and an indication of the fiducial limits of a measurement are now part of the equipment standard.

### 3.7.5.5 Detection limit in theory and practice

A) **Definition of detection limit and recognition limit**

If, as is the case with measurement methods for determining activity in water, "detection limit" values are required or indicated, misunderstandings can only be avoided by coming to prior agreement as to the meaning of this term.

DIN 25482, Part 1 (1985) stipulates that a strict distinction must be made between two different characteristic values which are frequently given the same name, namely "detection limit".

The detection limit defined by the DIN standard is the characteristic value of a measurement process which decides whether the measurement process satisfies a specific requirement, such as the detection of 0.1 Bq/l.

On the other hand, the second characteristic value, called the "recognition limit", is used to decide the value above which the measured pulse rate for a given individual measurement differs significantly from the background rate, i.e. it is used to decide whether the measurement sample is making a contribution or not.

The reliability of the statement, i.e. the statistical error probability, must of course also be indicated in both cases. The detection limit of a method defined in this manner can be calculated in accordance with the approximate formula

$$n_{limit} = 2.3 \frac{n_o}{t}$$

where $n_o$ is equal to the background rate measured over an extremely long period of time and t is equal to the measurement period for the sample. The factor of 2.3 recommended by DIN corresponds to a statistical error probability of 5 %.

B) **The detection limit for the evaporation method: a practical example**

In order to arrive at the limit activities in Bq/l from $n_{limit}$ in pulses per unit of time, allowance must be made for the individual calibration data of the measurement. The following example of a typical total beta activity determination using a large-surface counter serves to illustrate this.

A solid residue of 3 g remains when 1 litre of water is evaporated in the 200 mm diam. dish. The chosen calibration factor of the measurement set-up for medium beta energy (Tl204, 760 $keV_{max}$) with this residual quantity was 0.28 pulses per second/decays per second. The background rate is 0.12 $s^{-1}$ the counting time 1 h = 3600 s. $n_{limit}$ is thus 0.019 $s^{-1}$ or, in line with the calibration factor 0.28, 0.068 decays per second per litre = 0.068 Bq/l.

C) **The detection limits: Theory and reality**

It should be noted that the detection limit calculated in this manner represents a theoretical value which is not always attained under real conditions. There are two main reasons for this: on the one hand, a typical calibration value has to be formulated which may be far from correct for unknown radionuclides and, on the other hand, the background counting rate is often far less constant than is assumed in the calculation. The causes of this inconsistency, which are described comprehensively by Maushart, cannot be dealt with in detail here.

### 3.7.5.6 Notes on measurement

A) **Constancy of background rate**

As has already become apparent from the example relating to the detection limit, two values play a crucial role when measuring the activity in a sample: the background rate and the calibration factor.

The background rate is a characteristic value of the measurement set-up and ought to be constant on a long-term average. There are, however, factors which can lead to a change in the rate. The most common factor is the radioactive contamination of the detector by dust or vapours from the sample. The next most frequent cause is changes in the detector, such as a change of counting gas, contamination of the counting gas or internal contamination, and finally interference stemming from the electronics. Fluctuations in the background rate are also observed as a function of atmospheric pressure or the radon content of the ambient air. Thus, at least as regards the measurement of activity in the detection limit range, background measurements should always be taken between the sample measurements. To be absolutely certain, measurements should not be taken with empty sample dishes in the measurement position, but rather in the same way as the actual measuring samples.

Modern systems store the last background value in each case automatically and subsequently subtract it from the measured value. Such systems even compare the value with the previous value and indicate any statistically significant change.

**B) Optimization of calibration factor**

The calibration factor is not only a function of the detector - sample geometry and the efficiency of the counter for the respective type of emitter and radiation energy; it is influenced above all for alpha and beta measurements by the more or less uniform distribution of the residue in the counting dish. The calibration factor must therefore be determined under conditions which come as close as possible to those of the actual measurement.

Furthermore, the calibration factor must be determined for various residual quantities in the dish. With a 200 mm diam. dish the difference for example in the calibration factor between 0 g residue and 3 g residue for C 14 with 158 $keV_{max}$ is 3.2, whereas for Tl 204 with 760 $keV_{max}$. the difference is 1.3.

These values have been available for a long time in tabular form. Nevertheless it is advisable to produce one's own calibration standards for the specific analytical conditions, as the surface structure, the residue material and the uniformity or otherwise of its distribution over the surface of the dish can influence the calibration factor. Here again, advanced electronics can be a valuable aid, as the calibration factors assigned to the various samples can then be stored and applied automatically to the calculation of the results.

**C) Realistic error appraisal**

The main uncertainty factor when measuring total activity nevertheless remains the unknown radiation energy which makes the choice of the correct calibration factor into an educated guess.

The smaller the density of the residue per unit area, the smaller the error - hence the large dish diameters. Nevertheless the error is there and, despite all the progress which has been made in terms of equipment, it is still up to the judgement and experience of the expert concerned to give appropriate consideration to this error when stating the reliability of his results.

**Note:**

Details of further, more specific literature are obtainable from R. Maushart at the Prof. Dr. Berthold Laboratory, Wildbad, FRG.

Table. Radionuclides in water
Determination of total activity
**Measurement methods and detection limits for the various types of rays and radiation energy**

| Type of radiation or nuclide group | Typical nuclides | Sample quantity | Treatment of sample | Measurement method and measuring instruments | Detection limit in Bq/l |
|---|---|---|---|---|---|
| Alpha emitters | Ra-226<br>Pu-239<br>Am-241 | At least 1 litre<br><br>Max. residue 0.3 g/l | Transfer of radioactive substance to a 200 mm diam. sample dish by way of precipitation or evaporation (as solid residue). Evaporation is preferred, but may cause loss of readily volatile compounds | Measurement of alpha or beta radiation from solid residue with a low-level, large-surface proportional counter for 200 mm diam. sample dishes. For precise evaluation, the weight of the residue per unit area and the radiation energy must be known | 0.05 - 0.5; highly dependent on residual quantity |
| Pure beta emitters $E_{max.}$ > 150 keV | C-14<br>P-32<br>S-35<br>Sr-90<br>Pm-147 | At least 1 litre<br><br>Max. residue 3 g/l | | | 0.05 - 0.1 |
| Beta/gamma emitters | Co-60<br>I-131<br>Cs-137<br>numerous fission products | At least 1 litre<br><br>Max. residue 10 g/l | Transfer of solid residue from a precipitation reaction to a test tube; need not be dry | Measurement of gamma radiation of residue in 3" x 3" well-type NaI scintillation counter. Radiation energy can be determined by gamma spectroscopy | 0.05 - 0.1 |
| Pure gamma emitters | Cr-51<br>Co-57<br>Se-75<br>Tc-99m<br>I-125 | 1.5 ltrs<br><br>Solids content insignificant | None | Measurement of gamma radiation directly from volume of water with a 3" x 3" NaI scintillation counter for ring beakers with a volume of 1.5 litres. Radiation energy can be determined by gamma spectroscopy | 0.5 - 1 |
| Pure beta emitters $E_{max.}$ < 150 keV | H-3<br>Ni-63<br>Pb-210 | 5 - 10 ml | Usually none; otherwise distillation (H-3 only) or decolorization necessary depending on condition of water | Measurement of soft beta radiation directly from water at liquid-scintillation measuring station. Quench correction normally necessary. For precise evaluation the radiation energy must be known | 5 - 10 |

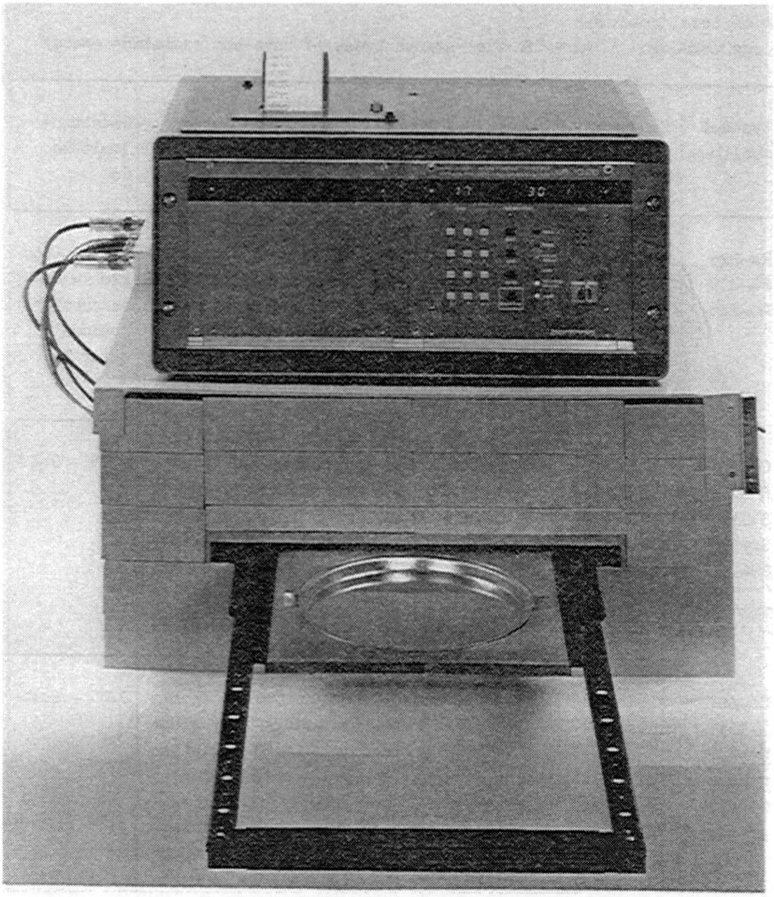

Fig. 115. Low-level measuring station for simultaneous, separate measurement of alpha and beta emitters directly in sample dishes with a diameter of 200 mm following evaporation of water sample. Microprocessor electronics with background subtraction and calculation in Bq/l

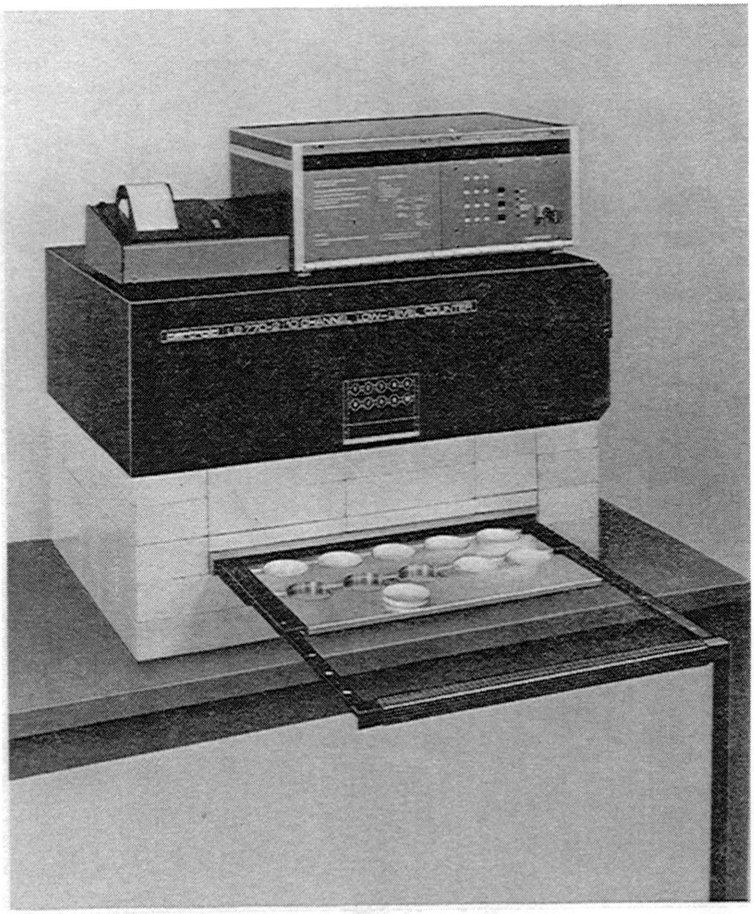

Fig. 116. Low-level measuring station for simultaneous measurement of 10 alpha or beta emitters in sample dishes with a diameter of 60 mm following chemical separation of individual nuclides or nuclide groups from water sample

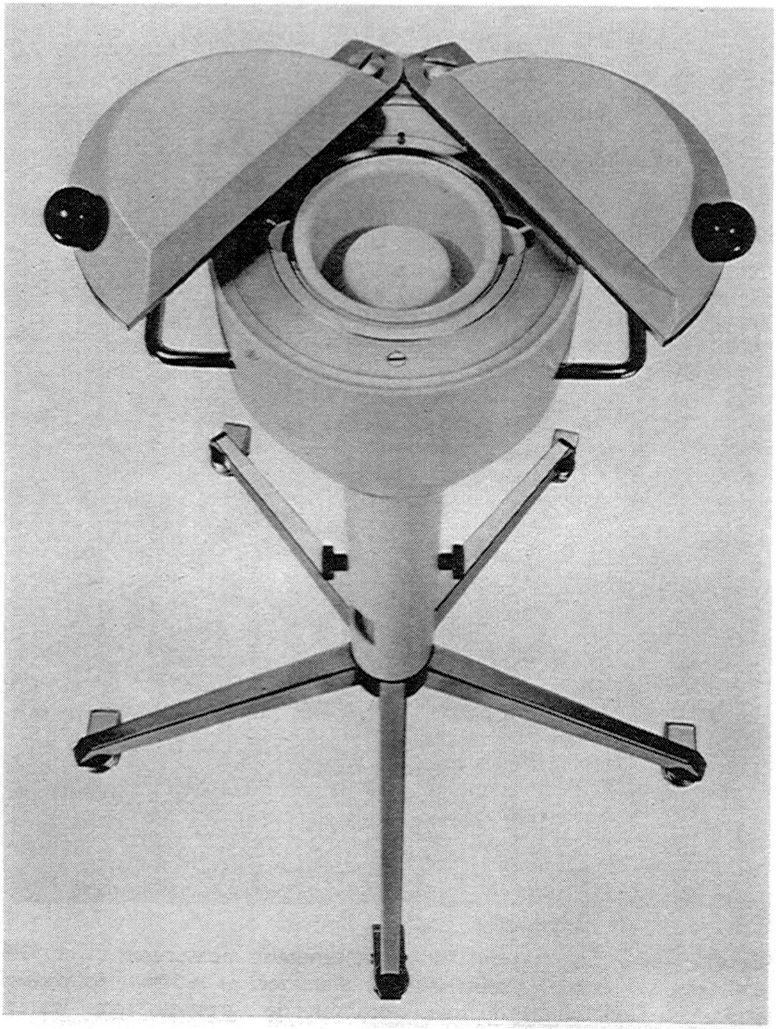

Fig. 117. Sodium iodide scintillation counter with ring-beaker arrangement for measurement of gamma emitters directly from the water sample

# 4 Organic Parameters

## 4.1 Overall organic parameters

### 4.1.1 TOC/DOC (Total Organic Carbon/Dissolved Organic Carbon)

Parameters such as biochemical oxygen demand (e.g. $BOD_5$) and chemical oxygen demand (COD) cannot simply be converted into a figure for the total content of organic matter. For this reason, additional overall parameters have been introduced to establish the total content of organically bound carbon (TOC) and the content of dissolved organic carbon DOC. TOC and DOC are precisely defined by the amount of organically bound carbon they contain and can be measured relatively accurately using modern instrumental analysis. However, these two parameters only represent part of the total organic matter and cannot be stoichiometrically converted to give the total figure because the composition of the chemical combination is unknown.

These methods can be used principally to measure concentrations in the range up to 10 mg/l and they produce a clear demarcation particularly in relation to COD. When measuring comparatively high TOC/DOC levels, for example in the case of sewage, the suspended matter is either homogenized (TOC) or separated by filtration (DOC).

In order to determine TOC/DOC, the organic substances contained in the water sample are oxidized with UV rays or by wet-chemical means, or burnt at high temperature. The amount of carbon dioxide released can then be measured. Commercial apparatus often uses a non-dispersive $CO_2$ infrared analysis device. If a flame ionization detector (FID) is used for gas analysis, the $CO_2$ released must first be converted into methane in a catalyst zone. Neither system of measurement is absolute, and so the equipment must be calibrated using standard solutions.

A constant check on the accuracy of the TOC/DOC analysis is kept by taking measurements of potassium hydrogen phthalate (PHP) which can be used both as an external and an internal standard. Any matrix disturbances which occur when measuring the TOC/DOC can be investigated by means of dilution analysis. If filtration through a glass-fibre filter or membrane filter is necessary, ensure that the first fraction of a few ml is rejected. Redistilled water is used for diluting, since de-ionized water has proved less suitable. The dilution water can be treated for a period of time (hours) with a low-pressure UV lamp in order to reduce any traces of organic matter which may be present.

Since most water samples contain inorganically bound as well as organically bound carbon ($CO_2$, $HCO_3^-$, $CO_3^{2-}$), the former must be removed by liberation before the TOC/DOC is determined. If volatile organic substances are present, the TOC/DOC level can be calculated from the difference between the total carbon content and the inorganically bound carbon content. But where a large proportion of inorganically bound carbon is present, this method entails a wide spread of errors.

These methods can be used with all types of water with TOC and DOC levels of approx. 0.1 mg/l to over 1 g/l.

In the case of sewage with high TOC/DOC levels, methods and types of equipment are usually different from those used for water with a low TOC/DOC content.

Under certain circumstances dilution is necessary to bring the sample within the measurement range of the method or equipment employed.

**Equipment**

Homogenization facilities, for example ultrasonic equipment, adequate for dealing with water samples containing suspended matter.

100 ml measuring flasks, 10 ml bulb pipettes, 0.1 - 10 ml graduated pipettes

Only use chemicals of "reagent purity". Follow the instructions of the apparatus manufacturer regarding the use of other chemicals which may require pretreatment.

Phthalate stock solution 1000 mg/l:
Dissolve 2.125 g potassium hydrogen phthalate in approx. 700 ml water in a measuring flask with a nominal volume of 1000 ml; the solution is then topped up with water to the 1000 ml mark. It can be kept for approx. 4 weeks at +4 °C.

Phthalate standard solution 100 mg/l:
Pipette 100 ml of the phthalate stock solution into a measuring flask with a nominal volume of 1000 ml and top up with water to the 1000 ml mark. This solution can be kept for approx. one week at +4 °C.

Control solution for determining the content of inorganically bound carbon:
1000 mg/l

Dissolve 4.404 g sodium carbonate (dried for 1 hour at 255 °C) in approx. 500 ml water in a measuring flask with a nominal volume of 1000 ml. Then add 3.497 g sodium hydrogen carbonate (dried using silica gel) and top up the flask with water to the 1000 ml mark. The solution can be kept at +4 °C for approx. 4 weeks.

Gases: Air, nitrogen, oxygen (free from $CO_2$ and organic impurities).

**Preliminary Treatment**

Put the samples into clean glass jars and, if they are not to be examined immediately, store in the refrigerator at +4 °C.

All undissolved matter must be removed on sampling when determining the amount of dissolved organic carbon (DOC). Filtration through a membrane filter has not proved successful in practice since carbon particles may be released by the filter material.

Glass-fibre filters which have first been washed with redistilled water are suitable for filtration. Alternatively, solid particles can be separated off by using a centrifuge for a sufficient period of time.

Method

Calibration

Calibration is necessary when comparative methods are used. Calibration curves can be drawn up by preparing calibration solutions from the phthalate stock solution or phthalate standard solution with suitable TOC/DOC concentrations so that the expected spread of measurements does not exceed the range of the apparatus or method used. If, for example, the TOC/DOC concentration range is from 10 to 100 mg/l, the procedure to be followed is:

Pipette 1.0 ml, 2.5 ml, 5.0 ml, 7.5 ml and 10.0 ml of the phthalate solution into five separate 100 ml measuring flasks and top them up to the 100 ml mark with water. Fill a sixth flask with plain water as a control. Analyze all the solutions, including the control, at least three times following the apparatus manufacturer's instructions.

A system of co-ordinates is used in which the TOC/DOC concentrations of the individual calibration samples are entered on the abscissa. These concentrations are calculated using the following equation:

$$K = \frac{V_K \cdot s}{V_o}$$

where:

$K$ = TOC/DOC mass concentration of the particular calibration sample in mg/l
$V_k$ = Volume of phthalate stock solution (or phthalate standard solution) used, in ml
$V_o$ = Maximum volume of the calibration sample, in this case $V_o$ = 100 ml
$s$ = TOC/DOC mass concentration of the phthalate stock solution (or phthalate standard solution) in mg/l.

The values to be entered on the ordinate depend on the apparatus used and must be taken from the manufacturer's instructions.

The compensating straight line for the series of measurement values obtained is then established. The reciprocal of the gradient of the straight line gives the factor f expressed in the unit required for the chosen method.

Measurement

Observe the manufacturer's instructions for the apparatus used. If necessary, dilute the water sample so that the resulting concentration of organically bound carbon falls within the range of measurement of the apparatus.

Before carrying out the TOC/DOC test, make sure that the prescribed performance checks are carried out with the apparatus at least as frequently as recommended by the manufacturer. The complete system should be regularly checked for leaks.

The sort of measurement values obtained will depend on the TOC/DOC measuring apparatus used, and these values must then be used to calculate the TOC/DOC level of the samples analyzed. The values may be obtained by intermittent measurement, e.g. peak heights (only where the shape of the peak is independent of the matter being measured) or peak areas. With continuous TOC/DOC measurement the $CO_2$ concentration in the carrier gas stream is recorded (e.g. as a line on a graphic recorder) for a particular water. The distance between this line and the datum line is proportional to the TOC/DOC concentration.

### 4.1.2 Oxidizable Organic Substances

#### General remarks

In order to evaluate water, it is important to know the amount of oxidizable organic substances it contains. This is determined either via the "total organic carbon (TOC)" or via the "dissolved organic carbon (DOC)" or, indirectly, via the quantity of oxygen which is needed to oxidize these chemically oxidizable organic substances as completely as possible. The total amount of needed to do this is termed the "chemical oxygen demand (COD)".

The latter is determined by adding to a measured volume of water sample a known quantity of oxygen in the form of an oxidizing agent, which carries out oxidation by chemical means, and determining the amount of oxidizing agent consumed, the result being expressed in mg/l $O_2$. Account has to be taken of the fact that oxidizable inorganic substances will also be oxidized at the same time. The amount of oxygen consumed as a result of this must be taken into consideration in the evaluation.

Potassium permanganate and potassium dichromate are the oxidizing agents most frequently used in determining the chemical oxygen demand of water. The chemical oxygen demand is defined as the mass of $KMnO_4$ or $K_2Cr_2O_7$, recalculated in terms of **oxygen**, which is needed to oxidize under defined conditions the organic substances contained in the water.

Summarizing, the following definition can be given:

#### Oxidizability

The amount of potassium permanganate consumed in the chemical oxidation of organic substances contained in water, this amount being quoted in "mg/l $KMnO_4$" as well as in "mg/l" as the corresponding oxygen equivalent.

#### Chemical oxygen demand (COD)

The amount of potassium dichromate consumed in the chemical oxidation of organic substances contained in water, the amount to be quoted in "mg/l" as the oxygen equivalent.

Oxidizability and chemical oxygen demand, like biological oxygen demand, do not provide an exact figure for the total amount of organic substances in water. However, they are used as characteristic values, whose usefulness has been proved in practice, for the evaluation of the total organic substances contained in a sample of water.

The oxidizability or the chemical oxygen demand is determined on untreated water, water which has been allowed to settle out or water which has been filtered, depending on the purpose of the investigation. The results should state under which of the above conditions the investigation was carried out.

### 4.1.2.1 Oxidizability

#### General remarks

Potassium permanganate has an oxidizing effect on many organic and a number of inorganic substances under acidic, neutral and alkaline conditions.

Analysis is always carried out with an excess of potassium permanganate. Under acidic conditions, the permanganate ion is reduced to the manganese (II) ion:

$$MnO_4^- + 5\ e^- + 8\ H^+ \longrightarrow Mn^{2+} + 4\ H_2O$$

The excess quantity of potassium permanganate is determined by titration with oxalic acid:

$$2\ MnO_4^- + 5\ C_2O_4^{2-} + 16\ H^+ \longrightarrow 2\ Mn^{2+} + 10\ CO_2 + 8\ H_2O$$

Under alkaline conditions, the permanganate ion is only reduced to quadrivalent manganese, which precipitates as brown manganese (IV) oxide.

$$MnO_4^- + 3\ e^- + 2\ H_2O \longrightarrow MnO_2 + 4\ OH^-$$

Since, after acidification, both the manganese (IV) oxide and the excess permanganate ions are reduced by the oxalic acid to manganese (II) ions:

$$MnO_2 + C_2O_4^{2-} + 4\ H^+ \longrightarrow Mn^{2+} + 2\ CO_2 + 2\ H_2O,$$

the nature of the medium (acidic or alkaline) is not of significance in determining oxygen consumption.

The method is suitable for determining the oxidizability of all water samples which have a potassium permanganate consumption of at least 1 mg/l (corresponding to an oxygen equivalent of 0.25 mg/l $O_2$).

Since the purpose is to record only the oxidizable organic substances, the influence of the oxidizable inorganic substances has to be excluded as far as possible or the content of the latter must be determined separately.

Chloride ions affect the analysis under acidic conditions if their concentration exceeds 300 mg/l. At higher $Cl^-$ concentrations, the investigation must be carried out under alkaline conditions.

Hydrogen sulphide, sulphide and nitrite ions can be removed when the analysis is carried out under acidic conditions by boiling out the sample with 5 ml of sulphuric acid (1.27 g/ml). When working in an alkaline medium, these substances as well as any ferrous ions must be determined separately and taken into account in calculating the results.

If the sample contains volatile organic compounds, it should be cold when the potassium permanganate solution is added to it. This fact should be noted in the results.

#### Equipment

Erlenmeyer flask 300 ml, with condenser

Potassium permanganate solution, 0.02 mol/l

Potassium permanganate solution, 0.002 mol/l:

This is prepared by diluting the 0.02 m potassium permanganate solution. The titre of the 0.002 m potassium permanganate solution only remains constant for a short time. It must be determined again every day before use.

Oxalic acid solution 0.05 mol/l

Oxalic acid solution 0.005 mol/l:

100 ml 0.05 m oxalic acid solution is made up to 1 litre with freshly boiled water and 50 ml sulphuric acid (1.84 g/ml). This solution can be kept for 2 - 3 weeks.

Sulphuric acid (1.27 g/ml):

Gradually add 1 part by volume of sulphuric acid (1.84 g/ml) to 3 parts by volume of distilled water, stirring all the time. Add to the former while it is warm until there is a permanent slightly pink coloration.

Sulphuric acid (1 + 2):

Gradually add 1 part by volume of sulphuric acid (1.84 g/ml) to 2 parts by volume of distilled water, stirring all the time. To this mixture, which should be at a temperature of approx. 40 °C, add 0.002 m potassium permanganate solution until there is a permanent slightly pink coloration.

Sodium hydroxide solution (1.36 g/ml):

Dissolve 330 g reagent-purity NaOH in 670 ml distilled water under cool conditions.

Diluting water:

Add 0.002 m potassium permanganate solution to boiling distilled water which has been made acidic with sulphuric acid, until there is a permanent slightly pink coloration.

Method

**Preparation of the samples**

If there are coarsely dispersed materials in the sample and if these are also to be determined, the sample must be homogenized for 3 minutes, for example, in a mixer.

If the oxidizability of the substances which are capable of settling out is not to be taken into account, the sample is to be allowed to settle out in an Imhoff funnel for two hours, and the water is then decanted off and analyzed.

If only the oxidizability of the dissolved substances is to be determined, a filtered sample of the water is used.

**Determining oxidizability under acidic conditions**

5 ml sulphuric acid (1.27 g/ml) is added to 100 ml of the sample in an Erlenmeyer flask and the mixture is heated rapidly to boiling. 15.0 ml of 0.002 m potassium permanganate solution is added immediately to the boiling solution. After the condenser has been placed on top, the solution is kept boiling for exactly 10 minutes from the time at which boiling commenced.

If, during the boiling process, the colour of the solution becomes brownish or if the coloration disappears altogether, the analysis must be repeated with a smaller amount of the sample made up with dilution water to 100 ml. This diluting of the water sample is to be taken into account in the calculation.

After boiling for 10 minutes, 15.0 ml of 0.005 m oxalic acid solution is added. If the solution does not become colourless immediately, it should be heated up again for a short time. The hot solution is then back-titrated with 0.002 m potassium permanganate solution until a just visible pink coloration appears and remains for at least 30 seconds (value a). Between 5 and 12 ml of the potassium permanganate solution should be consumed for this.

**Determining oxidizability under alkaline conditions**

0.5 ml sodium hydroxide (1.36 g/ml) is added to 100 ml of the water sample and the mixture is heated rapidly to boiling. 15.0 ml of 0.002 m potassium permanganate solution is added to the boiling solution. After attaching the condenser (or a reflux condenser), the solution is kept boiling for ten minutes from the time it starts to boil again.

After boiling for 10 minutes, 5.0 ml sulphuric acid (1.27 g/ml) and 15.0 ml 0.005 m oxalic acid solution are added. If the solution does not become colourless immediately, it is heated up again for a short time. The hot solution is then back-titrated with 0.002 m potassium permanganate solution until a just visible pink coloration appears and remains for at least 30 seconds (value a). Between 5 and 12 ml of the potassium permanganate solution should be consumed for this.

Calculation

The results are quoted as oxidizability (potassium permanganate consumption) in mg/l $KMnO_4$ and additionally in mg/l (oxygen equivalent) in a similar manner to the chemical oxygen demand. On the basis of the stoichiometric relationships whereby:

1 ml 0.002 m $KMnO_4$ corresponds to 0.316 mg $KMnO_4$ and
1 ml 0.002 m $KMnO_4$ corresponds to 0.08 mg oxygen,

$$\text{mg/l } KMnO_4 = \frac{(a) \cdot F \cdot 316}{V}$$

$$\text{mg/l (oxygen equivalent)} = \frac{(a) \cdot F \cdot 80}{V}$$

where:

a = Consumption of 0.002 m $KMnO_4$ solution in ml (sample)
F = Factor for the 0.002 m $KMnO_4$ solution
V = Volume of the water.

When oxidizable inorganic substances are present, the following corrections have to be taken into account:

The following have to be deducted from the result:

per 1 mg/l $Fe^{2+}$ : 0.57 mg/l $KMnO_4$ or 0.14 mg/l (oxygen equivalent)

per 1 mg/l $NO_2^-$ : 1.37 mg/l $KMnO_4$ or 0.35 mg/l (oxygen equivalent) per 1 mg/l $S^{2-}$ : 1.97 mg/l $KMnO_4$ or 0.50 mg/l (oxygen equivalent)

Rounding off of the results

$KMnO_4$ consumption (oxygen equivalent):

| | | |
|---|---|---|
| | up to 10 mg/l, | round to 0.1 mg/l |
| 10 | up to 100 mg/l, | round to 1 mg/l |
| 100 | up to 1000 mg/l, | round to 10 mg/l |
| | over 1000 mg/l, | round to 100 mg/l |

Example

Oxidizability of the filtered sample in alkaline or acid solution: 19 mg/l $KMnO_4$,corresponding to 5.0 mg/l $O_2$.

### 4.1.2.2 Chemical oxygen demand (COD)

General remarks

Method with potassium dichromate

Under hot, acidic conditions, potassium dichromate oxidizes practically all organic substances as well as a number of inorganic components and ions. The level of oxidation depends on the type and concentration of the organic materials, the concentration of the potassium dichromate and sulphuric acid, and the reaction temperature and time. For this reason it is important that the conditions of the investigation are always strictly maintained.

The analysis is carried out with an excess of potassium dichromate, whereby a part of the dichromate is reduced to the chromium (III) ion:

$$Cr_2O_7^{2-} + 6e^- + 14\ H^+ \rightarrow 2\ Cr^{3+} + 7\ H_2O$$

The excess of potassium dichromate ions is back-titrated against a ferrous solution using Ferroin as redox indicator:

$$Cr_2O_7^{2-} + 6\ Fe^{2+} + 14\ H^+ \rightarrow 2\ Cr^{3+} + 6\ Fe^{3+} + 7\ H_2O$$

This method is suitable for determining chemical oxygen demand in all types of water and waste water which have a potassium dichromate consumption of over 40 mg/l when an 0.0415 m potassium dichromate solution is used, or a potassium dichromate consumption of 10 - 40 mg/l when an 0.0083 m potassium dichromate solution is used. In the latter case, the results are generally 10 % lower than when using the 0.0415 m potassium dichromate solution. This shows that the results obtained when using different concentrations of the oxidizing agent cannot simply be compared with one another.

Certain organic substances which may be present in the water, e.g. benzene or pyridine, are not completely oxidized by the method quoted above. For this reason the COD value only reaches the 100 % limit of the TOD value (total oxygen demand) in specially favourable cases and may be considerably below this limit in waters which have an unfavourable make-up.

The method used to try and exclude the disturbances arising from the oxidation of the chloride ions is to add mercuric sulphate. The mercuric chloride which results is highly poisonous and must be made harmless in the proper manner (see note) in order to avoid subsequent contamination of the environment. Sulphur compounds which contain sulphur at an oxidation stage lower than +6 (sulphate sulphur) are oxidized almost completely to the sulphate level. Their influence can be calculated where it is possible to determine these compounds by means of a selective method.

Reduced nitrogen (as in ammonium ions, amino compounds, amides, nitriles) normally retains its oxidation stage of -3; exceptions are possible.

Oxidized nitrogen (as in nitrites, nitroso and nitro compounds) is almost completely oxidized to nitrate.

Ferrous ions are oxidized completely to ferric ions. Their influence can be determined by means of a separate iron (II) analysis and taken into account in the calculations.

Nevertheless, in spite of what has been said above, the fact remains that disturbing influences must be taken into account in determining the COD value. However, by adding silver sulphate as an oxidation catalyst, an average degree of oxidation of 95 - 98 % of the TOD value can be achieved with the majority of water samples which are not extremely heavily contaminated.

The COD method given here represents an attempt to offer a rapid method for daily laboratory practice which can be carried out quickly with simple laboratory equipment and which leads to useful rates of breakdown and good reproducibility.

It can be used between 15 and 200 mg/l COD and at $Cl^-$ ion contents of not more than 1000 mg/l.

If the upper limits are exceeded, the sample has to be diluted. The degree of dilution must be taken into account in the calculation.

**Equipment**

Erlenmeyer flask, 300 ml with standard ground-in stopper (29 mm diam./32 mm long)

Reflux condenser with standard ground-in stopper (29 mm diam./32 mm long)

Heater

Metering devices, resistant to the reagents employed

Boiling chips

Burette 50 ml

Stopwatch

Preparation water: dist. or redist. water with only traces of organic compounds.

Mercuric sulphate solution:
700 ml preparation water is added to 200 g mercuric sulphate and stirred for 15 minutes. Then 200 ml of diluted sulphuric acid, consisting of 100 ml

preparation water and 100 ml sulphuric acid (95 to 97 %), is added and the mixture is made up to 1 litre with preparation water. The mixture must remain colourless.

Potassium dichromate solution (0.0166 mol/l, corresponding to 0.1 n):
Approximately the required quantity of potassium dichromate is dried for 2 h at 105 °C and then cooled in a desiccator. 4.9031 g of the substance is dissolved in preparation water in a 1000 ml volumetric flask and made up to 1000 ml with preparation water. This solution is stable.

Silver sulphate solution:
80 g silver sulphate is dissolved in 1 litre of 95 to 97 % sulphuric acid.

Ferroin solution:
1.485 g 1,10-phenanthroline and 0.695 g ferrous sulphate is dissolved in preparation water and made up to 100 ml. The solution is stable.

Ferrous solution:
Approx. 1 litre preparation water is made largely free of oxygen by boiling for 5 minutes followed by cooling. 20 ml of 95 to 97 % sulphuric acid is added to 500 ml of the above and the mixture is cooled down. 10 g ammonium ferrosulphate is dissolved in this and the solution is transferred to a 1000 ml volumetric flask. The solution is then made up to 1000 ml with preparation water which has been boiled off.

Adjustment of the ferrous solution:
10 ± 0.02 ml 0.0166 m potassium dichromate solution is poured into a 300 ml Erlenmeyer flask. Then 100 ml preparation water and 40 ml of 95 to 97 % sulphuric acid are added and the mixture is cooled to room temperature. After adding 2 to 3 drops of ferroin solution, the solution is titrated against the ferrous solution. The end-point is a change from blue-green to reddish brown and takes place within approx. 2 drops.

**Note:**

The end-point can also be determined using a suitable electrochemical method.

Calculation of the factor:

$$f = \frac{A \cdot N_{Cr}}{B \cdot N_{Fe}}$$

where:

$f$ = factor of the ferrous solution (must be determined at least once a day)
$A$ = volume of the potassium dichromate solution used (10 ml)
$N_{Cr}$ = molar concentration of potassium dichromate solution, e.g. 0.0166 mol/l
$B$ = volume of the ferrous solution consumed in ml
$N_{Fe}$ = molar concentration of the ferrous solution (1 ml 0.0166 m potassium dichromate solution is equivalent to 5.5847 mg Fe)

**Procedure**

Measure out 5 ml mercury sulphate solution into the reaction flask.

Add a measured 20 ml of the sample to be tested and mix carefully.

Measure out and add 10 ml of 0.0166 m potassium dichromate solution.

Add 2 - 3 boiling chips.

Connect the reaction flask to condenser (do not grease the ground joints).

Add 40 ml of 95 to 97 % sulphuric acid through the condenser and mix together (observe safety rules).

Heat to boiling and start the stopwatch when boiling commences.

After boiling for 5 minutes, add 5 ml silver sulphate solution to the reaction flask through the condenser.

Stop the boiling after a further 10 minutes.

Allow to cool in air for 5 minutes.

Carefully add 50 ml preparation water through the condenser, shaking all the time (to avoid delayed boiling).

Remove the flask (rinsing off the ground joint with preparation water).

Allow to cool down to room temperature in a water bath.

Add 2 - 3 drops ferroin solution.

Titrate with the ferrous solution to the end point (blue-green to reddish brown).

The end point can also be determined by a suitable electro-chemical method.

**Blank test**

Carry out using 20 ml preparation water instead of the water sample to be investigated.

**Evaluation**

$$COD = \frac{(C - D)\ m_{Fe} \cdot f_e \cdot \frac{O}{2} \cdot 1000}{E} \quad (mg/l)$$

where:

C = volume in ml of the ferrous solution consumed for titration of the blank test value.
The arithmetic mean of at least two blank test results should be entered here.
D = volume in ml of the ferrous solution consumed for titration of the water sample.
E = volume of the test sample used (20 ml).
$\frac{O}{2}$ = equivalent weight of oxygen (= 8).
$m_{Fe}$ = 1ml 0,0166 m $K_2Cr_2O_7$ solution = 5,5847 mg Fe.
fe = Titer of the ferrous solution (daily control)

This gives the following simplified evaluation formula:

$$COD = (C - D) \cdot f_e \cdot 10\ (mg/l)$$

Calculation of the blank test value

$$b = 100 \cdot \frac{C \cdot m_{Fe} \cdot f_e \cdot 100}{F \cdot m_{Cr}} \quad (\%)$$

b = blank test value as a percentage of the quantity of oxidizing agent used.
F = volume of potassium dichromate solution used (10 ml).
mCr = mol/litre $K_2Cr_2O_7$ solution

The blank test value b as a percentage of the quantity of oxidizing agent used permits a simple check to be made on the cleanliness of the devices used, the reagents and the preparation water. It must not exceed 10 %.

**Note:**

**Elimination of Hg and Ag from the residual solutions of COD analysis.**
(Fresenius-Schneider)

This method, which renders the Hg and Ag residues from the COD residual solutions to a large extent harmless, is based on the fact that alkaline thiosulphate solutions react with acidic solutions containing mercury, silver sulphide and chromic hydroxide. The above mentioned compounds are separated off by filtration.

Method

After determining the chemical oxygen demand (COD), add 30 g sodium thiosulphate, followed by 300 ml of 30 % sodium hydroxide for every 500 ml of the collected reaction solutions containing mercury, silver and chromium ions. Mercury and silver are removed from the solution as sulphide while chromium is removed as chromic hydroxide. $Hg^{2+}$ ions are no longer detectable in the filtrate of the remaining colourless liquid (by cold vapour atomic-absorption spectro-photometry; detection limit: 0.001 mg/l). The residual concentration of silver is less than 0.1 mg $Ag^+$/l, while that of chromium (VI) is below 0.5 mg/l.

### 4.1.3 Biochemical oxygen demand (BOD)

General

The biochemical oxygen demand ($BOD_n$) is the mass of dissolved molecular oxygen which is needed by microorganisms for the oxidation (and also conversion) of organic substances in a sample (20 °C) of water under defined conditions and within a defined period of time (index n in days or hours).

In order to determine the BOD, the bacterial degradation process is allowed to proceed under controlled conditions in test flasks and the quantity of oxygen consumed is determined. The amount of dissolved molecular oxygen which is consumed for plain chemical oxygen processes - i.e. ones in which microorganisms are not involved and above all for inorganic substances in water - must be taken into account in determining BOD. It is assumed in general that these plain chemical oxidation processes will have proceeded to completion within 2 hours of taking the water sample.

The biochemical oxygen demand, which is quoted in mg/l water and is often designated as the "oxygen consumption" within the period of investigation, serves as a unit for evaluating the influence of organic contamination on the level of oxygen in a particular water. It has to be remembered that the biochemical breaking down or converting of organic substances proceeds via two stages which are not sharply separated from one another. In the first stage organic compounds are largely broken down to inorganic compounds. In the second stage, which is also designated the nitrification stage, the main process is the oxidizing of the ammonium created from the nitrous organic compounds to form nitrite and nitrate. However, this second stage does not usually start until after a particular period of time so that in determining $BOD_5$ it is generally insignificant. In special cases it can be determined separately.

However, only the oxygen consumption of the first stage of breaking down is desired for the BOD, and here nitrification is regarded as a disturbing effect. It will be clear from the above that the BOD in a defined, narrow sense is not a measure of the totality of the organic substances. It is not directly comparable with the chemical oxygen demand (COD) obtained using potassium dichromate or with the oxidizability obtained using potassium permanganate. After all, there are experimental values relating to waters of medium levels of contamination in which the oxidizability with potassium permanganate test gives 25 %, the $BOD_5$ test 70 % and the COD under optimum conditions up to 98 % of the oxygen needed for the complete oxidation of the organic substances (W. Leithe).

The magnitude of the BOD is influenced by many factors: nature and concentration of the organic substances in the water to be broken down, nature, number and adaptation of the microorganisms, nature and quantity of nutrients for the microorganisms, incubation time (period during which oxygen is consumed), temperature, effect of light and influencing of the biological and/or biochemical processes by substances having a toxic effect.

An incubation time of 5 days ($BOD_5$) has proved to be useful. Since the time bacterial processes require to start up varies, a shortening of this period could lead to incorrect values or ones having a low level of reproducibility. On the other hand, the reproducibility is also impaired if the period for the BOD analysis is too long, as the bacterial flora which bring about the breaking down process and which constitute the most important "reagent" cannot be defined in terms of quality and output. According to W. Leithe, the reproducibility (amount of variation arising with a number of analyses on the same sample), lies between 20 and 50 % of the value found, in particular when different workers or different laboratories are involved in the work.

$BOD_5$ is the quantity of dissolved molecular oxygen which is used by the microorganisms during an incubation period of 5 days in order to break down by oxidation the organic substances contained in the water at 20 °C. The $BOD_2$ may also be determined for surface waters, while incubation periods of 24 hours, 10 or 20 days can be selected for particular purposes.

Determining the $BOD_5$ is carried out on untreated water, water which has been allowed to settle out or water which has been filtered, depending on the purpose of the investigation. The results should make clear under which of these conditions the $BOD_5$ was determined.

The BOD5 is determined using oxygen flasks. When making up the test sample, care must be taken to ensure that there are adequate quantities of oxygen, nutrients and suitable microorganisms for optimal breaking down of the organic substances.

The BOD can be determined in accordance with one of two different principles:

1) The amount of a defined quantity of dissolved oxygen consumed is measured in a closed, calibrated vessel by determining the difference between the oxygen content of the sample at the beginning and at the end of incubation. The $O_2$ is determined volumetrically or electrometrically.

2) $O_2$ is supplied continuously from the gas space to the water sample in a closed set-up. The consumption of $O_2$ is followed during the incubation period manometrically (e.g. after Warburg) as the oxygen content in the gas space changes.

   The method employed in this standard method is based on principle 1.

Determination of $BOD_5$ after dilution

General

Substances which can be oxidized by oxygen without any help from microorganisms influence the results. This disturbing effect can be eliminated by stipulating that the start of the incubation period for the $BOD_5$ should be 2 hours after dilution of the sample.

Substances which act in a biologically inhibiting or toxic manner must be rendered inactive. If the elimination of such substances also brings about a significant alteration in the composition of the water sample, the $BOD_5$ test will give incorrect results.

Free chlorine is removed by addition of an equivalent quantity of sodium thiosulphate solution (1 ml 0.01 m solution approx. 0.4 m $Cl_2$).

Water which reacts in a strongly acidic or alkaline manner is set to a pH of between 7 and 8.

The sodium hydroxide contains sodium azide in order to eliminate the disturbing effect caused by nitrite, some of which does not form until during the incubation period. In the case of water samples in which the breaking down of organic substances to form inorganic compounds (first decomposition stage) proceeds very rapidly, the second decomposition stage (designated as "nitrification") can set in during the first 5 days of the consumption process. Such nitrification can be suppressed to a large extent by the addition of 0.5 mg n-allylthiourea per litre of dilution water, as this inhibits oxidation of the $NH_4^+$.

Additional bacteria have to be injected into samples of water which contain too few bacteria. This is carried out by adding domestic waste water which has been allowed to settle, biologically treated waste water or heavily contaminated river water. In each case the intrinsic BOD content of these additions must be taken into account.

Diluting the water sample can lead to a shortage of the necessary nutrient salts. This effect can be avoided by adding solutions of nutrient salts.

Equipment

Oxygen flasks (glass); 110 - 130 ml or 250 - 300 ml with flasks and stoppers having the same number.

Measuring pipettes, 2 ml, calibrated in 0.1 ml.

Dilution water:
The dilution water is brought to 20 °C and must be saturated with oxygen at this temperature. Its $BOD_5$ may not be above 1 mg/l ("exhausted dilution water" containing no more biodegradable substances). It is either taken from the water serving as the receiving body of water or is made from distilled water to which nutrient salts have been added. Preparation: see below.

Nutrient salt solutions:

Solution 1:
Dissolve 8.5 g $KH_2PO_4$, 21.75 g $K_2HPO_4$, 33.4 g $Na_2HPO_4 \cdot 2\,H_2O$ and 1.7 g $NH_4Cl$ in distilled water and make up to 1 litre. pH = 7.2.

Solution 2:
22.5 g $MgSO_4 \cdot 7\,H_2O$ in distilled water and make up to 1 litre.

Solution 3:
Dissolve 27.5 g $CaCl_2$ in distilled water and make up to 1 litre.

Solution 4:
Dissolve 0.25 g $FeCl_3$ in distilled water and make up to 1 litre.

Preparation of the dilution water:
The water used for this must be free of silver and copper ions (<0.005 mg/l) and active chlorine. Add 1 ml of each of solutions 1, 2, 3 and 4 to 1 litre distilled water. The dilution water obtained in this way must be repeatedly aerated until it is exhausted and saturated with oxygen. It is then stored in the dark.

Manganese (II) chloride solution:
Dissolve 800 g $MnCl_2 \cdot 4\,H_2O$ in 1 litre distilled water.

Sodium hydroxide, containing $KI/NaN_3$:
360 g NaOH (free of nitrite), 200 g KI (free of iodate) and 5 g $NaN_3$ are dissolved carefully (protective glasses to be worn) in distilled water and made up to 1 litre. The solution is filtered through glass wool.

Phosphoric acid (1.70 g/ml)

Sodium thiosulphate solution, 0.01 m

Zinc iodide starch solution:
Triturate 4 g starch with a little distilled water and add the resulting paste to a boiling solution of 20 g zinc (II) chloride in 100 ml dist. water. Boil the mixture until it becomes clear, replacing the water that evaporates. After dilution, add 2 g zinc iodide make up to 1 litre with distilled water. Then filter and store in a brown bottle.

## Determining $BOD_5$

### General remarks

Where the water to be tested contains only a low level of germs, 0.3 ml of waste water which has been allowed to settle or 5 - 10 ml river water are

**Table.** Determining $BOD_5$ by dilution

| Oxidizability ($O_2$) | Expected $BOD_5$ ($O_2$) | ml water sample, which must be diluted to 1000 ml |
|---|---|---|
| mg/l | mg/l | ml |
| up to 4 | up to 10 | 250 and 150 |
| 4 to 10 | 10 to 30 | 100 and 75 |
| 10 to 15 | 20 to 50 | 50 and 40 |
| 15 to 30 | 40 to 100 | 30 and 20 |
| 30 to 60 | 80 to 200 | 15 and 10 |
| 60 to 90 | 160 to 300 | 10 |

added to each litre of the diluted sample. The inoculating liquid is added after approximately half the envisaged quantity of dilution water has been added to the sample.

Depending on the BOD expected, **two or more different dilutions** must be prepared from the water sample (which has to be pretreated if necessary) and the dilution water and each of these must be inoculated. When testing domestic and municipal waste water for which the potassium permanganate consumption ("oxidizability") is known, work can be carried out in accordance with the table.

If the $BOD_5$ is expected to be over 300 mg/l, dilute the sample 1 + 10 with dilution water and use the table appropriately

**Method**

The diluting is carried out in 500 ml or 1 litre volumetric flasks in the following way. The measuring flask is partially filled with dilution water. After adding the measured water sample, the flask is filled up to the mark and mixed. Three oxygen bottles are filled bubble-free from each dilution. Manganese (II) chloride solution and the sodium hydroxide containing $KI/NaN_3$ are added immediately to one of the three samples. The quantities of these reagent solutions added are in each case 0.5 ml when the small oxygen bottles (110 - 130 ml) are being used and 1.0 ml when the larger bottles (250 - 300 ml) are being used. All three bottles are then immediately closed, care being taken to ensure that there are no air bubbles.

The sample to which the reagents have been added - hereafter designated sample (1) - is shaken well. The second and third samples are stored in the dark at 20 °C during the incubation period (consumption of oxygen period). After the expiration of this time (5 days), the two above-mentioned reagent solutions are added to the water in bottles 2 and 3 in the same manner as described above. The contents of each of the bottles are then treated as follows (oxygen determination after Winkler, see also Chapter 1).

The precipitate is dissolved by adding 2 ml phosphoric acid. The iodine released in this way is titrated against 0.01 m sodium thiosulphate. Towards the end of titration, 1 ml zinc iodide starch solution is added to the pale yellow coloured solution, and the resulting blue solution is titrated until it is colourless. The BOD of the dilution water is determined at the same time and taken into account in evaluation.

Only those samples whose oxygen consumption, excluding the consumption of the dilution water, is greater than 1 mg/l and whose residual oxygen content is at least 2 mg/l, are evaluated.

1 ml 0.01 m sodium thio sulphate solution corresponds to 0.08 mg $O_2$. Thus the oxygen content in the individual samples is calculated in accordance with the following equation:

$$G = \frac{a \cdot f \cdot 80}{V - V_R} \quad (\text{mg } O_2/\text{l})$$

where:

$a$ = consumption of 0.01 m sodium thio sulphate solution, in ml
$f$ = factor of the 0.01 m sodium thio sulphate solution
$V$ = volume of the oxygen bottle in ml
$V_R$ = quantities of reagent solutions added, in ml
$G$ = oxygen content, in mg $O_2$/l

Determining the $O_2$ content before and after the incubation period can also be carried out electrometrically. Evaluation is carried out in a similar manner.

Oxygen consumption in the dilutions is calculated from

$$G_1 - \frac{G_2 + G_3}{2} = Z \quad (\text{mg/l})$$

where:

$G_1$ = $O_2$ content of sample 1 in mg/l
$G_2$ = $O_2$ content of sample 2 in mg/l
$G_3$ = $O_2$ content of sample 3 in mg/l
$Z$ = $O_2$ consumption in mg $O_2$/l

$$BOD_5 = \frac{V_a}{V_b} (Z_p - Z_v) + Z_v \ (\text{mg/l})$$

where:

$V_a$ = total volume after dilution, in ml
$V_b$ = volume of the undiluted sample, in ml
$Z_p$ = oxygen consumption of the dilution in 5 days, given in mg $O_2$/l
$Z_v$ = oxygen consumption of the dilution water (or of the inoculated dilution water) in 5 days, given in mg $O_2$/l
$BOD_5$ = biochemical oxygen consumption of the water in 5 days, given in mg $O_2$/l

## Rapid Methods

### General

In addition to the dilution method, manometric methods (rapid methods which can also be carried out by semi-skilled personnel) are also employed for determining $BOD_5$. In general the waste water sample to be analyzed is put into a vessel having an airspace which can be closed off from the athmosphere. The water is kept in motion by stirring, vibrating or shaking so that the entire sample is continuously supplied with oxygen from the air above it. In order to be able to measure the amount of oxygen consumed by determining the drop in pressure in the vessel, the $CO_2$ produced, insofar as this is not dissolved or bonded in the sample, must be removed from the enclosed airspace. This is done by absorption with alkalis, primarily potassium hydroxide. Since there is considerably more oxygen available in the enclosed airspace than in the dilution water, the range of expected $BOD_5$ for which this method can be employed is much greater than with the dilution method (irrespective of whether similarly diluted or undiluted starting formulations are used with the latter).

The limits of this method can be pushed upwards by using devices with which measurement is carried out in accordance with the same principle but with which the drop in pressure only serves to control the feeding-in of pure oxygen to the enclosed airspace. The quantity of oxygen is then measured and registered in the latter.

These devices have an advantage over the $BOD_5$ dilution method in that the breaking-down processes can be followed and recorded continuously so that there are numerous opportunities for recording and recognizing various influences such as temperature, toxic effect of substances and deficiency of mineral salts as well as for investigating the effects of dilution and the adaptation of biocenoses to substrates.

If the biochemical oxygen demand of the settleable solids is to be determined, it may in the case of primary sludge be calculated from the loss on ignition of the settleable solids. According to Imhoff, 30 g organic substances cause 20 g $BOD_5$ in primary sludge.

### 4.1.4 UV absorption

#### General

As a complement to the equally unspecific overall parameters for determining organic substances dissolved in water such as oxidizability with potassium permanganate (COD) or also for organic dissolved carbon (TOC or DOC), measurement of UV absorption has become established as a rapid method for determining the spectral absorption coefficients and thus as an indirect way of measuring the level of dissolved organic materials. For this purpose spectral absorption is used at 254 nm (Sontheimer, Wagner).

For this method care must be taken to ensure that the water sample is not cloudy since this can give erroneous results. Any cloudiness which may be present must therefore be eliminated by filtration, e.g. using a membrane filter with an average pore size of 0.45 µm, before determination is carried out. When comparative measurements are carried out, centrifuging can also be carried out before determination.

It should be borne in mind that nitrate ions also show absorption in the UV range.

The further below 260 nm analysis is carried out, the stronger this influence will be. For this reason it is recommended that determination be carried out at the mercury line of 254 nm. The recording of an absorption spectrum can also be used to obtain further information. This method can be supplemented by determining absorption in the visible wavelength range at the mercury line of 436 nm.

### Equipment

Equipment for filtration via membrane filter and under vacuum.

Spectrophotometer or filter photometer with which the 254 nm mercury line and also, if necessary, the 436 nm mercury line can be utilized for measurement.

Quartz cuvettes of 1 cm, 5 cm and 10 cm path length

### Method

The water sample, which has been filtered through a membrane filter, is put into the quartz cuvette of the spectrophotometer and its spectral absorption measured. The decadic absorption coefficient is the quotient of the decadic absorption measured and the path length of the cuvetted. Normally extinction is used as a measure of absorption (see Chapter 2).

### Measured value

$$\alpha(\lambda) = \frac{A(\lambda)}{d}$$

When evaluating the results, the calculation of the spectral absorption coefficient is rounded to 0.1 $m^{-1}$. The result should be accompanied by a statement as to whether the water sample was filtered and at what wavelength the measurement was carried out.

### For example

Spectral absorption coefficient 254 nm 1.5 $m^{-1}$.

### Note:

Where continuous checks are made on particular waters, statistical comparative values can be obtained which permit speedy recognition of freak values resulting from, for example, increased pollution with dissolved organic substances.

If sufficient numbers of measured values are available for other overall parameters for dissolved materials, an attempt can be made to establish a correlation. If this is successful, the control measurement can be restricted to determining the spectral absorption coefficient at 254 nm.

## 4.1.5 Determination of organically bound halogens as overall parameters (EOX / AOX)

Recent years have increasingly seen the emergence of methods of analysis in which group or overall parameters are determined. The level of extractable organic halogens (EOX) or adsorbable organic halogens (AOX) is such an overall parameter. What is understood by the content of extractable or adsorbable organic halogen compounds in a water sample is the volume-related mass of chlorine, bromine and iodine found under the relevant conditions and quoted as EOX or AOX.

To separate the organic halogen compounds from the matrix of the water or waste water, physico-chemical methods of separation are used, these being based either on the solubility of the substances in organic solvents or on their adsorbability on active charcoal. As a result of the great variety of substances that can be contained in water and waste water, no single one of the methods used can fully record all the organic halogen compounds. For this reason different methods of concentrating such compounds must inevitably lead to different results.

## A) Extractable organic halogen (EOX)

When recording the EOX, the water sample is extracted with n-pentane and diisopropylether. Despite the influence of factors such as the pH and salt content of the water sample being investigated, it is necessary that the test conditions for carrying out the extraction are maintained exactly. The extracts obtained are then combusted using, for example, a Wickboldt apparatus, in which the solvent can be sucked in from an open vessel. In view of the relatively high volatility of the solvents and the occurrence of volatile halogen compounds, it is recommended that the apparatus be modified in order to exclude the possibility of errors arising as a result of evaporation. The modification suggested by Fischer and Rump is shown in Fig. 119.

After combustion has been carried out in an oxygen-hydrogen flame, the inorganic mineralization products are determined as chloride ions. This can be carried out by a number of different methods, among which microcoulometry has been proved especially successful.

### Equipment

Volumetric flasks 50 ml, 100 ml, 1000 ml

Magnetic stirrer

Pipettes 25 ml, 50 ml

pH measuring device

Wickbold combustion apparatus (combustion in an oxygen-hydrogen flame DIN 51400 Part I) (Fig. 118)

Microcoulometer with potentiometric indication

Oxygen

Hydrogen

Acetone

Distilled water (EOX content less than 20 µg/l)

n-pentane

Diisopropylether (store over NaOH)

Sulphuric acid, 96 %

Sodium hydroxide solution: 25 %

Sodium sulphate (calcinated at 600 °C)

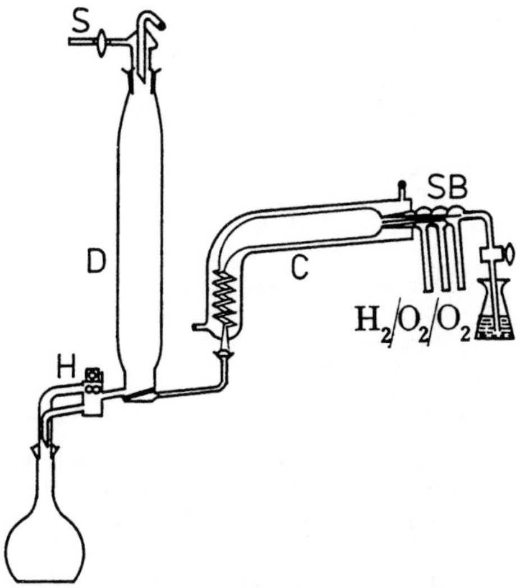


Fig. 118. Wickbold apparatus, type 5; SB) = Suction burner; C) = Combustion chamber with $H_2$ $O_2$ flame followed by condenser; D) = Absorption receiver for washing out gases; H) = Stopcock to round flask for condensate; S) = Feed stopcock for absorption liquid and rinsing liquid;

Pentachlorophenol

Halogen stock solution (100 mg/l Cl):

Weigh out 15 mg pentachlorophenol ($C_6Cl_5OH$) into a 100 ml volumetric flask. After adding approx. 50 ml n-pentane, make up the solution to the mark with diisopropylether.

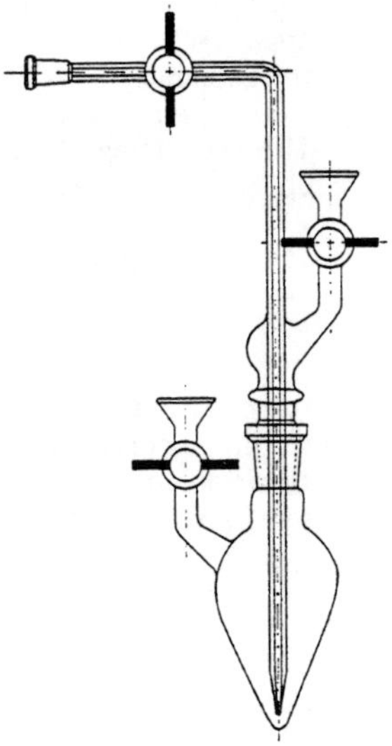

Fig. 119. Improved apparatus for EOX determination

Halogen standard solution (10 mg/l Cl):

Add 45 ml n-pentane to 10 ml of halogen stock solution in a 100-ml volumetric flask. Make up the solution to the mark with diisopropylether. The solution is stable for approx. 1 week.

Chloride stock solution (100 mg/l $Cl^-$):

Put 0.1648 g sodium chloride (dried at 180°C) into a 1 litre volumetric flask and make up to the mark with water.

Chloride standard solution (1 mg/l $Cl^-$):

Put 10 ml of chloride stock solution into a 1 litre volumetric flask and make up to the mark with water.

## Execution

The tests should be carried out as soon as possible (within 2 days) after the samples have been taken. If this is not possible, the samples should be kept at 4 °C. All the tests are carried out in glass vessels.

Put 20 g sodium sulphate and 950 ml of the water sample into a 1 litre volumetric flask. The pH is adjusted to between 6 and 8 with sulphuric acid or sodium hydroxide solution.

Add 25 ml n-pentane to the water sample prepared in the above manner. Then seal the flask and stir the mixture on a magnetic stirrer at approx. 1100 rpm for 10 minutes. After the phases have separated, the extract is sucked off and transferred to a 50 ml volumetric flask. A further 25 ml diisopropylether is now added to the water sample remaining in the 1 litre volumetric flask. The mixture is then stirred for a further 10 minutes and the extract transferred as above to the 50 ml volumetric flask. After making up to the mark, add between 1 and 4 g sodium sulphate depending on the moisture content of the solution. Then shake the mixture and allow the salt to settle out.

An aliquot part of the extract, whose size depends on the expected content of organically bound halogens, is combusted in the Wickboldt apparatus, care being taken to ensure no solid sodium sulphate gets into the combustion chamber. After this, 2 ml of acetone or one of the extraction agents are combusted, and this process is repeated at least once. The resulting condensate is put into a 50 ml volumetric flask and made up to the mark.

The coulometric measurement is carried out in this solution (which can be diluted if necessary).

Two portions, each of 10 ml, of the standard halogen solution are used to check that the complete apparatus and the measuring device are functioning correctly. The recovery rate of organically bound, extractable chlorine must be at least 90 % for each sample. Two blanks, consisting of distilled water, are tested in the same way.

In order to improve the determination limit, the sample volume and the volumes of extract to be combusted and aqueous solution to be titrated can be selected freely according to the expected organic halogen content. These volumes are then taken into account in the evaluation. The 20 : 1 ratio of water volume to particular extraction volume must not be altered. As a further means of improving the determination limit, the aqueous solution of the mineralization products can be restricted in a defined manner.

### Evaluation

The mineralization products recorded by the coulometer are shown as chloride. The content of "extractable organic halogen compounds (EOX)" is calculated in µg/l EOX and quoted as a chloride concentration.

$$EOX(\mu g/l) = \frac{f \cdot e_1 \cdot m_1 \cdot (n - n_o)}{u \cdot e_2 \cdot m_2}$$

where:

$f$ = equivalence factor (here: 35.5 µg/µg-mol)
$e_1$ = extraction volume (ml)
$m_1$ = total volume of the aqueous sample after mineralization (if necessary after boiling down by a defined amount) (ml)
$n$ = number of micromoles chloride in the analysis sample
$n_o$ = number of micromoles chloride in the blank
$u$ = volume of the initial sample (litre)
$e_2$ = extraction volume that is mineralized (ml)
$m_2$ = volume used for the coulometric titration (ml)

## B) Adsorbable organic halogens (AOX) (Sontheimer)

It is preferable to use highly active, pulverized active charcoal for the determination of adsorbable organic halogen compounds. As a rule at least 90 % of all the organic material in a water sample is adsorbed during the adsorption process. Since organic compounds in the water sample which do not contain halogens are also adsorbed and since their adsorption competes with the adsorption of the halogen-containing compounds, the level of dissolved organic carbon compounds (DOC) should not exceed 10 mg/l DOC if possible. The sample must be diluted if necessary. The chloride contained in the water sample is held in part on the active charcoal, but is then removed by elution with a nitrate solution. The charged active charcoal is then mineralized by being combusted in a stream of oxygen and the combustion products are then investigated by the appropriate method for the type of device being used.

### Equipment

Combustion apparatus (tube furnace)

Apparatus for the determination of chloride (e.g. microcoulometer)

Filtration device consisting of 50 ml suction flask, 25 mm diameter suction filter

Polycarbonate membrane filter 0.45 µm

Erlenmeyer flask 250 ml

Mechanical shaker

pH measuring device

Active charcoal:
Special active charcoal is used for determination. Not all active charcoals can be used, so that one has to rely on experience and the details quoted by the manufacturers. The blank value of the washed active charcoal must be less than 20 µg chloride per g active charcoal.

Nitric acid, 65 %

Hydrochloric acid, 0.1 mol/l

Oxygen

Nitrate stock solution:
17 g sodium nitrate is dissolved in water, 2 ml of nitric acid added and the mixture made up to 1000 ml with water.

Nitrate washing solution:
50 ml of the nitrate stock solution is made up to 1 litre with water.

Pentachlorophenol stock solution and pentachlorophenol standard solution:
Solutions as described under EOX.

**Execution**

Five ml nitrate stock solution is added to 100 ml of the water sample in a 250 ml Erlenmeyer flask. The pH of the solution should be between 1 and 3.5 if possible and nitric acid is added if it is too high. 50 mg pulverized active charcoal is now added. After having been shaken for at least one hour, the suspension is filtered off through an 0.45 µm membrane filter. In the case of chloride contents in excess of 1 g/l, the filter cake is shaken for a further hour with 50 ml of the washing solution. After filtration, the moist filter cake together with the membrane filter are put into a quartz boat.

The combustion temperature should be at least 950 °C and the boat should be carefully inserted into the furnace in such a way that no losses occur through spillage. The further course of the determination should be carried out in accordance with the operating instructions for the particular device being used. The halide-containing combustion gases pass into an absorber charged with sulphuric acid and then flow through the measuring device (e.g. a microcoulometer). The electrolyte and sulphuric acid must be replaced at regular intervals.

For calibration, the appropriate pentachlorophenol solution is used instead of the water sample. A blank determination should also be carried out with 50 mg uncharged active charcoal and 50 ml nitrate washing solution.

**Evaluation**

The content of "adsorbable organic halogen compounds (AOX)" is calculated in "µg/litre AOX" and quoted as a chloride concentration:

$$\text{AOX } (\mu g/l) = n - n_0 \cdot \frac{M \cdot n}{f \cdot v}$$

where:

$n$ = Measured value
$n_o$ = Blank value
$M$ = Molar mass of chlorine (35.5 · µg/µg-mol)
$f$ = Faraday constant (96.487 coulomb/mol)
$v$ = Sample volume used (litre)
$n$ = Current efficiency of electrolysis.

### 4.1.6 Cyanide

#### 4.1.6.1 General

Water is sometimes contaminated with cyanide compounds contained in the discharge from factories of the electroplating industry, in the washings from gas works and from blast furnace gas cleaning, and in the waste from coking ovens, hardening shops and certain chemical production processes.

The methods used for cyanide analysis must be suitable for these types of water. But they must also be sensitive enough to detect traces of cyanide occurring in concentrations of less than 0.05 mg/l in drinking water as a result of the contamination of surface water and ground water by discharged waste.

When describing such methods, a distinction is made between the detection of "easily released cyanide" and "total cyanide". No clear-cut chemical definition exists of the term "easily released cyanide", for which the analogous terms "easily decomposed" and - in a medico-toxicological context - "directly toxic" can be used. General experience indicates that the "easily released" and "immediately toxic" forms of cyanide are both characterized by the fact that hydrogen cyanide is easily split off in a slightly acid environment.

If "easily released cyanide" is to be a useful concept in analytical chemistry, an agreed definition must be found. This parameter is termed "easily released cyanide" and covers cyanide ions which become separated and determinable under precisely defined conditions. The term "total cyanide", on the other hand, is used to describe all compounds which contain at least one cyano-group which may be split off as cyanide ions or hydrogen cyanide under certain unfavourable environmental water conditions.

From the point of view of water analysis, "total cyanides" include all those compounds which - in contrast to the "easily decomposed" cyanide compounds - can release even very firmly bound (complexed) cyanide under the right conditions, such as those provided by decomposition distillation in a highly acid environment.

When analyzing sewage it is sometimes necessary to identify those cyanides or cyanide compounds which can be destroyed by technical means such as oxidation with chlorine or by catalytic methods and which can therefore be separated from the sewage. This group includes the simple cyanides (alkali cyanides) and cyanide complexes of zinc, cadmium, copper, silver and nickel, but not complexes of iron(II) and iron(III), unless they are present in concentrations of over 1000 mg/l.

These "destructible cyanides" (also described imprecisely as "cyanides which can be destroyed by chlorine") are by definition not identical to "easily decomposed cyanides". It is necessary to identify them when assessing the feasibility of, or need for, a decontamination plant to deal with this type of cyanide.

### 4.1.6.2 Total cyanide

#### General remarks

To determine the level of total cyanide in a water sample, all compounds which contain cyano groups and which give off cyanide ions into the distillate under the stated test conditions (decomposition distillation within a highly acid environment) are detected: hydrocyanic acid, cyanide ions and cyanide complexes ($Cd(CN)_4^{2-}$, $Zn(CN)_4^{2-}$, $Cu(CN)_4^{3-}$, $Ag(CN)_4^{3-}$, $Ni(CN)_4^{2-}$, $Fe(CN)_6^{4-}$, $Fe(CN)_6^{3-}$. The cobalt complex $Co(CN)_6^{3-}$, which is very difficult to destroy, can only be partially decomposed and therefore cannot be detected entirely. But this is of little importance for analyzing the water's toxicity since, being difficult to destroy, the cobalt complex is only very slightly toxic.

Under the distillation conditions used here some organic cyano compounds, e.g. cyanohydrins give off hydrogen cyanide and are detected as total cyanide. This does not, however, apply to the simple nitriles such as acetonitrile and benzonitrile, nor to cyanate ions and thiocyanate ions or cyanogen chloride.

**Quantitative analysis** is carried out by means of distillation from a mineral acid solution. The absorbed cyanide in the distillation receiver can then be subjected to:

1) spectrophotometric analysis with barbituric acid-pyridine
2) titration with silver nitrate using p-dimethylaminobenzylidene-rhodamine as an indicator.

#### Distillation method

The purpose of the distillation method is to release the total cyanide from the water sample under controlled conditions and to convert it into a form in which it can be measured using one of the methods described. This means the total cyanide which occurs in organic and inorganic cyanide compounds or complexes and which can be split up under the same conditions as those prevailing in the environment (temperature up to 30 °C, exposure to light, aeration, shift in pH, breakdown by microbes). The differentiation between "total cyanide" and "easily released cyanide" ("easily decomposed cyanide") results from the choice of decomposition conditions in the distillation flask.

Hydrochloric acid is used to bring about a pH below 1 for the decomposition of the total cyanide (add 10 ml of 25 % hydrochloric acid solution to 100 ml of the sample). In this case all complex cyanides (with the exception of the only partially decomposable cobalt complex) are correctly detected in their entirety.

Using air as the carrier gas, the cyanide is transferred as hydrogen cyanide into the absorption vessel where it is absorbed in 1 m sodium hydroxide.

## Equipment

Distillation apparatus (see Fig. 120):
Three-necked flasks (250 to 1000 ml, depending on the volume of the water sample to be distilled), reflux condenser, filling funnel, absorption vessel (with non-return arrangement), water jet pump.

Flow meter for an airflow of 20 to 60 l/h

Washing flask for pre-cleaning the air, 250 ml

Measuring flask, 25 ml

Hydrochloric acid, concentrated and 25 % solution

Sodium hydroxide, 5 m and 1 m

Tin(II)chloride solution:
Dissolve 500 g $SnCl_2 \cdot 2\ H_2O$ and 200 ml conc. hydrochloric acid with dist. water to make a total of 1 litre.

Cadmium acetate solution:
Dissolve 300 g $Cd\ (CH_3COO)_2 \cdot 2\ H_2O$ in 1 litre distilled water.

Buffer solution, pH 5.4:
Dissolve 60 g sodium hydroxide in approx. 500 ml distilled water and while the solution is still warm stir in 118 g reagent-purity succinic acid, $C_4H_6O_4$. After cooling, top up to 1 litre with distilled water.

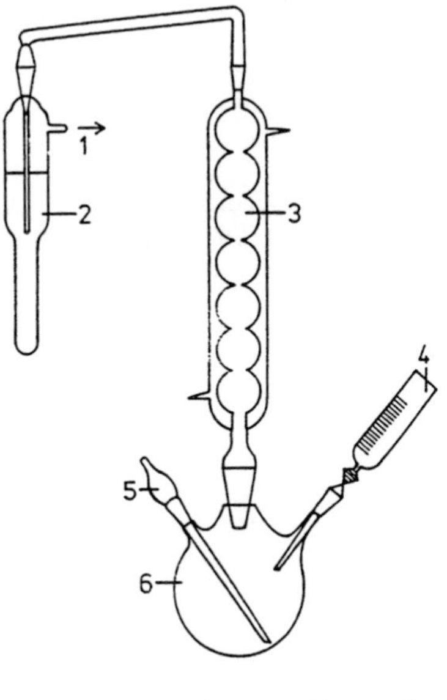


**Fig. 120.** Distillation Apparatus; 1) = Water jet pump; 2) = Absorption vessel; 3) = Reflux condenser; 4) = 50 ml dropping funnel; 5) = Capillary tube for admitting air; 6) = 500 ml three-necked spherical flask

Phenolphthalein solution:
Dissolve 1 g phenolphthalein in 100 ml ethanol.

### Area of use

The distillation method of determining total cyanide content is suitable for levels of between 0.02 and 100 mg $CN^-$/l, if 100 ml of water sample is used. If 500 ml sample is used, levels as low as 0.005 mg $CN^-$/l can be detected. If the total cyanide content is around or above 100 mg $CN^-$/l, the water sample must be diluted.

It can be assumed that the samples received for analysis from water supply authorities will mostly be free of sulphide and thiocyanate and will not contain any oxidizing agents which would interfere with the analysis. However, with chlorinated or otherwise partly contaminated water, an excess of oxidizing agents is to be expected. Samples taken from this latter type of water must have an excess of reducing agents added to them immediately after sampling. For this purpose first add 5 ml of 5 m sodium hydroxide for every litre of the water sample (in order to avoid any loss of free hydrocyanic acid) and then 5 ml of an approx. 50 % solution of tin(II)chloride.

Samples taken from water containing no oxidizing agents are first neutralized against phenolphthalein by adding 1 m hydrochloric acid or 1 m sodium hydroxide, and then slightly alkalized by adding 1 ml of 1 m sodium hydroxide for every litre of the sample.

### Method

After stabilizing the sample with sodium hydroxide solution and tin (II) chloride solution, pour 100 to 500 ml into the three-necked flask of the distillation apparatus. The volume of sample fluid used should be such that the expected absorption solution will contain approx. 2.5 - 50 µg $CN^-$. Pour 10 ml of 1 m sodium hydroxide solution into the absorption vessel. Using a flow meter, adjust the velocity of the air drawn in by the water jet pump to 20 l/h. This air must first pass through a wash bottle containing 100 ml of 1 m sodium hydroxide solution before entering the flask.

After adding 10 ml of 25 % hydrochloric acid solution for every 100 ml of the sample, switch on the heater and bring the contents of the flask to the boil for approx. 45 mins. (Reflux: approx. 1 to 2 drops per second). At the end of this time all decomposable cyanides have been transferred in the form of hydrogen cyanide via the carrier gas stream into the absorption vessel.

End the distillation process. If the absorption solution is turbid or if it is suspected that any substances which would interfere with the analysis still remain, distil the absorption solution a second time. For this purpose transfer the contents of the absorption vessel into a second distillation flask, into which 10 ml cadmium acetate solution and 40 ml buffer solution have first been poured. Rinse the absorption vessel with 40 ml distilled water and transfer this wash liquid to the flask as well. Then proceed as described for the first distillation.

In order to measure the quantity of cyanide ions, transfer the contents of the absorption vessel (after the first or second distillation) into the 25 ml measuring flask, rinse three times with 3 ml dist. water on each occasion and then top the measuring flask up to the 25 ml mark with distilled water.

### 4.1.6.3 Spectrophotometric method of analysis using barbituric acid-pyridine in the distillate

This method can be used after preliminary distillation to measure cyanide ion concentrations of between 0.005 and 0.05 mg $CN^-$/l in surface water and sewage. In this range, which corresponds to a cyanide content of 2.5 to 25 µg in 10 ml absorption solution (assuming that 500 ml of the water sample was used), Lambert's and Beer's laws are borne out.

Where the total cyanide concentration in the water sample is higher, it is necessary to dilute the sample or use smaller quantities of sample, or to use volumetric determination techniques.

Nearly all interfering substances, such as nitrogen-oxygen compounds, hydrogen sulphide, sulphur dioxide or sulphide, are virtually eliminated after the second distillation at the latest (see "Distillation Method"). If the distillate is still coloured even then, or if it contains organic compounds which together with the reagents used produce colouring, this method of cyanide analysis cannot be employed.

**Equipment**

Cuvettes, path length 1 cm to 5 cm

Measuring flasks: 25 ml, 50 ml, 250 ml

Hydrochloric acid (1.19 g/ml)

Hydrochloric acid, 1 m

Sodium hydroxide, 0.4 m

Buffer solution, pH 5.4: see p. 503

Chloramine T solution:
Dissolve 1 g analytically pure chloramine T (sodium p-toluene-sulphochloramide) in 100 ml distilled water. The solution can be kept for approx. 1 week, but its effectiveness should be checked iodometrically.

Barbituric acid-pyridine reagent:
Put 3 g analytically pure barbituric acid into a 50 ml measuring flask. Swirl with a little distilled water and then add 15 ml freshly distilled pyridine (b.p. 115 - 118 °C). Dilute with distilled water, shaking the flask constantly, until the barbituric acid is almost completely dissolved and then add 2.5 ml analytically pure hydrochloric acid (1.19 g/ml). Allow to cool to 20 °C before topping up to the 50 ml mark with distilled water. The reagent can be kept in a brown glass flask for 1 day or for 1 week in the refrigerator.

Cyanide reference solution:
Dissolve 25 mg analytically pure potassium cyanide in 1 litre of 0.4 m sodium hydroxide. 1 ml of this solution contains approx. 0.01 mg cyanide ions and its factor is determined by titration with 0.001 m silver nitrate solution (see "Volumetric Analysis using Silver Nitrate").

## Method

### Calibration

Using a pipette transfer 2, 5, 10, 20 and 25 ml of cyanide reference solution into five 250 ml measuring flasks and top each one up to the 250 ml mark with 0.4 m sodium hydroxide. Transfer 10 ml from each flask to five 25 ml measuring flasks. To each of these add 1 m hydrochloric acid, buffer solution, chloramine T solution and barbituric acid-pyridine reagent, as described under "Measurement". Then carry out photometric analysis, in which the measured values must lie on a straight line. Check the calibration curve from time to time.

### Measurement

Transfer 10 ml of the absorption solution from the 25 ml measuring flask to another 25 ml measuring flask and add in turn exactly 2 ml buffer solution, 4 ml 1 m hydrochloric acid and 1 ml chloramine T solution, shaking the flask constantly. Close the flask and leave to stand for at least 1 min. but not more than 5 min.. Now add 3 ml barbituric acid-pyridine reagent, top up to the 25 ml mark with distilled water and shake well. After 20 min. carry out photometric analysis at 578 nm using a reference solution which is prepared as follows:

Put 10 ml 0.4 m sodium hydroxide, 2 ml buffer solution, 4 ml 1 m hydrochloric acid, 1 ml chloramine T solution and 3 ml barbituric acid-pyridine reagent into a 25 ml measuring flask, shaking constantly, and top up to the 25 ml mark with distilled water.

For the purpose of evaluation a blank value must be established by subjecting a blank sample to the entire analytical process, including distillation. Use distilled water instead of the water sample and again carry out photometric analysis against the reference solution.

## 4.1.6.4 Volumetric analysis using silver nitrate

The method can be used for all distillates (see "Distillation Method") which contain more than 0.05 mg $CN^-$; this corresponds to a cyanide content in the water sample of over 0.1 mg/l, assuming that 500 ml of the sample was distilled.

The preceding single or double distillation eliminates virtually all interfering factors.

### Equipment

Glass beakers, magnetic stirrer with magnetic bar

Burette, 10 ml, or flask burette

Silver nitrate solution, 0.001 m:
The solution must be prepared freshly from 0.01 m silver nitrate solution each time it is used and its content checked.

Indicator solution:
Dissolve 20 mg p-dimethylaminobenzylidene-rhodamine, $C_{12}H_{12}N_2OS_2$, in 100 ml analytically pure acetone. The solution can be kept in a dark bottle for at least 1 week.

Sodium hydroxide solution, 1 m

## Method

### Measurement

The blank value is measured first, since the colour tone obtained is later used as a guide for the sample solution containing cyanide at the end of the titration process. Put 10 ml of 1 m sodium hydroxide solution into a glass beaker, add 20 ml distilled water and 0.1 ml indicator solution, then set the magnetic stirrer in motion. Immerse the tip of the burette in the distilled water solution and release enough silver nitrate solution to cause a change in colour from yellow to yellow-red or salmon-pink. The colouring only remains for a short time.

In order to analyze the sample containing cyanide, transfer the absorption solution resulting from the distillation process (contained in a 25 ml measuring flask) into another glass beaker and add 0.1 ml indicator solution. Carry out titration in the same way as for the analysis of the blank value and until the same colouring is obtained.

## 4.1.6.5 Easily Released Cyanide

### Separation of Easily Released Cyanide (Distillation Method)

The purpose of separation is to expel that part of the cyanide contained in the water sample which can be released at room temperature and in a slightly acid environment (pH 3.9 - 4.0) in the form of hydrogen cyanide in the air flow and to catch it in an alkaline absorption solution. The correct pH is obtained by adding potassium hydrogen phthalate buffer solution. The detection of iron-cyanide complexes is prevented by adding zinc sulphate, while the addition of zinc powder causes faster decomposition of the copper (II)-cyanide complexes. EDTA is added as a complex-building agent in order to largely prevent the catalytic-decomposing effect of heavy metal ions, in particular the copper ions. Cadmium sulphate checks any interference by sulphides or hydrogen sulphide.

The method can be used with all types of water whose content of easily released cyanide does not exceed 50 mg $CN^-$/l. If the cyanide content is higher than this, the water sample must be diluted. The method described for use with "total cyanide" (distillation method) also applies in virtually all respects to the method for separating easily released cyanide.

### Equipment

Distillation apparatus, flow meter, wash bottle and 25 ml measuring flasks are the same as described in the section "Total Cyanide" (Distillation Method).

Hydrochloric acid, concentrated and 25 % solution

Sodium hydroxide solution 5 m and 1 m

Tin(II)chloride solution:
Dissolve 500 g $SnCl_2 \cdot 2\ H_2O$ and 200 ml concentrated hydrochloric acid with distilled water to give a total of 1 litre.

Chloroform-phenolphthalein solution:
Dissolve 0.3 g phenolphthalein in 900 ml ethanol and add 100 ml $CHCl_3$.

Zinc sulphate/cadmium sulphate solution:
Dissolve 100 g $ZnSO_4 \cdot 7\ H_2O$ and 100 g $CdSO_4 \cdot 8\ H_2O$ with 1 litre distilled water.

Buffer solution, pH 4:
Dissolve 80 g analytically pure potassium hydrogen phthalate, $C_8H_5KO_4$, in 920 ml warm distilled water.

EDTA solution:
Dissolve 100 g of the disodium salt of analytically pure ethylenediamine tetra-acetic acid in 940 ml warmed up distilled water.

Analytically pure zinc powder.

**Method**

**Sample preparation**

When analyzing the water sample for easily released cyanide the method of treatment is particularly important right from the moment of sampling if the cyanide content is to be determined with accuracy.

Therefore 5 ml of 5 m sodium hydroxide solution, 10 ml chloroform-phenolphthalein solution and 5 ml tin(II)chloride solution is added for each litre of the water sample (after preliminary dilution in the case of sewage known to have a high cyanide content. If a red colouring results, this is removed by adding 1 m hydrochloric acid a drop at a time. If no red colouring occurs, add 1 m sodium hydroxide solution until a slight suggestion of redness begins to appear. The coloured solutions should have a pH of approx. 8 measured by electrometric means or with indicator paper.

Then add 10 ml zinc sulphate/cadmium sulphate solution and store in a cool, dark place. Carry out the analysis as soon as possible.

Pour 10 ml of 1 m sodium hydroxide into the adsorption vessel and connect it to the reflux condenser. Set the air flow speed initially at 30 to 60 l/h and pour 10 ml zinc sulphate/cadmium sulphate solution, 10 ml EDTA solution, 50 ml buffer solution and 100 ml of the pre-treated water sample into the three-necked flask. If the concentration of easily released cyanide is expected to be less than 0.1 mg/l, a larger volume of sample water (up to 500 ml) can be used. However, the quantities of zinc, zinc sulphate/cadmium sulphate solution and buffer solution must also be correspondingly increased.

Fit a pH electrode through the side nozzle so that it is immersed in the solution and add enough hydrochloric acid or sodium hydroxide solution through the filler nozzle to obtain a pH of $3.9 \pm 0.1$. After removing the electrode, add approx. 0.3 g zinc powder through the side nozzle and close the opening with a glass stopper. Connect the filler nozzle to the cleaning flask containing 1 ml 1 m sodium hydroxide and raise the air flow to 60 l/h.

End this process after 4 hours and transfer the absorption solution into the 25 ml measuring flask. Rinse three times, using 3 ml dist. water on each occasion, and add these rinsing solutions to the measuring flask. Then top up to the mark with distilled water.

### 4.1.6.6 Spectrophotometric analysis using barbituric acid-pyridine

The method of carrying out spectrophotometric measurement of "free cyanide" is identical in every respect with the procedure described in the section "Total Cyanide". (4.1.6.3)

When calculating the results take the calibrations as a basis and read off the values from the relevant calibration curve. Take blank tests and dilutions into account.

The results should be given to 0.01 mg/l for concentrations of less than 1 mg/l, and to 0.1 mg/l in the case of higher concentrations.

## 4.1.7 Detergents (surfactants)

### General Remarks

Surface-active substances are known as detergents or surfactants. They are frequently used in cleansing agents, washing agents, cosmetics and also as auxiliary agents in industrial products. The field of applications is widespread.

In a number of cleansing agents or technical products, not only anionic detergents are used but also other types such as cationic detergents, ampholytic detergents and nonionic detergents.

The following methods are described below:

4.1.7.1 Determination of total detergents (surfactants) for orientation purposes (ethanol extract of the dry residue).

4.1.7.2 Photometric determination of anionic surfactants with methylene blue (MBAS).

4.1.7.3 Potentiometric determination of nonionic surfactants.

4.1.7.4 Cationic detergents (surfactants) after ion exchange.

### 4.1.7.1 Determination of total detergents (surfactants)

Most detergents (surfactants) are soluble in ethanol. The content of ethanol-soluble substances in the evaporation residue of a water sample therefore provides an indication of the scale on which these substances are present. It should be noted that other ethanol-soluble substances have a disturbing effect and that inorganic substances contained in the water may also pass into the ethanol extract. It is therefore advisable to re-extract the first extract with ethanol.

Most detergents are substances which are sensitive to drying, so the drying time of the ethanol extract should be limited to 2 hours at 85 °C.

### Equipment

Ethanol, absolute

Strainer

Pleated filter

Platinum or glass dish

Drying cabinet

Analytical balance

Desiccator

**Method**

Add approximately 50 ml of hot ethanol to the evaporation residue of 1 or 2 litres water and use a glass rod to loosen and disperse any residues attached to the dish. After approximately 30 minutes on a boiling water bath, filter hot through a pleated filter into a glass dish. Then evaporate this first ethanol extract. Extract the residue once more with hot ethanol, and after filtration evaporate in a weighed dish (second extract).

Dry the second extraction residue for 2 hours at 85 °C, cool in the desiccator and weigh.

Convert the amount weighed to mg/l of water. This is used as a guide value for detergents (surfactants) in the water sample.

### 4.1.7.2 Photometric determination of anionic detergents (surfactants) with methylene blue (Methylene blue active substances, MBAS)

**General remarks**

Methylene blue has the property of a cationic detergent and forms a coloured complex with an ionic surfactant. This complex is extracted and photometrically evaluated.

The content of ionic substance should be between 0.1 and 1.5 mg/l of sample. Water containing higher concentrations must be diluted accordingly.

**Equipment**

Photometer (650 nm)

Cuvettes, 1, 2 and 5 cm

Separating funnel, volume approx. 250 ml

Measuring flask, volume 50 ml

Phosphate solution:
Dissolve 12.52 g of $Na_2HPO_4 \cdot 2H_2O$ (Sörensen's buffering agent) in 500 ml of distilled water. Set the solution to pH 10 with 0.5 m sodium hydroxide solution and make up to 1 litre with distilled water. If the solution is to be stored for a long period, the pH must be checked and adjusted as necessary.

Methylene-blue solution, neutral:
Dissolve 0.35 g of methylene blue in distilled water to 1 litre. The freshly prepared solution must be left to stand for 24 hours at 20 - 25 °C

before the calibration curve is plotted or the measurement taken. If a new batch of methylene blue is used, the calibration curve must be replotted.

Methylene-blue solution, acid:
Dissolve 0.35 g of methylene blue in 500 ml of distilled water and add 6.5 ml of $H_2SO_4$ (1.84 g/ml). Dilute the solution to 1 litre with distilled water. The freshly prepared solution must be left to stand for at least 24 hours at 20 - 25 °C before the calibration curve is plotted or the measurement taken. If a new batch of methylene blue is used, the calibration curve must be replotted.

Chloroform, reagent purity, freshly distilled

Cotton wool filter

Calibration substances:
Sodium salt of dodecane-1-sulphonic acid or (e.g. 0.1 g/litre in distilled water) tetrapropylene benzene sulphonate (sodium salt) or sodium lauryl sulphate or another anionic surfactant. The results should state which substance was used for calibration.

## Method

Filter turbid water and discard the first 100 ml of the filtrate. Take 100 ml of the filtered water in a separating funnel.

If less than 100 ml of water is used, make up to 100 ml with distilled water.

A blank test with 100 ml of distilled water must always be subjected to the entire procedure.

Add 10 ml of alkaline phosphate solution, 5 ml of neutral methylene-blue solution and 15 ml of chloroform. Seal the separating funnel and shake evenly but thoroughly for approximately 1 minute. Allow the chloroform and the water layer to separate. The chloroform phase settles to the base of the separating funnel; drain the chloroform off into a second separating funnel containing 100 ml of distilled water and 5 ml of acid methylene-blue solution. Shake the second separating funnel for about 1 minute. After separation of the phases, filter the chloroform through a cotton wool filter moistened with chloroform into a 50-ml measuring flask. Repeat the entire procedure twice more, with extraction of the water sample in separating funnel 1, and separation and extraction in separating funnel 2, using 10 ml of chloroform each time. No further phosphate solution or methylene-blue solution should be added. Rinse the cotton wool filter and top up the measuring flask with chloroform.

The blue complex of the anionic detergent (surfactant) with methylene blue is in the chloroform solution. Measure the extinction of the chloroform phase in a photometer at 650 nm in cuvettes with 1, 2 or 5 cm path length against chloroform. Treat and measure a blank test with 100 ml of distilled water in exactly the same way (with a 1-cm cuvette the measured value should be below 0.02).

## Evaluation

Evaluate the extinction on the basis of a calibration curve, taking the blank test into account and the volume of water or degree of dilution used.

Give the results in mg/l.

Since the various calibration substances have different molecular weights, the reference substance for calibration should be specified when the content is given in mg/l.

Calibration or plotting of the calibration curve must be conducted in precisely the same way as the procedure described. Take suitable concentrations of calibration solution, make up to 100 ml with distilled water and proceed with the method.

Note:

If disturbances are detected, the detergents (surfactants) can be separated by blowing out and the anionic substances determined in the way described.

### 4.1.7.3 Determination of nonionic detergents (surfactants)

(By potentiometry after precipitation with $KBiI_4$.)

General remarks

In the procedure described, Dragendorff's reagent, potassium iodide bismuthate $KBiI_4$, forms a deeply orange-red-coloured adduct with many nonionic detergents (surfactants) from aqueous solutions. The adduct is filtered off and dissolved, and the bismuth contained in the adduct is determined by potentiometric titration or by AAS or ICP.

The following types of nonionic detergents can be determined with the aid of this method:

Oxalkylated fatty acids, fatty alcohols, fatty amines, alkylphenols and fatty acid amides, and polythylene glycols and polypropylene glycols down to pentaethylene glycol and pentapropylene glycol.

The content of nonionic detergents (surfactants) is ideally measured between 0.2 and 0.8 mg/l.

Low-molecular alkylene glycols such as tetra-, tri- and di-alkylene glycol are no longer capable of precipitation. Protein and its breakdown products including the amino acids and other surface-active agents with cationic or anionic reactivity do not give a positive reaction unless ethylene oxide groups are present in the molecule.

Anionic detergents have no disturbing effect in a range of concentration up to 5 mg/l. However, anionic and cationic detergents which display polyethylene oxide functions in the molecule also form the adduct. Detergents of this type may be separated off by ion exchange.

Equipment

Ethyl acetate, reagent purity, freshly distilled

Sodium hydrogen carbonate, $NaHCO_3$, reagent purity,

1% hydrochloric acid, prepared from concentrated reagent-purity hydrochloric acid with redistilled water

Methanol, reagent purity, freshly distilled, stored in glass bottles

Bromocresol purple solution:
Dissolve 0.1 g of bromocresol purple in 100 ml of freshly distilled reagent-purity methanol.

Precipitating agent:
The precipitating agent is a mixture of 2 parts by volume of Solution a) and 1 part by volume of Solution b). The mixture is to be kept in an amber flask in a dark place and may be stored for about 1 week.

Solution a:
Dissolve 1.7 g of basic bismuth(III)nitrate ($BIONO_3 \cdot H_2O$) (superpure quality) in 20 ml of reagent-purity glacial acetate acid and make up with redistilled water to 100 ml. Dissolve 65 g of reagent-purity potassium iodide in about 200 ml of distilled water. Combine the two solutions in a 1000-ml measuring flask, add 200 ml of reagent-purity glacial acetic acid and make up to the mark with distilled water.

Solution b:
Dissolve 290 g of reagent-purity barium chloride ($BaCl_2 \cdot 2\ H_2O$) in distilled water and make up to 1000 ml.

Glacial acetic acid, 100 %, purest quality available

Ammonium tartrate solution:
Combine 12.4 g of reagent-purity tartaric acid and 18 ml of 25 % ammonium hydroxide solution, reagent purity, and make up to 1000 ml with distilled water.

Ammonium hydroxide solution:
1 %, prepared from 25 % ammonium hydroxide solution, reagent purity, by dilution with distilled water.

Standard acetate buffer:
Dilute or dissolve 120 ml of 100 % glacial acetic acid and 40 g of reagent-purity hydroxide (separately) in distilled water and mix the solutions. Make up to 1000 ml.

Pyrrolidine dithiocarbamate solution (carbate solution):
Dissolve 103.0 mg of pyrrolidine dithiocarboxylic acid, sodium salt, in distilled water, add 10 ml of reagent-purity n-amyl alcohol and 0.5 g of sodium hydrogen carbonate, and make up to 1000 ml with distilled water.

Copper sulphate solution:
Solution I:
Make up 1.249 g of reagent-purity copper(II)sulphate ($CuSO_4 \cdot 5\ H_2O$) and 50 ml of 0.5 m sulphuric acid to 1000 ml with distilled water.
Solution II:
Make up 50 ml of Solution I and 10 ml of 0.5 m sulphuric acid to 1000 ml with distilled water.

Magnetic stirrer with magnetic bar (25 to 30 mm)

Separating funnel, 250 ml

Porous porcelain crucible, size 2 (upper diameter 40 mm, height 42 mm, porosity 2)

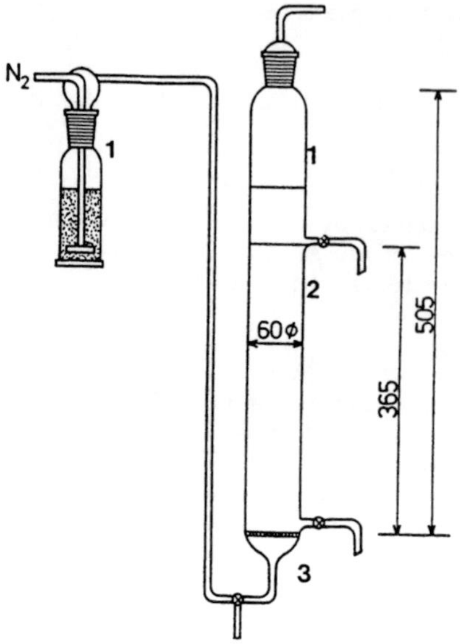


Fig. 121. 1-litre apparatus for surfactant concentration by blowing out, after Wickbold (Dimensions in mm); 1) = Ethyl acetate; 2) = Water sample; 3) = Glass filter frit

2 suction bottles with adapter for filtering crucible (500 and 250 ml) (suction systems)

Round glass-fibre-paper filters, 27 mm diameter, fibre diameter 0.5 to 1.5 µm

Polyethylene washing bottle, volume 500 ml, for 100 % glacial acetic acid

Recording potentiometer with platinum/calomel measuring cell or platinum-silver chloride measuring cell, measuring range 250 mV with automatic burette, volume 20 to 25 ml. Alternatively, a manual potentiometric apparatus. A potentiograph is recommended.

Methanolic hydrochloric acid, 10 %

AAS or ICP apparatus

Method

Isolation of detergents (surfactants)

In order to concentrate the surfactants and separate off disturbing substances, approximately 1 litre of the water sample, filtered through a paper filter, is required. Dilute water samples having higher concentrations of nonionic detergents than 0.8 mg/l as appropriate. Neutralize acid and alkaline water samples. Dissolve 100 g of sodium chloride and 5 g of sodium hydrogen carbonate in the water sample of the dilution. Fill the surfactant-blowout device, see diagram 121, with water sample to the upper discharge cock. On top of this, carefully add a layer of 100 ml of ethyl acetate. Fill the frit washing bottle in the gas feed line to about 2/3 with ethyl acetate. Pass a current of gas (nitrogen or air) through the blowout apparatus at a rate of approximately 50 to 60 l/h. The flow should be set such that there is no turbulence at the phase boundary. The phases should remain clearly separate. Mixing of the phases, leading to emulsification, is thus avoided. Stop the gas flow after 5 minutes.

Drain off the ethyl acetate phase into a 250-ml separating funnel. Return any water which accompanies the ethyl acetate to the blowout apparatus.

Again add a layer of 100 ml of ethyl acetate to the surfactant-blowout device and feed through nitrogen or air for a further 5 minutes. Once again drain off the organic phase into the separating funnel. Discard the remaining water in the separating funnel and filter the ethyl acetate phase through a pleated filter. Rinse the separating funnel and filter with about 20 ml of ethyl acetate. Combine the ethyl acetate extracts and the washing solution in a 250-ml beaker.

As a result of the procedure of blowing through gas, the detergents are concentrated at the surface of the aqueous phase and are taken up quantitatively in the ethyl acetate phase.

Carefully evaporate the combined extracts in a 250-ml beaker on a water bath. A light current of air over the beaker accelerates the evaporation process and reduces the risk of sensitive detergents being converted as a result of heat treatment.

Take up the evaporation residue in 5 ml of methanol, 40 ml of distilled water and 0.5 ml of 0.5 m sulphuric acid. Stir with a magnetic stirrer. Add 3 to 5 drops of bromocresol purple solution; the indicator should change to yellow. Add 30 ml of precipitating agent, and continue stirring. Cease stirring after 10 minutes and leave to stand for at least 5 minutes.

If the water sample contains nonionic detergents, an orange-red precipitate will have formed. Filter off the precipitate through a porous porcelain crucible by suction with a water-jet pump. Rewash the beaker, magnetic rod and crucible thoroughly with 100 % glacial acetic acid; approximately 150 to 200 ml should be used. Washing is performed more easily if a polyethylene washing bottle is used. It is not necessary to transfer the precipitate quantitatively from the beaker to the filter crucible since the beaker is later used to hold the solution of the precipitate.

Mount the filter crucible with the washed precipitate on another suction system than that used for filtration. Dissolve the precipitate by adding hot ammonium tartrate solution in 3 portions of 10 ml each. Pour a further 20 ml of ammonium tartrate solution into the precipitation beaker and dissolve any remnants of the precipitate. Use the ammonium tartrate solution from the suction system for rinsing and return the wash liquid to the precipitation beaker. The total volume of solution in the precipitation beaker should be between 150 and 200 ml.

Add a few drops of bromocresol purple solution, stir with the magnetic stirrer and adjust with 1 % ammonium hydroxide solution until the colour of the indicator changes to violet. Then add 10 ml of standard acetate buffer, after which a pH value of 4.6 is established. Place the beaker on the potentiograph and immerse the electrodes in the solution. Conduct potentiometric titration with the pyrrolidine dithiocarbamate solution beyond the potential jump. The rate of titration should be set at 2 ml/min. and the paper feed at 2 cm/ml.

The intersection of the tangents of the two branches of the potential curve is taken as the end point of the titration. A levelling off of the potential jump may occasionally be observed; this phenomenon may be eliminated by careful polishing of the platinum electrode with emery.

The bismuth forms a sparingly soluble metal complex with the pyrrolidine dithiocarbamate. The potentiometric method of titration is highly sensitive for small quantities of bismuth. An evaluable jump is still obtained with the standard solution. However, no explanation has yet been given as to why there is a potential jump at all. During potentiometric titration there is presumably a small quantity of a reversible oxidation product present, with the result that a measurable redox potential may be formed between the electrodes, which changes when the excesss dithiocarbamate appears in the solution as the equivalence point is exceeded.

Blank test

A blank test must be conducted using the same procedure as that described above. Take 5 ml of methanol and 40 ml of distilled water for the blank test and then continue as described under "Method". Conduct filtration, dissolution and titration even if there is no visible precipitate. The consumption of pyrrolidine dithiocarbamate solution should be below 1 ml in the blank test; if this is not the case, the reagents used should be prepared afresh.

(The purity of the reagents can be checked by emission spectrum analysis of the evaporation residue. No appreciable heavy-metal components should be detected).

Establishing the factor of the pyrrolidine dithiocarbamate solution: To 10 ml of copper sulphate calibration solution add 100 ml of distilled water and 10 ml of standard acetate buffer solution. Titrate as described above.

The factor is calculated thus:

$f = 10/a$
a = consumption of pyrrolidine dithiocarbamate solution in ml

The factor of the solution must be taken into account when evaluating the results of titration and those of the blank test.

Evaluation

Since there are many different types of nonionic detergents and the individual types may also have different ethylene oxide chain lengths, reference must be established to a standard substance with a known molecular weight; this is, of course, not necessary if it is known which nonionic surfactant is present in the water sample.

Nonylphenol decaglycol ether, for example, may be used as the standard substance. A conversion factor of 54 has been determined empirically for this nonionic surfactant.

The content of nonionic detergent (surfactant) in the water sample, in mg, is calculated according to the following formula:

$$g = (b - c) \cdot f \cdot 0.054$$

g = Content of nonionic detergent (surfactant) of the water sample in mg, referred to nonylphenol decaglycol ether
b = Consumption, in ml, of pyrrolidine dithiocarbamate solution during titration of the precipitate of the solution under analysis
c = Consumption of pyrrolidine dithiocarbamate solution in ml during the blank test
f = Factor of the pyrrolidine dithiocarbamate

The results are given in mg/l. A note should be made of which nonionic standard substance has been used for reference.

An advantageous and highly sensitive method of determining the bismuth content is to use ICP-AES (Chapters 2 and 3) on a separate concentration. The result may then be referred to nonionic surfactant, taking into account the calibration substance etc.

### 4.1.7.4 Cationic detergents (surfactants) after ion exchange

In the determination of nonionic surfactants according to the method described above (4.1.7.3), cationic surfactants are also detected. The content of nonionic detergent as measured is consequently too high. In order to check this, a second series of tests as described in 4.1.7.3 may be carried out. In this second analysis, the evaporation residue of the ethyl acetate extract is dissolved in 20 ml of methanol. An exchanger column should be prepared, suitable for filling with 10 ml of cation exchanger in the $H^+$ form. The particle size of the cation exchanger should be between 0.15 and 0.30 mm. The cation exchanger is again converted to the $H^+$ form with methanolic hydrochloric acid (11 ml of hydrochloric acid (1.125 g/ml) made up to 100 ml with methanol) and rewashed with methanol until the discharge is no longer acid against methyl red. The flow rate is set at a fairly rapid drip. The methanolic solution of the evaporation residue of the ethyl acetate is then put through the cation exchanger column. About 50 ml of methanol is used for rewashing.

Determination of the nonionic surfactants passing through the cation exchanger is performed exactly as described in Method 4.1.7.3.

If there are reproducible differences between the results of determination of the nonionic surfactants without cation exchange in comparison with those subjected to cation exchange, this means the water under analysis also contains cationic substances in addition to nonionic detergents (surfactants).

Direct determination of the cationic surfactants (detergents) is advisable only if considerable differences arise in the determination of nonionic surfactants (detergents) with and without ion exchange. The method using bromophenol blue may be employed as a more semi-quantitative determination of cationic surfactants (detergents) in water.

#### Equipment

Photometer (416 nm)

Calibration solution:
Cetyltrimethylammonium bromide (CTAB, molecular weight: 364.5) may be used as a cationic calibration surfactant. Dissolve 1 g of this substance

in distilled water and make up to 1 litre. This stock solution should be diluted in order to plot the calibration curve. Ten ml, for example, may be taken and diluted to 1000 ml with distilled water. One ml of this solution contains 0.01 mg of CTAB. Increasing quantities of this solution are used for preparation of the calibration curve according to the method described.

Bromophenol blue solution:
Dissolve 0.15 g of bromophenol blue in 200 ml of 0.01-m sodium hydroxide solution. When dissolved, add 42 ml of 0.1-m hydrochloric acid (e.g. Merck, Darmstadt, FRG).

Chloroform with approx. 1 % ethanol

Buffer solution:
Dissolve 21 g citric acid in 200 ml of 1-m sodium hydroxide solution, and make up to 1 litre with distilled water. Take 309 ml of this solution and make up to 1 litre with 0.1 m hydrochloric acid.

### Method

Pour 100 ml of water into a 250-ml glass separating funnel and add 10 ml of buffer solution, 5 ml of 0.1 m hydrochloric acid, 2 ml of bromophenol blue solution and 50 ml of chloroform. Shake for 3 minutes. After separating of the layers, pass the lower chloroform phase through a cotton wool filter moistened with chloroform. Discard the first 5 or 10 ml of the filtrate and subsequently measure the coloured solution against chloroform at 416 nm. Evaluate according to the calibration curve plotted.

Round off the results to 0.1 mg/l and mention the fact that determination was carried out according to the semi-quantitative method with bromophenol blue.

## 4.1.8 Determination of hydrocarbons (Oil and greaselike extractable substances)

### General

Water, and in particular surface water and waste water, can be polluted by mineral, vegetable and animal oils and greases, waxes etc. These can arise on the one hand in very small quantities in the water in either dissolved or dispersed form but can also, on the other hand, be present as massive contamination in the form of a two-phase system.

However, natural products such as humic materials may also be determined in part by extraction.

The formation of emulsions is aided by the presence of surface-active materials (detergents or surfactants).

Contamination with oil and greaselike substances affects the smell and taste of water and can cause technological problems. Such contamination can also lead to problems relating to health.

### Methods

A variety of methods are known for the testing of samples of water for oils and greases. Four methods are mentioned here:

4.1.8.1 Gravimetric determination following extraction with n-hexane:

Low-boiling substances are only recorded in part with this method; in addition errors can also arise as a result of the co-extraction of surface-active substances and in some cases natural waxes, humic materials etc.

4.1.8.2 Determination of hydrocarbons by infrared intensity spectroscopy:

Following extraction of the hydrocarbons with 1,1,2-trichlorotrifluoroethane, measurements are carried out in the wavelength range from 3.2 to 3.6 µm.

4.1.8.3 Measurement of UV fluorescence following separation by thin-layer chromatography:

The fluorescence spectrum is recorded following extraction with carbon tetrachloride. This method can be considered as augmenting the infrared intensity spectroscopy method.

4.1.8.4 Gas chromatography head-space analysis:

In this method of investigation, volatile organic compounds (e.g. fuels, organic solvents), which go over into the gas phase after the vapour pressure has reached equilibrium, are removed in gaseous form and subjected to analysis by gas chromatography.

### 4.1.8.1 Gravimetric determination following extraction with n-hexane

General remarks

By extracting a sample of water with n-hexane in the pH range usual for water, namely > pH 6, it is possible to detect the following main groups:

Mineral oils and mineral greases
Vegetable and animal oils and greases (triglycerides)
Free fatty acids

Other organic constituents, where these can be extracted with a non-polar solvent, e.g. certain surface-active substances (detergents, tensides), waxes etc., are also recorded.

The following are not recorded:

Fatty acids which are present as fatty acid compounds (soaps). These can be recorded following extraction with n-hexane and acidification of the aqueous solution to a pH range of pH 1 to pH 2 by means of an additional extraction process.

For this reason the pH of the water sample should always be quoted with the results of the analysis.

The method can be used down to concentrations of some 0.1 mg/l depending on the size of the sample used; this can be between 1 litre and about 10 litres of water.

Readily volatile oils and greases, which evaporate during drying depending on their vapour pressure and the drying time, are not recorded.

The extraction can be disturbed if emulsions are present. An emulsion can be broken down by being salted out (with sodium chloride or sodium sulphate).

**Taking of samples** (see also Chapter 1)

In addition to the instructions given in Chapter 1 on the taking of samples of water, full details of how to take samples are included here.

Only glass bottles with ground glass stoppers should be used for the taking of water samples to determine oil and greaselike substances.

These bottles, including the ground glass seat and stopper, are thoroughly cleaned as usual and are then extracted to exhaustion with n-hexane and dried. Add 50 ml n-hexane to bottles prepared in the above-mentioned way with a volume of up to 5 litres, or 100 ml n-hexane to bottles with a volume of up to 10 litres. At the place where the sample is to be taken, the stopper is removed, taking care not to touch the ground glass joints with the hands, and the water is poured into the bottle. The quantity of water can either be measured on the spot with a grease-free measuring cylinder or any desired quantity of water can be poured into the bottle containing the n-hexane and the water level marked. The quantity of water added can then be determined later in the laboratory with account being taken of the volume of the n-hexane. Finally it is also possible to weigh the sample bottles containing the n-hexane, and then weigh them again in the laboratory after the quantity of water has been added.

When taking samples, it should be kept in mind that oil- and greaselike substances frequently accumulate at the surface of the water as a thin film and that where emulsions are present, separation of the phases may already have started. Finally it has to be taken into account that oil- and fatlike substances can accumulate in sediments as the result of adsorptive processes. Sampling must therefore be geared to the particular task specified and it is important to state whether the sample was taken from the surface or at a particular depth. Samples of sediment are to be removed in an analogous manner but should be tested separately from the water.

After the samples have been taken, the ground-glass stoppers should be secured in such a way that there is no chance of them opening during transportation.

Equipment

Sampling bottle, glass with ground glass stopper, 5 or 10 litres

Separating funnel, capacity approx. 1 litre

Glass funnel, diameter approx. 7 cm

Filter paper

Water bath with thermostat, 80 °C

Drying cabinet, 80 °C

Desiccator with silica gel

Platinum or glass dishes, greasefree

n-hexane, reagent purity

Alcoholic potassium hydroxide, 0.1 m

Sulphuric acid or hydrochloric acid, 2 m

Burette

Erlenmeyer flask with ground glass joint and stopper, 250 ml

Reflux condenser

Centrifuge

Semimicro balance

**Procedure**

The pH of the water to be investigated is determined by electrometric means using a separate sample.

Extraction is carried out in the sampling bottle by means of thorough shaking (1 minute). After the phases have separated, the aqueous phase is transferred with the aid of a siphon into a second glass sampling bottle which also contains 50 or 100 ml n-hexane. Here the second extraction is carried out. After the phases in this second bottle have separated, the water is siphoned off and discarded or is used for determining the soaps. The residues of the water sample and the hexane phases are now in both bottles. The contents of both sampling bottles are transferred to a grease-free 1 litre separating funnel and both bottles are washed out with 2 portions, each of 20 to 50 ml, of n-hexane. The total extraction solutions and the residual water are vigorously shaken (1 minute) in the separating funnel and, after the phases are separated, the aqueous phase is run off. The n-hexane phase, which contains all the organic substances capable of being extracted at the pH in question, is transferred through a greasefree filter into a constant weight platinum or glass dish which has been previously weighed. When there is reason to expect that the amount to be subsequently weighed out will be more than 5 mg, constant weight glass dishes can also be used.

The n-hexane in the platinum dish or glass dish is vaporized off on the water bath at a maximum of 80 °C. Then the dish with the extractable substances is dried at 80 °C in the drying cabinet and is then weighed out on a semimicro balance after a definite cooling time in a desiccator. The precautions usual when working in the semimicro range are to be observed.

The extract which has been weighed out can be processed further. It is possible to differentiate between mineral oils and greases on the one hand and triglycerides and/or fatty acids on the other hand by determining the content of saponifiable and non-saponifiable components.

The saponifiable part can be investigated by gas chromatography following the formation of the methyl esters in order to determine which fatty acids are present.

The non-saponifiable part can be further investigated by means of infrared spectroscopy or UV fluorescence. This can give indications as to which types of hydrocarbons are present (paraffins or aromatic substances or mixtures of both).

(See the appropriate section of this chapter for the methods of carrying out these additional analysis).

If the fatty acids present as fatty acid compounds (soaps) in water are also to be recorded by this method, then the water samples extracted with n-hexane and the rinsing water (see method above) should not be discarded but collected in a sampling bottle and acidified with sulphuric acid or hydrochloric acid to a pH of 1 to 2. This process causes the soaps to be broken down and the fatty acids released. These should then be extracted with n-hexane and further processed in the same way as described above.

When the quantities weighed out following direct extraction and following extraction after acidification are below 1 mg, then the investigation should be repeated with a larger quantity of water. But experience has shown that 10 litres water should be regarded as the upper limit.

Blanks are to be carried out with distilled water and the results taken into account. (The blanks must be tested in a fully analogous manner, i.e. the appropriate quantity of distilled $H_2O$ should be extracted with the appropriate quantity of n-hexane in a sampling bottle and the extract subjected to the full series of analyses.)

Evaluation

The quantity of extractable substances weighed out, corrected for the blank value and taking account of the quantity of water used, is quoted in mg/l (the pH at which the extraction was carried out should be quoted and it should also be noted that the result was obtained using the n-hexane extraction method).

If the fatty acid compounds have been recorded as well following extraction in acidic solution, then this should also be noted as an additional result.

Appendix 1 to Method 4.1.8.1 (Extraction with n-hexane)

Determination of the saponifiable and non-saponifiable parts

For the separation of the saponifiable and non-saponifiable parts, ethanol is used to transfer the quantity weighed out from the platinum or glass dish into a glass flask with ground glass stopper. The part which is not soluble in ethanol should also be washed over into the flask. After the addition of alcoholic potassium hydroxide, the saponification is carried out by boiling with a reflux condenser.

The contents of the flask are allowed to cool down and the ethanolic potassium hydroxide is transferred into a separating funnel. The saponification flask is washed out with n-hexane (some 10 to 50 ml). These n-hexane portions are also transferred into the separating funnel and the extraction is then carried out by shaking the flask thoroughly. The ethanolic potassium hydroxide is separated from the n-hexane phase and is then extracted again with n-hexane. The cleaned n-hexane extracts are transferred via a greasefree filter into a constant-weight platinum dish which has been previously weighed, and are then further treated as described for the main process.

The saponified parts in the ethanolic potassium hydroxide solution are set to pH 1 to 2 with sulphuric acid and then extracted twice with n-hexane. Both n-hexane extracts separated off in this way are filtered into a platinum dish which has been weighed and are then further processed in an analogous manner to that described for the main process. The following data are obtained from the quantities weighed out:

1. Non-saponifiable mineral oils and greases. The result is quoted in mg/l.

2. The fatty acids of the saponifiable parts. The result is quoted in mg/l.

These two fractions can be further analyzed if necessary using GC, GC-MS, IR or UV analysis.

## Appendix 2 to Method 4.1.8.1 and to Appendix 1

### Infrared analysis

When the non-saponifiable part is to be evaluated by means of infrared analysis, then this should always be carried out in accordance with Method 2 of this section.

In order to carry out infrared analysis on the quantity of non-saponifiable residue obtained, the latter is dissolved in n-hexane. The n-hexane solution is transferred into a greasefree burette. Some 300 mg potassium bromide is placed on a watchglass under the tap of the burette and the n-hexane solution with the non-saponifiable part dissolved in it is allowed to drip slowly onto the potassium bromide. At the same time the potassium bromide mixture is heated with an infrared lamp placed at a suitable distance, and the rate of the drops and the distance between the tip of the burette and the potassium bromide are selected in such a way that the n-hexane evaporates soon after it has dropped down onto the potassium bromide. This method enables the complete n-hexane extract to be applied in a rapid and simple manner to the potassium bromide while the n-hexane is given the opportunity to evaporate off at the same time. The burette should then be washed out with a few millilitres of n-hexane and the latter should also be allowed to drop down and evaporate off in an analogous manner. After mixing, a potassium bromide pellet is made from the potassium bromide prepared in this way and this is used for the infrared analysis. It is important that the blank which has been obtained in a fully analogous manner is also investigated in a fully analogous manner and that this result is taken into account since the n-hexane itself can also cause the result to be influenced.

Some relevant valuation bands are summarized below, for these non-saponifiable, extractable substances of low volatility:

a) Paraffins show the following characteristic absorption bands
   C-H stretching vibrations between 3.33 - 3.57 µm (3000 - 2800 $cm^{-1}$)
   C-H bending vibrations $CH_2$ and $CH_3$ 6.89 - 7.29 µm (1450 - 1350 $cm^{-1}$)
   For long-chain paraffins with more than four $CH_2$ elements:
   Skeletal vibrations at about 13.89 µm (approx. 720 $cm^{-1}$)

b) Double bonds are principally indicated by:
   C=C stretching vibration: 6.25 µm (1600 $cm^{-1}$)

as a result of C-H stretching vibration: approx. 3.29 µm (3040 $cm^{-1}$)

c) Aromatic substances show characteristic absorption bands
C-H stretching vibrations between 3.23 - 3.33 (3100 - 3000 $cm^{-1}$)
Vibrations of the aromatic skeleton at approx.
6.18 µm (1600 $cm^{-1}$)
6.66 µm (1500 $cm^{-1}$)
6.89 µm (1450 $cm^{-1}$)

Attention should be paid to the aromatic C-H bending vibrations in the range 11.1 - 15.4 µm (900 - 650 $cm^{-1}$). Their spectral position is characteristic of the type of substitution and of any substituents present.

Appendix 3 to Method 4.1.8.1 and Appendix 1

Analysis by gas chromatography (see also Chapter 2)

If the saponifiable part is to be investigated in terms of, for example, its fatty acid spectrum, esterification to the methyl esters is to be recommended. For this the weighed out, saponifiable part is dissolved in some 50 to 60 ml methanol containing some 10 % by weight of sulphuric acid. This solution is transferred into a glass flask and boiled with a reflux condenser for about 1 hour. This operation causes the fatty acids to be transformer into their methyl esters (the esterification can also be carried out with diazomethane).

The methanolic solution is exhaustively extracted with n-hexane after having been allowed to cool down, and the n-hexane extract is then concentrated to a volume of some 5 ml by evaporation on a water bath at 80 °C. This residual volume is transferred into a 10 ml graduated flask and made up to the mark at 20 °C with n-hexane. Aliquot parts are removed using a microlitre syringe and analyzed by gas chromatography. For a packed column the following programme can be recommended.

| | |
|---|---|
| Column: | Glass column 2 m long, 1/8 inch with 10 % diethylene glycol succinate on Chromosorb 80 (80 to 100 mesh) |
| Carrier gas: | Nitrogen (20 to 25 ml/min.) |
| Injector: | Temperature 210 °C |
| Column temperature: | 190 °C |
| Detector: | Flame ionization detector (FID), 250 °C |

Blank values are to be obtained in an analogous manner. The retention times are determined by calibration with mixtures of the methyl esters of, for example, palmitic acid, oleic acid and stearic acid (see also Chapter 2 and Section 4.2).

### 4.1.8.2 Determination of hydrocarbons by infrared intensity spectroscopy

General remarks (see also Chapter 2)

The energy of most molecular vibrations lies in the infrared range of the electromagnetic spectrum. The characteristic vibrations of particular functional groups are to be found in defined regions of the IR range. The stretching vibrations of O-H, N-H and C-H are to be found in the range from 3.2 to 3.6 µm. The stretching vibrations of CH-, $CH_2$- and $CH_3$-groups produce characteristic absorption signals in the infrared spectral range between 3.57 µm and 3.23 µm (2800 to 3100 $cm^{-1}$). Extraction agents such as 1,1,2-trichlorotrifluoroethane do not possess significantly strong absorption bands in this spectral range, which means they hardly affect the quantitative determination at all. For this reason infrared spectroscopy can be used for quantitative determinations. The Lambert-Bouguer-Beer law applies here too:

$$c_i = \frac{E_{c,i,\lambda}}{e_{i,\lambda} \cdot d \cdot k_{c,i,\lambda}}$$

$E_{c,i,\lambda}$ = Extinction of the substance i at wavelength $\lambda$ and concentration c
$e_{i,\lambda}$ = Specific extinction of substance i at wavelength $\lambda$
$c_i$ = Concentration of substance i
$d$ = Thickness of the solution being irradiated in the cuvette (path length)
$k_{c,i,\lambda}$ = Constant for the arrangement of the device at wavelength $\lambda$ ; the measurement and evaluation factors are covered by this constant.

The quantitative determination of hydrocarbons in water and waste water with infrared intensity spectroscopy is technically a form of IR photometry with a large path length performed in the absence of water. For this reason the hydrocarbons have to be extracted before being determined. Quartz as the cuvette material (Infrasil cuvettes) is transparent in this wavelength range. As a result the extinction of hydrocarbons can be measured to a large extent without any interference between 3.6 and 3.2 µm. However, regardless of the particular arrangement selected, such extinction measurements are always less accurate than wavelength measurements; this is a fundamental disadvantage of intensity measurements.

## Sampling

The procedure is the same as for method 4.1.8.1, except that 1,1,2-trichlorotrifluoroethane must be used instead of n-hexane.

For this method of analysis, the level of hydrocarbons in the water should be over 0.01 mg/l or better still over 0.1 mg/l.

The trichlorotrifluoroethane used for the extraction must be of a very high degree of purity. For this reason it is recommended that it be tested for purity in the quoted range before being used for IR spectroscopy. If it is found to be contaminated, it must be distilled via a column before being used.

Other extractable substances such as triglycerides, fatty acids, surfactants etc. are also recorded at the same time, some completely, some only in part.

If the trichlorotrifluoroethane-water-emulsion is reluctant to separate, the addition of a neutral salt (sodium chloride, sodium sulphate) can help.

Incorrect adjustment of the spectrograph as well as measurements of very high or very low transmissions can falsify the results.

## Equipment

Infrared spectrophotometer, if necessary with Fourier transformation

Quartz cuvettes (Infrasil) with path lengths varying from 0.2 to 5 cm

Glass bottles, 1 litre or 2 litres (to be cleaned as for Method 4.1.8.1)

Separating funnel, 300 ml

Agitator motor with folding blade agitator for 3000 to 4000 rpm or turbo-agitator

Erlenmeyer flasks, 50 to 200 ml with ground glass stopper

1,1,2-trichlorotrifluoroethane

Florisil (Mg-Al-silicate) 60 to 100 mesh

Sodium sulphate, anhydrous, reagent purity, extracted with trichlorotrifluoroethane and then roasted at approx. 600 °C.

## Method

The pH of the water to be investigated is determined electrometrically using a separate sample.

The volume of the water sample added to the trichlorotrifluoroethane on the spot is marked; in addition it is either measured later or the quantity of water is determined as the difference in weight. The water sample together with the extraction agent is then stirred in the sampling bottle with an agitator for 30 seconds at between 3000 and 4000 rpm; a turbo-agitator can also be used. Then the aqueous phase and the organic phase are allowed to separate. (This process can take up to several hours).

Water is siphoned off from the upper phase until a residual volume of approx. 200 ml is left. The water that has been siphoned off is discarded. The remainder of the water phase and the organic phase are transferred to a separating funnel. The sampling bottle is washed out with a small quantity of trichlorotrifluoroethane and the solution obtained in this way is also added to the separating funnel. After the organic phase and the aqueous phase are separated, the trichlorotrifluoroethane is filtered off through sodium sulphate into an Erlenmeyer flask. The sodium sulphate is then washed off into the same Erlenmeyer flask with a little trichlorotrifluoroethane.

If the emulsion of organic extraction agent in water fails to break down after agitation in the sampling bottle, centrifuging must be carried out at 6000 rpm.

1 g Florisil is added to the filtered trichlorotrifluoroethane solution to separate off polar substances and the mixture is shaken well for 2 minutes. The Florisil is then allowed to settle out and the clear solution is made up to a definite volume. e.g. 50 ml.

Before the extinction measurement is carried out, the IR spectrophotometer must be accurately set.

Work is carried out with medium recorder speed, minimal damping and high resolution. The recorder should be fitted with a 0.25 mm cannula as its pen. The absorption curve of the cuvettes to be used for the measurement is checked against air in the range from 2.8 to 3.8 µm (see IR spectrum 1).

A sample of the trichlorotrifluoroethane to be used for the extraction is added to one of the cuvettes to be used for measurement of the extinction and checked for purity.

After the trichlorotrifluoroethane has been checked for purity, one of the empty cuvettes to be used for the measurement of the transmission is brought into the measuring beam while another one is brought into the reference

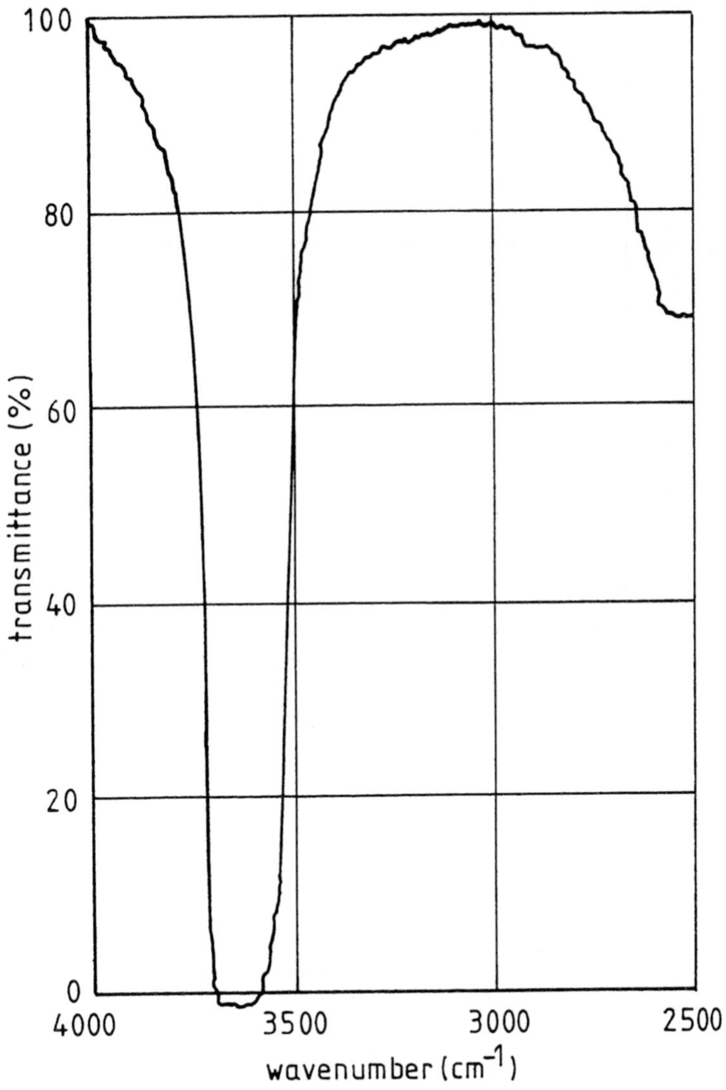


Fig. 122. IR spectrum 1. Empty 10 mm quartz cuvette against air

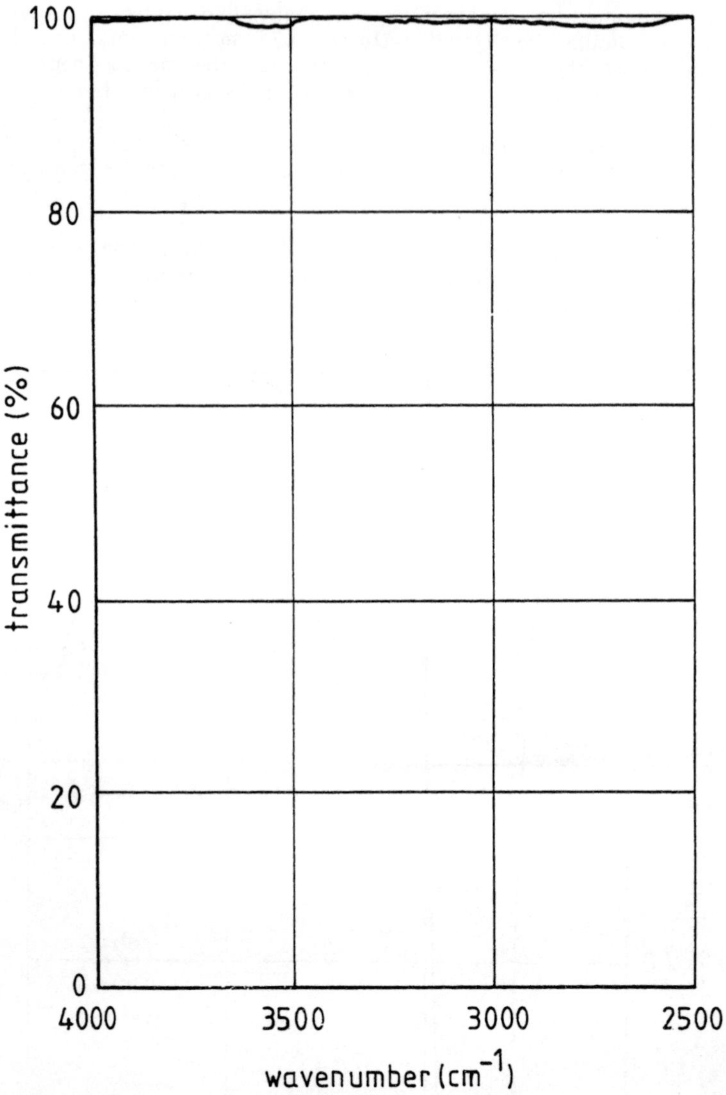


Fig. 123. IR spectrum 2. Compensated spectrum; 2 empty quartz cuvettes

beam. The cuvette in the measuring beam and the cuvette in the reference beam are then turned relative to one another around the vertical axis until optimum compensation has been reached (see IR spectrum 2).

Finally the two cuvettes are filled with the trichlorotrifluoroethane as used for the extraction, the spectrum is set to 100 % transmission and the quality of compensation is checked.

After these preparations, the trichlorotrifluoroethane extract is put into the cuvette in the measuring beam and the transmission is determined in the range from 3.2 to 3.6 µm.

## Carrying out the measurement

First of all cuvettes with only a short path length, e.g. 0.2 cm, are used in the measuring beam and in the reference beam. If it is found during the measurement of the transmission spectrum that the path length is too short, cuvettes with the next larger path length are used following the same methods for setting and adjustment as already described. This process is continued until the transmission spectrum obtained covers approx. 1/3 to 2/3 of the complete transmission range.

The scattering of the IR radiation increases with increasing path length of the solution being irradiated. If it is found that cuvettes with a path length in excess of 2 cm are necessary, a series of dilutions of trichloro-trifluoroethane should be used for spectroscopy. In this way errors from the scattering of the IR radiation arising from the use of cuvettes with long path length can be avoided.

## Evaluation

There is a relationship between the amount by which the transmission of the extraction agent is reduced as a result of the presence of the hydrocarbons in it and the concentration of these hydrocarbons in the water which it is desired to determine. This concentration can be determined by two evaluation methods:

1. The transmission of reference samples having known but different concentrations is determined and calibration curves are drawn. They provide information on the relationship between wavelength transmission and concentration. The concentration in the sample can then be determined with the aid of a series of curves obtained in this way.

2. The measured transmission of the sample being investigated is put into an empirically determined formula and the desired concentration is obtained from this.

Evaluation method 1 is demanding and takes a lot of time because it requires a knowledge of the substances being extracted. Method 1 can be used directly in those cases where the contamination substance or mixture of substances is adequately known. However, the evaluation can also be carried out via the absorption of the $CH_2$ bands at 3.45 µm. Here the setting is made using squalene ($C_{30}H_{62}$) and results must be related to squalene.

Evaluation method 2 is less time consuming. It is valid in each case for just one class of substances and can be used for 2 groups of mineral oil products:

a) "Motor gasolines with a high $CH_3$-group content and an aromatic substance content of up to 25 % by volume", e.g. gasoline;

b) "All other mineral oil products having a predominant proportion of $CH_2$ groups and an aromatic substance content of less than 10 % by volume", e.g. heating oil.

Evaluation method 2 cannot be used for mineral oil products having a predominantly aromatic substance content because the extinction in the spectral range of the aromatic substance CH stretching vibrations is too low.

Paraffinic hydrocarbons give a characteristic transmission minimum in the region of 3.4 μm, aromatic hydrocarbons in the region of 3.3 μm and naphthenic hydrocarbons in the region of 3.4 to 3.5 μm.

The method of evaluation in accordance with the semi-empirically obtained formulas for gasoline and heating oil are described below.

$$G_V = \frac{1.3a}{b \cdot d} (1.1 \cdot E_1 + 0.12 \cdot E_2 + 0.19\ E_3)\ \text{mg/l}$$

where:

$G_V$ = Content of extractable motor gasoline in the water sample
a = Volume of the solvent used for the extraction in ml
b = Volume of the water sample used in litres
d = Path length used in the measurement in cm
$E_1$ = Extinction at 3030 $cm^{-1}$ (3.30 μm); absorption of the aromatic CH groups
$E_2$ = Extinction at 2959 $cm^{-1}$ (3.38 μm); absorption of the $CH_3$ groups
$E_3$ = Extinction at 2924 $cm^{-1}$ (3.42 μm); absorption of the $CH_2$ groups

The evaluation is explained using IR spectrum 3 (Fig.124) as an example.

Evaluation of IR spectrum 3:

$$G_V = \frac{1.3a}{b \cdot d} (1.1 \cdot E_1 + 0.12E_2 + 0.19E_3)\ \text{mg/l}$$

$$E_1 = \log \frac{1}{T_1} = \log \frac{100}{79} = 0.102$$

$$E_2 = \log \frac{1}{T_2} = \log \frac{100}{31} = 0.509$$

$$E_3 = \log \frac{1}{T_3} = \log \frac{100}{40} = 0.398$$

a = 25 ml
b = 1.003 litres
d = 1 cm

$G_V$ = 32.403 · 0.249 = 8.1 mg/l gasoline
= result approx. 8 mg/l gasoline-type hydrocarbons following extraction and IR analysis.

The following formula is used for the recording of heating oil:

$$G_M = \frac{1.3 \cdot a}{b \cdot d} (0.12 \cdot E_2 + 0.19 \cdot E_3)\ \text{mg/l}$$

where:

$G_M$ = Content of extractable heating oil in the water sample
a = Volume of the solvent used for the extraction in ml
b = Volume of the water sample in litres

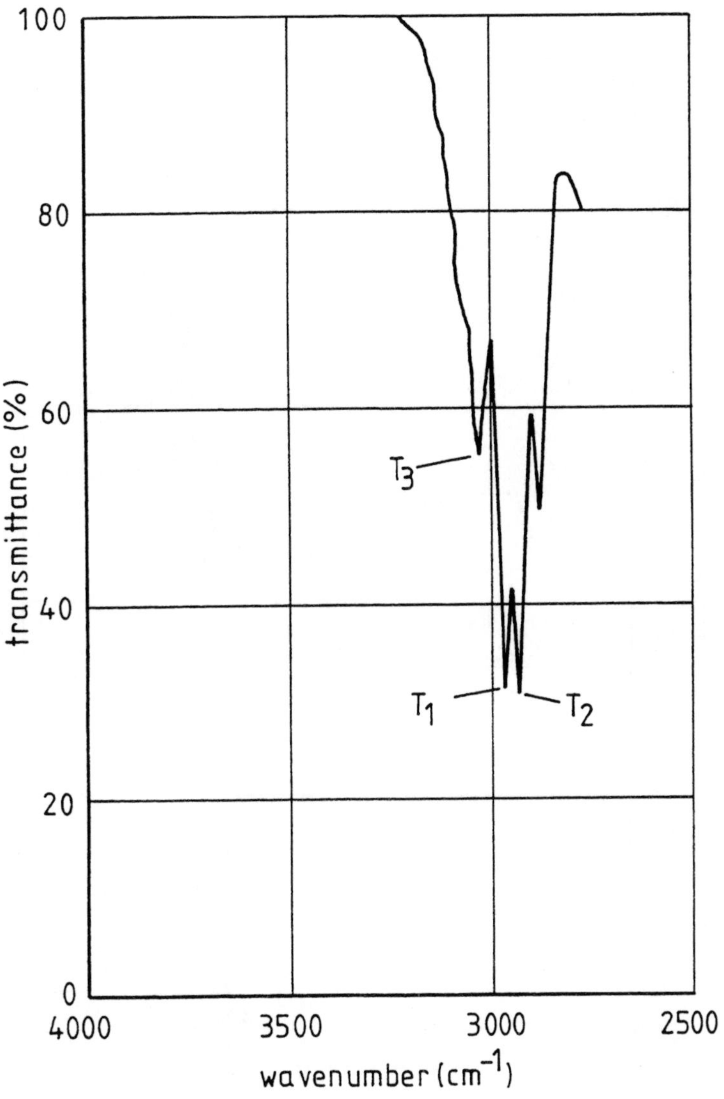


Fig. 124. IR spectrum 3. Gasoline in water

d = Path length used in the measurement in cm
$E_2$ = Extinction at 2959 $cm^{-1}$ (3.38 μm); absorption of the $CH_3$ groups
$E_3$ = Extinction at 2924 $cm^{-1}$ (3.42 μm); absorption of the $CH_2$ groups

$$G_M = \frac{1.3 \cdot a}{b \cdot d} (0.12 \cdot E_2 + 0.19 \cdot E_3) \text{ mg/l}$$

$$E_2 = \log \frac{1}{T_2} = \log \frac{100}{38} = 0.420$$

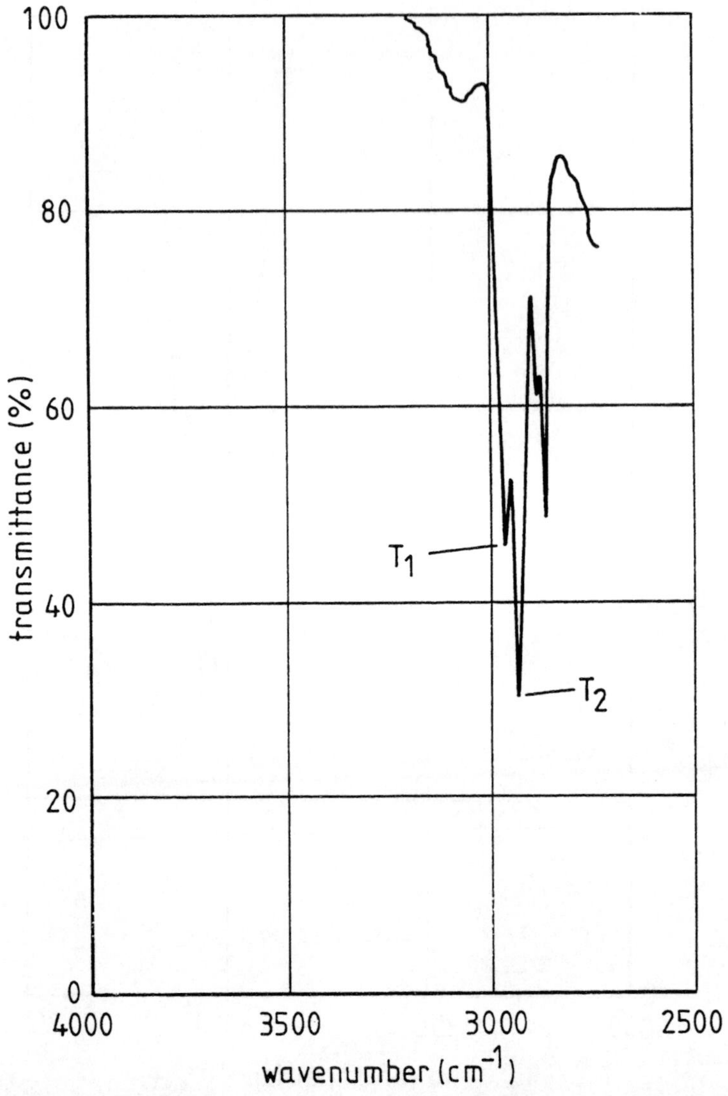


Fig. 125. IR spectrum 4. Heating oil in water

$$E_3 = \log \frac{1}{T_3} = \log \frac{100}{22} = 0.658$$

a = 25 ml
b = 0.274 litre
d = 1 cm

$G_V$ = 127.737 · 0.175 = 22.4 mg/l heating oil
= result 22 mg/l heating-oil type hydrocarbons after extraction and IR analysis.

### 4.1.8.3 Determination of hydrocarbons (measurement of UV fluorescence)

#### General remarks

The method of ascending two-dimensional thin layer chromatography permits aromatic hydrocarbons to be separated from the other accompanying substances which are also present after extraction with, for example, carbon tetrachloride or trichlorotrifluoroethane. Volatile aromatic hydrocarbons are lost during the drying processes in the method. By selecting the excitation wavelengths and the wavelengths for the measurement of the fluorescence, it can be arranged that aromatic hydrocarbons with more than four fused rings are not recorded in this method.

Electrons are excited as a result of the high energy of UV radiation. In the longer wavelength range of UV light, electrons are excited out of $\pi$ orbitals and in particular from $\pi$ conjugated systems. The conjugations of double bonds and substitution by residues with free electron pairs in the $\alpha$ position causes the absorption to be displaced to longer wavelengths. This can be demonstrated by using aromatic substances as an example:

| | |
|---|---|
| Benzene | Absorption maxima at 184 nm, 203.5 nm and 254 nm |
| Phenol | Absorption maxima at 210.5 nm and 270 nm |
| Naphthalene | Absorption maximum at 215 nm |
| Anthracene | Absorption maximum at 265 nm |

Emission of the absorbed radiation is by fluorescence or phosphorescence. Fluorescence is classified as being the emission of longer wavelengths within $10^{-5}$ seconds. Radiation which is emitted more than $10^{-5}$ seconds after absorption is termed phosphorescence.

Here fluorescence is used for the measurements. Fluorescence takes place over a relatively large range of wavelengths so that the fluorescence of different chemical compounds can be measured together by using a carefully selected wavelength.

The separating of the aromatic hydrocarbons from the other compounds is carried out by means of two-dimensional thin layer chromatography using n-hexane and benzene as solvents.

The water sample should contain between 1 and 5 mg/l of substances extractable with carbon tetrachloride. A suitable volume of water is selected for the extraction.

The n-hexane and benzene used as solvents should be of suitable purity for spectroscopy. If necessary, they should be distilled using a column.

#### Equipment

Same devices and chemicals as for Method 4.1.8.2 - Determination of hydrocarbons using infrared intensity spectroscopy.

In addition:

Carbon tetrachloride
n-hexane for spectroscopy
Benzene for spectroscopy

Caution! Observe safety regulations

Plates for thin layer chromatography
UV fluorescence spectrophotometer
Circulating air drying cabinet 60 °C
Microlitre syringe, 50 µl

Taking of samples: As recommended for Method 4.1.8.1

## Method

The carbon tetrachloride and water are mixed with a high speed agitator or a turbo-agitator in the sampling bottle. Then the phases are allowed to separate. The aqueous phase is discarded. Working with a fume hood, the carbon tetrachloride extract is concentrated carefully by evaporation to approx. 5 ml on a water bath and is then brought to exact volume in a 5 ml graduated flask.

The TLC plate is cleaned with chloroform and dried. Then 10 µl amounts are taken approx. 5 times from the carbon tetrachloride solution and dropped on the plate at one point, the carbon tetrachloride being blown off with a hairdryer between the individual applications. The point should be at the left hand bottom end of the plate.

After this has been done, the plate is placed in a chamber for ascending thin layer chromatography and developed with n-hexane as vehicle up to a 10 cm solvent front. The plate is taken out and dried at 60 °C in a circulating air drying cabinet. After drying, the plate is developed in the same direction using benzene or toluene with a solvent front up to 8 cm and is then dried again in a circulating air drying cabinet.

The n-hexane as a non-polar solvent first of all separates the paraffinic hydrocarbons from the other hydrocarbons. They migrate within the solvent front from bottom to top. The benzene or toluene brings primarily the aromatic substances as a group to a height of 8 cm. The plate is now turned through 90 °C and is first of all again developed with n-hexane up to 10 cm.

The paraffinic hydrocarbons are now to be found at the top end of the solvent front. The aromatic hydrocarbons are separated in accordance with their polarity relative to the n-hexane, being distributed over the path of the solvent.

Usually the degree of separation is still not quite adequate so that a fourth developing run is made in the second direction again using a mixture of 80 % n-hexane by volume and 20 % benzene or toluene by volume. The solvent front should reach 10 cm. Drying is then carried out again.

The position of the substance groups is quoted in $R_f$ units. The starting point is given the $R_f$ value 0, the solvent front the $R_f$ value 1. The starting point of the second running direction is indicated by the blanks of the runs in the first dimension.

After development in the second dimension, the paraffins are to be found just beneath the solvent front with an $R_f$ value of approx. 0.95, the naphthenic hydrocarbons in a similar position with an $R_f$ value of 0.9. The aromatic hydrocarbons are distributed having $R_f$ values between 0.8 and 0.1. Strongly polar substances remain near the starting zone.

The fluorescence measurement is carried out with a chromatogram UV fluorescence spectrophotometer. An analysis width of approx. 1 cm is recommended.

Two excitation wavelengths are used to excite the fluorescence:

a) 313 nm
b) 360 nm.

The analyzer is set to

a) 365 nm
b) 445 nm.

Identification

a) When the spectrophotometer is set with the exciter wavelength at 313 nm and the receiver wavelength at 360 nm, experience has shown that the 1-ring and 2-ring and, to a small extent too, the 3-ring aromatic hydrocarbons are recorded. Naphthenic and paraffinic hydrocarbons are not excited. Aromatic hydrocarbons with more than 4 fused rings are also not indicated either.

b) When the exciter wavelength of 365 nm is combined with a fluorescence wavelength of 445 nm, the aromatic hydrocarbons, which are less polar relative to n-hexane, are recorded as well as reaction products resulting from the action of natural photochemical processes on oils and greases in water and other polar substances in the start region of the thin layer plate.

Evaluation

The strength of the fluorescence signals in relation to the $R_f$ values can be used for the semi-quantitative estimation of the level of aromatic hydrocarbons in the water sample.

The following measures should be observed:

A thin layer plate which has been previously cleaned with chloroform is developed using the same solvents but without application of the test solution. The method described above is followed. The developed plate is used to estimate the background fluorescence caused by the solvent itself.

For calibration purposes, gasoline, heating oil, lubricating oil, heterocyclic aromatic substances and also aromatic substances with fused rings are separated using the method described above. The test solution must be applied quantitatively, i.e. carbon tetrachloride solutions having a known content must be applied.

Benzene and naphtalene derivatives show fluorescence maxima between $R_f$ values of 0.6 and 0.9; they are found as the result of contamination from lubricating oils and middle distillates.

Aromatic hydrocarbons of natural origin are found to be between $R_f$ values of 0.2 to 0.8. If only "biogenic contamination" is present, the range with $R_f$ values > 0.8 remains free.

4.1.8.4 For gas chromatography head-space analysis see Section 4.2.11

### 4.1.9 Phenol (phenol index)

#### General remarks

Phenols are aromatic compounds with one or more hydroxyl groups bonded directly to the benzene ring. Surface waters in general possess a natural phenol content, which lies in the region of a few µg/l and is made up of higher molecular weight compounds; these compounds can arise in the course of biological transformation (humification) of vegetable material (foliage, algae) in the ground and in the water.

Anthropogenic phenols are often superimposed on this natural phenol level. They get into surface and ground water as a result of, for example, industrial, domestic, or communal waste water having been inadequately cleaned.

Phenols pollute untreated water that is to be turned into drinking water. Whereas phenol as such ($C_6H_5OH$) can be detected by its smell or taste in water at levels of 0.01 to 0.1 mg/l, this effect is increased by one to two orders of magnitude to some µg/l as a result of the effect of chlorine during the purification of drinking water which causes the chlorine phenols, which have a more intense smell and taste, to be formed. Phenols are volatile with steam to varying degrees; they dissolve in ethanol and diethyl ether and form phenolates with alkalis. As far as evaluation is concerned, it is important to classify phenols into those which are distillable with steam and those which are not.

The group of phenols which are distillable with water vapour include phenol, the cresols, the xylenols, guaiacol, thymol, the main fraction of 1,2-dihydroxybenzene and a small fraction of 1-naphthol. Those which are not distillable with water vapour include 2-naphthol, hydroquinone, resorcinol, pyrogallol and phloroglucinol, and part of 1-naphthol and of 1,2-dihydroxybenzene.

#### Methods

(The first three methods as overall parameters)

4.1.9.1 Phenol index, with p-nitroaniline and extraction, and also with additional steam distillation

4.1.9.2 Phenol index, with 4-aminoantipyrine without extraction but with steam distillation

4.1.9.3 Phenol index, with 4-aminoantipyrine after extraction with and without steam distillation

4.1.9.4 Phenols, see Section 4.2 for the gas chromatography method (individual determination of the phenol-type substances)

In the case of methods 4.1.9.1 - 4.1.9.3 the overall parameter (phenol index) is determined by photometry. Either p-nitroaniline or 4-amino-antipyrine can be used; these give coloured coupling compounds which can be evaluated by photometry under the stated conditions.

### 4.1.9.1 Determination of phenol-type substances which are capable of coupling, in natural waters, with p-nitroaniline

**General**

For extraction with n-butanol, the volume of the water or the distillate used is such that the concentration of phenols in the butanol extract is equivalent to a level of 0.001 to 0.4 mg/l phenol in the water sample.

Equations for reaction:

$$O_2N-\!\!\!\bigcirc\!\!\!-NH_2 + 2\,HCl + NaNO_2 \longrightarrow$$

$$O_2N-\!\!\!\bigcirc\!\!\!-N=NCl + NaCl + 2\,H_2O$$

$$O_2N-\!\!\!\bigcirc\!\!\!-N=NCl + \bigcirc\!\!\!-OH \longrightarrow$$

$$O_2N-\!\!\!\bigcirc\!\!\!-N=N-\!\!\!\bigcirc\!\!\!-OH + HCl$$

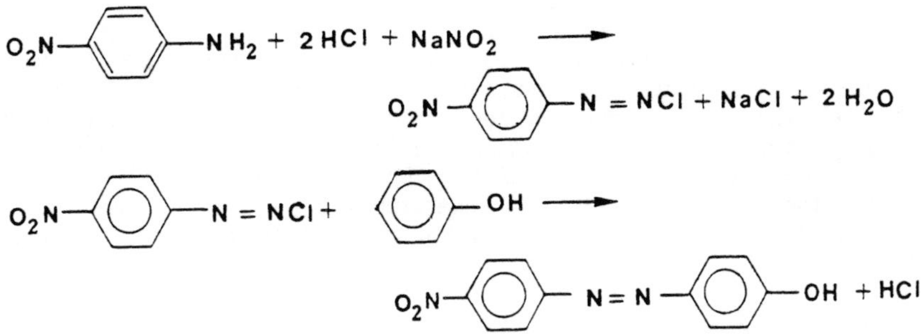

Interference arising from sulphide or cyanide ions is eliminated by adding $CuSO_4$ or $CoSO_4$ solution.

**Equipment**

Photometer

Cuvettes

Distillation apparatus consisting of 500-ml round-bottom flask, spherical distilling head and descending Liebig condenser

Separating funnel, 1 litre

Measuring flask, 100 ml

pH meter

Phosphoric acid (1.70 g/ml)

Copper(II)sulphate solution:
    Dissolve 10 g $CuSO_4 \cdot 5\,H_2O$ in 100 ml $H_2O$

Cobalt(II)sulphate solution:
    Dissolve 10 g $CoSO_4 \cdot H_2O$ in 100 ml $H_2O$

p-nitroaniline solution:
    0.69 g p-nitroaniline is dissolved in 155 ml 1 m hydrochloric acid and made up to 1 litre with distilled water

Sodium nitrite solution, saturated

Sodium carbonate solution, 1 m

n-butanol

Phenol stock solution:

1.000 g reagent-purity phenol (freshly distilled) is made up to 1 litre with distilled water. This solution is used to prepare suitable dilutions for calibration purposes.

Method

200 ml of the water sample (or an appropriately smaller volume which is then made up to 200 ml with distilled water) is treated with 30 ml 1 m sodium carbonate solution in a 500-ml separating funnel and the pH brought to approx. 11.5 (electrometric check) by (if necessary) adding 30 % sodium hydroxide drop by drop. After adding 20 ml diazotized p-nitroaniline solution (drops of saturated sodium nitrite solution are added to the p-nitroaniline solution until the latter becomes colourless), the mixture is allowed to stand for 20 min. and the coloured substance which has been formed is then extracted by shaking with 50 ml n-butanol. After a further 10 min., the aqueous phase is separated off and the extinction of the butanol extract is measured against a blank test which is carried out in parallel.

Calibration

To produce the calibration curve, suitable dilutions of the phenol stock solution having phenol contents of 0.001 - 0.4 mg are prepared, these being made up in each case to 200 ml with distilled water. The standard solutions are then subjected to the same analysis process as described above for the water sample.

**Distillation**

200 ml of the water sample (or an appropriate smaller volume which is then made up to 200 ml with distilled water) is treated in the distillation apparatus with 1 ml $CuSO_4$ solution and/or $CoSO_4$ solution as necessary (in order to bind sulphide and/or cyanide). After adding 10 ml phosphoric acid (1.70 g/ml), the mixture is distilled into a flask containing 30 ml 1 m sodium carbonate solution until just 20 ml of the water sample remains in the distillation flask.

The rest of the method is as given above.

Evaluation

The phenol content is read off from the calibration curve against the measured extinction value from which the blank value has been subtracted. After taking into account the volume of water used, the content of phenols is then expressed in mg/l and stated as the phenol index measured in accordance with this method.

### 4.1.9.2 Phenol index with 4-aminoantipyrine without extraction after steam distillation

**General**

Phenol-type substances, and also certain other compounds capable of coupling react with 4-aminoantipyrine, to produce - in alkaline reaction and under oxidizing conditions - antipyrine dyes which can be extracted with organic solvents, e.g. chloroform (4.1.9.3). Under particular conditions, these extracted colour complexes can be evaluated by photometry to yield information on phenol-type substances in water samples. In the case of the method described here, steam distillation is carried out at a pH of around 4 in a similar manner to that described for method 4.1.9.1, but without extraction of the colour complex.

Care should be taken to ensure that chlorinated water is treated with ascorbic acid; all other possible causes of interference are eliminated by the distillation method described under method 4.1.9.1.

**Equipment**

As for method 4.1.9.1 plus ascorbic acid.

Buffer solution pH 10:
Dissolve 34 g ammonium chloride and 200 g sodium potassium tartrate in approx. 700 ml distilled water, add 150 ml ammonium hydroxide solution (0.91 g/l) and make up to 1 litre.

4-amino-antipyrine solution:
Dissolve 2 g of 4-amino-2,3-dimethyl-1-phenyl-3-pyrazoline-5-one in 100 ml water. This solution must be freshly prepared every day.

Potassium hexacyanoferrate (III) solution:
8 g in 100 ml

**Procedure**

Distillation as for 4.1.9.1. The pH is set to 4 with phosphoric acid, and a water sample volume of 500 ml is recommended.

Some 400 ml is distilled over. This distillate is then treated with 20 ml of the buffer solution and made up to 500 ml. The pH (determined electrometrically) should then be 10 $\pm$ 0.2.

Three ml of the aminoantipyrine solution is now added, the mixture is shaken and finally 3 ml of the potassium hexacyanoferrate solution is added. After the mixture has been allowed to stand for 10 min., it is evaluated by photometry at 510 nm. A blank sample is tested in the same way. In addition it is recommended that aqueous solutions containing known amounts of phenol are subjected to the same process of analysis, i.e. including distillation.

### 4.1.9.3 Phenol index with 4-aminoantipyrine after extraction and also if necessary after distillation

**General remarks**

Where determination is to be preceded by distillation, the distillate described in 4.1.9.2 is used; otherwise the water sample, which should be investigated as soon as possible after the sample has been taken, is acidified to a pH of less than 4 and then 0.5 g of copper sulphate is added. The water sample to be investigated is transferred to a separating funnel; 20 ml of the buffer solution is added and finally the pH is checked electrometrically; it should be 10 ± 0.2. 3 ml of the aminoantipyrine solution and 3 ml of the potassium peroxodisulphate solution (0.65 % in water) are added, the mixture is shaken and then allowed to stand for between 30 and 60 min. in the dark. This solution is then extracted with 2 times 15 ml chloroform. The two chloroform extracts are purified and made up to 50 ml at 20 °C. They are then measured at 460 nm. The photometer is set at 460 nm against chloroform. A blank sample is tested in the same way. It is also recommended that known quantities of phenol be added to the water sample being investigated after the admixture operation and that the retrieval rate then be determined using this method. This process yields information on possible interference. Evaluation is carried out using a previously prepared calibration curve and taking account of the quantity of water used, the dilutions, the extraction volume, the cuvette path length etc.

When quoting phenol index figures, it should be observed that the determination limit with this method is around 2 µg/l.

#### 4.1.9.4 Separation by gas chromatography of phenol-type substances including halogenated phenols (see Section 4.2)

It is frequently observed in analytical practice that the phenol index as a total parameter gives positive measured values whereas the analysis for individual phenols using gas chromatography does not yield positive results. In such cases a check should always be made to see whether the gas chromatography program covers phenol-type substances capable of coupling - or substances reacting as such - which can arise in nature or anthropogenically. It must also be noted that humic matter contains phenol-type groups, which means that discrepancies between the total parameter of the phenol index and the determination of individual phenols obtained by gas chromatography also appear possible for this reason, too.

### 4.1.10 Nitrogen compounds

#### General remarks

The nitrogen contained in organic substances in water is converted to ammonium ions by Kjeldahl's decomposition method. The ammonium ions are distilled from alkaline solution as ammonia, collected in boric acid solution and determined acidimetrically or photometrically in the receiver. Ammonium ions originally contained in the sample are separated, identified and deducted, or separated off by distillation before carrying out the Kjeldahl decomposition. Nitrite and nitrate are volatilized by $H_2SO_4$ in the Kjeldahl decomposition process.

The method is suitable for the determination of organically bonded nitrogen in concentrations greater than about 0.2 mg/l. Aromatic and heterocyclic compounds with N atoms in the ring are only partially detected.

#### Total nitrogen

The sum of nitrate nitrogen, nitrite nitrogen, and ammonium nitrogen and organically bonded nitrogen, each determined separately, is designated total nitrogen. Elemental nitrogen dissolved in the water is not contained in this total.

## Calculation

Determine separately the content of nitrate, nitrite and ammonium ions in the water, convert the results to mg/l N and add to the content of organically bonded nitrogen in mg/l N:

$$\text{Total nitrogen N mg/l} = A \cdot 0.23 + B \cdot 0.30 + C \cdot 0.89 + D$$

A = Nitrate concentration in mg/l $NO_3^-$
B = Nitrite concentration in mg/l $NO_2^-$
C = Ammonium concentration in mg/l $NH_4^+$
D = Content of organically bonded nitrogen in the sample in mg/l N

## Conversion factors

| mg | | N mg | $NH_4^+$ mg | $NO_2^-$ mg | $NO_3^-$ mg |
|---|---|---|---|---|---|
| N | corresponds to | 1.00 | 1.29 | 3.28 | 4.43 |
| $NH_4^+$ | " | 0.78 | 1.00 | 2.55 | 3.44 |
| $NO_2^-$ | " | 0.30 | 0.39 | 1.00 | 1.35 |
| $NO_3^-$ | " | 0.23 | 0.29 | 0.74 | 1.00 |

## Equipment

Distillation apparatus with receiver

Sulphuric acid, conc., (1.84 g/ml)

Sulphuric acid (1.27 g/ml):
Carefully add 100 ml of conc. sulphuric acid to 300 ml of dist. water.

Sodium sulphite solution:
$Na_2SO_3$, reagent purity, 5 % in dist. water

Iron (III) chloride solution:
$FeCl_3 \cdot 6\ H_2O$, reagent purity, 10 % in dist. water

Copper sulphate solution:
$CuSO_4 \cdot 5\ H_2O$, reagent purity, 10 % in dist. water

Potassium sulphate, $K_2SO_4$, reagent purity

Wieninger's selenium reaction mixture

a) Merck, Art. No. 8030:
97 % $Na_2SO_4$ (anhydrous)
1.5 % $CuSO_4 \cdot 5\ H_2O$
1.5 % Selenium (elemental)

b) or Art. No. 15 348:
3 % $CuSO_4 \cdot 5\ H_2O$
3 % $TiO_2$
94 % $K_2SO_4$ (anhydrous)

Phenolphthalein solution:
Dissolve 1 g phenolphthalein in 100 ml ethanol

Boric acid solution:
40 g of reagent-purity $H_3BO_3$ diluted to 1 litre in ammonium-free water (1 ml of this solution absorbs about 2 mg of ammonium ions)

Sodium hydroxide solution:
NaOH 30 % in dist. water

Reagents for acidimetric titration or photometric analysis of the ammonium-ion content in the distillate: see "Acidimetric determination following distillation" or "Photometric determination as indophenol" in Section 3.3.5 "Ammonium".

Method

Filter and, if necessary to remove ammonium ions, distill the water sample following alkalization. Transfer up to 500 ml of the sample to an appropriate Kjeldahl flask.

Treat the sample with 10 ml of sulphuric acid, (1.27 g/ml), 10 ml of sodium sulphite solution and 5 drops of iron(III)chloride solution, and evaporate to a volume of about 20 ml. When cool, add 10 ml of conc. sulphuric acid, 1 ml of copper sulphate solution and between 1 and 3 g of selenium reaction mixture. Heat the sample until white $SO_3$ vapour appears, then place a cooling bulb in position. Decomposition is complete when the solution is completely clear, which generally takes 20 - 30 minutes. If decomposition is difficult, for example in the case of waste waters, the boiling point of the sulphuric acid may be raised by adding 5 g of potassium sulphate to the reaction mixture.

When the solution is cool, add 250 ml of distilled water, some boiling chips and a few drops of phenolphthalein solution. Add sodium hydroxide solution until a pink coloration appears and then immediately distill into a receiver containing about 50 ml of boric acid solution. The end of the condenser must dip into the boric acid solution. Continue until about 200 ml of distillate has passed over.

a) **Titration.** Treat the distillate with 5 drops of methyl red solution and titrate from yellow to red with 0.025 m $H_2SO_4$. Boil up briefly to expel $CO_2$, allow to cool to room temperature and titrate again to the same shade of colour. For the blank test, carry out decomposition with ammonium-free distilled water and all the reagents in the same way, subsequently distill and titrate the distillate or measure photometrically according to b).

b) **Photometric determination.** Transfer the distillate to a 200-ml measuring flask and make up to the mark with ammonium-free water. Determine aliquot parts according to the method described under "Photometric determination of ammonium as indophenol" (Section 3.3.5.1).

Calculation

Titration method

1 ml of 0.025 m $H_2SO_4$ is equivalent to 0.70 mg of N. The content of organically bonded nitrogen can be calculated according to the following formula:

$$(a - b) \cdot F \cdot 0.70 \cdot \frac{1.000}{V} = \text{mg/l} \quad \text{Organically bonded nitrogen}^{1}$$

a = Consumption of 0.025 m $H_2SO_4$ in ml for the distillate of the sample
b = Consumption of 0.025 m $H_2SO_4$ in ml for the blank
F = Factor of the 0.025 m $H_2SO_4$
V = Volume of the water sample in ml

Calculation

Photometric determination

Take the average of several individual measurements, deduct the blank reading and read off the content of ammonium ions from the calibration curve. Convert the result to nitrogen ("organically bonded nitrogen") by multiplying by 0.78[1].

Up to 10 mg of N/l, the values should be rounded off to 0.1 mg/l. Above 10 mg of N/l round off to whole numbers.

## 4.1.11 Organic Acids

General remarks

Organic acids frequently occur in polluted water, sometimes in considerable concentrations, e.g. in leakage water from rubbish tips, household and industrial sewage etc. They are formed in especially large quantities when "acid fermentation" of organic substances takes place under anaerobic conditions (stage preceding the methane phase in sludge and waste materials). If the organic acids are not recorded separately for this type of polluted water, they can cause spurious results, e.g. when measuring the level of hydrogen carbonate, and in general can lead to an incorrect assessment, e.g. when drawing up "ion balances". The method selected for the analysis of organic acids must always be stated in the report.

### 4.1.11.1 Determination of organic acids which are volatile with steam

General remarks

Organic acids which are volatile with water vapour and whose low level of dissociation can be further suppressed by adding phosphoric acid, are distilled from the water sample and measured in the distillate by titration with sodium hydroxide solution against phenolphthalein.

This method can be used with leakage water from rubbish tips and polluted surface water containing volatile fatty acids in concentrations of approx. 0.1 mmol/l and above. Almost without exception carbon dioxide present in such water samples must be expelled from the distillate by boiling and using a backflow condenser before titration is carried out. If hydrogen sulphide is present in quantities large enough to be a problem, it must be combined by adding copper sulphate to the water sample.

---

1 If $NH_4^+$ ions have not been previously separated off they should be determined separately, converted to N and deducted from the results of the measurement.

The influence of nitrates is eliminated by adding approx. 100 ml of amidosulphonic acid or urea to the sample. If the sample contains phenols which may be transferred into the distillate, titration should not be carried out against phenolphthalein, but instead against a mixed indicator (methylene blue/neutral red).

Equipment

Apparatus for steam distillation with receiver and flask with reflux condenser, boiling off $CO_2$

Analytically pure phosphoric acid (1.70 g/ml)

Sodium hydroxide: 0.1 m

Phenolphthalein 1 % solution in ethanol

Mixed indicator:
0.05 g methylene blue and 0.2 g neutral red are dissolved in 100 ml 70 % ethanol.

Method

Transfer 0.1 to 1 litre of the homogenized or filtered (depending on purpose of test) sample into the distillation flask, acidify with 5 ml phosphoric acid and distill with steam. If sludge is to be analyzed, first dilute the sample with sufficient distilled water to make it capable of distillation. Ensure that the volume in the flask remains virtually constant throughout the process. Stop distillation once the quantity of distillate has reached 500 ml, then boil the latter for 10 min. using the reflux condenser, in order to expel any gases dissolved (e.g. $CO_2$). Leave to cool to room temperature, add 5 drops of phenolphthalein solution and titrate with 0.1 N sodium hydroxide solution until a permanent pink colour is reached.

If phenols are present, titrate until the colour of the mixed indicator changes to bluish-green.

Calculation

1 ml 0.1 m NaOH corresponds to 0.1 mmol volatile organic acids. The quantity is calculated in mmol/l, taking into account the volume of the sample used.

### 4.1.11.2 Quantitative analysis of organic acids after separation by column chromatography

General remarks

Organic acids are absorbed from an acidified aqueous solution by being passed through a chromatographic column of silica gel. After elution with a chloroform/butanol mixture, titration is carried out in the eluate itself with methanolic sodium hydroxide solution. This method is only recommended for dissolved organic acids after preliminary filtration of the water sample, where necessary.

Dissolved organic acids in concentrations of 0.5 mmol/l and above can be detected. Apart from organic acids, alkyl sulphates, alkylaryl sulphonates and other chemically related surface-active substances are absorbed by the silica gel column and are eluted with the organic mixture of solvents. However, these generally occur in such low concentrations that they can be ignored for the purpose of evaluation. In exceptional cases they have to be measured separately and taken into account in the calculation.

## Equipment

Centrifuge

Glass filter crucible for 20 g silica gel

1000 ml shaking funnel

Silica gel for chromatography, 50 - 200 mesh. To separate off any excessively fine material, elutriate the preparation with distilled water, let it settle for 15 min., and decant the excess water. Then dry in the drying oven at 100 - 105 °C and store in the dessiccator.

Chloroform/butanol mixture:
Mix 300 ml analytically pure chloroform, 100 ml n-butanol and 80 ml of 0.25 m sulphuric acid in the shaking funnel and once the phases have separated, allow the bottom, organic phase to run off through a folded filter into a dry flask.

Thymol blue indicator solution:
Dilute 80 mg thymol blue in 100 ml abs. methanol.

Phenolphthalein indicator solution:
Dissolve 80 mg phenolphthalein in 100 ml abs. methanol.

Concentrated, analytically pure sulphuric acid (= 1.84 g/ml).

Methanolic sodium hydroxide solution, 0.02 m:
Dilute 20 ml 1.0 m sodium hydroxide solution with abs. methanol to make 1000 ml. The normality of the solution must be checked frequently, and absorption of $CO_2$ into the solution must be prevented.

## Method

Filter or centrifuge the water sample and transfer 20 ml of the clear sample to a flask. Add a few drops of thymol blue indicator solution, then add concentrated sulphuric acid one drop at a time until the indicator colour changes (pH 1.0 - 1.2).

Place a glass filter crucible on a suction flask, add 20 g silica gel, connect up to the water jet pump and tap the apparatus so that the silica gel settles uniformly. Then add 10 ml of the acidified water sample using a pipette to give a uniform distribution of the liquid over the entire surface of the silica gel layer. As soon as the latter has absorbed all the liquid, immediately elute in the water-jet vacuum with a total of 100 ml chloroform/butanol mixture added in stages. Stop elution as soon as the last of the mixture has been absorbed into the silica gel layer.

Now transfer the eluate from the suction bottle into the titration vessel and drive off any $CO_2$ by briefly passing hydrogen into the vessel. Add a few drops of the phenolphthalein indicator solution and titrate in a $CO_2$-free atmosphere using 0.02 m methanolic sodium hydroxide solution until the colour changes to pink.

Carry out the same procedure with a blank solution of 10 ml distilled water.

### Calculation

1 ml 0.02 m NaOH is equivalent to 0.02 mmol of organic acids whose quantity can thus be calculated in mmol/l, taking into account the sample volume used.

### Note:

If the term "organic acids" is purposely modified to read "titratable organic substances" it is possible, in overall titration to pH 4.3 (m value; see Chapter 1), to take account of the latter when calculating the level of hydrogen carbonate ions.

The sum of the titration value in mmol/l for total titration to pH 4.3 is calculated as mg hydrogen carbonate ions (a).

The free dissolved carbon dioxide level is measured in the field (sampling location) by means of back titration (Chapter 1) (b).

In addition both the free and bound carbon dioxide are fixed with CaO or NaOH and the total carbonic acid level is determined in the laboratory by means of "gravimetric measurement after distillation" (as described in Section 3.2. "Total Carbon Dioxide") (c).

If the value for free dissolved carbon dioxide as per (b) is subtracted from the result obtained as per (c) the remainder is part of the titration obtained as per (a), which can be classified with the hydrogen carbonate ions found to be present. The difference between the value (c) minus (b) and the value (a) (all in mmol/l) gives the level of "titratable organic substances", a category not defined in any greater detail.

This method is also suitable for correcting the m-value measurement obtained, for example, with samples of sewage and leakage water from refuse tips containing unidentified organic substances which are otherwise incorrectly taken as "hydrogen carbonate ions". The latter error frequently causes discrepancies in the figures obtained for cation and anion equivalents in so-called full analyses. Using the method described above, this type of error can be identified and possibly reduced.

It must be pointed out that this method of testing and calculation is only to be regarded as giving general information. It is, however, important to add that if large discrepancies occur between the levels of hydrogen carbonate ions measured by direct titration and by subtracting the free $CO_2$ from the total carbon dioxide, this is in all probability attributable to the presence of other titratable substances.

### 4.1.11.3 Analysis of organic acids using gas chromatography

Gas chromatography is also suitable (as is HPLC) for separating volatile and non-volatile organic acids. There is no standard method and it should be noted that each different type of water or sewage sample must be individually prepared for this sort of testing.

In the method described below, samples are first passed through an anion exchanger and then esterified before the GC analysis takes place.

### Test procedure

For the exchange resin put 2 - 3 g XAD-4 into a glass column (12 cm · 0.8 cm) and pass 200 ml of the sample (waste water) or 1 - 4 litres (surface water) through the exchanger at a rate of 4 - 5 ml/min.

Wash through with 25 ml diethyl ether, then with 25 ml methanol and finally again with 25 ml diethyl ether.

Extract the sorbed anions with 20 ml diethyl ether saturated with HCl. Return this eluate to the column and wash through with ether-HCl.

Boil this new eluate down until it is dry and use a stream of $N_2$ to remove any last remaining liquid. Add 2 - 3 ml diazomethane until a constant yellow colour is obtained, then top up with diethyl ether to a defined volume. Inject an aliquot quantity of the resulting solution into a gas chromatograph.

Example of test conditions:

Type of column:
e.g. 30 m SE 30 glass capillary tube

Temperature of injector and F.I.D.: 275 °C

Gas 1 ml/min. Split: 1 : 25

Oven: 50 °C - 250 °C

Temperature program: 6 - 30 °C/min.

## 4.1.12 Isolation and measurement of humic substances*

### General remarks

Over 50 % of dissolved organic carbon (DOC) in surface water is classified under the category of humic substances (HUS). These are relatively high-molecular compounds with a complex structure which still remains to be clarified accurately. For this reason it is especially important to characterize HUS by determining their physico-chemical properties.

In order to characterize and quantify aquatic HUS specifically, they must first of all be isolated. The method most widely used at present is sorption with synthetic resins at a low pH followed by elution with alkaline solution (i.e. at a high pH). Mantoura and Riley used amberlite on a polystyrene base (XAD-2) whilst Thurman and Malcolm used polyacrylamide (XAD-8). The method used by Frimmel and Niedermann on numerous bodies of water is shown in the chart below.

---

* F. H. Frimmel

Chart

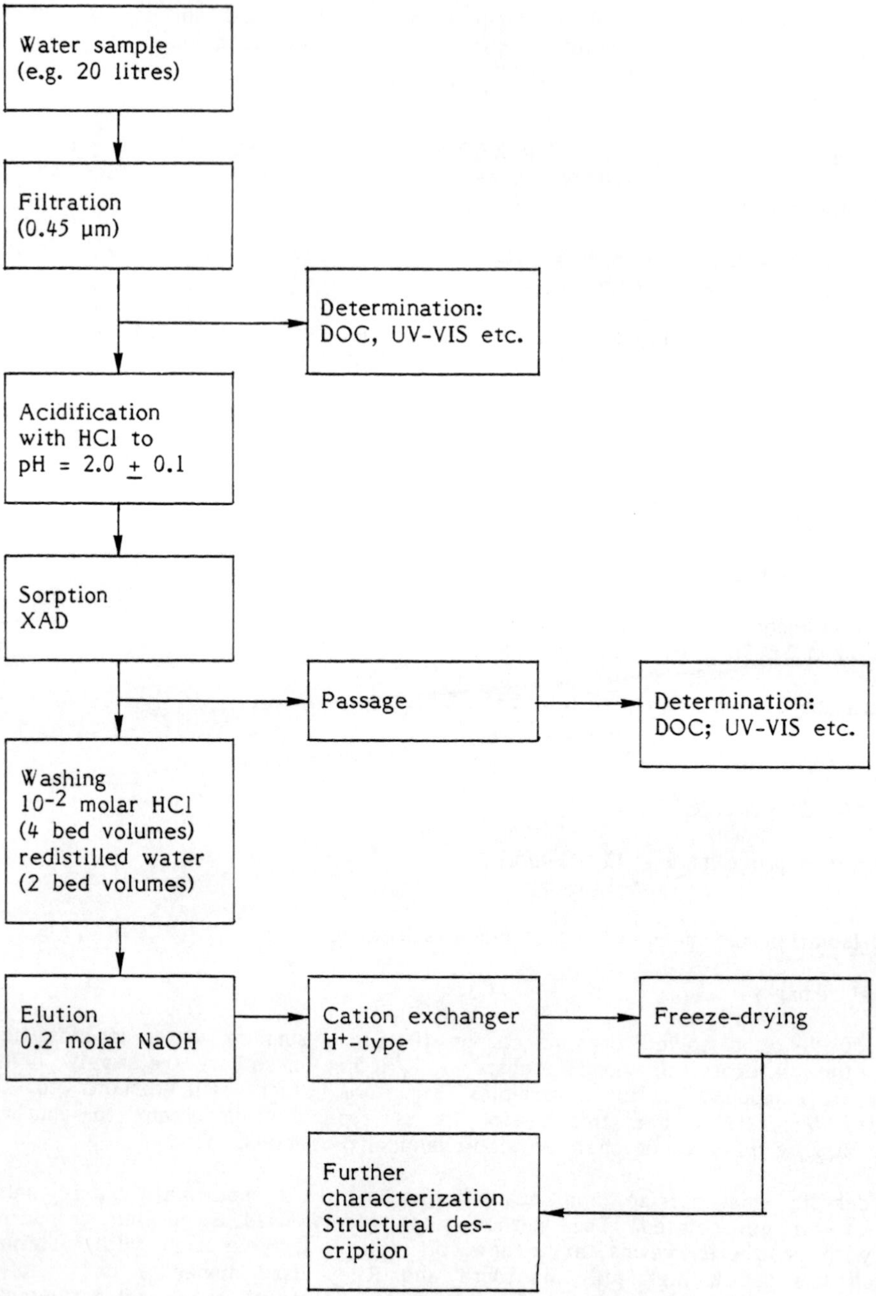


**Fig. 126.** Sequence of steps for isolating and characterizing aquatic humic substances

| Parameter | Type of sample | Method | Units |
|---|---|---|---|
| Elementary analysis (C, H, O, N, S) | s | Element | relative % |
| DOC | d | DOC analyzer | mg/l |
| A (254 nm) | d | Spectrometer | $cm^{-1}$ |
| A (436 nm) | d | Spectrometer | $cm^{-1}$ |

**Fig. 127.** Characterization of HUS (HA, FA) in either solid (s) or dissolved (d) form

A further division into humic acids (HA) and fulvic acids (FA) can be made. According to the working definition, HAs are only soluble in an alkaline environment, whereas FAs are soluble in both acidic and alkaline media. To differentiate between the two, add HCl to the alkaline eluate in the XAD column until pH 2.0 ± 0.1 is reached. Filter off the precipitated HA through a 0.45 µm membrane, wash with approx. $10^{-2}$ molar HCl and, after drying, keep available for further analysis. Concentrate the acidic solution containing FA again in a small XAD column (approx. 10 cm long) and wash with redistilled water until the eluate is free of chloride. Then elute with 0.2 molar NaOH and, in order to avoid denaturation as far as possible, pass the eluate through a highly acidic ion exchanger (Lewatit S 1080) and either analyze immediately or freeze-dry.

Characterization of the HUS, HA or FA should be conducted following the criteria given in Fig. 127.

The values obtained are especially informative if they are made comparable by being expressed in relation to 1 mg $C_{org}$. The resulting quantities A (254 nm)/mg DOC and A (436 nm)/mg DOC are also called specific UV absorption or specific yellowing.

It is usual to determine further characteristic quantities in addition to the basic parameters. The interaction of HUS with metal ions can be described in terms of complexation capacity (CC), in relation to a reference metal, e.g. Cu (II). The available methods of determination are polarography and fluorimetric investigation.

Fractionation using various methods of separation (gel permeation chromatography; reversed-phase chromatography) allows conclusions to be drawn concerning the molecular size and polarity of the substances contained in the HUS.

### 4.1.13 Urochrome

#### General remarks

Urochromes are decomposition products of blood and bile pigments. They are contained in urine and faeces, and may find their way into ground and surface

face waters, for example via leaking liquid manure cisterns or through natural fertilization of grassland and fields. Chemically, they probably constitute derivatives of porphyrine, though precise clarification of their chemical constitution has so far proved impossible.

The urochrome component in the form of a yellow urine pigment may be used in the detection of urine in swimming-bath waters. Since, however, the urea test is about 50 times more sensitive, the urochrome test is of no significance in this connection.

Urochrome determination is important as an indicator of faecal contamination of waters. Its primary application is in the hygienic monitoring of such waters or of waters which are used for bathing purposes.

Urochrome is determined by adsorptive concentration with aluminium hydroxide and photometric determination in a solution containing formic acid. Since urochromes present no pressingly important problem with regard to water hygiene and water treatment, differentiation between urochrome A and urochrome B is unnecessary when determining urochrome in water. The result is influenced by the content of humic acids, which often accompany urochromes; this may be taken into account by using a differential method.

### Colorimetric determination (hydroxide method)

The urochromes dissolved in water are precipitated by coprecipitation with aluminium hydroxide. The precipitate is decanted or centrifuged off and dissolved in formic acid. After adding phosphoric acid to eliminate iron coloration, the urochrome content is determined by measurement of the extinction at 380 nm. If humic acids are present, a second extinction measurement is taken at 530 nm and taken into account in the calculation. Instead of calibration with a urochrome reference solution, the urochrome content is calculated by multiplying the "colour value" by an empirical factor.

The method is suitable for the determination of urochromes in drinking water, surface water and ground water. The lack of specificity of this method dictates that the results should be used to evaluate waters only in combination with other chemical and microbiological data.

### Equipment

Photometer with filters (380 nm and 530 nm)

Cuvettes of suitable light path length

Glass flask, volume 1000 ml

Measuring flask, volume 50 ml

Potassium aluminium sulphate solution 0.1 m:
Dissolve 4.74 g $KAl(SO_4)_2 \cdot 12\ H_2O$ in dist. water to 100 ml

Formic acid, 85 %

Phosphoric acid, 85 %

Ammonium hydroxide solution, 5 % (0.977 g/ml)

Phenolphthalein solution, 1 % in ethanol

### Procedure

Take 500 ml of the water sample in the 1000 ml glass flask and treat with 20 ml of 0.1 m potassium aluminium sulphate solution and two drops of phenolphthalein solution. Add about 2 - 4 ml of 5 % ammonium hydroxide solution dropwise, shaking constantly, until just past the point where the red coloration disappears (or leaving a slight pink coloration); if possible, the pH should be 7.8. An excess of $NH_4OH$ is to be avoided since this reduces the absorption of the urochrome.

After about 30 minutes, half of the supernatant clear liquid may be siphoned off. The remainder is centrifuged or left to stand overnight. Dissolve the precipitate, which should be isolated from the supernatant liquid as far as possible, with 5 ml of 85 % formic acid, add 0.5 ml of 85 % phosphoric acid and dilute to the mark in the 50-ml measuring flask with distilled water at 20 °C.

After 30 minutes, measure the extinction of the clear formic acid-urochrome solution at 380 nm. First filter the solution if it is not completely clear. If humic acids can be expected to be present, conduct a second measurement at 530 nm.

### Calculation

Multiply the extinction value, measured at 380 nm, by the number of ml of solution for measurement (50 ml). Multiply the "colour value" obtained in this way by the empirical factor 1.9, established by gravimetric urochrome determination by H. O. Hettche. We therefore obtain

$$\text{mg/l urochrome} = E_{380} \cdot V_F \cdot 1.9$$

$E_{380}$ = Extinction in 1-cm cuvettes at 380 nm
$V_F$ = Number of ml of formic acid colour solution

To estimate the proportion of humic acid, take the logarithm of the extinction values of the measurements at 380 nm and 530 nm and multiply by 1000:

$$\log E_{380} \cdot 1000 \quad - \quad \log E_{530} \cdot 1000 \quad = \quad Q$$

The "Q value" is 0.903 for pure urochrome solutions, and 0.573 for pure humic acid solutions.

It may be concluded that no humic acid is present in the sample solution if the Q value is around 0.9 or above. Q values between 0.9 and 0.8 indicate a considerable portion of urochromes, mixed, however, with a more than negligible quantity of humic acids. In the case of Q values of 0.6 and below, only humic acid is present.

## 4.2 Organic compounds

### 4.2.1 Polycyclic Aromatic Hydrocarbons

**General remarks**

Among the harmful substances which can be detected in drinking water, polycyclic aromatic hydrocarbons (PAH) are considered to be particularly important because of their potential carcinogenic properties. The circumstances in which these substances occur include the incomplete burning of organic material. They also appear in substantial concentrations in cigarette smoke, car exhaust fumes and other cases of pyrolysis. The following figures are given by Kunte and Borneff as a guideline for the level of selected polycyclic aromatic hydrocarbons in the water:

| | |
|---|---|
| Ground water | 10 - 50 ng/l |
| Slightly polluted surface water | 50 - 250 ng/l |
| Heavily polluted surface water | up to 1000 ng/l |
| Wastewater | up to 100,000 ng/l |

(ng/l = Nanogram/l)

A routine analysis of all compounds in this group of substances is scarcely practicable. However, methods of fluorescent spectrometry have proved useful and comparatively simple for analyzing representative PAHs. These methods are used after enrichment and separation using thin-layer chromatography and after high pressure liquid chromatography. The 1986 German Drinking Water Regulations lay down a maximum permissible level of 0.00025 mg/l (calculated as C) for total polycyclic aromatic hydrocarbons. The following substances are included (designated in accordance with IUPAC rules):

fluoranthene, benzo(b)fluoranthene, benzo(k)fluoranthene, benzo(a)pyrene, benzo-(ghi)perylene and indeno(1,2,3-cd)pyrene.

The substances listed are not all equally carcinogenic. Fluoranthene, which can also occur naturally, must be regarded as scarcely if at all carcinogenic, whereas benzo(a)pyrene is extremely so.

#### 4.2.1.1 Detection of polycyclic aromatic hydrocarbons using thin-layer chromatography

**Extraction of the water samples**

Two-litre wide-necked reagent bottles with conical shoulders (brown glass with ground glass stoppers) are used for the water sampling. After adding 60 ml cyclohexane, the test sample is extracted for 5 minutes using a mixer such as the Ultra-Turrax. Alternatively it can be mechanically shaken for an hour. When the sample has stood for a sufficient length of time, the remaining cyclohexane solution is removed using a separator consisting of two ascending pipes. For relatively clean water, such as ground water, 2 to 2.5 litres should be used, but with waste water and seepage water from refuse dumps 100 ml is sufficient. The latter quantity of water is diluted to approx. 2 litres and is then treated in the same way as the other types of water. With heavily polluted water, emulsions can easily form. In such cases 100 ml water is evaporated in the rotary vaporizer at 40 °C in the water-jet vacuum. The residue is extracted using cyclohexane and then sampling continues as described below. Under certain circumstances the emulsions which form can be dispersed using a centrifuge.

## Enrichment

The extract is dried with sodium sulphate, then transferred to a 100 ml conical flask (brown glass) and concentrated to approx. 1 to 2 ml in the rotary vaporizer. The resulting solution can be transferred into a smaller conical flask and further concentrated.

## Preparatory Cleaning

With ground water and drinking water, preparatory cleaning is only necessary in rare cases, but with waste water and seepage water it is essential. The following procedure has given good results for cleaning and recovery:

1 g Florisil (magnesium silicate) is put into a small 1.6-cm diam. brown glass column approx. 10 cm long. Cyclohexane is added for preliminary cleaning, and then 1 to 2 ml of the extract for analysis is added. A TLC measurement is then taken. If difficulties arise during the separation process, the quantity of the sample used in the TLC measurement may have to be varied. A further 3 ml cyclohexane is then added for secondary cleaning. Finally 4 elutions follow using 1 ml benzene cyclohexane mixture (1 + 1) each time. The combined samples are then reduced in the rotary vaporizer to between 100 and 200 µl and the concentrate is transferred to a thin-layer plate.

## Thin-layer Chromatography

Thin-layer chromatography (separations) can be made using either plates prepared in the laboratory or ready-prepared TLC plates. Plates with reversed-phase material have also proved successful. The procedure for preparing plates in the laboratory is as follows:

To make 5 plates (20 cm x 20 cm), a mixture of 20 g aluminium oxide G (suitable for gas chromatography) and 10 g cetyl cellulose is thoroughly homogenized with 65 ml ethanol and then immediately spread on the plates in a layer 0.3 mm thick. Once they are dry on the surface, the plates are activated for 30 minutes in the drying cabinet at 110 to 130 °C. They are then stored in the desiccator until needed.

A microlitre syringe is used to deposit a spot of the previously cleaned extract onto one corner of the plate. This initial spot should not be larger than 5 mm in diameter. The supply vessel is then rinsed twice with cyclohexane and the resulting liquid is also applied to the plate.

The plates are developed in the first direction with n-hexane/benzene (176 : 24, v/v). After 30 minutes the plates are dried and developed at 90 °C to the first direction using methanol/diethyl ether/water (80 : 80 : 20, v/v). The plates must be protected from light throughout the developing process, and are finally dried using an electric hair-dryer (cold).

For reference and calibration chromatograms, suitable quantities of the standard solution are applied and developed in the appropriate way.

## Evaluation

The position of the spots and the colour of their fluorescence can be observed by stimulating with a UV lamp at 365 nm. Measurements are taken with a suitable densitometer. The peaks correspond to different intensities of fluorescence, and are proportional to the amounts of substance. The fluorescence wavelengths should be set as follows:

fluoranthene 462 nm, benzo(b)fluoranthene 452 nm, benzo(k)fluoranthene 431 nm, benzo(a)pyrene 430 or 405 nm, benzo(ghi)perylene 419 or 407 nm, indeno-(1,2,3-cd)-pyrene 500 nm.

The surface area of the intensity curve is proportional to the amount of substance in a spot on the plate. The surface area is calculated either with an electronic integrator or using the normal formula height x width half way up. The mass concentration of the individual substance in the water is calculated using the following equation:

$$M = \frac{m \cdot a_b \cdot v_c}{v_w \cdot a_a \cdot v_e}$$

$M$ = mass concentration of the individual substance
$m$ = mass of the individual substance on the reference plate
$a_b$ = surface area of the intensity curve for the individual substance on the sample plate
$v_c$ = volume of cyclohexane added for extraction
$a_a$ = surface area of the intensity curve for the individual substance on the reference plate
$v_w$ = volume of extracted water sample
$v_e$ = volume of cyclohexane extract used.

In order to establish the total quantity of polycyclic aromatic hydrocarbons, the concentrations of the 6 individual substances are added together.

## Statement of Results

For drinking water, the carbon content of the 6 polycyclic aromatics is calculated by multiplying their total quantity by 0.95.

### 4.2.1.2 Determination Using High Pressure Liquid Chromatography

For carrying out routine high pressure liquid chromatography separation, it is the reversed-phase method which is most suitable, and in particular using RP 8 or RP 18 columns. These materials are based on silica gel with a chemically bonded, non-polar stationary phase of differing polarity.

The various commercially available ready-made columns do not always give the same separation results and so no general statement can be made about the success of separation. For elution, an approx. 85 : 15 (v/v) methanol/water mixture is used. Under these conditions it often proves impossible to separate benzo(b)fluoranthene and benzo(k)fluoranthene. Preliminary treatment of the water samples is the same as described above.

### 4.2.2 Determination of phenols in water (see also Section 4.1.9)

General remarks

Extraction of the water sample with n-hexane. Derivatization with pentafluorobenzoyl chloride. GC measurement with double ECD and two capillary columns of different polarity.

Equipment

100 ml separating funnel

100 ml glass flasks

Funnel

100 ml round-bottomed flask with 1-ml tip

Injection syringes: 100 µl, 250 µl, 500 µl

1-litre wide-necked reagent bottles with conical shoulders, with tilting pipettes: 10 ml, 20 ml, 50 ml

GC with double ECD and 2 capillary columns of different polarity

Plotter etc.

Water, for example from Millipore superpure water analysis

$Na_2SO_4$ ignited at 450 °C

1 m NaOH

1 m $NaHCO_3$

Hexane, nanograde

Decane

Pentafluorobenzoyl chloride (10 % solution in nanograde toluene) (e.g. Fluka, Art. No. 76733)

Standard phenol solution:
Approx. 5 µg/ml (see list below) in nanograde acetone

Internal standard solution:
2,3-dichlorophenol, approx. 5 µg/ml, in nanograde acetone

Sample preparation

| | |
|---|---|
| Reference: | 80 ml of superpure water + 100 µl of internal standard + 50 µl of standard phenol solution |
| Blank test: | 80 ml of superpure water + 50 µl of internal standard |
| Sample: | 80 ml of water sample + 50 µl of internal standard |

Pour 80 ml of water into a 100-ml separating funnel, add 10 ml of 1 m NaOH and 20 ml of n-hexane and shake for 2 minutes. Drain off the aqueous phase into a 100-ml glass flask; discard the organic phase. Return the aqueous phase into the separating funnel, and after the addition of 20 ml of 1 m NaOH, 20 ml of nanograde n-hexane and 200 µl of pentafluorobenzoyl chloride solution, shake vigorously for 5 minutes.

Subsequently discard the aqueous phase. Add 50 ml of 1 m NaOH to the organic phase and shake again for 1 minute. Discard the aqueous phase. Filter the organic phase through $Na_2SO_4$ into a 100-ml round-bottomed flask with 1-ml tip and carefully concentrate by evaporation at 40 °C with 0.5 ml of decane as a keeper.

**Measurement**

| | |
|---|---|
| Temperature program: | Initial temperature 140 °C, 10 minutes isothermal prerun, then heat to 280 °C at 5 °C/min., 10 min. isothermal tail run |
| Split: | 30 ml/min. |
| Injection temperature: | 280 °C |
| Detector temperature (ECD): | 350 °C |
| Injection volume: | 2 µl |
| Capillary columns: | A column: 30 m DB1 0.25 µm film thickness<br>B column: 30 m DB1701 0.25 µm film thickness |

**Evaluation**

Internal standard method

Peak height evaluation

**Phenolic substances, by gas chromatography**

| (Standard solution approx. 5 µg/ml) | Determination limit µg/l |
|---|---|
| Phenol | 0.5 |
| 2-chlorophenol | 0.1 |
| 4-chlorophenol | 0.1 |
| 2,4-dichlorophenol | 0.1 |
| 3,5-dichlorophenol | 0.1 |
| 2,3,5-trichlorophenol | 0.1 |
| 2,4,6-trichlorophenol | 0.1 |
| 2,3,4,6-tetrachlorophenol | 0.1 |
| Pentachlorophenol | 0.1 |
| o-cresol | 0.5 |
| m-cresol | 0.5 |
| p-cresol | 0.5 |
| 4-chloro-m-cresol | 0.5 |
| 2,4-dimethyl phenol | 0.1 |
| 3,4-dimethyl phenol | 0.1 |
| 2,3,5-trimethyl phenol | 0.1 |
| o-phenyl phenol | 0.5 |

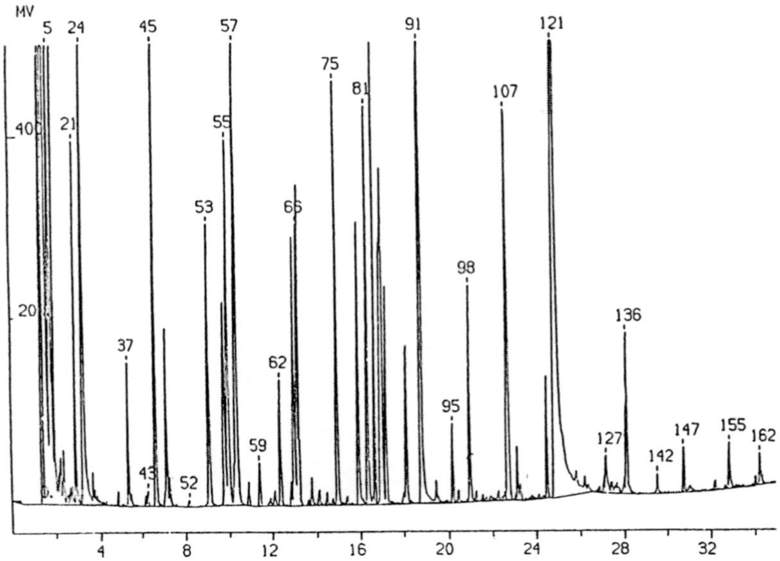


Sample: Phenols DB5

| No. | Min. | Area-% | Name |
|---|---|---|---|
| 45 | 6.68 | 7.49 | phenol |
| 53 | 9 14 | 6.01 | o-cresol |
| 55 | 10.02 | 8.15 | m-cresol |
| 57 | 10.39 | 14.1 | p-cresol |
| 62 | 12.42 | 2.38 | 2-chlorophenol |
| 65 | 13.04 | 5.05 | 4-chlorophenol |
| 66 | 13.20 | 5.18 | 2,4-dimethyl phenol |
| 75 | 14.99 | 7.29 | 3,4-dimethyl phenol |
| 81 | 16.38 | 6.69 | 4-chloro-m-cresol |
| 83 | 16.95 | 4.09 | 3,5-dichlorophenol |
| 84 | 17.03 | 5.36 | 2,3,5-trimethyl phenol |
| 85 | 17.21 | 3.72 | 2,4-dichlorophenol |
| 89 | 18.15 | 2.77 | Rf:2,3-dichlorophenol |
| 95 | 20.22 | 1.26 | 2,4,6-trichlorophenol |
| 98 | 21.01 | 3.42 | 2,3,5-trichlorophenol |
| 107 | 22.70 | 5.58 | o-phenyl phenol |
| 119 | 24.53 | 1.85 | 2,3,4,6-tetrachlorophenol |
| 136 | 28.15 | 2.60 | pentachlorophenol |

%-threshold: 0.30 %
60 peaks out of 78 (total area percentage = 7.04 %) are below threshold.
109 peaks had been suppressed.

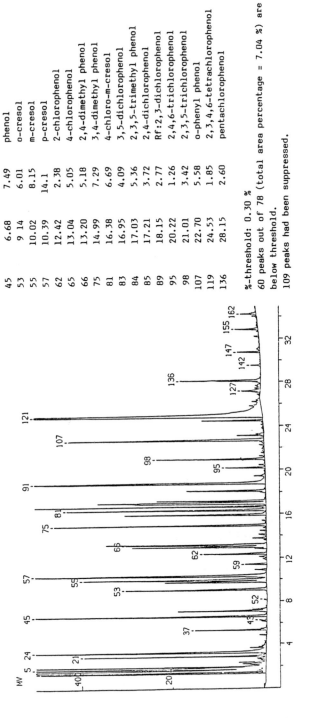

Fig. 128. Example of a chromatogram for phenols

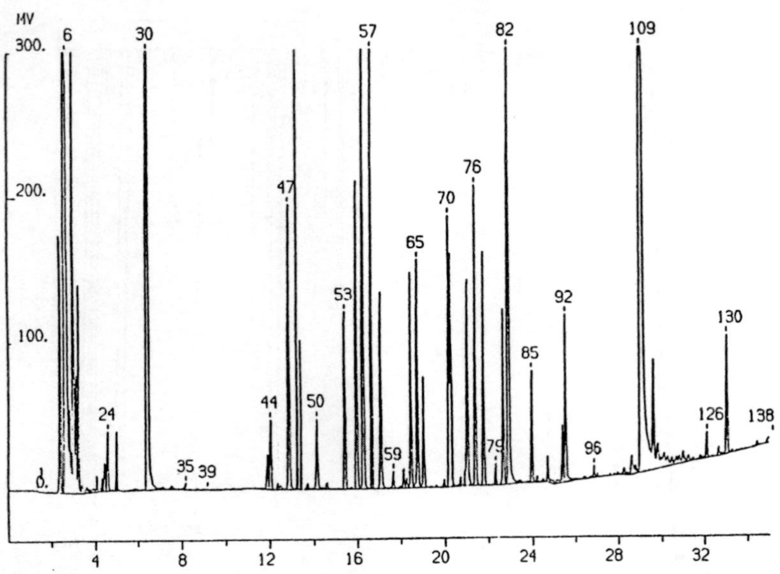

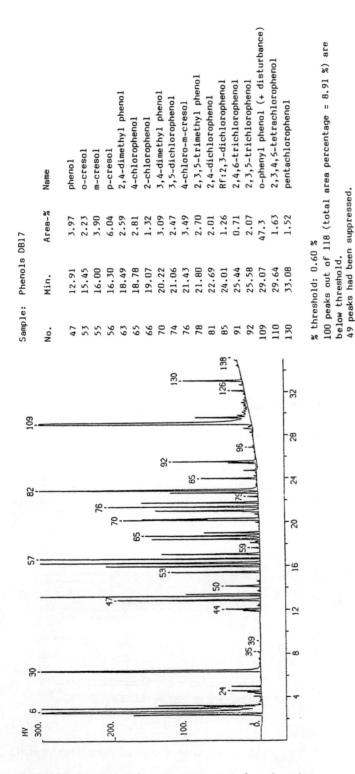

Sample: Phenols DB17

| No. | Min. | Area-% | Name |
|---|---|---|---|
| 47 | 12.91 | 3.97 | phenol |
| 53 | 15.45 | 2.23 | o-cresol |
| 55 | 16.00 | 3.90 | m-cresol |
| 56 | 16.30 | 6.04 | p-cresol |
| 63 | 18.49 | 2.59 | 2,4-dimethyl phenol |
| 65 | 18.78 | 2.81 | 4-chlorophenol |
| 66 | 19.07 | 1.32 | 2-chlorophenol |
| 70 | 20.22 | 3.09 | 3,4-dimethyl phenol |
| 74 | 21.06 | 2.47 | 3,5-dichlorophenol |
| 76 | 21.43 | 3.49 | 4-chloro-m-cresol |
| 78 | 21.80 | 2.70 | 2,3,5-trimethyl phenol |
| 81 | 22.69 | 2.01 | 2,4-dichlorophenol |
| 85 | 24.01 | 1.26 | Rf:2,3-dichlorophenol |
| 91 | 25.44 | 0.71 | 2,4,6-trichlorophenol |
| 92 | 25.58 | 2.07 | 2,3,5-trichlorophenol |
| 109 | 29.07 | 47.3 | o-phenyl phenol (+ disturbance) |
| 110 | 29.64 | 1.63 | 2,3,4,5-tetrachlorophenol |
| 130 | 33.08 | 1.52 | pentachlorophenol |

% threshold: 0.60 %
100 peaks out of 118 (total area percentage = 8.91 %) are below threshold.
49 peaks had been suppressed.

Fig. 129. Example of a chromatogram for phenols

### 4.2.3 Determination of aromatic hydrocarbons, kerosene, heating oil, diesel oil or petrol (gasoline) etc. in water

General remarks

Extraction with pentane

GC measurement with double FID on two capillary columns of different polarity

Equipment

100-ml measuring flask

2-ml bulb pipette

10-µl syringe

Mechanical shaker

Gas-chromatographic apparatus

Standard solution:
a) for petrol(gasoline): standard petrol solution (regular grade) 1 g/50 ml nanograde pentane
b) for kerosene: standard kerosene solution, 1 g/50 ml nanograde pentane
c) for fuel oil: standard fuel oil (extra-light heating oil) solution or diesel oil solution, 1 g/50 ml nanograde pentane
d) aromatics: standard aromatics solution, approx. 1 mg/ml in nanograde pentane

Benzene

Toluene

Ethylbenzene

o-xylene

m-xylene

p-xylene

Internal standard solution of 1-chloroheptane, 20 µl/l in nanograde pentane

Water, e.g. from a Millipore superpure water system

Sample preparation

| | |
|---|---|
| Reference: | 100 ml of superpure water + 10 µl of appropriate standard solution + 2 ml of internal standard solution |
| Blank test: | 100 ml of superpure water + 2 ml of internal standard solution |
| Sample: | 100 ml of water sample + 2 ml of internal standard solution |

## Procedure

Fill a 100-ml glass flask with 100 ml of water sample, add 2 ml of internal standard solution (see "Sample preparation"), and shake for 30 minutes on the mechanical shaker. Transfer the organic phase to GC sampler flasks.

## Measuring conditions

| | |
|---|---|
| GC: | e.g Varian 3700 |
| Capillary columns: | |
| Column A: | 30 mµ DB1, film thickness 0.25 µm |
| Column B: | 30 m DB17, film thickness 0.25 µm |
| Temperature program: | Initial temperature 40 °C, 4 minutes isothermal prerun, heat to 300 °C at 6 °C/min.; 5 minutes isothermal tail run |
| Injector temperature: | 300 °C |
| Detector temperature(ECD): | 330 °C |
| Injection volume: | 2 µl |
| Carrier gas: | Helium, 1.1 bar |

## Evaluation

Internal standard solution
Peak height evaluation

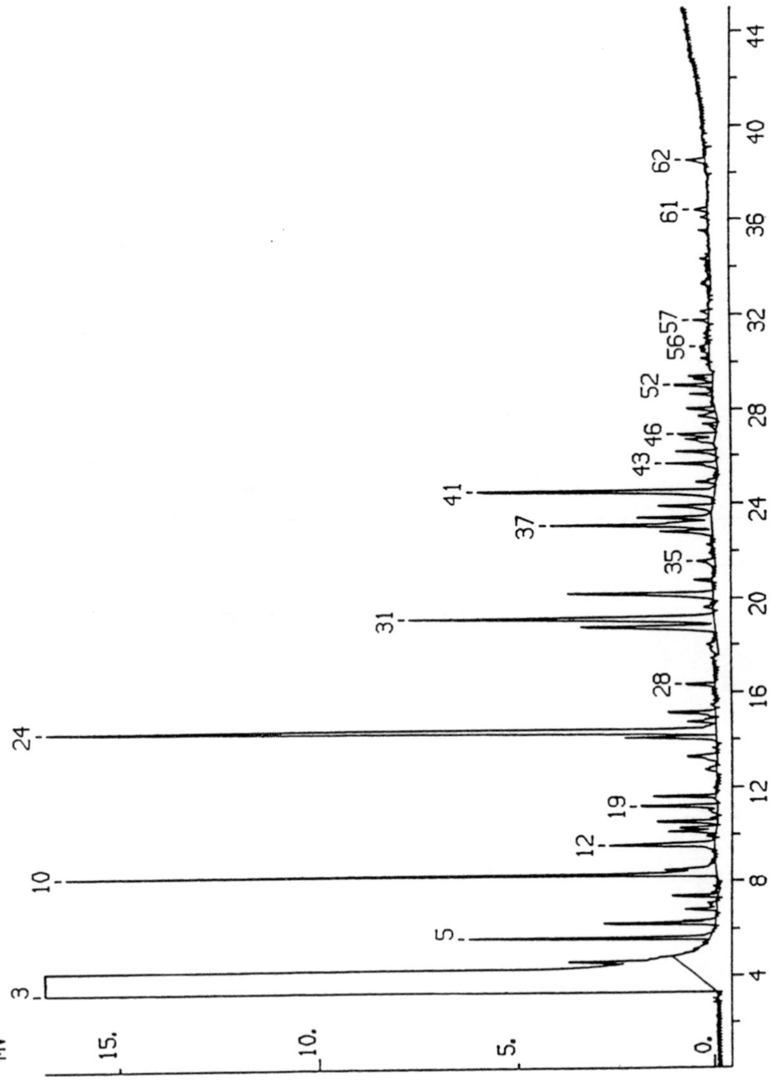


Fig. 130. Super grade petrol

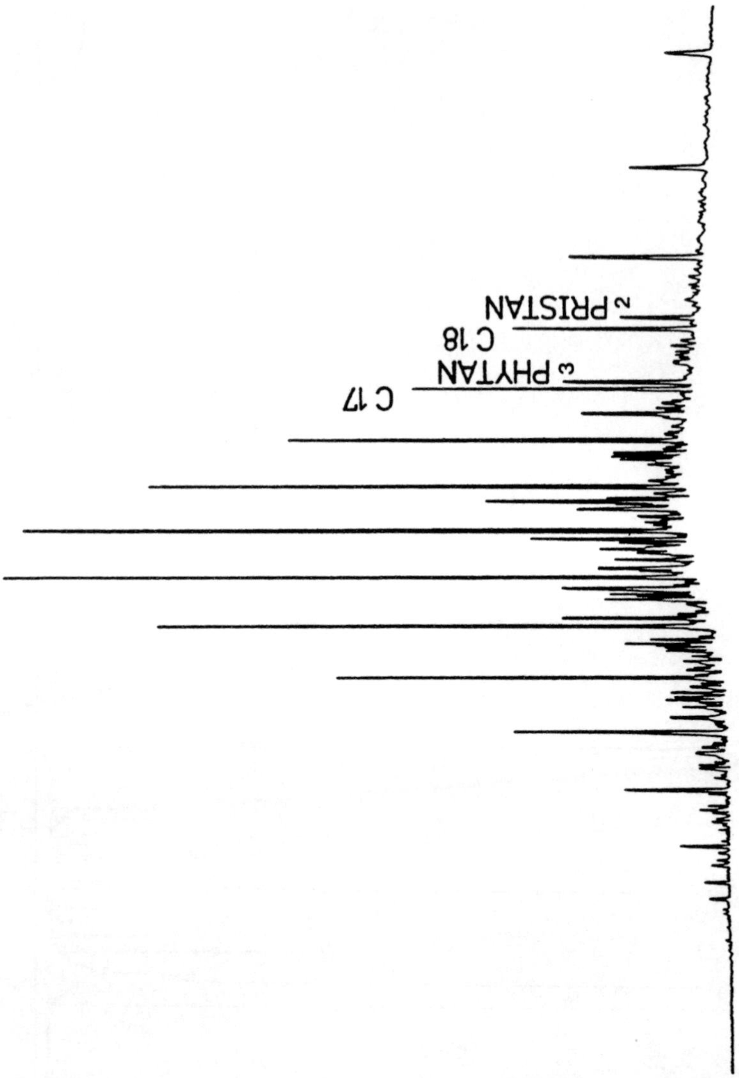


Fig. 131. Diesel fuel

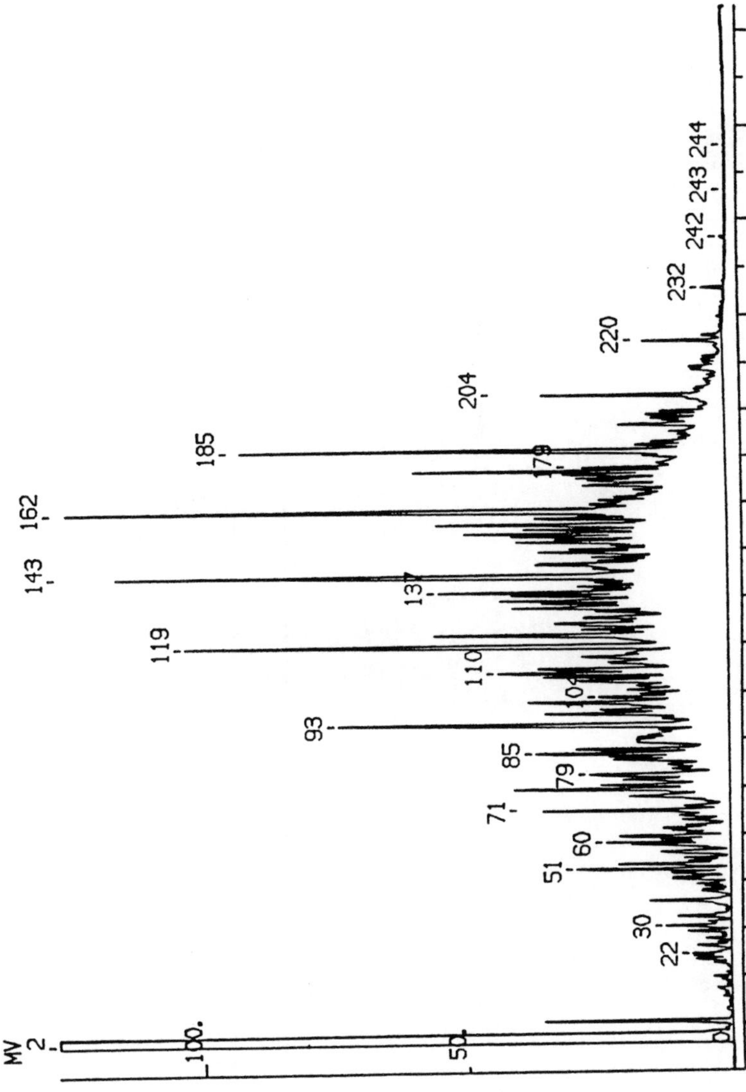


Fig. 132. Kerosene DB1

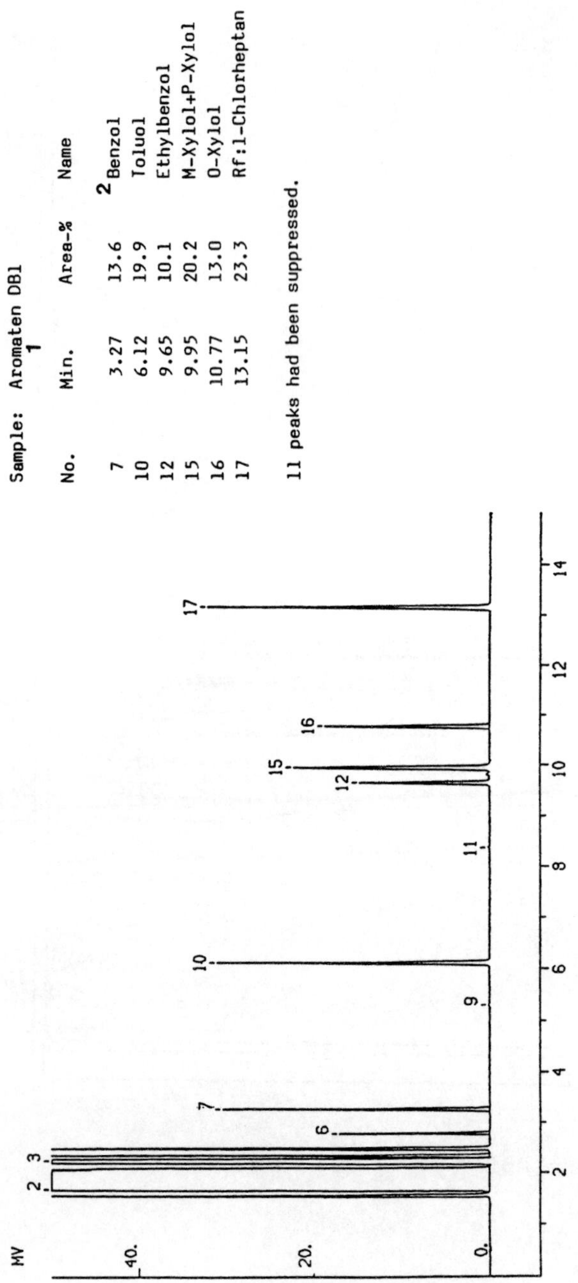


Fig. 133. 1) = Aromatics; 2) = Benzene, Toluene, Ethylbenzene, m-xylene + p-xylene, o-xylene, Rf: 1-chloroheptane

## 4.2.4 Systematic determination of highly volatile halogenated hydrocarbons (HHC) in water samples using gas chromatography

Stages of the determination process

A Preparation of the samples
B Calibration
C Gas-chromatographic measurements and quantification of the individual components
D Qualitative assignment of the compounds

### A) Preparation of the samples

AI Samples with low loadings:

Take the sample of water in a weighed 100-ml amber glass flask with ground-glass stopper. Determine the quantity of sample by weighing the sample flask before and afterwards.

For extraction, take between 2 and 5 ml of the sample with a pipette, and add the same quantity of pentane and 5 µl of internal standard. Subsequently shake for 5 minutes (mechanical shaker).

Draw off the pentane solution into sample flasks. (It is generally not necessary to use a phase separator.)

AII Samples with higher loadings:

Reduce the ratio of aqueous sample to pentane so that the concentration ranges of the compounds under analysis are within the range of the calibration curve.

Prepare as for samples with low loadings

AIII Analogous preparation etc. of a blank test with 100 ml of superpure water (e.g. Millipore).

**Equipment**

Microlitre syringes, 5, 10, 25, 50, 100 µl

Measuring flasks, 10, 20, 50, 100 ml

Pasteur pipettes

Pipettes, 2, 5, 10 ml

Mechanical shaker

Pentane (suitable for the analysis of highly volatile halogenated hydrocarbons)

Polyethylene glycol 400

Dichloromethane, superpure quality

Carbon tetrachloride, superpure quality

1,1,1-trichlorethane, superpure quality

Chloroform, superpure quality

Trichloroethene, superpure quality

Tetrachloroethene, superpure quality

Bromodichloromethane, superpure quality

Dibromochloromethane, superpure quality

Bromoform, superpure quality

1-bromo-3-chloropropane, superpure quality

Gas chromatograph with 2 ECDs and simultaneous splitter

Two-channel intergrator, recorder

Capillary column, 25 m, coated with OV 225 (DB 225)

Capillary column, 25 m, coated with OV 210 (DB 210)

or other capillary columns of equivalent separation efficiency.

Internal standard solution:
e.g. 5 µl of a solution of 200 mg of 1-bromo-3-chloropropane in 50 ml of polyethylene glycol 400

**B) Calibration**

Weigh and mix thoroughly several grams of the pure substances to be determined (see equipment).

Add aliquot masses between 50 and 500 mg of the mixture to between 10 and 50 ml of polyethylene glycol 400 (PEG 400) (multi-component calibration solution).

Internal standard solution:
200 mg of 1-bromo-3-chloropropane in 50 ml of PEG 400

Per calibration point:
Add a set volume (e.g. 5 µl) of the multi-component calibration solution and of the internal standard solution to 100 ml of superpure water (100-ml amber glass flask with ground-glass joint).

Multi-point calibration:
Various aliquot volumes of the multi-component calibration solution provide different concentrations per 100 ml of superpure water.

**C) Gas-chromatographic measuring conditions**

| | |
|---|---|
| Gas chromatograph: | e.g Varian, Model 3700 |
| Columns: | 50 m capillary columns<br>OV 225 OV 1 |

| | |
|---|---|
| Oven temperature: | 35 °C, 5 min. isothermal<br>35 °C to 150 °C - 6 °C/min.<br>possibly: 150 °C, 20 min., isothermal |
| Injector: | e.g. Gerstel, FRG -<br>simultaneous splitter on 2 capillary columns<br>Temperature: 150 °C |
| Detector: | ECD, temperature: 300 °C |
| Carrier gas: | Helium |
| Purge gas: | Argon/methane 95 : 5 (v/v) |
| Split: | e.g. Gerstel, approx. 1 : 10 |

**General remarks on specific gas-chromatographic techniques for these halogenated hydrocarbons**

Fig. 135 shows the calibration curves of various highly volatile halogenated hydrocarbons in a very low concentration range. The calibration curves are linear, but do not all pass through zero (Fig. 135). If the concentration range is extended, the corresponding calibration curves are no longer linear, as shown in Fig. 136. This is particularly clear in the case of the compound 1,1,1-trichloroethane. The period of validity of the calibration curves measured for the calculation of substances in samples can be only a matter of days, depending on the operating state of the substance measured with the detector. Except for dichloromethane, the calibration curves in the higher and lower concentration ranges shown in Fig. 137 were all measured with the same detector within a few days. It can be seen that the calibration curves in the higher concentration range are generally not a direct extension of the calibration curves in the lower concentration range. This is particularly distinct in the case of the compounds carbon tetrachloride and 1,1,1-trichloroethane. This means that the use of a measured calibration function to determine the contents of sample solutions is somewhat problematical over a long period, and frequent control measurements are necessary.

**D) Qualitative assignment of the compounds**

The measured signal of the compound must be present on 2 capillary columns of different polarity (with as great a polarity difference as possible).

Assignment is by comparison of retention (comparison of relative retentions - the reference substance is the internal standard). Reproducibility: ± 0.02 - 0.03 min. (applies to isothermal and programmed-temperature work).

If the relative retention times agree: identity of the substance is **highly probable** (criterion is sufficient for routine analyses).

The identity of the substance may be considered as **certain** following confirmation by GC/MS analysis.

For further examples see Figs. 134 -137 (Scholz-Betz-Rump 1984).

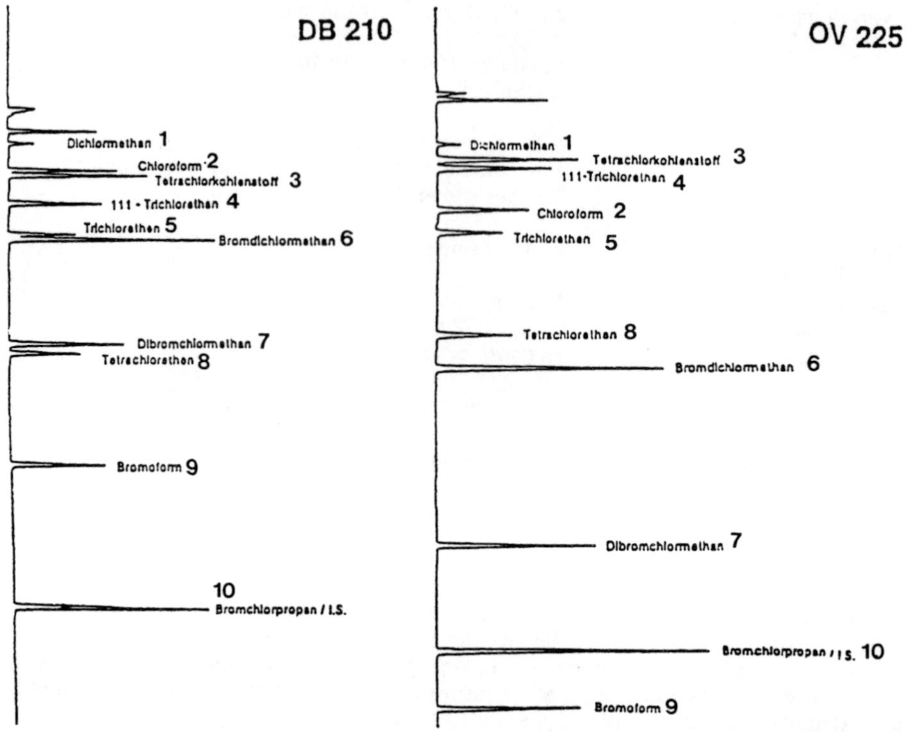


Fig. 134. Chromatograms of highly volatile halogenated hydrocarbons on two columns with different polarity; 1) = Dichloromethane; 2) = Chloroform; 3) = Carbon tetrachloride; 4) = 1,1,1-trichloroethane; 5) = Trichloroethene; 6) = Bromodichloromethane; 7) = Dibromochloromethane; 8) = Tetrachloroethene; 9) = Bromoform; 10) = Bromochloropropane/I.S.

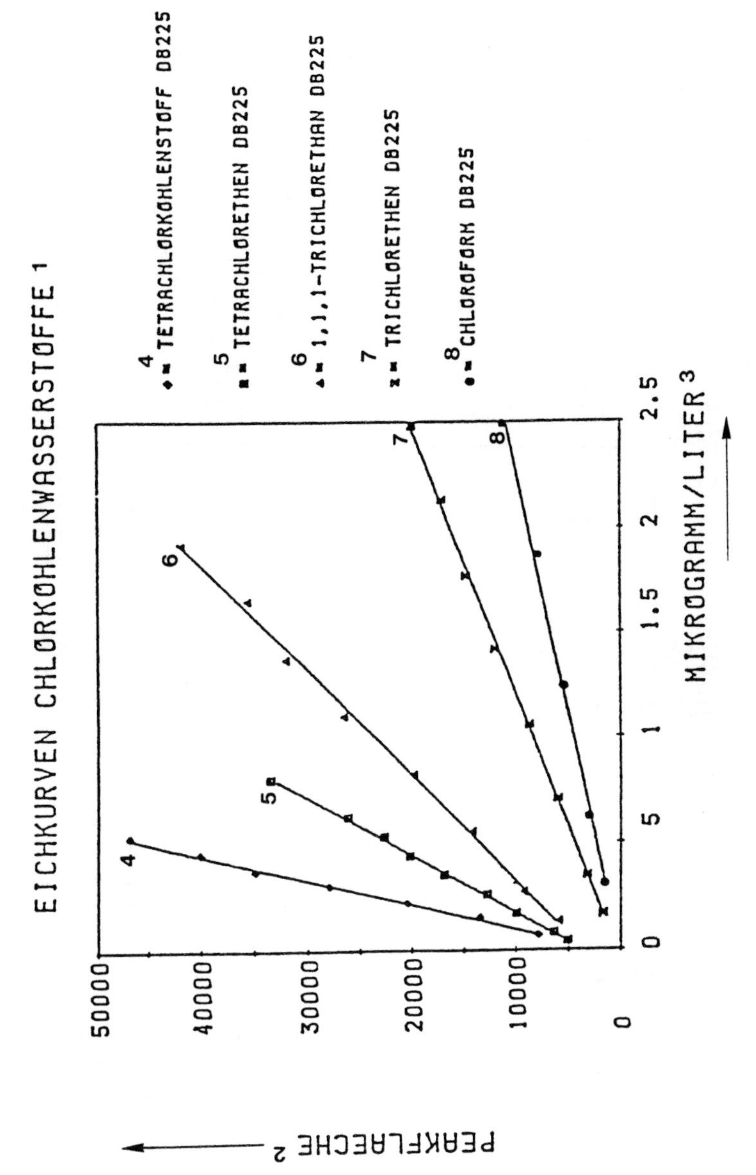


Fig. 135. 1) = Calibration curves of chlorinated hydrocarbons; 2) = Peak area; 3) = Micrograms/litre; 4) = Carbon tetrachloride; 5) = Tetrachloroethene; 6) = 1,1,1-trichloroethane; 7) = Trichloroethene; 8) = Chloroform

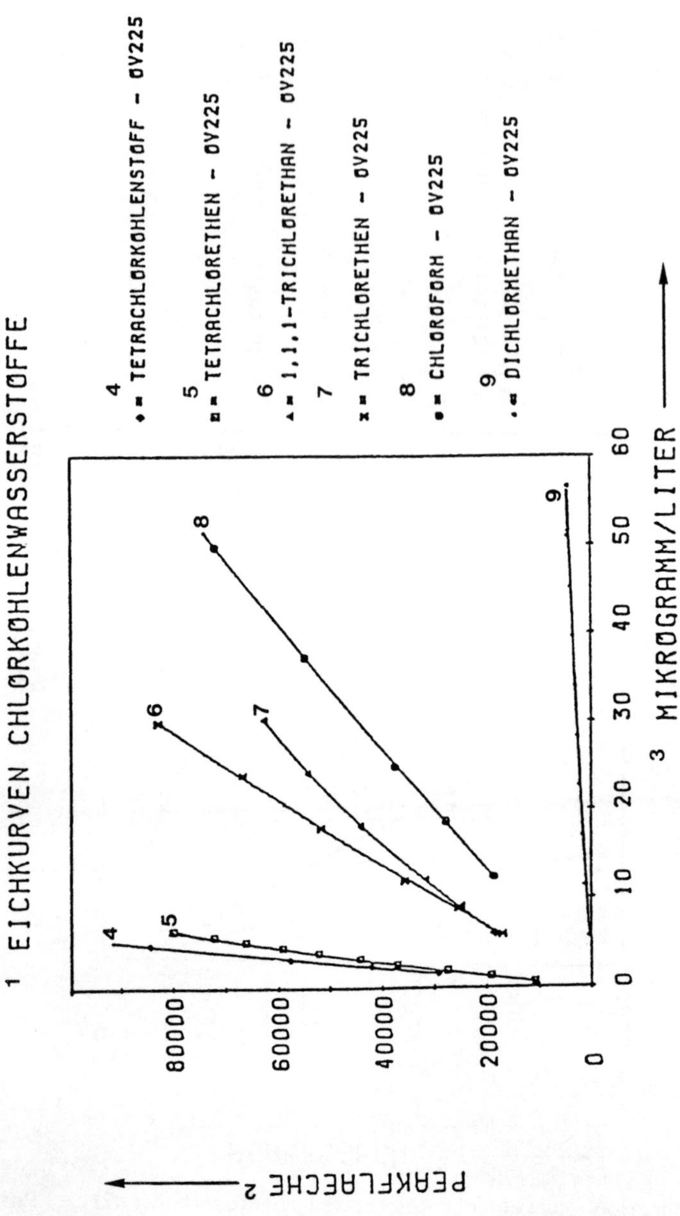


Fig. 136. 1) = Calibration curves of chlorinated hydrocarbons; 2) = Peak area; 3) = Micrograms/litre; 4) = Carbon tetrachloride; 5) = Tetrachloroethene; 6) = 1,1,1-trichloroethane, 7) = Trichloroethene; 8) = Chloroform; 9) = Dichloromethane;

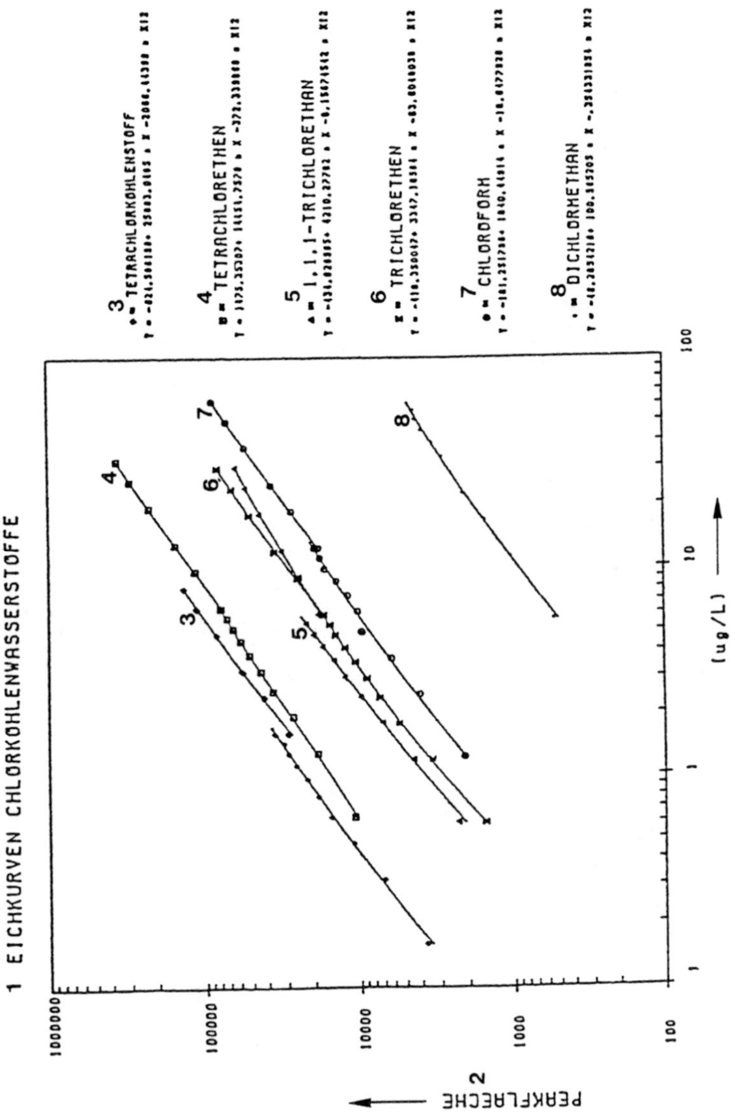


Fig. 137. 1) = Calibration curves of chlorinated hydrocarbons, 2) = Peak area; 3) = Carbon tetrachloride; 4) = Tetrachloroethene; 5) = 1,1,1-trichloroethane, 6) = Trichloroethene; 7) = Chloroform; 8) Dichloromethane

### 4.2.5 Determination of nitroaromatics and higher-boiling halogenated compounds in water

General remarks

Extraction with toluene

GC measurement with double ECD on two capillary columns of different polarity

Equipment

100-ml measuring flask

Mechanical shaker

2-ml pipette

500 µl syringe

25 µl syringe

10 µl syringe

GC with capillary columns, integrator or recorder

Water, e.g. from a Millipore superpure water system

Nanograde toluene

Standard solutions:
- Nitroaromatics, approx. 0.1 - 6 µg/ml in nanograde acetone (Table A)
- Higher-boiling halogenated compounds, approx. 0.5 - 15 µg/ml in nanograde acetone (Table B)

Internal standard solution: 1,1,2,2-tetrabromomethane, approx. 1 µg/ml in nanograde acetone

**Table A**

List of selected nitroaromatics with determination limits in water (empirical):

| Nitroaromatics | Determination limit µg/l |
|---|---|
| Nitrobenzene | 10.0 |
| 2-nitrotoluene | 20.0 |
| 3-nitrotoluene | 20.0 |
| 4-nitrotoluene | 20.0 |
| 1-chloro-2-nitrobenzene | 2.0 |
| 1-chloro-3-nitrobenzene | 2.0 |
| 1-chloro-4-nitrobenzene | 2.0 |
| 2-chloro-4-nitrotoluene | 2.0 |
| 4-chloro-2-nitrotoluene | 2.0 |
| 6-chloro-2-nitrotoluene | 2.0 |
| 1-chloro-2,4-dinitrobenzene | 2.0 |
| 2,6-dinitrotoluene | 2.0 |
| 3,4-dinitrotoluene | 2.0 |

Table B

List of selected higher-boiling organohalogen compounds with determination limits in water (empirical):

| | µg/l |
|---|---|
| 1,1,1,2-tetrachloroethane | 1.0 |
| Hexachloroethane | 1.0 |
| Hexachloro-1,3-butadiene | 1.0 |
| 1,2-dichlorobenzene | 10.0 |
| 1,4-dichlorobenzene | 10.0 |
| 1,2,3-trichlorobenzene | 1.0 |
| 1,2,4-trichlorobenzene | 1.0 |
| 1,3,5-trichlorobenzene | 1.0 |
| 1,2,3,4-tetrachlorobenzene | 1.0 |
| 1,2,3,5-tetrachlorobenzene | 1.0 |
| Pentachlorobenzene | 1.0 |
| 2,4-dichlorotoluene | 15.0 |
| 2,6-dichlorotoluene | 10.0 |
| 3,4-dichlorotoluene | 15.0 |

**Sample preparation**

| | |
|---|---|
| Reference 1: | 100 ml of superpure water + 10 µl of internal standard solution + 50 µl of nitroaromatics mixture |
| Reference 2: | 100 ml of superpure water + 10 µl of internal standard solution + 50 µl of higher-boiling halogenated compounds mixture |
| Blank test: | 100 ml of superpure water + 10 µl of internal standard solution |
| Sample: | 100 ml of water + 10 µl of internal standard solution |

**Procedure**

Take 100 ml of water, reference solution and blank in separate 100-ml measuring flasks and add 2 ml of nanograde toluene by pipette to each. Seal each measuring flask firmly and shake vigorously for 10 minutes. Draw off the organic phase with a Pasteur pipette and fill into GC sampler flasks.

**Measuring conditions**

| | |
|---|---|
| Temperature program: | Initial temperature 80 °C, 5 minutes isothermal prerun, heat to 280 °C at 5 °C/min., 5 minutes tail run (isothermal) |
| Injection temperature: | 280 °C |
| Detector temperature: | 350 °C |
| Split: | 30 ml/min. |
| Helium: | 1.0 bar |
| Injection volume: | 2 µl |

Capillary columns

Column A: 30 m DB1 (e.g. 0.25 μm film thickness)

Column B: 30 m DB170 (e.g. 0.25 μm film thickness)

**Evaluation**

Internal-standard method
Peak height evaluation

A number of examples of chromatograms and measuring conditions are given in Figs. 138 - 141.

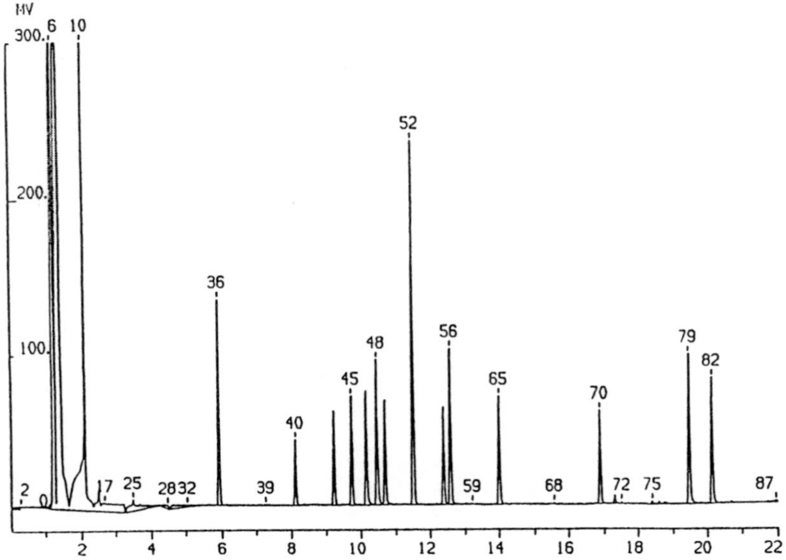

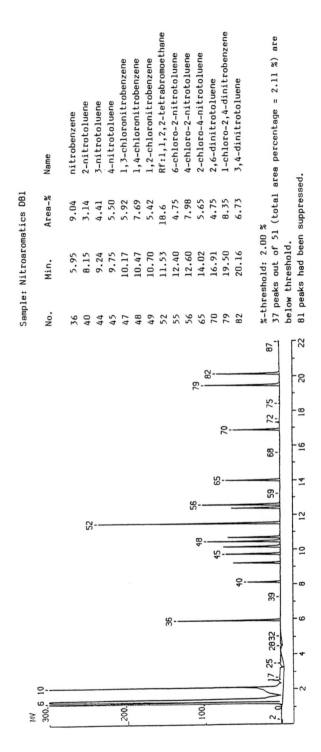

Sample: Nitroaromatics DB1

| No. | Min. | Area-% | Name |
|---|---|---|---|
| 36 | 5.95 | 9.04 | nitrobenzene |
| 40 | 8.15 | 3.14 | 2-nitrotoluene |
| 44 | 9.24 | 4.41 | 3-nitrotoluene |
| 45 | 9.75 | 5.50 | 4-nitrotoluene |
| 47 | 10.17 | 5.92 | 1,3-chloronitrobenzene |
| 48 | 10.47 | 7.69 | 1,4-chloronitrobenzene |
| 49 | 10.70 | 5.42 | 1,2-chloronitrobenzene |
| 52 | 11.53 | 18.6 | Rf:1,1,2,2-tetrabromoethane |
| 55 | 12.40 | 4.75 | 6-chloro-2-nitrotoluene |
| 56 | 12.60 | 7.98 | 4-chloro-2-nitrotoluene |
| 65 | 14.02 | 5.65 | 2-chloro-4-nitrotoluene |
| 70 | 16.91 | 4.75 | 2,6-dinitrotoluene |
| 79 | 19.50 | 8.35 | 1-chloro-2,4-dinitrobenzene |
| 82 | 20.16 | 6.73 | 3,4-dinitrotoluene |

%-threshold: 2.00 %
37 peaks out of 51 (total area percentage = 2.11 %) are below threshold.
81 peaks had been suppressed.

**Fig. 138.** Example of a chromatogram of nitroaromatics

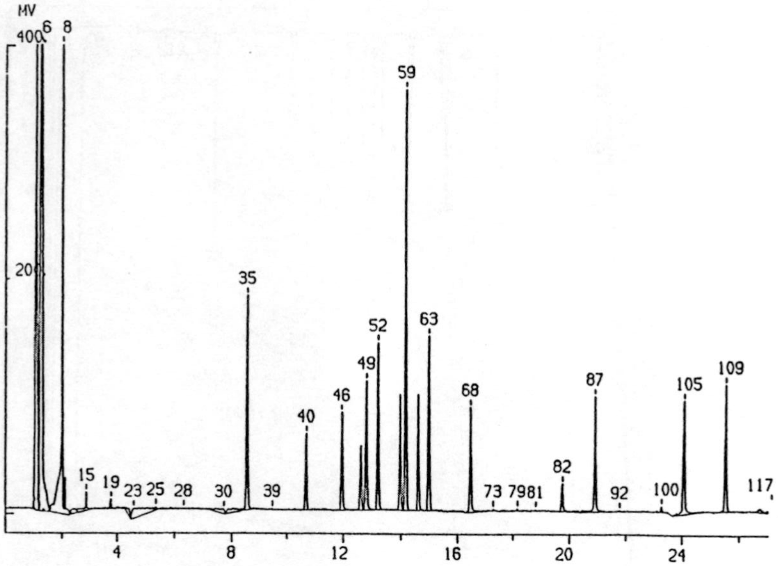


Sample: Nitroaromatics DB1701

| No. | Min. | Area-% | Name |
|---|---|---|---|
| 35 | 8.57 | 9.86 | nitrobenzene |
| 40 | 10.68 | 3.34 | 2-nitrotoluene |
| 46 | 11.97 | 4.64 | 3-nitrotoluene |
| 48 | 12.63 | 3.11 | 4-nitrotoluene |
| 49 | 12.83 | 6.24 | 1,3-chloronitrobenzene |
| 52 | 13.22 | 7.43 | 1,4-chloronitrobenzene |
| 58 | 14.00 | 5.50 | 1,2-chloronitrobenzene |
| 59 | 14.20 | 19.0 | Rf:1,1,2,2-tetrabromoethane |
| 61 | 14.65 | 5.18 | 6-chloro-2-nitrotoluene |
| 63 | 15.02 | 8.16 | 4-chloro-2-nitrotoluene |
| 68 | 16.51 | 5.41 | 2-chloro-4-nitrotoluene |
| 87 | 20.92 | 5.37 | 2,6-dinitrotoluene |
| 105 | 24.07 | 6.47 | 1-chloro-2,4-dinitrobenzene |
| 109 | 25.54 | 6.22 | 3,4-dinitrotoluene |

%-threshold: 1.00 %
64 peaks out of 78 (total area percentage = 4.06 %) are below threshold.
79 peaks had been suppressed.

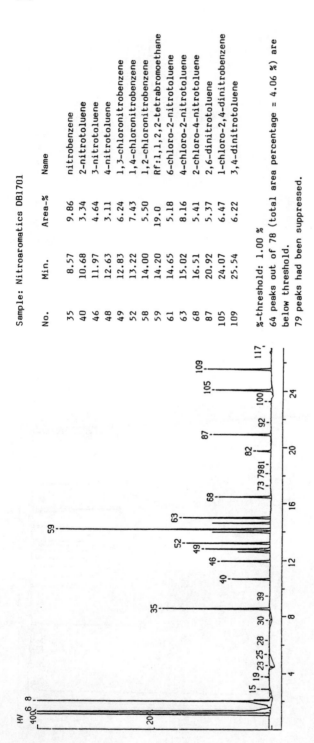

Fig. 139. Example of a chromatogram of nitroaromatics

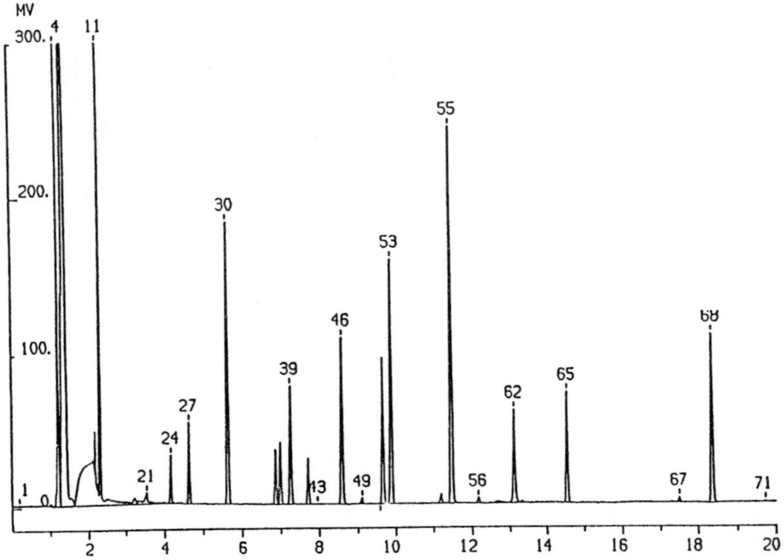


Sample: Higher-boiling halogen compounds DB1

| No. | Min. | Area-% | Name |
|---|---|---|---|
| 11 | 2.32 | 12.4 | 1,1,1,2-tetrachloroethane |
| 24 | 4.17 | 1.37 | 1,4-dichlorobenzene |
| 27 | 4.65 | 2.64 | 1,2-dichlorobenzene |
| 30 | 5.64 | 10.3 | hexachloroethane |
| 36 | 6.89 | 2.08 | 2,4-dichlorotoluene |
| 37 | 7.02 | 2.51 | 2,6-dichlorotoluene |
| 39 | 7.28 | 4.80 | 1,3,5-trichlorobenzene |
| 41 | 7.74 | 1.81 | 3,4-dichlorotoluene |
| 46 | 8.61 | 7.07 | 1,2,4-trichlorobenzene |
| 52 | 9.69 | 6.04 | 1,2,3,-trichlorobenzene |
| 53 | 9.91 | 10.5 | hexachloro-1,3-butadiene |
| 55 | 11.50 | 16.0 | Rf:1,1,2,2-tetrabromoethane |
| 62 | 13.12 | 4.33 | 1,2,3,5-tetrachlorobenzene |
| 65 | 14.55 | 4.71 | 1,2,3,4-tetrachlorobenzene |
| 68 | 18.39 | 7.88 | pentachlorobenzene |

%-threshold: 0.50 %
46 peaks out of 61 (total area percentage = 5.52 %) are below threshold.
56 peaks had been suppressed.

Fig. 140. Example of a chromatogram of higher-boiling halogen compounds

Sample: Higher-boiling halogen compounds DB170

| No. | Min. | Area-% | Name |
|---|---|---|---|
| 11 | 2.36 | 13.2 | 1,1,3,2-tetrachloroethane |
| 20 | 4.59 | 1.88 | 1,4-dichlorobenzene |
| 22 | 5.37 | 12.3 | hexachloroethane+1,2-dichlorobenzene |
| 28 | 7.36 | 4.55 | 2,4-dichlorotoluene+2,6-dichlorotoluene |
| 29 | 7.44 | 4.83 | 1,3,5-trichlorobenzene |
| 34 | 8.70 | 1.87 | 3,4-dichlorotoluene |
| 37 | 9.31 | 7.01 | 1,2,4-trichlorbenzene |
| 38 | 9.44 | 10.2 | hexachloro-1,3-butadiene |
| 45 | 10.74 | 6.05 | 1,2,3-trichlorobenzene |
| 61 | 13.52 | 3.87 | 1,2,3,5-tetrachlorobenzene |
| 65 | 14.17 | 13.8 | Rf:1,1,2,2-tetrabromoethane |
| 67 | 15.45 | 4.73 | 1,2,3,4-tetrachlorobenzene |
| 81 | 18.83 | 6.59 | pentachlorobenzene |

%-threshold: 0.50 %
113 peaks out of 126 (total area percentage = 9.14 %) are below threshold.
16 peaks had been suppressed.

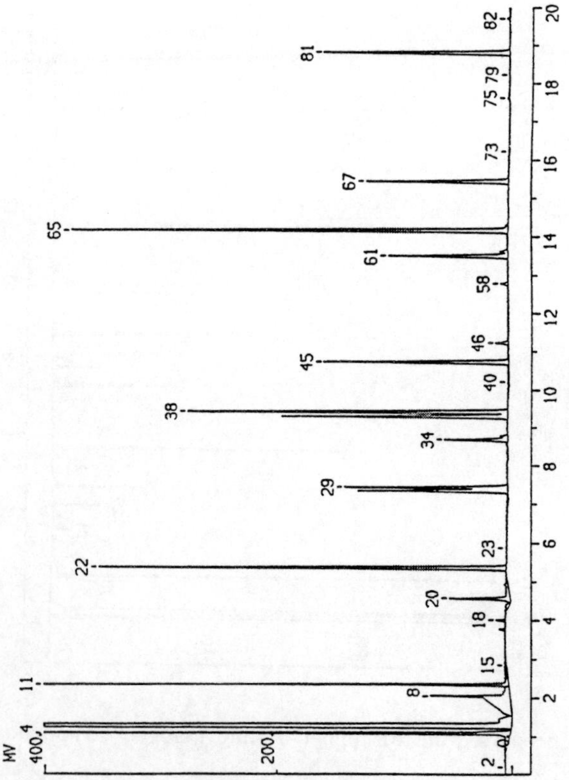


Fig. 141. Example of a chromatogram of higher-boiling halogen compounds

## 4.2.6 System of determination of organochlorine pesticides, organophosphorus pesticides and triazines in water

### General remarks

Extraction with dichloromethane from the water sample. GC measurement with double ECD and double TSD with two capillary columns of different polarity.

### Equipment

1-l separating funnel

Funnel

250-ml round-bottomed flask with 1-ml tip (at lower end for concentration of extract)

1-litre measuring cylinder

500-ml Erlenmeyer flask with 50-ml tilting pipette

1-ml pipettes

Water, e.g. from Millipore superpure water system

NaCl ignited at 450 °C

$Na_2SO_4$ ignited at 450 °C

Nanograde toluene

Nanograde dichloromethane

Standard solutions: (See empirical list)
Organochlorine pesticides, approx. 0.3 µg/ml
Organophosphorus pesticides, approx 0.3 µg/ml
Triazine, approx. 0.5 µg/ml
Chlordane, approx. 5 µg/ml
Arochlor 1254, approx. 5 µg/ml

Internal standard solutions:
Hexabromobenzene, approx. 0.5 µg/ml
Bromophosmethyl, approx. 0.5 µg/ml

All standard solutions in nanograde acetone

Gas-chromatographic apparatus

### Sample preparation

Reference 1:
1 litre of superpure water + 1 ml of internal standard solution + 1 ml each of standard solutions of organochlorine pesticide, organophosphorus pesticide and triazine

Reference 2:
1 litre of superpure water + 1 ml each of internal standard solutions + 1 ml each of arochlor 1254 + chlordane standard solutions

Blank test:
1 litre of superpure water + 1 ml each of internal standard solutions

Sample:
1 litre of water + 1 ml each of internal standard solutions

**Procedure**

Take 1 litre of water (sample) in a 1-litre separating funnel, add 30 g of NaCl and shake twice for 5 minutes, each time with 50 ml of nanograde dichloromethane. Transfer the organic phases via $Na_2SO_4$ to a 250-ml round-bottomed flask with 1-ml tip; concentrate to 1 ml with 1 ml of toluene (water bath 40 °C).

**Measuring conditions at the ECD (example):**

| | |
|---|---|
| Temperature program: | initial temperature 140 °C, 10 minutes prerun (isothermal), heat to 270 °C at 5 °C/min., 15 minutes tail run (isothermal) |
| Split: | 30 ml/min. |
| Injector temperature: | 280 °C |
| Detector temperature: | 350 °C |
| Helium: | 1.8 bar |
| Injection volume: | 2 µl |
| Capillary columns: | Column A: DB1 (e.g. from HP) 25 m, 0.17 µm film thickness<br>Column B: DB1701 (e.g. from HP) 25 m, 0.17 µm film thickness |

**Measuring conditions for TSD (example):**

| | |
|---|---|
| Temperature program: | initial temperature 180 °C, 6 minutes isothermal prerun, heat to 270 °C at 8 °C/min.; 5 minutes isothermal tail run |
| Split: | 15 ml/min. |
| Helium: | 2 bar |
| Injector temperature: | 250 °C |
| Detector temperature: | 300 °C |
| Injection volume: | 2 µl |

| | A | B |
|---|---|---|
| Column | 60 m DB1 | 60 m DB1701 |
| $H_2$ | 1.2 bar | 1.8 bar |
| Attenuation (example) | $1 \cdot 10^{-12}$ | $2 \cdot 10^{-12}$ |
| Make up | 250 scale units | 292 scale units |

**Evaluation**

Internal-standard method
Peak height evaluation

**Empirical list**

| Organochlorine pesticides | Determination limit µg/l |
|---|---|
| Aldrin | 0.05 |
| Chlordane | 1.0 |
| Dieldrin | 0.05 |
| Endrin | 0.05 |
| Heptachlorepoxide | 0.05 |
| p,p'-DDE | 0.1 |
| p,p'-DDD | 0.1 |
| p,p'-DDT | 0.1 |
| Hexachlorobenzene | 0.05 |
| Quintozene | 0.05 |
| -HCH (lindane) | 0.05 |
| -Endosulfan | 0.05 |
| PCBs (Arochlor 1254) | 1.0 |
| **Organophosphoros pesticides** | |
| Dimethoate | 0.05 |
| Parathionmethyl | 0.05 |
| Parathionethyl | 0.05 |
| Bromophosethyl | 0.05 |
| Diazinon | 0.05 |
| Mevinphos | 0.05 |
| Malathion | 0.05 |
| **Phenylurea herbicides** | |
| Buturon | 0.5 |
| Fenuron | 0.5 |
| Linuron | 0.5 |
| Methoxuron | 0.5 |
| **Triazine herbicides** | |
| Atrazine | 0.1 |
| Propazine | 0.1 |
| Simazine | 0.1 |
| Terbutryn | 0.1 |
| **Carbamate herbicides** | |
| Carbetamide | 0.5 |
| Chlorbufam | 0.5 |
| Propham | 0.5 |
| **Phenoxyalkane carboxylic acid herbicides** | |
| 2,4-D | 1.0 |
| 2,4,5-T | 1.0 |
| MCPA | 1.0 |

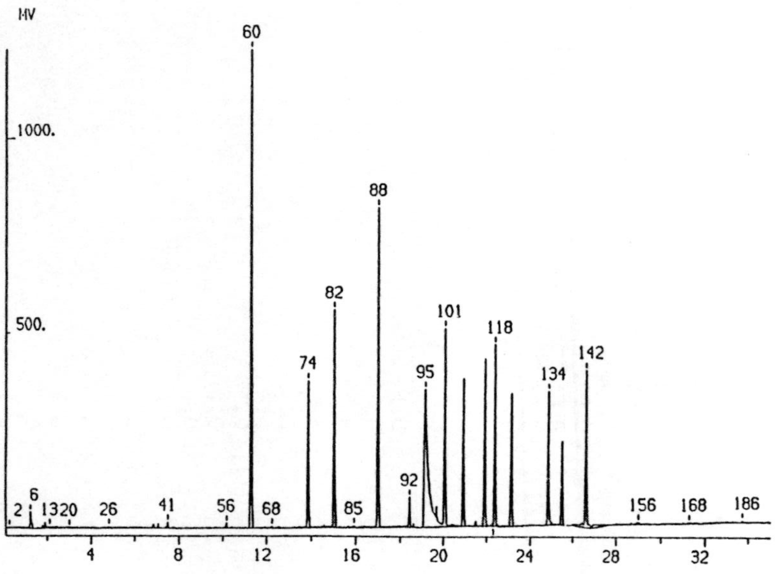

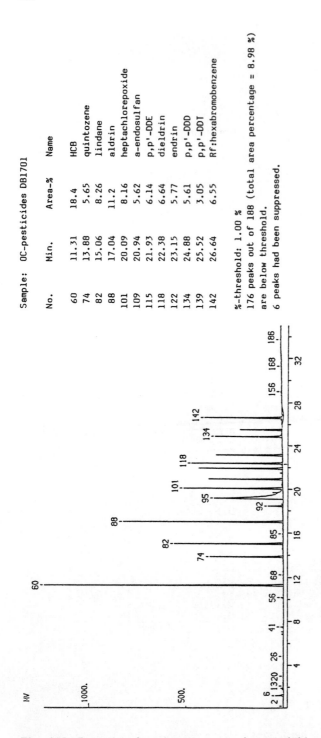

Sample: OC-pesticides DB1701

| No. | Min. | Area-% | Name |
|---|---|---|---|
| 60 | 11.31 | 18.4 | HCB |
| 74 | 13.88 | 5.65 | quintozene |
| 82 | 15.06 | 8.26 | lindane |
| 88 | 17.04 | 11.2 | aldrin |
| 101 | 20.09 | 8.16 | heptachlorepoxide |
| 109 | 20.94 | 5.62 | a-endosulfan |
| 115 | 21.93 | 6.14 | p,p'-DDE |
| 118 | 22.38 | 6.64 | dieldrin |
| 122 | 23.15 | 5.77 | endrin |
| 134 | 24.88 | 5.61 | p,p'-DDD |
| 139 | 25.52 | 3.05 | p,p'-DDT |
| 142 | 26.64 | 6.55 | Rf:hexabromobenzene |

%-threshold: 1.00 %
176 peaks out of 188 (total area percentage = 8.98 %)
are below threshold.
6 peaks had been suppressed.

Fig. 142. Example of a chromatogram for pesticides

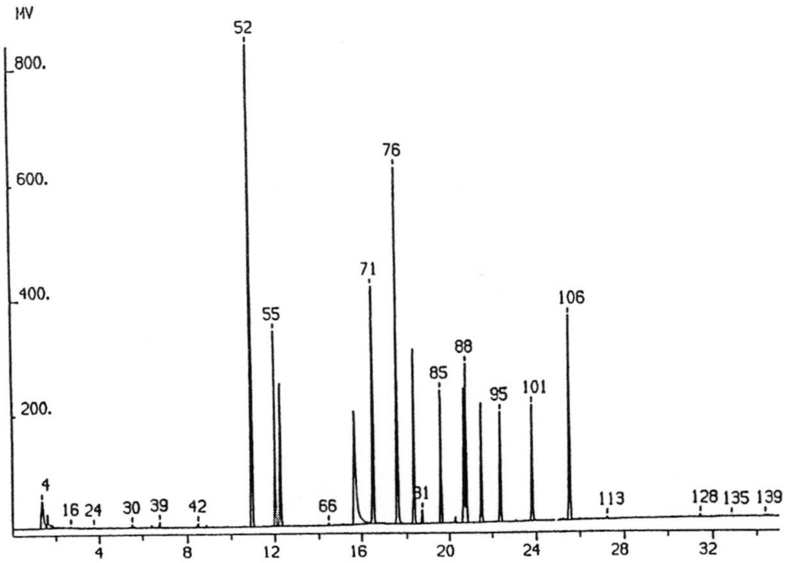


Sample: OC-pesticides DB5

| No. | Min. | Area-% | Name |
|---|---|---|---|
| 52 | 10.96 | 19.1 | HCB |
| 55 | 12.05 | 8.26 | lindane |
| 56 | 12.30 | 6.21 | quintozene |
| 71 | 16.54 | 10.7 | aldrin |
| 79 | 18.39 | 7.92 | heptachlorepoxide |
| 85 | 19.65 | 5.76 | a-endosulfan |
| 87 | 20.69 | 6.03 | p,p'-DDE |
| 88 | 20.80 | 6.84 | dieldrin |
| 90 | 21.52 | 5.31 | endrin |
| 95 | 22.41 | 4.89 | p,p'-DDD |
| 101 | 23.85 | 5.08 | p,p'-DDT |
| 106 | 25.56 | 9.51 | Rf:hexabromobenzene |

%-threshold: 0.50 %
123 peaks out of 135 (total area percentage = 4.44 %)
are below threshold.
5 peaks had been suppressed.

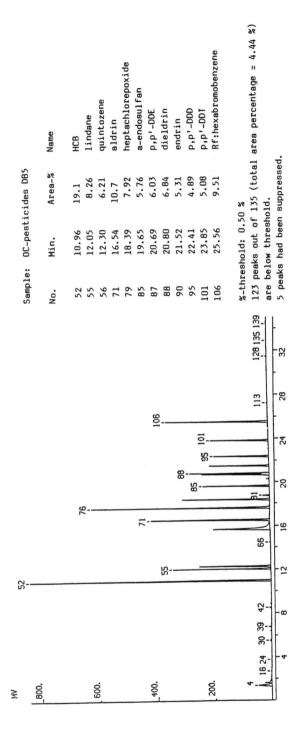

Fig. 143. Example of a chromatogram for pesticides

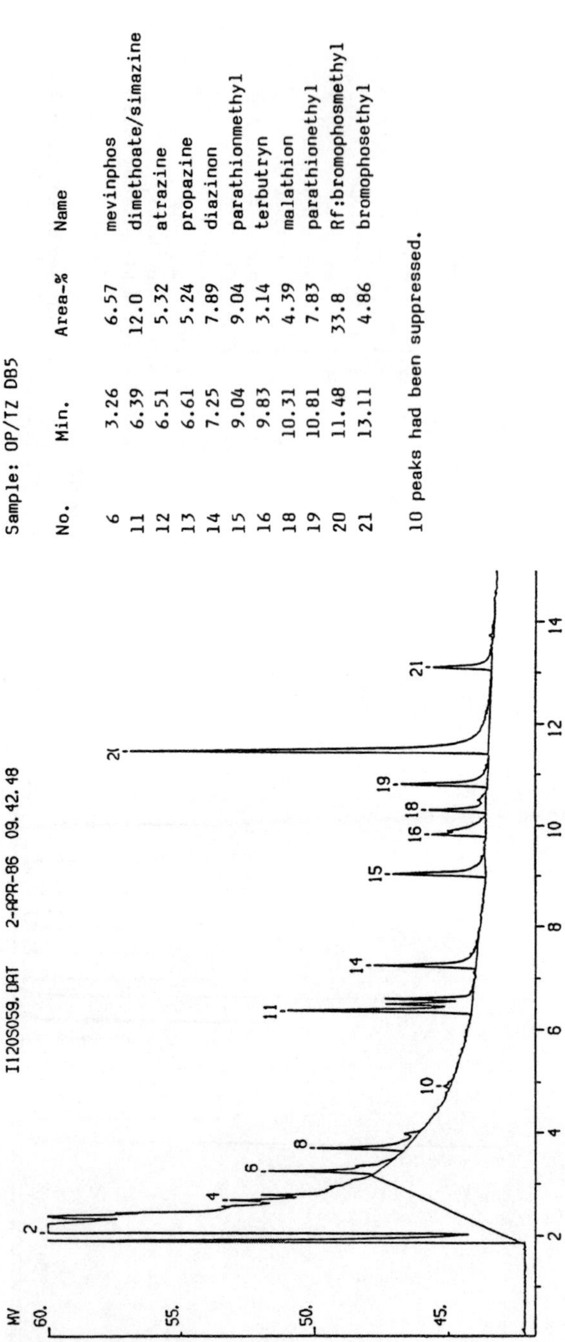

I120S059.DAT 2-APR-86 09.42.48
MV
60.
55.
50.
45.
2
4
6
8
10
11
14
15
16
18
19
20
21
2
4
6
8
10
12
14
Sample: OP/TZ DB5
No. Min. Area-% Name
6 3.26 6.57 mevinphos
11 6.39 12.0 dimethoate/simazine
12 6.51 5.32 atrazine
13 6.61 5.24 propazine
14 7.25 7.89 diazinon
15 9.04 9.04 parathionmethyl
16 9.83 3.14 terbutryn
18 10.31 4.39 malathion
19 10.81 7.83 parathionethyl
20 11.48 33.8 Rf:bromophosmethyl
21 13.11 4.86 bromophosethyl
10 peaks had been suppressed.


Fig. 144. Example of a chromatogram for organophosphorus pesticides and triazine herbicides

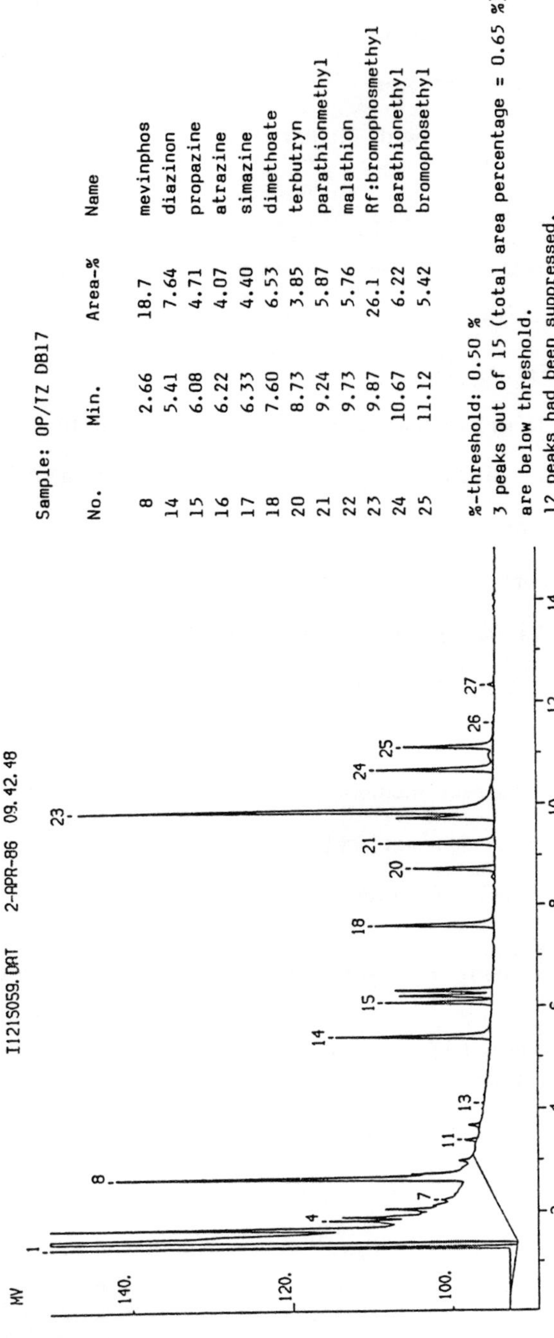

I121S059.DAT 2-APR-86 09.42.48
MV
140.
120.
100.
1
4
7
8
11
13
14
15
18
20
21
23
24
25
26
27
2
4
6
8
10
12
14
Sample: OP/TZ DB17
No. Min. Area-% Name
8 2.66 18.7 mevinphos
14 5.41 7.64 diazinon
15 6.08 4.71 propazine
16 6.22 4.07 atrazine
17 6.33 4.40 simazine
18 7.60 6.53 dimethoate
20 8.73 3.85 terbutryn
21 9.24 5.87 parathionmethyl
22 9.73 5.76 malathion
23 9.87 26.1 Rf:bromophosmethyl
24 10.67 6.22 parathionethyl
25 11.12 5.42 bromophosethyl
%-threshold: 0.50 %
3 peaks out of 15 (total area percentage = 0.65 %)
are below threshold.
12 peaks had been suppressed.


**Fig. 145.** Example of a chromatogram for organophosphorus pesticides and triazine herbicides

### 4.2.7 System for the determination of phenylurea and herbicidal carbamates in water

**General remarks**

Extraction with dichloromethane, HPCL measurement on RP-C18 column and UV detection

**Equipment**

HPLC device with UV detector

1-litre separating funnel

Funnel

250-ml round-bottomed flask with 1-ml tip

500-ml Erlenmeyer flask with 50-ml tilting pipette

1-litre measuring cylinder

1-ml pipettes

Water, e.g. from Millipore superpure water system

NaCl ignited at 450 °C

$Na_2SO_4$ ignited at 450 °C

Nanograde dichloromethane

Nanograde methanol

Nanograde acetonitrile

Standard solution:
Phenylureas and herbicidal carbamates, approx. 1 µg/ml in nanograde methanol (see empirical list 4.2.6)

Internal standard solution:
Chlorpropham (CIPC), approx. 1 µg/ml in nanograde methanol

**Sample preparation**

Reference: 1 litre of superpure water + 1 ml of internal standard + 1 ml of standard solution of phenylurea and herbicidal carbamates

Blank test: 1 litre of superpure water and 1 ml of internal standard

Sample: 1 litre of sample + 1 ml of internal standard

**Procedure**

Take 1 litre of water sample in a 1-litre separating funnel, add 30 g of NaCl and shake twice for 5 minutes, each time with 50 ml of dichloromethane. Pass the organic phases through a filter with $Na_2SO_4$ into a 250-ml round-bottomed flask with 1-ml tip, and carefully evaporate to dryness at 40 °C. Take up the residue with 1 ml of nanograde methanol and subsequently blow off with nitrogen to approx. 200 µl.

Measuring conditions

| | |
|---|---|
| Column: | Latek 250 · 4 C18-2/5µm |
| Mobile phase: | $CH_3CN/H_2O$ (acetonitrile/water) = 40/60 (v/v) |
| Flow: | 1.5 ml/min. |
| Injection volume: | 35 µl |
| | $\lambda$ = 235 nm |

Evaluation

Internal-standard method
Peak height evaluation

### 4.2.8 System for the determination of phenoxyalkane carboxylic acids in water

General remarks

Extraction in an acid medium with dichloromethane, HPLC measurement on RP-C18 column with UV detection.

Equipment

HPLC device with UV detector

1-litre separating funnel

Funnel

250-ml round-bottomed flask with 1-ml tip

1-litre measuring cylinder

1-ml pipettes

500-ml Erlenmeyer flask with 50-ml tilting pipette

Water, e.g. from Millipore superpure water system

NaCl ignited at 450 °C

$Na_2SO_4$ ignited at 450 °C

2 m $H_2SO_4$

Nanograde dichloromethane

Nanograde methanol

KCl, reagent purity

2 ml HCl

Tetrahydrofuran, reagent purity

Standard solution:
Phenoxyalkane carboxylic acids, approx. 10 µg/ml in nanograde methanol (see empirical list, 4.2.6)

Internal standard solution:
2,4,5-trichlorophenoxypropionic acid, approx. 20 µg/ml in nanograde methanol

**Sample preparation**

| | |
|---|---|
| Reference: | 1 litre of superpure water + 1 ml of internal standard solution + 1 ml of standard phenoxyalkane carboxylic acid solution |
| Blank test: | 1 litre of superpure water + 1 ml of internal standard solution |
| Sample: | 1 litre of sample + 1 ml of internal standard solution |

**Procedure**

Take 1 litre of water in a 1-litre separating funnel, add 30 g of NaCl and acidify to pH 3 with 2 m $H_2SO_4$. Then shake twice for 5 minutes, each time with 50 ml of nanograde dichloromethane. Filter the organic phases through $Na_2SO_4$ into a 250-ml round-bottomed flask with 1-ml tip and carefully evaporate to dryness at 40 °C. Take up the residue with 1 ml of nanograde methanol and blow off with nitrogen to approx. 200 µl.

**Measuring conditions**

| | |
|---|---|
| Column: | Latek 250 · 4 C18-2,5 µm |
| Mobile phase: | 0.75 g of KCl + 900 ml of $H_2O$ + 1200 ml of $CH_3OH$ + 10 ml of tetrahydrofuran + 2 m HCl to pH 3 |
| Flow: | 1.3 ml/min. |
| Injection volume: | 35 µl |
| | $\gamma$ = 280 nm |

**Evaluation**

Internal-standard method
Peak-height evaluation

## 4.2.9 Determination of pesticides in water by gas chromatography according to the methods of the Deutsche Forschungsgemeinschaft (German Research Society GRS)

### General remarks

Gas chromatography plays a dominant role compared to other methods in the detection and determination of very small quantities of residues of crop protection products in water. The necessity for the analysis of residues of crop protection products arises from the increasing use of pesticides, which then occur in toxic concentrations, usually in production waste waters, leading to the pollution of receiving waters and other bodies of water. Sudden epidemics of fish mortality in the waters of highly industrialized countries have resulted. It should also be noted that while natural waters are welcome, "cheap" means of transporting waste, they also provide raw water for drinking water supplies. The testing of pesticide residues therefore deserves a place alongside the analysis of other pollutants within the framework of general drinking-water analysis. The methods of analysis described below originate primarily from the collection of methods for the analysis of residues of pesticides by the Deutsche Forschungsgemeinschaft (GRS).

### Notes on the collection and preparation of samples

Water samples should generally be taken in quantities of 1 or 2 litres, or occasionally up to 20 litres in several glass bottles with ground-glass stoppers. Before use, the bottles should be carefully cleaned with a mixture of concentrated sulphuric acid and concentrated nitric acid (1 + 1), then with distilled water and finally with an organic solvent (e.g. acetone), and dried at 200 °C in a drying cabinet.

Bottles or canisters made of polyethylene or aluminium may not be used for collecting samples, since uncontrollable substance losses may occur as a result of diffusion or adsorption processes. As a general rule, the water samples are examined in an unfiltered state in order to detect the substances adsorbed on suspended material. It may be advisable to process the filter residue separately when examining waters with high loadings of undissolved substances. In contrast to most organochlorine compounds, the organophosphorus insecticides are frequently highly unstable in water. For this reason, the water sample should be treated at the sampling point with n-hexane for the analysis of halogenated organic compounds or with 30 ml of dichloromethane per litre of sample for the analysis of organophosphorus compounds. The sample should then be stored in a cool and dark place.

After thorough shaking, the extract is separated at the latest within 12 hours. Organophosphorus insecticides may be determined directly by gas chromatography in the dichloromethane extract. If necessary, primary treatment may be carried out with florisil or silica gel; in the latter case, prefractionation is possible at the same time. The hexane extract is used for the test for organohalogen compounds.

### Organohalogen compounds

#### Equipment

Separating funnels, 5000 to 250 ml

Shaking cylinders (mixing cylinders) or test tubes, graduated, with stoppers, 30 - 50 ml

Vacuum evaporator (if necessary, without condensor) or

Kuderna-Danish evaporator

Chromatographic columns, 12 mm diam., height 15 cm,

Glass frit 5 mm diam., height 30 cm

Mechanical shaker with attachment in which the separating funnel can be inserted such that its discharge tube points diagonally upwards

Ultra Turrax (e.g. Jahnke and Kunkel, FRG)

Gas chromatograph with suitable detectors. Packed separating columns or capillary columns

Extreme purity of the solvents is a prerequisite for the detection of very small residues of insecticides. Purification is by distillation through a Snyder column. Discard the first 10 % of the distillate, and continue distillation until about 20 % of the original volume remains in the distillation flask. Test the purity of the distillate by preparing an aliquot portion after 100 : 1 concentration in a Kuderna-Danish evaporator and checking that there are no measurable interfering peaks after injection into the gas chromatograph. If such peaks do appear, repeat the purification process. ECD-pure solvents are obtained by boiling hexane which has been prepurified by distillation, with finely dispersed sodium in the reflux.

To prepare the sodium suspension, heat sodium disks under kerosene in ground-glass-stoppered flasks until they are seen to melt (120 °C, caution when working with an oil bath), immediately seal the flasks and shake vigorously.

Dichloromethane

Purified toluene

Purified n-hexane

Purified isooctane

Florisil deactivated with 8 % water

Silica gel for adsorption, activity level 1 (e.g. Woelm, FRG; Alcoa, USA)

Sodium sulphate, anhydrous, heated for at least 2 hours at 400 - 500 °C

Silica gel

Argon

Helium

## Procedure

### A) General method

Take a maximum of 4 litres of water sample in a 5-litre separating funnel. On the mechanical shaker, make successive additions of 15, 10 and 10 ml of n-hexane (with one litre of sample or less) or of 25, 15 and 15 ml of n-hexane (with 3 to 4 litres of sample), shaking for 3 minutes each time. Leave to stand for 5 minutes to allow the phases to separate (10 - 15 minutes after the third shaking), transfer most of the aqueous phase to a second 5-litre separating funnel and drain off the n-hexane with the remaining water into a 250-ml separating funnel. Rinse out the first large funnel each time, particularly following the third extraction, with pure water using a washing bottle. Collect the rinsing water in the 250-ml funnel. In this funnel, separate off the water, dry the hexane phase with sodium sulphate and transfer through the neck of the vessel into a graduated container (mixing cylinder, or if evaporation takes place, a large test tube) and rerinse twice with a few ml of n-hexane. The extract in each case (25 ml or 50 ml) may be analyzed by gas chromatography immediately afterwards.

### B) Processing of small volumes of samples (100 to 150 ml)

Add 10 ml of n-hexane to the unfiltered water sample in a wide-necked bottle with conical shoulder and homogenize for 2 minutes with, for example, an ultrastirrer. In order to accelerate the subsequent phase separation, transfer the sample to a centrifuge beaker and centrifuge for 15 minutes at 9000 rpm. Then separate the aqueous phase in a 250-ml separating funnel, collecting the water once more in the wide-necked bottle as it drains off. Repeat extraction, centrifugation and phase separation with a further 10 ml of n-hexane, using the same apparatus. Dry the fully separated n-hexane phase while still in the separating funnel with sodium sulphate and transfer through the neck of the vessel into a graduated tapered flask, rinsing twice with a few ml of n-hexane.

### C) Processing of larger volumes of samples (1 to 20 litres)

Add 100 ml of n-hexane to the unfiltered water sample in one or more wide-necked bottles with conical shoulders, as appropriate to the quantity of sample, and homogenize for 15 minutes as above. Allow to stand for 2 hours, by which time the phases are separated to the extent that the solvent can be siphoned off and transferred to a 1000-ml separating funnel. Return any aqueous phase in the separating funnel to the wide-necked bottle. Transfer the n-hexane phase through the neck of the separating funnel into a 300-ml tapered flask through a glass funnel with $Na_2SO_4$. Proceed with a second extraction in the same way, using the same apparatus. If the water sample contains perceptible quantities of suspended matter, relatively persistent gels may form at the phase boundary layer, which then remain after separation of the aqueous and hexane phases. The remaining hexane can be isolated from the gels by shaking after the addition of sodium sulphate or by stirring up the gel with sodium sulphate. (For the analysis of organophosphorus compounds the treatment is the same, using dichloromethane.)

## Extract purification when using n-hexane for the analysis of halogenated compounds

Purification of the extract is necessary if traces are to be detected or if the water samples have a high content of accompanying substances.

Concentrate the extract in a rotary evaporator or in a Kuderna-Danish evaporator to approx. 1 ml (without an air flow). Inject the extract into a chromatographic column of 12 mm diameter and 15 cm height, which has previously been packed to half height with n-hexane, then to a height of 7 cm with florisil and above that 1 cm of hexane-toluene mixture 1 : 1, maintaining a drop rate of 1 to 2 drops/second (3 to 5 ml/min). For the separation of non-polar interfering substances (in particular polychlorinated biphenyls (PCBs)) and for preseparation, it is advisable to use a silica gel column instead of or in addition to the above:

For this purpose pack a chromatographic column of 5 mm diameter and 30 cm length to a height of 10 cm with silica gel and activate for approximately 15 hours at 110 °C. Adjust the raw extract to a set volume and then inject 1 ml into the column, previously rinsed with 15 ml of isooctane (prefractionation). Elution is with isooctane and toluene in various mixture ratios.

Apply 15 ml of isooctane-toluene 95 : 5 (fraction I), then 10 ml of isooctane-toluene 70:30 (fraction II) and subsequently 10 ml of pure toluene (fraction III). The fractions obtained are divided as follows:

1 ml of prefraction, containing the main proportion of impurities
Fraction I: containing aldrin, DDE and DDT, heptachlor and PCB,
Fraction II: containing lindane and other HCH isomers,
Fraction III: containing dieldrin $\alpha$- and ß-endosulfan and methoxychlor.

The runoffs, which contain the active substances, should again be concentrated by evaporation to an appropriate volume in order to achieve high detection sensitivity.

Gas-chromatographic measurement

Use a gas chromatograph with packed or capillary columns and an electron capture detector. The following stationary phases are recommended: one non-polar (e.g. Dow 11, SE 30, OV 1) and one weakly polar (e.g. QR-1, XE-60, OV-225) or a mixed phase.

| | |
|---|---|
| Carrier gas: | After- purified nitrogen (through molecular sieve 5 A), additionally dried |
| Temperatures: | Injection block 200 to 220 °C<br>Column oven 185 to 195 °C<br>Detector oven 190 to 205 °C |

Any gas chromatograph which fulfils the above conditions may be used for gas-chromatographic determination of pesticides.

Compare the chromatogram obtained with external standards. In a standard solution, which may conveniently be prepared with toluene or hexane, the following active substances (in picograms per µl) give approximately the same peak heights: lindane and other HCH isomers 30, heptachlor 30, HCB 40, aldrin 40, heptachlorepoxide 80, dieldrin 80, $\alpha$-endosulfan 100, DEE 120, ß-endosulfan 200, DDD 250, DDT 500, methoxychlor 500. The peak area is used as the basis for evaluation.

Information on the sensitivity of the determination of various pesticides on the basis of peak heights can only serve as a guide. The peak areas are used for actual evaluation, and are compared with corresponding standard substances for calibration. Calibration must be performed every day.

The following minimum quantities can be detected from a 4-litre sample without extract purification (in ng/l):

Aldrin, HCH, heptachlor 10 - 5, DDD, DDT, methoxychlor 200 - 100, the other organochlorine insecticides 40 - 20.

Depending on the method used, and the quantity and type of active substance, the yields are between 80 and 100 %.

Blank tests must be conducted according to the same method using superpure water, and their results taken into account.

For calibration, the compounds to be tested should be added to water samples (e.g. in acetonic solution), and the samples processed in the same way as other samples and the extracts used for calibration.

In order to be certain that no substances are present which may influence the pesticide determination, it is necessary to test the purity of the reagents before use. This is required in order to utilize the lower gas-chromatographic detection limit when analyzing concentrated water extracts.

**Note:**

The water solubility of certain active substances is of the following order of magnitude: (in µg/l)

Aldrin 50, dieldrin 200, PP-, DDT 5, heptachlor 50, endosulfan ( + isomers) 500, endrin 250, hexachlorocyclo-hexane ( + -isomers) 10000, methoxychlor 100.

### 4.2.10 Organophosphorus pesticides, thiophosphoric acid esters, chlorinated hydrocarbons and triazine herbicides

**General remarks**

If the aim is also to detect organophosphorus insecticides, the water is extracted with dichloromethane instead of hexane. The following is a description of the method of the Deutsche Forschungsgemeinschaft used for foodstuffs of vegetable origin, but which may also be applied to water. The separation of interfering substances is achieved with an active charcoal - silica gel column. The retained active substances are eluted with a mixture of dichloromethane-toluene-acetone, the eluate is evaporated off and the residue picked up with toluene. This solution is subjected to gas-chromatographic determination, using an electron capture detector and phosphorus detector.

**Equipment**

Funnel with flat perforated bottom

Beakers, 1000 ml, 50 ml

Separating funnels, 1000 ml and 5000 ml (ground-glass stoppers)

Chromatographic tubes with fritted glass filter plate and stopcock, length 40 cm, diameter 2.5 cm

Vacuum rotary evaporator

Gas chromatograph with electron capture detector ECD and specific phosphorus detector (flame ionization detector AFID)

Packed or capillary columns

Acetone, chemically pure, distilled in rotary evaporator at 40 °C

Toluene, superpure quality, additionally distilled in fractionating attachment

Dichloromethane, superpure quality

Elution mixture:
1000 ml of dichloromethane, 200 ml of toluene, 200 ml of acetone
Sodium chloride solution, saturated
Sodium chloride, reagent purity
Sodium sulphate, anhydrous, heated for at least 2 hours at 400 to 500 °C

Active charcoal, reagent purity, e.g. Merck No. 2186, Chemviron etc.

Celite 545, e.g. C. Roth, Karlsruhe, FRG

Silica gel for column chromatography, 0.05 to 0.2 mm, e.g. Merck No. 7734

Qualitative round filter, diameter 11 cm

Argon/methane

Helium

Air, synthetic

Hydrogen

**Procedure**

Take up to a maximum of 4 litres of unfiltered water sample in a 5-litre separating funnel. Extract on a mechanical shaker, successively adding 25, 15 and 15 ml of dichloromethane, 3 minutes each time. Wash the combined dichloromethane extracts with 250 ml of distilled water and 25 ml of saturated solution of common salt, drain off the dichloromethane and dry for 30 minutes with 10 g of ignited sodium sulphate. Filter the dried extract through a fluted filter. Rinse the vessel and filter with 30 ml of dichloromethane in three portions. Evaporate the filtrate and the rinse to about 2 ml in the rotary evaporator and remove the remainder of the solvent by turning the tilted flask, held in the hand. Take up the residue in 10 ml of dichloromethane.

**Preparation of the purification column**

Wash 5 g of silica gel with 15 ml of elution mixture into a column containing dichloromethane to a height of 1 cm. Discard the supernatant liquid. Thoroughly mix a further 15 g of silica gel and 1 g of active charcoal in a 50 ml beaker and slowly mix with 35 ml of elution mixture. Stirring constantly, pour the charcoal-silica gel mixture into the column over the silica gel substructure via a funnel, initially pouring slowly then all at once. Leave the stopcock of the column open. Liquid which has passed through is used to rinse the pouring vessel. Drain off the elution mixture down to 2 cm above the edge of the packing and coat the packing in small portions with a total of 5 g of sodium sulphate. Prewash the column with 50 ml of eluting agent. The column is now ready for operation.

### Purification of the extract

Feed the residue taken up in 10 ml of dichloromethane quantitatively into the prepared column, rinsing with dichloromethane. Collect any liquid passing through at this stage and the later eluate in a 250 ml round-bottomed flask. Use 140 ml of elution mixture for elution of the column. Concentrate the eluate to approximately 2 ml in the rotary evaporator and remove the remaining solvent by turning the tilted flask, held in the hand. Remaining traces of dichloromethane lead to undesirable broadening of the solvent peak in the ECD and may be eliminated by repeating evaporation. Take up the residue in 10 ml of toluene and transfer to a graduated ground-glass tube. The residue is now ready for gas chromatography. In routine cases, doses of 2 µl should be injected to determine the organophosphorus compounds and the thiophosphoric acid esters. The measurements may also be conducted with a 2-column gas chromatograph with a phosphorus detector and an electron capture detector if a test is also to be carried out for chlorinated hydrocarbons.

### Chlorinated hydrocarbons

| | |
|---|---|
| Detector: | ECD |
| Column: | 183 cm/6.35 mm glass column with 3.8 % SE-30 on Diatoport S-80 to 100 mesh, or capillary column |
| Carrier gas: | Helium |
| Purge gas: | Argon/methane, 50 ml per minute |
| Temperature: | Oven 210 °C (isothermal)<br>ECD 300 ° - 380 °C |
| Injection block: | 230 °C |

### Thiophosphoric acid esters

| | |
|---|---|
| Detector: | AFID |
| Column: | 183 cm/6.35 mm glass column with 2 % FS1265 on Chromosorb WAW DMCS, 60 to 80 mesh or capillary column |
| Carrier gas: | Helium |
| Fuel gases: | |
| Hydrogen: | Approx. 40 ml per minute |
| Air: | 380 ml per minute |
| Temperatures | Oven 210 °C (isothermal)<br>AFID 220 °C |
| Injection block | 230 °C |

### Equipment

Rotary evaporator

Chromatographic tube, diameter 18 mm, length at least 200 mm

Gas chromatograph with a detector specific to each of nitrogen, chlorine and sulphur

Packed or capillary columns

Microinjection syringes, 10 and 50 µl

Ethanol, distilled or nanograde

Ether, distilled or nanograde (diethylether)

Toluene, distilled or nanograde

Chloroform, distilled or nanograde

n-hexane, distilled or nanograde

Active-substance standard solution:
10 µg/ml of each in hexane - ethanol 1 : 1

Aluminium oxide, e.g. Woelm or Alcoa 200, basic, activity level V (19 % water addition)

Nitrogen, after-purified

Helium

Hydrogen, after-purified

Oxygen, after-purified

### Extraction

Extract the water sample (500 ml, pH 6 - 9) twice, each time with one tenth of its volume, with chloroform. Drain off the chloroform via a wad of cotton wool and evaporate in the rotary evaporator at a water-bath temperature of 40 °C (use fume hood).

Pack a chromatographic tube with n-hexane and trickle in aluminium oxide, activity level V, to a height of 7 cm. Drain off the hexane to the height of the aluminium oxide. Dissolve the residue in the rotary evaporator in 5 ml of toluene, feed the solution onto the column and allow to seep in. Rewash the flask twice, each time with 5 ml of n-hexane, feed this solution successively onto the column with 80 ml of n-hexane and discard these first runnings. Then elute with 75 ml of a mixture of n-hexane/ether 2 : 1. Collect the eluate and evaporate in the rotary evaporator at a water-bath temperature of 40 °C.

### Gas-chromatographic measurement

First prepare a general chromatogram with a selective alkali flame ionization detector using a carbowax 20 M column. This column separates all listed triazine herbicides with the exception of the two pairs terbuthylazine/isobumetone and simazine/terbutryn. If peaks appear at the points at which terbuthylazine, isobumetone, terbutryn and simazine are to be expected according to the standard chromatogram, re-inject the sample using a flame-photometric detector with S filter. If peaks then appear at the same points, they may be evaluated quantitatively. Calculate the content of methoxtriazine by taking the difference.

The analysis may also be performed using capillary columns in the device.

**Calibration**

Commencing with a standard at the detection limit, inject a calibration series of 4 points up to the quantity corresponding to the maximum triazine content in the samples under analysis. For example: detection limit 20 nanograms; maximum quantity of triazine injected with one sample 400 nanograms; calibration series 20, 50, 150 and 400 nanograms. Peak height or peak area is used as the measure.

**Yield**

If active substances are added to untreated water samples, in a range between 0.05 and 1 ppm 80 to 120 % of the added active substances are regained. Prepare a double logarithm calibration curve from the standard injections, either graphically or using an electronic computer. Take the quantities, in nanograms, corresponding to the peak heights of the sample from the calibration curve. Calculate the content of active substance referred to the water sample in mg/l or µg/l, based on the injected sample aliquot.

**Note:**

Column chromatographic purification of water samples is in many cases unnecessary; instead, it is enough to dissolve the residue from dichloromethane or chloroform extraction with hexane/ethanol and to inject it directly. As an alternative to a nitrogen-selective alkali flame ionization detector a nitrogen-specific conductivity detector after Coulson may also be used.

**Appendix to 4.2.9 and 4.2.10**

The test for triazine herbicides is based on the gas-chromatographic detection of undecomposed active substances with the assistance of various molecule-specific detectors. This method combines high sensitivity and good selectivity. The active substances which may be determined using this method are listed in the table below.

| Name | Chemical notation |
|---|---|
| Ametryne | 2-Methylthio-4-ethyl-amino-6-isopropylamino-s-triazine |
| Atratone | 2-Methoxy-4-ethylamino-6-isopropylamino-s-triazine |
| Atrazine | 2-Chloro-4-ethylamino-6-isopropylamino-s-triazine |
| Desmetryne | 2-Methylthio-4-methylamino-6-isopropylamino-s-triazine |
| Isobumetone | 2-Methoxy-4-sec.butylamino-6-ethylamino-s-triazine |
| Methoprotryne | 2-Methylthio-4-isopropylamino-6-γ- methoxypropylamino-s-triazine |
| Prometone | 2-Methoxy-4,6-bis-isopropylamino-s-triazine |
| Prometryne | 2-Methylthio-4,6-bis-isopropylamino-s-triazine |

| | |
|---|---|
| Propazine | 2-Chloro-4, 6-bis-isopropylamino-s-triazine |
| Simazine | 2-Chloro-4,6-bis-ethylamino-s-triazine |
| Terbuthylazine | 2-Chloro-4-ethylamino-6-tert.butylamino-s-triazine |
| Terbutryne | 2-Methylthio-4-ethylamino-6-tert.butylamino-s-triazine |

### 4.2.11 Gas chromatography head-space analysis

**General remarks** (see also Section 2.13)

The sampler must describe the results of his odour tests at the sampling point. Select the size of flask according to odour intensity.

**Method A: GC with packed columns**

Fill 3 glass flasks of equal volumes of the water under investigation. If possible, a further 3 flasks should be filled with the untreated water in addition, as blank tests or for calibration purposes.

**Examples of quantities per flask**

| | |
|---|---|
| 250-ml flasks: | 200 ml of water |
| 500-ml flasks: | 450 ml of water |
| 1-litre flasks: | 950 ml of water |
| 2-litre flasks: | 1900 ml of water |

N.B.: The glass flasks must be absolutely odour-free.

For each analysis, the same size of flask is to be used for the various samples under analysis and for the blank or calibration readings.

For the samples under analysis, the headspace equilibrium is established at, for example, 20 °C, 30 °C or 50 °C. The water sample may either be shaken by hand and then headspace equilibrium is established at various temperatures, or a mechanical shaker may be used for a defined period, following which headspace equilibrium is established at the specified temperature. Next withdraw various volumes of gas (1 to 5 ml) with a gas syringe through the septum from the gas space of the flask, and inject into a gas-chromatograph.

**Equipment**

Glass flasks, from 0.25 to approx. 2 litres

Ground-glass attachment with septum, to fit glass flasks

Gas syringes, 0.5 to 10 ml

Gas chromatograph with accessories and various packed columns

Flame ionization detector (FID)

Drying cabinet

Mechanical shaker

Columns

a) Non-polar column with SE 30 on Varoport (100 mesh) 2-m steel column, diameter 1/8 inch

b) Non-polar separation column, 3 % Apiezon M on Chromosorb W-AWS, 80 - 100 mesh
2-m glass column or 2-m steel column, 1/8 inch

c) Polar column, Carbowax 20M on Chromosorb G-DMCS, 80-100 mesh,
2-m steel column, 1/8 inch

d) Polar column, 10 % phthalic acid -bis- (-3,3,5-trimethylcyclohexyl)-ester or Chromosorb W-AW, 80 - 100 mesh, 2-m steel column, 1/8 inch

e) Polar column with silver nitrate on diethylene glycol on Chromosorb R, NAW, 60 - 80 mesh,
2-m steel column, 1/8 inch

f) Polar column, nitrile silicone oil XE 60 on kieselguhr 60 - 100 mesh, 2-m steel colum, 1/8 inch

g) Capillary columns, see Method B.

Carrier gas: Nitrogen or helium, 20 ml/min.

Injector temperature: 250 °C

Detector (flame ionization detector) 250 °C

It is suggested that a temperature programme is used in which the temperature of the various columns is raised to maximum temperature at a rate of 10 °C per minute, beginning at 70 °C.

Following each gas chromatographic analysis, the column and equipment used should be thoroughly baked out. It is recommended that the gas space of at least two sample flasks is examined. If calibrations are carried out according to the admixture method, the same water, but unpolluted, should if possible be charged with known quantities of the suspected volatile organic substances and subjected to further analysis in the same way. This makes it possible to draw quantitative conclusions as well concerning the concentration of certain volatile organic compounds in the water.

### Method B. Example of a head-space analysis using capillary columns

(Determination of diisopropyl ether (DiPE) in water)
(Can be used similarly for other substances or mixtures)

#### General remarks

The water sample is thermostatically conditioned until equilibrium is established in the head-space. A gas sample from the head-space is then injected into the gas chromatograph.

## Equipment

5-ml septum bottles with beaded caps or head-space bottles of different volume

Closure tool

2-ml bulb pipette

10-µl syringe

Head-space sampler, eg DANI SPT 3750

Gas chromatography unit

Capillary columns etc

Standard solution:
Dissolve approx. 50 mg of diisopropyl ether in 50 ml of polyethylene glycol 400.

Water, e.g. from a Millipore superpure water system

## Preparation of sample

| | |
|---|---|
| Reference: | e.g. 2 ml superpure water + 1 µl standard solution |
| Blank test: | 2 ml superpure water |
| Sample: | 2 ml sample, or some other volume (if necessary diluted, if the concentration to be expected in the vapour phase is very high) |
| Admixture: | e.g. 2 ml sample + standard solution (the amount of standard solution added should be based on the concentration of DiPE expected in the sample). |
| Example: | DiPE concentration in sample = 1 mg/l<br>Additional quantity of DiPE in admixture = 1 - 2 µl of standard solution (for example) |

## Conditioning

Pipette 2 ml of sample into a 5-ml septum bottle and seal with a beaded cap. Condition the samples, calibration and blank tests for at least 30 min. at 90 °C, thereby establishing head-space equilibrium.

## Measuring conditions

50 m Carbowax 400, capillary columns

Temperatures:

$T_i$ = 200 °C
$T_s$ = 30 °C
$T_D$ = 290 °C

Admission pressure 2.8 bar helium, split 15 ml/min.

Injection methods:

1. Manual: Inject 0.5 ml of gas phase with a gas syringe conditioned to 90 °C.

2. Injection using P.E. HS-6' semi-automatic system

3. Injection using DANI SPT 3750 Autosampler

## Evaluation

The results are obtained by direct comparison of the DiPE peak areas for the sample and the calibration. The augmentation of the sample by admixture is necessary as a qualitative check.

$$\frac{\text{DiPE peak area}_{\text{Sample}} \cdot \text{Conc.}_{\text{Calibration}}\ (\text{mg/l})}{\text{DiPE peak area}_{\text{Calibration}}} = \text{Concentration}_{\text{Sample}}$$

Theoretical result: 0.14 mg/l

Example of GC conditions:

50 m CP sil-5 CB 1.0 µm
$T_i$ = 220 °C
$T_s$ = 40 °C, 6 min. | 6°/min. | 240 °C
$T_D$ = 300 °C FID
1 bar helium, split 30 ml/min.
AT = $2 \cdot 10^{-12}$
GC 3400 Varian
Injector volume 100 µl manual

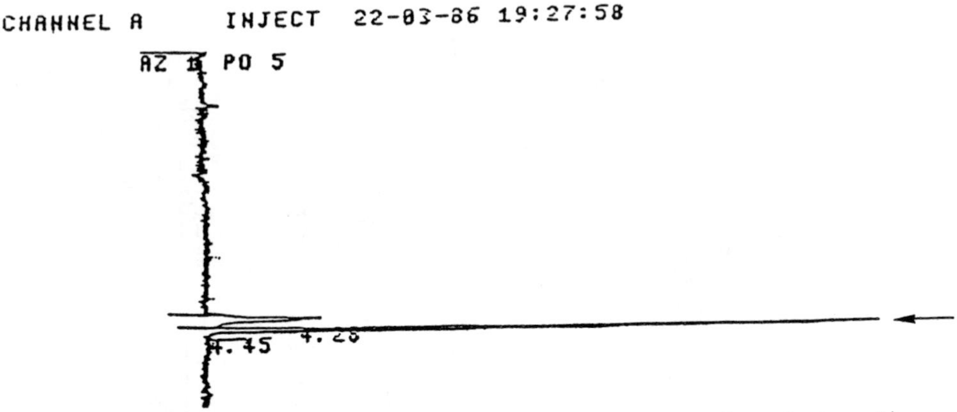


Fig. 146 a. Example of chromatogram of head-space analysis (practical work)

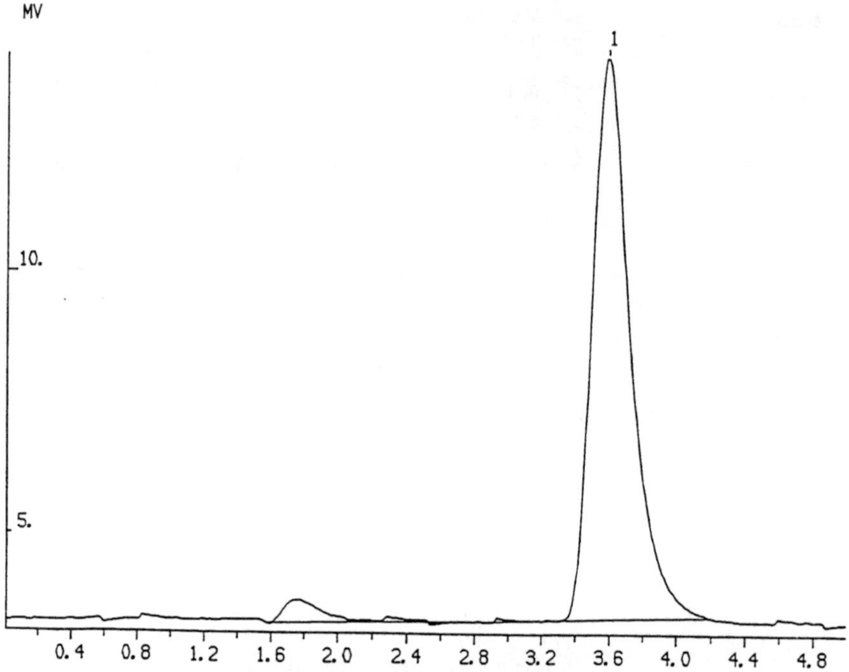


**Fig. 146 b.** Example of chromatogram of head-space analysis (practical work); 1) = Vinyl chloride

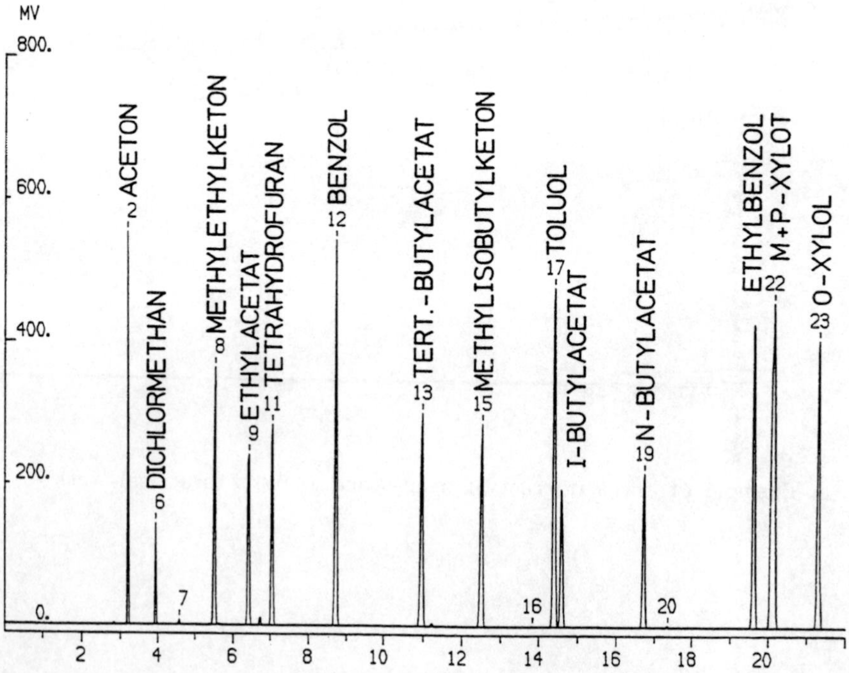


**Fig. 147.** Mixture of organic solvents Tetrahydrofurane, Acetone, Dichloromethane, Methyl ethyl ketone, ethylacetate, butyl-acetate, Benzene, Isobutyl, methyl ketone, Toluene, ethylbenzenė, m+p-xylene, o-xylene

### 4.2.12 Determination of plasticizers in water

**General remarks**

Plasticizers may enter waste water from the production of plastics and may also enter bottled water from seals.

Extraction of the water with n-hexane
GC analysis with double FID and two capillary columns of different polarity

**Equipment**

1-litre separating funnel

1-litre measuring cylinder

500-ml Erlenmeyer flask with 50 ml tilting pipette

100-ml beakers

10-µl syringe

100-ml round-bottomed flask with 1-ml tip

GC with double FID and two capillary separation columns of different polarity

Recorder etc.

Water, e.g. from a Millipore superpure water system

Nanograde hexane

Nanograde toluene

Standard plasticizer solution:
e.g. approx. 10 µg DOP/ml in nanograde acetone (see empirical list)

Internal standard solution:
Docosane, approx. 1 mg/ml in nanograde hexane

**Sample preparation**

| | |
|---|---|
| Reference: | 1 litre of superpure water + 10 µl of internal standard + 10 ml of standard plasticizer solution |
| Blank test: | 1 litre of superpure water + 10 µl of internal standard |
| Sample: | 1 litre of water sample + 10 µl of internal standard |

**Procedure**

Take 1 litre of water in a 1-litre separating funnel and shake with 50 ml of nanograde hexane for 5 minutes. Discard the aqueous phase. Transfer the organic phase to a 100-ml beaker, subsequently decant into a 100-ml round-bottomed flask with 1-ml tip and evaporate in a water bath at 40 °C with 1 ml of toluene as a keeper.

**Measurement**

| | |
|---|---|
| Temperature program: | Initial temperature 200 °C, heat to 280 °C at 5 °C/min., 2 min. isothermal tail run |
| Injector temperature: | 280 °C |
| Detector temperature (FID): | 330 °C |
| Split: | 20 ml/min. |
| Helium: | 1.45 bar |
| Injection volume: | 2 µl |
| Capillary columns: | A column: 30 m DB5, film thickness 0.25 µm<br>B column: 30 m DB1701, film thickness 0.25 µm |

**Evaluation**

Internal-standard method
Peak height evaluation

| Empirical list of plasticizers in water | Empirical determination limit in mg/l |
|---|---|
| Di-(2-ethylhexyl)-phthalate (DOP) (Dioctylphthalate) | 0.001 |
| Dibutylphthalate (DBP) | 0.001 |
| Di-(2-ethylhexyl)-adipate (DOA) | 0.001 |
| Dibutylsebacate (DBS) | 0.001 |
| Tricresylphosphate (TCP) | 0.001 |

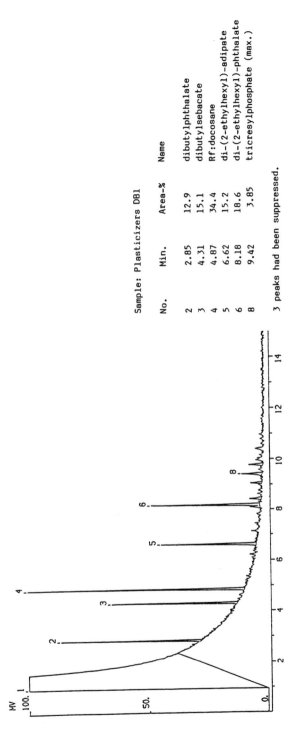

Sample: Plasticizers DB1

| No. | Min. | Area-% | Name |
|---|---|---|---|
| 2 | 2.85 | 12.9 | dibutylphthalate |
| 3 | 4.31 | 15.1 | dibutylsebacate |
| 4 | 4.87 | 34.4 | Rf:docosane |
| 5 | 6.62 | 15.2 | di-(2-ethylhexyl)-adipate |
| 6 | 8.18 | 18.6 | di-(2-ethylhexyl)-phthalate |
| 8 | 9.42 | 3.85 | tricresylphosphate (max.) |

3 peaks had been suppressed.

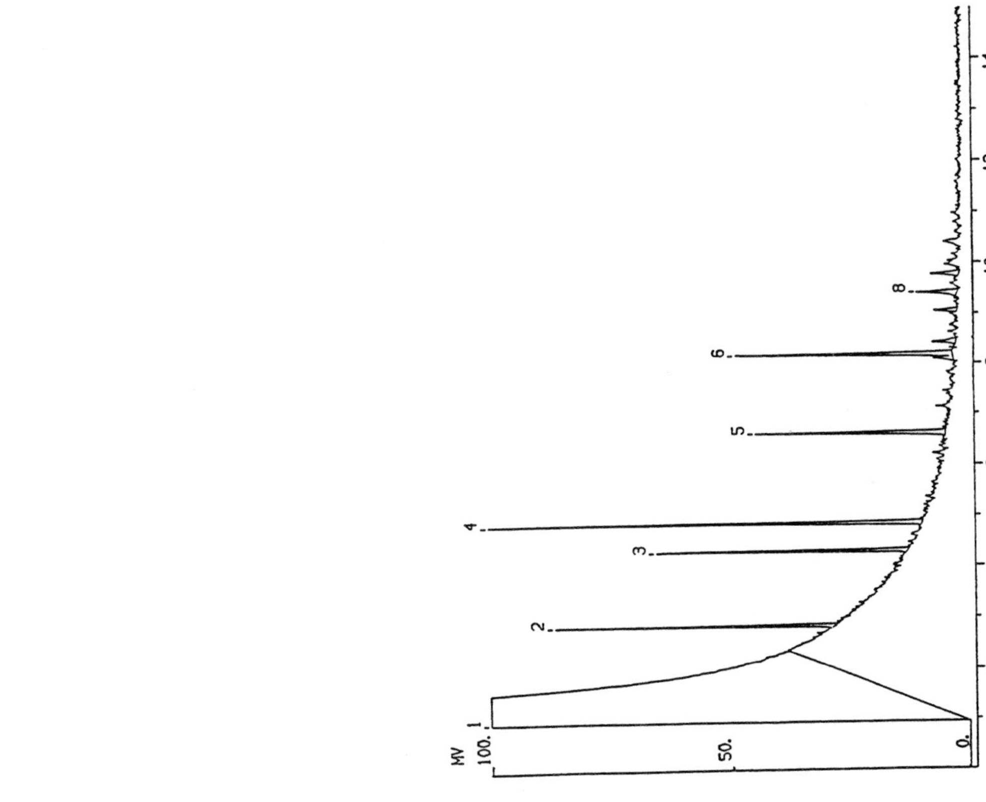


Fig. 148. Example of a chromatogram for plasticizers

Sample: Plasticizers DB1701

| No. | Min. | Area-% | Name |
|---|---|---|---|
| 1 | 2.96 | 11.7 | dibutylphthalate |
| 2 | 3.09 | 29.8 | Rf:docosane |
| 4 | 3.99 | 13.1 | dibutylsebacate |
| 5 | 5.98 | 13.9 | di-(2-ethylhexyl)-adipate |
| 7 | 8.03 | 16.2 | di-(2-ethylhexyl)-phthalate |
| 20 | 11.25 | 2.10 | tricresylphosphate (max.) |

%-threshold: 2.00 %
19 peaks out of 25 (total area percentage = 13.11 %)
are below threshold.

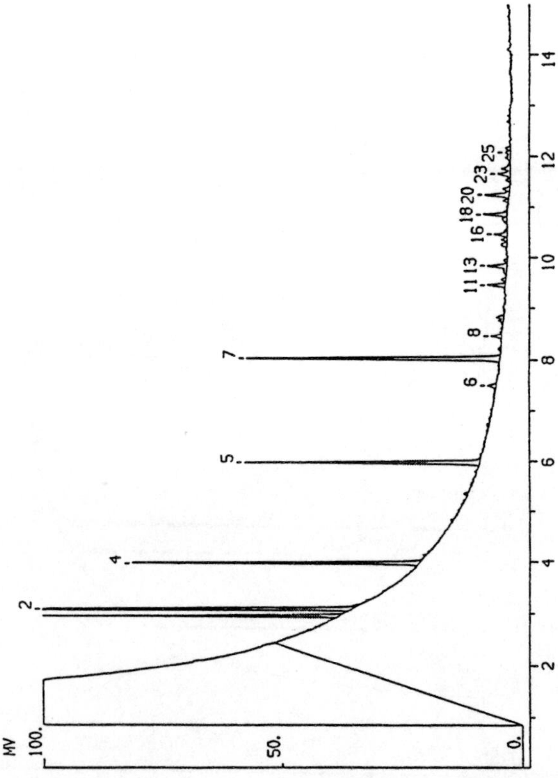


Fig. 149. Example of a chromatogram for plasticizers

### 4.2.13 Determination of antioxidants in water

**General remarks**

Antioxidants occur only rarely in water.
Extraction with dichloromethane.

Silylation

GC measurement with double FID on two capillary columns of different polarity

**Equipment**

100-ml separating funnel

500-ml Erlenmeyer flask with 5-ml tilting pipette

Syringes, 200 µl, 500 µl

Pipettes: 1 ml, 2 ml, 5 ml, 10 ml

5-ml septum flasks with beaded cap

Gas-chromatographic apparatus

Water, e.g. from a Millipore superpure water system

2 m HCl

Nanograde dichloromethane

Trimethylchlorosilane (TMCS)

MSHFBA (N-Methyl-N-trimethylsilyl-heptafluorobutyramide)

Pyridine, reagent purity

Standard antioxidant solution:
Approx. 1 mg/ml in nanograde acetone (propylgallate = 2 mg/l)

| **Antioxidants** (empirical list) | Determination limit in mg/l |
|---|---|
| 2,6-Di-tert.-butyl-p-cresol | 0.05 |
| 3-tert.-Butyl-4-hydroxy-anisole | 0.05 |
| Propylgallate | 0.2 |
| Octylgallate | 0.05 |
| Dodecylgallate | 0.05 |

Internal standard solution, 2,3,4,6-tetrachlorophenol, approx. 1 mg/ml in nanograde acetone

**Sample preparation**

Reference: 100 ml of superpure water + 500 µl of internal standard + 200 µl of standard antioxidant solution

Blank test: 100 ml of superpure water + 500 µl of internal standard

Sample: 100 ml of water + 500 µl of internal standard

**Procedure**

Take 100 ml of sample in 100-ml separating funnel and acidify with 2 ml of 2 m HCl. Shake for 5 minutes with 5 ml of nanograde dichloromethane. Drain off the organic phase into a 5-ml septum flask and blow off with nitrogen until dry. Add 1 ml of reagent solution to the residue, seal the flask with the beaded cap and heat in a drying cabinet at 100 °C for 45 minutes.

Reagent solution: Pyridine/TMCS/MSHFBA = 2/5/10 (v/v/v)

**Measuring conditions**

| | |
|---|---|
| Temperature program: | Initial temperature 180 °C, 4 minutes isothermal prerun, heat to 290 °C at 8 °C/min., 3 minutes isothermal tail run |
| Carrier gas: | Helium 1.5 bar |
| Split: | 20 ml/min. |
| Injector temperature: | 280 °C |
| Detector temperature: | 330 °C |
| Injection volume: | 2 µl |
| Capillary columns: | Column A: 30 m DB5, film thickness 0.25 µm<br>Column B: 30 m DB1701, film thickness 0.25 µm |

**Evaluation**

Internal standard solution
Peak height evaluation

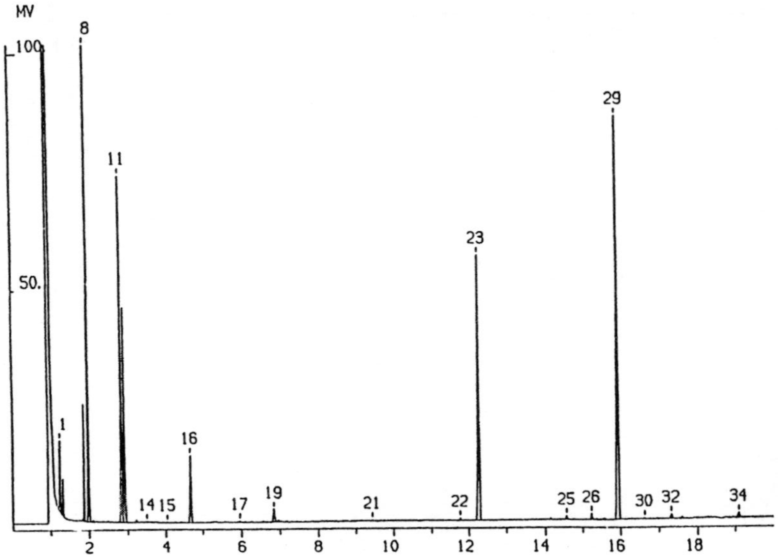

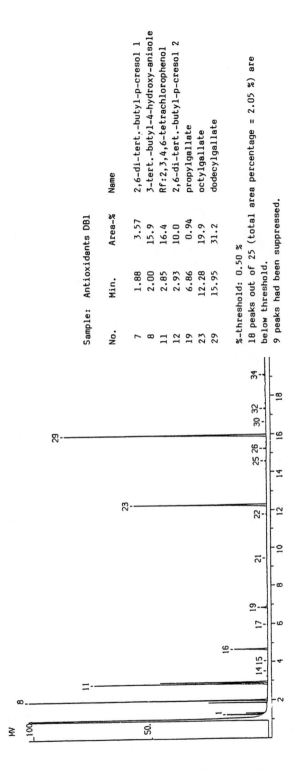

Sample: Antioxidants DB1

| No. | Min. | Area-% | Name |
|---|---|---|---|
| 7 | 1.88 | 3.57 | 2,6-di-tert.-butyl-p-cresol 1 |
| 8 | 2.00 | 15.9 | 3-tert.-butyl-4-hydroxy-anisole |
| 11 | 2.85 | 16.4 | Rf:2,3,4,6-tetrachlorophenol |
| 12 | 2.93 | 10.0 | 2,6-di-tert.-butyl-p-cresol 2 |
| 19 | 6.86 | 0.94 | propylgallate |
| 23 | 12.28 | 19.9 | octylgallate |
| 29 | 15.95 | 31.2 | dodecylgallate |

%-threshold: 0.50 %
18 peaks out of 25 (total area percentage = 2.05 %) are below threshold.
9 peaks had been suppressed.

Fig. 150. Example of a chromatogram for antioxidants

Sample: Antioxidants DB1701

| No | Min. | Area-% | Name |
|---|---|---|---|
| 6 | 1.55 | 3.89 | 2,6-di-tert.-butyl-p-cresol 1 |
| 7 | 1.60 | 17.1 | 3-tert.-butyl-4-hydroxy-anisole |
| 9 | 2.04 | 10.3 | 2,6-di-tert.-butyl-p-cresol 2 |
| 11 | 2.25 | 16.8 | Rf:2,3,4,6-tetrachlorophenol |
| 20 | 5.63 | 0.90 | propylgallate |
| 26 | 10.98 | 18.8 | octylgallate |
| 33 | 14.54 | 29.2 | dodecylgallate |

%-threshold: 0.50 %
30 peaks out of 37 (total area percentage = 3.04 %) are below threshold.
2 peaks had been suppressed.

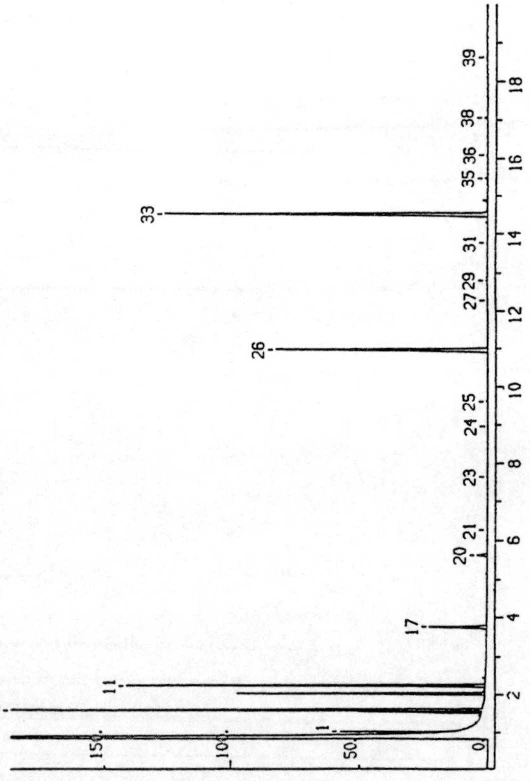


Fig. 151. Example of a chromatogram for antioxidants

## 4.2.14 Enzymatic determination of urea

### Fundamentals

Urea in aqueous solutions is broken down quantitatively into ammonium in the neutral range (pH at approximately 7) by the enzyme "urease". The reaction can be represented by the following equation:

$$(NH_2)_2CO + 2\ H_2O \xrightarrow{\text{urease}} 2\ NH_4^+ + CO_3^{2-}$$

$NH_4^+$ is then subsequently determined analytically. The determination limit and accuracy are not dependent on the enzymatic transformation but on the analytical detectability of the ammonium formed.

It is advisable to carry out the process of determination photometrically, employing Berthelot's colour reaction. This reaction is based on the formation of the indole dye indophenol blue. Firstly, hypochlorite is used to convert $NH_4^+$ into chloramine under alkaline conditions.

$$NH_4^+ + NaOH \longrightarrow NH_3 + Na^+ + H_2O$$

$$NH_3 + NaOCl \longrightarrow NH_2Cl + NaOH$$

$$NH_4^+ + NaOCl \longrightarrow NH_2Cl + Na^+ + H_2O$$

Chloramine reacts with phenolate in the presence of sodium nitroprusside as catalyst and in several reaction steps forms the dye indophenol blue, which is yellow in the non-dissociated and blue in the dissociated state. This reaction can be represented by the following empirical formula:

$$2\ C_6H_5{\cdot}ONa + NH_2Cl + 2\ NaOCl \longrightarrow \underset{\text{(indophenol blue)}}{O{\cdot}C_6H_4{\cdot}N{\cdot}C_6H_4{\cdot}ONa} + 3\ NaCl + 2\ H_2O$$

### Conducting the process for the enzymatic determination of urea

Quantity of sample:
Twice 2 ml water

### Equipment

Photometer

Urease solution:
2 mg/ml urease + 50 mmol/l phosphate buffer, pH = 6.5

Phenol solution:
106 mmol/l phenol + 0.17 mmol/l sodium nitroprusside

Alkaline sodium hypochlorite solution:
11 mmol/l sodium hypochlorite in 0.125 m sodium hydroxide solution

Urea standard:
0.5 mmol/l urea (= 30 mg/l)

Demineralized water

The reagents mentioned above can be acquired ready-prepared as a complete test set, for example from Boehringer Mannheim GmbH, Mannheim, FRG. It is imperative that attention be paid to the printed expiry dates.

**Process of Analysis**

The following batches should be prepared:

| | |
|---|---|
| Sample: | 2 ml of the water to be analyzed + 0.1 ml urease solution |
| Sample blank reading: | 2 ml of the water to be analyzed + 0.1 ml demineralized water |
| Standard: | 0.2 ml urea standard + 1.8 ml demineralized water + 0.1 ml urease solution |
| Standard blank reading: | 0.2 ml urea standard + 1.9 ml demineralized water |
| Urease blank reading: | 0.1 ml urease solution + 12 ml demineralized water |

All batches are cultivated for 15 minutes at 37 °C after thorough mixing. Subsequently, 5 ml phenol solution and 5 ml sodium hypochlorite solution are added to batches 1 to 4 which are then mixed well. All batches - including batch 5 - are then cultivated for a further 30 minutes at 37 °C.

Depending on the urea content or the original ammonium content, the cultivated batches are coloured different shades of blue. For evaluation, the extinctions are now measured at a wavelength of 546 nm in cuvettes with a path length of 4 mm. The intensity of the colour of indophenol blue formed is constant for several hours at room temperature.

Batch 5 must be uncoloured and display no extinction, with demineralized water and urease solution of perfect quality. The same applies to the standard blank reading (batch 4). If blue colorations are displayed in batches 4 and 5, the reagents are contaminated and should as far as possible not be used. Blue coloration in batch 4 indicates that ammonium ions are present in the urea solution. If the standard solution cannot be produced to be free of ammonium, the extinction of the standard blank reading must be subtracted from the extinction of the standard. This corrected standard value is then included in the calculation of the urea content.

**Calculation**

The urea contained in the water to be analyzed is calculated as follows from the extinction measurements E:

$$\left(\frac{E_{\text{Sample}}}{E_{\text{Standard}}} - \frac{E_{\text{Sample blank reading}}}{E_{\text{Standard}}}\right) \cdot 6 = \text{mg/l urea}$$

**Interference**

Heavy metal ions, in particular copper, inhibit enzyme activity in the urease. In the case of waters containing heavy metals, e.g. bathing water treated with copper, add 0.1 % EDTA (ethylene diamine tetraacetic acid) to complex the metal ions.

In waters containing chlorine, urea may be destroyed; chlorine is therefore removed from the water at the time of sampling by adding sodium thiosulphate.

The sequence of operations in enzymatic determination of urea is summarized in outline in the following pipetting chart.

| Solutions | Sample | Sample blank reading | Standard | Standard blank reading | Urease blank reading |
|---|---|---|---|---|---|
| Demineralized water | - - | 0.1 ml | 1.8 ml | 1.9 ml | 12.0 ml |
| Urease solution | 0.1 ml | - - | 0.1 ml | - - | 0.1 ml |
| Sample solution | 2.0 ml | 2.0 ml | - - | - - | - - |
| Standard solution | - - | - - | 0.2 ml | 0.2 ml | - - |
| | Mix.<br>Seal reaction vessels.<br>Cultivate for 15 min. at 37 °C | | | | |
| Phenol solution | 5.0 ml | 5.0 ml | 5.0 ml | 5.0 ml | - - |
| Hypochlorite solution | 5.0 ml | 5.0 ml | 5.0 ml | 5.0 ml | - - |
| | Mix immediately.<br>Cultivate for 30 min at +37 °C.<br>Measure extinction. | | | | |
| | | | Once per series | | Once per solution used. Measurement against demineralized water $\Delta$ E must be equal to 0. |

# 5 Biological Analysis

## 5.1 Significance of biological and ecological investigations when evaluating the quality of flowing water

When evaluating the quality of water which is flowing, it is essential to carry out a biological analysis as well as to determine the hydrographic, microbiological, chemical, physical and toxicological state of the water. Unless all forms of life have been made impossible for particular reasons, all waters represent habitats for organisms which differ to a greater or lesser extent depending on the overall state of the water. The particular biocenosis and the population density of the organisms are dependent on the conditions of the water as well as on external factors such as water movements, velocity, etc.

When biological and ecological conditions of a specific water are known, quite definite deductions can be made with regard to its state as well as changes in position and time.

According to the literature, the biological methods for analyzing water can be classified in two groups:

Ecological Methods
and
Physiological Methods

The "system of animal and vegetable saprobes" by Kolkwitz and Marsson (which has been revised several times) serves as the basis for biological and ecological analysis aimed at determining the degree of organic contamination of water or the stage of self-cleaning this water has reached. In this system there are four levels of saprobidity.

Special biological knowledge is necessary to be able to carry out the appropriate investigations and thus to be able to correctly classify the trophic level and degree of saprobidity of a section of water. When employing the biological method, it is essential that the species of organisms present at the particular locality are identified and determined accurately and that account is taken of the number of species, the individual density of each species and the particular ecological valency, i.e. the proven dependence on one particular state of the milieu. In the case of the saprobic system listed below, one must consider the limitations on the distribution of individual organisms to particular areas of the earth. This applies particularly to the macroorganisms.

**Polysaprobic zone (ps-Z)**

Sections of water with extremely high levels of organic contamination and in which ample quantities of nutrients for microorganisms are to be found, are classified in this level. Bacteria can develop en masse. The number of types is small, the density of individual types is often high. Decomposers, which are capable of splitting any organic materials present into simpler compounds, dominate; producers are almost completely absent. Organisms with high oxygen requirements are not present. Macroorganisms are not found.

**alpha-mesosaprobic zone (ams-Z)**

This level relates to a section of water with high organic contamination. In addition to numerous types of microorganisms, macroorganisms are frequently found in this zone. Decomposers still dominate, but producers and animal consumers are increasing. The total number of types is greater than in the polysaprobic zone.

**beta-mesosaprobic zone (bms-Z)**

This consists of a section of water with moderate organic loading. Such a section of water offers most organisms ideal conditions for survival. In these zones one finds a large increase in the number of producers and consumers as well as in the number of different species with, at the same time, a large decrease in the number of decomposers. The biocenoses are characterized by high fixity and presence of species.

**Oligosaprobic zone (os-Z)**

This term describes a section of water in which the water is hardly organically loaded at all and in which macroorganisms clearly dominate. In these zones, it is primarily producers which occur while, at the same time, there is a decrease in the number of consumers. The number of species is large and the density of the individuals in each species is generally small.

### 5.1.1 Significance of worm eggs in waste water, sewage sludge, surface water, swimming-pool water, drinking water and process water

Waste water, above all in its untreated state, is to be regarded as a special source of danger for the transmission and propagation of worm parasites among human beings and animals. This is because the eggs and larvae of various worm parasites (helminthes) are passed into waste water along with the faeces of human beings and domestic animals. The level of worm eggs can vary from day to day and with the time of day depending on the quantity and composition of the waste water.

The introduction of untreated or inadequately treated waste water into rivers, lakes, ponds and areas of the sea can lead to transmission and propagation of these parasite eggs. In order to monitor the destruction or the retention of worm eggs in a sewage treatment plant, it is necessary to determine the worm egg content before and after the treatment of the water. In many cases during the treatment of waste water, the worm eggs pass into the sewage sludge and survive the putrefaction process so that a danger of infection also exists if the sewage sludge is used as a fertilizer in agriculture.

Proven sources of danger for the transmission of worm eggs are therefore fruits and vegetables which have been fertilized with untreated waste water or with faecal matter. In addition, water for irrigation purposes should always be analyzed for worm eggs before being used. In this connection too, the watering or spraying of fruit, vegetables and other cultures with surface water which contains waste water or untreated or inadequately treated waste water must be rejected.

Furthermore, there is always the possibility of worm eggs being transmitted in swimming-pools.

In the case of waterworks, and especially ones which are fed with surface water, it is advisable to carry out monitoring for parasite eggs in addition to the usual chemical, physical and bacteriological analyses, as well as, where necessary, testing for any pathogenic germs and viruses. (See also section 5.2)

5.1.2 A clear classification of water quality was published in 1985 by the West German organization "Länderarbeitsgemeinschaft Wasser" - LAWA - (Water Study Group of the German Federal States)

A) Quality of flowing waters

Quality grade I:

Unpolluted to very slightly polluted

Stretches of water with pure water which is low in nutrients and always virtually saturated with oxygen; low bacterial content; moderately dense colonization, mainly by algae, mosses, turbellarians and insect larvae; if cool in summer, spawning ground for Salmonidae.

Quality grade I - II:
Slightly polluted

Stretches of water with slight supply of organic or inorganic nutrients without appreciable oxygen consumption; dense colonization, usually with great variety of species; if cool in summer, water for Salmonidae.

Quality grade II:
Moderately polluted

Stretches of water with moderate pollution and oxygen supply; very great variety of species and density of individuals of algae, snails, small crustaceans, insect larvae; aquatic plant stands cover considerable areas; high-yield fish water.

Quality grade II - III:
Critically polluted

Stretches of water in which the pollution load of organic, oxygen-consuming substances produces a critical state; fish mortality possible as a result of shortage of oxygen; decline in the number of species of macroorganisms; certain species show tendency to mass development; algae frequently form considerable stands covering large areas. Usually still high-yield fish water.

Quality grade III:
Heavily polluted

Stretches of water with heavy organic, oxygen-consuming pollution and usually low oxygen content; local deposits of digested sludge; filamentous sewage bacteria and sessile ciliates in colonies covering large areas exceed the extent of algae and higher plant life; only a few macroorganisms of types not sensitive to oxygen shortage, such as sponges, leeches, water-lice, are present, sometimes in massive quantities, low fish-yields; periodic fish mortality must be expected.

**Quality grade III - IV:**
**Very heavy polluted**

Stretches of water with substantially restricted living conditions as a result of very heavy pollution with organic, oxygen-consuming substances, often aggravated by toxic influences; turbidity due to suspended effluent matter; extensive deposits of digested sludge, densely colonized by bloodworms (Chironomus larvae) or sludge worms (Tubificidae); decline in filamentous sewage bacteria; fish not present to any permanent extent and then only on a locally restricted basis.

**Quality grade IV:**
**Excessively polluted**

Stretches of water with excessive pollution as a result of organic, oxygen-consuming effluent; putrefaction processes predominate; over long periods oxygen is present in very low concentrations or totally absent; colonization primarily by bacteria, flagellates and free-living ciliates; fish absent; biological depopulation in the presence of heavy toxic load.

**B) Criteria for assessing the quality grades of flowing waters**

In addition to the characteristic features listed for the quality grades under A) above, further identification criteria are provided below, including those relating to specialized techniques of biological status analyses (eg saprobic index, doublets/triplets) and also characteristic chemical values.

**Quality grade I:**
**Unpolluted to very slightly polluted**

This quality grade generally comprises headwater areas and only very slightly polluted upper reaches of flowing waters which are cool in summer. The water is clear and low in nutrients, the bed usually stony, gravelly or sandy; if sludge occurs it is of a mineral character. Stretches which are cool in summer are spawning grounds for Salmonidae.

Moderately dense colonization, especially by red algae (Batrachospermum), diatoms (Meridion, Diatoma hiemale, Achnanthes minutissima), mosses, turbellarians, stone fly larvae, caddis fly larvae and beetles.

The saprobic index is less than 1.5.

Reliable classification is possible if even only one of the following species is present: Polycelis felina, Crenobia alpina, Elmis latreillei, Esolus angustatus, Leuctra (nigra and related species), Agapetus. The $O_2$ content is close to (approx. 95 to 105% of) saturation level and not less than 8 mg/l. The five-day biochemical oxygen demand ($BOD_5$) is usually around 1.0 mg/l. $NH_4$-N is present in traces, if at all.

**Quality grade I - II:**
**Slightly polluted**

This group comprises slightly polluted flowing water, usually upper reaches. The water is still clear, the nutrient content low.

Stretches which are cool in summer are waters for Salmonidae. Characteristic fish: bullhead (Cottus gobio).

Dense colonization especially by algae (Ulothrix), mosses, Spermatophyta (Berula, Callitrich), turbellarians, stone-fly, may-fly and caddis-fly larvae, and beetles (Elminthidae, Hydraenidae).

The saprobic index is between 1.5 and 1.8.

Reliable classification is possible if more than one of the following species are present: Dugesia gonocephala, Amphinemura, Brachyptera, Perla marginata, Silo, Hydraena belgica, H. gracilis, Limnius perrisi, Oreodytes rivalis.

The $O_2$ content is still high (as a rule over 8 mg/l) but frequently there is already a detectable deficit ($O_2$ content approx. 85 to 95 % of saturation). $BOD_5$ is generally between 1.0 and 2.0 mg/l. $NH_4$-N is only present in small concentrations (averaging 0.1 mg/l).

**Quality grade II:**
**Moderately polluted**

This group comprises stretches of water with moderate pollution due to organic substances and their decomposition products. At times of heavy algal development there is appreciable turbidity. The bed is stony, gravelly, sandy or sludgy; even if the undersides of stones are blackened as a result of bacterial formation of iron sulphide, there is not yet any formation of digested sludge. High-yield fish waters.

Very dense colonization by algae (all groups), Spermatophyta (often covering considerable areas), snails, small crustaceans and insects of all groups and their larvae.

The saprobic index is between 1.8 and 2.3.

Reliable classification on the basis of individual abundant species is only possible in comparatively rare cases, for example Anabolia, Athripsodes, Atherix, Oulimnius tuberculatus and Orectochilus villosus; usually such classifications is only safe if certain pairs of species (doublets) are present. Examples of significant doublets include: Polycentropus flavomaculatus **with** Ecdyonurus venosus **or with** Riolus cupreus **or with** Hydropsyche **or with** Rhyacophila **or with** Baetis; Ranunculus fluitans **with** Ancylus **or with** Rhyacophila; Navicula gracilis **with** Dendrocoelum lacteum **or with** Elmis maugei (sensu lato) **or with** Bithynia tentaculata; Rhycophila **with** Hydropsyche; Gammarus roeseli **with** Glossiphonia complanata (typica).

The $O_2$ content displays substantial fluctuations (deficit to supersaturation) as a result of sewage pollution and algal development, but is high enough for no fish mortality to occur, ie it is always over 6 mg/l. $BOD_5$ is frequently from 2 to 6 mg/l. $NH_4$-H is frequently below 0.3 mg/l.

**Quality grade II - III:**
**Critically polluted**

As a result of the considerable pollution load of organic substances, the water is always slightly turbid; digested sludge may occur locally.

Usually still high-yield fish water (without high-class fish).

Dense colonization by algae and Spermatophyta (Potamogeton, Nuphar), sponges, bryozoa, small crustaceans, snails, shellfish, leeches and insect larvae (with the exception of stone-flies). Usually colonial mass development of several species. Sewage fungi are often visible to the naked eye - albeit not on a massive scale as yet. Greatest variety of species of ciliates.

The saprobic index is between 2.3 and 2.7.

Reliable classification on the basis of individual abundant species is hardly ever possible (with the exception of Potamanthus luteus, for example). Examples of significant pairs of species include Helobdella stagnalis with Rivulogammarus pulex or with Planaria torva or with Potamogeton natans or with Radix peregra or with Dendrocoelum lacteum; Erpobdella octoculata with Navicula rhynchocephala; Bithynia tentaculata with Nitzschia palea; Fontinalis antipyretica with Stentor roeseli; and Campanella umbellaria with Vorticella campanula; and also the triplet Gammarus pulex with Asellus and Cladophora.

The $O_2$ content often drops to half the saturation level. Frequently $BOD_2$ (two-day biochemical oxygen demand) is 2 to 5 mg/l and $BOD_5$ between 5 and 10 mg/l. $NH_4$-N is frequently below 1 mg/l.

**Quality grade III:**
**Heavily polluted**

The water is turbid as a result of sewage outfall. Stony to sandy bed is generally blackened by iron sulphide. Where the current is weak, digested sludge is deposited. Fish-yields low; periodic fish mortality due to lack of oxygen is to be expected.

Colonization by macroscopic organisms displays little variety of species, but there is mass growth of certain types (water-lice, leeches, sponges). Colonies of sessile ciliates (Carchesium, Vorticella) and sewage bacteria (Sphaerotilus) covering considerable areas are conspicuous: algae and Spermatophyta, by contrast, diminish sharply in importance.

The saprobic index is between 2.7 and 3.2.

Reliable classification on the basis of individual abundant species is never possible; even significant doublets are rare: Chironomus thummi with Helobdella stagnalis; Gomphonema olivaceum with Tubificidae; Erpobdella octoculata with Tetrahymena pyriformis. Triplets of highly abundant species are often reliable indicators, for example: Erpobdella octoculata with Tubificidae and Rotaria rotatoria; Chironomus thummi with Erpobdella octoculata and Carchesium polypinum.

$O_2$ is still always present, but drops at times to levels around 2 mg/l. $BOD_2$ is frequently 4 to 7 mg/l and $BOD_5$ 7 to 13 mg/l. $NH_4$-N is usually in excess of 0.5 mg/l and not infrequently reaches several milligrams per litre.

**Quality grade III - IV:**
**Very heavily polluted**

The water is turbid as a result of sewage outfall and the bed is usually covered in sludge (digested sludge). Fish are only encountered locally and then not on any permanent basis.

Colonization almost exclusively by microorganisms, especially ciliates, flagellates and bacteria. The only macroorganisms still present are bloodworms (Chironomus larvae) and sludge worms, but these often occur in massive quantities.

The saprobic index is between 3.2 and 3.5.

The only significant doublets are those of microorganisms, for example: Colpidium colpoda with Beggiatoa; Nitzschia palea with Stigeoclonium; Paramecium trichium with Navicula accommoda.

The $O_2$ content is sometimes less than 1 mg/l and as a rule does not exceed a few milligrams per litre. $BOD_2$ is frequently 5 to 10 mg/l and $BOD_5$ 10 to 20 mg/l. The concentration of $NH_4$-N is usually several milligrams per litre.

Even if the chemical situation is otherwise favourable, toxic influences may be the cause of serious depletion of the biocenosis.

**Quality grade IV:**
**Excessively polluted**

The water is highly turbid as a result of sewage outfall and the bed is usually characterized by thick deposits of digested sludge. In many cases the water smells of hydrogen sulphide. Fish are absent.

Colonization almost exclusively by bacteria, fungi and flagellates; of the ciliates, only a few free swimming species are present, often in massive quantities.

The saprobic index is over 3.5.

Only a very small number of doublets are significant, for example: Paramecium caudatum with Zoogloea; Colpidium campylum with Paramecium trichium or with Fusarium or with Stigeoclonium. Mass development of Colpidium campylum in the absence of other ciliates is also significant.

Such stretches of water are so heavily polluted by inputs of organic effluent that the $O_2$ content of the water is either very low or totally non-existent. Putrefaction processes predominate. $BOD_2$ is usually in excess of 8 mg/l and $BOD_5$ over 15 mg/l. The concentration of $NH_4$-N is usually several milligrams per litre. Where toxic pollution is heavy, biological depopulation can occur.

**Classification of the quality of flowing water**

The classification of flowing water quality is summarized in the table below. It is important to note that if this table is used blindly, incorrect classification of water quality may result. In particular, it can occur that the chemical characteristics, saprobic stages and saprobic indices assigned to the individual quality grades in the table do not agree, in which case it will not always be possible to draw reliable conclusions about chemical concentrations on the basis of biological findings and vice versa.

| Quality grade | Degree of organic pollution | Saprobic stage | Saprobic index | Chemical parameters $BOD_5$[a] (mg/l) | $NH_4$-N[a] (mg/l) | $O_2$ minima[ab] (mg/l) |
|---|---|---|---|---|---|---|
| I | Unpolluted to very slightly polluted | Oligo-saprobic | 1.0 - 1.5 | 1 | traces | 8 |
| I - II | Slightly polluted | Transitional from oligosaprobic to beta-mesosaprobic | 1.5 - 1.8 | 1-2 | around 0.1 | 8 |
| II | Moderately polluted | Betameso-saprobic | 1.8 - 2.3 | 2-6 | 0.3 | 6 |
| II - III | Critically polluted | **Borderline between beta- and alpha-meso-saprobic** | 2.3 - 2.7 | 5-10 | 1 | 4 |
| III | Heavily polluted | Alphameso-saprobic | 2.7 - 3.2 | 7-13 | 0.5 to several mg/l | 2 |
| III - IV | Very heavily polluted | Transitional from alpha-mesosaprobic to polysaprobic | 3.2 - 3.5 | 10-20 | several mg/l | 2 |
| IV | Excessively polluted | Polysaprobic | 3.5 - 4.0 | 15 | several mg/l | 2 |

[a] The chemical data given in this table are intended merely as a guide to concentrations frequently encountered.

[b] The fast-flowing mountain and upland streams the oxygen minima for the quality grades II to IV are frequently higher than the figures quoted in the table.

## 5.2 Microbiological water analysis

### 5.2.1 General remarks

Water is not an especially good culture medium for microorganisms. Nevertheless, microorganisms can live and increase in water and be transmitted by water. Of particular significance is waste water which usually contains a high level of germs, especially pathogens which can then find their way into surface and/or underground water and thus be transmitted by the water.

The main risk is from drinking infected drinking water. For this reason drinking water must be free from pathogens.

Three measures are necessary in order to fulfil this basic requirement:

1. Hygienically safe collecting, transporting, storing and distribution of the water to the consumers.

2. Protecting the ground water in the tapping zone and catchment area from influences which are detrimental to health and which reduce the quality of the water (marking and maintaining protection zones).

3. Where the untreated water is infected, measures must be taken at the purification stage to eliminate the germs, e.g. by the use of filters through which bacteria cannot pass or by means of chemical disinfectants approved for the treatment of water (chlorine gas, hypochlorite compounds, chlorine dioxide, ozone, silver preparations).

Microbiological analysis of water is used to monitor the microbiological quality and safety of water used as drinking water, process water, water for swimming pools and in other relevant areas. In general, such analysis includes determining the total number of germs capable of multiplying (total colony count) as well as detecting special types of germ which are considered indicators as to the possible presence of hygienically unacceptable contamination or even pathogenic germs.

### 5.2.2 Microorganisms in water

Microorganisms are minute organisms which are invisible to the naked eye and frequently consist of just one single cell. The cell of a bacterium measures, as a rule, only 1 to 2 µm (one to two thousandth of a mm); it can have a spherical, rodlike, hook-shaped or screwlike form. **Bacteria** are the most important microorganisms in water microbiology. Considerably smaller than bacteria are viruses, which are of significance for waste water and surface waters in which they may be present to a more or less numerous extent. Viruses can be pathogens and can also be transmitted to human beings. **Virological analyses of water** are time-consuming and often troublesome and are not carried out in the usual microbiological analyses of water. For this reason, they will not be discussed here.

**Yeasts and moulds** occur only rarely in ground water, so that they only play a subordinate role in drinking water microbiology. They are frequently detected together with bacteria when analyses are being carried out with cultures. In special cases, however, separate analyses should be carried out to detect these microorganisms. If necessary, these should be carried out with larger quantities of water.

**Algae** can also be considered as belonging to the group of microorganisms living in water. However, they can generally only be detected and counted by microscopic means. They will therefore only be included here as belonging to the group of microorganisms to be found in water and will not be treated further in this text. The same applies to the single-cell animal organisms known as **protozoa.**

Bacteria are just visible under an optical microscope with thousandfold magnification, especially if the cells have been dyed by a suitable method. If necessary, they can be counted under the microscope by the direct method.

### 5.2.3 Direct counting of germs

A counting chamber can be used to determine the number of bacterial cells in water. In such a chamber, the number of bacterial cells in a particular, small quantity of water (e.g. a cube with 0.1 mm sides = 0.001 $mm^3$) can be counted under a microscope. Each bacterial cell determined in such a cube represents $10^6$ germs in 1 ml. Thus, in order to determine a bacterial count at all with this method, the germ content of the water must be very high. This direct counting method cannot be used for water containing low levels of germs.

A second possible method for the direct determination of the number of bacteria in water would be to filter a definite quantity of water through a fine membrane filter (pore size 0.2 or 0.45 µm), on which a grid has been marked. The bacteria are dyed on the filter with suitable dyes, the filter is illuminated, and the bacteria are counted under a microscope. This method certainly allows the bacteria in large quantities of water containing low levels of germs to be concentrated. However, the accuracy of counting is relatively low, resulting in a relatively high error rate. Similarly, it is also impossible to distinguish between living and dead bacterial cells with this method.

Direct methods of counting germs are very time-consuming and inexact and are therefore as a rule not employed in microbiological water analyses.

### 5.2.4 Indirect methods of counting germs

The indirect method of determining the germ level has been used since the first beginnings of microbiological water analysis. In this method, a definite quantity of water is added to a sterile culture medium, which is then incubated under definite conditions. After the incubation period, the colonies which have formed are counted and quoted in the analysis report as the bacterial count. It is assumed that each colony can be related to a bacterial cell which was originally present in the water and therefore one also speaks of the number of colony-forming units in the water being analyzed. However, colony-forming units may be both individual bacterial cells and small accumulations of cells which were present in the water prior to the incubation process.

In the case of the indirect method of determining the number of germs, only those germs which have developed into colonies under the conditions employed are detected. Comparison of the results from the direct and indirect methods of determining the numbers of germs shows that the indirect method falls a long way short of detecting all the microorganisms which are actually present. Nevertheless, when the analytical conditions are always kept the same, the indirect method yields useful results and permits sufficiently reliable statements to be made about the germ content of a particular water sample.

### Standard value

In satisfactory ground water the colony number will not exceed the standard value of 100 per ml at an incubation temperature of 20 °C. In the case of disinfected drinking water, the colony number should be less than 20 per ml after completion of the treatment processes.

The bacterial count is defined as the number of colonies visible under 6 to 8-fold magnification which form in pour-plate cultures with nutrient-rich, peptonic culture mediums (1 % meat extract, 1 % peptone) from 1 ml of the water being analyzed after incubation at 20 ± 2 °C for 44 ± 4 hours.

## 5.2.5 Indicator germs

Drinking water and process water must not contain living pathogens such as salmonella, shigella, cholera vibrios and other organisms capable of causing epidemics. Since, however, the presence of pathogens which are capable of causing epidemics cannot always be detected even when large quantities of water are used, and since these pathogens can only be detected with difficulty on cultures, the approach adopted is to analyze water for bacteria whose number and identity can be more easily determined in cultures and which have the same natural habitat as these pathogens. The most important examples of such bacteria are the intestinal bacteria because these, like the pathogens which inhabit the intestine, are excreted, albeit in much greater numbers than the latter. Thus the presence of these bacteria in water indicates that faecal contamination is present so that the presence of pathogens from the intestinal tract cannot be ruled out.

The germs generally used to indicate faecal contamination in water are Escherichia coli, faecal streptococci and sulphite-reducing, anaerobic, spore-forming organisms. In addition, the coliform bacteria and Pseudomonas aeruginosa are used as indicator germs to determine whether water stocks have been contaminated to a hygienically significant degree.

### Escherichia coli

Escherichia coli normally lives in intestinal tracts and, in particular, in the large intestine of human beings and warm-blooded animals. Outside the intestinal tract, it can only live for a short time in water and in the ground and therefore indicates relatively fresh faecal contamination. Thus, when Escherichia coli is present in water, one also has to reckon on the presence of pathogenic intestinal bacteria such as Salmonella, Shigella and cholera vibrios. Escherichia coli is easy to cultivate on culture mediums and can, by reason of its metabolic characteristics in the so-called "coloured series", be identified relatively easily by checking its biochemical characteristics. As a rule Escherichia coli is not itself a pathogen. However, rare cases are known of enteropathogenic Escherichia coli types, which can cause diarrhoeal illnesses in babies and small children.

If the enteropathogenicity of coliform strains which have been detected is of interest from a clinical point of view, the serological analyses necessary to permit an exact determination of type must be carried out in a laboratory equipped for this purpose.

The rule for satisfactory drinking water is that Escherichia coli must not be detectable in 100 ml of drinking water.

(For mineral water in Europe, there must be no Escherichia coli in 250 ml.)

### Faecal streptococci

Here we are talking about gram-positive streptococci, which are classified as belonging to serological group D. The normal habitat of these germs, which are also called enterococci, is the intestinal tract of human beings and warm-blooded animals. Thus their presence in water also indicates faecal contamination. In addition, they are frequently more resistant to environmental influences than Escherichia coli.

The principal representatives of this group, **Streptococcus faecalis** and **Streptococcus faecium,** can also be detected easily using cultures under aerobic conditions and can be identified more exactly with relatively simple methods of analysis.

**Rule:**

Faecal streptococci must not be detectable in 100 ml of drinking water or 250 ml of mineral water.

### Sulphite-reducing, spore-forming Anaerobes

The spore-forming organisms of the genus Clostridium live in the intestinal tract of human beings and warm-blooded animals. However, they also remain capable of reproducing outside the intestinal tract in water and in the ground for a long time, in particular when they are in their sporulated state. Their presence in water may therefore have been brought about by faecal contamination which occurred a long time before, because they are considerably more resistant than Escherichia coli and faecal streptococci and, in unfavourable situations, can outlive the latter many times over. Of particular importance in this group is **Clostridium perfringens,** which counts as a facultative pathogen and the cause of gaseous gangrene. For the purpose of analysis, these microorganisms must be cultivated under anaerobic conditions. Cultivation should only be carried out in laboratories having appropriately trained personnel, so that no danger to health can be caused by the laboratory work. In operating water works, consideration has to be given to the fact that clostridium spores are very resistant and are frequently not killed by the chlorination treatment usually used in disinfecting water.

**Rule:**

There should not be more than 1 sulphite-reducing, spore-forming anaerobe in 20 ml of drinking water. They must not be detectable at all in 50 ml of mineral water.

### Coliform bacteria

A significant characteristic of coliform bacteria, which belong to the family of enterobacteria, is their ability to cause lactose to ferment with the formation of gas and acid. The types in question here are Enterobacter, Klebsiella and Citrobacter. They are frequently to be found in waste water, surface water, cultivated ground and other substrates and also live in the intestinal tract of warm-blooded animals.

They can also multiply outside the intestinal tract and are indicator germs for hygienically significant impairment of water quality, since they are not present in clean ground water. The designation "coliform" shows that these germs possess morphological and biochemical similarities to Escherichia coli. For this reason, colonies suspected of containing coliform bacteria should always be exactly identified.

Rule:

Coliform bacteria should not be detectable in 100 ml of drinking water. In the case of mineral water, they must not be present in 250 ml.

Pseudomonas aeruginosa

Pseudomonas aeruginosa rates as a problem germ from a sanitary point of view and is frequently to be found in waste water and surface water contaminated with waste water. It is classified as a facultative pathogen, since it infects wounds and causes the formation of pus. It can cause inflammations in the human ear, for example, which are known as otitis. In addition, this bacterium can become a problem germ not only in human and veterinary medicine, the food sector and the manufacture of cosmetic and pharmaceutical products, but also, above all, in the swimming pool sector. Pseudomonas aeruginosa is not to be found in clean ground water and spring water and when found in such water is an indicator of contamination caused by waste water or contamination caused by human beings. This germ has the ability to survive for a long time in water and often shows a high resistance to disinfectants. It can also adapt to higher water temperatures and colonize tanks and pipelines having a temperature of around 50 °C.

Rule:

Pseudomonas aeruginosa should not be detectable in 100 ml drinking water. There must be no Pseudomonas aeruginosa present in 250 ml mineral water.

Other hygienically significant germs

**Proteus vulgaris** is frequently to be found as a putrefactive agent in rotting food and stagnant water as well as on organic substances which are in a state of decomposition. It can also occur in the intestinal tracts of warm-blooded animals and in the urinary passage of human beings. It spreads on the surface of the culture medium of plate cultures and spoils in this way the course of the culture analysis, in particular when identifying Escherichia coli and coliform bacteria. In such cases it is necessary to form cleaning passages on the culture medium by the addition of a wetting agent, since the spreading of the proteus bacteria is prevented by wetting agents.

**Serratia marcescens**, which is to be found in surface water and layers of earth near the surface, indicates that the ground water has been influenced by surface water. However, these bacteria can also cause the spoilage of food and, like Pseudomonas aeruginosa, have been found to be a cause of hospital cross-infection.

**Aeromonades** are frequent inhabitants of the intestinal tract of cold-blooded animals and are almost always to be found in surface waters and in waste water. These bacteria are to be found relatively rarely in the intestinal contents of human beings and warm-blooded animals. Some representatives of the aeromonas genus can also cause lactose to ferment at 37 °C and can, in certain circumstances, simulate the presence of coliform bacteria.

The presence of **flavobacteria, achromobacter types** and the apathogenic representatives of the genus **Pseudomonas** indicate water from layers of soil near the surface.

### 5.2.6 General requirements for microbiological work

Risks both for the analyst as well as for the environment are associated with work on microorganisms which are inevitably pathogenic. In order to prevent infection, this work must only be carried out by experts with the appropriate special knowledge, observing the necessary precautions. In many countries precise legal regulations exist for work on pathogens.

Work on germs which are not inevitably pathogenic also requires trained personnel as well as rooms and equipment which are suitably equipped for microbiological work.

An important requirement for proper microbiological work is the avoidance of secondary infections both in the handling of the sample material prior to actual analysis and in the analysis itself.

#### Sterilization of apparatus

Where glass or metal equipment which can be used several times is employed instead of commercially available sterile disposable plastic equipment, then the former must be carefully cleaned and sterilized each time before it is used.

#### Cleaning

All apparatus must be mechanically cleaned using cleaning agents, brushes, etc. They should be rinsed first with clean tap water, to which under certain circumstances 1 % hydrochloric acid has been added, and then with distilled or demineralized water. In order to prevent infection in the course of the rinsing process, vessels and equipment containing germs should be autoclaved at approx. 120 °C for 30 minutes before the cleaning process.

#### Drying and sterilization

After cleaning, the devices are first dried and then sterilized for 2 hours at 180 °C to 200 °C in a hot-air sterilizer, remembering to allow for the warming up time. Glass vessels which have been sealed with cotton wool or cellulose should be heated for 2 hours at 160 °C so that the plugging material is not caused to go too brown. In the case of bottles with glass stoppers, a strip of filter paper approx. 6 cm long and 1 cm wide is laid between the ground surfaces prior to sterilization. This strip is then not removed until the flask is filled when the sample is taken. After sterilization, the glass stopper and neck of the flask are carefully protected against secondary infection with the aid of sterile aluminium foil. Where pipettes are sterilized in metal tins, care must be taken that the air holes in the tins are open during the sterilization process and are then immediately closed after sterilization when the tins have cooled down to 60 °C.

In order to avoid the glass breaking, the door of a hot-air sterilizer should not be opened after sterilization until the temperature has dropped to 60 °C. It is recommended that the effectiveness of a sterilizer should be checked at regular intervals using either melting point or colour indicators or by carrying out a bacteriological check using packets of spores which contain the thermoresistant spores Bacillus subtilis or Bacillus stearothermophilus. The indicators that are used should be placed at different points in the sterilizer both inside and outside the objects to be sterilized in order to verify the uniformity of sterilizing effectiveness inside the sterilizer.

The baking of platinum needles and eyes, etc. as well as the immersing of pincers, scissors, spatulas, etc. in alcohol followed by flaming can be regarded as **fast sterilization methods** in microbiological work. Care must be taken that the heat is allowed to act for long enough to kill all the micro-organisms which might be on the item in question.

In addition care should be taken that items of equipment which have been treated in the above-mentioned way are allowed to cool down sufficiently before being used in the course of an analysis so that any microorganisms that may be present are not damaged by heat.

### Sterilization of culture media

The most reliable method for the sterilization of culture media and culture solutions has proved to be that of treatment with superheated steam in an autoclave (superheated steam sterilizer) at 121 °C, which corresponds to 1 bar over atmospheric pressure, for 20 to 30 minutes. The operating conditions quoted by the manufacturer must be observed when using an autoclave.

Thermally labile culture media whose composition and properties are impaired by heating at high temperatures should be fractionally sterilized in circulating steam in a steam autoclave. This process is also described as tyndallizing.

### Sterilization in stages

The culture medium is heated up in circulating steam at 100 °C for a sterilizing time of 30 minutes on three consecutive days. Between the individual heating periods, the culture medium is incubated at about 25 °C. When the culture medium is heated up to 100 °C, the vegetative cells are killed. Spores of bacteria and moulds, however, survive a single treatment at 100 °C. This is why the intermediate periods of incubation are introduced. Vegetative forms develop from the steam-resistant spores and these are then killed by the subsequent heating period. The heat treatment is carried out three times to increase the reliability of the process.

## 5.2.7 Taking and transporting water samples for microbiological investigations

Secondary infection or technical errors in sampling can falsify the accuracy of the entire micro-biological investigation; it is therefore of decisive importance that sampling should be carried out expertly.

As a rule, sterile glass-stoppered bottles with a capacity of 250 to 500 ml are used for sampling. These are sterilized in the laboratory and safeguarded against secondary infection with aluminium foil. Bottles used for sampling chlorinated water must be treated with sodium thiosulfate before sterilization. A 250 ml bottle should be charged with 0.25 ml of a solution of 0.01 m sodium thiosulfate and a 500 ml bottle with 0.5 ml of the same solution in order to bind immediately any chlorine or chloramine which may be present in water.

The sampler should first convince himself at the sampling location of the suitability of the proposed sampling point. If he finds the sampling point suitable in principle, he should take care that the water can be sampled without negative influences. Taps must be initially cleansed mechanically and rinsed so as to be free of particles where necessary. The tap should subsequently be flamed until it is completely dry and a hissing noise is clearly discernible when opened. After flaming, the water should be allowed to flow for 5 to 10 minutes without changing the position of the cock during this time. It is advisable to measure the water temperature during the period of flow and not to take the sample until a constant temperature has been achieved. In the case of wells with manual pumps, the pump must be operated evenly for approximately 10 minutes before sampling. The discharged water may not be allowed to flow back into the inside of the well or infiltrate the direct vicinity of the well.

Water can be taken from dug wells and water containers with the aid of a sterile sampling advice. The sample should be taken from approximately 30 cm below the surface of the water. If no sampling device is available, a sample can be taken by immersing the sample bottle in the water with the aid of sterilized crucible tongs and slowly moving the bottle through the water approximately 20 to 30 cm below the water surface with the opening pointing diagonally upwards. Care must be taken that germs do not enter the sampling bottle from the hands of the sampler.

The aluminium foil should be removed before filling the bottle; the strips of filter paper remaining from sterilization should be carefully shaken out when opening the bottle, and the water then allowed to flow into the bottle. Any contamination from touching the bottle opening or the stopper with fingers, from breathing, from contact with clothing or with the water outlet, etc., should be avoided.

After filling the bottle, this should be sealed immediately and the bottle neck protected with the aluminium foil.

### Transport and storage of the sample

In order to prevent changes in the microbiological quality of the water, the filled sampling bottles must be transported in boxes insulated against heat. Where necessary, they should be cooled in the case of high outside temperatures. The sampling bottles should also not be subjected to direct sunlight for any length of time because germs could then be killed by UV radiation.

The samples taken should be examined immediately after arrival at the laboratory. If this is not possible in exceptional circumstances, the samples should be stored in a refrigerator at +4 °C. The period of storage should nevertheless be kept to a minimum. If the time span between sampling and investigation is too long, the bacteriological test will have to be carried out on site. A record should be made in the examining report of any considerable intervals between sampling and examination.

## 5.2.8 Performing the microbiological analysis of the water

### 5.2.8.1 Determining the total colony counts

In order to determine the colony count, 1 ml of water in each case is pipetted into a sterile Petri culture dish and mixed with sterile nutrient gelatine or sterile nutrient agar. Nutrient gelatine is liquefied in a water bath at 35 °C and cooled to about 30 °C before pouring into the culture dish. Nutrient agar is liquefied in boiling water and cooled to 46 ° ± 2 °C before use. Before liquefying, a visual inspection should be carried out to check whether the nutrient medium contained in the tube is free of secondary infection, i.e. there are no indications of stored colonies of bacteria. 10 ml of the liquefied nutrient medium are added free of air bubbles to the pipetted water in the culture dish. Secondary infection from bumping or from water droplets on the outside of the nutrient medium glass should be avoided when filling the nutrient medium into the culture dish. Before pouring the nutrient medium, the tube edge should be flamed. Immediately after pouring, the nutrient medium and the water are mixed well by carefully swirling the culture dish sealed with a lid, using a motion in the shape of a "figure 8". The prepared culture must then be allowed to solidify in a horizontal position.

Nutrient gelatine solidifies at temperatures below 25 °C and cooling is therefore necessary in certain circumstances.

Note:

If high counts are expected in the water to be examined, it is advisable to prepare series of dilutions with sterile water and then to test the dilution stages 1 : 100, 1 : 1000, etc.

The culture with the solidified layer of nutrient medium is incubated at the prescribed temperature in the incubator or incubating chamber, whereby a maximum of 4 to 6 plates should be stacked one above the other. Plates and nutrient agar medium should be turned over after solidification and incubated with the layer of nutrient medium upwards in order to avoid precipitation of condensed water, particularly at higher incubation temperatures.

After the prescribed incubation period has elapsed, the visible colonies are counted with the aid of a magnifying glass with 6x to 8x magnification. In order to facilitate counting, a Wolfhügel counting plate or any other suitable counting device can be used. Only cultures with a count not exceeding 300/ml should be used to determine the count. If dilution series were prepared of contaminated waters, those plates should be counted on which between 30 and 300 colonies have grown. If more than 1/4 of the surface of the nutrient medium is overgrown with spreading colonies, the plate should be discarded.

Gelatine cultures can only be incubated at 20 ° ± 2 °C, as they liquefy at higher temperatures. Liquefication of gelatine is also possible as a result of microorganisms with proteolytic enzymes. It is therefore recommended that an agar culture be prepared in addition to gelatine cultures, in which the germ yield is as a rule higher than in agar cultures, so that figures for the numbers of germs can still be given even when gelatine liquefiers are present. Gelatine liquefiers frequently occur in surface water. The figures obtained from dilution series should be multiplied according to the stage of dilution.

Rule:

If the colony count determined lies above 100, the figures are rounded down to complete tens, in the case of values over 1000 to complete hundreds, etc.

It is usual to indicate the nutrient media used and the length and temperature of incubation in the analysis report.

The **membrane filter process** and the **dip slide process** are unofficial methods of determining the colony count.

In the **membrane filter process**, larger quantities of water can be pressed or sucked through a sterile membrane filter inserted in the sterile filtering device. The filter is then stretched free of bubbles on the surface of the soldified nutrient medium in a Petri culture dish and the culture thus prepared then incubated. The nutrient substances in the medium migrate through the layer of the membrane filter to the germs on its upper surface

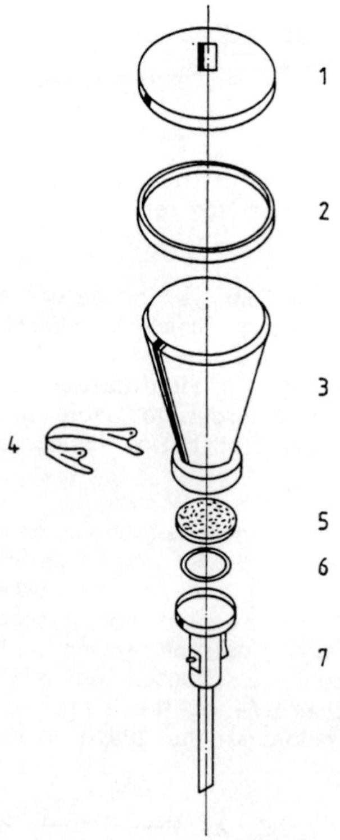


Fig. 152. Suction membrane filtration unit of stainless steel; 1) = Lid with handle; 2) = Silicone seal for the lid; 3) = Funnel-shaped upper section, capacity 500 ml; 4) = Clip (is secured to the funnel-shaped upper section and locks into the metal pins on the lower part when the unit is coupled together); 5) = Metal frit (the membrane filter disc is placed on this); 6) = Teflon ring between frit and lower section; 7) = Lower section with tap (when filtering, the spout of the lower section extends through the hole in the plug into the suction flask)

so that they can form colonies there. It is important in this context that the surface of the filter does not remain excessively damp because otherwise the germs float away in the residual water and no countable individual colonies are formed. In general, fewer germs are recorded by the membrane filter method than by the pour-plate method which is officially prescribed in many cases.

Instead of the nutrient agar in Petri culture dishes, cardboard culture discs may also be used. The manufacturers of cardboard culture discs have incorporated soluble nutrient substances into the cardboard disc and sterilized them. The analyst must moisten the cardboard culture disc with sterile water according to the manufacturer's instructions and then place the filter disc on the cardboard, also making sure that there are no air bubbles.

The **dip slide process** is based on the membrane filter method. It uses commercially available test kits, comprising a nutrient medium carrier in which the medium part is inserted in a beaker. This beaker serves as dipper for taking the water sample and also protects the nutrient medium part from secondary infection and from drying out during the incubation period.

The medium carrier can be held by a grip on the non-sterile part and its sterile nutrient medium part projects into the beaker. The surface of the medium part is designed as a membrane filter layer, and when the medium part is immersed in the water to be examined in the full beaker, the inside of the medium part sucks water through the medium layer. In this way, the germs present in the water which has been sucked through the medium layer remain adhering to the surface of the filter layer. It is important to observe strictly the period for which the medium part is immersed in the beaker part completely filled with the water to be examined, as prescribed by the manufacturer. Only if this is done can it be guaranteed that the colony count determined later can really be referred to 1 ml water. After the prescribed exposure time, the medium part is removed from the water to be examined, the beaker is emptied and then slipped over the medium part and pressed down. The usual period of incubation and the determining process for the colony count can then follow after the prescribed incubation period.

Colony counts can also be made using the **surface method.** In this 0.1 to 0.3 ml water is spread out on the surface of the solidified, sterile agar culture medium in the petri dish with the aid of a sterile spatula (Drigalski spatula). However, this method also results in certain differences from the pour-plate method.

### 5.2.8.2 Detection of Escherichia coli and coliform bacteria

The cultivation of these germs is carried out in one process. Two methods can be employed:

Liquid enrichment with lactose-peptone solution

Membrane-filter method using endoagar or endo-nutrient cardboard disks

**Liquid enrichment method**

Where the only question to be decided is whether or not Escherichia coli and/or coliform bacteria are present in 100 ml water, it is sufficient to mix 100 ml of the water being analyzed with 100 ml lactose-peptone solution of double concentration. After an incubation period of 20 ± 4 hours at 37 °C, an examination is made to see whether acid and gas are being formed or not. If they are not being formed, the composition of the water in terms of its levels of Escherichia coli and coliform bacteria meets the drinking water requirements and the analysis can be stopped. If, however, it is established that the lactose is fermenting with the formation of gas and acid, one has to establish whether the germs causing this are Escherichia coli or coliform bacteria, or whether the germs do not belong to the group of coliforms. For this, a small quantity of the lactose-peptone solution, which has become turbid as a result of the growth of germs, is removed with a sterile platinum loop and fractionated on the endoagar (fractionated means that the spreading is not carried out with the loop over the entire endo plate but that instead just one single strip of germs is applied to the surface of the culture medium at the edge of the Petri dish.)

With the second loop, part of the material is now spread out on a third of the culture medium surface at right angles to the first. Then the dish is turned through a further 20° and part of the material is applied with a third sterile loop to the part of the culture medium surface which has not yet been coated. It is possible in this way to obtain individual colonies, which can then be identified in the so-called "colour series".

Moist, dark red colonies with a gold, iridescent metallic sheen can be suspected of being Escherichia coli. Coliform bacteria grow as moist, red colonies with a continuous or discontinuous metallic sheen, with or without the formation of slime.

Generally speaking, colour series work today is carried out using the prepared systems or kits which are commercially available, such as API, Enterotube, Titertek, etc. These consist of prepared culture medium systems which are inoculated with the cell material from one single colony and then incubated. Handling and incubation must be carried out in accordance with the manufacturer's instructions. Evaluation is frequently carried out by determining a number code on the basis of positive or negative metabolic reactions which take place with the individual culture mediums used. After determining the number code, one can read off the type of germ the suspect colony consists of in the catalogue supplied.

When such prepared or ready-to-use systems are not available, the identification culture mediums must be prepared in accordance with the recipes given in the section on culture media. These are then inoculated and evaluated in accordance with the characteristics obtained.

### Determining biochemical characteristics

For the colour series, the culture media listed below are inoculated and incubated at the stated temperatures for 20 ± 4 hours:

If the colony which was injected onto the identifying culture medium was a pure culture, then only colonies which are typical for the particular nutrient agar plate will grow. These colonies will have a uniform appearance and will not show any formation of pigment. If different types of colony are present, then the colony which was transferred must have consisted of a mixed colony of different types of germs. Such mixed colonies cannot be used for differentiating between different types of germs. Further subcultures must be started on the endoagar. When a pure

| Culture Medium | Incubation Temperature | Positive Reaction | Negative Reaction |
|---|---|---|---|
| Nutrient agar plate | 37 °C | Exlusively uniform, typical colonies | Morphologically different colonies |
| Simmons citrate agar (sloping culture) | 37 °C | Growth with colour change from green to blue | No growth, no alteration in colour |
| Koser citrate medium | 37 °C | Turbidity as the result of bacterial growth | No growth, clear, no turbidity |
| Glucose-peptone broth<br>Culture A)<br>Culture B) | <br>37 °C<br>44 °C | Turbidity, gas formation, colour indicator changes from purple to yellow | No growth, no gas formation, no colour change |
| Lactose-peptone broth | 44 °C | Turbidity, gas formation, colour change from purple to yellow | No growth, no gas formation, no colour change |
| Mannitol-neutral red broth | 44 °C | Turbidity, gas formation, colour change from red to yellow | No growth, no gas formation, no colour change |
| Kligler urea agar | 37 °C | Sloping surface:<br>colour change from red to yellow as a result of acid formation<br>Stab: gas formation, colour change to black as a result of $H_2S$ formation, $NH_4$ formation as a result of decomposition of urea | No gas formation, no colour change as a result of acid formation, no turning black as a result of $H_2S$, no $NH_4$ formation |
| Tryptophane-tryptone broth | 37 °C | Growth with turbidity, red coloration when indole reagent is added | No growth, no red coloration with indole reagent |
| Buffered nutrient broth | 37 °C | Divide solution into 2 sterile test tubes: | |
| | | a) Methylene red test | |
| | | Colour change from yellow to red | Indicator remains yellow |
| | | b) Voges-Proskauer reaction (Addition of potassium hydroxide and creatine) | |
| | | Red colouration after 1 - 2 minutes | No colour formation after 2 minutes |
| Nutrient gelatine (stab culture) | 20 - 22 °C | Liquefaction in the region of the stab | No liquefaction |

Biochemical characteristics of Escherichia Coli and Coliform Bacteria

| | Escherichia Coli | Entero-bacter | Klebsiella | Citro-bacter |
|---|---|---|---|---|
| Glucose fermentation | | | | |
| at 37 °C | + | + | + | + |
| at 44 °C | + | +/- | +/- | +/- |
| Lactose fermentation | | | | |
| at 44 °C | + | +/- | +/- | +/- |
| Mannitol fermentation | | | | |
| at 44 °C | + | +/- | +/- | +/- |
| Citrate decomposition | - | + | + | + |
| Indole formation | + | - | - | +/- |
| Methyl red test | + | - | - | + |
| Voges-Proskauer reaction | - | + | + | - |
| Urea decomposition | - | +/- | + | +/- |
| $H_2S$ formation | - | - | - | +/- |
| Gelatine liquefaction | - | - | - | - |

\+ = positive
\- = negative
+/- = various strains behave in different ways

culture is obtained, the cytochromoxidase reaction is carried out on the nutrient agar plate. Here 2 to 3 drops of Nadi reagent are dropped onto the colonies with the aid of a dropping bottle. In the case of a positive reaction, the colonies turn blue-violet within 1 to 2 minutes. No colour change takes place in the case of a negative reaction.

When the cytochromoxidase reaction is positive, Escherichia coli and coliform bacteria are not present.

When the cytochromoxidase reaction is negative, the colour series which has been prepared is evaluated in accordance with the table above.

### Important

Escherichia coli grows at 44 °C, fermenting glucose, lactose and mannitol with the formation of gas. It forms indole and has a positive methyl red test, while the Voges-Proskauer reaction, decomposition of urea and the hydrogen sulphide formation tests are negative. Citrate is not decomposed.

Coliform bacteria frequently do not grow at 44 °C but do at 37 °C. All coliform bacteria can cause citrate to decompose. It is possible to differentiate between the different coliform bacteria on the basis of the $H_2S$ formation, urea decomposition, indole formation, methyl-red reaction and Voges-Proskauer reaction tests.

### Coli titre

The liquid concentration method can be arranged in a more differentiated manner by preparing cultures not only with 100 ml of water but also with 10 ml, 1 ml, 0.1 ml, etc. In this way one can determine the smallest quantity of water in which Escherichia coli or coliform bacteria are still detectable.

The coli titre gives the smallest quantity of water in which at least one Escherichia coli or coliform bacterium capable of multiplication is present.

The following is then stated in the analysis report:

Escherichia coli is detectable in ... ml.

Coliform bacteria (possibly with details of the type) are detectable in ... ml.

One inserts in this text the smallest quantity of water giving a positive result in the coli titre test; the presence of Escherichia coli and/or coliform bacteria must also be confirmed by the colour series. If it is certain that Escherichia coli and coliform bacteria are not present in 100 ml of water, the following is stated:

Escherichia coli and coliform bacteria are not detectable in 100 ml of water.

### Membrane-filter method

To detect Escherichia coli and coliform bacteria in 100 ml of water and in larger quantities of water, the sample to be analyzed is filtered under sterile conditions through a membrane-filter with a pore width of 0.45 µm in a membrane-filter device. The membrane filter is then incubated in single-strength lactose-peptone solution at 37 °C. The same method is used as for the liquid concentration method. Alternatively the membrane-filter disk is placed onto endoagar or onto an endonutrient cardboard disk (taking care to avoid air bubbles) and is then incubated at 37 °C ± 1 °C for 20 ± 4 hours. The final diagnosis is carried out by checking the suspect colonies which have formed with the aid of the colour series.

The membrane-filter method is not suitable for turbid water samples because such water frequently blocks the filters or the build-up of foreign substances on the filters causing the growth of the germs to be disturbed.

### 5.2.8.3 Detection of faecal streptococci

Faecal streptococci are able to grow in the presence of sodium azide ($NaN_3$), while many other types of bacteria are inhibited or killed by azide. Thus, by using culture mediums containing azide, it is possible to increase the number of faecal streptococci while, at the same time, suppressing the growth of other germs.

#### Liquid enrichment

The quantity (100 ml or 250 ml) of water to be analyzed is mixed with double-strength azide-dextrose broth and incubated at 37 ° ± 1 °C for at least 20 ± 4 hours. The incubation time can also be extended to 44 ± 4 hours if it is suspected that insufficient enrichment will have taken place after just one day's incubation. After incubation, the enriched culture is spread onto blood agar (nutrient agar with the addition of approx. 1 % of defibrinated sheep's blood). Faecal streptococci are capable of haemolysis and grow in blood agar into colonies which are surrounded with a yellow ring. Such colonies must be subsequently identified more accurately.

#### Membrane-filter method

The quantity of water to be tested is filtered through a membrane filter with a pore size of 0.45 µm. The membrane filter is placed on a special culture medium, which contains triphenyl-tetrazoliumchloride and sodium azide. The suspect colonies are counted after an incubation time of 20 ± 4 hours or 44 ± 4 hours at 37° ± 1 °C. Faecal streptococci grow in red or red-brown colonies which, after they have been counted, are differentiated further to permit final diagnosis.

#### Diagnosis

a) Gram staining

Faecal streptococci are gram-positive oval to lance-shaped diplococci or short-chain streptococci.

b) Analysis with cultures

Faecal streptococci must grow both in a nutrient broth whose pH has been set to 9.6, as well as in a nutrient broth which contains 6.5 % common salt (normal nutrient groth contains 0.5 % common salt). Incubation temperature in both cases: 37° ± 1 °C; incubation time: 20 ± 4 hours. In addition, faecal streptococci are capable of causing Aesculin to decompose to Aesculetin. Aesculin broth is injected and incubated at 44 ± 4 hours at 37 °C. Aesculin has decomposed when there is a colour change from yellow-brown to olive-green to black after the addition of a freshly prepared, 7 % aqueous solution of ferric chloride ($FeCl_3$).

If necessary, determination of the exact type can be carried out with the diagnosis strips (e.g. API-Strep) which are commercially available, or in accordance with the following table.

c) Serological analysis

The majority of the streptococci possess group-specific antigens which can be extracted and determined with the appropriate antiserum. If strepto-

Extracts from Hahn and Tolle

| | Streptococcus faecalis var faecalis | Streptococcus faecium |
|---|---|---|
| Growth at 45 °C | positive | positive |
| Growth with 6.5 % sodium chloride | positive | positive |
| Bile 40 % | positive | positive |
| Litmus milk | reduction and coagulation | reduction and coagulation |
| Reduction of potassium tellurite | positive | negative |
| Splitting of: | | |
| Arginine | positive | positive |
| Mannitol | positive | positive |
| Lactose | positive | positive |
| Glycerine | positive | negative |
| Gelatine | negative | negative |
| Starch | negative | negative |

cocci colonies are present, it is therefore possible to determine in a rapid microscopic slide test whether these colonies belong to Lancefield group A, B, C, D, F, or G. Faecal streptococci belong to Lancefield group D.

The serological test can be carried out with the commercially available test kits in accordance with the instructions supplied.

### 5.2.8.4 Detection of sulphite-reducing, spore-forming anaerobes

#### Liquid enrichment

20 to 50 ml of water is mixed with the same quantity of a double-strength dextrose-iron-citrate-sodium-sulphite medium and added to a sterile glass flask with a groundglass stopper, taking care not to introduce air bubbles. If necessary, the flask must be topped up with single-strength medium. If the groundglass stopper is not an absolutely tight fit, the neck of the flask can be dipped in hot paraffin wax and covered with a paraffin wax layer. This is necessary because the incubation must take place under anaerobic conditions (attention: do not grease the groundglass stopper in order to make it tight, because the grease may contain anaerobic, sulphite-reducing spore-forming agents).

The culture is incubated for up to 44 ± 4 hours at 37 ° ± 1 °C. A positive reaction is indicated by the liquid culture medium turning black.

Membrane-filter method

The water to be analyzed is filtered through a membrane filter. The filter disk is placed with the side containing the layer downwards onto a dextrose/iron sulphate/sodium sulphite agar. Incubation is carried out for up to 44 ± 4 hours at 37 ° ± 1 °C in an anaerobic pot. Other methods common in microbiology for anaerobic cultivation of germs (Fortner culture, Pyrogallol culture) can also be employed if such a pot is not available. Alternatively the membrane-filter disk which has been laid on the culture medium can simply be covered over again with dextrose/iron citrate/sodium sulphite agar. The thickness of the culture medium covering layer should be at least 5 mm.

Fortner culture

Instead of the lid of the Petri dish, a second Petri dish is placed on top of the first. This second dish contains a nutrient medium layer which has been intensively inoculated with Serratia marcescens; these bacteria consume oxygen rapidly. The two dish halves are joined to one another in an airtight manner (using several layers of adhesive tape, insulating tape or plasticine). The Serratia germs, which multiply rapidly, consume practically all the oxygen in the system leaving only very small residual quantities and thus create conditions which are very largely anaerobic. This means that even anaerobic germs are provided with the most favourable environment and can grow into colonies capable of being counted.

Pyrogallol culture

Approx. 5 g pyrogallol is thoroughly moistened with 8 to 10 ml of a 10 % solution of sodium carbonate ($Na_2CO_3$) in the lower part of a Petri dish. The dish with the membrane-filter culture to be analyzed is immediately placed on top and the two dish halves are sealed to one another in an airtight manner. Pyrogallol binds the oxygen present in the system. Care must be taken that the culture medium layer with the membrane filter is not wetted by the liquid in the lower dish part as a result of careless handling.

The cultures are incubated at 37 ° ± 1 °C and evaluated after not longer than 44 ± 4 hours. Black colonies, which can if necessary be counted and related to the quantity of water used, are positive.

5.2.8.5 Detection of Pseudomonas aeruginosa

Liquid enrichment

100 ml or 250 ml of the water to be analyzed is mixed with double-strength malachite-green broth and incubated for up to 44 ± 4 hours at 37 ° ± 1 °C. The subcultures are then placed on a suitable selective culture medium or on endoagar and the colonies formed after an incubation time of 20 ± 4 hours at 37 °C are analyzed in terms of their ability to form fluorescin and pyocyanin. For this, the suspect colonies are inoculated onto culture mediums A and B (after King) and incubated for 44 ± 4 hours at 37 °C. In addition, a check should be carried out as to whether ammonia is formed from acetamide. For this purpose a solution of ammonium-free acetamide is injected and checked with Nessler's reagent for the presence of ammonia after an incubation time of 20 ± 4 hours at 37 °C.

### Membrane-filter method

The quantity of water to be analyzed is filtered through a membrane filter. The membrane filter with the layer side upwards is placed on malachite-green agar or cetrimide agar, taking care not to include air bubbles. Incubation is carried out at 37 ° ± 1 °C, and the evaluations are carried out after 20 ± 4 hours and 44 ± 4 hours. Colonies showing a blue-green pigment are suspect. In addition the cultures possess a characteristic sweet-aromatic smell. The acetamide decomposition test and the tests for the formation of fluorescin and pyocyanin are used for the final diagnosis.

## 5.2.8.6 Occurrence, significance and detection of sulphate-reducing bacteria

Sulphate-reducing bacteria are to be found in the ground and in so-called reduced waters (waters with a redox potential in the negative mV range).

In the absence of oxygen, they reduce sulphate to sulphide. As a result of this capability, they play an important role in, amongst other things, the recovery of mineral oil. In an anaerobic environment, they form considerable quantities of hydrogen sulphide, which leads to significant changes in the biocenosis of surface waters as a result of its poisonous effect on other microorganisms. The sulphate-reducing bacteria are to be regarded as noxious as far as supplies of drinking water and process water are concerned, because the hydrogen sulphide they form produces an unacceptable smell in the water and the activities of these bacteria in pipelines give rise to biological corrosion.

The two most important sulphate-reducing bacterial genera are Desulfovibrio and Desulfotomaculum.

### Genus Desulfovibrio

This consists of gram-negative rods or spirally-curved rods with a half turn and polar flagella. Pleomorphic forms can occur frequently. Growth takes place strictly anaerobically, the optimum temperature being 44 °C. Gelatine is not liquefied, nitrates are not reduced and as a rule hydrogenase is present. No growth takes place in a strongly acid environment (the pH must not fall below 5.0). The most well-known representative of this genus is Desulfovibrio desulfuricans.

### Desulfotomaculum

The genus is represented exclusively by the type Desulfotomaculum nigrificans, a gram-negative, peritrichally flagellated rod with rounded corners. The cells are straight to slightly curved. It grows in a temperature range from 45° to 70 °C, the optimum temperature being around 55 °C. It can, however, become acclimatized to temperature ranges around 37 °C by a process of adaptation. Desulfotomaculum nigrificans is fundamentally to be regarded as thermophilic.

### Detection of sulphate-reducing bacteria

For detection and/or quantitative determination, 1 to 2 litres of water or graduated quantities thereof are filtered through a membrane-filter with grid. The filter is then placed with the coated side down on the special culture medium, taking care to avoid air bubbles. It is then covered over with about 5 mm of the same culture medium. Two cultures are always prepared in the same way, one culture being incubated at 28 ° to 30 °C, the other at 55 °C, in both cases under anaerobic conditions.

The cultures are examined after 2 days and 4 - 5 days incubation. In the case of a negative result, the incubation time should be extended to 10 to 14 days in order to be quite certain. The sulphate-reducing bacteria appear as black colonies as a result of the formation of ferrous sulphite.

The same principle is used for cultivation in liquid culture media. Here the same selective media but without the addition of agar are sterilized in glass-stoppered, 50 ml flasks. The membrane filter with the germs isolated from the water to be analyzed is rolled up and put into the culture medium in such a way that it can be closed without there being any air bubbles. Care must be taken that the glass stopper is an absolutely tight fit.

### 5.2.8.7 Autotrophic microorganisms in water

In terms of obtaining drinking and process water, it is not only saprophytic bacteria which are of significance. In addition the so-called iron and manganese bacteria and sulphur bacteria can be of special importance as autotrophic bacteria.

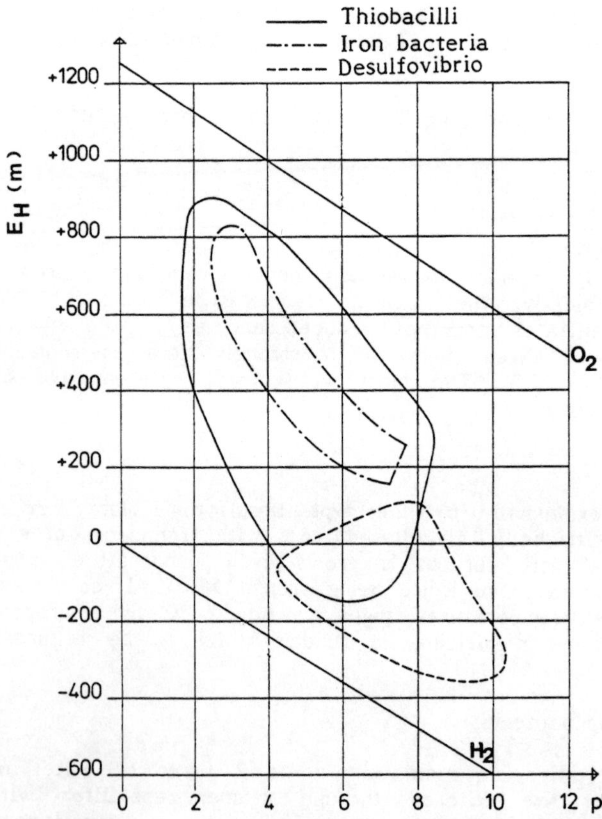


**Fig. 153.** G. Rheinheimer: Microbiology of Waters 3rd edition 1981, p. 110, Gustav Fischer Verlag, Stuttgart

## Iron and manganese bacteria

Iron and manganese bacteria live primarily in waters containing iron and manganese. They oxidize bivalent iron and manganese compounds in these waters to higher levels of oxidation. For this purpose these bacteria need oxygen, albeit only in very small quantities.

As a rule metallic iron is not attacked directly but, instead, only ferrous ions are utilised and for this their concentration must lie between 0.2 and 12 mg/l. Cultures stop growing when the iron concentration is more than 14 mg/l. A redox potential of at least + 10 mV is necessary for the large scale development of iron and manganese bacteria in water. There is a relationship between the pH and the necessary redox potential. The following graph shows the growth field.

In practice, iron and manganese bacteria are only of significance when they develop in large quantities, in which case they can lead to the ferric incrustation of springs and wells and the blocking of pipes. This biological ferric incrustation occurs in springs and wells when the following preconditions exist:

1. Presence of iron and manganese bacteria.

2. Presence of ferrous ions or manganese (II) ions in the necessary concentrations.

3. Redox potential and pH in the growth field for iron and manganese bacteria.

4. Increased flow velocity of the water relative to the natural conditions underground.

Biological ferric incrustation in springs and wells can only be combatted with difficulty. Certain success can be achieved by chlorinating the spring or well at high levels at regular intervals. The chlorine levels for this should be between 50 and 100 mg/l.

To establish the presence of iron and manganese bacteria, microscope slides are attached at intervals of 2 to 3 m to a suitable line with a weight at the end. The line is then hung in the centre of the well and left there for 4 weeks. The microscope slides are then removed and examined under the microscope for the presence of iron and manganese bacteria.

## Microscopic detection (Daubner staining technique)

The sample material to be analyzed is filtered onto a membrane filter. The membrane-filter disk is then dried and laid on filter paper which has been saturated with a 3 % solution of potassium hexacyanoferrate (yellow potassium ferrocyanide = $K_4Fe(CN)_6 \cdot 3\ H_2O$). After 15 minutes, the filter is transferred to a filter paper saturated with 5 % hydrochloric acid and, after completion of the oxidation, is dried again. The membrane-filter disk is then laid for 30 minutes on a filter paper which has been soaked with carbolic erythrosine. After a further drying process, the membrane filter can be lightened with cedar oil and examined under the microscope. The bacteria which have been caught are stained red, the caps and threads of the iron bacteria blue.

For waterworks practice, the following iron and manganese bacteria are of significance as causing problems:

1. **Gallionella ferruginea (Genus Gallionella Ehrenberg)**

   These iron bacteria show a characteristic morphological structure and can be detected easily under the microscope. The bacterial cell of Gallionella ferruginea is kidney or bean shaped. It has a length of 1.5 μm and a diameter of 0.5 to 0.6 μm. It is attached to a long, flat, spirally wound stem. The front of the cell is convex, the back concave. Ferric hydroxide is deposited on the concave side of the cell and, together with organic material, forms the strip which is wound in the form of a screw and which can be seen well under the microscope. The strips divide when the bacterial cell multiplies by transverse division. Since, however, these strips are very fragile, one can usually only find fragments of them during microscopic examination. Gallionella ferruginea is microaerophilic and prefers neutral waters which contain low levels of nutrient and have only low levels of organic substances. This bacterium is only capable of precipitating iron and cannot utilize manganese. The deposits formed by Gallionella ferruginea have a conspicuously light ochre colour.

   Gallionella **minor** is morphologically similar. However, its strips are somewhat wider than those of Gallionella ferruginea.

2. **Leptothrix ochracea (Genus Leptothrix)**

   Leptothrix ochracea belongs to the group of sheath bacteria. The cylindrical individual cells of these are lined up in rows one after the other and are surrounded by a gelatinous sheath of varying thickness, so that a filament appears to be of different lengths. The sheaths are coloured ochre-yellow to dark-brown as a result of the inclusion of iron or manganese compounds. When the sheaths become thicker and thicker as a result of the iron and manganese deposits, the bacteria can wander outwards and form new sheaths. Thus, under the microscope, there are always numerous empty sheaths to be observed as well as the living filaments. Leptothrix ochracea is facultatively autotrophic and can also utilize organic substances. The filaments are never branched or attached.

3. **Crenothrix polyspora (Genus Crenothrix Cohn)**

   This type is also a sheath bacterium, in which the individual cells are connected to one another to make filaments of different lengths and lie in a gelatinous sheath. The length and the width of the cells vary. In general, the sheath is thinner and colourless in the upper, younger part, while in the older, basal part, it is often thickened and coloured brown as a result of iron or manganese deposits. Individual bacterial cells or cell groups can emerge from the sheath and form new filaments. Crenothrix polyspora, which is also known as "spring thread", form unbranched but attached filaments. It is also facultatively autotrophic so that it can grow well in still and flowing waters regardless of the ferrous iron concentration, provided organic substances are present. It is generally not possible to cultivate it in pure cultures.

4. **Genus siderocapsa**

   The generally coccoid, but occasionally rod shaped cells of the bacteria belonging to this genus are imbedded in more or less clearly formed capsule-like sheaths and can occur individually or in small or large accumulations. According to present knowledge, Siderocapsa types are not capable of precipitating iron but do oxidize manganese. In contrast to the previously named types, detection of these bacteria under the microscope is very difficult. It is usually only possible by microscopic evaluation after enrichment of a culture.

5. Ferrobacillus ferrooxidans

This representative of the genus Thiobacillus also belongs in the wider sense of the word to the group of iron bacteria. It grows in a strongly acid environment at pH values from about 2.5 to 4.4 and dies in a short time in the neutral range. In the case of Ferrobacillus, the so-called mineralization products are deposited freely in the medium. This type of germ plays only a subordinate role in waterworks operations, since the strongly acid conditions needed for growth do not as a rule prevail in waterworks.

Cultivation of cultures of iron and manganese bacteria belonging to the genera Gallionella, Leptothrix, Siderocapsa and Ferrobacillus can be carried out when suitable culture mediums are used, the composition of which is given in the section on culture mediums.

The following diagrams serve as an aid to microscopic differentiation of iron and manganese bacteria.

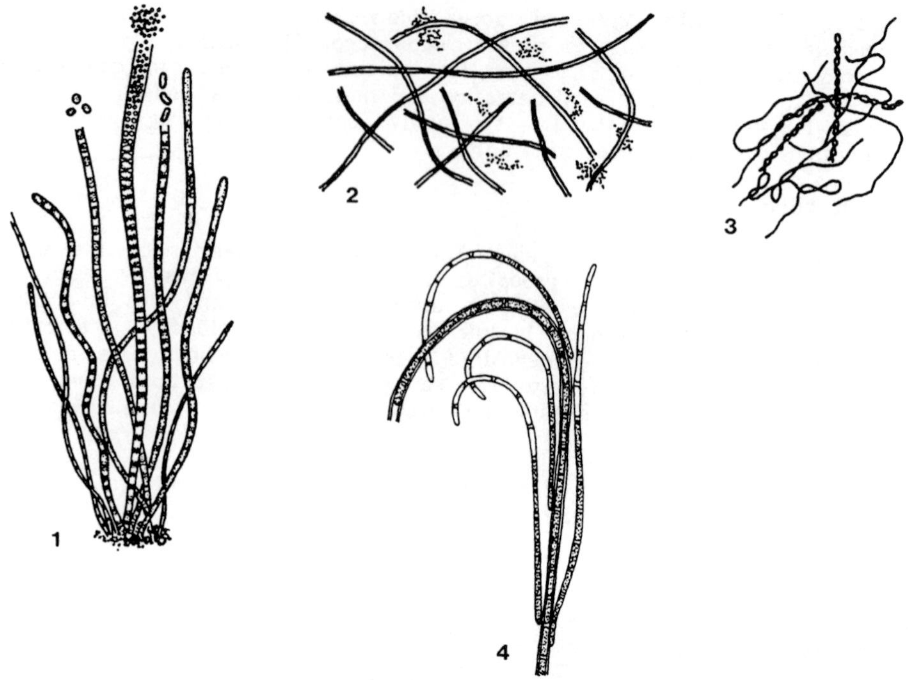


Fig. 154 - 157. Iron bacteria; 1) = Crenothrix polyspora, "spring filament": Filaments several mm long, at the bottom 1.5 - 5 µm, at the top 6 - 9 µm thick with sheaths. Short cells are deposited in sheaths. In the course of multiplication, the cells of the threads simply emerge from the sheaths or they first decompose to form smaller spherules (spores). Iron is deposited in the form of $Fe^{2+}$ salts. Rust formation as a result of these salts oxidizing and forming oxides. Oligosaprobic. In springs and water pipes; 2) = Clonotrix fusca, branched iron and manganese bacterium: Clear, pronounced, heavily encrusted sheaths, in which iron and manganese are stored; old sheaths up to 24 µm thick; otherwise the filaments are 5 - 7 µm thick at the base, narrowing down to 2 µm approaching the top. Cells 2 µm thick. Floccules of a few millimetres in length with a brown to black appearance. In springs and water towers. Often in the company of Crènothrix; 3) = Gallionella ferruginea, "twisted iron bacterium": Forms very thin, only 1 µm thick solid filaments, these are dichotomously branched, the two parts being wound around each other. The size of the kidney or coccus-shaped cells at the ends of the branches is only about 0.5 - 1.2 µm. Generally contains hydrated ferric oxide. Oligosaprobic. In springs and wells containing iron. 4) = Leptothrix ochracea, "common iron bacterium": Sheath when young thin and only slightly coloured; thickens later and becomes yellow to brown as a result of iron or manganese deposits. Cells without sheath only 1 µm thick. Eggshaped gonidia. In ground water and ditches.

### 5.2.8.8 Sulphur Bacteria

Of the sulphur bacteria, which oxidize hydrogen sulphide, the genera Beggiatoa and Thiothrix are of significance for drinking water and process water supplies.

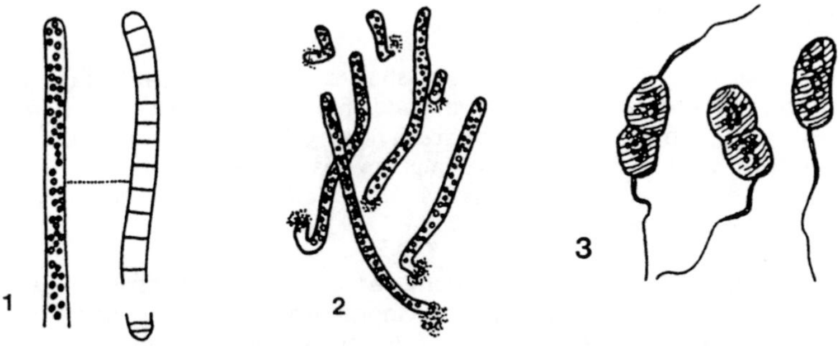


Figs. 158 - 160. 1) = Beggiatoa alba ("White Sulphur Bacteria"): These are fairly long filaments, up to 5 μm thick, composed of individual cells, approx. 3 μm thick. They are unbranched and may either be attached or move by means of sliding or oscillating movements. They do not have a clearly defined sheath. They are often packed with grains of sulphur, which refract light considerably. These release carbon disulphide when the sample is dried. Multiplication through decomposition into smaller filament pieces. When it occurs en masse, Beggiatoa, which can be recognized macroscopically, frequently covers objects (mud, leaves, etc.) like a white veil, which can be easily torn when touched. The white colour comes from the sulphur. The transverse walls in the filaments usually only appear after the sulphur has been removed by natural or artificial means; 2) = Thiothrix nivea ("White Sulphur Filament"): Similar to Beggiatoa, but non-mobile, 1.5 - 2.4 μm thick. Usually heavily filled with sulphur grains. Forms white coatings, similar to Beggiatoa. $\alpha$-mesosaprobic. 3) = Chromatium okenii ("Red Sulphur Bacteria, Little Sulphur Barrels"): Some 8 μm long and 5 μm thick, flagellated, roundish cells which move fairly rapidly (the movements cease for a while when the viewing field is suddenly darkened). Red in colour, usually containing a distinct concentration of sulphur grains. The organism is to be found in plankton, often in company with species of Oscillatoria and other similar species. Often forms red spot-like structures, or colours all the water red in uncleaned water. Polysaprobic to mesosaprobic.

The best-known representative of the genus Beggiatoa is Beggiatoa alba. This bacterium is a colourless, unbranched, gram-negative filament bacterium. The filaments have a diameter of 2.5 to 5 μm and are composed of cell segments having a length of more than 3 μm. Beggiatoa alba can move freely by means of sliding, crawling movements. In the presence of hydrogen sulphide, grains of sulphur are deposited inside the cell. When a shortage of hydrogen sulphide occurs, these grains are further oxidized to form sulphuric acid. The sulphuric acid reacts with the carbonates in the surrounding medium, producing a rearrangement of these carbonates to form sulphates accompanied by carbon dioxide. The presence of Beggiatoa alba in water counts as a biological indicator of the presence of hydrogen sulphide in ground water and surface water.

**Thiothrix types**

These sulphur bacteria form colourless filaments. In contrast to the Thiorhodaceae, which like high hydrogen sulphide levels, the Thiothrix types prefer habitats with hydrogen sulphide concentrations below 4 mg/l.

Spherical sulphur bacteria

In addition to the filament-shaped sulphur bacteria, there are coccoid forms, which as a rule form reddish pigments. The most well-known representative of these sulphur bacteria is Chromatium okenii, also known as "red sulphur bacterium", and Lamprocystis roseo-persicina. The latter is often to be found on water plants and forms a pink-coloured covering when it develops en masse.

The pH, redox potential, light and oxygen conditions as well as temperature, presence of organic and inorganic substances and the movement of the water, all play a role in the mass development of sulphur bacteria. In water supply systems, sulphur bacteria can cause or favour corrosion and its occurrence always indicates the presence of hydrogen sulphide in the system. Mass development of sulphur bacteria is always undesirable, even if sulphur bacteria themselves do not play a particular role in epidemic-hygiene terms.

The **genus Thiobacillus** is of significance in terms of corrosion. Thiobacillus thiooxidans is capable of oxidizing sulphur and sulphur compounds to form sulphuric acid. This can bring about localised concentrations of sulphuric acid, which attack metals and acid-sensitive materials substantially. In waterworks operations, the growth of Thiobacillus thiooxidans is rare because it prefers a strongly acid environment (ideal pH between 2.0 and 3.5), which as a rule is not present in water supply systems. Thiobacillus thiooxidans is closely related to Ferrobacillus ferrooxidans; both are gram-negative short rods and require strictly autotrophic conditions.

**Detection of sulphur bacteria**

Sulphur bacteria are generally detected by means of microscopic investigation. Their cell forms are shown in the following diagram. They can be cultivated in special culture mediums, the composition of which is given in the section on culture mediums.

### 5.2.8.9 Distribution of worm parasites

The spectrum of zooparasites involved here varies from country to country and from climatic region to climatic region.

The following table by Jawetz, Melnick and Adelberg provides information on the occurrence of a variety of worm parasites.

In this connection special mention should be made of the disease bilharzia. In the case of bilharzia, the human being is infected by the free cercaria penetrating directly into the human skin. The human being can be infected by standing or bathing in water or by drinking infected water.

Further details for the identification of tapeworms, threadworms and trematodes can be taken from L. Hallmann's tables (Klinische Chemie und Mikroskopie 1980, p. 488 to 497) which are reproduced below.

| Species | Occurrence | Intermediate host | Final host apart from human beings |
|---|---|---|---|
| Cestoda or tapeworms (Taeniasis, Diphyllobothriasis): Taenia saginata (cattle tapeworm) | Worldwide | Cattle, camels | -- |
| Taenia solium (pork tapeworm): | Worldwide | Swine | -- |
| Diphyllobothrium latum (fish tapeworm): | Worldwide, temperate zones | From crustaceans to fish | Dogs, cats and carnivores |
| Trematodes: Leaches (Schistosomiasis) Schistosoma mansoni (intestinal): | Africa-Arabia South America | Snails | Hardly any |
| Schistosoma haematobium (urinary passage, sometimes intestinal): | Africa-Europe Middle East West Indies | Snails | Hardly any |
| Schistosoma japonicum (intestinal): | China, Japan the Philippines | Snails | Many, but only human beings significant |
| Intestinal flukes Fasciolopsis buski (large intestinal fluke): | China-Indonesia | Snails; Metacercaria on edible water plants | Swine, dogs |
| Heterophyes heterophyes (small intestinal fluke) and Metagonimus yokogawai: | Egypt Far East and Japan | Snails and fish | Dogs, cats and carnivores |
| Liver fluke Clonorchis sinensis (East Asian or oriental liver fluke): | Far East | From snails to fish | Dogs, cats |
| Lung fluke (Paragonimiasis) Paragonimus westermani | Worldwide especially Far East | From snails to various crustaceans | Dogs, cats |

Tapeworms = Cestodes

| Parasite | Worm | | | | |
|---|---|---|---|---|---|
| | Length | Segments | Head | Uterus after Lightening with Dilute Acetic Acid | Sexual Opening |
| **Taenia solium** Pork tapeworm | 2 - 3 m occassionally longer up to 8 m | Mature segments: pumpkin pip shape | Spherical, 1 mm dia. 4 suckers and a double ring of teeth with 25 to 50 hooks | Treelike branching on each side, each with 7 - 10 thick branches | Irregularly alternating right or left at the side |
| **Taenia saginata** Cattle tapeworm | 4 - 10 m | Mature segments: pumpkin pip shape, but wider than T. solium | Pear shaped, 1 to 2 mm dia., 4 round, hemispherical suckers, without ring of hooks | Abundant branching (20 - 30) on both sides, but narrow and dichotomous | Irregularly alternating right or left at the side |
| **Diphyllobothrium latum** Bothriocephalus latus Pike (fish) tapeworm Wide tapeworm | 2 - 8 m occasionally up to 20 m | Wide, very short | 2 - 5 mm long, almond shaped, two shallow elongated suction depressions, one ventral and one dorsal | Rosette shaped, in the middle of each segment. Uterus opening is the lower genital opening. | Medial on the stomach side |
| **Hymenolepis nana** Dwarf tapeworm | 10 - 25 mm long 0.55 - 0.9 mm wide | Wide, very short | Spherical up to 3 mm dia. Rostellum with single ring of hooks. Hooks 14 - 18 µm long | Not branched | Unilateral |

| Final Host | Intermediate Host | Eggs or embryophores | | | Detection, Secondary Findings |
|---|---|---|---|---|---|
| | | Size | Appearance | Colour of embryonic skin | |
| Human beings | Swine, occasionally human beings | Roundish, approx. 31 - 50 µm in diameter | Like "spherical bubbles". Radially striped shells; less transparent than T. saginata. Embryo with 3 pairs of hooks[a)] less clearly recognizable | Yellowish to light brown | Detection in stool: mature proglottides. Eggs only occasionally when the proglottides decompose in the intestine. Charcot-Leyden crystals frequent. |
| Human beings | Cattle, rarely human beings | Egg shaped 30 - 40 µm long 20 - 30 µm wide | Oval to egg-shaped. Thick skinned. Skin radially striped. Inside, 6 embryonic small hooks[a] more clearly recognizable than with T. solium | Dark brown | Detection in stool: mature proglottides. Eggs only in the case of the decomposition of the proglottides in the intestine. Charcot-Leyden crystals frequent. |
| Human beings, dogs, cats, foxes | Two intermediate hosts: 1. very small crabs 2. various fish | 68 - 71 µm dia. 45 - 54 µm | Ellipsoidal with small cap on the upper pole. Yolk cells and a still uncleaved germ cell. - Similarity to eggs of the large liver fluke | Pale yellowish to yellow-brown | Detection in stool: tapeworm eggs. The tapeworm segments usually decompose in the intestine - in the blood often severe anaemia of the pernicious type. |
| Human beings, possibly dogs, rodents | Insects, e.g. fleas, but not inevitable | Outer membrane 45 - 60 µm to 34 - 45 µm Inner membrane 29 - 30 µm | Longitudinally elliptical eggs with 2 thin membranes, between which liquid and convoluted threads spread out. | Very transparent, matt pink to colourless | Occurrence: cysticeroid in the small intestine villi of human beings, later also as tapeworm. Also autoinfection. Rare, sometimes present in large |

**Tapeworms = Cestodes** (cont.)

| **Parasite** | Worm | | | | |
|---|---|---|---|---|---|
| | Length | Segments | Head | Uterus after Lightening with Dilute Acetic Acid | Sexual Opening |
| **Hymenolepis nana** (cont.) | | | | | |
| **Dipylidium caninum** Taenia cucume-rina Cucumber seed tapeworm | 15 - 40 cm long 2 - 3 mm wide | Stretched in length, look like "cucum-ber seeds". Immature segments reddish. Mature seg-ments con-tain indi-vidual, roundish sacs, which contain eggs | Small Ros-tellum with 4 rings of hooks. Hooks have rose thorn shape, 5-15 µm | Not branched | Immature seg-ments, on each side two opposed genitalpori |
| **Echinococcus Cysticus** Dog tapeworm | 2 - 5 mm | 3 Proglo-ttides | Ring of hooks 4 suckers | Not branched | Alternately right or left at the side |
| **Echinococcus alveolaris** | 1.9 - 2.5 mm | 3 - 5 Pro-glottides | Double ring of hooks 4 suckers | | |

From: L. Hallmann (1980), Klinische Chemie und Mikroskopie, 11th edition, Published by: Verlag Georg Thieme, Stuttgart-New York

| Final Host | Intermediate Host | Eggs or embryophores | | | Detection, |
|---|---|---|---|---|---|
| | | Size | Appearance | Colour of embryonic skin | Secondary Findings |
| **Hymenolepis nana** (cont.) | | | Threads (3 - 4) are attached like tufts to the two poles of the inner membrane. 6 clear embryonic hooks. | | numbers, primarily in children, also autoinfection. Detection in stool: eggs transparent, difficult to recognize. |
| Dogs, cats rarely human beings | Insects, e.g. fleas | 25 - 40 µm in diameter | Spherical, embryonic skin, thin, hyaline, not radially striped, embryonic hooks recognizable | Pale brick red | Occurrence: tapeworms in the intestine of domestic dogs, cats and also children (from swallowing infected fleas). Detection in stool: Proglottides. Occasionally also as packets of 8 - 15 eggs surrounded by a sac-like membrane |
| Dogs | Sheep, cattle, swine, goats, horses, human beings | 32 - 36 µm dia. 21 - 30 µm | Rather egg-shaped to oval, thick skinned (similar to Taenia saginata eggs). Inside, likewise 3 pairs of small hooks. | | |
| Foxes, dogs, domestic cats | Field mice, human beings | | | | |

[a] These 3 embryo pairs of small hooks have nothing to do with the small hooks which form later at the head of the tapeworms.

Round (Thread) Worms = Nematodes

| Parasite | Worm | | Eggs |
|---|---|---|---|
| | Size in mm | Appearance | Size in µm |
| **Enterobius** (Oxyuris) **vermicularis**, Threadworm Pinworm Seatworm | ♂ 2 - 3 dia. 0.2 ♀ 9 - 12 dia. 0.5 | Whitish, almost colourless, roundish, maggot shaped | 50 - 60 dia. 30 - 32 |
| **Ascaris lumbricoides** Eelworm | ♂ 140 - 200 dia. 3.2 - 4 ♀ 200 - 400 dia. 5 - 6 | Grey-pink or yellowish-pink, roundish, like an earthworm, thick as a pencil | 45 - 78 dia. 35 - 60 Unfertilized eggs, see note (last column) |
| **Trichuris trichiura** Trichocephalus dispar Whipworm | ♂ and ♀ 30 - 50 dia. 1 | Roundish, like a maggot head whip-shaped and thin. In the case of thicker, spirally rolled in with Spiculum; in the case of ♀ curved like a scythe | 50 - 55 dia. 22 - 25 |
| **Ancylostoma duodenale** **Necator americanus** Hookworm | ♂ 8 - 11 ♀ 10 - 18 | Reddish white, thread shaped; slightly thinner at the front; mouth capsule with teeth which are bell-shaped in ♂, pointed in ♀ | 60 dia. 30 - 40 |
| **Strongyloides stercoralis** Dwarf thread worm | Rhabditiform larva 200 - 300: 14 - 16 µm Filariform larva double length | Larva similar to the larva of hookworms: three-part oesophagus, club-shaped front section; postmedian constriction followed by short bulb. Short, wide mouth cavity | 60 - 80 to 40 - 45 |
| **Filariasis Group** **Trichinella spiralis** Trichinae | These consist of Nematodes which live as parasites primarily in the blood or tissue. | | |

| Eggs | | Detection in stool; secondary findings |
|---|---|---|
| Appearance | Colour | |
| Oval, asymmetrical, seen from above: with one flat and one rounded side. Shell thin, double contoured. Contents fine-grained, with nucleus and all stages of development from yolk up to developed, tadpole-shaped embryo. | Transparent, almost colourless | Detection: worms in stool; detection of eggs using adhesive tape method is better. Charcot-Leyden crystals frequent. |
| Oval with undulating egg-white sac or without sac but with thin, smooth, double-contoured skin. Large round nucleus in the centre with numerous refractive lecithin granules. No yolk cleavage. Occasionally also unfertilized eggs (see remarks in last column). | a) with egg-white sac: brownish-yellow due to colouring by bile pigment<br>b) without sac: colourless | Detection: eggs (after enrichment if necessary). Ocassional worm in stool. Charcot-Leyden crystals frequent. N.B. The eggs are very resistant to drying out (risk of infection). Remarks: Unfertilized eggs are more elongated, 88 - 94 : 39 - 60 µm. Their interior is filled with numerous highly refractive droplets. Egg-white sac is lost if exposed to hydrochloric acid (Telemann enrichment technique). |
| Lemon-shaped. Skin double-contoured, with a light-coloured slimy plug at each pole. In stool generally uncleaved, granular mass in interior. | Yellowish-brown | Detection: Virtually only worm eggs, best detected after enrichment. Anaemia only in cases of severe infestation with worms. Charcot-Leyden crystals rare. |
| Skin thin, smooth, single-contoured, in fresh stool usually 4 - 8 cleavage globules, in older stool more cells or even mobile larvae. | Colourless, diphanous, highly refractive. | Detection: eggs in untreated preparation; if negative: enrichment. In blood: anaemia of the post-haemorrhagic type. Remarks: epidemics formerly common in mining operations. |
| Ellipsoidal, surrounded by fragile envelope similar to hookworm eggs, but morula and larval stages even in fresh stool. | Colourless, diaphanous | Detection: eggs rare; but highly mobile rhabditiform larvae (which emerge within the intestinal tract, unlike hookworm). Charcot-Leyden crystals always present. |

From: L. Hallman (1980), Klinische Chemie und Mikroskopie;
Publ. by: Verlag Georg Thieme; Stuttgart-New York

**Trematodes** (flukes)

| Species | Parasite | | Eggs | Infection path - Occurrence<br>Diagnostic peculiarities |
|---|---|---|---|---|
| | Size | Appearance | | |
| **Fasciolopsis buski**<br>Large intestinal fluke | 30 - 70 mm long<br>14 - 15 mm wide | Flat, grey with lateral pigmented areas. No intestinal branching | 130 - 140 µm : 80 - 85 µm<br>Similar to Fasc. hepatica: oval with cap, dark-coloured, numerous yolk cells in addition to egg-cell. | Intermediate host: species of snail which are penetrated by the miracidium developing in the water. Cercaria encyst on aquatic plants (water chestnut). Oral infection of humans through eating infected plants (= intestinal distomiasis with diarrhoea). Occurrence: India, South-East Asia, China, Thailand.<br>Diagnosis: eggs in stool; usually the eggs are abundant, so no need for enrichment. |
| **Heterophyes heterophyes** | 1 - 2 mm long<br>0.5 - 0.7 mm wide | Oval. Reddish coat of anterior part of body covered with extremely fine hairs. Ventral suction pad and separate genital suction pad. | 30 - 70 µm<br>Oval with cap, light brown with thick skin | Infection path unknown. - Occurrence: in small intestine of dogs, cats, foxes, occasionally humans. Egypt, Asia, Orient.<br>Diagnosis: eggs in stool. |
| **Metagonimus yokogawai** | 1 - 2.5 mm long<br>425 - 730 µm wide | Pear-shaped, brown. Large, irregularly rounded or elliptical testes. | 27 - 30 µm : 15 - 17 µm<br>Similar to egg of Clonorchis: very small, oval with thick skin and cap, yellow-brown. Small spine on pole without cap. | Metacercaria development in various fish and freshwater snails. - Occurrence: in small intestine of cats, dogs, swine, very rarely in humans. Japan, China, Korea. - Diagnosis: eggs in stool. |

**Trematodes** (flukes) (cont.)

| Species | Parasite | | Eggs | Infection path - Occurrence<br>Diagnostic peculiarities |
|---|---|---|---|---|
| | Size | Appearance | | |
| **Echinostoma ilocanum** (lindoensis)<br>Small intestinal fluke | 0.75 - 1.35 mm long,<br>0.5 - 1 mm wide | Transparent, pinkish | 88 - 111 µm : 47 - 53 µm<br>Oval with cap | Metacercaria development in various fish and freshwater snails. - Occurrence: in upper intestinal regions in humans. Philippines, Java, Celebes. - Diagnosis: untreated stool preparation for detection of eggs, which are usually abundant. |
| **Watsonius watsoni** | 4 - 5 mm long,<br>8 - 10 mm wide | 4 mm thick, pear-shaped, reddish colour | 130 µm : 75 µm<br>Oval with cap | Infection path unknown. - Occurrence: in upper intestinal regions in humans. Especially Africa. - Diagnosis: eggs in stool. |
| **Gastrodiscus hominis** | 6 - 8 mm long<br>3 - 4 mm wide | 2 mm thick. Reddish colour. Body consists of conical anterior and discus-shaped posterior part. | 150 µm : 72 µm<br>Oval with cap | Infection path unknown. - Occurrence: detected in humans in colon and ileo-caecal region. Especially India. - Diagnosis: eggs in stool. |
| **Fasciola hepatica**<br>Large liver fluke | 20 - 30 mm long | Thin, flattened, light-coloured in centre, darker at edge (intestinal branching) | 140 µm : 80 µm<br>Oval with cap, light brown. | Intermediate host: small mud-snail. Cercaria encyst in water or on plants. Infection through intake of encysted metacercaria with drinking water or plants as food. - Occurrence: flukes in bile ducts of sheep, cattle, occasionally humans. - Diagnosis: detection of eggs in |

Trematodes (flukes) (cont.)

| Species | Parasite: Size | Parasite: Appearance | Eggs | Infection path - Occurrence<br>Diagnostic peculiarities |
|---|---|---|---|---|
| **Fasciola hepatica** (cont.) | | | | stool or duodenal fluid (but eggs detected in human faeces are usually not from parasite in the human, but from eating infected sheep or cattle liver, i.e. only "passing through"). Also: skin test and complement fixation reaction (Hamburg Institute of Tropical Medicine). Secondary finding in infected humans: severe eosinophilia. |
| **Dicrocoelium dendriticum** (lanceolatum) Small liver fluke | 5 - 12 mm long 1.5 - 2.5 mm wide | Lancet shaped, tapered at the ends, transparent, so uterus filled with dark eggs is recognizable. | 38 - 45 µm : 22 - 30 µm Oval with cap and thick skin to egg, usually flattened on one side. The eggs contain an embryo when they are laid. | Intermediate hosts: snails. 1st to 2nd order egg sporocysts with cercaria. Excreted in form of slime globules adhering to one another and to plants. Infection through eating plants. - Occurrence: as for Fasciola hepatica. - Diagnosis: detection of eggs in untreated preparation of stool or duodenal fluid. |
| **Clonorchis sinensis** Chinese liver fluke | 10 - 20 mm long 2 - 4 mm wide | Longish and flattened, reddish colour, almost transparent. At the rear end 2 heavily branched testes lying next to one another. | 27 - 35 µm : 14 - 19 µm Oval, bottle-shaped with very clear small cap and with ring shaped bulge. Dark colour. Shell double contoured. Miracidium detectable at laying. | Two intermediate hosts: first Japanese pond snails, then fish (carp types), the cercaria encyst and become capable of infection as encysted metacercaria. Infection orally by the eating of raw, parasited fish. Immigration of the metacercaria through the Ductus chole- |

**Trematodes** (flukes) (cont.)

| Species | Parasite | | Eggs | Infection path - Occurrence Diagnostic peculiarities |
|---|---|---|---|---|
| | Size | Appearance | | |
| **Clonorchis sinensis** (cont.) | | | | dochus into the bile ducts of the liver, and also into the pancreas ducts. Of considerable human-pathological significance (liver distomatoses, opisthorchiosis) in China, Indochina, Japan, Tongking. - Diagnosis: detection of eggs in native stool or preparation of duodenal juice; enrichment is usually necessary. |
| **Opisthorchis felineus** Cat's liver fluke | 7 - 12 mm long 2 - 2.5 mm wide | Lancet shaped, transparent, of reddish colour. Characteristicly there is clear lapping of the testes. | 26 - 30 µm : 11 - 15 µm (somewhat thinner than Cl. sinensis eggs). Oval, with less clearly visible cap and with small spine on the pole without a cap. | Two intermediate hosts: as for Clonorchis sinensis. - Occurrence: especially in cats and dogs in the bile and pancreas ducts, also in human beings (liver distomatosis), Memel delta, Russia, Tongking, France, Italy. - Diagnosis: Detection of eggs in native stool or preparation of duodenal juice: enrichment may be necessary. |
| **Paragonimus westermani** (rings) Lung fluke | 8 - 20 mm long 4 - 9 mm wide | 2 - 5 mm thick egg-shaped, plump, reddish-brown | 80 - 118 µm : 48 - 65 µm Oval with gold-brown, relatively thick skin and clear cap, which appears to have sunk in a little; the opposite poles have among other things a | Two intermediate hosts: black snails - then small types of crustacea and crabs (Wollhand crabs). Cercaria encyst in the muscles of the torso. Infection through the eating of raw prawns. Metacercaria migrate through the intestinal wall and midriff into the thoracic cavity and then |

Trematodes (flukes) (cont.)

| Species | Parasite: Size | Parasite: Appearance | Eggs | Infection path - Occurrence Diagnostic peculiarities |
|---|---|---|---|---|
| **Paragonismus Westermani** (cont.) | | | small protuberance. In the inside often one striated germ cell with 5 - 10 vitelline cells. | the bronchioles (formation of cysts). - Occurrence: in dogs, cats, etc. and also in human beings. - Diagnosis: detection of eggs in the sputum (prefer brownish-bloody parts; enrichment if necessary, also the stool (swallowed sputum). - Secondary findings: rust brown discharge; often Charcot-Leyden crystals; eosinophilia. |
| **Schistosoma** | | | | |
| a) **Sch. haematobium** Exciter of urogenital schistosomiasis | 9 - 15 mm long 1 mm wide | Separate sexes; at front end 2 suckers, light colour, significantly thicker than with 4 large testes and abdominal channel (Canalis gynaecophorus) for the reception of the , thread-shaped, of dark colour, since the black intestine shows through. | Large, oval (spindle shaped), transparent with pale yellowish skin without a cap. Present on the inside as a fully formed miracidium in faeces and urine.<br><br>Size of the eggs:<br><br>ad a) 120 - 160 µm : 46 - 50 µm with terminal spine.<br><br>ad b) 110 - 162 µm : 60 - 70 µm with lateral spine | Intermediate hosts are different types of snail. The cercaria penetrate through the skin into the blood vessels, where the sexually mature schistosoma live together in pairs. Eggs (first of all without a miracidium) are laid in the capillaries<br><br>ad a) Breakthrough of the eggs (with miracidium) through the capillaries of the bladder. Excretion in the urine: hematurial schistosomiasis (quantities of blood particularly in the last portion of the urine).<br><br>ad b) and c) Sexually mature pairs of worms live predominantly in the veins of the abdominal organs. Breakthrough |
| b) **Sch. mansoni** Exciter of intestinal schistosomiasis | 12 - 22 mm long, at front 0.1 mm wide, at rear 0.2 mm | | | |
| c) **Sch. japonicum** Exciter of the Katayama illness, vascular schistosomiasis, Chinese-Japanese schistosomiasis | | | | |

Trematodes (flukes) (cont.)

| Species | Parasite | | Eggs | Infection path - Occurrence<br>Diagnostic peculiarities |
|---|---|---|---|---|
| | Size | Appearance | | |
| Schistosoma (cont.) | | | ad c) 70 - 100 µm : 50 - 65 µm with small, often rudimentary spine. | of the eggs (insofar as they have a miracidium) through the capillaries of the intestinal wall. Excretion in stool: Intestinal schistosomiasis.<br><br>Secondary findings: severe eosinophilia and clear leucocytosis in the acute stage. |

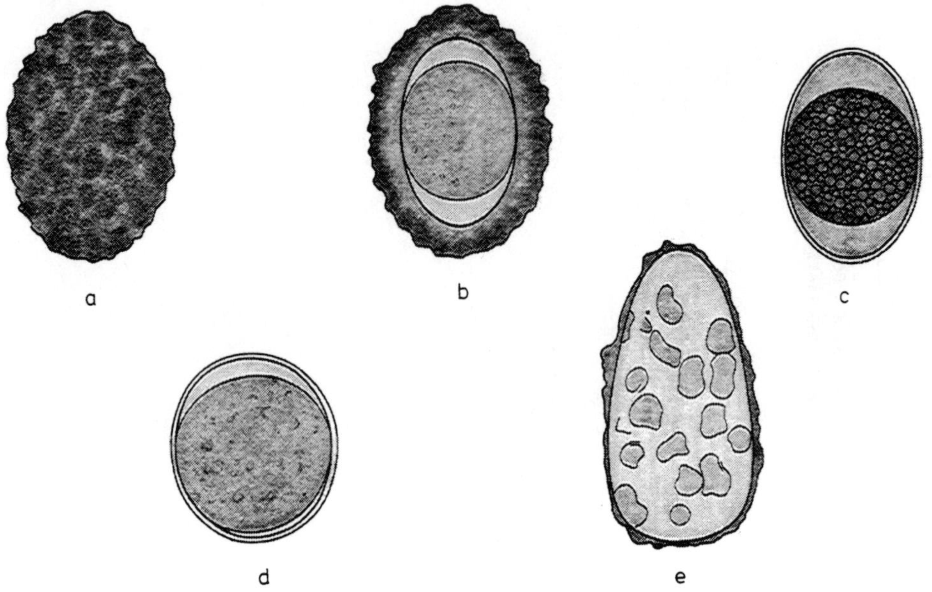


Fig. 161 a - e. Eggs of Nematodes (Threadworms) and Cestodes (Tapeworms). Magnification approx. 400 fold; a = Egg with protein sheath in view; b = The same at the optical centre; c - d = Eggs without sheaths; e = Unfertilized egg (rare)

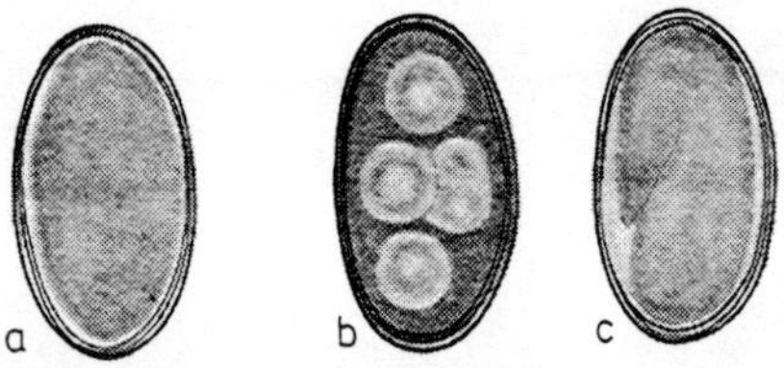


Fig. 162. Eggs of Enterobius (Oxyuris) vermicularis (threadworm) in different stages of development

Fig. 163. Egg of Trichuris trichiura = Trichocephalus dispar (whipworm)

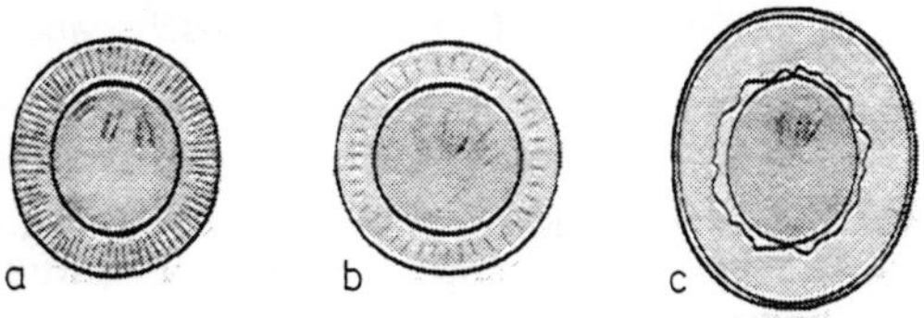


Fig. 164 a - c. a = Egg of Taenia saginata (cattle tapeworm); b = Egg of Taenia solium (swine tapeworm); c = Egg of Hymenolepis nana (dwarf tapeworm)

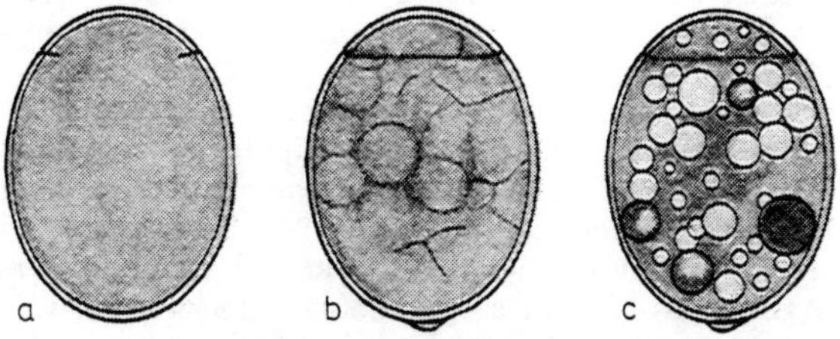


Fig. 165. Eggs of Diphyllobothrium latum (fish or pike tapeworm) in different stages of development

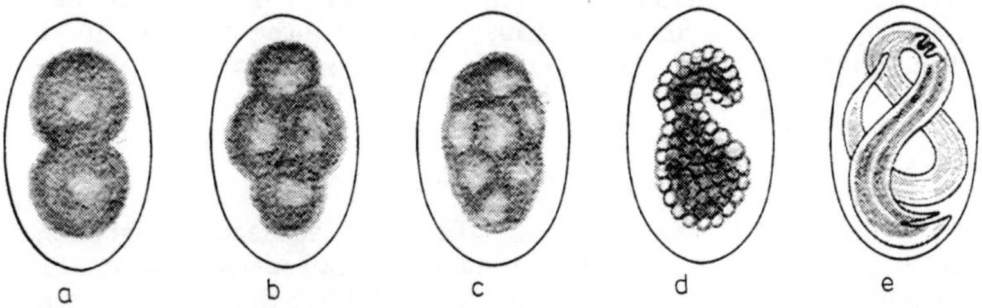


Fig. 166 a - e. Eggs of Ancylostoma duodenale (hookworm) in different stages of development. In practical terms only stages a and b are detectable in fresh (!) stool. - The egg of Necator americanus is somewhat longer (70 : 40 µm), but has the same appearance.

From. Lothar Hallmann,
Klinische Chemie und Mikroskopie
11th edition, 1980
Publ. by: Verlag Georg Thieme,
Stuttgart - New York

## 5.2.9 Preparation of culture solutions and culture media

### General

For colony counts in the microbiological analysis of water, both culture media having a gelatine base and culture media with an agar base are used. In addition, liquid culture solutions are used as enrichment media and, in the case of the "colour series", for differentiating between different entero bacteria.

Gelatine and agar-agar are nutrient carriers. They bring about the solidification of the culture medium. This enables the germs which are present to be held fixed and isolated from one another, so that when they multiply accumulations of daughter cells and hence colonies come into being. Gelatine is a commercially available protein of high molecular weight. Gelatine culture media become liquid at temperatures above 25 °C, so that they can only be incubated at temperatures below 25 °C. It is usual to carry out the incubation at between 20 °C and 22 °C. Agar-agar is a polysaccharide ester of sulphuric acid and is obtained from certain seaweeds (red algae). Agar-agar is also available commercially. Agar culture media liquefy at temperatures of approx. 100 °C; they solidify again when the temperature falls below 45 °C. When agar culture media have solidified they can be incubated at temperatures above 45 °C.

In the case of gelatine culture media, consideration should be given to the fact that certain bacteria and moulds form enzymes which break down protein and are able to cause the culture medium to liquefy by breaking down the gelatine. Proteolitic bacteria (primarily Pseudomonas types) live particularly in surface water, so that the increased occurrence of bacteria able to liquefy gelatine in samples of water from great depths can indicate influence by surface water.

Of considerable significance for the subsequent growth of germs in a culture medium is the maintenance of an optimum pH. For this reason it is necessary that the pH is regulated when the culture medium is being prepared. To lower a value which is too high, 1 m-HCl is used; to raise a value which is too low, 10 % soda solution ($Na_2CO_3$) or 1 m NaOH.

The following instructions apply to culture media used for the microbiological analysis of water:

### Recipes

**Nutrient gelatine:** (Recipe 1)

**Composition**

| | |
|---|---|
| Meat extract | 10 g |
| Peptone | 10 g |
| Sodium chloride (NaCl) | 5 g |
| Gelatine | 120 - 150 g |

(At warm times of the year, the gelatine content must be somewhat higher than at cooler times of the year).

| | |
|---|---|
| Demineralized water | 1000 ml |

**Preparation**

1000 ml demineralized water is poured over the above-mentioned quantities of meat extract, peptone, sodium chloride and gelatine in a 2-litre Erlenmeyer flask. The gelatine is allowed to swell for 1 hour at about 25 °C. It is then dissolved on a hot-water bath at about 50 °C. Excessive heating before the pH is set should be avoided, since gelatine reacts as a rule in an acid manner, so that protein coagulates at high temperatures. After the gelatine has dissolved, the culture medium is set to a pH of 7.2. To clear the solution, the whites of 2 hens' eggs, which have been beaten to a foam, are added.

After mixing well, the solution is heated in a steam bath to 100 °C for 30 - 45 minutes, and the coagulated egg white sinks to the bottom. The clear culture medium is poured through a previously moistened folded filter in a hot-water filter funnel or in a steam bath. The first amounts of the filtrate passing through are refiltered again and this is repeated until the media comes out clear. Nutrient gelatine should be completely clear and possess a yellow colour. 10 ml of the filtered culture medium is put into each test tube. Each test tube is closed with a plug of cotton wool or cellulose or with a metal cap and is then sterilized in a steam bath (3 times 20 minutes with intervals of 24 hours between each sterilizing process).

In the case of nutrient gelatine, too long a period of heating must be avoided since otherwise the gelatine will lose its ability to solidify.

**Nutrient agar:** (Recipe 2)

**Composition**

| | |
|---|---|
| Meat extract | 10 g |
| Peptone | 10 g |
| Sodium chloride (NaCl) | 5 g |
| Agar-agar | 30 g |
| Demineralized water | 1000 ml |

**Preparation**

1000 ml of demineralized water is poured over the above-mentioned quantities of meat extract, peptone, sodium chloride and agar-agar in a 2-litre Erlenmeyer flask. The mixture is allowed to swell at about 25 °C and is then dissolved by heating in a steam bath. The pH is set to between 7.2 and 7.5 by the careful addition of soda solution or sodium hydroxide (NaOH) to the hot and liquid culture medium. If the culture medium is turbid, it is cleared with egg white using the same method as for nutrient gelatine. 10 ml of the liquid culture medium is then put into each test tube and these are closed with cotton wool or cellulose plugs or with metal caps. Sterilization is carried out either in an autoclave for 15 minutes at 121 °C or by sterilization in a steam bath (3 times 30 minutes with intervals of 24 hours between each sterilizing process).

Nutrient agar melts at 100 °C and solidifies at 45 °C. Thus nutrient agar can also be used for the cultivation of thermotolerant or thermophilic microoganisms.

Special Forms

a) Surfactant-agar (to prevent spreading in the case of proteus bacteria)

10 ml of 10 % aqueous anionic surfactant solution is added to 1000 ml nutrient agar. After sterilization, the agar is poured out into Petri dishes as soon as it has cooled down to 44 °C.

b1) Blood agar (for the examination of haemolytic properties)

10 ml of sterile, defibrinated sheep's blood (or another type of blood) is added to 1000 ml culture medium after the latter has been sterilized and allowed to cool down to 46 °C. The culture medium is then poured out into plates.

b2) For the cultivation of enterococci, the culture medium can be made selective as a blood-azide agar by adding 0.3 g/l sodium azide.

c) Nutrient broth

The composition of nutrient broth is the same as for nutrient agar except that it does not contain agar-agar. Once again, the pH is set to between 7.2 and 7.5. Filtration is necessary if there is turbidity. Sterilization is carried out in an autoclave (15 minutes at 121 °C).

**Lactose-peptone broth:** (Recipe 3)

**Composition**

| | |
|---|---|
| Peptone | 20 g |
| Sodium chloride (NaCl) | 10 g |
| Lactose | 20 g |
| Demineralized water | 1000 ml |
| Bromocresol purple indicator (Stock solution) | (dissolve 1 g bromocresol purple in 100 ml demineralized water) |

2 ml of the above is used for Recipe 3.

**Preparation**

The given quantities of peptone and sodium chloride are dissolved in 1000 ml demineralized water by heating in a steam bath. After a dwell time in the steam bath of about 1 hour, the prescribed quantity of lactose is added and the mixture is heated for a further 20 minutes. The pH is set to 7 by the addition of soda solution or sodium hydroxide solution and 2 ml of the above-mentioned bromocresol purple indicator solution is added. Quantities of 100 ml or 10 ml of this solution, which is designated **double-strength,** is poured into the culture vessels for determining the coli titre in accordance with the liquid enrichment method and is then sterilized in an autoclave for 20 minutes at 121 °C.

To prepare the **single-strength** lactose-peptone broth, the nutrient broth described above is diluted with the same volume of demineralized water prior to the addition of the bromocresol purple indicator. The pH is then set and the indicator solution added. Test-tubes are then each filled with 10 ml of this solution; a Durham tube is added to each test tube and the test tubes are then sterilized in an autoclave for 30 minutes at 121 °C.

Durham tubes are approx. 4 - 5 cm long, similar to test tubes, with a diameter of 6 - 8 mm. One Durham tube is put into each test tube with its opening pointing downwards. During sterilization, the air in the Durham test tube escapes so that it is completely filled with liquid after the sterilization process. If the inoculated solution forms gas later during incubation, the gas collects in the Durham test tube.

Endoagar (Lactose-fuchsin-sulphite agar): (Recipe 4)

Composition

| | |
|---|---|
| Nutrient agar | 1000 ml |
| Lactose | 15 g |
| Concentrated, alcoholic fuchsin solution | 5 ml (dissolve 10 g diamond fuchsin in 90 ml ethanol) |
| 10 % sodium sulphite solution | approx. 25 ml (dissolve 10 g $Na_2SO_3 \cdot 7\ H_2O$ sodium sulphite in 90 ml demineralized water) |

Preparation

The above-mentioned quantities of lactose and fuchsin solution are added to 1000 ml nutrient agar (Recipe 2), which is in a 2000 ml Erlenmeyer flask and which has been liquefied by heating on a steam bath. The mixture is thoroughly mixed and the culture medium acquires an intense red colour. The culture medium is then decolorized by the addition of sodium sulphite solution. The sodium sulphite solution must be added very carefully, until the hot culture medium is still a weak pink colour (in general about 25 ml sodium sulphite solution will be needed for this). The culture medium will then be almost colourless when cold. This fact should be tested by pouring part of the culture medium into a test tube and allowing it to solidify by cooling it under a stream of water.

The endoagár prepared in this way contains 3 % agar and is suitable for streaking subcultures. The culture medium is sensitive to light and must be kept cool and in the dark.

For starting endoagar cultures with membrane filters, a lower level of agar-agar is desirable in order to allow better diffusion of the nutrient and indicator. Thus, for the preparation of endoagar for membrane-filter cultures, one starts from a nutrient agar that contains only 1 % agar-agar (10 g in 1 litre). Thus an appropriate nutrient agar has to prepared before this special endoagar is prepared. The endoagar with 1 % agar-agar is not suitable for streaking subcultures, because this culture medium is very soft so that the surface of the solidified culture medium is easily damaged when touched by the platinum loop or needle.

The endoagar culture medium is not poured into test tubes. Instead it is sterilized in stages (3 times 20 minutes with intervals of 24 hours between each stage) in an Erlenmeyer flask and is then poured directly from this flask into sterile Petri dishes. The pouring out operation must be carried out in a room that can be blacked out, so that the culture medium can be allowed to solidify in the Petri dishes in the dark. If such a room is not available, the pour plates must be covered with a material through which light cannot pass (e.g. several layers of cellulose) while the pour plates are solidifying. The endoagar is kept cool and in the dark until it is needed. Prior to use, the pour plates are predried at 37 °C in the incubator. For this, both parts of the Petri dish (lower part with culture medium and lid) are placed in the incubator at 37 °C for some 30 minutes with their insides facing down.

**Culture media of the "Colour Series"**

**Glucose-peptone broth:** (Recipe 5)

This culture medium is prepared in exactly the same way as the preparation of the lactose-peptone broth (Recipe 3), the only difference being that glucose is used instead of lactose (same quantity).

Since only the "single-strength" medium is used, the double-strength medium must first be diluted with the same volume of demineralized water. 5 ml of the diluted medium is poured into test tubes. After the addition of Durham tubes, this is sterilized in an autoclave at 121 °C for 20 minutes.

**Tryptophane-tryptone broth:** (Recipe 6)

**Composition**

| | |
|---|---|
| Tryptone | 10 g |
| DL-Tryptophane | 1 g |
| Sodium chloride | 5 g |
| Demineralized water | 1000 ml |

**Preparation**

The stated quantities of the above-mentioned components are dissolved in 1000 ml demineralized water in an Erlenmeyer flask by heating in a steam bath. The pH is set to 7.2 ± 0.1 by the addition of sodium hydroxide and the broth is then filtered through a folded filter. 5 ml of the filtered culture medium is placed in test tubes which are then sterilized in an autoclave for 20 minutes at 121 °C.

**Buffered nutrient broth:** (Recipe 7)

**Composition**

| | |
|---|---|
| Peptone | 11 g |
| Glucose | 5 g |
| Dipotassium hydrogen phosphate ($K_2HPO_4$) | 5 g |
| Demineralized water | 900 ml |

**Preparation**

The components listed above are heated in a flask for 20 minutes in a steam bath, filtered, cooled down and then made up to 1000 ml with demineralized water. 10 ml quantities of the broth are poured into test tubes and then sterilized in stages (3 times 20 minutes) on a steam bath.

**Koser citrate medium:** (Recipe 8)

**Composition**

| | |
|---|---|
| Sodium chloride (NaCl) | 5 g |

| | |
|---|---|
| Magnesium sulphate ($MgSO_4 \cdot 7\ H_2O$) | 0.2 g |
| Ammonium phosphate ($(NH_4)_3PO_4$ | 1 g |
| Dipotassium hydrogen phosphate ($K_2HPO_4$) | 1 g |
| Trisodium citrate | 0.277 g |
| Demineralized water | 1000 ml |

**Preparation**

The constituents listed above are dissolved in water by warming and the pH is set to 6.8. 5 ml quantities of this solution are poured into test tubes and sterilized in an autoclave for 20 minutes at 121 °C.

**Simmons citrate agar:** (Recipe 9)

**Composition**

| | |
|---|---|
| Magnesium sulphate ($MgSO_4 \cdot 7\ H_2O$) | 0.2 g |
| Sodium ammonium hydrogen phosphate ($NaNH_4HPO_4 \cdot 4\ H_2O$) | 0.8 g |
| Ammonium dihydrogen phosphate ($NH_4H_2PO_4$) | 0.2 g |
| Trisodium citrate | 2 g |
| Sodium chloride (NaCl) | 5 g |
| Agar-agar | 15 g |
| Bromothymol blue indicator solution | 40 ml of the solution prepared as described below |
| Demineralized water | 1000 ml |

**Preparation of the indicator solution**

1 g bromothymol blue is dissolved in 25 ml of 1 m sodium hydroxide and this solution is then made up to 500 ml with demineralized water.

**Preparation of the culture medium**

The stated quantities of salts and agar-agar are heated in 1000 ml of demineralized water in a steam bath until they have dissolved completely. While it is still hot, the pH of the solution is set to 6.9 ± 0.1 and 40 ml of the indicator solution is then added. After mixing thoroughly, 6.5 ml quantities of this solution are poured into test tubes, which are sterilized in an autoclave for 20 minutes at 121 °C. Before the culture medium solidifies, the test tubes are placed at an angle to the vertical, so that there will be an angled surface in the test tube. However, about 3 cm of the test tube should still be completely filled.

**Kligler urea agar:** (Recipe 10)

**Composition**

| | |
|---|---|
| Meat extract | 3 g |
| Peptone | 20 g |
| Sodium chloride | 5 g |
| Lactose | 10 g |
| Glucose | 10 g |
| Ferrous sulphate ($FeSO_4 \cdot 7\ H_2O$) | 0.2 g |
| Sodium sulphite ($Na_2SO_3 \cdot 7\ H_2O$) | 0.8 g |
| Sodium thiosulphate ($Na_2S_2O_3 \cdot 5\ H_2O$) | 0.125 g |

| | |
|---|---|
| Urea | 20 g |
| Agar-agar | 20 g |
| Phenol red | 0.025 g |
| Demineralized water | 1000 ml |

**Preparation**

The stated quantities of meat extract, peptone, sodium chloride and agar-agar are dissolved in 1000 ml demineralized water and filtered. The stated quantities of lactose, glucose, ferrous sulphate, sodium sulphite and sodium thiosulphate are added to the hot filtered solution. After these components have dissolved, the pH is set to 7.4. Then the stated quantities of phenol red and urea are added and dissolved. 6.5 ml quantities of the culture medium are put into test tubes. These are then sterilized in stages (3 times 20 minutes with intervals of 24 hours between each sterilization process). After the third sterilization process, the test tubes are placed at an angle as already described for Simmons citrate agar (Recipe 9).

**Mannitol-neutral red broth (after Bulin):** (Recipe 11)

**Composition**

| | |
|---|---|
| Meat extract | 10 g |
| Peptone | 25 g |
| Sodium chloride (NaCl) | 15 g |
| Mannitol | 30 g |
| 0.1 % neutral red solution | 20 ml (0.1 g neutral red dissolved in 100 ml demineralized water) |
| Demineralized water | 1000 ml |

**Preparation**

The stated quantities of meat extract, peptone and common salt are dissolved in 350 ml demineralized water, warming gently, and filtered. The filtrate is then made up to 1000 ml with demineralized water. The stated quantity of mannitol is then added to this solution and dissolved by vigorous shaking. The pH is then set to 7.0. 20 ml of the 0.1 % neutral red solution is now added. After mixing thoroughly, 10 ml quantities of the solution are put into test tubes. A Durham tube is placed in each test tube. The test tubes are then sterilized in stages (3 times 20 minutes with 24 hours between each sterilizing process).

**Selective culture mediums for the detection of Enterococci**

**Burkwell Hartmann culture medium:** (Recipe 12 a)

**Composition**

| | |
|---|---|
| Soya peptone | 5 g |
| Yeast extract | 5 g |
| Trypticase | 15 g |
| Potassium dihydrogen phosphate ($KH_2PO_4$) | 4 g |
| Sodium azide ($NaN_3$) | 0.4 g |
| Triphenyl tetrazolium chloride (TTC) | 0.1 g |

| | |
|---|---|
| Tween 80 | 0.5 g Tween 80 = polyoxyethylene sorbitane-mono-oleate |
| Sodium carbonate | 2.0 g |
| Agar-agar | 15 g |
| Demineralized water | 1000 ml |

**Preparation**

The stated quantities of soya peptone, yeast extract, trypticase, potassium dihydrogen phosphate, Tween 80 and agar-agar are dissolved in the demineralized water by heating. The stated quantity of sodium carbonate is then added. As a result of the buffer effect of the sodium carbonate, it is not necessary to set the pH. The solution is sterilized by being heated in an autoclave for 20 minutes at 121 °C. The culture medium is then cooled down to 46 °C. Then 10 ml of a 1 % sterile-filtered aqueous solution of TTC and 4 ml of 10 % sterile-filtered aqueous solution of sodium azide solution are added. After mixing well, the culture medium is poured out into sterile Petri dishes.

**Slanetz + Bartley culture medium:** (Recipe 12 b)

**Composition**

| | |
|---|---|
| Tryptose | 20 g |
| Yeast extract | 5 g |
| Glucose | 4 g |
| Disodium hydrogen phosphate ($Na_2HPO_4 \cdot 2\ H_2O$) | 4 g |
| Sodium azide ($NaN_3$) | 0.4 g |
| Triphenyl tetrazolium chloride (TTC) | 0.1 g |
| Agar-agar | 10 g |
| Demineralized water | 1000 ml |

**Preparation**

The specified quantities with the exception of sodium azide and TTC are dissolved in the stated quantity of water by careful heating to boiling point. Then, after cooling to 46 °C, 10 ml of 1 % sterile-filtered aqueous solution of TTC and 4 ml of 10 % sterile-filtered aqueous sodium azide solution are added. No further sterilization is carried out. The culture media are poured immediately into sterile Petri dishes and allowed to solidify on a horizontal base. This culture medium is particularly suitable for water analyses using the membrane-filter method. However, it can only be kept for a limited time.

**Azide dextrose broth:** (Recipe 12 c)

**Composition (double-strength)**

| | |
|---|---|
| Tryptone | 30 g |
| Meat extract | 10 g |
| Dextrose (glucose) | 15 g |
| Sodium chloride (NaCl) | 15 g |
| Sodium azide ($NaN_3$) | 0.4 g |
| Demineralized water | 1000 ml |

**Preparation**

The stated components with the exception of sodium azide are dissolved in the demineralized water by heating. After being allowed to cool down, the solution is filtered through a folded filter and 4 ml of 10 % sterile-filtered aqueous solution of sodium azide is added. The pH is set to 7.2. The necessary quantities are poured into the appropriate culture vessels and sterilization is carried out in an autoclave for 15 minutes at 121 °C.

A single-strength solution is obtained by diluting the double-strength solution with an equal quantity of demineralized water prior to the sterilization stage.

**Aesculin bile broth:** (Recipe 13)

**Composition**

| | |
|---|---|
| Peptone | 5 g |
| Meat extract | 3 g |
| Ox bile | 40 g |
| Aesculin | 1 g |
| Demineralized water | 1000 ml |

**Preparation**

The stated quantities of the components are dissolved in the demineralized water and the pH is set to 6.6. 5 ml quantities of this solution are then poured into test tubes which are then sterilized in an autoclave for 15 minutes at 121 °C.

The growth of enterococci can be optimized by the addition of 3 % - 5 % of horse serum.

**Special culture medium for the detection of Clostridium perfringens**

**Sulphite polymixine sulfadiazine agar (Angelotti SPS agar):** (Recipe 14)

**Composition**

| | | |
|---|---|---|
| Tryptone | 15 | g |
| Yeast extract | 10 | g |
| Ferrous citrate | 0.5 | g |
| Sodium sulphite ($Na_2SO_3$) | 0.5 | g |
| Sodium thioglycolate | 0.1 | g |
| Tween 80 | 0.05 | g |
| Sodium sulfadiazine | 0.12 | g |
| Polymyxine-B sulphate | 0.01 | g |
| Agar-agar | 15 | g |
| Demineralized water | 1000 | ml |

**Preparation**

The stated quantities of the components listed above are suspended in 1000 ml demineralized water and dissolved by heating to boiling point. The pH is then set to 7.0 and the culture medium is put into small bottles for storage. These are then sterilized in an autoclave for 15 minutes at 121 °C. For use, the agar in the storage bottles is dissolved and then poured into sterile Petri dishes. Incubation is carried out under anaerobic conditions.

**Selective agar for the detection of Pseudomonas aeruginosa**

**Malachite green broth (double-strength):** (Recipe 15 a)

**Composition**

| | | |
|---|---|---|
| Peptone | 15 g | |
| Meat extract | 9 g | |
| Malachite green solution | 4 ml | (0.75 g malachite green dissolved in 100 ml demineralized water) |
| Demineralized water | 1000 ml | |

**Preparation**

Peptone and meat extract are dissolved in water by heating. The pH is then set to between 7.3 and 7.4 and the malachite green is added.

100 ml quantities of this solution are put into 250 ml Erlenmeyer flasks or 250 ml quantities are put into 750 ml Erlenmeyer flasks. The flasks are plugged with cotton wool or cellulose and are then sterilized in an autoclave for 20 minutes at 121 °C.

**Malchite green agar:** (Recipe 15 b)

**Composition**

| | | |
|---|---|---|
| Peptone | 7.5 g | |
| Meat extract | 4.5 g | |
| Malachite green solution | 4 ml | (composition as for Recipe 15 a) |
| Agar-agar | 15 g | |
| Demineralized water | 1000 ml | |

**Preparation**

The peptone, meat extract and agar-agar are allowed to swell for 15 minutes and are then dissolved by warming to 20 °C - 25 °C. The pH is the set to 7.3 - 7.4 and the malachite green solution is added. The solution is then sterilized in an autoclave for 15 minutes at 121 °C. After being allowed to cool down to 46 °C, the solution is poured out into sterile Petri dishes.

**Cetrimide agar:** (Recipe 16)

**Composition**

| | |
|---|---|
| Peptone | 20 g |
| Magnesium chloride ($MgCl_2$) | 1.4 g |
| Potassium sulphate ($K_2SO_4$) | 10 g |
| N-cetyl-N,N,N-trimethyl-ammonium bromide | 0.5 g |
| Agar-agar | 13.6 g |
| Glycerine, bidistilled | 10 ml |
| Demineralized water | 1000 ml |

Preparation

The stated quantities of the components with the exception of the glycerine are placed in a glass flask with 1000 ml demineralized water and distributed by vigorous shaking. The components are allowed to swell for between 15 and 30 minutes at 20 °C - 25 °C, during which time the nutrient mixture is allowed to stand. 10 ml of bidistilled glycerine is then added to the solution, which is heated to boiling with frequent swirling. The pH is then set to between 7.3 and 7.4. The selective agar is either put into test tubes (10 ml in each test tube) which are then sterilized in an autoclave for 15 minutes at 121 °C, or the entire quantity of the culture medium is sterilized in the flask and then the culture medium is poured out into sterile Petri dishes before it solidifies.

**Medium King A for the detection of pyocyanine:** (Recipe 17 a)

Composition

| | | |
|---|---|---|
| Peptone | 20 | g |
| Glycerine, bidistilled | 10 | ml |
| Potassium sulphate, anhydrous ($K_2SO_4$) | 10 | g |
| Magnesium chloride, anhydrous ($MgCl_2$) | 1.4 | g |
| Agar-agar | 15 | g |
| Demineralized water | 1000 | ml |

Preparation

The stated quantities of the components are dissolved in the water by heating in a steam bath. The pH is then set to 7.2. The culture medium is either sterilized in stages (heating three times for 20 minutes in a steam bath with intervals of 24 hours between each heating process) or in an autoclave for 20 minutes at 121 °C. The culture medium is then poured out into Petri dishes.

**Medium King B for the detection of fluorescein:** (Recipe 17 b)

Composition

| | | |
|---|---|---|
| Peptone | 20 | g |
| Glycerine, bidistilled | 10 | ml |
| Dipotassium hydrogen phosphate, anhydrous ($K_2HPO_4$) | 1.5 | g |
| Magnesium sulphate ($MgSO_4 \cdot 7\ H_2O$) | 1.5 | g |
| Yeast extract | 10 | g |
| Saccharose | 50 | g |
| Agar-agar | 15 | g |
| Demineralized water | 1000 | ml |

Preparation

The stated quantities of the components listed above are dissolved in the demineralized water by heating in a steam bath. The pH is then set to 7.2. The culture medium is then either sterilized in stages (heating three times for 20 minutes in a steam bath with an interval of 24 hours between each heating process), or sterilized by autoclaving for 20 minutes at 121 °C. The culture medium is then poured into sterile Petri dishes.

**Acetamide culture broth:** (Recipe 18)

**Composition**

| | |
|---|---|
| Solution A: | |
| Dipotassium hydrogen phosphate ($K_2HPO_4$), anhydrous | 1.0 g |
| Magnesium sulphate ($MgSO_4 \cdot 7\ H_2O$) | 0.2 g |
| Acetamide ($CH_3CONH_2$) | 2.0 g |
| Sodium chloride (NaCl) | 0.2 g |
| Demineralized water | 900 ml |
| Solution B: | |
| Sodium molybdate ($Na_2MoO_4 \cdot 2\ H_2O$) | 0.5 g |
| Ferrous sulphate ($FeSO_4 \cdot 7\ H_2O$) | 0.05 g |
| Demineralized water | 100 ml |

**Preparation**

The pH of solution A is set to 7.0. 1 ml of solution B is added to solution A, which is then made up to 1000 ml with demineralized water. 5 ml quantities of the broth are put into test tubes. Sterilization is then carried out in an autoclave for 20 minutes at 121 °C.

**Culture media for the cultivation of iron bacteria**

**a) Lieske nutrient fluid:** (Recipe 19)

**Composition**

| | |
|---|---|
| Ammonium sulphate $(NH_4)_2SO_4$ | 1.5 g |
| Potassium chloride (KCl) | 0.05 g |
| Magnesium sulphate ($MgSO_4 \cdot 7\ H_2O$) | 0.05 g |
| Dipotassium hydrogen phosphate ($K_2HPO_4$) | 0.05 g |
| Calcium nitrate ($Ca(NO_3)_2 \cdot 4\ H_2O$) | 0.01 g |
| Demineralized water | 1000 ml |

Iron fillings or freshly precipitated ferrous sulphide (FeS)

**Vitamin mixture:**

| | |
|---|---|
| Aneurin ($B_1$) | 0.2 µg |
| Riboflavin | 0.2 µg |
| Calcium salt of pantothenic acid | 0.2 µg |
| Pyridoxin | 0.2 µg |
| p-aminobenzoic acid | 0.6 µg |
| Biotin | 0.002 µg |
| Nicotinic acid | 0.04 µg |
| Folic acid | 0.2 µg |
| Vitamin $B_{12}$ | traces |

**Trace-element mixture**

| | |
|---|---|
| Zinc sulphate ($ZnSO_4 \cdot 7\ H_2O$) | 4.4 mg |
| Calcium chloride ($CaCl_2 \cdot 6\ H_2O$) | 1.0 mg |
| Molybdenum trioxide ($MoO_3$) | 3.0 mg |

| | |
|---|---|
| Barium carbonate ($BaCO_3$) | 15.0 mg |
| Cobalt nitrate ($Co(NO_3)_2 \cdot 6\ H_2O$) | 2.5 mg |
| Boric acid ($H_3BO_3$) | 56.0 mg |
| Potassium iodide (KI) | 1.3 mg |
| Aluminium sulphate ($Al_2(SO_4)_3 \cdot 6\ H_2O$) | 250 mg |
| Potassium silicate ($K_4SiO_4$) | traces |

The stated quantities of the above salts are dissolved in 1000 ml of demineralized water.

**Preparation**

The stated quantities of the components of the culture medium are first of all dissolved in the demineralized water without the addition of the vitamin and trace element mixtures. The medium is then poured into a 100 ml flask and sterilized in an autoclave for 15 minutes at 121 °C. Before inoculation, the vitamin mixture and 1 ml of the sterile-filtered trace element mixture are added, as are also 2 - 3 g of iron filings or a small quantity of freshly precipitated ferrous sulphide.

Lieske nutrient fluid is especially suitable for the cultivation of Gallionela ferruginea. The inoculated culture is incubated at 6 - 10 °C. After some days, the culture medium is examined with a magnifying glass to see whether light brown floccules have formed. The evaluation is then carried out by analyzing the floccules that have formed under the microscope.

b) **Silvermann/Lundgren nutrient fluid:** (Recipe 20)

**Composition**

| | |
|---|---|
| Ammonium sulphate $(NH_4)_2SO_4$ | 3.0 g |
| Dipotassium hydrogen phosphate ($K_2HPO_4$) | 0.5 g |
| Potassium chloride (KCl) | 0.1 g |
| Magnesium sulphate ($MgSO_4 \cdot 7\ H_2O$) | 0.5 g |
| Calcium nitrate ($Ca(NO_3)_2 \cdot 4\ H_2O$) | 0.01 g |
| Ferrous sulphate ($FeSO_4 \cdot 7\ H_2O$) | 52 g |
| 0,05 m sulphuric acid | 1.0 ml |
| Demineralized water | 1000 ml |

**Preparation**

1 ml of the 0,05 m sulphuric acid is added to 300 ml demineralized water. The stated quantity of ferrous sulphate is added to this acidified water. The other salts listed are dissolved in the remaining quantity (700 ml) of demineralized water. Then the two solutions are combined and mixed well together. 8 - 10 ml quantities of this nutrient fluid are put into test tubes, which are then sterilized in an autoclave for 10 minutes at 121 °C. The Silvermann/Lundgren nutrient fluid is especially suitable for the cultivation of Ferrobacillus ferro-oxidans. The growth of the bacteria can be intensified if the culture which has been started is aerated during the incubation period.

c) **Special culture medium according to Winogradsky:** (Recipe 21)

**Composition**

| | |
|---|---|
| Ammonium nitrate ($NH_4NO_3$) | 0.5 g |

| | |
|---|---|
| Sodium nitrate ($NaNO_3$) | 0.5 g |
| Dipotassium hydrogen phosphate ($K_2HPO_4$) | 0.5 g |
| Magnesium sulphate ($MgSO_4 \cdot 7\ H_2O$) | 0.5 g |
| Calcium chloride ($CaCl_2 \cdot 6\ H_2O$) | 0.2 g |
| Ferric ammonium citrate (commercial quality) | 10 g |
| Agar-agar | 15 - 20 g |
| Demineralized water | 1000 ml |

**Preparation**

The stated quantities of the components are dissolved in 1000 ml of demineralized water by heating. The solution is filtered while hot. 8 - 10 ml quantities of the culture medium are poured into test tubes, which are then closed with cotton wool or cellulose plugs. The test tubes are then sterilized in an autoclave for 15 minutes at 121 °C.

For the cultivation of iron bacteria, the culture medium is liquefied in the normal way by heating and is then poured out into sterile Petri dishes. This recipe can also be used to give a liquid culture medium instead of a solid culture medium by leaving out the agar.

**Special culture media for the detection of sulphate-reducing bacteria**

a) **Culture medium according to Sokolova/Sorokin:** (Recipe 22)

**Composition**

| | |
|---|---|
| Dipotassium hydrogen phosphate ($K_2HPO_4$) | 0.5 g |
| Disodium hydrogen phosphate ($Na_2HPO_4$) | 0.3 g |
| Sodium sulphate ($Na_2SO_4 \cdot 10\ H_2O$) | 0.5 g |
| Magnesium sulphate ($MgSO_4 \cdot 7\ H_2O$) | 0.1 g |
| Ammonium sulphate $(NH_4)_2SO_4$ | 0.2 g |
| Sodium or calcium lactate | 2.0 g |
| Agar-agar | 15 g |
| Tap water | 50 ml |
| Demineralized water | 950 ml |
| Ferrous sulphate solution | 0.1 % |

**Preparation**

The tap water and demineralized water are mixed. The other components listed above are dissolved in this mixture. Sterilization is carried out in an autoclave for 15 minutes at 121 °C. Then the ferrous sulphate solution is added carefully until the culture medium turns dark. 15 ml quantities of the culture medium are poured into test tubes.

b) **Culture medium according to Fresenius:** (Recipe 23)

**Composition**

| | |
|---|---|
| Components A: | |
| Sodium lactate | 5 g |
| Ammonium chloride ($NH_4Cl$) | 1 g |
| Dipotassium hydrogen phosphate ($K_2HPO_4$) | 0.5 g |
| Ferrous sulphate ($FeSO_4 \cdot 7\ H_2O$) | 0.3 g |
| Sodium sulphite ($Na_2SO_3 \cdot 7\ H_2O$) | 1 g |
| Peptone | 5 g |

| | | |
|---|---|---|
| Demineralized water | 1000 | ml |

Components B:

| | | |
|---|---|---|
| Sodium lactate (50 % solution) | 5.4 | ml |
| Dipotassium hydrogen phosphate ($K_2HPO_4$) | 0.67 | g |
| Asparagine | 1.34 | g |
| Ferrous sulphate ($FeSO_4 \cdot 7\ H_2O$) | 0.3 | g |
| Sodium sulphite ($Na_2SO_3 \cdot 7\ H_2O$) | 0.67 | g |
| Sodium thiosulphate ($Na_2S_2O_3 \cdot 5\ H_2O$) | 0.34 | g |
| Peptone | 5 | g |
| Demineralized water | 1000 | ml |

**Preparation**

The two component groups are prepared separately, the stated quantities of the named components being dissolved in the demineralized water in each case. The two component group solutions are then each set to a pH of 7.3 to 7.5. 50 ml quantities of component group A solution are then put into 100 ml flasks with tightly sealing glass stoppers. The same procedure is repeated with the component group B solution. Both sets of flasks are then sterilized in an autoclave for 15 minutes at 121 °C.

When starting liquid cultures for the detection of sulphate-reducing bacteria, a component group A flask is first inoculated and the flask is then filled up with component group B without any air bubbles. When larger quantities of water are being analyzed, membrane filtration is carried out. The membrane-filter disk is then rolled up and put into the flask with component group A.

The culture medium can also be used as a solid culture medium if 2 % agar (20 g agar-agar to 1000 ml culture liquid) is added to the components. In addition, 2 % (20 g/l) sodium chloride (NaCl) can be added for halophilic strains.

c) **LS agar according to Baars (lactate-sulphate agar):** (Recipe 24)

**Composition**

| | | |
|---|---|---|
| Sodium lactate (50 % solution) | 7.5 | ml |
| Ammonium chloride ($NH_4Cl$) | 1 | g |
| Calcium chloride ($CaCl_2 \cdot 6\ H_2O$) | 0.1 | g |
| Dipotassium hydrogen phosphate ($K_2HPO_4$) | 0.5 | g |
| Ferrous sulphate ($FeSO_4 \cdot 7\ H_2O$) | 3 | mg |
| Magnesium sulphate ($MgSO_4 \cdot 7\ H_2O$) | 1.5 | g |
| Sodium sulphate ($Na_2SO_4 \cdot 10\ H_2O$) | 3.4 | g |
| Demineralized water | 1000 | ml |

**Preparation**

The above components are dissolved in demineralized water. The pH is set to 7.5 by the addition of 1 or 2 m NaOH. Then 50 ml or 100 ml quantities of this solution are put into small flasks and sterilized for 15 minutes at 121 °C. When used as a solid culture medium, 2 % (20 g/litre) agar-agar is added to this culture medium.

d) YLS agar according to Starkey: (Recipe 25)
(Yeast-extract-lactate-sulphate agar)

This culture medium has the same composition as the LS agar according to Baars. The only difference is that 1.0 g yeast extract is added to 1000 ml of the liquid nutrient solution. The addition is carried out after the other components have been dissolved in the demineralized water and immediately before the pH is set.

This culture medium too can be used as a solid culture medium if 2 % (20 g/litre) agar-agar is added.

e) GPS agar according to Starkey: (Recipe 26)
(Glucose-peptone-sulphate agar)

Composition

| | | |
|---|---|---|
| Glucose | 10 | g |
| Peptone, tryptically digested | 5 | g |
| Calcium chloride ($CaCl_2 \cdot 6\ H_2O$) | 0.1 | g |
| Dipotassium hydrogen phosphate ($K_2HPO_4$) | 0.5 | g |
| Magnesium sulphate ($MgSO_4 \cdot 7\ H_2O$) | 1.5 | g |
| Sodium sulphate ($Na_2SO_4 \cdot 10\ H_2O$) | 3.4 | g |
| Ferrous sulphate ($FeSO_4 \cdot 7\ H_2O$) | 3 | mg |
| Demineralized water | 1000 | ml |

Preparation

The peptone and the inorganic salts are dissolved in the demineralized water and the pH set to between 7.3 and 7.5. The stated quantity of glucose is now added. After the glucose has dissolved, 50 ml quantities of this medium are poured into 100 ml glass flasks, which are then sterilized in an autoclave for 15 minutes at 121 °C. This culture medium too can be used as a solid culture medium by the addition of 2 % (20 g/l) agar-agar.

Special culture medium for the cultivation of sulphur bacteria

a) Beggiatoa culture medium according to Scotten/Stokes: (Recipe 27)

Composition

| | | |
|---|---|---|
| Calcium chloride ($CaCl_2 \cdot 6\ H_2O$) | 0.2 | g |
| Dipotassium hydrogen phosphate ($K_2HPO_4$) | 0.5 | g |
| Magnesium sulphate ($MgSO_4 \cdot 7\ H_2O$) | 0.5 | g |
| Yeast extract | 1.0 | g |
| Meat extract | 1.0 | g |
| Peptone | 1.0 | g |
| Demineralized water | 1000 | ml |

Preparation

The individual components are dissolved in water by heating. The medium is autoclaved for 20 minutes at 121 °C. Separately, a vitamin mixture is prepared and germs are removed from it by means of membrane filtration. 0.4 ml of this vitamin mixture is added to 100 ml of the culture medium. The concentrations of the various vitamins per 1 ml of demineralized water are as follows:

| | | |
|---|---|---|
| Pantothenic acid | 25 | µg |
| Nicotinic acid | 100 | µg |
| Biotine | 0.5 | µg |
| p-aminobenzoic acid | 10 | µg |
| Riboflavin | 100 | µg |
| Thiamine | 100 | µg |
| Pyridoxin | 100 | µg |
| Inositol | 500 | µg |
| Vitamin $B_{12}$ | 100 | µg |
| Folic acid | 1 | µg |

**b) Beggiatoa culture medium according to Cataldi:** (Recipe 28)

10 g of dried hay is boiled for 10 minutes in 250 ml demineralized water. The water is then poured off, a fresh quantity is added and the hay is again boiled for 10 minutes. This procedure is repeated a third time. Then the hay is dried at 37 °C and cut into approx. 1 cm lengths. 8 g of this pretreated hay is put into a 150 ml Erlenmeyer flask with 100 ml of tap water or, better still, with the water which is to be analyzed itself. To this mixture 0.5 g sodium sulphide ($Na_2S$) is added. The culture medium prepared in this way is sterilized in an autoclave for 15 minutes at 121 °C.

**c) Nutrient broth according to Starkey:** (Recipe 29)

**Composition**

| | | |
|---|---|---|
| Ammonium sulphate $(NH_4)_2SO_4$ | 0.3 | mg/l |
| Potassium dihydrogen phosphate ($KH_2PO_4$) | 3.5 | g |
| Magnesium sulphate ($MgSO_4 \cdot 7\ H_2O$) | 0.5 | g |
| Calcium chloride ($CaCl_2 \cdot 6\ H_2O$) | 0.25 | g |
| Ferric sulphate ($Fe_2(SO_4)_3 \cdot 9\ H_2O$) | 0.01 | g |
| Sulphur, amorphous | 10 | g |
| Demineralized water | 1000 | ml |

5 g sodium thiosulphate ($Na_2S_2O_3 \cdot 5\ H_2O$) can also be used instead of the amorphous sulphur.

**Preparation**

The stated components are dissolved in the demineralized water and the pH is set to 7.5. For the cultivation of Thiobacterium thiooxidans, the pH is set to 3 with the aid of 0.5 m sulphuric acid. The culture medium is sterilized in an autoclave for 15 minutes at 121 °C.

### 5.2.10 Prepared culture media

The specialist trade also supplies dry culture media in powder form, which contain all the components which are necessary for the cultivation of micro-organisms on the particular culture medium. In accordance with the recipes which are given, a definite quantity of the powder must be weighed out. A specified quantity of demineralized water, tap water or the water which is to be analyzed is then poured over this powder. This is then generally dissolved with heating and the pH is set to the desired value. After the solution has been put into test tubes or flasks, it is sterilized either in an autoclave or by sterilization in stages in a steam bath. It is always to be recommended that the pH is set or that it is checked. Some manufacturers also offer the prepared culture medium to the trade in the form of tablets. Each tablet is related to a definite quantity of water, so that the weighing-out process is not necessary.

In addition to numerous, often small manufacturers of dry culture media, the following manufacturers are known around the world:

Difco (Difco Laboratories, Detroit, Michigan/USA)

Oxoid (Oxoid Limited, London S.E.1, England or Oxoid Deutschland GmbH, 4230 Wesel, FRG)

Merck (E. Merck AG, 6100 Darmstadt, FRG)

Nutrient cardboard disks are supplied by:

Sartorius GmbH, 3400 Göttingen, FRG
(with subsidiaries in USA, France, United Kingdom, the Netherlands, Austria).

Millipore GmbH, Hauptstraße, 6236 Eschborn, FRG

### 5.2.11 Detection reagents for biochemical reactions

**Indole reagent (according to Kovacs):** (Recipe 30)

**Composition**

| | |
|---|---|
| p-dimethyl aminobenzaldehyde | 5 g |
| Amyl alcohol | 75 ml |
| Hydrochloric acid (HCl), (1.18 g/ml) | 25 ml |

**Preparation**

The p-dimethyl aminobenzaldehyde is dissolved in the amyl alcohol by warming in a water bath for 5 minutes at 60 °C. After the solution has been cooled down, the specified quantity of hydrochloric acid is added, stirring all the time. This gives the solution a red colour. This lasts about 6 to 7 hours and the solution then turns yellow. The reagent is ready to use when the solution has taken on the yellow shade.

**Carrying out the reaction**

5 to 10 drops of the reagent are added and mixed in by swirling the tryptophane-tryptone broth (Recipe 6), which has been inoculated and incubated for 20 ± 4 hours. If indole is present (positive reaction), a pink to red coloured ring will form on the surface of the liquid culture after 1 to 2 minutes. In the case of a negative reaction, a yellow-brown ring will form.

**Nadi reagent for the detection of cytochromoxidase:** (Recipe 31)

**Composition**

| | |
|---|---|
| Alpha-naphthol (1) | 1 g |
| Ethanol, 95 % | 100 ml |

| | |
|---|---|
| N,N-dimethyl-p-phenylene diammonium dichloride | 1 g |
| Demineralized water | 100 ml |

Preparation

The specified quantity of alpha-naphthol (1) is dissolved in the specified quantity of ethanol, the stated quantity of dimethyl-p-phenylene diammonium dichloride is dissolved in the stated quantity of demineralized water. Both solutions are light-sensitive; they must not be warmed either. It is therefore recommended that not too large quantities of the two solutions are mixed together. The reagent can be kept as a mixture for a short time (some days) in a refrigerator. The naphthol solution can no longer be used if it is coloured red, and the dimethyl p-phenylene diammonium dichloride solution cannot be used if it has taken on a violet to brown colour.

**Reagent for the methyl red reaction:** (Recipe 32)

40 mg of solid methyl red is dissolved in 100 ml of 60 % ethanol.

Carrying out the reaction

A few drops of the methyl red reagent are added to the buffered nutrient broth of the "colour series" in which growth has taken place. In a positive case, a red coloration occurs when the reagent is added, in a negative case a permanent, yellowish coloration takes place.

Gram staining

Certainly the most important method for differentiating between microorganisms is the gram-staining method. A distinction is made between gram-positive and gram-negative bacteria. When the stained preparation is examined under the microscope, gram-positive bacteria appear as cells which have been stained blue-violet, while gram-negative bacteria appear as cells which have been stained magenta red.

**Preparation of the colour solutions**

Stock solutions are prepared from the two basic aniline dyes, aniline violet and diamond fuchsine, by pouring over the dyes 9 times their weight of ethanol. The mixtures are allowed to stand for approx. 1 week at approx. 20 °C with frequent intermittent shaking. The mixtures are then filtered to remove the parts which have not dissolved. Stock solutions must be kept cool and in the dark.

The following dyestuff solutions are used for the gram-staining test:

1. Phenolic aniline violet (10 ml of aniline-violet stock solution, 1 ml of liquefied phenol, 100 ml of demineralized water)

2. Phenolic fuchsine (10 ml of fuchsine stock solution, 5 ml of liquefied phenol, 100 ml of demineralized water)

3. Lugol iodine solution (2 g potassium iodide, 1 g iodine, 300 ml demineralized water)

Carrying out of the gram staining

1 drop of the bacterial suspension to be investigated is applied to a dry, grease-free microscope slide, thinly and evenly distributed and allowed to dry in the air. The air-dried preparation is heat-fixed by heating it in the oxidizing flame of a bunsen burner. For this, the microscope slide is drawn three times slowly through the flame with the aid of a pair of pincers with the layer side upwards. After allowing the microscope slide to cool down, it is placed with the layer side upwards on a colour bench. This consists of 2 horizontal metal strips laid parallel to one another, on which the microscope slide can be placed and under which a collecting dish is placed for the quantities of dyestuff which flow off. The phenolic aniline violet dyestuff solution is now applied drop by drop to the heat-fixed bacterial suspension, the dyestuff solution being filtered through a paper filter. After allowing the dyestuff solution to act for 3 minutes, it is poured off and rinsed away with a little Lugol iodine solution. Immediately thereafter, Lugol iodine solution is applied to the microscope slide and allowed to act for 1.5 minutes. This solution too is then poured off and the preparation is then rinsed by dripping ethanol onto it until no more clouds of colour are visible (maximum 1.5 minutes). The preparation is then rinsed off with distilled water and the phenolic fuchsine colour solution is applied with filtration to give the contrast coloration. The phenolic fuchsine colour solution should be allowed to act for 30 seconds. After this, the colour solution is rinsed off with water and the preparation is dried by placing a filter paper on it. The preparation is then examined under the microscope with oil immersion. The immersion oil is applied directly to the stained preparation; it is not necessary to use a cover glass.

## 5.2.12 Work sheets for microbiological water analysis

### 5.2.12.1 Sheet 1 - Taking and handling samples

Apparatus

1. Sterilized glass bottles 250 ml, 500 ml (if required 1 litre, 2 litres, 5 litres) containing sodium thiosulphate (0.5 - 5 ml 0.1 m) for dechlorination, if the water contains chlorine.

2. Chlorine meter

3. Thermometer

4. Blow lamp or "Camping GAZ" burner for flaming taps

5. Crucible tongs and scoop device for taking samples

6. Adhesive labels, notebook, writing implements

Taking the sample

Before taking the sample, check whether the sampling point is suitable.

1. Has the water for analysis been correctly located?

2. Has the water stagnated in the pipes before sampling?

3. Are there any filters, dosing devices, storage tanks, boilers or other installations before the sampling point?

4. Are the pipes made from a material which has an oligo-dynamic effect (copper, lead, zinc), with the result that changes in bacterial content occur?

5. Are any other factors present which can have an effect on the microbiology of the water?

6. In the case of drawn water: does the construction of the well or spring-intercepting structure present any problems, and are there any special features in the catchment area?

Sampling from the tap

1. The tap is cleaned mechanically, hoses, strainers, aerators, etc. are removed.

2. The tap is briefly turned on full several times in order to rinse away any particles.

3. The tap is heated with a flame so that when turned on a hissing noise is heard.

4. The tap is allowed to run for 5 - 10 minutes with a steady stream (until the temperature is constant) and the sample is taken. The sterilized bottle is carefully opened, the sterilization strip is shaken off and the bottle filled with water (approx. 4/5 full, so that an air space is left). Secondary infections from contact with objects or hands, coughing, etc. must be avoided. Close the bottle immediately.

5. The bottle is labelled. The tap is returned its original state.

Sampling from hand pumps

1. Check the condition of the hand pump! Ensure that work has not been carried out on the pump immediately before sampling and that the pump has not been filled with water from a different source for priming purposes.

2. Pump the water for approx. 10 minutes and ensure that the pumped water does not flow back or seep away in the immediate vicinity of the well.

3. Take the sample in the same fashion as described for taps.

Scoop samples

Scoop samples are taken 20 - 30 cm below the water surface either with a sampling device or with a sterilized bottle and flamed crucible tongs. Ensure that no secondary infection is passed to the water from the sampler. When sampling with crucible tongs, the mouth of the bottle is held upwards at an angle and the bottle is moved slowly through the water with the mouth pointing forwards. In flowing water, samples should be taken in midstream wherever possible, not less than 1 m from the bank and without stirring up sediment.

Sample handling

Once taken and appropriately labelled, the sample is protected from sunlight and taken to the laboratory in insulated transportation cases, and if necessary under refrigeration. It is essential to keep the sample cool if it cannot be analyzed within 3 hours. It can be stored in the refrigerator at + 4 °C, for not more than 30 hours. Long delays between sampling and analysis must be recorded in the analysis report.

Special considerations

A) When testing for suspected pathogenic germs, samples from taps and pumps should be taken once before letting the water run and a second time after it has run for 10 minutes. At least 5 litres water must be taken.

B) If water is taken for chemical analyses at the same time, microbiological samples are taken first in the case of scoop sampling, but last in the case of sampling from taps and pumps.

5.2.12.2 Sheet 2 - Determining the colony count

Apparatus and materials

1. Sterilized Petri dishes (plate diameter 90 mm)
2. Sterilized pipettes, 1 ml capacity (if necessary plugged with cotton wool)
3. Burner for flaming
4. 10 ml sterilized culture media in test tubes
5. Writing implement for labelling the plates
6. Membrane-filter unit complete with suction device (water-jet pump, electric pump)
7. Vacuum hose, Wulff bottle
8. Sterilized membrane filters, to fit the filter unit, pore size 0.45 µm or 0.6 µm with mesh screen
9. Forceps
10. Incubators for the required incubation temperatures (generally 20 °C - 22 °C, 28 ° - 30 °C, 37 °C), equipped with calibrated thermometers for temperature control
11. Colony counter or Wolfhügel counting plate with magnifying glass, magnification 6 - 8x
12. Laboratory journal or analysis report sheets.

Sample preparation

1. Check the label on the sample. Are there any comments from the sampler which must be taken into account for the analysis?

2. Label the Petri dishes for the analysis so that they can be uniquely related to the results when the sample is being evaluated. Other project workers will thus be able to carry out the evaluation and collation of results for the sample (label the plates with precise information!).

3. Shake the sample bottle well before opening, so that the germs are as evenly distributed throughout the sample as possible.

4. Thoroughly flame the neck and mouth of the bottle immediately after opening, taking care that the glass does not crack or break.

A) Pour-plate method

1 ml water from the sample bottle is transferred to each of at least 3 Petri dishes using a pipette. If the water has a very high microbiological content, it is first diluted to make a decimal dilution series, and at least 3 Petri dishes are prepared from each stage as described. Then 10 ml of the liquefied culture medium cooled to 45 °C is added after first flaming the mouth of the test tube. For this purpose the lid of the Petri dish is raised but not completely removed. The culture medium and water are mixed by circular movements of the dish, then the dish is stood on a horizontal surface and the layer of culture medium is left to set. The plates are later stacked not more than 10 high in the incubator.

B) Membrane-filter method

The prepared and sterilized membrane-filter unit is connected to the suction device and fitted with a sterilized membrane-filter disk. The machine is not under suction during this procedure. The coated side of the filter disk with the mesh screen faces upward. The funnel is put in place and firmly locked with the bayonet locking device. The water to be analyzed is poured into the funnel and is drawn through the filter. Approx. 5 seconds later the suction is stopped and the funnel removed. The filter disk is placed on the set culture medium in the Petri dish using forceps which have been flamed and then cooled. The filter disk must have been flamed and then cooled. The filter disk must have the mesh-screen side upward and there must be no air bubbles. The culture medium must be dried in advance and show no visible signs of moisture on the surface. On the other hand it must not have dried out.

The membrane-filter method is only suitable for water with a very low germ content and no turbidity or sediment.

Incubation

The cultures are grown in incubators pre-heated to the required temperatures - gelatine culture media at 20 - 22 °C only. The incubation times specified must be observed.

Evaluation

When the incubation time has elapsed, the magnifying glass is used to count the colonies which have grown. The Wolfhügel counting plate, or some other counting device, can be used for this purpose. If the number of colonies on the plate is over 100, they are counted on zones and multiplied by the appropriate number (area of the Petri dish = 56 $cm^2$).

Example: 8 zones of 1 $cm^2$ each counted · 7 = colony count.

When counting zones, care must be taken to check that the growth of colonies is spread evenly over the plate and that the zones counted are similarly evenly spread over the surface and so are representative.

Statement of results

The number of colonies counted is entered in the laboratory journal or analysis report, stating the culture medium used, the cultivation temperature and time, and the quantity of water analyzed. With dilution series the result has to be multiplied by the number of dilution stages.

Example: number of colonies on 1 ml (culture agar at 20 °C after 44 ± 4 hours) = 85.

### 5.2.12.3 Sheet 3 - Detection of Escherichia coli and coliform bacteria

Apparatus and materials

A) Titre method

1. Graduated pipettes, 10 ml and 1 ml
2. Sterile lactose-peptone broth in suitable vessels, double-strength for 100 ml and 10 ml water (broth I), single-strength for 1 ml and 0.1 ml water (broth II) with Durham tubes (see Recipe 3)
3. Incubator 37 °C and 44 °C

B) Membrane-filter method

1. Complete membrane-filter unit; as working sheet 2
2. Sterile membrane filters, pore size 0.6 µm
3. Endo-agar in Petri dishes (see Recipe 4)
4. Forceps and pipettes, 100 ml

C) Germ differentiation

1. Platinum loop, platinum needle, burner for flaming and annealing
2. Endo-agar for cleaning passages or for producing individual colonies (see Recipe 4)
3. Culture media of the "Multicoloured Series" or normal commercially available identification systems (see Recipe 5)
4. Detection reagents for oxidases and indole (see Recipe 30)
5. Writing implements for labelling the cultures

Analysis procedure

A) Titration method

The cultures are labelled to correspond to the sample:

| | | |
|---|---|---|
| 200 ml vessel | with | 100 ml broth I |
| 25 ml vessel | with | 10 ml broth I |
| 1 test tube | with | 10 ml broth II and Durham tubes |
| 2 test tubes | with | 10 ml broth II and Durham tubes |

After checking the labels, the appropriate amount of water (100 ml, 10 ml, 1 ml, 0.1 ml) is poured into the culture vessels from the sample bottle. All the vessels must be carefully flamed on opening, and secondary infections must be avoided.

Once filled and properly closed, the nutritive vessels are incubated at 37 °C. After 20 ± 4 hours, or at the latest after 44 ± 4 hours incubation, the results are evaluated. The findings are positive if the culture medium displays a yellow colouring, turbidity and gas formation. These cultures are smeared onto endo-agar with the platinum loop or needle for further identification. These sub-cultures are incubated for approx. 24 hours at 37 °C.

The result is negative if none of the treated culture displays any colour change, turbidity or gas formation. In the latter case an entry must be made in the laboratory ledger or analysis report: Escherichia coli and coliform bacteria cannot be detected in 100 ml.

B) Membrane-filter method

The analysis is conducted in the way described in work sheet 2. The membrane-filter disk is laid on endo-agar, taking care to avoid air bubbles, and the culture is incubated for 20 ± 4 hours at 37 °C.

After incubation, suspected colonies (red, with fuchsine lustre, usually with dark patches on the underside of the filter disk) are counted and more closely identified.

C) Identification

With suspected colonies, the oxidase reaction is tested with Nadi reagent. Colonies which react positively (blue-violet colouring after 1 - 2 minutes) are not coliform bacteria and no further identification is necessary.

Colonies which react negatively (no colour change) must be suspected of containing coliform bacteria, since Enterobacteriaceae do not form cytochrome oxidase.

Cell material from the suspect, oxidase-negative colonies is injected into the culture media of the "Coloured Series" using the platinum needle. The injected cultures are incubated at 37 °C or 44 °C and evaluated after 20 ± 4 hours.

Commercially available identification systems can be used instead of "Coloured Series" prepared in the laboratory.

Important features:

| | E. coli | Citro-bacter | Entero-bacter | Klebsiella |
|---|---|---|---|---|
| Growth at 44 °C | + | d | d | d |
| Indole formation | + | - | - | d |
| Methyl-red test | + | + | - | - |
| Voges-Proskauer reaction | - | - | + | + |
| Citrate break-down | - | + | + | + |
| $H_2S$ formation | - | + | - | - |
| Mobility | d | + | + | - |

+ = positive; - = negative; d = variable behaviour

Statement of results

With the titre method, one states the smallest amount of water in which Escherichia coli and coliform bacteria were detected, e.g.

> Escherichia coli detected in 10 ml
> coliform bacteria detected in 0.1 ml

With the membrane-filter method, the number of colonies in relation to the amount of water used is recorded, e.g.:

in 100 ml water 6 colonies of Escherichia coli
8 colonies of coliform bacteria

### 5.2.12.4 Sheet 4 - Detection of faecal streptococci Pseudomonas aeruginosa and sulphite-reducing Clostridia

Growing and identification of Pseudomonas aeruginosa and sulphite-reducing, spore-forming anaerobes (Clostridia) should only be carried out by persons with appropriate training.

Apparatus and materials

1. Complete membrane-filter unit; as work sheet 2

2. Culture media[1]

| Liquid (in suitable culture flasks) | Solid (in Petri dishes) |
|---|---|
| **Faecal streptococci**<br>Azide-dextrose broth<br>a) Double-strength<br>b) Single-strength | Tetrazolium-sodium azide agar |
| **Pseudomonas aeruginosa**<br>Malachite green medium<br>a) Double-strength<br>b) Single-strength | Cetrimide agar |
| **Clostridia** | SPS agar |

3. Pipettes, 100 ml and 50 ml, sterilized membrane filter, pore size 0.6 µm, forceps, burner for flaming, writing instruments

4. Anaerobic jar with accessories (for growing Clostridia)

5. Culture media used for identification
   Faecal streptococci:
   Aesculin bile broth, 7 % ferrous chloride solution
   Nutrient broth with pH = 9.6
   Nutrient broth with 6.5 % common salt (NaCl)

---

[1] Other suitable culture media can also be used instead of those named above.

Pseudomonas aeruginosa:
King F medium, King P medium, Acetamide culture broth

Clostridia:
Commercially available identification systems (if more specific identification is required at all)

### Analysis procedure

#### A) Qualitative analyses

1. Liquid enrichment

The double-strength culture solution suitable for the species of germ under examination is sterilized and put into culture vessels (bottles, Erlenmeyer flasks). These vessels must also be able to hold the same quantity of the substance being investigated (water from the sample flask). The required quantity of water is removed from the sample flask under sterile conditions and mixed with the culture medium in the culture vessel. It is then incubated for 20 ± 4 hours or 44 ± 4 hours at 37 °C. If turbidity occurs in the cultures owing to the growth of germs, these must be more closely identified.

2. Membrane filter enrichment method

The quantity of water to be examined (250 ml or 100 ml) is filtered through the membrane filter under sterile conditions. The filter disk is then put into 50 ml sterile, single-strength culture medium, and incubated at 37 °C. Evaluation follows after 20 ± 4 hours or 44 ± 4 hours. If turbidity occurs owing to the growth of germs, these must be more closely identified.

#### B) Quantitative analysis

The quantity of water to be examined (250 ml, 100 ml, 50 ml or 20 ml) is filtered through a membrane filter under sterile conditions. Then the filter disk is laid onto the appropriate solid special medium and the culture is incubated at 37 °C for 20 ± 4 hours or 44 ± 4 hours.

For detecting Clostridia, cultures must be incubated under anaerobic conditions in the anaerobic jar, following the manufacturer's instructions.

Suspected colonies are counted and more closely identified after the incubation time is ended.

### Identification

When carrying out liquid enrichments, the cultures with bacterial turbidity are smeared onto the appropriate special media using a platinum loop or platinum needle. The subcultures are then incubated for 20 ± 4 hours at 37 °C, and suspect colonies are further identified.

In the case of suspect colonies, cell material is removed with the platinum needle and transferred to the identification media listed in Sheet 5 under "Apparatus and Materials". The injected cultures are incubated at 37 °C and evaluated after 20 ± 4 hours.

Evaluation

Faecal streptococci

These grow with a sodium chloride content of 6.5 % and with pH = 9.6. They break down aesculin.

Pseudomonas aeruginosa

forms fluorescein and pyocyanin and produces ammonia from acetamide.

Clostridia

Under anaerobic conditions sulphite is reduced to hydrogen sulphide, which causes black colouring of the medium with iron ions as a result of FeS or $FeS_2$.

Statement of results

When carrying out liquid enrichments, it must be reported whether the species of germ looked for was present in the water examined.

Example:
Faecal streptococci are not detectable in 100 ml.

When making the quantitative analysis, the number of positive colonies is given in relation to the quantity of water examined.

Example:
27 faecal streptococci were detected in 100 ml.

### 5.2.12.5 Sheet 5 - Cleaning with sterilization

Apparatus and materials

1. Sink with hot and cold taps or apparatus washing machine.
2. Bottle brushes of various sizes, scrapers, scratchers, handbrush.
3. 2 - 3 buckets, approx. 10 litre capacity, rubbish containers, towels, plastic bowls.
4. Commercially available cleaning agents, disinfectants, hydrochloric acid.
5. Distilled or demineralized water.
6. Drying cabinet, hot-air sterilization cabinet heatable to 200 °C.
7. Steam bath; autoclave

Cleaning

New glass apparatus is treated with hydrochloric acid and rinsed twice, first with alkaline cleaning agents, then with demineralized water.

Used glass apparatus is first heat-sterilized or left overnight in a disinfectant solution. The apparatus is then mechanically cleaned, rinsed off with tap water and treated with acid and alkaline cleaning agents.

It is important to carry out adequate final rinsing with demineralized water in order to remove any remaining traces of cleaning agents which would inhibit growth.

## Drying

After the cleaned apparatus has drained, it is dried for 1 hour in the drying cabinet at 100 °C. Vessels with covers and bottles with glass stoppers are left open. Care must be taken to ensure that parts which belong together e.g. bottles and stoppers (unless all of standard size) are not confused, so that they fit together and are germ-tight.

The ventilation openings of the drying cabinet are left open during drying so that the water vapour which is formed can escape.

At the end of the drying time, the heater is switched off and the door opened to produce a rapid exchange of hot, damp atmosphere in the cabinet. This prevents condensation moisture on the dried apparatus.

After cooling to approx. 40 °C, the dried apparatus can be removed from the cabinet.

## Sterilization

### A) Sterilization in the hot-air cabinet

The dried apparatus is sterilized in the hot-air cabinet for 2 hours at 180 °C. Flasks, test tubes, etc. are first closed with cotton wool or cellulose plugs or caps, and for bottles with glass stoppers the sterilizing strip is inserted and the stopper is loosely fitted. Pipettes are sterilized in sterilizing boxes made of tin. The ventilation holes must be left open during sterilization.

When sterilization is complete, the cabinet heater is switched off and the door left closed. When the interior temperature has fallen to 60 °C, the sterile apparatus is taken out, stoppers are pushed in and the ventilation holes of the sterilizing boxes are closed.

### B) Sterilization in the steam bath

Sterilization takes place in freely flowing steam which is not under pressure. The steam bath must be used in accordance with the manufacturer's instructions. Care must be taken to ensure that the steam bath always contains a sufficient quantity of water. If water with a significant calcium content is used, the steam bath must be delimed at regular intervals.

The steam bath is usually used for sterilizing sensitive culture media containing sugar by means of so-called "fractionated sterilization". The substance to be sterilized is heated to 100 °C for 30 minutes on each of 3 consecutive days. It is kept at room temperature for 24 hours between the sterilizations.

Agar and gelatine media are also filtered in the steam bath to prevent the medium setting during filtration.

C) Sterilization in the hot-steam sterilizer (Autoclave)

The autoclave must always be used in accordance with the manufacturer's operating instructions. Demineralized water is recommended, and care must be taken to ensure that the machine always contains sufficient water. Sterilization is usually at 121 °C (1 bar gauge), and lasts for 20 - 30 minutes. The autoclave must be correctly ventilated during heating, and this can best be done using the method described in the operating instructions. After sterilization, the pressure inside the autoclave must be carefully reduced. If the pressure is released too rapidly, the vessels may shatter or their tops (stoppers, caps) may fly off.

The autoclave must only be opened once the excess pressure has been released completely and the pressure inside the autoclave has been equalized with the air pressure outside.

Checking the sterilizers

As with all technical equipment, malfunctions can occur with sterilizers. These must be immediately recognized and remedied. Temperature control is the most important precaution, and every sterilizer must be equipped with a thermometer in an easily visible position. It is recommended (especially with autoclaves) that the temperature actually reached inside the sterilizer be checked at least once a day, by putting a maximum thermometer in the machine during operation.

Sterilizers should be biologically tested at least every 6 months, or preferably quarterly. For this purpose use packets of spores (Bac. subtilis, Bac. stearothermophilus) or spore strips available from specialist dealers.

## 5.3 Biological toxicity tests

### 5.3.1 Bacterial inhibiting tests

Minimum requirements and limits which often have no toxicological basis are used for evaluating the harmful effects of chemical substances on the biocenosis of water and sewage, and in considering unfavourable after-effects. However, even with adequate knowledge of the effects of individual substances on microorganisms, it is not always possible to apply laboratory results to the complex conditions which exist in areas of water or in sewage treatment plants. Analyses of toxic substances and inhibitors in sewage have been concerned primarily with establishing upper limits of toxic effects on selected microorganisms such as pseudomonads and blue algae. For this purpose the toxicity of substances is often determined individually, which means that synergistic effects (which have an additive and multiplicative effect) have not been investigated. The findings therefore often have little bearing on the practical assessment of sewage. The following factors play a part in determining toxicity:

pH, incubation temperature, concentration of the active substance, quantity of the substrate added, type of substrate, duration of exposure, adsorption characteristics of the active substance and the inoculated material. Whereas pH, incubation temperature, concentration of the active substance and quantity of the substrate added can be laid down by convention, the choice of type of substrate gives rise to difficulties. If sewage is used partly or exclusively as the substrate, the measurement results will vary according to the source of sewage. Using a single defined substrate, such as glucose or peptone, does not solve the problem because the development of the biocenosis can be significantly different from that of the sewage. Besides, with certain active substances the toxicity measurements are also affected by the amount of substrate. The choice of duration of exposure also makes a great difference to the inhibiting effects, since time-dependent changes in the biocenosis often provoke non-linear inhibiting effects.

Adsorption and the inoculated material are two factors which have a great influence on the toxic effects of substances. If activated sludge is used as the inoculum in large quantities, the question arises whether toxicity can be correctly determined at all using this method. Depending on the condition of the sludge (e.g. age or pollution), the adsorptive effects may be modified to the extent that it is scarcely possible to achieve standardized conditions.

### 5.3.1.1 Respirometric measurements

A frequently used parameter for determining the toxicity of household and industrial sewage and of chemicals which affect the environment, is the change in rate of gas metabolism by microorganisms. Two basic variants are possible:

- Measurement of the respiration rate of a sludge-water suspension in a system (duration of measurement e.g. 15 minutes). With this method adaptations of the mixed bacteria population used are not taken into account.

- Measurement of the respiration rate in the closed system over a longer period of time (e.g. 5 days) and using small quantities of sewage as the inoculum.

This method is usually used as a direct method of establishing the biochemical oxygen demand (BOD). In spite of all reservations, the first method is the only practicable one for protecting sewage-treatment plants from sudden increases in load, since it provides information rapidly. If toxicity measurements are taken on the basis that they are largely unaffected by adsorption, changes in the test biocenosis with time must be taken into account. In this case incubation times of 5 days or more may be necessary to permit summarization of the kinetics of reactions which depend on decay products of biochemical decomposition. Normal BOD analysis deals only with metabolism which is influenced by substrate, but respirometric measurements of toxicity also take into account the respiration of endogenous bacteria and the respiration of protozoa.

The oxygen demand of microorganisms is reduced by the presence of toxic or inhibiting substances. The reduction serves as a means of measuring inhibiting effects, and for this purpose reference cultures must be prepared concurrently to which the substance under examination is not added.

All respirometric measurements can be made with simple, barometric BOD measuring instruments.

## Equipment

Apparatus for respirometric measurements

Beakers

Pipettes

Dilution water:
1 ml samples of the solutions A, B, C and D are each diluted to 1000 ml immediately before use.

A: 8.5 g $KH_2PO_4$, 21.75 g $K_2HPO_4$, 33.4 g $Na_2HPO_4 \cdot 2\ H_2O$, 2.5 g $NH_4CL$ - dissolved in 1000 ml demineralized water; pH should be 7.2.

B: 22.5 g $MgSO_4 \cdot 7\ H_2O$ - dissolved in 1000 ml demineralized water.

C: 27.5 g $CaCl_2 \cdot 2\ H_2O$ - dissolved in 1000 ml demineralized water.

D: 0.25 g $FeCl_2 \cdot 6\ H_2O$ - dissolved in 1000 ml demineralized water.

Inoculation solution:
Discharge from a domestic-sewage treatment plant; addition of 10 drops to the trial culture.

1 % Peptone solution:
Dissolved in demineralized water.

Sewage:
Coarsely filtered sewage is used from the pre-sedimentation discharge of the relevant sewage-treatment plant.

## Method

A known quantity of the substrate (sewage, peptone, sewage + peptone) is taken. The volume is adjusted by the addition of dilution water so that the range of measurement is fully utilized. A 500 ml sample bottle is filled. The pH is checked and, if necessary, adjusted to pH 7 using 0.5 m $H_2SO_4$ and 1 m NaOH. 4 drops of 45 % potassium hydroxide solution are put into the plastic insert. The sealing surfaces are smeared with high-viscosity silicone grease and then screwed up. Readings are taken 4 to 5 times daily and the experiment lasts 5 or more days. No nitrification inhibitors are added.

## Evaluation

Graphs are drawn from the results obtained; the area underneath the BOD curve is proportional to the amount of oxygen. The area representing a preparation of the sewage under examination plus peptone additive is compared with the area for a pure peptone solution. In this way any inhibition of respiration can be recognized. The comparison is usually carried out using a basic computer program, the individual measurement points being joined to make a single curve which enables the area below the curve to be numerically integrated. The following diagram makes this clear. The individual curves for the decomposition of sewage and peptone are depicted as well as the cumulative curve for the sewage plus peptone. The difference between the curve which is theoretically to be expected and the actual BOD decomposition curve is given by the area of broken lines representing

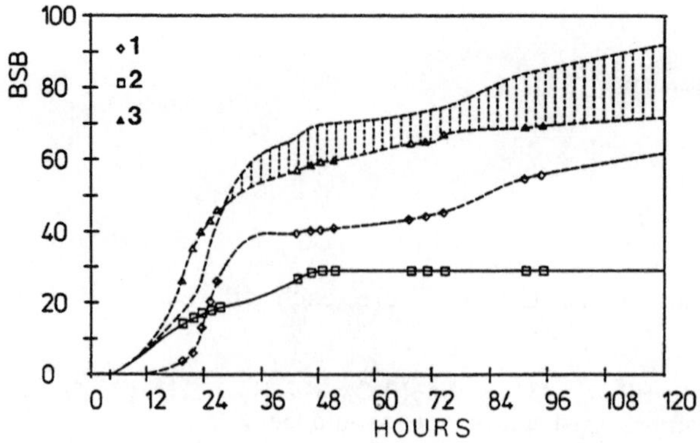


Fig. 167. BOD curves for various substrates: 1) = Peptone; 2) = Sewage; 3) = Sewage + peptone

inhibiting effects. This area can be integrated and used as a measure of toxic inhibition.

### 5.3.1.2 Test using Pseudomonas fluorescens

Dissolved toxic substances contained in sewage inhibit the ability of the bacterium Pseudomonas fluorescens to produce organic acids from glucose.

Inhibition of acid production can be quantitatively determined by measuring the pH. For this purpose the pH and acid consumption of the sewage must be adjusted to match the corresponding values for the receiving water before testing begins.

The ratio of volume of sewage to total volume of receiving water plus sewage must then be found for the level of dilution which produces the smallest deviation from the pH of the receiving water sample by the end of the test period. This ratio provides a measure of the biological damage caused by a particular toxic sewage.

The method can be used with all types of industrial effluent.

Equipment

Incubator, steam bath, vibrator, pressure-filtering device for membrane filters, membrane-filters pore size 0.2 µm (1 µm = 0.001 mm), photometer with nephelometer attachment, filter Hg 436 nm, cuvettes (path length 10 mm), electric pH gauge, culture tubes with caps, Erlenmeyer flasks (300 ml), graduated pipettes (1 ml), graduated pipettes (10 ml), measuring cylinders (100 ml), glass beads (diameter approx. 2 mm), inoculation loop, aqueous sodium hydroxide, hydrochloric acid.

Culture media

Medium for parent and preliminary cultures:

1.060 g Sodium nitrate, $NaNO_3$, reagent purity

| | | |
|---|---|---|
| 0.600 | g | Anhydrous di-potassium hydrogen phosphate, $K_2HPO_4$. |
| 0.300 | g | Potassium hydrogen phosphate, $KH_2PO_4$, reagent purity |
| 0.200 | g | Magnesium sulphate, $MgSO_4 \cdot 7\ H_2O$, reagent purity |
| 10.000 | g | D(+)-glucose |
| 18.000 | g | Bacto-agar (Difco) |
| 1.5 | ml | Trace element solution |

Glucose and nutrient salts are dissolved separately in 500 ml redistilled water, sterilized for 30 minutes in steam bath and the separate solutions combined after cooling.

Trace element solution (given in grams per litre of redistilled water):

| | |
|---|---|
| 0.055 | $Al_2(SO_4)_3 \cdot 18\ H_2O$ |
| 0.028 | KI |
| 0.028 | KBr |
| 0.055 | $TiO_2$ |
| 0.028 | $SnCl_2 \cdot 2\ H_2O$ |
| 0.028 | LiCl |
| 0.389 | $MnCl_2 \cdot 4\ H_2O$ |
| 0.614 | $H_3BO_3$ |
| 0.055 | $ZnSO_4 \cdot 7\ H_2O$ |
| 0.055 | $CuSO_4 \cdot 5\ H_2O$ |
| 0.059 | $NiSO_4 \cdot 6\ H_2O$ |
| 0.055 | $Co(NO_3)_2 \cdot 6\ H_2O$ |

The culture tubes are each filled with 6 ml of the prepared culture medium and then subjected to fractional sterilization (three times 30 minutes). The medium is then left to set in a slanting position.

Stock solution I:
20.000 g D(+)-glucose
4.240 g $NaNO_3$
2.400 g Anhydrous $K_2HPO_4$
1.200 g $KH_2PO_4$
and 30 ml trace element solution
are dissolved in 1000 ml sterile redistilled water.

Stock solution II:
0.200 g $FeSO_4 \cdot 7\ H_2O$
4.000 g $MgSO_4 \cdot 7\ H_2O$
are dissolved in 1000 ml sterile redistilled water.

NaCl solution:
0.500 g NaCl are dissolved in 1000 ml bidistilled water. The solution is sterilized for 30 minutes in the steam bath.

**Working instructions**

Parent cultures of a test strain of Pseudomonas (suggested: American Type Culture Collection - ATCC -, Rockville, USA, culture No. 13525) are kept in slanting agar culture tubes on the culture medium for parent and preliminary cultures. Further supplies of the test strain are grown by preparing new parent cultures at intervals of one week. The inoculated parent cultures are incubated for 20 ± 4 hours at 25 °C and then kept in stock.

Preliminary cultures are prepared as required from parent cultures on the medium described above in slanting agar culture tubes, and incubated for 20 ± 4 hours at 25 °C. The cell material is then washed away with sterile NaCl solution. The level of turbidity of the bacterial suspension is determined by photoelectric measurement of the transmission ratio of the monochromatic test beam Hg 436 nm. This measured value is used to adjust (by diluting with sterile NaCl solution) the final level of turbidity of the bacterial suspension for a path length of 10 mm to the transmission factor T = 37 % of the monochromatic test beam Hg 436 nm (nm = nanometre).

Before preparing the test cultures, the sewage for analysis is neutralized. This involves choosing the concentration of acid or lye which will permit the smallest possible volume to be added. The sewage is then filtered through a membrane filter (pore size 0.2 μm) by means of a pressure-filter machine.

The sewage prepared in the way described above is used to make up four parallel series of dilutions in 300 ml Erlenmeyer flasks closed with cotton-wool-filled plastic caps. The dilutions each contain (by volume) one part of sewage to 2, 4, 8, 16, 32, 64, then 100, 200, 400, etc. parts of diluent. When preparing the series of dilutions, 20 ml less of redistilled water than corresponds to the theoretical dilution ratio of sewage in redistilled water is added to each test flask. Accordingly, each test flask first receives 80 ml liquid.

Each test flask in the three series of sewage dilutions which are to be inoculated is then topped up to the set level of 100 ml. This means adding to each flask 5 ml of stock solution I, 5 ml of stock solution II and 10 ml of the prepared bacterial suspension from the preliminary culture which has a known adjusted transmission factor (T = 37 %). The initial level of turbidity of the test cultures after injection - for a path length of 10 mm - corresponds to a transmission factor of T = 89 % of the monochromatic test beam (Hg 436 nm).

Each test flask in the one sewage dilution series not injected is also topped up to the set level of 100 ml. Each flask receives 5 ml of stock solution I, 5 ml of stock solution II and 10 ml of NaCl solution.

At the same time control cultures are prepared in order to examine the standard biological reaction of the test organisms in the test medium when the latter is free of sewage. For this purpose 80 ml redistilled water, 5 ml of stock solution I, 5 ml of stock solution II and 10 ml of the prepared bacterial suspension from the preliminary culture which has a known adjusted transmission factor (T = 37 %) are put into each of five Erlenmeyer flasks.

All sewage dilution series and control cultures are kept at 25 °C for 16 hours. Then 1 ml hydrochloric acid (1.125 g/ml) and glass beads are added before they are shaken for 30 minutes in a vibrator at approx. 250 rotary oscillations per minute. The transmission factor of the monochromatic test beam (Hg 436 nm) is measured in the control cultures and the sewage dilution series for a path length of 10 mm.

Colouring may occur in the sewage dilution series or chemico-physical turbidity may appear after acidification. In this case the analogous concentration levels of the uninjected dilution series are utilized as nephelometric blank values for measuring the turbidity of the inoculated dilution series.

Evaluation

The lowest sewage dilution level whose average transmission factor is not above the average transmission factor of the control cultures at the end of the test period, is regarded as non-toxic. The dilution factor in question is stated in the results.

### 5.3.2. Fish test for sewage

When investigating the toxic effect of sewage on organisms, it is always necessary to make a compromise between obtaining as many useful results as possible and the feasibility of tests in practice.

One of the problems of ecotoxicology is deciding whether an observed change is ecologically significant. Past experience shows that biotic systems have the ability to adapt to changed conditions and to regenerate to a certain extent. Only irreversible changes can be unequivocally attributed to toxicological factors. For this reason, attention must be given primarily to persistent and/or irreversible effects when conducting tests with organisms.

The aim of toxicological experiments is to establish the relationship between dose and effect. With organisms which do not live in water, analysis of the substance in question can, for example, be conducted by administering it orally. However, this is not possible in aquatic ecotoxicology, since here the substance is in the water. This is the reason for measuring the effects of the concentration. In fish tests, the physico-chemical behaviour of the substances contained in the water, such as volatility and solubility, is particularly important with regard to effect on the test organisms. The dissolved substances can also be influenced by other factors such as oxidation, hydrolysis, adsorption, etc., before they have an effect on the organisms. Frequently, however, the substances contained in sewage are not known, and so one has to proceed without this advance information. In such cases the relationships between dose and effect cannot be established, and the test is instead purely empirical.

In general fish tests can be conducted in three different ways:

- static testing, for which the water being examined remains in the treatment tank for the entire duration of the tests

- semi-static testing, in which the test medium is renewed periodically

- flow testing, in which the test medium comes into contact with the test organisms at a constant level of concentration.

In the Federal Republic of Germany a test is performed in accordance with DIN 38412, Part 20, using the golden orfe (Leuciscus idus), but it has not been possible to establish any world-wide uniformity in the use of fish species. For the purpose of legislation relating to chemicals the zebra fish (Brachydanio rerio) is favoured in many countries but if we wish to compare findings acquired using different species of fish, new standard procedures must be developed both for the static method and for the semi-static and flow methods. The reason is that the static method can only be used for substances which remain stable for the duration of experiments lasting 48 to 96 hours. The semi-static method is used primarily for substances which do not change in the space of 24 hours. The flow method, which is also the most expensive, has to be used for unstable substances in order to guarantee that at least 80 % of the substance is constantly present in the system.

The following description is of the test used in the Federal Republic of Germany with the golden orfe.

In order to determine the toxic effect of the sewage, the sample is diluted with diluting water in integral volume ratios. After the tests are completed, the toxic effect of the sewage on fish is indicated by applying a dilution factor $G_F$. The dilution factor G indicates the ratio of one part by volume of sewage to the total volume of the test water produced by adding diluting water to the sewage.

Example:
1 part sewage + 4 parts diluting water; G = 5

The lowest value of G for the test water in which all the fish survive is called dilution factor $G_F$.

Equipment

Aquariums, capacity 10 litres

Graduated pipettes

Bulb pipettes

Measuring flasks

Measuring cylinders

Erlenmeyer flasks 250 ml

Glass beakers 250 ml

Aeration stones

Thermometers

Oxygen meter

pH meter with electrodes

De-ionized water.

0.5 mol/l calcium chloride solution:
109.55 g $CaCl_2 \cdot 6\ H_2O$ is dissolved in de-ionized water; the solution is made up to 1 litre with de-ionized water. 1 ml of the solution contains 0.5 mmol calcium ions, equivalent to 20 mg $Ca^{2+}$/l.

0.5 mol/l magnesium sulphate solution:
123.25 g $MgSO_4 \cdot 7\ H_2O$, pure, crystallized, is dissolved in de-ionized water; the solution is made up to 1 litre with de-ionized water. 1 ml of the solution contains 0.5 mmol magnesium ions, equivalent to 12 mg $Mg^{2+}$/l.

0.1 mol/l sodium hydrogen carbonate solution:
8.401 g $NaHCO_3$, superpure, is dissolved in de-ionized water; the solution

is made up to 1 litre with de-ionized water. 1 ml of the solution added to 1 litre water increases the acid capacity $C_A$ 4.3 of the water by 0.1 mmol/l.

1 mol/l hydrochloric acid:
1 ml of this solution added to 1 litre water reduces the acid capacity $C_A$ 4.3 of the water by 1 mmol/l.

Diluting water consisting of chlorine-free drinking water:
For diluting the sewage sample, chlorine-free drinking water with a concentration of calcium ions of (2.2 ± 0.4) mmol/l (equivalent to 88 ± 16 mg $Ca^{2+}$/l) and of magnesium ions (0.5 ± 0.1) mmol/l (equivalent to 12 ± 2 mg $Mg^{2+}$/l) can be used. (The mol ratio of calcium ions to magnesium ions should be approx. 4 : 1).

Synthetic dilution water:
22 ml calcium chloride solution, 5 ml magnesium sulphate solution and 5.0 ml sodium hydrogen carbonate solution are diluted with de-ionized water to make a volume of 5 litres. The diluted water is aerated until the pH remains constant.

**Test fish and their treatment**

The golden orfe (Leuciscus idus (L.), golden variety = golden orfe) is used as the test fish. They should have an overall length of 5 to 8 cm and a corpulence factor K of 0.8 to 1.1 g/cm$^3$. Only healthy fish may be used.

The corpulence factor K is calculated using the following equation:

$$K = \frac{100\ m}{l^3} \quad (g/cm^3)$$

in which:

K = corpulence factor, in g/cm$^3$
m = live weight of fish, in g
l = length of fish, measured from tip of mouth to end of caudal fin, in cm.

The test fish are not sexually mature and so cannot be distinguished as male or female. The sex of the fish is therefore not taken into account in this experiment.

The fish are kept in aerated, chlorine-free drinking water, preferably flowing, at 10° to 20 °C. Not more than 5 fishes should be kept in 1 litre water. If standing water is used, it must be adequately circulated and filtered, and frequently replaced.

The fish should be fed with a suitable dry food (e.g. dry food for fry, grain-size 0) for as long as they are kept in tanks.

If the temperature of the tank water is below 18 °C, an adaptation period of at least 28 hours at 20 °C must be allowed.

The fish should be kept for at least 1 week; the mortality rate thereafter does not usually exceed 1 % per week.

## Procedure

The aquarium is filled with 5 litres test water. The pH of the test water is adjusted to pH = 7.0 ± 0.2 by adding hydrochloric acid or sodium hydroxide. 5 golden orfes are then placed in each tank. No more food is given to the fish during the test.

The temperature of the test water is (20 ± 1) °C. During the experiment the oxygen concentration in the test water must be maintained at not less than 4 mg/l $O_2$. This is possible without aeration in many cases for the whole duration of the experiment. The experiment lasts 48 hours. A fish is regarded as being dead if no spontaneous movement can be detected even when touched.

A control test using 5 litres diluting water is made in the same way. If one or more fish die in the control test, the experiment cannot be regarded as valid.

Fish which have survived the test must not be used a second time.

## Evaluation

The lowest dilution factor $G_F$ of the test water in which all fish survive is taken as the result of the experiment.

Dilution factors are only given in whole numbers.

# 6 Evaluation of Analysis Data

## 6.1 Introduction to statistical evaluation

The quality of an analysis result is the basis for the overall evaluation of analytical data. **Accuracy** - approximation to the true value, which is a theoretical, unknown quantity - is therefore of prime importance. **Correctness**, i.e. the difference between the set value and the average value of the analysis results for a concentration level, can be verified using round-robin experiments and recovery experiments.

Accuracy is determined not only by correctness, but also by **precision:** systematic errors affect correctness and random errors affect precision.

Random errors can be revealed by repeating the analysis under the same conditions or by comparing analyses made at different times under different circumstances (i.e. technicians, apparatus, laboratories).

Systematic errors may be constant throughout the range of application and are only discovered by comparing with a different method of analysis. Proportional systematic errors are independent of concentration and can only be revealed by recovery experiments. A detailed description of the terms used and their statistical formulation is given by Massart et al, Elsevier 1978.

Fig. 168 shows the dispersion of test results caused by random and systematic errors. An interesting point to observe is the overall error affecting an analysis result. There are various definitions of what should be classed as part of the overall error, e.g. the systematic error (resulting from the method) plus the standard deviation resulting from repeat experiments in the laboratory.

Errors can also arise in the evaluation of analysis results if a substance not present in the sample is erroneously identified (error Type 1), or vice versa i.e. a substance which is present is not identified (error Type 2).

Other sources of errors can arise from inexpert use of statistical methods. For example the use of statistical tests based on the assumption of normal distribution can lead to significant errors if the population from which the random sample was taken is not based on normal distribution.

Fig. 169 shows the problem of time-dependent data when statistical data are presented in a simplified form. Examples a and b in Fig. 169 are based on the same empirical frequency distribution. Their statistical parameters (arithmetical mean, median, geometrical mean, standard deviation) are therefore identical. However, they conceal an essential piece of information, namely the time dependence factor.

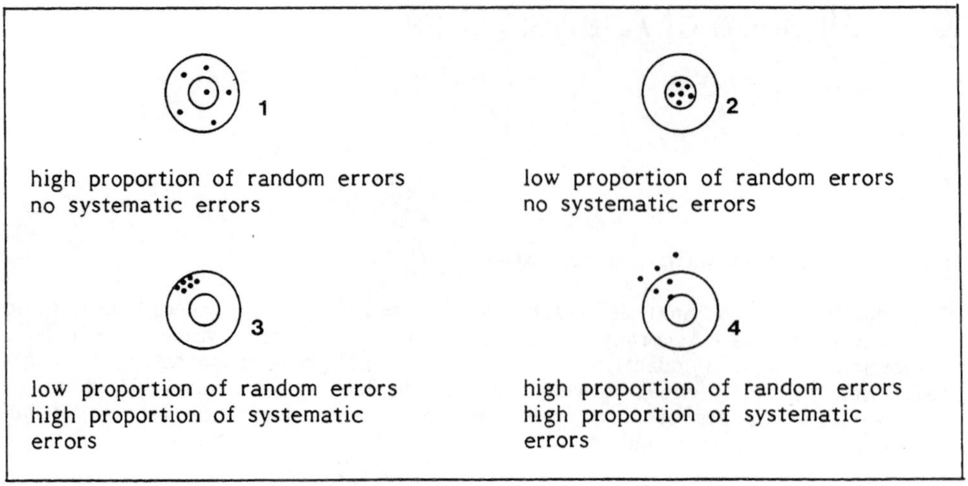


Fig. 168. Systematic and random errors (Kirchmer)

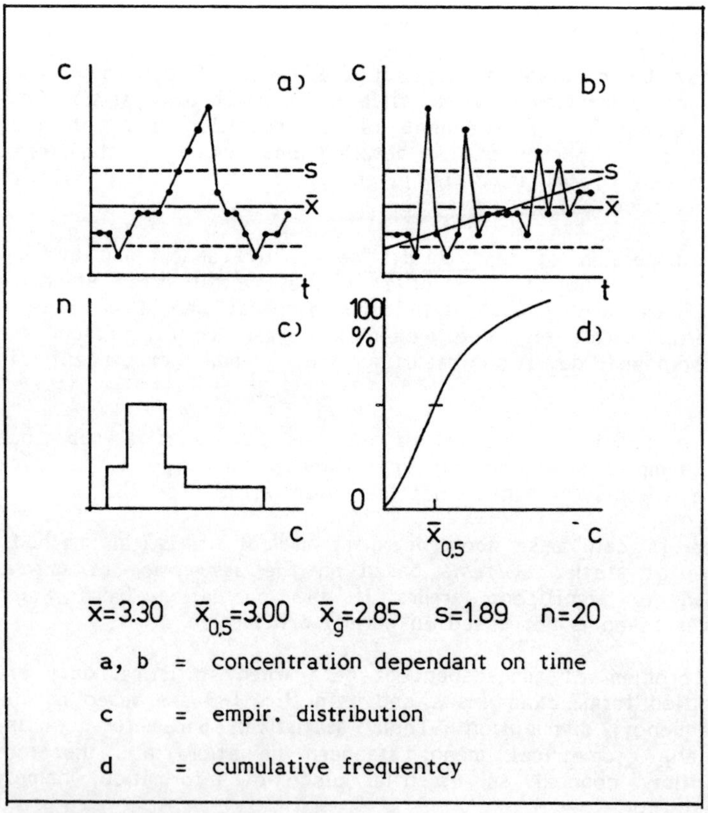


Fig. 169 a - d. Problems in the statistical presentation of time-dependent data on water quality

## 6.2 Applications of statistical methods in water analysis

The following pages describe selected statistical methods suitable for tackling individual problems arising from routine analysis with the aim of being able to evaluate and compare test results.

### 6.2.1 Calibration

Every analytical method is based on instructions for working, measuring, calibration and evaluation. Whereas working and measuring instructions are given in each individual method or in the corresponding standard method, instructions for calibration and evaluation apply to all methods of analysis that are capable of being calibrated. The latter instructions should be drafted so as to enable analytical results based on them to be more easily compared. A description is given of how a calibration function is established from given standard concentrations ($x_i$) and measured information values ($y_i$), which can then be used for converting information values ($y_i$) from future analyses into concentrations. On the basis of these characteristic data it is possible to state the degree of sensitivity and precision of particular analytical methods. A comparison can then be made with the characteristic data of standardized analytical methods in order to provide a basis for judging the quality of the analysis results.

It is necessary to differentiate between establishing and familiarizing oneself with new methods of analysis, and establishing a calibration function in routine analysis.

### 6.2.2 Calibration when using new methods of analysis

When preparing a method of analysis, extensive calibration should be carried out which will enable parameters to be given for the method in question. By using linear regression analysis, for example, a calibration curve can be expressed as a function. The linear calibration function is as follows:

$$y = a_0 + a_1 x$$

However, it may only be used if:

- the measured information values ($y_i$) are normally distributed throughout every standard concentration ($x_i$), and

- all calibration samples are independent of each other.

The latter condition means that standard solutions must be taken not from a previous step in the analysis but from separate ones. Alternatively, if this is not possible, independent dilutions from a single parent solution must be prepared. Dilutions made in succession lead to systematic errors.

In addition steps must be taken to ensure:

- the homogeneity of the variances, and

- the linearity of the function in the working range (linearity test).

In the preparation and familiarization stages of a new analytical method, DIN 38402 recommends that the information values $y_i$ be established for N = 10 calibration samples. In so doing, the intervals of the individual concentration levels should be equidistant, and so equally distributed over the working range. In order to verify that the variance of the information values $y_i$ is independent of the concentration, a statistical test of the homogeneity of the variance can be used. For this purpose one analyses n = 10 standard samples of the lowest ($y_{j,1}$) and n = 10 standard samples of the highest concentration level ($y_{j,10}$).

The variances $S_1{}^2$ and $S_{10}{}^2$ are calculated as follows:

$$s_1^2 = \frac{\sum_{j=1}^{10} (y_{j,1} - \bar{y}_1)^2}{n_1 - 1} \qquad s_{10}^2 = \frac{\sum_{j=1}^{10} (y_{j,10} - \bar{y}_{10})^2}{n_{10} - 1}$$

With the mean value:

$$\bar{y}_1 = \frac{\sum_{j=1}^{10} y_{j,1}}{n_1} \qquad \bar{y}_{10} = \frac{\sum_{j=1}^{10} y_{j,10}}{n_{10}}$$

The homogeneity of the variances is checked using the F-test, for which the test quantity PG is determined by

$$PG = \frac{s_{10}^2}{s_1^2} \quad \text{mit } s_{10}^2 > s_1^2$$

or

$$PG = \frac{s_1^2}{s_{10}^2} \quad \text{mit } s_1^2 > s_{10}^2$$

and is compared with the value in the F-table.

If PG < F ($f_1$, $f_2$, 95 %), then the difference of the variances is random. If PG > F ($f_1$, $f_2$, 95 %), then the difference is significant and the working range must be narrowed down until the difference between the variances is random.

Apart from homogeneity of the variance, the linearity test, as already mentioned, is also a precondition for being able to use linear regression as a calibration function.

For this purpose, it is necessary to test whether the function expressed by 10 equidistant calibration samples can be better adapted to a linear ($y_a$) or a quadratic calibration function ($y_b$).

$y_a = a_o + a_1 x$ Calcuiation of the residual dispersion $S_{ya}$

$y_b = a_o + a_1 x + a_2 x^2$ Calculation of the residual dispersion $S_{yb}$

From this follows the difference $DS^2$ of the variances, calculated from:

$DS^2 = (N - 2) \cdot S_{ya}{}^2 - (N - 3) \cdot S_{yb}$

If the calculated test quantity PG

$$PG = \frac{DS^2}{S_{yb}^2}$$

is smaller than the F-value (F-test) given in the table, linearity can be assumed to exist.

If PG > F, then the working range must be narrowed down until PG < F, i.e. linearity is assured.

If the conditions for linear regression as a calibration function now exist, the procedural parameters can be determined. These include:

- the **residual standard deviation** ($s_y$):

  This is the standard deviation of the differences between the information values and the values of the associated calibration curve, and therefore also a precision measurement for calibration.

$$a_1 = \frac{\sum_{i=1}^{N} (x_i - \bar{x}) \cdot (y_i - \bar{y})}{\sum_{i=1}^{N} (x_i - \bar{x})^2}$$

- the **sensitivity** of a method is reflected in the gradient of the calibration function:

$$s_y = \sqrt{\frac{\sum_{i=1}^{N} (y_i - \bar{y}_i)^2}{N-2}} = \sqrt{\frac{\sum_{i=1}^{N} (y_i - (a_0 + a_1 x_i))^2}{N-2}}$$

- the **standard deviation** of the method ($S_{xo}$):

  This is the quotient of the residual standard deviation ($s_y$) and the sensitivity ($a_1$) of the analytical method (gradient of the calibration function) and is a measure of the quality of the analytical method in the working range under consideration.

$$S_{xo} = \frac{s_y}{a_1}$$

  The ordinate intercept can be interpreted as a calculated blank value.

An unknown concentration $\hat{x}$ is calculated from the measured information value as follows:

$$\hat{x} = \frac{y1 - a_o}{a_1}$$

This calculation is, however, affected by an error, which results from the uncertainty in the determination of the information value y and the uncertainty of the calibration function. Therefore, for every calculated concentration x, there is a confidence interval $\bar{x}$, which includes the true information value y.

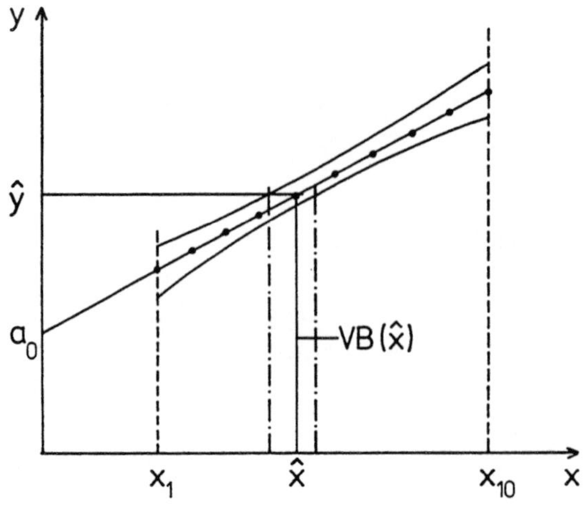


Fig. 170. Confidence range VB ($\hat{x}$) surrounding an analysis value

This confidence range is calculated using the following approximation equation:

$$\hat{x}_{1,2} = \hat{x} \pm VB(\hat{x})$$

$$\hat{x}_{1,2} = \frac{\hat{y} - a_0}{a_1} \pm \frac{s_y}{a_1} \cdot t \sqrt{\frac{1}{N} + \frac{1}{\hat{n}} \frac{(\hat{y} - \bar{y})^2}{a_1^2 \sum_{i=1}^{N} (x_i - \bar{x})^2}}$$

n = number of parallel analyses of a concentration level
$\hat{x}$ = analysis result calculated using the calibration function

The true analysis value is therefore enclosed by the confidence range VB ($\hat{x}$) with a confidence level given by Student's t-factor.

N.B.: The confidence range VB ($\hat{x}$) is to a large extent governed by the number of parallel conditions n, the resulting information value y, and the procedural parameters $s_y$ (residual standard deviation) and $a_1$ (sensitivity).

The quality of the analytical method therefore increases as sensitivity rises and residual dispersion falls.

### 6.2.3 Calibration in routine analysis

The previous section described how the conditions for postulating a linear regression function as a calibration function are checked, and how the parameters of an analytical method are determined. In routine analysis, the scientist now has the task of proving that the procedural characteristic data which he has obtained are not significantly different from the prescribed data. For this purpose the information values ($y_i$) and the procedural standard deviation $s_{xo(R)}$ are determined from 10 equally spaced concentration levels ($x_i$).

A simple variance test (F test) shows whether the procedural standard deviation $s_{xo}(R)$ determined in routine analysis differs significantly from the procedural standard deviation $s_{xo}$ prescribed in the standard.

If the first calibration function appears non-linear, the simplest solution is to attempt to position the working range so that the calibration function is linear within the working range.

If this is not possible, parameters of the calibration function $a_o$, $a_1$ and $a_2$ can be determined according to the rules of polynomial approximation of the 2nd degree:

$$y = a_o + a_1 x + a_2 x^2$$

The residual standard deviation $s_y$ is calculated as follows:

$$s_y = \sqrt{\frac{\sum_{i=1}^{N} (y_i - (a_0 + a_1 x_1 + a_2 x_i^2))^2}{N-3}}$$

Only the sensitivity in the middle of the working range can be given as a procedural parameter ($\bar{x}$):

$$E = a_1 + 2a_2\bar{x}$$

The procedural standard deviation ($s_{xo}$) in this case is:

$$s_{xo} = \frac{s_y}{E}$$

The confidence range encompassing an analysis value is calculated from:

$$VB(\hat{x}) = \frac{s_y \cdot t}{(a_1 + 2a_2\hat{x})} \cdot \sqrt{\frac{1}{N} + \frac{1}{\hat{N}} + \frac{1}{Q_x \cdot Q_{xx} - (Q_x)^2} \cdot \left\{ (\hat{x}-\bar{x})^2 Q_x + \left(\hat{x}^2 - \frac{\sum_{i=1}^{N} x_i^2}{N}\right)^2 Q_{xx} - 2(\hat{x}-\bar{x}) \cdot \left(\hat{x}^2 - \frac{\sum_{i=1}^{N} x_i^2}{N}\right) \cdot Q_x \right\}}$$

$$Q_{xx} = \Sigma x_i^2 - ((\Sigma x_i)^2/N)$$

$$Q_x^3 = \Sigma x_i^3 - ((\Sigma x_i) \cdot (\Sigma x_i^2)/N)$$

$$Q_x^4 = \Sigma x_i^4 - ((\Sigma x_i^2)^2)/N$$

If there is a non-linear concentration of analysis data, this is usually caused by additional factors. If it is possible to establish what these factors are, they can be included in the calibration function, e.g. in the form

$$y_i = f_i \ (x_1, x_2, \ldots x_{n-1}, x_n)$$

Then, however, deviations in the form of partial sensitivities should be given only for the individual factors. A detailed description of this procedure is given in D. L. Massart, A. Dijkstra, L. Kaufman: Evaluation and Optimization of Laboratory Methods and Analytical Procedures; Elsevier Scientific Publishing Company, Amsterdam - Oxford - New York 1978. A confidence range can no longer be derived in the simple form.

### 6.2.4 Detection limit and determination limit

Detection and determination limits should not be established from the dispersion of the blank tests, but from calibration experiments (1,5). This gives the definition of the detection limit (XN) as the smallest quantity or concentration of a substance which can be **qualitatively** detected or quantitatively estimated by a single analysis with the required statistical certainty (level of significance $\alpha$ = 5 %). This definition therefore represents the upper confidence-range limit of a concentration $x_C$ whose lower confidence-range limit is concentration zero. The determination limit (XB) represents the smallest quantity or concentration of a substance which can be quantitatively determined by a single analysis with the required certainty ($\alpha$ = 5 %).

For all concentrations falling below the detection limit, the risk of erroneously identifying a substance not present in a sample is greater than 50 %.

Actual presence of the substance cannot therefore be distinguished from zero with sufficient statistical certainty.

Analysis results between the detection limit XN and the determination limit XB can only be distinguished qualitatively from zero. The statistical risk of error resulting in a substance which is present not being identified will only be 5 % or less when the analysis results lie on or above the determination limit.

In draft DIN norm 1983 two procedures are suggested for establishing the detection and determination limits. The following requirements apply jointly to both:

- independent calibration-standard solutions

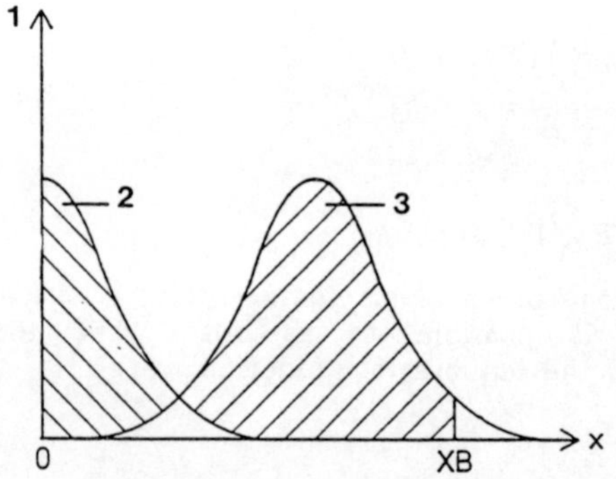


Fig. 171. Distribution of blank and analysis values; 1) = Frequency; 2) = Blank-value distribution; 3) = Analysis-value distribution

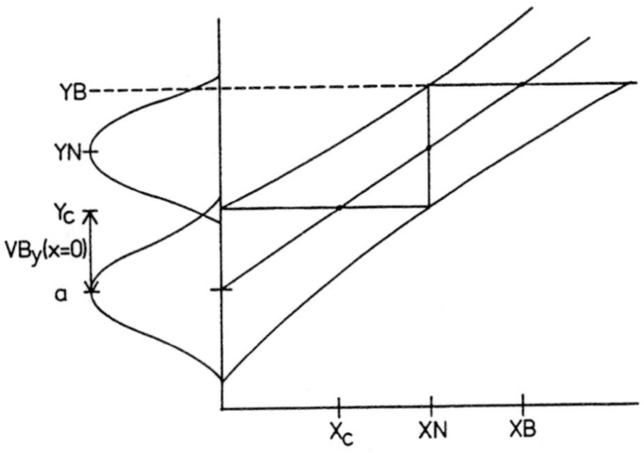


**Fig. 172.** Detection limit and determination limit (XN, XB), establishing the subsidiary value $x_c$

- linear calibration function
- homogenous variances

a) **Statistical procedure**

The standard deviation of the calibration curve to be determined from a known calibration curve is used with the concentration value x = 0 as a subsidiary quantity.

$$s_0 = s_{x0} \cdot \sqrt{\frac{1}{N} + 1 + \frac{\bar{x}^2}{\sum_{i=1}^{N} (x_i - \bar{x})^2}}$$

The necessary values are determined from 10 equidistant calibration solutions with the concentrations

$$x_1 = s_o$$
$$x_2 = 2s_o$$
$$x_{10} = 10s_o$$

b) **Dynamic procedure**

This must be used if no values in the proximity of the blank value can be established with the analytical methods employed. The aim is to measure standard calibration solutions of ever decreasing concentrations until the calculated test quantity $x_p$ corresponds approximately to the lowest calibration concentration of the lowest selected working range. If this aim is achieved, then

$$x_p \simeq x_1 = XB$$

The detection limit is calculated using

$$XN = 2 \cdot \frac{s_y \cdot t}{b} \sqrt{\frac{1}{N} + 1 + \frac{(y_c - \bar{y})^2}{b^2 \sum_{i=1}^{N} (x_1 - \bar{x})^2}}$$

The confidence range of a concentration $y_c$ corresponds to

$$y_c = a + VB_y\,(x=0)$$

$$y_c = a + s_y \cdot t \sqrt{\frac{1}{N} + 1 + \frac{(0-\bar{x})^2}{\sum_{i=1}^{N}(x_i-\bar{x})^2}}$$

The following applies to the determination limit:

$$XB = \frac{y_h - a}{b} + \frac{s_y \cdot t}{b} \sqrt{\frac{1}{N} + 1 + \frac{(y_h-\bar{y})^2}{b^2 \sum_{i=1}^{N}(x_i-\bar{x})^2}}$$

where

$$y_h = a + 2 \cdot VB_y\,(x=x_c)$$

$$y_h = a + 2 \cdot s_y \cdot t \sqrt{\frac{1}{N} + 1 + \frac{(x_c-\bar{x})^2}{\sum_{i=1}^{N}(x_i-\bar{x})^2}}$$

and

$x_c = VB_x\ (x = 0)$ extrapolated confidence range of the concentration $x = 0$

$$x_c = \frac{s_y \cdot t}{b} \cdot \sqrt{\frac{1}{N} + 1 + \frac{\bar{x}^2}{\sum_{i=1}^{N}(x_i-\bar{x})^2}}$$

Calculation examples are given in draft DIN norm 1983.

### 6.2.5 Blank values

The dispersion of blank values can give indications of random and systematic deviations, e.g. in order to

- uncover carry-over errors
- describe the influence of reagents and individual workers
- detect drift caused by apparatus, etc.

For this reason blank values should be analyzed both before and after a series of analyses. They can be experimentally determined from separately conducted multiple analyses. The mean blank value is subtracted from the information value $y_i$ and this corrected information value gives the concentration.

Another possibility for determining the blank value consists in extrapolating the calibration curve. There the blank value is already recorded by the ordinate intercept $a_o$ (UBA research report 10205114, 1982, Verlag Chemie, Weinheim, FRG).

### 6.2.6 Standard addition

Matrix effects influence the correctness of a determined value. They can take the form of constant, systematic errors or of proportional errors which are dependent on concentration. The latter can be detected by volume-increment experiments.

For example, iron causes proportional errors when determining the nitrite content in water. The procedure described in the following paragraphs is taken from, where it is described in detail. It only applies to the linear range and the incremental concentrations must be chosen accordingly.

Furthermore, it is recommended that 4 equidistant incremental concentrations be used, whose maximum concentration corresponds to that of the original sample.

**Example**

Concentration of the original sample: 8 mg/l

$x_1 = 2$ mg/l
$x_2 = 4$ mg/l
$x_3 = 6$ mg/l
$x_4 = 8$ mg/l

The incremental volume should be as small as possible in comparison to the sample volume and should be taken in aliquot steps.

**Example**

Incremental volume VS — Concentration $x_{A1}$
Incremental volume 2 · VS — Concentration $x_{A2}$

In this way the additional volume VS changes, but the total volume remains constant and is topped up with a solvent (e.g. distilled water) if necessary.

The sample volume and incremental volume are thoroughly analyzed following the method laid down. The calibration function and the confidence range are calculated from the 5 pairs of values obtained. The blank value YB corresponds to the axis intercept $a_o$ of the calibration function. A line is drawn through YB parallel to the abscissa, and this line gives the required value x at the intersection with the extrapolated calibration function. It also gives its confidence bands.

### 6.2.7. Matrix effects

Water samples, in particular sewage samples, always display matrix effects to a greater or lesser extent. A measure for assessing these effects statistically is provided by the recovery rate. For this purpose the original sample (concentration $a_o$) plus 9 incremental volume samples are analyzed and a linear calibration function is created:

$$x_g = a_o + a_1 x_A$$

The recovery rate WFR is given by:

$$WFR = \frac{x_g - a_o}{x_A} \cdot 100\ \%$$

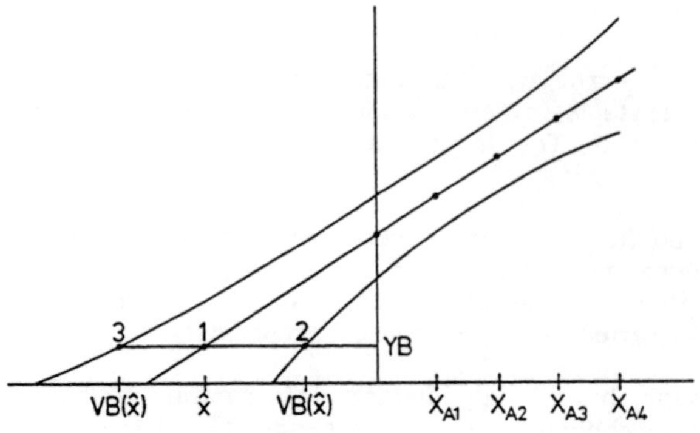


Fig. 173. Determining the sample content and the associated confidence range by means of volume-increment experiments

The different types of variance analyses offer another possibilty for revealing matrix effects.

Multivariate approaches are based on the fact that various possible factors are detected quantitatively in all samples. Using the data matrix so obtained, it is possible to identify the factors controlling the analysis results, since the peripheral information which accompanies a measuring result is regarded as of equal value for calculation purposes.

If the factors which cause the matrix dependence are known, they can be described as a function, e.g. in the form of a multiple linear regression. An introduction to this can be found in D. L. Massart et al.

### 6.2.8 Comparison of analytical methods

For both methods (A, B), a calibration should be carried out in matrix-free solution over a defined working range, in order to be able to establish the procedural characteristic data ($s_{xoA}$, $s_{xoB}$) and also the determination limits.

The F-test is used to check whether the two procedural standard deviations differ significantly from each other:

$$PG = \left(\frac{s_{xoA}}{s_{xoB}}\right)^2$$

They differ if the test quantity PG is greater than the comparative value listed in the table for the F distribution (P = 99 %).

If the influence of matrices is to be examined, 10 parallel analyses should be made from 3 real samples with different concentration levels. This should be done for both methods for the matrix-free calibration function and for the calibration function of the standard addition, making a total of 4 series.

The multiple analyses are checked for mavericks using the Grubbs technique. The homogeneity of the variances is also tested. Finally, the mean values

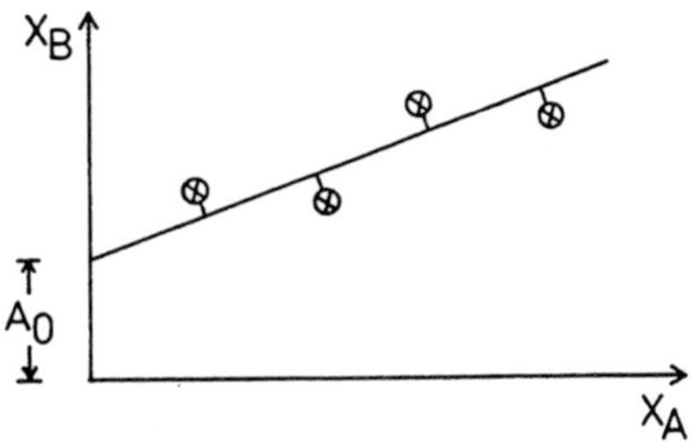


Fig. 174. Orthogonal regression for comparing analytical methods

of the 4 series of analyses are compared two at a time using the mean-value-t-test. If the test quantity so calculated is greater than or equal to the value given in the table, this means that there are systematic differences between the mean values.

A further possible method of comparison consists in subjecting 30 pairs of values, obtained from double analyses by means of both methods, to a linear orthogonal regression. For this purpose the compensating straight line is assessed so as to minimize the deviations from the measuring point in the horizontal and vertical directions.

Systematic errors are present if the straight line of the calibration function does not cut the system of co-ordinates at the origin, i.e. $a_0 \neq 0$ (constant systematic deviation) and the gradient does not equal 1 i.e. ($a_1 \neq 1$).

The differences t-test can be used to discover whether the two methods differ significantly from each other:

$$D = X_{Ai} - X_{Bi}$$

$$PG = \frac{\frac{1}{N}\sum_{i=1}^{N} X_{Ai} - \frac{1}{N}\sum_{i=1}^{N} X_{Bi}}{\sqrt{\frac{\sum_{i=1}^{N} D_i^2 - \frac{1}{N}\cdot\left(\sum_{i=1}^{N} D_i\right)^2}{(N-1)\cdot N}}} \quad \text{(mit } f = N-1)$$

If the test quantity PG is greater than or equal to t (f, P = 99 %), then the two methods differ significantly from each other.

The statistical tests described so far (t-test, F-test) are classified as parametric tests and require a normally distributed population. If this does not exist, non-parametric tests should be used instead, i.e. tests which are not dependent on distribution.

These include: the Wilcoxon test, the Kruskal-Wallis test or the Kolmogoroff-Smirnov test.

Variance analyses can be used instead of the orthogonal regression for comparing methods.

A typical example of the sort of question which should be asked when conducting univariate variance analysis is: do the two methods to be compared differ significantly from each other in respect of their share of systematic errors-assuming that the random deviations are of equal size?

Bivariate and multivariate variance analyses permit several test series and alternative sequences of samples to be compared for significant differences between the series. It is also possible to investigate whether, for example, the difference between the series is attributable to the sequence in which the samples were made (D. L. Massart et al).

### 6.2.9 Optimizing analytical methods

Questions frequently arise, especially during the development phase of analytical methods, about how to optimize parameters, such as reaction time, concentration of reagents, etc. and how to optimize analysis procedure, e.g. extraction method, choice of reagent, etc. Questions constantly crop up, e.g. "What effect do pH and EDTA level have on the extraction of a metal ion by means of a chelating agent?" - or "What significance do pH and redox potential have in the desorption of heavy metals from sands?"

Parametric tests (t-test, F-test) are not really adequate for answering the sort of question mentioned above, since they require a standard distribution of the population and can only compare 2 components with each other at any one time. A better solution is offered above all by variance analyses or factorial experiment planning and evaluation.

Alternatively one can fall back on methods used in operations research, such as the simple or revised Simplex method, the Uniplex method, etc. A more detailed description with practical examples and suggestions for further reading is given in (D. L. Massart et al).

## 6.3 Quality control

Analysis results must be reliable and comparable. This requires both internal quality control conducted in the laboratory, and also external quality control of analysis results in order to be able to make comparisons with other laboratories.

### 6.3.1 Internal quality control

The quality of an analytical method can be expressed by the procedural characteristic data ($s_{xy}$, $s_y$, $a_o$, $a_1$), as already explained in Section 6.2.1.

On this basis various analytical methods can be compared with each other (see Section 6.2.8). For example, when a method is used for the first time in a laboratory, the results (in the form of parameters) can be assessed in comparison with the parameters of the person who devised the method.

If this method is used by various workers in the same laboratory, it is possible to judge whether there is a significant difference between the results they achieve.

Internal quality control should aim to document the reliability of analytical methods. This can be achieved by first adopting a familiarization phase with the purpose of being able to adhere to the characteristic data of the prescribed analytical method. When there is no longer any significant difference between the results achieved in the laboratory and the prescribed characteristic data, it can be assumed that the method has been mastered.

The next step should be to investigate whether over a period of time, e.g. 1 week or 1 month, and with various workers involved, the analysis results deviate from given standards. Calibration solutions, blank samples (real or incremental) or real samples can be used as standards.

The analysis results are entered on so-called "control cards" or "quality regulation cards" after each investigation. In this way routine errors may be revealed by time discrepancies, trends and regular fluctuations.

There are different types of control card, such as

- the mean value control card

- the standard deviation card

- range card

- the recovery rate card

- the drift card, etc.

Graphs can be used to give a clear picture of the point at which newly obtained analysis values overstep the tolerable levels of discrepancies and an "out-of-control situation" arises. It is a prerequisite for this decision that limit values must be laid down, beyond which the "out-of-control situation" is held to exist. A preliminary training phase is necessary in order to have an idea of the proportion of tolerable random errors, which as a rule affect all methods. The level of precision to be maintained is established during this familiarization phase.

The calibration standard, for example, can be monitored using quality regulation cards. Ageing of the reagents then causes the analysis results to drift until they go out of control.

Variances, differences, ranges of multiple analyses can also be monitored, as can blank values or recovery rates. In all cases an acceptable tolerance

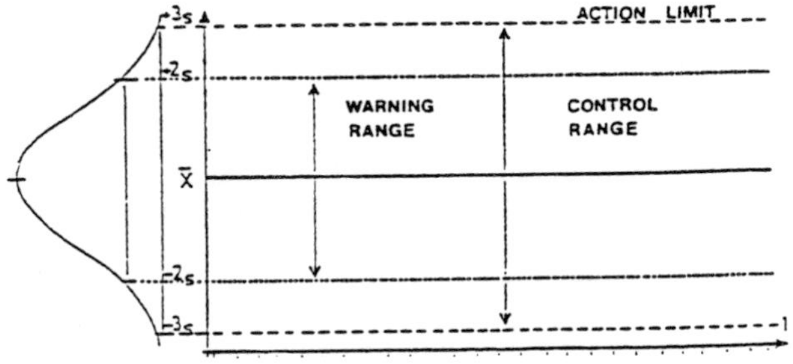


Fig. 175. Mean-value control card (quality regulation card)

range (precision) forms the basis of the monitoring. This tolerance range is calculated during a preliminary phase by means of analytical investigation and various statistical methods. Checks are then made during routine analyses to see whether the tolerance range is being kept to.

The quality regulation cards enable a laboratory to determine how precisely it is working and how reliable its analysis results are.

### 6.3.2 External quality control

Internal quality control should be conducted to establish whether analysis results are precise. But to prove whether they are correct is only possible in the long run by means of a round-robin test, in which the results are compared with those of other laboratories.

Only the evaluation of round-robin tests can reveal systematic discrepancies in the results of individual laboratories. The original research conducted by these laboratories, for example, can lead to inexact description of the different steps in the analysis method. In this case corrective measures could be instigated, which would allow correctness limits to be established at a later date.

The planning, organization and evaluation of round-robin tests are covered by DIN or ISO standards.

Apart from the description mentioned above, a supplementary analysis of significant differences, e.g. between laboratories, between different methods and within separate series of analyses carried out by different laboratories, is given by D. L. Massart et al. Here multivariate variance analysis is used to clarify whether quantifiable factors influence the results of the co-operative test, and if so which factors these are.

## 6.4 Data evaluation

In the following pages further statistical methods of evaluation are set out, which permit temporal and spatial fluctuations of substances contained in water to be described and analyzed.

In contrast to chemical analysis, in which standardization is already well advanced, statistical methods used for evaluating water quality data are, with few exceptions, not standardized but mostly a matter of tradition. The basis of all statistical evaluations should be an investigation plan which encompasses formulation of problems, data acquisition and an evaluation strategy. When drawing up plans for random sampling, all information which is already available about possible fluctuations of the system under examination should be taken into account.

The requirement for statistical independence demands that two areas of data evaluation be kept separate from each other.

- The area of analysis in order to guarantee individual measured values

- The evaluation of analytical data already obtained.

In the first case the total volume of the sample under examination can be designated as the population, from which several random samples are taken for analysis, which are independent of each other. As a rule dispersion of the results follows a normal distribution, if there are no temporal or peripheral effects which cause possible discrepancies.

If, on the other hand, various time-dependent and/or location-dependent data are collected, the task of describing the population represented by these data can be more problematical. With surface water all substances contained can be defined as belonging to the parent population, but the relationship of the substances to each other - the matrix composition - depends on external factors such as water runoff or sewage discharge, etc. As a model, it could be assumed that a "stratified" population has to be dealt with.

The need for sampling to be representative and for the individual random samples to be independent, e.g. in all correlation calculations, must not be overlooked. It is necessary to check whether a measurement being made is influenced by the size of the previous measurement. This applies in particular when taking continuous measurements of temperature, oxygen or discharge values. These so-called "autocorrelated" measured values can lead to over-interpretation, if they are evaluated using statistical methods which do not take the temporal and spatial sequence of the measured values into account.

The aim of more detailed statistical evaluation is to build up a body of information which can be used to describe individual measured quantities and to understand how they are related to each other. Depending on the number of measured quantities under consideration, it is possible to distinguish between univariate, bivariate and multivariate time-dependent or non-time-dependent statistical evaluation.

### 6.4.1. Univariate statistical evaluation

The empirical distribution of measured values can be described in terms of

- positional measurements e.g. arithmetical mean, median, mode
- dispersion measurements, e.g. variance, standard deviation, variation coefficient, etc.

As a rule the substances contained in water are not symmetrically distributed since, as already mentioned, they consist of various "strata" of a population, which are not always equally represented. Time-dependent fluctuations in a univariate, time-dependent approach entail a loss of information, since the temporal dynamics cannot be adequately described.

### 6.4.2 Bivariate statistical evaluation

Trends and fluctuations can have different causes. These can be simply understood through their relationship to other water quality parameters by means of correlation calculation. The correlation coefficient r is a standardized measure of the linear correlation between two randomly distributed characters x and y (e.g. water quality parameters) The correlation coefficient r always lies between -1 and +1. Where $r = 0$ there is no interrelationship. Since the number of random samples has a very significant effect on the degree of correlation, the significance of the correlation should always be tested and examined on the basis of the relevant table values. The correlation calculation requires standard distribution of the populations of both variables x and y. Furthermore, the samples should be randomly independent, i.e. the quantity of the previous x value must not influence the following x value (no autocorrelation). It is clear from the method of calculating the correlation coefficient r that the variables are compared in pairs and that the time aspect is not taken into account.

$$r = \frac{\sum_{i=1}^{n}(x_i - \bar{x})(y_i - \bar{y})}{\sqrt{\sum_{i=1}^{n}(x_i - \bar{x}) \cdot \sum_{i=1}^{n}(y_i - \bar{y})^2}}$$

n = number of measured values
$x_i$ = i-th measured value of the variable A, i = 1.2, ..., n
$y_i$ = i-th measured value of the variable B, i = 1.2, ..., n

Extreme values make a great difference to correlation coefficients. The linear interrelationship between two variables can be described functionally by a straight-line regression curve. However, this type of description does not always lead to satisfactory results, because deviations from the straight line are frequently large. But precisely these deviating measurements can provide interesting clues for investigating what exterior peripheral conditions obtained during the sampling and how the concentration of the remaining substances in the water was distributed in these measurement results. The multivariate statistical methods described below are helpful in this respect.

### 6.4.3 Multivariate statistical investigations

If the pair of values x and y do not fit satisfactorily on a linear or non-linear regression line, this is a fundamental indication that other influencing factors are involved. The basis of all multivariate statistical methods is a data matrix which not only deals with all directly relevant variables, such as the concentrations of different substances contained in the water, but also uses possible indirect influencing factors, such as weather, run-off, quantity of sewage, etc. as additional variables. Such data are usually been recorded at every sampling.

Interrelationships or dependences between different substances in the water and their presumed influencing factors are analyzed on the basis of correlations or absolute distance measurements. They are further structured according to the method used, categorized or described as functional interrelation-ships.

#### 6.4.3.1 Principal component analysis

This involves describing linear interrelationships on the basis of correlation calculations and co-variance calculations. Groups are formed such that within each group the variables belonging thereto are highly correlated, but between the groups of individual variables there is no linear interrelationship.

Fig. 176 shows how the feed to sewage treatment plants can be characterized by co-ordinating individual variables. Principal component analysis of sewage from a residential area indicates that the substances contained originate essentially from diffuse sources. Sewage from an industrial area shows a clear correlation of heavy metals with other heavy metals, turbidity and content of solid matter. This means it is probable that heavy metals are predominantly bound to solid particles.

The diagram depicts the correlation of variables with different principal components (factors), expressed as principal component load. As has already been mentioned, variables that are correlated with one principal component are independent of variables that are co-ordinated with other principal components. If a sharp division is not possible, the variable with the smaller principal component load appears in two or more principal components.

### 6.4.3.2 Cluster analyses

This group of statistical methods permits variables (substances contained in the water and additional quantifiable influences) to be grouped according to their degree of similarity. The individual methods differ from each other in the criteria of similarity applied, e.g. absolute or relative differences or correlation coefficients. They can be used for categorization and classification, etc. in the same way as principal component analyses.

Standard distribution of the variables is taken as the basis for both statistical groups of methods. Transformations and standardization of variables may therefore become necessary, and this makes interpretation more difficult. Practical examples of these methods are widely published.

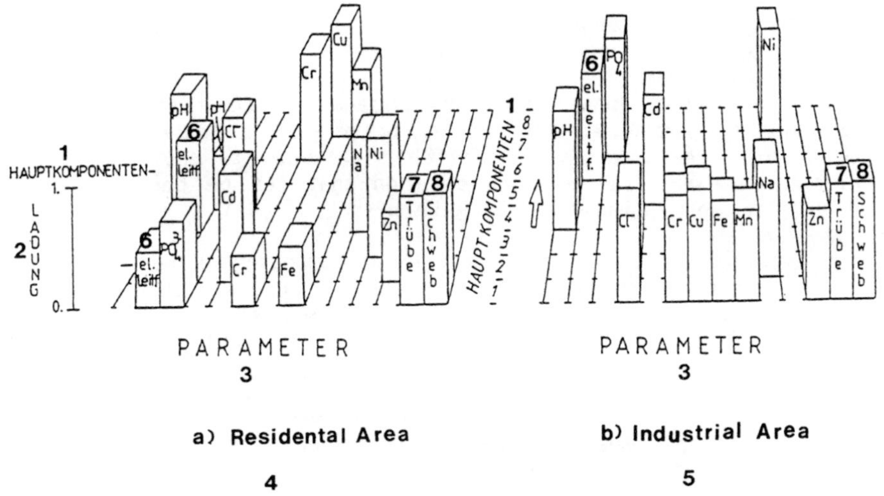


Fig. 176. Example: Comparison of feeds to 2 sewage treatment plants; 1) = Principal components; 2) = Load; 3) = Parameters; 4) Residental area; 5) = Industrial area; 6) = Electrical conductivity; 7) = Turbidity; 8) = Suspended matter

### 6.4.3.3 Multidimensional discriminatory analysis

With this method a linear compilation of independent variables - the discriminant function - is made, which is characteristic of optimum differentiation in given groups. The division into groups can be made subjectively or on the basis of a cluster analysis.

Using the discriminant function, it is possible to work out for individual samples separation values which determine which group these results belong to. Fig. 177 shows how a preliminary approximate structural analysis of the relationship between water quality parameters can be used to assign a water sample to given water types. Three different groundwaters and subterranean waters were analyzed for 10, 25 and 30 variables. In this example, the variables sulphate and boric acid proved the most useful for separation purposes.

## 6.5 Time-series analyses

Time-series analyses are distinguished from all evaluation methods so far discussed in that they take account of the temporal sequence of the observations. In contrast to the previously described methods, it is assumed in this case that dependences will exist. On the other hand, sophisticated time-sequence analyses impose substantially stricter conditions as regards the data itself:

- the observations must be made at equal intervals

- missing observation values must be estimated, thus raising the question of the method of interpolation

- under certain circumstances data must be adjusted every working day.

The aim of time-series analysis of water quality data is to discover whether a trend exists and, if so, whether there are fluctuations around this trend or periodical fluctuations, and how great a part is played by random influences. In addition, by juxtaposing the temporal order of two water-quality parameters the aim is to provide a scale for comparing the para-meters. The examples (Fig. 178) is intended to show the results of a bivariate analysis. The extra information that can be gained from this type of analysis can be summarized as follows:

- A cycle of 59 weeks (equivalent to approximately 1 year) for temperature and oxygen is revealed by mathematical analysis. There is a seasonal fluctuation of 3 to 4 months, but this is somewhat shorter for oxygen and nitrite. The phase displacement by -3.1 is confirmed in the negative correlation.

- Nitrite only coincides with the other variables of the first principal components in respect of the yearly cycle. It displays additional periodicity in the 6 to 8 week range.

- Water delivery and phosphate are affected not only by the yearly cycle, but also by the summer high-water in the half-yearly cycle. In contrast to the yearly rhythm, the half-yearly cycle in the case of ammonium is closely connected with a stochastic 2.5-month cycle.

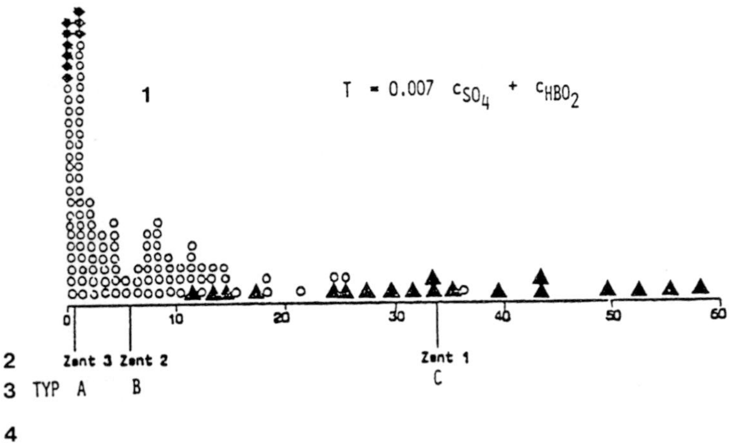


Fig. 177. Discriminatory analysis; assignment of a water sample to three specified water types (A, B, C) as applied to groundwater and subterranean water; 1) = Discriminant function: $T = 0.007\ C_{SO_4} + C_{HBO_2}$; 2) = Centre 3, 2, 1; 3) = Type A, B, C; 4) = Condition of the water in uni-dimensional discrimination field

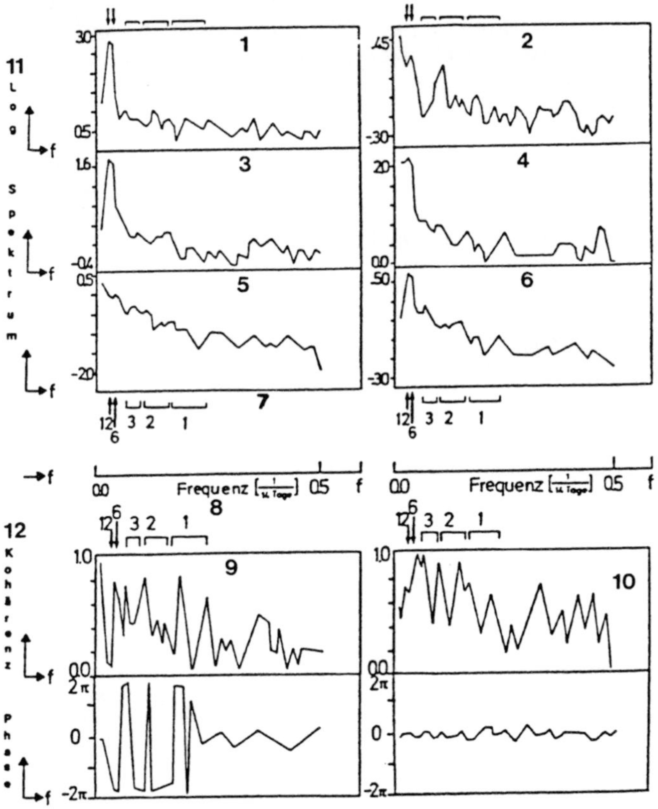


Fig 178. Analysis Surface Water Ruhr Villigst 1976 - 1982; 1) = W. temperature; 2) = Ammonium; 3) = Oxygen; 4) = Nitrate; 5) = Phosphate; 6) = Nitrite; 7) = Periodicity in months; 8) = Frequency (1/14 days); 9) = Ammonium/phosphate; 10) = Nitrite/phosphate; 11) = Log spectrum; 12) = Coherence phase

## 6.6 Technical advice

All the statistical methods described in this chapter are available from a wide range of manufacturers and other suppliers as computer programs. The multivariate statistical methods and the time-series analyses require office-oriented data processing technology, if a large quantity of measurement results and variables are involved. The methods mentioned here are currently offered in the English language by extensive computer program libraries. They are accompanied by a continuous scientific back-up and technical program maintenance service. The programs available include BMDP, SPSS, SAS, IMSL, NAG and CLUSTAN. The ARTHUR and SIMCA programs are specially designed for chemical analysis applications. Descriptions are given in Kowalski, Anal. Chem. 1980 and 1982.

Seemingly objective evaluation strategies can nevertheless produce spurious results if the restrictions set out in the method are not observed or if the evaluation is not carried out in a way which is adequate for the problem.

## 6.7 Assessment of water analysis findings

How the findings of physico-chemical, chemical and microbiological water analysis, and where appropriate the question of biological indicators, are to be assessed and weighted, depends on the use to which the water is to be put or on its relative importance in the environment. For example: whether it is untreated water for purification to obtain drinking water, whether it is wastewater or water from flowing or standing bodies of water. Many countries have their own rules regarding quality specifications for water that is to be supplied as drinking water or treated to obtain drinking water, some of them in the form of statutory provisions. On a worldwide scale the World Health Organization's Guidelines for Drinking-Water Quality are important, as set out, for example, in Volume 1: Recommendations, Geneva 1984. This English-language compilation of guidelines on drinking-water quality is a fundamental work which deserves close attention.

Wherever an assessment of water for drinking purposes is required, water analysts are urgently recommended to make use of the WHO booklet. The basic requirements are summarized in 5 tables which are reproduced below.

### 6.7.1 WHO-Guidelines for drinking water quality (1984)

Table 1. Microbiological and biological quality

| Organism | Unit | Guideline value | Remarks |
|---|---|---|---|
| **I. Microbiological quality** | | | |
| **A. Piped water supplies** | | | |
| A.1 Treated water entering the distribution system | | | |
| faecal coliforms | number/100 ml | 0 | turbidity < 1 NTU; for disinfection with chlorine, pH preferably < 8.0; free chlorine residual 0.2 - 0.5 mg/litre following 30 minutes (minimum) contact |
| coliform organisms | number/100 ml | 0 | |
| A.2 Untreated water entering the distribution system | | | |
| faecal coliforms | number/100 ml | 0 | |
| coliform organisms | number/100 ml | 0 | in 98 % of samples examined throughout the year - in the case of large supplies when sufficient samples are examined |
| coliform organisms | number/100 ml | 3 | in an occasional sample, but not in consecutive samples |
| A.3 Water in the distribution system | | | |
| faecal coliforms | number/100 ml | 0 | |
| coliform organisms | number/100 ml | 0 | in 95 % of samples examined throughout the year - in the case of large supplies when sufficient samples are examined |
| coliform organisms | number/100 ml | 3 | in an occasional sample, but not in consecutive samples |
| **B. Unpiped water supplies** | | | |
| faecal coliforms | number/100 ml | 0 | |
| coliform organisms | number/100 ml | 10 | should not occur repeatedly; if occurrence is frequent and if sanitary protection cannot be improved, an alternative source must be found if possible |

Table 1. Microbiological and biological quality (cont.)

| Organism | Unit | Guideline value | Remarks |
|---|---|---|---|
| **C. Bottled drinking-water** | | | |
| faecal coliforms | number/100 ml | 0 | source should be free from faecal contamination |
| coliform organisms | number/100 ml | 0 | |
| **D. Emergency water supplies** | | | |
| faecal coliforms | number/100 ml | 0 | advise public to boil water in |
| coliform organisms | number/100 ml | 0 | case of failure to meet guideline values |
| Enteroviruses | -- | no guideline value set | |
| **II. Biological quality** | | | |
| protozoa (pathogenic) | -- | no guideline value set | |
| helminths (pathogenic) | -- | no guideline value set | |
| free-living organisms (algae, others) | -- | no guideline value set | |

Table 2. Inorganic constituents of health significance

| Constituent | Unit | Guideline value | Remarks |
|---|---|---|---|
| arsenic | mg/l | 0.05 | |
| asbestos | -- | no guideline value set | |
| barium | -- | no guideline value set | |
| beryllium | -- | no guideline value set | |
| cadmium | mg/l | 0.005 | |
| chromium | mg/l | 0.05 | |
| cyanide | mg/l | 0.1 | |
| fluoride | mg/l | 1.5 | natural or deliberately added; local or climatic conditions may necessitate adaptation |
| hardness | -- | no health-related guideline value set | |
| lead | mg/l | 0.05 | |
| mercury | mg/l | 0.001 | |
| nickel | -- | no guideline value set | |
| nitrate | mg/l(N) | 10 | |
| nitrite | -- | no guideline value set | |
| selenium | mg/l | 0.01 | |
| silver | -- | no guideline value set | |
| sodium | -- | no guideline value set | |

Table 3. Organic constituents of health significance

| Constituent | Unit | Guideline value | Remarks |
|---|---|---|---|
| aldrin and dieldrin | µg/l | 0.03 | |
| benzene | µg/l | 10[a] | |
| benzo[a]pyrene | µg/l | 0.01[a] | |
| carbon tetrachloride | µg/l | 3[a] | tentative guideline value[b] |
| chlordane | µg/l | 0.3 | |
| chlorobenzenes | µg/l | no health-related guideline value set | odour threshold concentration between 0.1 and 3 µg/l |
| chloroform | µg/l | 30[a] | disinfection efficiency must not be compromised when controlling chloroform content |
| chlorophenols | µg/l | no health-related guideline value set | odour threshold concentration 0.1 µg/l |
| 2,4-D | µg/l | 100[c] | |
| DDT | µg/l | 1 | |
| 1,2-dichloroethane | µg/l | 10[a] | |
| 1,1-dichloroethene[d] | µg/l | 0.3[a] | |
| heptachlor and heptachlor epoxide | µg/l | 0.1 | |
| hexachlorobenzene | µg/l | 0.01[a] | |
| gamma-HCH (lindane) | µg/l | 3 | |
| methoxychlor | µg/l | 30 | |
| pentachlorophenol | µg/l | 10 | |
| tetrachloroethene[d] | µg/l | 10[a] | tentative guideline value[b] |
| trichloroethene[d] | µg/l | 30[a] | tentative guideline value[b] |
| 2,4,6-trichlorophenol | µg/l | 10[a,c] | odour threshold concentration 0.1 µg/l |
| trihalomethanes | | no guideline value set | see chloroform |

[a] These guideline values were computed from a conservative hypothetical mathematical model which cannot be experimentally verified and values should therefore be interpreted differently. Uncertainties involved may amount to two orders of magnitude (i.e. from 0.1 to 10 times the number).

[b] When the available carcinogenicity data did not support a guideline value, but the compounds were judged to be of importance in drinking-water and guidance was considered essential, a tentative guideline value was set on the basis of the available health-related data.

[c] May be detectable by taste and odour at lower concentrations.

[d] These compounds were previously known as 1,1-dichloroethylene, tetrachloroethylene, and trichloroethylene, respectively.

Table 4. Aesthetic quality

| Constituent or characteristic | Unit | Guideline value | Remarks |
|---|---|---|---|
| aluminium | mg/l | 0.2 | |
| chloride | mg/l | 250 | |
| chlorobenzenes and chlorophenols | -- | no guideline value set | these compounds may affect taste and odour |
| colour | true colour units (TCU) | 15 | |
| copper | mg/l | 1.0 | |
| detergents | -- | no guideline value set | there should not be any foaming or taste and odour problems |
| hardness | mg/l (as $CaCO_3$) | 500 | |
| hydrogen sulfide | -- | not detectable by consumers | |
| iron | mg/l | 0.3 | |
| manganese | mg/l | 0.1 | |
| oxygen - dissolved | -- | no guideline value set | |
| pH | -- | 6.5-8.5 | |
| sodium | mg/l | 200 | |
| solids - total dissolved | mg/l | 1000 | |
| sulfate | mg/l | 400 | |
| taste and odour | -- | inoffensive to most consumers | |
| temperature | -- | no guideline value set | |
| turbidity | nephelometric turbidity units (NTU) | 5 | preferably < 1 for disinfection efficiency |
| zinc | mg/l | 5.0 | |

Table 5. Radioactive constituents

| Constituent or characteristic | Unit | Guideline value | Remarks |
|---|---|---|---|
| gross alpha activity | Bq/l | 0.1 | a) If the levels are exceeded more detailed radionuclide analysis may be necessary.<br><br>b) Higher levels do not necessarily imply that the water is unsuitable for human comsumption |
| gross beta activity | Bq/l | 1 | |

The significance of the Guidelines is described as follows:

## Introduction

### Consumer perception of drinking-water quality

In assessing the quality of drinking-water, the consumer relies completely upon his senses. Water constituents may affect the appearance, smell, or the taste of the water and the consumer will evaluate the quality and the acceptability essentially on these criteria. Water that is highly turbid, highly coloured, or has an objectionable taste will be regarded as dangerous and will be rejected for drinking purposes. However, we can no longer rely entirely upon our senses in the matter of quality judgement. The absence of any adverse sensory effects does not guarantee the safety of water for drinking.

The primary aim of the **Guidelines for drinking-water quality** is the protection of public health and thus the elimination, or reduction to a minimum of constituents of water that are known to be hazardous to the health and wellbeing of the community.

### Priorities as regards water quality

The relative priorities assigned to the many substances for which guideline values are given later in this book will depend on local circumstances. Some guideline values, e.g., for colour and pH, are not related directly to health, but have been applied widely and successfully over many years to ensure the wholesomeness of water.

The microbiological quality of drinking-water is of the greatest importance, however, and must never be compromised in order to provide aesthetically pleasing and acceptable water.

A higher quality may be required for some special purposes, such as renal dialysis.

When a guideline value is exceeded this should be a signal: (i) to investigate the cause, with a view to taking remedial action; (ii) to consult with authorities responsible for public health for advice.

Although the guideline values describe a quality of water that is acceptable for lifelong consumption, the establishment of these guidelines should not be regarded as implying that the quality of drinking-water may be degraded to the recommended level. Indeed, a continuous effort should be made to maintain drinking-water quality at the highest possible level.

The guideline values specified have been derived to safeguard health on the basis of lifelong consumption. Short-term exposures to higher levels of chemical constituents, such as might occur following accidental contamination, may be tolerated but need to be assessed case by case, taking into account, for example, the acute toxicity of the substance involved.

Short-term deviations above the guideline values do not necessarily mean that the water is unsuitable for consumption. The amount by which, and the period for which, any guideline value can be exceeded without affecting public health depends on the specific substance involved.

It is recommended that, when a guideline value is exceeded, the surveillance agency (usually the authority responsible for public health) should be consulted for advice on suitable action, taking into account the intake of the substance from sources other than drinking-water (for chemical constituents), the likelihood of adverse effects, the practicability of remedial measures, and similar factors.

In developing national drinking-water standards based on the these guidelines, it will be necessary to take account of a variety of local geographical, socioeconomic, dietary, and industrial conditions. This may lead to national standards that differ appreciably from the guideline values.

In the case of radioactive substances, the term guideline value is used in the sense of "reference level" as defined by the International Commission on Radiological Protection (ICRP).

In arriving at the guideline values for various substances in water, the total intake from air, food and water for each substance is taken into consideration, as far as possible from the information available; it is assumed that the daily **per capita** consumption of water is 2 litres.

For the majority of the substances for which guideline values are proposed, the toxic effect in man is predicted from studies with laboratory animals. The accuracy and reliability of a quantitative prediction of toxicity in man from animal experimentation depend upon a number of factors, e.g., choice of animal species, design of the experiment and, not least, extrapolation methods. However, for most of the organic compounds considered, the difference in chemical pathogenesis between animals and man is mainly quantitative, although qualitative differences also exist.

Data on the toxicity of chemicals are obtained from experiments in which the adverse effect occurs at considerably higher dosages than would be experienced in man. When extrapolating from such animal data to man, therefore, a safety factor must be introduced to provide for the unknown factors involved. The current doubts concerning both the biological and the mathematical reliability of methods of extrapolating from high doses to low doses necessitate the use of somewhat arbitrary safety factors, such as reduction by a factor of 100 or 1000.

These uncertainties arise from the nature of the toxic effects and the quality of the toxicological information. Other considerations are the size and type of the population to be protected, and thus under certain conditions safety factors (or uncertainty factors) as high as 1000 may be necessary.

However, assessment of the health risk to the population involves more than routine application of safety factors, and it must be emphasized that strictly speaking the extrapolation from animal experimentation applies only to the conditions of the particular experiment.

The existing methods of extrapolation from animal data to man deal with exposures to single substances, whereas in the human environment a large number of hazardous chemicals and other factors may interact. In the special case of substances possessing carcinogenic properties, this book illustrates the rationale of using a risk factor in arriving at the proposed guideline value. Owing to the considerable uncertainties in the available evidence, the proposed guideline values are in many cases deliberately cautious in character and therefore must not be interpreted as standards.

A judgement about safety - or what is an acceptable risk level in particular circumstances - is a matter in which society as a whole has a role to play. The final judgement as to whether the benefit from adopting any of these proposed guidelines does or does not justify the risk is for each country to decide. What must be re-emphasized is that the guideline values proposed are not strict standards that must be adhered to, but are subject to a wide range of flexibility and are provided essentially in an endeavour to protect public health and enable a judgement to be made regarding the provision of drinking-water of acceptable quality.

**Table 6.** Inorganic constituents of potential health significance

| Constituent | Guideline values set | Background document drafted | Referred for consideration of aesthetic and organoleptic aspects | No action required |
|---|---|---|---|---|
| aluminium | | | x | |
| antimony | | | | x |
| arsenic | x | x | | |
| asbestos | | x | | |
| barium | | x | | |
| beryllium | | x | | |
| boron | | | | x |
| cadmium | x | x | | |
| chromium | x | x | | |
| cobalt | | | | x |
| copper | | | x | |
| cyanide | x | x | | |
| ferrocyanide | | | | x |
| fluoride | x | x | | |
| hardness (calcium and magnesium) | | x | x | |
| iron | | | x | |
| lead | x | x | | |
| lithium | | | | x |
| magnesium | | | x | |
| manganese | | | x | |
| mercury | x | x | | |
| molybdenum | | | | x |
| nickel | | x | | |
| nitrate | x | x | | |
| nitrite | | | | x |
| selenium | x | x | | |
| silver | | x | | |
| sodium | | x | x | |
| tellurium | | | | x |
| thallium | | | | x |
| thiocyanate | | | | x |
| tin | | | | x |
| titanium | | | | x |
| tungsten | | | | x |
| uranium | | | | x |
| vanadium | | | | x |
| zinc | | | x | |

Table 7. Guideline values for health-related inorganic constituents

| Constituent | Guideline value (mg/litre) |
|---|---|
| arsenic | 0.05 |
| cadmium | 0.005 |
| chromium | 0.05 |
| cyanide | 0.1 |
| fluoride | 1.5[a] |
| lead | 0.05 |
| mercury | 0.001 |
| nitrate (as N) | 10.00 |
| selenium | 0.01 |

[a] Guideline value may vary depending upon climatic conditions and water consumption

Table 8. Groups of organic compounds of potential health significance

| Contaminant | Detailed examination required | No further action required[a] |
|---|---|---|
| 1. Source contaminants | | |
| humic substances | | x |
| chlorinated alkanes and alkenes | x | |
| nitrosamines | | x |
| polynuclear aromatic hydrocarbons (PAH)[b] | x | |
| nitrilotriacetic acid (NTA) | | x |
| phenols | x | |
| synthetic detergents | | x |
| pesticides[b] | x | |
| polychlorinated biphenyls (PCB) | | x |
| phthalate esters | | |
| petroleum oils, including gasoline | | x |
| chlorobenzenes | x | |
| chlorinated phenols | x | |
| benzene and alkylaromatics | x | |
| carbon tetrachloride | x | |
| 2. Introduced during treatment | | |
| carbon tetrachloride | x | |
| acrylamide | | x |
| trihalomethanes | x | |
| 3. Introduced during distribution | | |
| vinyl chloride monomer | x | |
| polynuclear aromatic hydrocarbons (PAH)[b] | x | |

[a] Substances did not comply with criteria enumerated on the previous page
[b] Mentioned in International standards for drinking water.

**Table 9.** Organic compounds for which no guideline value is recommended

| | |
|---|---|
| Chlorinated alkanes and alkenes:<br>dichloromethane<br>1,1,1-trichloroethane<br>1,2-dichloroethene<br>vinyl chloride | Chlorinated phenols:<br>2-chlorophenol<br>4-chlorophenol<br>2,4-dichlorophenol<br>2,6-dichlorophenol<br>2,4,5-trichlorophenol |
| Pesticides:<br>alpha-HCH<br>beta-HCH<br>triazine herbicides | Others:<br>trihalomethanes other than chloroform |
| Chlorobenzenes:<br>chlorobenzene<br>1,2-dichlorobenzene<br>1,4-dichlorobenzene<br>trichlorobenzenes | |

**Table 10.** Guideline values for health-related organic contaminants

| **Contaminant** | **Guideline value (μg/litre)** |
|---|---|
| aldrin and dieldrin | 0.03 |
| benzene | 10 |
| benzo[a]pyrene[a] | 0.01 |
| chlordane (total isomers) | 0.3 |
| chloroform[a,d] | 30 |
| 2,4-D | 100 |
| DDT (total isomers) | 1 |
| 1,2-dichloroethane[a] | 10 |
| 1,1-dichloroethene[a,e] | 0.3 |
| heptachlor and heptachlor epoxide | 0.1 |
| hexachlorobenzene[a] | 0.01[c] |
| gamma-HCH (lindane) | 3 |
| methoxychlor | 30 |
| pentachlorophenol | 10 |
| 2,4,6-trichlorophenol[a,b] | 10 |

[a] The guideline values for these substances were computed from a conservative, hypothetical, mathematical model that cannot be experimentally verified and therefore should be interpreted differently. Uncertainties involved are considerable and a variation of about two orders of magnitude (i.e. from 0.1 to 10 times the number) could exist.

[b] The threshold taste and odour value for this compound is 0.1 g/litre.

[c] Since the FAO/WHO conditional ADI of 0.0006 mg/kg body weight has been withdrawn, this value was derived from the linear multi-stage extrapolation model for a cancer risk of less than 1 in 100,000 for a lifetime of exposure.

[d] The microbiological quality of drinking-water should not be compromised by efforts to control the concentration of chloroform.

[e] Previously known as 1,1-dichloroethylene

Table 11. Organic substances for which tentative guideline values are recommended

| Contaminant | Tentative guideline value µg/litre |
|---|---|
| carbon tetrachloride | 3 |
| tetrachloroethene [a] | 10 |
| trichloroethene[a] | 30 |

a Previously known as tetrachloroethylene and trichloroethylene, respectively

Table 12. Guideline values and ADIs for certain pesticides

| Compound or group of isomers | Guideline value (µg/litre) | ADI (mg/kg body weight) |
|---|---|---|
| DDT (total isomers) | 1 | 0.005 |
| aldrin and dieldrin | 0.03 | 0.0001 |
| chlordane (total isomers) | 0.3 | 0.001 |
| hexachlorobenzene | 0.01[a] | --- |
| heptachlor and heptachlor epoxide | 0.1 | 0.0005 |
| gamma-HCH (lindane) | 3 | 0.01 |
| methoxychlor | 30 | 0.1 |
| 2,4-D | 100 | 0.3 |

a Since the FAO/WHO conditional ADI of 0.0006 mg/kg body weight has been withdrawn, this value was derived from the linear multi-stage extrapolation model for a cancer risk of less than 1 in 100,000 for a lifetime of exposure.

## 6.7.2 Directive of the Council of the European Communities on the quality of water for human consumption

As an example of comprehensive specifications for drinking water, the European Communities' requirements, which have to be translated into national legislation by the individual member states of the EEC, are reproduced below (from the Official Journal of the European Communities, No. L 229 of July 15, 1980).

List of Parameters

**Table A.** Organoleptic Parameters

| | Parameters | Expression of the results[a] | Guide level (GL) | Maximum admissible concentration (MAC) | Comments |
|---|---|---|---|---|---|
| 1 | Colour | mg/l Pt/Co scale | 1 | 20 | |
| 2 | Turbidity | mg/l $SiO_2$<br>Jackson units | 1<br>0.4 | 10<br>4 | - Replaced in certain circumstances by a transparency test, with a Secchi disc reading in meters:<br>GL: 6 m<br>MAC: 2 m |
| 3 | Odour | Dilution number | 0 | 2 at 12 °C<br>3 at 25 °C | - To be related to the taste tests |
| 4 | Taste | Dilution number | 0 | 2 at 12 °C<br>3 at 25 °C | - To be related to the odour tests |

[a] If, on the basis of Directive 71/354/EEC as last amended, a Member State uses in its national legislation, adopted in accordance with this Directive, units of measurement other than these indicated here, the values thus indicated must have the same degree of precision.

**Table B.** Physico-Chemical Parameters (in relation to the water's natural structure)

| | Parameters | Expression of the results | Guide level (GL) | Maximum admissible concentration (MAC) | Comments |
|---|---|---|---|---|---|
| 5 | Temperature | °C | 12 | 25 | |
| 6 | Hydrogen ion concentration | pH unit | $6.5 \leq pH \leq 8.5$ | | - The water should not be aggressive<br>- The pH values do not apply to water in closed containers<br>- Maximum admissible value: 9.5 |
| 7 | Conductivity | $\mu S\ cm^{-1}$<br>at 20 °C | 400 | | - Corresponding to the mineralization of the water<br>- Corresponding relativity values in ohms/cm: 2500 |

Table B. Physico-Chemical Parameters (in relation to the water's natural structure) (cont.)

| Parameters | Expression of the results | Guide level (GL) | Maximum admissible concentration (MAC) | Comments |
|---|---|---|---|---|
| 8 Chlorides | Cl mg/l | 25 | | - Approximate concentration above which effects might occur: 200 mg/l. |
| 9 Sulphates | $SO_4$ mg/l | 25 | 250 | |
| 10 Silica | $SiO_2$ mg/l | | | |
| 11 Calcium | Ca mg/l | 100 | | |
| 12 Magnesium | Mg mg/l | 30 | 50 | |
| 13 Sodium | Na mg/l | 20 | 175 (as from 1984 and with a percentile of 90) | - The values of this parameter take account of the recommendations of a WHO working party (The Hague, May 1978) on the progressive reduction of the current total daily salt intake to 6 g. |
| | | | 150 (as from 1987 and with a percentile of 80) | - As from 1st January 1984 the Commission will submit to the Council reports on trends in the total daily intake of salt per population. |
| | | | (these percentiles should be calculated over a reference period of three years) | - In these reports the Commission will examine to what extent the 120 mg/l MAC suggested by the WHO working party is necessary to achieve a satisfactory total salt intake level, and, if appropriate, will suggest a new salt MAC value to the Council and a deadline for compliance with that value. |
| | | | The WHO (Guidelines 1984): At present there is insufficient evidence to justify a guideline value for sodium in water based on health-risk considerations. | Before 1 January 1984 the Commission will submit to the Council a report on whether the reference period of three years for calculating these percentiles is scientifically well founded. |

Table B. Physico-Chemical Parameters (in relation to the water's natural structure) (cont.)

| Parameters | Expression of the results | Guide level (GL) | Maximum admissible concentration (MAC) | Comments |
|---|---|---|---|---|
| 14 Potassium | K mg/l | 10 | 12 | |
| 15 Aluminium | Al mg/l | 0.05 | 0.2 | |
| 16 Total hardness | | | | - See Table F. |
| 17 Dry residues | mg/l after drying at 180 °C | | 1500 | |
| 18 Dissolved oxygen | % $O_2$ saturation | | | - Saturation value > 75 % except for underground water. |
| 19 Free carbon dioxide | $CO_2$ mg/l | | | - The water should not be aggressive |

Table C. Parameters concerning substances undesirable in excessive amounts[a]

| Parameters | Expression of the results[a] | Guide level (GL) | Maximum admissible concentration (MAC) | Comments |
|---|---|---|---|---|
| 20 Nitrates | $NO_3$ mg/l | 25 | 50 | |
| 21 Nitrites | $NO_2$ mg/l | | 0.1 | |
| 22 Ammonium | $NH_4$ mg/l | 0.05 | 0.5 | |
| 23 Kjeldahl Nitrogen (excluding N in $NO_2$ and $NO_3$) | N mg/l | | 1 | |
| 24 (K Mn $O_4$) Oxidizability | $O_2$ mg/l | 2 | 5 | - Measured when heated in acid medium |

[a] Certain of these substances may even be toxic when present in very substantial quantities.

Table C. Parameters concerning substances undesirable in excessive amounts[a] (cont.)

| Parameters | Expression of the results[a] | Guide level (GL) | Maximum admissible concentration (MAC) | Comments |
|---|---|---|---|---|
| 25 Total organic carbon (TOC) | C mg/l | | | - The reason for any increase in the usual concentration must be investigated |
| 26 Hydrogen sulphide | S µg/l | | undetectable organoleptically | |
| 27 Substances extractable in chloroform | mg/l dry residue | 0.1 | | |
| 28 Dissolved or emulsified hydrocarbons (after extraction by petroleum ether); Mineral oils | µg/l | | 10 | |
| 29 Phenols (phenol index) | $C_6H_5OH$ µg/l | | 0.5 | - Excluding natural phenols which do not react to chlorine |
| 30 Boron | B µg/l | 1000 | | |
| 31 Surfactants (reacting with methylene blue) | µg/l (lauryl sulphate) | | 200 | |
| 32 Other organochlorine compounds not covered by parameter No 55 | µg/l | 1 | | - Haloform concentrations must be as low as possible. |

[a] Certain of these substances may even be toxic when present in very substantial quantities.

Table C. Parameters concerning substances undesirable in excessive amounts[a] (cont.)

| Parameters | Expression of the results[a] | Guide level (GL) | Maximum admissible concentration (MAC) | Comments |
| --- | --- | --- | --- | --- |
| 33 Iron | Fe µg/l | 50 | 200 | |
| 34 Manganese | Mn µg/l | 20 | 50 | |
| 35 Copper | Cu µg/l | 100<br>- at outlets of pumping and/or treatment works and their substations<br>3000<br>- after the water has been standing for 12 hours in the piping and at the point where the water is made available to the consumer | | - Above 3000 µg/l astringent taste, discoloration + corrosion may occur. |
| 36 Zinc | Zn µg/l | 100<br>- at outlets of pumping and/or treatment works and their substations<br>5000<br>- after the water has been standing for 12 hours in the piping and at the point where the water is made available to the consumer | | - Above 5000 µg/l astringent taste, opalescence and sand-like deposits may occur. |
| 37 Phosphorus | $P_2O_5$ µg/l | 400 | 5000 | |
| 38 Fluoride | F µg/l<br>8 - 12 °C<br>25 - 30 °C | | <br>1500<br>700 | - MAC varies according to average temperature in geographical area concerned. |

[a] Certain of these substances may even be toxic when present in very substantial quantities.

Table C. Parameters concerning substances undesirable in excessive amounts[a] (cont.)

| Parameters | Expression of the results[a] | Guide level (GL) | Maximum admissible concentration (MAC) | Comments |
|---|---|---|---|---|
| 39 Cobalt | Co μg/l | | | |
| 40 Suspended solids | | None | | |
| 41 Residual Chlorine | Cl μg/l | | | |
| 42 Barium | Ba μg/l | 100 | | |
| 43 Silver | Ag μg/l | | 10 | If, exceptionally, silver is used non-systematically to process the water, a MAC value of 80 μg/l may be authorized. |

[a] Certain of these substances may even be tooxic when present in very substantial quantities.

Table D. Parameters concerning toxic substances

| Parameters | Expression of the results | Guide level (GL) | Maximum admissible concentration (MAC) | Comments |
|---|---|---|---|---|
| 44 Arsenic | As μg/l | | 50 | |
| 45 Beryllium | Be μg/l | | | |
| 46 Cadmium | Cd μg/l | | 5 | |
| 47 Cyanides | CN μg/l | | 50 | |
| 48 Chromium | Cr μg/l | | 50 | |
| 49 Mercury | Hg μg/l | | 1 | |
| 50 Nickel | Ni μg/l | | 50 | |

Table D. Parameters concerning toxic substances (cont.)

| Parameters | Expression of the results | Guide level (GL) | Maximum admissible concentration (MAC) | Comments |
|---|---|---|---|---|
| 51 Lead | Pb µg/l | | 50 (in running water) | Where lead pipes are present, the lead content should not exceed 50 µg/l in a sample taken after flushing. If the sample is taken either directly or after flushing and the lead content either frequently or to an appreciable extent exceeds 100 µg/l, suitable measures must be taken to reduce the exposure to lead on the part of the consumer. |
| 52 Antimony | Sb µg/l | | 10 | |
| 53 Selenium | Se µg/l | | 10 | |
| 54 Vanadium | V µg/l | | | |
| 55 Pesticides and related products | µg/l | | | "Pesticides and related products" means: |
| - substances considered separately | | | 0.1 | - insecticides: - persistent organochlorine compounds - organophosphorus compounds - carbamates |
| - total | | | 0.5 | - herbicides - fungicides - PCBs and PCTs |
| 56 Polycyclic aromatic hydrocarbons | µg/l | | 0.2 | - reference substances: - fluoranthene/benzo 3,4 - fluoranthene/benzo 11,12 - fluoranthene/benzo 3,4 - pyrene/benzo 1,12 - perylene/indeno (1,2,3-cd) pyrene |

Table E. Microbiological parameters

| Parameters | Results: volume of the sample in ml | Guide level (GL) | Maximum admissible concentration (MAC) | |
|---|---|---|---|---|
| | | | Membrane filter method | Multiple tube method (MPN) |
| 57 Total coliforms[a] | 100 | -- | 0 | MPN < 1 |
| 58 Faecal coliforms | 100 | -- | 0 | MPN < 1 |
| 59 Faecal streptococci | 100 | -- | 0 | MPN < 1 |
| 60 Sulphite-reducing Clostridia | 20 | -- | -- | MPN $\leq$ 1 |

Water intended for human consumption should not contain pathogenic organisms.

If it is necessary to supplement the microbiological analysis of water intended for human consumption, the samples should be examined not only for the bacteria referred to in Table E but also for pathogens including:

- salmonella,
- pathogenic staphylococci,
- faecal bacteriophages,
- entero-viruses;

not should such water contain:

- parasites,
- algae
- other organisms such as animalcules.

[a] Provided a sufficient number of samples is examined (95 % consistent results).

Table E. Microbiological parameters (cont.)

| | Parameters | | Results: size of the sample (in ml) | Guide level (GL) | Maximum admissible concentration (MAC) | Comments |
|---|---|---|---|---|---|---|
| 61 | Total bacteria counts for water supplied for human consumption | 37 °C | 1 | 10[a,b] | -- | |
| | | 22 °C | 1 | 100[a,b] | -- | |
| 62 | Total bacteria counts for water in closed containers | 37 °C | 1 | 5 | 20 | On their own responsibility and where parameters 57, 58, 59 und 60 are complied with, and where the pathogen organisms given on page 22 are absent, Member States may process water for their internal use the total bacteria count of which exceeds the MAC values laid down for parameter 62. MAC values should be measured within 12 hours of being put into closed containers with the sample water being kept at a constant temperature during that 12-hour period. |
| | | 22 °C | 1 | 20 | 100 | |

[a] For disinfected water the corresponding values should be considerably lower at the point where it leaves the processing plant.

[b] If, during successive sampling, any of these values is consistently exceeded a check should be carried out.

Table F. Minimum required concentration for softened water intended for human consumption

| | Parameters | Expression of the results | Minimum required concentration (softened water) | Comments |
|---|---|---|---|---|
| 1 | Total hardness | mg/l Ca | 60 | Calcium or equivalent cations. |
| 2 | Hydrogen ion concentration | pH | | The water should not be aggressive |
| 3 | Alkalinity | mg/l $HCO_3$ | 30 | |
| 4 | Dissolved oxygen | | | |

NB: - The provisions for hardness, hydrogen ion concentration, dissolved oxygen and calcium also apply to desalinated water.

- If, owing to its excessive natural hardness, the water is softened in accordance with Table F before being supplied for consumption, its sodium content may, in exceptional cases, be higher than the values given in the "Maximum admissible concentration" column. However, an effort must be made to keep the sodium content at as low a level as possible and the essential requirements for the protection of public health may not be disregarded.

**Table of correspondence between the various units of water hardness measurement**

| | French degree | English degree | German degree | Milligrams of Ca | Millimoles of Ca |
|---|---|---|---|---|---|
| French degree | 1 | 0.70 | 0.56 | 4.008 | 0.1 |
| English degree | 1.43 | 1 | 0.80 | 5.73 | 0.143 |
| German degree | 1.79 | 1.25 | 1 | 7.17 | 0.179 |
| Milligrams of Ca | 0.25 | 0.175 | 0.140 | 1 | 0.025 |
| Millimoles of Ca | 10 | 7 | 5.6 | 40.08 | 1 |

**Patterns and Frequency of Standard Analyses**

**A. Table of standard pattern analyses** (parameters to be considered in monitoring)

| Standard analyses / Parameters to be considered | Minimum monitoring (C 1) | Current monitoring (C 2) | Periodic monitoring (C 3) | Occasional monitoring in special situations or in case of accidents (C 4) |
|---|---|---|---|---|
| A Organoleptic parameters | - odour[a]<br>- taste[a] | - odour<br>- taste<br>- turbidity (appearance) | | The competent national authorities of the Member States will determine the parameters[e] according to circumstances, taking account of all factors which might have an adverse affect on the quality of drinking water supplied to consumers. |
| B Physico-chemical parameters | - conductivity or other physicochemical parameter<br>- residual chlorine[c] | - temperature[b]<br>- conductivity or other physico-chemical parameter<br>- pH<br>- residual chlorine[c] | | |
| C Undesirable parameters | | - nitrates<br>- nitrites<br>- ammonia | | |
| D Toxic parameters[d] | | | | |
| E Micro-biological parameters | - total coliforms or total counts at 22 °C and 37 °C<br>- faecal coliforms | - total coliforms<br>- faecal coliforms<br>- total counts at 22 °C and 37 °C | | |

**Note:** An initial analysis, to be carried out before a source is exploited, should be added. The parameters to be considered would be the current monitoring analyses plus inter alia various toxic or undesirable substances presumed present. The list would be drawn up by the competent national authorities.

[a] Qualitative assessment

[b] Except for water supplied in containers

[c] Or other disinfectants and only in the case of treatment

[d] These parameters will be determined by the competent national authority, taking account of all factors which might affect the quality of drinking water supplied to users and which could enable the ionic balance of the constituents to be assessed.

[e] The component national authority may use parameters other than those mentioned in Annex I to this Directive.

B. Table of minimum frequency of standard analyses[c]

| Volume of water produced or distributed in $m^3$/day | Population concerned (assuming 200 l/day per person) | Analysis C 1 | Analysis C 2 | Analysis C 3 | Analysis C 4 |
|---|---|---|---|---|---|
| | | Number of samples per year | Number of samples per year | Number of samples per year | |
| 100 | 500 | a | a | a | Frequency to be determined by the competent national authorities as the situation requires |
| 1 000 | 5 000 | a | a | a | |
| 2 000 | 10 000 | 12 | 3 | a | |
| 10 000 | 50 000 | 60 | 6 | 1 | |
| 20 000 | 100 000 | 120 | 12 | 2 | |
| 30 000 | 150 000 | 180 | 18 | 3 | |
| 60 000 | 300 000 | 360[b] | 36 | 6 | |
| 100 000 | 500 000 | 360[b] | 60 | 10 | |
| 200 000 | 1 000 000 | 360[b] | 120[b] | 20[b] | |
| 1 000 000 | 5 000 000 | 360[b] | 120[b] | 20[b] | |

[a] Frequency left to the discretion of the competent national authorities. However, water intended for the food-manufacturing industries must be monitored at least once a year.

[b] The competent health authorities should endeavour to increase this frequency as far as their resources allow.

[c] (1) In the case of water which must be disinfected, microbiological analysis should be twice as frequent.

(2) Where analyses are very frequent, it is advisable to take samples at the most regular intervals possible.

(3) Where the values of the results obtained from samples taken during the preceding years are constant and significantly better than the limits laid down in Annex I, and where no factor likely to cause a deterioration in the quality of the water has been discovered, the minimum frequencies of the analyses referred to above may be reduced:

- for surface waters, by a factor of 2 with the exception of the frequencies laid down for microbiological analyses;
- for ground waters, by a factor of 4, but without prejudice to the provisions of point (1) above.

Reference methods of analysis

A. Organoleptic parameters

| | |
|---|---|
| 1 Colour | Photometric method calibrated on the Pt/co scale. |
| 2 Turbidity | Silica method - Formazine test - Secchi's method. |
| 3 Odour | Successive dilutions, tested at 12 °C or 25 °C. |
| 4 Taste | Successive dilutions, tested at 12 °C or 25 °C. |

B. Physico-chemical parameters

| | |
|---|---|
| 5 Temperature | Thermometry. |
| 6 Hydrogen ion concentration | Electrometry. |
| 7 Conductivity | Electrometry. |
| 8 Chlorides | Titrimetry - Mohr's method. |
| 9 Sulphates | Gravimetry - complexometry - spectrophotometry. |
| 10 Silica | Absorption spectrophotometry. |
| 11 Calcium | Atomic absorption - complexometry. |
| 12 Magnesium | Atomic absorption. |
| 13 Sodium | Atomic absorption. |
| 14 Potassium | Atomic absorption. |
| 15 Aluminium | Atomic absorption - absorption spectrophotometry. |
| 16 Total hardness | Complexometry. |
| 17 Dry residue | Dessication at 180 °C and weighing. |
| 18 Dissolved oxygen | Winkler's method - Specific electrode method. |
| 19 Free carbon dioxide | Acidimetry. |

C. Parameters concerning undesirable substances

| | |
|---|---|
| 20 Nitrates | Absorption spectrophotometry - Specific electrode method. |
| 21 Nitrites | Absorption spectrophotometry. |
| 22 Ammonium | Absorption spectrophotometry. |
| 23 Kjeldahl Nitrogen | Oxidation with Titrimetry or Absorption spectrophotometry. |
| 24 Oxidizability | Boiling for 10 minutes with $KMnO_4$ in acid medium. |
| 25 Total organic carbon (TOC) | --- |
| 26 Hydrogen sulphide | Absorption spectrophotometry. |
| 27 Substances extractable in chloroform | Liquid/liquid extraction using purified chloroform at neutral pH, weighing the residue. |
| 28 Hydrocarbons dissolved or in emulsion); Mineral oils | Infra-red absorption spectrophotometry. |
| 29 Phenols (phenol index) | Absorption spectrophotometry, paranitroaniline method and 4-aminoantipyrine method. |
| 30 Boron | Atomic absorption - Absorption spectrophotometry. |
| 31 Surfactants (reacting with methylene blue) | Absorption spectrophotometry with methylene blue |
| 32 Other organo-chlorine compounds | Gas-phase or liquid-phase chromatography after extraction by appropriate solvents and purification - Identification of the constituents of mixtures if necessary. Quantitative determination. |
| 33 Iron | Atomic absorption - Absorption spectrophotometry. |
| 34 Manganese | Atomic absorption - Absorption spectrophotometry. |
| 35 Copper | Atomic absorption - Absorption spectrophotometry. |
| 36 Zinc | Atomic absorption - Absorption spectrophotometry. |
| 37 Phosphorus | Absorption spectrophotometry. |
| 38 Fluoride | Absorption spectrophotometry - Specific electrode method. |
| 39 Cobalt | --- |

C. Parameters concerning undesirable substances (cont.)

| | |
|---|---|
| 40 Suspended solids | Method of filtration on to µ 0.45 porous membrane or centrifuging (for at least 15 minutes with an average acceleration of 2,800 to 3,200 g) dried at 105 °C and weighed. |
| 41 Residual chlorine | Titrimetry - Absorption spectrophotometry. |
| 42 Barium | Atomic absorption. |

D. Parameters concerning toxic substances

| | |
|---|---|
| 43 Silver | Atomic absorption. |
| 44 Arsenic | Absorption spectrophotometry - Atomic absorption. |
| 45 Beryllium | --- |
| 46 Cadmium | Atomic absorption. |
| 47 Cyanides | Absorption spectrophotometry. |
| 48 Chromium | Atomic absorption - Absorption spectrophotometry. |
| 49 Mercury | Atomic absorption. |
| 50 Nickel | Atomic absorption. |
| 51 Lead | Atomic absorption. |
| 52 Antimony | Absorption spectrophotometry. |
| 53 Selenium | Atomic absorption. |
| 54 Vanadium | --- |
| 55 Pesticides and related products | See method 32. |
| 56 Polycyclic aromatic hydrocarbons | Measurement of intensity of fluorescence ultra-violet after extraction using hexane - gas-phase chromatography or measurement in ultraviolet after thin layer chromatography - Comparative measurements against a mixture of six standard substances of the same concentration[a] |

[a] Standard substances to be considered: fluoranthene/benzo-3,4-fluoranthene/benzo-11,12-fluoranthene/benzo-3,4-pyrene/benzo-1,12-perylene and indeno (1,2,3-cd)pyrene.

E. Microbiological parameters

| | | |
|---|---|---|
| 57[b] | Total coliforms | Fermentation in multiple tubes. Subculturing of the positive tubes on a confirmation medium. Count according to MPN (most probable number)<br>or<br>Membrane filtration and culture on an appropriate medium such as Tergitol lactose agar, endo agar, 0.4 % Teepol broth, subculturing and identification of the suspect colonies -<br>Incubation temperature for total coliforms: 37 °C<br>Incubation temperature for fecal coliforms: 44 °C |
| 58[b] | Fecal coliforms | |
| 59[b] | Fecal streptococci | Sodium azide method (Litsky). Count according to MPN -<br>Membrane filtration and culture on an appropriate medium. |
| 60[b] | Sulphite-reducing Clostrida | A spore count, after heating the sample to 80 °C by:<br>- seeding in a medium with glucose, sulphite and iron, counting the black-halo colonies;<br>- membrane filtration, deposition of the inverted filter on a medium with glucose, sulphite and iron covered with agar, count of black colonies;<br>- distribution in tubes of differential reinforced clostridial medium (DRCM), subculturing of the black tubes in a medium of litmus-treated milk, count according to MPN. |
| 61/62[b] | Total counts | Inoculation by placing in nutritive agar. |

Additional tests

| | |
|---|---|
| Salmonella | - Concentration by membrane filtration. Inoculation on a pre-enriched medium. Enrichment, subculturing on isolating agar. Identification. |
| Pathogenic staphylococci | - Membrane filtration and culture on a specific medium (e.g. Chapman's hypersaline medium). Test for pathogenic characteristics. |
| Fecal bacteriophages | - Guelin's process. |
| Enteroviruses | - Concentration by filtration, flocculation or centrifuging, and identification. |
| Protozoa | - Concentration by filtration on a membrane, microscopic examination, test for pathogenicity. |
| Animalcules (worms - larvae) | - Concentration by filtration on a membrane. Microscopic examination. Test for pathogenicity. |

F. Minimum required concentration

| | |
|---|---|
| Alkalinity | Acidimetry with Methyl orange. |

### 6.7.3 Federal Republic of Germany: Recommendation on halogenated hydrocarbons

In the Federal Republic of Germany, volatile halogenated hydrocarbons in ground water and drinking water are covered by a recommendation issued by the Federal Health Ministry (Federal Health Gazette 25 No. 3 of March 1982). This proposes a maximum concentration of 25 µg/l as a mean annual value for the total of the following four compounds:

Chloroform ($CHCl_3$)
Monobromo dichloromethane ($CHBrCl_2$)
Dibromo monochloromethane ($CHBr_2Cl$)
Bromoform ($CHBr_3$).

A mean annual value of 25 µg/l is also set for the following volatile halogenated organic solvents in drinking water:

Trichloroethylene ($C_2HCl_3$)
Tetrachloroethylene ($C_2Cl_4$)
1,1,1-trichloroethane ($C_2H_3Cl_3$)
Dichloromethane ($CH_2Cl_2$).

### 6.7.4 Treatment of surface water to obtain drinking water

As early as 1975 the European Community issued a directive on quality requirements for surface waters used for drinking-water supplies in the member countries. These requirements are reproduced below.

**Definition of the standard methods of treatment for transforming surface water of categories A1, A2 and A3 into drinking water**

Category A1

Simple physical treatment and disinfection, e.g. rapid filtration and disinfection.

Category A2

Normal physical treatment, chemical treatment and disinfection, e.g. prechlorination, coagulation, flocculation, decantation, filtration, disinfection (final chlorination).

Category A3

Intensive physical and chemical treatment, extended treatment and disinfection e.g. chlorination to break-point, coagulation, flocculation, decantation, filtration, adsorption (activated carbon), disinfection (ozone, final chlorination).

**Characteristics of surface water intended for the abstraction of drinking water**

| | Parameters | | A1 G | A1 I | A2 G | A2 I | A3 G | A3 I |
|---|---|---|---|---|---|---|---|---|
| 1 | pH | | 6.5 to 8.5 | | 5.5 to 9 | | 5.5 to 9 | |
| 2 | Coloration (after simple filtration) | mg/l Pt scale | 10 | 20[a] | 50 | 100[a] | 50 | 200[a] |
| 3 | Total suspended solids | mg/l SS | 25 | | | | | |
| 4 | Temperature | ºC | 22 | 25[a] | 22 | 25[a] | 22 | 25[a] |
| 5 | Conductivity | $\mu s/cm^{-1}$ at 20 ºC | 1000 | | 1000 | | 1000 | |
| 6 | Odour | (dilution factor at 25ºC) | 3 | | 10 | | 20 | |
| 7[b] | Nitrates | mg/l $NO_3$ | 25 | 50[a] | | 50[a] | | 50[a] |
| 8[c] | Fluorides | mg/l F | 0.7 to 1 | 1.5 | 0.7 to 1.7 | | 0.7 to 1.7 | |
| 9 | Total extractable organic chlorine | mg/l Cl | | | | | | |
| 10[b] | Dissolved iron | mg/l Fe | 0.1 | 0.3 | 1 | 2 | 1 | |
| 11[b] | Manganese | mg/l Mn | 0.05 | | 0.1 | | 1 | |
| 12 | Copper | mg/l Cu | 0.02 | 0.05[a] | 0.05 | | 1 | |
| 13 | Zinc | mg/l Zn | 0.5 | 3 | 1 | 5 | 1 | 5 |

**Characteristics of surface water intended for the abstraction of drinking water** (cont.)

| | Parameters | | A1 G | A1 I | A2 G | A2 I | A3 G | A3 I |
|---|---|---|---|---|---|---|---|---|
| 14 | Boron | mg/l B | 1 | | 1 | | 1 | |
| 15 | Beryllium | mg/l Be | | | | | | |
| 16 | Cobalt | mg/l Co | | | | | | |
| 17 | Nickel | mg/l Ni | | | | | | |
| 18 | Vanadium | mg/l V | | | | | | |
| 19 | Arsenic | mg/l As | 0.01 | 0.05 | | 0.05 | 0.05 | 0.1 |
| 20 | Cadmium | mg/l Cd | 0.001 | 0.005 | 0.001 | 0.005 | 0.001 | 0.005 |
| 21 | Total chromium | mg/l Cr | | 0.05 | | 0.05 | | 0.05 |
| 22 | Lead | mg/l Pb | | 0.05 | | 0.05 | | 0.05 |
| 23 | Selenium | mg/l Se | | 0.01 | | 0.01 | | 0.01 |
| 24 | Mercury | mg/l Hg | 0.0005 | 0.001 | 0.0005 | 0.001 | 0.0005 | 0.001 |
| 25 | Barium | mg/l Ba | | 0.1 | | 1 | | 1 |
| 26 | Cyanide | mg/l CN | | 0.05 | | 0.05 | | 0.05 |
| 27 | Sulphates | mg/l $SO_4$ | 150 | 250 | 150 | 250[a] | 150 | 250[a] |
| 28 | Chlorides | mg/l Cl | 200 | | 200 | | 200 | |

**Characteristics of surface water intended for the abstraction of drinking water** (cont.)

| | Parameters | | A1 G | A1 I | A2 G | A2 I | A3 G | A3 I |
|---|---|---|---|---|---|---|---|---|
| 29 | Surfactants (reacting with methylene blue) | mg/l (laurylsulphate) | 0.2 | | 0.2 | | 0.5 | |
| 30[b,d] | Phosphates | mg/l $P_2O_5$ | 0.4 | | 0.7 | | 0.7 | |
| 31 | Phenols (phenol index) paranitraniline 4 aminoantipyrine | mg/l $C_6H_5OH$ | | 0.001 | 0.001 | 0.005 | 0.01 | 0.1 |
| 32 | Dissolved or emulsified hydrocarbons (after extraction by petroleum ether) | mg/l | | 0.05 | | 0.2 | 0.5 | 1 |
| 33 | Polycyclic aromatic hydrocarbons | mg/l | | 0.0002 | | 0.0002 | | 0.001 |
| 34 | Total pesticides (parathion, BHC, dieldrin) | mg/l | | 0.001 | | 0.0025 | | 0.005 |
| 35[b] | Chemical oxygen demand (COD) | mg/l $O_2$ | | | | | 30 | |
| 36[b] | Dissolved oxygen saturation rate | % $O_2$ | > 70 | | > 50 | | > 30 | |
| 37[b] | Biochemical oxygen demand ($BOD_5$) (at 20 °C without nitrification) | mg/l $O_2$ | < 3 | | < 5 | | < 7 | |

**Characteristics of surface water intended for the abstraction of drinking water** (cont.)

| | Parameters | | A1 G | A1 I | A2 G | A2 I | A3 G | A3 I |
|---|---|---|---|---|---|---|---|---|
| 38 | Nitrogen by Kjeldahl method (except $NO_3$) | mg/l N | 1 | | 2 | | 3 | |
| 39 | Ammonia | mg/l $NH_4$ | 0.05 | | 1 | 1.5 | 2 | 4[a] |
| 40 | Substances extractable with chloroform | mg/l SEC | 0.1 | | 0.2 | | 0.5 | |
| 41 | Total organic carbon | mg/l C | | | | | | |
| 42 | Residual organic carbon after flocculation and membrane filtration (5μ)TOC | mg/l C | | | | | | |
| 43 | Total coliforms 37 ºC | /100 ml | 50 | | 5000 | | 50000 | |
| 44 | Faecal coliforms | /100 ml | 20 | | 2000 | | 20000 | |
| 45 | Faecal streptococci | /100 ml | 20 | | 1000 | | 10000 | |
| 46 | Salmonella | | Not present in 5000 ml | | not present in 1000 ml | | | |

G = guide
I = mandatory

a Exceptional climatic or geographical conditions.
b See Article 8 (d).
c The values given are upper limits set in relation to the mean annual temperature (high and low).
d This parameter has been included to satisfy the ecological requirements of certain types of environment.

It is nevertheless important to take a critical approach to the requirements outlined in the above tables prepared by the European Community for drinking water. This applies, for example, to cases where the available drinking water exceeds certain parameters, such as those for sulphate, sodium, potassium, dry matter and fluoride. In our opinion, exceeding these permissible maximum concentrations is perfectly tolerable if the condition of the water is otherwise acceptable and above all if there are no reservations from a microbiological point of view. The decision will depend on local conditions and on the possible uses to which the water supply can be put. The data quoted for example in the EEC Directive on drinking water should not therefore be indiscriminately applied to all other conditions in all other countries. From a purely technical point of view, it should also be noted that there is cause for criticism of certain methods. This is true, for example, of the permissible upper limit value for phenols which is to be determined using the so-called phenol index and which is given as 0.5 µg/l except for natural phenols which do not react with chlorine. This requirement is impossible to satisfy from the outset, since the maximum permissible concentration of 0.5 µg/l cannot be determined using the so-called phenol index, as the sensitivity of this method must be put at around 2 µg/l given an extremely good method of working. Moreover, there is no way of telling how natural phenols which do not react with chlorine are then to be distinguished at such concentration levels. As regards Item 27 involving substances which can be extracted with chloroform, the standard value of 0.1 mg/l (determined by way of the dry matter) is only valid for substances with a higher boiling point and can only be verified to a certain degree if at least 5 or 10 litres of water is used for analysis. The analytical method used to determine so-called Kjeldahl nitrogen (Item 23) would also appear to be highly problematical. When all is said and done, the opinion has been expressed in certain quarters that there is no objection to exceeding a barium content of for example 100 µg/l. The WHO Guidelines for Drinking Water Quality indicate for example that no definite influence can be established even at 10 mg/l. As regards Item 55 concerning pesticides and the like, it would also appear more or less impossible to realize the maximum permissible concentration of 0.1 µg/l for all conceivable pesticides. This not only applies to the analytical aspect, but also takes account of the wide range of substances which may be used and which may not be available as reference substances in a laboratory. These critical remarks indicate that analysts must view such methods with a certain amount of caution and should always choose the method which appears most favourable under the given circumstances.

This means, however, that a great deal of expert knowledge is required of evaluation or supervisory bodies when it comes to assessing the values determined by the analyst. The WHO Guidelines and the recommendations outlined in the Drinking Water Directive issued by the Council of the European Community on Drinking Water should aid this assessment. This expert knowledge of the assessing bodies is however always called for, in order to be able to achieve the primary aim of any analysis, namely that of providing users - under the given local conditions - with a standard of drinking water which does not pose health problems. Particularly stringent compliance with microbiological demands is required here. With the results of certain chemical analyses it is possible to consider tolerating values in excess of those listed in the above Standards. Under such circumstances it may be that individual decisions cannot be taken, but rather that a working party must establish whether or not water which exceeds the standard values can still be used as drinking water. This question is relatively easy to answer in areas where modern water treatment plants are available, but extremely complicated in regions in which the water may have to be used with a minimum of treatment, as otherwise no water would be available at all. Similar considerations must be applied as regards the demands to be made of the quality of surface water used to obtain drinking water. Here again an attempt must be made to employ the treatment stages, and where necessary to utilize even simple and primitive treatment methods, whilst bearing in mind the absolute priority of providing the population with usable water. Although assessments of course vary greatly from country to country, we would mention as an example a study conducted by Dr. Samia al Azharia Jahn of the GTZ entitled "Traditional Water Purification in Tropical Developing Countries" (GTZ 1981, 6236 Eschborn 1, FRG).

### 6.7.5 Waste water

Similar considerations apply to the assessment of waste water. As an example of the limiting conditions applying to waste water constituents and the requirements placed on waste water in industrial nations, the following list outlines the stipulations made for example in the Federal Republic of Germany:

#### 6.7.5.1 Stipulations applying to the discharge of industrial sewage

Sewage (examples):

| | |
|---|---|
| **1. Physical parameters** | |
| 1.1 Temperature | max. 35 °C |
| 1.2 pH | 6.5 - 9.0 |
| 1.3 pH (cyan. sewage) | 8.0 - 9.0 |
| **2. Settlable solids** | |
| Sludge-like substances and solids from industrial sewage treatment plants (e.g. neutralizing and detoxification plants) | 1 ml/l after 2 hours settling period in a settling glass |
| **3. Organic substances and solvents** | |
| 3.1 Organic solvents | 10 mg/l |
| 3.2 Halogenated hydrocarbons calculated as organically bonded chlorine | 5 mg/l |
| 3.3 Phenols (total) | 20 mg/l |
| 3.4 Mineral oils/fats Non-saponifiable substances extractable with petroleum ether | 20 mg/l |
| 3.5 Organic oils/fats Saponifiable substances extractable with petroleum ether | 50 mg/l |
| **4. Inorganic substances (dissolved)** | |
| 4.1 Cyanides (total) | 1 mg/l |

| | |
|---|---|
| 4.2 Cyanides which can be decomposed by chlorine | 0.2 mg/l |
| 4.3 Sulphates | 400 mg/l |

**5. Inorganic substances (total)**

| | |
|---|---|
| 5.1 Arsenic | 0.1 mg/l |
| 5.2 Lead | 2.0 mg/l |
| 5.3 Cadmium (if necessary, separate treatment of sewage containing cadmium is required). | 0.5 mg/l |
| 5.4 Chromium | 2.0 mg/l |
| 5.5 Chromium VI | 0.2 mg/l |
| 5.6 Iron | 20 mg/l |
| 5.7 Copper | 2.0 mg/l |
| 5.8 Nickel | 3.0 mg/l |
| 5.9 Mercury (if necessary, separate treatment of sewage containing mercury is required.) | 0.05 mg/l |
| 5.10 Silver | 0.5 mg/l |
| 5.11 Zinc | 5.0 mg/l |
| 5.12 Tin | 3.0 mg/l |

### 6.7.6 Requirements to be satisfied by bathing waters

Finally, reference is made to a recommendation of the European Community dated 1975, which is concerned with the quality of bathing waters. Its requirements are summarized below:

This Directive deals with the quality requirements to be satisfied by bathing waters with the exception of water for therapeutical purposes and water for swimming pools.

Within the meaning of this Directive, the following definitions apply:

a) "Bathing waters" refers to those flowing or standing inland waters or parts thereof and seawater in which bathing

- is expressly permitted by the competent authorities in a particular Member State or

- is not prohibited and in which large numbers of people tend to go bathing;

b) "Bathing area" refers to the location at which the above water is to be found;

c) "Bathing season" refers to the period in which a major influx of bathers can be expected taking into account local customs, any local bathing regulations and weather conditions.

Within the scope of the checks performed, the bathing waters are considered to comply with the relevant parameters if the samples taken at the sampling point with the frequency stipulated in the Appendix are found to conform with the values of the parameters for the respective water quality in 95 % of cases in respect of parameters coinciding with those given in Column I of the Appendix, and

in 90 % of cases in respect of all other parameters, with the exception of the parameters "total coliform bacteria" and "faecal coliform bacteria" where the percentage of samples may be 80 %,

and provided that in the case of the 5 %, 10 % or 20 % respectively of samples which do not conform

the measured values do not deviate by more than 50 % from the value of the parameters concerned, with the exception of the microbiological parameters, the pH value and the dissolved oxygen; and

consecutive water samples taken in a statistically meaningful time sequence do not deviate from the parameters concerned.

Samples are to be taken at those points where on average the greatest numbers of bathers per day are recorded. Samples are preferably to be taken 30 cm beneath the water's surface; this does not however apply to mineral oil samples which are taken on the surface. Sampling is to begin two weeks prior to the start of the bathing season.

In the case of flowing water the upstream conditions, and in the case of standing inland waters or seawater the conditions in the surrounding area, are to be closely investigated on site at regular intervals, in order to determine the geographical and topographical conditions, the amount and nature of all polluting and possibly polluting substances introduced into the water and their significance in relation to the distance from the bathing area.

Should testing of the water by the authorities or sampling and sample analysis reveal that substances which could reduce the quality of the water used for swimming are being introduced into the water, or if this is suspected, additional samples are to be taken. Additional samples are also to be taken if a drop in the quality of the water is suspected elsewhere.

The analytical procedures (reference methods) for the parameters in question are indicated in the Appendix. Laboratories using other methods must make sure that the results obtained are equivalent to or comparable with the results indicated in the Appendix.

Deviations from this Directive are permitted

a) For certain parameters marked with an (0) in the Appendix in the case of unusual metereological or geographical conditions;

b) If the bathing waters are subject to natural enrichment with certain substances over and above the limit values established in the Appendix.

Natural enrichment is taken to be the process in which a certain volume of water absorbs certain substances contained in the soil without human intervention.

6.7.6.1 Quality requirements for bathing water

| | Parameters | | G | I | Minimum sampling frequency | Method of analysis and inspection |
|---|---|---|---|---|---|---|
| | **Microbiological:** | | | | | |
| 1 | Total coliforms | /100 ml | 500 | 10000 | Fortnightly b | Fermentation in multiple tubes. Subculturing of the positive tubes on a confirmation medium. Count according to MPN (most probable number) or membrane filtration and culture on an appropriate medium such as Tergitol lactose agar, endo agar, 0.4 % Teepol broth, subculturing and identification of the suspect colonies. In the case of 1 and 2, the incubation temperature is variable according to whether total or faecal coliforms are being investigated. |
| 2 | Faecal coliforms | /100 ml | 100 | 2000 | Fortnightly b | |
| 3 | Faecal streptococci | /100 ml | 100 | -- | c | Litsky method. Count according to MPN (most probable number) or filtration on membrane. Culture on an appropriate medium. |
| 4 | Salmonella | /1 litre | -- | 0 | c | Concentration by membrane filtration. Inoculation on a standard medium. Enrichment - subculturing on isolating agar - identification. |
| 5 | Entero viruses | PFU/10 litres | -- | 0 | c | Concentrating by filtration, flocculation or centrifuging and confirmation. |

**Quality requirements for bathing water** (cont.)

| | Parameters | | G | I | Minimum sampling frequency | Method of analysis and inspection |
|---|---|---|---|---|---|---|
| | **Physico-chemical:** | | | | | |
| 6 | pH | | -- | 6 to 9[a] | c | Electrometry with calibration at pH 7 and 9. |
| 7 | Colour | | -- | No abnormal change in colour[a] | Fortnightly b | Visual inspection or photometry with standards on the Pt.Co scale. |
| | | | -- | -- | c | |
| 8 | Mineral oils | mg/litre | -- | No film visible on the surface of the water and no odour | Fortnightly b | Visual and olfactory inspection or extraction using an adequate volume and weighing the dry residue. |
| | | | $\leq 0.3$ | --- | c | |
| 9 | Surface-active substances reacting with methylene blue | mg/litre (lauryl-sulfate) | -- | No lasting foam | Fortnightly b | Visual inspection or absorption spectrophotometry with methylene blue. |
| | | | $\leq 0.3$ | -- | c | |

**Quality requirements for bathing water** (cont.)

| | Parameters | | G | I | Minimum sampling frequency | Method of analysis and inspection |
|---|---|---|---|---|---|---|
| 10 | Phenols (phenol indices) | mg/litre $C_6H_5OH$ | -- | No specific odour | Fortnightly b | Verification of the absence of specific odour due to phenol or absorption spectrophotometry 4-aminoantipyrine (4 AAP) method |
| | | | $\leq 0.005$ | $\leq 0.05$ | c | |
| 11 | Transparency | m | 2 | 1[a] | Fortnightly b | Secchi's disc. |
| 12 | Dissolved oxygen % saturation $O_2$ | | 80 to 120 | -- | c | Winkler's method or electrometric method (oxygen meter). |
| 13 | Tarry residues and floating materials such as wood, plastic articles, bottles, containers of glass plastic, rubber or any other substance. Waste or splinters. | | Absence | | Fortnightly b | Visual inspection. |
| 14 | Ammonia | mg/litre $NH_4$ | | | d | Absorption spectrophotometry, Nessler's method, or indophenol blue method. |
| 15 | Nitrogen Kjeldahl | mg/litre N | | | d | Kjeldahl method. |

**Quality requirements for bathing water** (cont.)

| | Parameters | | G | I | Minimum sampling frequency | Method of analysis and inspection |
|---|---|---|---|---|---|---|
| | **Other substances regarded as indications of pollution** | | | | | |
| 16 | Pesticides (parathion, HCH, dieldrin) | mg/litre | | | c | Extraction with appropriate solvents and chromatographic determination. |
| 17 | Heavy metals such as: | | | | | |
| | - arsenic | mg/litre As | | | c | Atomic absorption possibly preceded by extraction. |
| | - cadmium | Cd | | | | |
| | - chrome VI | Cr VI | | | | |
| | - lead | Pb | | | | |
| | - mercury | Hg | | | | |
| 18 | Cyanides | mg/litre CN | | | c | Absorption spectrophotometry using a specific reagent |
| 19 | Nitrates and phosphates | mg/litre $NO_3$<br>$PO_4$ | | | c | Absorption spectrophotometry using a specific reagent |

**Quality requirements for bathing water** (cont.)

I = mandatory

G = guide

a = Provision exists for exceeding the limits in the event of exceptional geographical or meteorological conditions.

b = When a sampling taken in previous years produced results which are appreciably better than those in this Annex and when no new factor likely to lower the quality of the water has appeared, the competent authorities may reduce the sampling frequency by a factor of 2.

c = Concentration to be checked by the competent authorities when an inspection in the bathing area shows that the substance may be present or that the quality of the water has deteriorated.

d = These parameters must be checked by the competent authorities when there is a tendency towards the eutrophication of the water.

### 6.7.7 Substances used for treating drinking water

An excerpt from the German Drinking Water Treatment Order (draft of January 1985) is enclosed as an example of substances which can be added to drinking water during treatment.

**Additives remaining in treated drinking water**

| No. | Substance | Usage conditions | | | |
|---|---|---|---|---|---|
| | Name | Purpose | Max. permitted admixture | Max. quantity in treated drinking water[a] | Further restrictions |
| a | b | c | d | e | f |
| 1 | Chlorine, sodium hypochlorite, calcium hypochlorite and magnesium hypochlorite Chlorinated lime | Disinfection | 34 $mmol/m^3$ corresponding to 1.2 mg/l of chlorine, in exceptional cases[b] 50 $mmol/m^3$ corresponding to 1.8 mg/l of chlorine | 8.5 $mmol/m^3$ corresponding to 0.3 mg/l of free chlorine, in exceptional cases[b] 17 $mmol/m^3$ corresponding to 0.6 mg/l of free chlorine | Max. values for trihalogen methanes as secondary reaction product; total of 0.025 mg/l in treated drinking water[a] |
| 2 | Chlorine dioxide | Disinfection | 6 $mmol/m^3$ corresponding to 0.4 mg/l of chlorine dioxide | 3 $mmol/m^3$ corresponding to 0.2 mg/l, in exceptional cases[b] 6 $mmol/m^3$ corresponding to 0.4 mg/l of chlorine dioxide | Maximum value for chlorite as reaction end product 3 $mmol/m^3$ corresponding to 0.2 mg/l in treated drinking water[a] |
| 3 | Ammonia, Ammonium chloride, Ammonium sulphate | | 30 $mmol/m^3$ corresponding to 0.5 mg/l calculated as ammonium | 30 $mmol/m^3$ corresponding to 0.5 mg/l calculated as ammonium | Addition only together with the substances listed under Nos. 1 and 2 (chlorine and chlorine compounds) |

**Additives remaining in treated drinking water** (cont.)

| No. | Substance<br>Name | Usage conditions<br>Purpose | Max. permitted admixture | Max. quantity in treated drinking water[a] | Further restrictions |
|---|---|---|---|---|---|
| a | b | c | d | e | f |
| 4 | Ozone | Disinfection<br>Oxidation | | 1 $mmol/m^3$ corresponding to 0.05 mg/l | Max. value for trihalogen methanes as secondary reaction product; total of 0.025 mg/l in treated drinking water[a] |
| 5 | Silver, silver chloride, Sodium/silver chloride complex, silver sulphate | Disinfection | 0.9 $mmol/m^3$ corresponding to 0.1 mg/l calculated as silver | 0.9 $mmol/m^3$ corresponding to 0.1 mg/l calculated as silver | |
| 6.1 | Sodium, potassium and calcium salts of mono and polyphosphoric acids | Inhibition of scale deposits | 50 $mmol/m^3$ corresponding to 4.7 mg/l calculated as phosphate ($PO_4^{3-}$) | 50 $mmol/m^3$ corresponding to 4.7 mg/l calculated as phosphate ($PO_4^{3-}$) | Maximum value for potassium 300 $mmol/m^3$ corresponding to 12 mg/l in treated drinking water[a] |

**Additives remaining in treated drinking water** (cont.)

| No. | Substance | Usage conditions | | | |
|---|---|---|---|---|---|
| | Name | Purpose | Max. permitted admixture | Max. quantity in treated drinking water[a] | Further restrictions |
| a | b | c | d | e | f |
| 6.2 | Sodium, potassium and calcium salts of mono and polyphosphoric acids, incl. such salts when alkalized with sodium carbonate, sodium hydroxide or sodium silicate | Inhibition of corrosion | 30 mm/m$^3$ corresponding to 2.8 mg/l calculated as phosphate ($PO_4^{3-}$) | 30 mm/m$^3$ corresponding to 2.8 mg/l calculated as phosphate ($PO_4^{3-}$) | Max. value for potassium 300 mmol/m$^3$ corresponding to 12 mg/l<br>Max. value for silicate 350 mmol/m$^3$ corresponding to 20 mg/l calculated as silicon dioxide in treated drinking water[a] |
| 7 | Calcium carbonate<br>Part-calcinated dolomite<br>Calcium oxide<br>Calcium hydroxide<br>Magnesium carbonate<br>Magnesium oxide<br>Magnesium hydroxide<br>Sodium carbonate<br>Sodium hydroxide<br>Sodium hydrogen carbonate | Increase in pH and increase in calcium content<br>Increase in acid capacity<br>Increase in pH<br><br>Increase in acid capacity | | | Max. pH of 9.5, max. value for sodium 6500 mmol/m$^3$ corresponding to 150 mg/l treated drinking water[a]<br><br>Max. value for magnesium 2000 mmol/m$^3$ corresponding to 50 mg/l |

**Additives remaining in treated drinking water** (cont.)

| No. | Substance | Usage conditions | | | |
|---|---|---|---|---|---|
| | Name | Purpose | Max. permitted admixture | Max. quantity in treated drinking water[a] | Further restrictions |
| a | b | c | d | e | f |
| 8 | Calcium sulphate<br>Calcium chloride | For setting calcium content of drinking water intended for production of beer and malt extract | | | Max. value for sulphate 2500 $mmol/m^3$ corresponding to 240 mg/l |
| 9 | Sulphuric acid<br>Hydrochloric acid<br>Magnesium hydrogen-sulphate<br>Calcium hydrogensulphate | Reduction of pH | | | Minimum pH of 6.5<br>Max. value for sulphate 2500 $mmol/m^3$ corresponding to 240 mg/l in treated drinking water[a]<br><br>Max. value for magnesium 2000 $mmol/m^3$ corresponding to 50 mg/l |
| 10 | Sulphur dioxide<br>Sodium sulphite<br>Calcium sulphite | Combining oxygen | 60 $mmol/m^3$ corresponding to 5 mg/l calculated as $SO_3^{2-}$ | 25 $mmol/m^3$ corresponding to 2 mg/l calculated as $SO_3^{2-}$ | - |

**Additives remaining in treated drinking water** (cont.)

| No. | Substance | Usage conditions | | | |
|---|---|---|---|---|---|
| | Name | Purpose | Max. permitted admixture | Max. quantity in treated drinking water[a] | Further restrictions |
| a | b | c | d | e | f |
| 11 | Hydrogen peroxide | Disinfection Oxidation | 500 mmol/$m^3$ corresponding to 17 mg/l | 150 mmol/$m^3$ corresponding to 5 mg/l | - |

[a] Including natural content and content resulting from other treatment stages

[b] The addition of greater amounts is only permitted if the microbiological requirements cannot be satisfied by other means

**Residual quantities and usage conditions for additives which, with the exception of residual amounts, are removed from the water again during the drinking water treatment process**

| No. | Substance | Usage conditions | | | |
|---|---|---|---|---|---|
| | Name | Purpose | Max. permitted admixture | Max. quantity in treated drinking water[a] | Further restrictions |
| a | b | c | d | e | f |
| 1 | Iron III chloride<br>Iron II sulphate<br>Iron III sulphate<br>Iron sulphate-chloride | Flocculation | - | 2 $mmol/m^3$ corresponding to 0.1 mg/l calculated as Fe | Max. value for sulphate: 2500 $mmol/m^3$ corresponding to 240 mg/l |
| 2 | Potassium permanganate | Oxidation | - | 1 $mmol/m^3$ corresponding to 0.05 mg/l calculated as Mn | Max. value for potassium: 300 $mmol/m^3$ corresponding to 12 mg/l in treated drinking water[a] |
| 3 | Aluminium sulphate<br>Aluminium chloride<br>Aluminium hydroxychloride<br>Aluminium hydroxysulphate<br>Sodium aluminate<br>Aluminium hydroxychloride-sulphate | Flocculation | - | 7 $mmol/m^3$ corresponding to 0.2 mg/l calculated as Al | Max. value for sodium: 6500 $mmol/m^3$ corresponding to 150 mg/l in treated drinking water[a]<br>Max. value for sulphate: 2500 $mmol/m^3$ corresponding to 240 mg/l |

**Residual quantities and usage conditions for additives which, with the exception of residual amounts, are removed from the water again during the drinking water treatment process** (cont.)

| No. | Substance | Usage conditions | | | |
|---|---|---|---|---|---|
| | Name | Purpose | Max. permitted admixture | Max. quantity in treated drinking water[a] | Further restrictions |
| a | b | c | d | e | f |
| 4 | Clays<br>Activated charcoal | Flocculation and adsorption | - | Total of 0.5 mg/l | - |
| 5 | Anionic and non-ionic polyacrylamides (monomeric acrylamide content 0.05 % referenced to the polymer) | Flocculation | 1 mg/l of polyacrylamide | 5 µg/l of polyacrylamide | - |
| 6 | Silicic acid and its sodium compounds | Flocculation | 700 $mmol/m^3$ corresponding to 40 mg/l calculated as silicon dioxide | - | Max. value for sodium 6500 $mmol/m^3$ corresponding to 150 mg/l in treated drinking water[a] |

[a] Including natural content and content resulting from other treatment stages

**Usage conditions when employing ion exchangers or sacrificial anodes for drinking water treatment**

| No. | Ion | Usage conditions | | | |
|---|---|---|---|---|---|
| | Name | Purpose | Max. permitted admixture | Max. quantity in treated drinking water[a] | Further restrictions |
| a | b | c | d | e | f |
| 1 | Hydrogen ion | Calcium removal<br>Reduction of pH or acid capacity | - | - | Min. value for calcium 1500 $mmol/m^3$ corresponding to 60 mg/l<br>Minimum pH of 6.5 in treated drinking water |
| 2 | Sodium ion | Calcium removal for adapting calcium content in drinking water on connection of central water supply systems or in the case of private water supply systems and individual water supply systems | - | 6500 $mmol/m^3$ corresponding to 150 mg/l of sodium | Min. value for calcium 1500 $mmol/m^3$ corresponding to 60 mg/l in treated drinking water[a] |

**Usage conditions when employing ion exchangers or sacrificial anodes for drinking water treatment** (cont.)

| No. | Ion | Usage conditions | | | |
|---|---|---|---|---|---|
| | Name | Purpose | Max. permitted admixture | Max. quantity in treated drinking water[a] | Further restrictions |
| a | b | c | d | e | f |
| 3 | Chloride ion<br>Hydrogen carbonate ion<br>Hydroxyl ion | Removal of humic substances<br>Nitrate ion removal<br>Sulphate ion removal<br>Selenium ion removal | - | - | Max. pH of 9.5 in treated drinking water[a] |
| 4 | Magnesium ion | Calcium removal | - | 2000 $mmol/m^3$ corresponding to 50 mg/l of magnesium | Min. value for calcium 1500 $mmol/m^3$ corresponding to 60 mg/l in treated drinking water[a] |
| | | Anodic corrosion protection | | 2000 $mmol/m^3$ corresponding to 50 mg/l of magnesium | - |

[a] Including natural content and content resulting from other treatment stages

**Treatment with chlorine tablets**

| Substance | Usage conditions | |
|---|---|---|
| | Purpose | Content in tablets for treatment of 10 litres of water[a] |
| Name | | |
| Sodium dichloroiso-cyanurate | Disinfection | Min. quantity 230 mg<br>Max. quantity 250 mg |
| Sodium chloride<br>Sodium carbonate<br>Sodium hydrogen carbonate<br>Tartaric acid<br>Adipic acid<br>Sodium benzoate<br>Polyoxyethylene/polyglycol waxes | Tablet aids | |

[a] In the case of tablets for disinfecting other amounts of water the permissible content is to be correspondingly converted

6.7.8

To sum up: as far as the assessment of water analysis findings is concerned, the reader's attention is once again drawn to the mathematical part of this Chapter (Section 6.1), which permits conclusions to be drawn about the reliability of the individual results obtained.

If the results are suitably reliable, or even if they are only of general information value, the above extracts from WHO, EEC and West German guidelines may serve as a useful guide in arriving at a more detailed assessment of water quality.

The final assessment, however, should always depend on the expert knowledge of the person making the assessment, and the application of such knowledge to the individual local situation. Where necessary, decisions may have to be taken by a committee of experts, but the deciding factor must always be the interest of man and animals, and also of the environment.

For example, where decisions have to be taken in emergencies or as a result of disasters, it is better to disinfect drinking water by heating it or using chemicals (and ignoring the chemical parameters etc.) than to die of thirst.

It is to be hoped, therefore, that those entrusted with the task of assessing water quality will see every new situation as a fresh challenge to their expert training and critical faculties.

1) Natural mineral waters see official Journal of the European Communities No L 229/1 from 30.08.80 "Council Directive of 15 July 1980 on the approximation of the laws of the Member States relating to the exploitation and marketing of natural mineral waters." (EC, Brussel's, Belgium)
2) Verordnung über natürliches Mineralwasser, FRG 1984 (German law of natural mineral water 1984, available: Verband Deutscher Mineralbrunnen e.V., Kennedyallee 28, D-5300 Bonn 2, FRG.

# 7 Subject Index